W0254326

EXOTIC NUCLEI AND NUCLEAR/PARTICLE ASTROPHYSICS (II)

EXOTIC NUCLEI AND NUCLEAR/PARTICLE ASTROPHYSICS (II)

Proceedings of the Carpathian Summer School of Physics 2007

Sinaia, Romania   20 – 31 August 2007

EDITORS

Livius Trache
Cyclotron Institute, Texas A&M University
College Station, Texas, U. S. A.

Sabin Stoica
Horia Hulubei National Institute of
Physics and Nuclear Engineering
Bucharest-Magurele, Romania

SPONSORING ORGANIZATION
Romanian National Authority for Scientific Research (ANCS)

Melville, New York, 2008
AIP CONFERENCE PROCEEDINGS ■ 972

Editors

Livius Trache
Cyclotron Institute
Texas A&M University
MS 3366
College Station, TX 77843-3366
U.S.A.

E-mail: livius_trache@tamu.edu

Sabin Stoica
Horia Hulubei National Institute
of Physics and Nuclear Engineering
407 Atomistilor str., POB MG-6
Bucharest-Magurele, Romania

E-mail: stoica@theory.nipne.ro

Cover design by Dr. Adriana Banu.

L.C. Catalog Card No. 2007942646
ISBN 978-0-7354-0490-8
ISSN 0094-243X

Printed in the United States of America

This volume is dedicated to the 60th anniversary of Professor Robert E. Tribble

Anniversary Symposium honoring Prof. R.E. Tribble
Carpathian Summer School of Physics, Aug. 20-31 2007, Sinaia, Romania (I)

Carpathian Summer School of Physics Aug. 20-31 2007, Sinaia, Romania (II)

CONTENTS

PART I: INVITED LECTURES

1. ANNIVERSARY MINI-SYMPOSIUM HONORING PROFESSOR ROBERT E. TRIBBLE

2. LOOK INTO THE FUTURE

3. EXOTIC NUCLEI—STRUCTURE AND REACTIONS

4. EXOTIC MATTER AND EXOTIC DECAYS

5. SUBNUCLEONIC DEGREES OF FREEDOM

6. NUCLEAR PHYSICS FOR ASTROPHYSICS

9. NEUTRINO PHYSICS

PART II: COMMUNICATIONS

PART III: OUTREACH SESSION AND CLOSING SESSION

CONCLUDING REMARKS—
NUCLEAR ASTROPHYSICS, NUCLEAR STRUCTURE,
PARTICLE ASTROPHYSICS

Organized by
"*Horia Hulubei*" National Institute of Physics and Nuclear Engineering, Bucharest
Cyclotron Institute, Texas A&M University, College Station TX

Sponsored by
Romanian National Authority for Scientific Research
Bucharest, Romania

Foreword

This book contains the invited lectures and contributions presented at the **Carpathian Summer School of Physics 2007** (CSSP07), held in Sinaia, Romania, August 20-31, 2007. In many ways, we, the organizers of this 22nd edition of the school, see it a direct continuation of its 20th edition CSSP05, in Mamaia. While looking for an appropriate title for this edition of the school, we found that the one we found in 2005 expresses best what we want to do and say and we kept it. Therefore, the title of the school became *"Exotic Nuclei and Nuclear/Particle Astrophysics (II)"* which is also the title of this book.

The lectures and contributions covered the topics announced in the school circulars:

- Structure of Exotic Nuclei
- Exotic Matter and Decays
- Phase transitions in hot nuclei
- Nucleosynthesis
- Nuclear Physics for Astrophysics
- High Energy Cosmic Rays
- Neutrino Physics
- Large Facilities

and in most cases these were the titles of the sessions during the schools, and are the titles of the chapters in the book.

The aim of this event, called a "school" mostly by tradition, was to offer a venue and occasion of interaction between scientists, physicists in particular, from many continents and countries, working in different or similar areas, but mainly to offer the possibility of such interactions for younger researchers, mostly from the host country and from the region, but open to the whole world. This was not a school in the sense of a limited number of "professors" teaching a large number of hours to "students". We had a rather large number of invited lectures (52 in total) addressing an attendance that was about 120 participants in total for the two weeks. Many of the "students" were very young or young; some were only young in spirit. In any case everybody had to gain from the arrangement, according to the opinions of the participants shared with us. We were very glad to hear that! From the students, 29 of them presented contributions of 20 minutes each. Most of the lectures and contributions are included in this volume.

It was particular at the school that we embedded a one-day symposium honoring professor Robert E Tribble at his 60th anniversary. It took place on Friday Aug. 24th and was a simple celebration of one of our ones, during which past or current collaborators talked about the physics research they did or do together with Bob or under his leadership. **We dedicate the present volume to Bob Tribble's anniversary**, and start the book with a chapter containing two specials from the symposium. We intended to bring in Sinaia professor Gerry Garvey, Bob's advisor from the time of his graduate studies in Princeton, but could only bring a recorded message that started the day. Second, Carl Gagliardi, a longtime collaborator of Bob's from the Texas A&M University, has reviewed his many years and many areas of work. The rest of the day, past or current students, and collaborators presented lectures or contributions on their work: X. Tang (Notre Dame), R.G. Pizzone (Catania), F. Carstoiu (Bucharest), S. Mioduszewski (Texas A&M), V. Kroha (Prague), A. Banu (Texas A&M), T. Al-Abdullah (Aman, Jordan), M. LaCognata (Catania), G. Tabacaru (Texas A&M). The affiliations give only some impression of the areas covered by all Bob's collaborators. The resulting papers are spread over the chapters of this book, according to their subject. We refer to the program of the school included in this volume to locate them. The day ended with a *symposion* at that spectacular location which is the restaurant "Cota 1400", anchored on the side of the mountain, half the way between the city and the top of Bucegi.

After the end of the conference, almost all of the presentations were posted on the website of the school: http://cyclotron.tamu.edu/cssp07 where you can find them. An overwhelming majority of the speakers at the school, lecturers and contributors, send their papers in time to be included in this volume. The volume itself speaks for the quality of their work, and implicitly, the quality of the school. Part I sums the invited lectures, Part II the contributions, and Part III the Outreach and the Closing Sessions.

We cannot end without expressing our many thanks to the members of the International Advisory Committee, and in particular to those that took time from their very busy schedule to attend, present lectures and send their contributions to this volume, enhancing considerably our stature and quality. Also, we thank to the members of the Organizing Committee that worked across two continents, with very little face-to-face contacts, to make the event happen. They worked before and during the event and some also after. It is our duty and pleasure to acknowledge here the contributions of our scientific secretaries, Iliana Brancus and Florin Carstoiu, of the webmasters Gabriel Tabacaru and Ed Pauna, of Mihai Horoi, Corina Simion and Adriana Banu. Adrian Sokolov provided the graphics for much of the material of the school. We all have to thank Aurora Anitoaie, school's secretary and Alexandra Olteanu, our financial director for making it run smoothly. In addition, we have to thank the management and staff of the "International" hotel in Sinaia for providing an excellent venue for the school and excellent services to us all.

Dr. Adriana Banu, to whom we are particularly indebted, handled a considerable part of the technical problems for editing this volume.

With this 22nd edition being a success, we are sure that future schools will be organized.

See you next time in the Carpathians!

Livius Trache
Sabin Stoica
Editors

Note: many of the graphs appearing in black-and-white in the printed version of the book are in color in the original pdf version and appear as such in the online version on the AIP website.

Carpathian Summer School of Physics 2007

Exotic Nuclei and Nuclear/Particle Astrophysics II

August 20-31, 2007, Sinaia, Romania

Program

Monday, August 20th, 2007	
Morning Session: Opening **Chair: L. Trache**	
9:00	*S. Lenzi*: Mirror nuclei: symmetry breaking and nuclear dynamics
9:55	*A. Bonaccorso*: Direct Reactions with Exotic Nuclei
10:50 Coffee break	
11:10	*S. Stoica:* Double beta decay and derivation of neutrino mass limits
12:05	*M. Horoi*: Correlations In Nuclei: Configuration Interaction, Coupled Clusters and Beyond
13:00 Lunch	
Afternoon Session **Chair: D. Delion**	
15:00	*A. Petrovici:* Beyond Mean Field Description of the Gamow-Teller β-decay
15:55 Coffee break	
16:10	*D. Poenaru*: Shell corrections stabilising superheavy nuclei and semispheroidal atomic clusters

Tuesday, August 21st, 2007	
Morning Session: **Chair: R.E. Tribble**	
9:00	*M. Teshima*: Very High Energy Gamma-Ray Astronomy
9:55	*C. Gagliardi*: pp Collisions at RHIC
10:50 Coffee break	
11:10	*R. Rapp*: Phase Transitions with Subnucleonic Degrees of Freedom
12:05	*M. Petrovici*: Selected aspects of flow phenomena in heavy ion collisions
13:00 Lunch	
Afternoon Session **Chair: A. Bonaccorso**	
15:00	*D. Delion*: Probing nuclear structure by cold emission processes
15:55 Coffee break	
16:10	*Communications: Laurent, Gherghescu, Neacsu, Peltonen*

Wednesday, August 22nd, 2007	
Morning Session: Nuclear Physics for Astrophysics I **Chair: P. J. Woods**	
9:00	*C. Rolfs*: Quests in experimental nuclear astrophysics
9:55	*M. Hass*: Low-energy nuclear astrophysics – the fascinating region of A = 7
10:50 Coffee break	
11:10	*M. Gai:* The Coulomb Dissociation of ^{8}B: A New Tool in Nuclear Astrophysics
12:05	*L. Trache*: Reaction rates for H-burning in stars from experiments with radioactive nuclear beams
13:00 Lunch	
Afternoon Session: Nuclear Physics for Astrophysics II **Chair: C. Rolfs**	
15:00	*K. Yoneda*: Coulomb dissociation of proton-rich sd-shell nuclei for astrophysical interest and related topics
15:55 Coffee break	
16:10	*M. Junker*: Nuclear Astrophysics Underground – Experiences and Results of the LUNA Experiment at Gran Sasso
17:05	Communications: *F. Negoita* (30 min) *Husti, Spiridon*

Thursday, August 23rd, 2007	
Morning Session: Exotic decays **Chair: A. Petrovici**	
9:00	*R.E. Tribble*: USA Nuclear Physics Long Range Plan 2007
9:55	*J.C. Hardy*: Superallowed nuclear β decay: a window on the weak interaction
10:50 Coffee break	
11:10	*J. Suhonen*: Recent Advances in Calculations of the Nuclear Matrix Elements for Double β Decay
12:05	*P.J. Woods*: The Structure Proton-rich Nuclei in Explosive Nuclear Astrophysical Scenarios
13:00 Lunch	
Afternoon Session **Chair: J.C. Hardy**	
15:00	*S. Shlomo*: Properties of finite nuclei and the nuclear matter equation of state
15:55 Coffee break	
16:10	*M. Harakeh*: Study of spin-isospin modes and their influence on astrophysical processes

17:05	Communications: *Stetcu, Kuz'min, Cozma*

Friday, August 24th, 2007	
Morning Session: Anniversary *mini-symposium* Robert E. Tribble **Chair: H. Rebel**	
9:00	*C.A. Gagliardi*: Fundamental Interactions, Nuclear Masses, Astrophysics, and QCD
9:55	*X. Tang*: The ^{16}N β-delayed α decay: Today and Tomorrow
10:50 Coffee break	
11:10	*R.G. Pizzone*: Trojan Horse Method: recent results
12:05	*F. Carstoiu*: Breakup of radioactive nuclei as indirect method in nuclear astrophysics
13:00 Lunch	
Afternoon Session: Anniversary *mini-symposium* Robert E. Tribble *(cnt'd)* **Chair: C. A. Gagliardi**	
15:00	*S. Mioduszewski*: An Overview of RHIC with a Look to the Future
15:55 Coffee break	
16:10	*V. Kroha*: Recent astrophysical experiments on cyclotron of NPI at Řež in collaboration with Texas A&M University
17:05	*A. Banu, T. Al-Abdullah, M. LaCognata, G. Tabacaru (20 min each)*

19:00 *Symposium* at restaurant "Cota 1400"

Saturday, Aug. 25th, 2007

School's excursion:
Sinaia - departure at 8:00am
Bran Castle ("Dracula's castle")
Brasov: 13h - lunch, visit town centre and old city
18:00 visit the Black Church, 18:30 organ concert
21h - return in Sinaia and dinner

Sunday Aug. 26th, 2007

Variae, in Sinaia, or TBD:
- hiking
- visit Peles Castle
- dinner at local restaurants

Monday, August 27th, 2007	
Morning Session: Nucleosynthesis **Chair: H. Ejiri**	
9:00	*B. Fields*: Primordial Nucleosynthesis in the New Cosmology
9:55	*K.-L. Kratz*: The astrophysical r-process: From B2FH via FK2L to present
10:50 Coffee break	
11:10	*H. Schatz*: X-ray bursts and superbursts
12:15 Lunch	
Round table: Science Education and the Public Understanding of Science ***Moderator: L. Trache***	
14:00	*A. Wolfendale*: The Public Understanding of Science: "The Wolfendale Report" *R. Tribble:* A view point from across the ocean *A. Haungs:* The road to KIT – Karlsruhe Institute of Technology
15:55 Coffee break	
16:10	*M. Miclea:* Some Psychological Remarks for Scientists' Use *P. Biermann:* From my teaching experience on 6 continents and 30 countries; *C. Diaconu:* A high energy perspective of science communication; *Guest interventions: A. Bararu, T. Suomijarvi, V. Popa*

Tuesday, August 28th, 2007	
Morning Session: High-Energy Cosmic Rays **Chair: Y.Y. Wong**	
9:00	*A. Wolfendale*: Cosmic Rays and Cosmology
9:55	*A. Bettini*: Highlights from Gran Sasso
10:50 Coffee break	
11:10	*T. Suomijarvi:* The Pierre Auger Observatory: study of the highest energy cosmic rays
12:05	*G. Navarra*: The physics and first results of KASCADE-Grande
13:00 Lunch	
Afternoon Session **Chair: I. Brancus**	
15:00	*A. Haungs*: Experimental Status of Air Shower Measurements with Radio Antennas: The LOPES project
15:55 Coffee break	
16:10	*Y. Wong:* Neutrinos in cosmology
17:10	*C. Diaconu*: Matter Structure from Deep Inelastic Scattering
18:00	Computation in Nuclear, Astro- and High-Energy Physics (*Diaconu-convenor)* Communications: *Rizea, L. Popa*

Wednesday, August 29th, 2007	
Morning Session **Chair: G. Navarra**	
9:00	*S. Grevy*: A "revisited view" of the shell structure and the magic number
9:55	*P. Biermann*: The nature of dark matter
10:50 Coffee break	
11:10	*A. Wolfendale*: Cosmic Rays and Global Warming
12:05	*H. Rebel*: The muon charge ratio in cosmic ray air showers
13:00 Lunch	
Afternoon Session **Chair: H. Rebel**	
15:00	*G. Wrochna*: Multimessenger observations of GRB and other rapid astrophysical processes
15:55 Coffee break	
16:10	Communications: *Garzelli, Toma, Meli, P.G. Isar, Saftoiu, Medinaceli, Mitrica*

Thursday, August 30th, 2007	
Morning Session **Chair: M. Horoi**	
9:00	*H. Ejiri*: Majorana Neutrinos by Double β Decays and MOON s
9:55	*J. Vary*: Realistic NN and NNN interactions and the nuclear single-particle basis
10:50 Coffee break	
11:10	*G. Giacomelli:* Atmospheric neutrinos and long baseline experiments
12:05	*J. Bonn*: The direct determination of the neutrino mass
13:00 Lunch	
Afternoon Session: **Chair: J. Vary**	
15:00	*T.S. Kosmas:* Neutrino-nucleus interactions at low and intermediate neutrino-energies
15:55 Coffee break	
16:10	*C. Borcea: Studies with exotic nuclei: two-proton radioactivity*
17:05	Communications: *A. Isar, Luczak, Pavalas, Mirea, Joulaeizadeh*

19:30 Banquet – restaurant "La Portile Regatului" in Pârâul Rece

Friday August 31st, 2007	
Morning Session: Closing session **Chair: S. Stoica**	
9:00	*N. Grama*: Exotic resonant states whose wave functions are localized in the barrier region
9:55	*J. Vary:* Effective interactions and operators + recent progress in the ab-initio No Core Shell Model
10:50 Coffee break	
11:10	**CLOSING Remarks:** *L. Trache, M. Horoi, H. Rebel, S. Stoica*
12:00 Lunch	
am: Departure from Sinaia to Bucharest	

PART I: INVITED LECTURES

1. ANNIVERSARY MINI-SYMPOSIUM HONORING PROFESSOR ROBERT E. TRIBBLE

Message of a (Proud) Mentor

Gerald T. Garvey[a,b]

[a]*then, Robert E. Tribble's PhD advisor at Princeton University*
[b]*now at Los Alamos National Laboratory, Los Alamos, New Mexico, USA*

Good morning,

I hoped to be with you all at this most interesting summer school in Romania, but my travel plans, unfortunately, would not allow it. I want to thank Livius Trache for the opportunity to present these personal remarks recalling my interaction with Bob Tribble over the years. First of all Bob, congratulations on having reached your 60th year, you have clearly made extraordinarily good use of your time.

I first encountered Bob as a young grad-student at Princeton in 1969. It was hard to evaluate how a young fellow from Missouri was going to fare in the heady environment of the Princeton Physics Department in those days. It did not appear to trouble Bob very much as he sailed through all the requirements and tackled a very difficult thesis problem. At that time, 1971, it was an accumulation of evidence pointing to the existence of large second-class currents. Bob's thesis work was one of the first experiments suggesting that this was unlikely, a point proven correct by later experimental work. Bob received his PhD in 1973.

By 1979, he had been made chair of the Physics Department at Texas A&M University. That surprised me a good deal, but it was a clear and early recognition of his good judgment and leadership qualities.

Our paths crossed again in the late '80's while I was the director of the Los Alamos Meson Physics Facility and Bob, along with Carl Gagliardi, came to Los Alamos to work on the MEGA project. This was an experiment to search for the standard model violating process where a muon decays to an electron plus gamma. MEGA proved to be an extremely difficult experiment, but Bob and Carl stuck it out and provided much of the talent required to bring MEGA to a useful conclusion. I was very impressed by their role.

I should add that with the lessons learned at MEGA, Bob and Carl joined the TWIST collaboration at the TRIUMF facility in Vancouver and carried off a precise measurement of the allowed muon decay where the muon goes in to electron plus 2 neutrinos.

Next, in the 90s, Bob joined our Drell-Yan experiment at Fermi Lab, E866, which was to measure the $\bar{d}/\bar{u}$ asymmetry in the nucleon sea. It was an extremely successful endeavor, and demonstrated the sizeable asymmetry in the nucleon sea. It was a surprise to most high-energy physicists, but quite natural to nuclear physicists who knew that a proton was often a neutron plus a positive pion. Bob's student, Eric Hawker won the 1999 dissertation prize from the Division of Nuclear Physics [of the

CP972, *Exotic Nuclei and Nuclear/Particle Astrophysics (II)—Proceedings of the Carpathian Summer School of Physics 2007,* edited by L. Trache and S. Stoica

American Physical Society] for the outstanding thesis of 1999 for his work in the write-up of E866.

More recently, Bob addressed the National Academy of Science's RISAC Committee, of which I was a member, on the work of his important committee called "Guidance for Implementing the NSAC Long Range Plan - 2005". This extremely sensitive report was occasioned by the precipitous decline of federal support for nuclear physics in the fiscal year 2006. His report to RISAC, and the physical report generated by the committee, played a significant role in helping RISAC reach its findings and conclusions.

Most recently, Bob has been made chair of NSAC (*Nuclear Science Advisory committee – editors note*), a high honor, and well deserved on his part. It reflects the judgment of the responsible federal officials and of those in the nuclear physics community, on the breadth of his knowledge, the fairness of his judgment and the soundness of his character. We are lucky to have Bob Tribble in our community and I feel honored to have played a role in his career!

Congratulations Bob!

Fundamental Interactions, Nuclear Masses, Astrophysics, and QCD

C. A. Gagliardi

Cyclotron Institute, Texas A&M University, College Station, TX, 77843, USA)

Abstract. During his long and varied career, Robert Tribble has made important contributions in many areas of nuclear physics. He has set new limits on the existence of second-class currents, lepton-flavor violation, and right-handed interactions. He optimized the use of the (^{4}He,^{8}He) reaction to determine nuclear masses and study charge-dependent effects in nuclei. He has developed a new indirect procedure to determine astrophysical reaction rates and applied it to study important nuclear reactions that occur in our sun, in massive stars, and in novae. He has explored anti-quark distributions in nucleons and nuclei, and the polarization of gluons in the nucleon. A brief overview of Bob Tribble's many accomplishments is presented.

Keywords: Second-class currents; right-handed interactions; nuclear masses; antiquark distributions; gluon polarization; asymptotic normalization coefficients.
PACS: 23.40.Bw, 13.35.Bv, 21.10.Dr, 24.85.+p, 26.20.Cd.

INTRODUCTION

During his long and varied career, Robert Tribble has made important contributions in many areas of nuclear physics. His dissertation research involved a search for second-class currents in nuclear beta decay, and he has continued to study fundamental interactions ever since then. He optimized the study of the (^{4}He,^{8}He) reaction as a tool to measure nuclear masses and structure. He has measured anti-quark distributions in nucleons and nuclei and gluon polarization in the nucleon. He proposed the asymptotic normalization coefficient approach to determine the rates of direct radiative capture reactions at astrophysical energies, and has applied it to important nuclear reactions that occur in many different astrophysical environments. As part of the celebration of Bob Tribble's 60th birthday, this article provides a brief overview of his many accomplishments.

FUNDAMENTAL INTERACTIONS

Study of the fundamental interactions has been a recurring theme over the course of Bob Tribble's career. In the early 1970's, his dissertation research involved a search for second-class currents in nuclear beta decay. During the 1980's and '90's, he searched for the lepton-flavor violating decay $\mu \to e\gamma$. For the past decade, he has performed a sensitive search for right-handed interactions in normal muon decay.

CP972, *Exotic Nuclei and Nuclear/Particle Astrophysics (II)—Proceedings of the Carpathian Summer School of Physics 2007,* edited by L. Trache and S. Stoica

The nucleon is a complex object. Therefore, the weak nucleon-nucleon current includes recoil-order contributions that are forbidden for free quarks. A well-known example is the weak magnetism term. Weinberg [1] characterized the recoil-order terms according to their behavior under *G*-parity. He called a vector (axial-vector) contribution that transforms in the same way as the normal Fermi (Gamow-Teller) current "first class", and one that transforms with the opposite sign as "second class". The most general nucleon-nucleon weak current includes a second-class, axial vector term, the "induced tensor" contribution. Renormalizable gauge theories have difficulty including second-class currents, so it is important to explore whether or not they exist in nature.

Bob Tribble's dissertation research was a search for second-class currents in the beta-alpha angular correlation in mass 8 [2,3]. The angular correlation takes the form $P(\theta) = 1 + a_{\pm} \cos(\theta) + p_{\pm} \cos^2(\theta)$, where the + or - sign denotes a β^+ or β^- decay. The difference $(p_- - p_+)$ between the correlations observed in analog beta decays is sensitive to the weak magnetism and second-class induced tensor terms. Prior to Bob Tribble's work, evidence had been mounting that second-class currents are present in beta decay. In contrast, his result, shown in Fig. 1, was consistent with the angular correlation expected if only weak magnetism contributed. Subsequent research [4] has limited the second-class induced tensor contribution in mass 8 to be no more than ~15% of weak magnetism. In later years, Bob Tribble also measured beta-gamma correlations [5,6] to set limits on second-class currents in the mass 20 system.

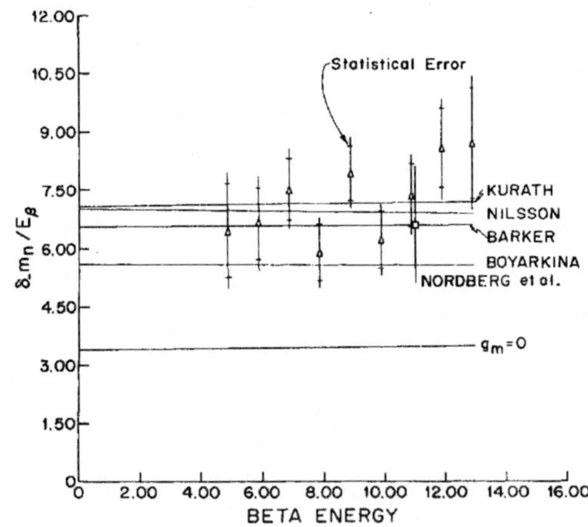

FIGURE 1. Measurement of the difference between the beta-alpha angular correlations in ^{8}Li and ^{8}B. The results are in good agreement with calculations that include weak magnetism, but no second-class induced tensor contribution. This figure is taken from Ref. [3].

The lepton-flavor violating decay $\mu \rightarrow e\gamma$ has been a benchmark for our understanding of the weak interaction. Prior to the discovery of neutrino oscillations, the Standard Model conserved lepton flavor and, thus, forbid $\mu \rightarrow e\gamma$ decay. However, $\mu \rightarrow e\gamma$ is permitted in essentially all extensions of the Standard Model, including most supersymmetric theories. Bob Tribble was one of the original members of the MEGA

Collaboration to search for $\mu \rightarrow e\gamma$ in an experiment at the Los Alamos Meson Physics Facility. The signature for $\mu \rightarrow e\gamma$ is the observation of a 52.83 MeV positron back-to-back and in time coincidence with a 52.83 MeV photon. The primary background arises from random coincidence of a positron from normal muon decay with a photon from a $\mu \rightarrow e\gamma\nu\nu$ decay. MEGA utilized a high-precision, low-mass positron spectrometer, surrounded by a set of three concentric, cylindrical photon pair spectrometers. Texas A&M and the University of Houston were responsible for design, construction, operation, simulation, and analysis of the photon detection system. Notably, Bob Tribble played a key role in the physics design of the MEGA photon pair spectrometers, he developed the pattern recognition algorithms that identified the high-energy photon events, and he performed the final-pass hand-scan of the entire candidate photon data set. He also served as the editor for the final MEGA archival publication [7]. MEGA found that the branching ratio for $\mu \rightarrow e\gamma$ is less than 1.2×10^{-11} with 90% confidence.

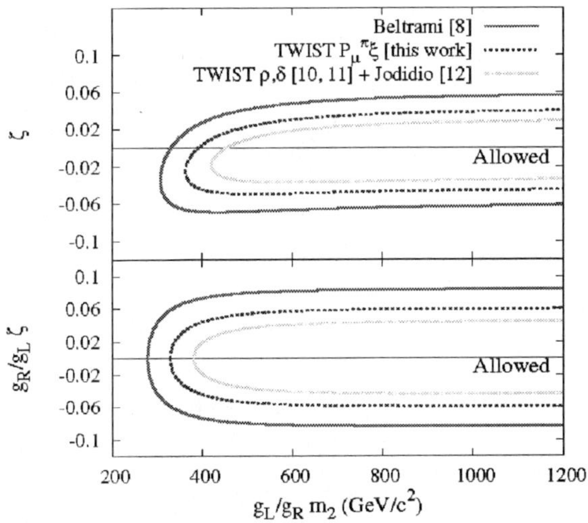

FIGURE 2. Limits on the parameters in left-right symmetric models derived from the TWIST Michel parameter measurements. The upper panel shows the limits for manifest left-right symmetric models, and the lower panel shows the limits for general left-right models. This figure is taken from Ref. [10].

The energy and angle distributions of the positrons following polarized muon decay may be characterized in terms of the "Michel parameters". The parameters ρ and η describe the momentum-dependence of the unpolarized spectrum, while ξ and δ describe the amplitude and momentum-dependence of the polarized spectrum. These parameters take on precise values in the Standard Model: $\rho = 3/4$, $\eta=0$, $\xi=1$, and $\delta=3/4$. ξ is always measured in the product $P_\mu\xi$, where P_μ is the polarization of the muon in pion decay and equals unity in the Standard Model. Deviations from these values require new physics, notably involving right-handed interactions. Bob Tribble was an original member of the TWIST Collaboration to provide order-of-magnitude improvements in our knowledge of ρ, δ, and $P_\mu\xi$ in an experiment at the TRIUMF cyclotron.

TWIST is the first experiment to ever measure the Michel parameters using a blind analysis. The TWIST spectrometer consists of a stack of 44 ultra-thin planar drift chambers in a uniform 2-T solenoidal magnetic field. Muons from pion decay are tracked to a thin (75-μm Al) target in the center of the detector stack, where they stop and decay. The outgoing positron helices are then measured. The measured 2-dimensional momentum-angle spectrum is then compared to expectations based on a detailed Monte Carlo simulation of the detector. To date, TWIST has published first results for ρ [8], δ [9], and $P_\mu\xi$ [10] that are factors of 2-3 more precise than the previous Particle Data Group evaluations. Figure 2 indicates the region of parameter space in left-right symmetric models that is allowed by the published TWIST results. Within the past month, TWIST has also announced new results for ρ and δ [11] that are another factor of 2 more precise than the values that were obtained in [8,9].

NUCLEAR MASSES

When Bob Tribble arrived at Texas A&M, he initiated a program of precision nuclear mass measurements utilizing the (^{4}He,^{8}He) reaction. He developed a detector system for the Enge spectrometer that achieved excellent background rejection. The background level in his first experiment was <0.5 nb/(sr MeV) [12], as shown in Fig. 3. Ultimately, he achieved a background level of <0.1 nb/(sr MeV).

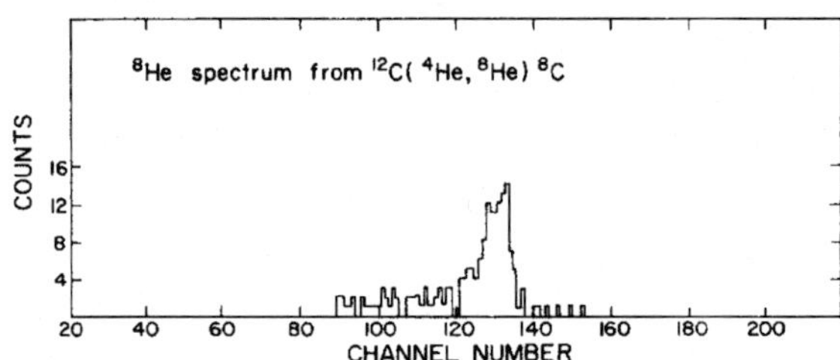

FIGURE 3. Measurement of the mass of ^{8}C, using the ^{12}C(^{4}He,^{8}He)^{8}C reaction. This figure is taken from Ref. [12].

Bob Tribble used the (^{4}He,^{8}He) reaction to study the T=2 nuclei ^{8}He [13], ^{8}C [12], ^{16}Ne [14], ^{20}Mg [15], ^{24}Si [16], and ^{36}Ca [17]. In all cases, he compared his mass measurement to predictions based on the quadratic isobaric multiplet mass equation (IMME) to search for evidence of nuclear charge-dependent interactions. The mass 8 system showed a ~2.5 sigma deviation from the quadratic IMME. Approximately 1 sigma effects were observed in masses 16 and 20. The mass 24 and 36 systems were consistent with the quadratic IMME. However, it was found that the observed Coulomb energies in mass 36 were better reproduced when a charge-dependent interaction that obeys the IMME was included.

At the same time that he measured the mass of ^{8}He, Bob Tribble studied the angular distribution of the ^{64}Ni(^{4}He,^{8}He)^{60}Ni reaction [18]. He found that it was better described by a four-nucleon cluster transfer calculation than by a ^{4}He → ^{6}He → ^{8}He two-step process. This indicates that the (^{4}He,^{8}He) reaction has a direct reaction component.

Bob Tribble did not restrict his mass measurements to the (^{4}He,^{8}He) reaction. In a foreshadowing of his future interests in nuclear astrophysics, he measured the masses of ^{61}Fe and ^{62}Fe [19] in order to gain a better understanding of *r*-process nucleosynthesis and the mass of ^{57}Cu [20] to predict the rate of the important *rp*-process reaction ^{56}Ni(p,γ)^{57}Cu.

It is striking that 8 of the 9 nuclear masses that Bob Tribble measured 20-30 years ago still fall within 1 sigma of the most recent accepted values. The only one that doesn't is ^{24}Si, which is only 1.2 standard deviations away. This is a clear testament to the care that Bob Tribble applies in his research.

QUANTUM CHROMODYNAMICS

For many years, people assumed that the quark-antiquark "sea" of the nucleon arises from gluon splitting ($g \rightarrow u\bar{u}, g \rightarrow d\bar{d}$), leading to flavor symmetry. In the early 1990's, the New Muon Collaboration reported measurements of deep-inelastic muon scattering off hydrogen and deuterium targets [21] that implied $\bar{d} > \bar{u}$ in the proton. Bob Tribble was an original member of the Fermilab E866/NuSea Collaboration that measured Drell-Yan dimuon production on hydrogen and deuterium targets to obtain the first determination of the momentum dependence of $\bar{d}/\bar{u}$.

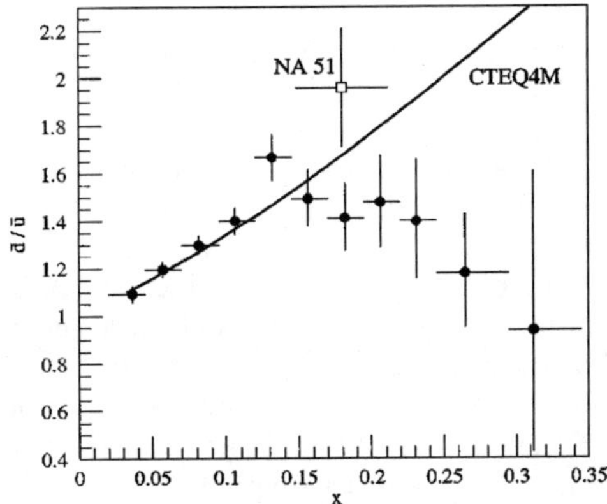

FIGURE 4. First determination of the $\bar{d}/\bar{u}$ ratio in the proton by FNAL E866/NuSea. This figure is taken from Ref. [22].

Eric Hawker, Bob Tribble's graduate student, performed the first FNAL E866/NuSea measurement of $\bar{d}/\bar{u}$ [22] for his dissertation research. Figure 4 shows some of the results that he found. The anti-quark asymmetry was found to increase as a function of anti-quark momentum fraction up to $x \sim 0.17$ and then decrease, in striking contradiction to the previous theoretical expectations that it would be monotonic. Eric Hawker received the 1999 APS Division of Nuclear Physics Thesis Prize in recognition for this work.

The modification of the anti-quark density in a nucleus and the energy loss of partons in cold nuclear matter can be probed by comparing the Drell-Yan yield per nucleon on heavy vs. light targets. Texas A&M led the FNAL E866/NuSea measurement of the nuclear dependence of Drell-Yan dimuon production on Be, Fe, and W targets [23]. The observed nuclear dependence was found to be consistent with expectations based on fits to shadowing in deep-inelastic lepton scattering off of nuclei, with little room for additional nuclear dependence due to parton energy loss.

More recently, Bob Tribble has joined the STAR Collaboration at Brookhaven National Laboratory, where one of the major goals is a measurement of the gluon polarization in the proton. The longitudinal double-spin asymmetry A_{LL} for $p + p \rightarrow \gamma$ + jet + X is the "golden channel" to determine gluon polarization. Over 90% of the cross section for this reaction arises from quark-gluon Compton scattering, which gives it a large and well-known gluon analyzing power. Furthermore, the observation of the γ + jet final state allows reconstruction of the incident parton kinematics at leading order. Texas A&M participated in the design, construction, and assembly of the STAR Endcap Electromagnetic Calorimeter (EEMC) [24]. The EEMC covers the kinematic region where detected photons provide the greatest sensitivity to gluon polarization, making it a critical component for the measurement, which will be performed over the next two years.

ASTROPHYSICS

Perhaps Robert Tribble's most notable contribution to nuclear physics, and certainly his contribution that is most familiar to many of the participants of this meeting, has been the asymptotic normalization coefficient technique. Nuclear reactions in astrophysical environments occur at very low energies. The cross sections are extremely small and dominated by the existence of the Coulomb barrier. This makes direct measurements in the laboratory very challenging, especially if one of the participating nuclei is radioactive. Hence, indirect techniques play a key role in determining the rates of nuclear reactions of astrophysical interest.

Due to the Coulomb barrier, direct proton radiative capture reactions typically occur at very large radii in astrophysical environments. For example, at solar energies over 99% of ^{7}Be(p,γ)^{8}B captures occur at radii greater than 5 fm. At low energies, the p + nucleus scattering wave function is a simple Coulomb scattering wave. Likewise, the shape of the tail of the nuclear overlap function in the classically forbidden region is well known; it follows the Whittaker function. Thus, if you know the amplitude of the tail of the nuclear overlap function, which is given by the asymptotic normalization coefficient (ANC), you can calculate the astrophysical S factor at low energies. The relationship between the ANC and the astrophysical S factor has been known for many years. However, it wasn't exploited because we didn't have a reliable general method to measure ANCs.

In 1994, Robert Tribble proposed that asymptotic normalization coefficients could be determined from the cross sections for peripheral nucleon transfer reactions [25] if the reactions and beam energies are chosen carefully. Initially, the nuclear astrophysics community was skeptical, emphasizing the need to test the ANC technique before it could be adopted [26]. The first test was performed in 1999 [27].

The angular distribution for the $^{16}O(^{3}He,d)^{17}F$ reaction was measured in a pair of experiments, one at the Nuclear Physics Institute of the Czech Republic and the other at Texas A&M. The ANCs for $^{17}F \leftrightarrow {}^{16}O + p$ were determined from a comparison of the cross section to DWBA calculations. Those ANCs were then used to calculate the astrophysical S factor for $^{16}O(p,\gamma)^{17}F$. Figure 5 shows the comparison of the predictions to direct measurements that had been performed independently [28]. The agreement demonstrates that the ANCs can be used to predict the direct capture rate to better than 10%. Several additional tests have now demonstrated the validity of the ANC technique in other nuclear systems.

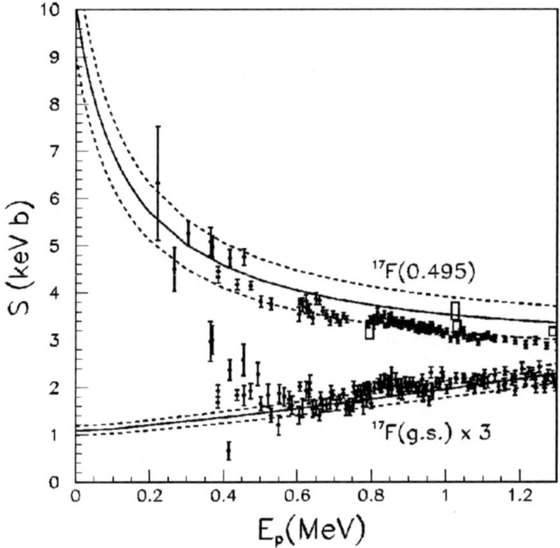

FIGURE 5. A comparison of the astrophysical S factor for $^{16}O(p,\gamma)^{17}F$ calculated from the measured ANCs (curves) to direct measurements (points). This figure is taken from Ref. [27].

Our group at Texas A&M then undertook a sequence of measurements to apply the ANC technique to determine the astrophysical S factor for $^{7}Be(p,\gamma)^{8}B$, a key reaction for solar neutrino studies. The initial determination focused on a measurement of the ANC for $^{8}B \leftrightarrow {}^{7}Be + p$ with the $^{10}B(^{7}Be,^{8}B)^{9}Be$ reaction, using an 84 MeV ^{7}Be radioactive beam that was produced using the Texas A&M recoil separator MARS [29,30]. Figure 6 shows the angular distribution that was obtained [31]. $^{10}B + {}^{7}Be$ elastic scattering data were measured simultaneously in order to verify the optical model parameters that were adopted for a DWBA calculation. The curve shows that DWBA calculation. It has been normalized to the experimental data to determine the ANC, $C^2 = 0.460 \pm 0.072$ fm^{-1}, which implies the astrophysical S factor for $^{7}Be(p,\gamma)^{8}B$ is $S_{17}(0) = 17.8 \pm 2.8$ eV b.

The ANC method has now become a standard tool utilized by nuclear astrophysicists in laboratories throughout the world. It has been used to explore astrophysical reactions that occur in the sun, in various explosive stellar environments, and in first-generation stars. Alternative techniques to determine ANCs using peripheral nuclear break-up reactions [32] and through measurements of mirror nuclear transitions [33] have been developed. The ANC has also been used to determine the radius of halo nuclei [34]. A sample of recent work involving the ANC technique may be found in several other talks presented at this meeting [35,36,37,38,39].

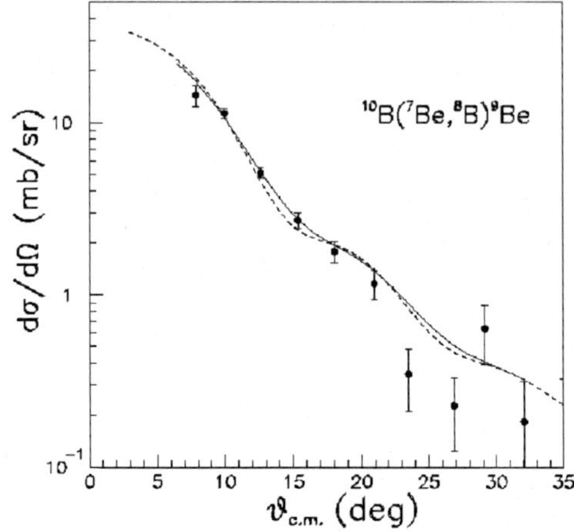

FIGURE 6. The angular distribution for $^{10}B(^{7}Be,^{8}B)^{9}Be$ that was used in the first determination of the ANC for $^{8}B \leftrightarrow {}^{7}Be + p$. This figure is taken from Ref. [31].

CONCLUSION

This review has focused on Robert Tribble's research activities. However, it is important to note that he has also played a major role as an administrator. He served has Head of the Texas A&M University Physics Department during the period 1979-87. Under his leadership, the Department experienced significant growth. More recently, he has served as Director of the Texas A&M Cyclotron Institute, which is undergoing a major upgrade at present [40]. Over the years, Bob Tribble has also made very significant contributions to the evolution of nuclear science in the United States. Currently, he is the Chair of the U.S. DOE/NSF Nuclear Science Advisory Committee (NSAC), and is leading the NSAC effort to develop a new 2007 Long-Range Plan for Nuclear Science.

During his career, Robert Tribble has made a major impact on several different areas of nuclear physics. We look forward to his continued accomplishments for many years to come.

ACKNOWLEDGMENTS

The author acknowledges Robert Tribble for over twenty-five years of close collaboration and friendship. This work was supported in part by the U.S. Department of Energy.

REFERENCES

1. S. Weinberg, *Phys. Rev.* **112**, 1375 (1958).
2. R.E. Tribble and G.T. Garvey, *Phys. Rev. Lett.* **32**, 314 (1974).
3. R.E. Tribble and G.T. Garvey, *Phys. Rev.* C **12**, 967 (1975).
4. R.D. McKeown, G.T. Garvey, and C.A. Gagliardi, *Phys. Rev.* C **22**, 738 (1980).
5. R.E. Tribble and D.P. May, *Phys. Rev.* C **18**, 2074 (1978).
6. R.E. Tribble, D.P. May, and D.M. Tanner, *Phys. Rev.* C **23**, 2245 (1981).
7. M. Ahmed *et al.* (MEGA Collaboration), *Phys. Rev.* D **65**, 112002 (2002).
8. J.R. Musser *et al.* (TWIST Collaboration), *Phys. Rev. Lett.* **94**, 101805 (2005).
9. A. Gaponenko *et al.* (TWIST Collaboration), *Phys. Rev.* D **71**, 071101 (2005).
10. B. Jamieson *et al.* (TWIST Collaboration), *Phys. Rev.* D **74**, 072007 (2006).
11. R. MacDonald (for the TWIST Collaboration), TRIUMF colloquium, Nov. 15, 2007.
12. R.E. Tribble, R.A. Kenefick, and R.L. Spross, *Phys. Rev.* C **13**, 50 (1976).
13. R.E. Tribble, J.D. Cossairt, D.P. May, and R.A. Kenefick, *Phys. Rev.* C **16**, 1835 (1977).
14. C.J. Woodward, R.E. Tribble, and D.M. Tanner, *Phys. Rev.* C **27**, 27 (1983).
15. R.E. Tribble, J.D. Cossairt, and R.A. Kenefick, *Phys. Lett.* **61B**, 353 (1976).
16. R.E. Tribble, D.M. Tanner, and A.F. Zeller, *Phys. Rev.* C **22**, 17 (1980).
17. R.E. Tribble, J.D. Cossairt, and R.A. Kenefick, *Phys. Rev.* C **15**, 2028 (1977).
18. R.E. Tribble, J.D. Cossairt, K.-I. Kubo, and D.P. May, *Phys. Rev. Lett.* **40**, 13 (1978).
19. R.E. Tribble, J.D. Cossairt, and R.A. Kenefick, *Phys. Rev.* C **15**, 1685 (1977).
20. C.A. Gagliardi, D.R. Semon, R.E. Tribble, and L.A. Van Ausdeln, *Phys. Rev.* C **34**, 1663 (1986).
21. P. Amaudruz *et al.* (New Muon Collaboration), *Phys. Rev. Lett.* **66**, 2712 (1991).
22. E.A. Hawker *et al.* (FNAL E866/NuSea Collaboration), *Phys. Rev. Lett.* **80**, 3715 (1998).
23. M.A. Vasiliev *et al.* (FNAL E866/NuSea Collaboration), *Phys. Rev. Lett.* **83**, 2304 (1999).
24. C.E. Allgower *et al.*, *Nucl. Instrum. and Meth.* **A499**, 740 (2003).
25. H.M. Xu *et al.*, *Phys. Rev. Lett.* **73**, 2027 (1994).
26. E.G. Adelberger *et al.*, *Rev. Mod. Phys.* **70**, 1265 (1998).
27. C.A. Gagliardi *et al.*, *Phys. Rev.* C **59**, 1149 (1999).
28. R. Morlock *et al.*, *Phys. Rev. Lett.* **79**, 3837 (1997).
29. R.E. Tribble, R.H. Burch, and C.A. Gagliardi, *Nucl. Instrum. and Meth.* **A285**, 441 (1989).
30. R.E. Tribble, C.A. Gagliardi, and W. Liu, *Nucl. Instrum. and Meth.* **B56-57**, 956 (1991).
31. A. Azhari *et al.*, *Phys. Rev. Lett.* **82**, 3960 (1999).
32. L. Trache, F. Carstoiu, C.A. Gagliardi, and R.E. Tribble, *Phys. Rev. Lett.* **87**, 271102 (2001).
33. L. Trache *et al.*, *Phys. Rev.* C **67**, 062801 (2003).
34. F. Carstoiu *et al.*, *Phys. Rev.* C **63**, 054310 (2001).
35. L. Trache, these proceedings.
36. F. Carstoiu, these proceedings.
37. V. Kroha, these proceedings.
38. A. Banu, these proceedings.
39. T. Al-Abdullah, these proceedings.
40. G. Tabacaru, these proceedings.

2. LOOK INTO THE FUTURE

The Frontiers of Nuclear Science: A New Long Range Plan for the United States

Robert E. Tribble

Cyclotron Institute, Texas A&M University, College Station, TX 77843, USA

Abstract. The U.S. Nuclear Science Advisory Committee has developed a new Long Range Plan to guide the program over the next decade. The new plan recommends a program of upgrades and new construction to move the U.S. nuclear physics program into the next decade. An overview of the plan is provided.

Keywords: nuclear physics planning
PACS: 20.

INTRODUCTION

The U.S. Nuclear Science Advisory Committee (NSAC) was formed in 1977 as an Advisory Committee to provide guidance to the Department of Energy (DOE) and the National Science Foundation (NSF) on the national nuclear science program. Prior to the formation of NSAC, there existed no formal mechanism to advise DOE and NSF on future developments in the field. Rather reviews by the National Academy of Science and panels set up by the agencies were used to make recommendations about future directions. Soon after NSAC was formed, it was asked to develop a long range plan for the field. The first plan, which was submitted in 1979, documented a number of new initiatives that involved upgrades of existing accelerators and the construction of a new facility. An outcome of the first plan was the construction of an electron accelerator and the development of a new national facility—The Thomas Jefferson National Accelerator Laboratory (JLab). A new Long Range Plan (LRP) was developed by NSAC a few years later and was submitted near the end of 1983. The new plan—the first of the modern LRPs—gave a comprehensive overview of the field of nuclear science in the U.S. It included a description of the science, an overview of the accelerator facilities and instrumentation, a discussion of the applications arising from nuclear science and the connections to other fields, information on graduate education and the workforce, an overview of international cooperation and a section discussing the budgets needed to implement the plan. The 1983 LRP recommended the construction of the Relativistic Heavy Ion Collider (RHIC) as its top priority.

In 1988, a request was submitted to NSAC to develop a new LRP. A series of Town Meetings were organized by NSAC in order to get community input into the plan. Summaries from the Town Meetings were generated and used by a Working Group—50 members of the nuclear science community, including the members of NSAC—to develop the recommendations. The top two recommendations from the

CP972, *Exotic Nuclei and Nuclear/Particle Astrophysics (II)—Proceedings of the Carpathian Summer School of Physics 2007,* edited by L. Trache and S. Stoica

1989 plan were to complete the construction of the Continuous Electron Beam Accelerator Facility at JLab and to initiate the construction of RHIC. The process used to develop the 1989 LRP was formalized for the 1996 plan. In preparation for the 1996 plan, the American Physical Society's Division of Nuclear Physics (DNP) sponsored Town Meetings for the different areas of nuclear science. White Papers were generated from the Town Meetings and were used as input to a Working Group of 65 members who developed the recommendations and wrote the LRP. The top recommendations in the 1996 plan were to complete the construction of RHIC and to utilize RHIC and CEBAF in order to exploit their scientific potential. The most recent LRP was completed in 2002. The process followed a similar path as for the 1996 plan with Town Meetings set up by the DNP and White Papers produced from those Town Meetings. A 56 member Working Group developed the recommendations for the 2002 plan and wrote the report. The top recommendation in the 2002 LRP was to effectively utilize the U.S. flagship facilities, CEBAF, NSCL and RHIC. The second recommendation was to construct the Rare Isotope Accelerator (RIA).

In a letter dated July 17, 2006, the Department of Energy (DOE) and the National Science Foundation (NSF) asked NSAC to 'conduct a study of the opportunities and priorities for United States nuclear physics research and recommend a long range plan that will provide a framework for coordinated advancement of the Nation's nuclear science research programs over the next decade.' This request set in motion a bottom up review and forward look by the nuclear science community following a format similar to that used in developing the 2002 plan. With input from the community-wide Town Meeting process, a 59 member working group, which included the present NSAC members, gathered at the beginning of May, 2007, for a week-long meeting to develop guidance on how to optimize the future research directions for the field based on the projected resources outlined in the charge letter from DOE and NSF. A new long range plan—The Frontiers of Nuclear Science—grew out of this meeting.

Building on the foundation of the past two decades, nuclear science in the U.S. today is focused on three broad but highly related research frontiers: (1) QCD and its implications and predictions for the origin of matter in the early universe, quark confinement, the role of gluons and the structure of the proton; (2) the structure of atomic nuclei and nuclear astrophysics, which addresses the origin of the elements, the structure and limits of nuclei, and the evolution of the cosmos; and (3) a new standard model, and its implications for the origin of matter and the properties of neutrinos and other subatomic particles.

Each nuclear science frontier is guided by overarching questions which serve to define it. These questions are listed here.

Quantum Chromodynamics

- **What are the phases of strongly interacting matter, and what roles do they play in the cosmos?**
- **What is the internal landscape of the nucleons?**
- **What does QCD predict for the properties of strongly interacting matter?**
- **What governs the transition of quarks and gluons into pions and nucleons?**

- **What is the role of gluons and gluon self-interactions in nucleons and nuclei?**
- **What determines the key features of QCD, and what is their relation to the nature of gravity and spacetime?**

Nuclei and Nuclear Astrophysics

- **What is the nature of the nuclear force that binds protons and neutrons into stable nuclei and rare isotopes?**
- **What is the origin of simple patterns in complex nuclei?**
- **What is the nature of neutron stars and dense nuclear matter?**
- **What is the origin of the elements in the cosmos?**
- **What are the nuclear reactions that drive stars and stellar explosions?**

Fundamental Symmetries and Neutrinos

- **What is the nature of the neutrinos, what are their masses, and how have they shaped the evolution of the universe?**
- **Why is there now more visible matter than antimatter in the universe?**
- **What are the unseen forces that were present at the dawn of the universe, but disappeared from view as it evolved?**

RECENT ACCOMPLISHMENTS

Since the last Long Range Plan was written, many significant new discoveries have been made in nuclear science. Below are just a few of those that are highlighted in the 2007 LRP.

Near-Perfect Liquid

In one of the most surprising discoveries of the past few years, the hot matter that is formed when gold nuclei accelerated by the Relativistic Heavy Ion Collider (RHIC) collide nearly head on seems to expand like an almost perfect liquid. A property used to characterize liquids—the ratio of the viscosity to entropy density—is smaller than for any other known liquid. It is so small, in fact, that it appears to be very close to a lower limit recently derived from string theory, indicating that the quark-gluon plasma formed in the collisions at RHIC is a strongly-coupled plasma and not a dilute gaseous plasma as originally expected. One of the most important scientific challenges for the next decade is a quantitative exploration of this new state of matter.

Spin of the Nucleon

In the early days of nuclear science, we found that protons and neutrons have an intrinsic angular momentum, or spin, of 1/2. Nearly forty years ago we discovered that our nuclear building blocks are themselves composite particles, made up of two types, or flavors, of quarks—up (u) and down (d)—that also have spin 1/2, and that are held together by spin-1 gluons. So how does the nucleon get its spin from its constituents? Recent measurements at the Thomas Jefferson Laboratory (JLab) and DESY in Germany have shown that one of the two flavors of quarks in the proton has its spin aligned along the polarization direction, while the other has its spin aligned in the opposite direction. When these quark spins are added together, however, , they account for only about 25% of the proton's total spin. Very recent measurements at RHIC indicate that gluons may also contribute very little to the spin. So the solution of the spin puzzle of the nucleon is still not known. Further measurements at RHIC and measurements following the upgrade at JLab will add more information to this fundamental issue.

Jet Quenching

QCD jets occur in all kinds of very-high-energy collisions, arising from the hard scattering of incoming quarks and gluons and their subsequent break up into a characteristic spray of particles that are measured in a detector. Jets have been observed at RHIC in proton-proton, deuteron-gold and gold-gold collisions. But a dramatic change occurs in the jet distributions from gold-gold collisions. When collisions are nearly head-on, the typical pattern of back-to-back jets, which are clearly seen in proton-proton and deuteron-gold collisions, disappears and only one jet is observed. This jet quenching indicates that the scattered quarks and gluons which would produce the second jet undergo large energy loss as they traverse the matter formed in the collision. Recent results further suggest that the energy lost by the high-energy jets may appear as a collective "sonic boom." This effect of jet quenching may open a new way for us to probe the gluon density distribution of the matter—the quark-gluon plasma—formed in these violent collisions.

Charge Distribution of the Neutron

The neutron, as its name implies, is an electrically neutral particle. But the neutron has magnetic properties that are similar to its electrically charged counterpart, the proton, which suggests that its internal charge distribution is quite complex. A sustained effort world-wide over the last decade (including experiments at JLab and the Bates Laboratory at the Massachusetts Institute of Technology) that utilized new polarized beams and targets, has resulted in a much clearer picture of the neutron's charge distribution. The core of the neutron is positively charged. The neutron becomes electrically neutral due to the significant cloud of negative charge produced by virtual mesons that surround the core.

Probing the Limits of Stability

How many neutrons can we add to a stable nucleus before it cannot hold any more? Today we know the answer to this question only for the lightest of the elements. As we push measurements to more neutron-rich isotopes, we learn new features about the strong interaction, which ultimately dictates where this limit, called the neutron drip-line, occurs in nature. Recent measurements of very neutron-rich isotopes of Al and Mg indicate that the drip line for these elements is likely further from the line of stability than previously thought. Mapping the neutron drip line will provide a wealth of information on how the strong interaction saturates. But it is also vital in order for us to fully understand the origin of the elements. Very heavy elements that exist in nature are thought to have formed during stellar explosions in an environment where rapid neutron capture occurs on an element until it approaches the neutron drip line. Through a sequence of neutron captures and nuclear beta decays, the very heaviest elements, such as uranium, can be made. This r-process nucleosynthesis is one of the key astrophysical processes that we believe must occur—but so far, we have little nuclear data to constrain it.

Stellar Reaction Rates

In recent measurements, the $^{14}N(p,\gamma)^{15}O$ reaction cross section that controls hydrogen burning in stars via the CNO cycle has been measured to be about a factor of 2 lower than previously thought. This increases the derived ages for globular clusters and related limits for the age of the Galaxy, by about a billion years. Meanwhile, pioneering measurements of reaction rates with the present generation of stable beam and rare isotope beam facilities have employed new techniques to constrain a number of key reaction rates in nova explosions. This has reduced uncertainties in modeling the hot CNO and NeNa cycles and thus the production of long-lived radioisotopes such as ^{18}F and ^{22}Na—nuclear decays that space-based gamma-ray telescopes look for in novae remnants. In a breakthrough for nuclear theory, ab-initio nuclear calculations are achieving sufficient precision to be able to determine nuclear reaction rates and resonance lifetimes.

Neutrino Oscillations and Neutrino Mass

Just before the 2002 Long Range Plan was published, the SNO collaboration completed the first measurements of both charged- and neutral-current neutrino scattering for solar neutrinos. These results verified, beyond doubt, that neutrinos produced in the sun were changing their character by oscillating into a different type of neutrino. Similar oscillation effects had been found a few years earlier for atmospheric neutrinos. Now recent observations from KamLAND have confirmed that neutrinos produced in a nuclear reactor also undergo oscillations. The fact that neutrino oscillations occur, and therefore neutrinos have mass, has had an enormous impact on both nuclear and particle physics. In just a few years, the neutrino mass mixing matrix has been established and many of the mixing angles are now known. Nuclear scientists are now working to determine the absolute mass scale for neutrinos and to ascertain if neutrinos, unlike any other particle in nature, may be their own anti-particle.

PLANNING FOR THE FUTURE

As nuclear scientists have pushed the boundaries of technology with new accelerators, more powerful computers, and new detectors, we have made major discoveries about the nature of the strong and electroweak interactions. The 2007 LRP focuses on the new science we must confront to improve our understanding of these basic interactions, as well as the tools needed to carry out this program and open up the path for even more discoveries. The plan has four recommendations to guide the future program.

Recommendation I

We recommend completion of the 12 GeV Upgrade at Jefferson Lab. The Upgrade will enable new insights into the structure of the nucleon, the transition be-

tween the hadronic and quark/gluon descriptions of nuclei, and the nature of confinement.

A fundamental challenge for modern nuclear physics is to understand the structure and interactions of nucleons and nuclei in terms of QCD. Jefferson Lab's unique Continuous Electron Beam Accelerator Facility has given the U.S. leadership in addressing this challenge. Doubling the energy of the JLab accelerator will enable three-dimensional imaging of the nucleon, revealing hidden aspects of its internal dynamics. It will complete our understanding of the transition between the hadronic and quark/gluon descriptions of nuclei, and test definitively the existence of exotic hadrons, long-predicted by QCD as arising from quark confinement. Through the use of parity violation, it will provide low-energy probes of physics beyond the Standard Model complementing anticipated measurements at the highest accessible energy scales.

Recommendation II

We recommend construction of the Facility for Rare Isotope Beams, FRIB, a world-leading facility for the study of nuclear structure, reactions and astrophysics. Experiments with the new isotopes produced at FRIB will lead to a comprehensive description of nuclei, elucidate the origin of the elements in the cosmos, provide an understanding of matter in the crust of neutron stars, and establish the scientific foundation for innovative applications of nuclear science to society.

We now have a roadmap to achieve the goal of a comprehensive and unified description of nuclei. New data on exotic isotopes, which only FRIB will provide, are essential as they will allow us to understand the nature of the forces that hold the nucleus together, to assess the validity of the theoretical approximations, and to delineate the path towards integrating nuclear structure with nuclear reactions. Advances in astrophysics and astronomy are driving the need for new and improved information on rare isotopes, including those near the very limits of nuclear stability that will be available for the first time with suitable rates at FRIB, to understand the chemical history of the universe and the synthesis of elements in stellar explosions. Rare isotopes are needed to test the fundamental symmetries of nature, and are essential for the many cross-disciplinary contributions they enable in basic sciences, national security, and societal applications. To launch the field into this new era requires the immediate construction of FRIB with its ability to produce ground-breaking research and effective utilization of current User facilities, NSCL, HRIBF and ATLAS.

Recommendation III

We recommend a targeted program of experiments to investigate neutrino properties and fundamental symmetries. These experiments aim to discover the nature of the neutrino, yet-unseen violations of time-reversal symmetry, and other key ingredients of the new standard model of fundamental interactions. Construction of a Deep Underground Science and Engineering Laboratory is vital to U.S. leadership in core aspects of this initiative.

The discovery of flavor oscillations in solar, reactor, and atmospheric neutrino experiments—together with unexplained cosmological phenomena such as the dominance of matter over anti-matter in the Universe—call for a new standard model of fundamental interactions. Nuclear physicists are poised to discover the symmetries of the new standard model through searches for neutrinoless double beta decay and electric dipole moments, determination of neutrino properties and interactions, and precise measurements of electroweak phenomena.

The Deep Underground Science and Engineering Laboratory (DUSEL) will provide the capability needed for ultra-low background measurements in this discovery-oriented program. Experiments also will exploit new capabilities at existing and planned nuclear physics facilities. Assembling the new standard model using the breadth of new experimental results will require enhanced theoretical efforts.

Recommendation IV
The experiments at the Relativistic Heavy Ion Collider have discovered a new state of matter at extreme temperature and density—a quark-gluon plasma that exhibits unexpected, almost perfect liquid dynamical behavior. We recommend implementation of the RHIC II luminosity upgrade, together with detector improvements, to determine the properties of this new state of matter.

The major discoveries in the first five years at RHIC must be followed by a broad, quantitative study of the fundamental properties of the quark-gluon plasma. This can be accomplished through a ten-fold increase in collision rate, detector upgrades, and advances in theory. The RHIC II luminosity upgrade, using electron cooling, enables measurements using uniquely sensitive probes of the plasma such as energetic jets and rare bound states of heavy quarks. The detector upgrades make important new types of measurements possible while extending significantly the physics reach of the experiments. Achieving a quantitative understanding of the quark-gluon plasma also requires new investments in modeling of heavy ion collisions, in analytic approaches, and in large scale computing.

FURTHER INTO THE FUTURE

Understanding the structure and properties of hadrons in terms of the fundamental theory of strong interactions, QCD, is one of the central thrusts of physics and seeks to provide the underpinning for all of nuclear science. Recent theoretical breakthroughs and experimental results suggest that both nucleons and nuclei, when viewed at high energies, appear as dense systems of gluons, creating fields whose intensity may be the strongest allowed in nature. The emerging science of this universal gluonic matter drives the development of a next-generation facility, the high-luminosity Electron Ion Collider, EIC. Polarized beams in the EIC will give unprecedented access to the spatial and spin structure of gluons in the proton.

An Electron Ion Collider (EIC) with polarized beams has been embraced by the U.S. nuclear science community as embodying the vision for reaching the next QCD frontier. EIC would provide unique capabilities for the study of QCD well beyond those available at existing facilities worldwide and complementary to those planned for the next generation of accelerators in Europe and Asia. While significant progress has been made in the design of EIC, substantial, open questions remain. Realization of an EIC will require advancements in accelerator science and technology, and detector R&D. The nuclear science community has recognized the importance of this future facility and makes the following recommendation.

- **We recommend the allocation of resources to develop accelerator and detector technology necessary to lay the foundation for a polarized Electron Ion Collider. The EIC would explore the new QCD frontier of strong color fields in nuclei and precisely image the gluons in the proton.**

INITIATIVES

As part of the planning process, a number of initiatives were discussed. Three of those merited special consideration by the working group and are considered to be very important to the future of the field.

Nuclear Theory

Experimental and theoretical developments in nuclear science go hand-in-hand, sometimes led by experiments and other times by theory. A very strong theory program is critical to the success of the field. Recent successes in areas such as the exploration of the internal structure and interactions of hadrons, the properties of hot and dense matter, the explosion mechanism of supernovae, the internal structure of nuclei, and the nuclear equation of state have come from the application of large-scale computing to problems in nuclear theory.

Beyond continued support for a healthy base program, we recommend the funding of finite-duration, multi-institutional topical collaborations initiated through a competitive, peer-review process. In addition to focusing efforts on important scientific problems, these initiatives are intended to bring together the best in the field, leverage resources of smaller research groups and provide expanded opportunities for the next generation of nuclear theorists.

Accelerator Research and Development

Much of the experimental research in nuclear science is based at particle accelerators. The future program is tied strongly to developments in accelerator technology. Accelerators also are used in many applications outside of nuclear science. We must support the development of new capabilities that will enable future discoveries. Advances in accelerator science and technology and the ability to attract students to this field will be strongly enhanced by **targeted support of proposal-driven accelerator R&D supported by DOE and NSF nuclear physics.**

Gamma-Ray Tracking

Gamma-ray detector arrays with high efficiency and high resolution have had a major impact on our understanding of nuclear structure over the past decade. But conventional arrays do not maintain their power when the particles undergoing decay move at high speeds. Now a new technology that allows for tracking of gamma rays promises to make the same high resolution, and even higher efficiency, possible for fast particles. A 1π detector array, GRETINA, is being built based on this new technology, and its first element has verified that the device will work as planned. The implementation of a 4π gamma-ray tracking detector, GRETA, based on this new technology will revolutionize gamma-ray spectroscopy and provide sensitivity improvements of several orders of magnitude for studies in nuclear structure, nuclear astrophysics, and weak interactions. Thus **the construction of GRETA should begin upon successful completion of GRETINA. This gamma-ray energy tracking array will enable full exploitation of compelling science opportunities in nuclear structure, nuclear astrophysics, and weak interactions.**

THE PROGRAM TODAY

Realization of the program that follows from implementing the principal recommendations of the 2007 plan will take nearly a decade. During this time, the U.S. must maintain an active program by effectively utilizing its existing facilities and those available elsewhere in the world.

Research and Operations

Progress in nuclear science depends critically on a healthy research program, and the operation of the broad range of facilities needed to carry out the research. Research is carried out by individual faculty members, large university groups and groups at national laboratories. Each of these components plays an important role in the research enterprise. In particular, the integration of research and teaching at universities provides a natural environment for the education and training of the future scientific workforce.

University-based accelerator facilities support a compelling and diverse portfolio of research in nuclear structure, nuclear reactions, nuclear astrophysics and fundamental symmetries. The infrastructure at universities, either in support of operations and research at local accelerator facilities or of participation in development of detectors, instrumentation and equipment for collaborative research projects, enables university scientists to contribute significantly to new initiatives in the field. Support for the university facilities and research infrastructure is essential to enable them to remain significant partners in the nation's scientific endeavor.

Operation of the major user facilities is clearly an essential component of the U.S. nuclear science program. Strategic investments in new and upgraded facilities have positioned the nation in a world leadership role in many areas of nuclear science. Exploiting the extraordinary opportunities for scientific discoveries made possible by these investments, preserving scientific output, and educating young scientists vital to meeting national needs, requires the effective operations of these facilities. Specifically, it is critical to sustain effective facility operations at CEBAF and RHIC in support of their world-leading research in QCD, at levels that also allow facility upgrades to proceed in a timely way, to operate NSCL at optimal levels, taking advantage of its world-leading capabilities with fast rare isotope beams, and to invest in the ATLAS and HRIBF user facilities, with their unique low-energy heavy-ion and rare-isotope beams.

International Collaboration

For many decades, nuclear scientists in the U.S. have sought the best facilities in the world to carry out their research. Some examples of this include measurements done by U.S. researchers studying QCD at high-energy physics facilities such as DESY, FERMILAB and SLAC, experiments carried out on nuclei far from stability at the first rare isotope beam facility, ISOLDE at CERN, experiments involving relativistic heavy-ion beams at the SPS at CERN, and the underground laboratories throughout the world that have housed neutrino detectors. Certainly the U.S. has benefited from the influx of researchers from abroad who have participated, and often led, experiments at U.S. facilities.

As nuclear science facilities become larger and more complex, the costs of construction and operation escalate. It is not only natural but necessary that nuclear scientists maintain a global view of the field. New facilities that will impact nuclear science are coming on-line in Europe and Japan. In Europe the LHC at CERN will soon be observing Pb-Pb collisions in a totally new energy regime. U.S. nuclear scientists already are involved in preparing several of the large detectors for these experiments. Construction has now begun at GSI on the Facility for Antiproton and Ion Research (FAIR). The FAIR facility will open new possibilities for research in rare isotopes and in QCD. Also underway in Europe is an upgrade of the facilities at GANIL in France which will greatly extend the rare isotope research capabilities there. In Japan, the RI Beam Factory (RIBF) at RIKEN, has begun initial operations. As the beam power at RIBF is increased, it will become the premier facility for producing nuclei far from stability until FRIB is completed. Also in Japan, the JPARC facility, which features high-power, high-energy proton beams, is nearing completion. These new facilities, along with existing ones such as ISAC at TRIUMF, provide opportunities for nuclear scientists in the U.S. to carry out research abroad that is not possible with our existing facilities.

Education

Education and outreach are fundamental underpinnings that support the mandate of DOE and NSF to advance the broad interests of society and to help ensure U.S. competitiveness in the physical sciences and technology. The role of educating the next generation workforce in nuclear science is crucial since the U.S. is facing a potentially serious shortage of qualified workers in pure and applied nuclear science research, nuclear medicine, nuclear energy and national security. A key to increasing the number of new nuclear science PhDs is to reach potential students while they are still undergraduates. Outreach by nuclear scientists also is vital. The public, and even scientists in other disciplines, are often uninformed or misinformed about nuclear science and its benefits. To increase the number and diversity of students who pursue a graduate degree in nuclear science and to affect a change in the understanding of the field by the public, we recommend (1) **the enhancement of existing programs and the inception of new ones that address the goals of increasing the visibility of nuclear science in undergraduate education and the involvement of undergraduates in research; (2)**

the development and dissemination of materials and hands-on activities that demonstrate core nuclear science principles to a broad array of audiences.

Applications

The world we live in today is a reflection of the enormous advances in technology of the past century. Nuclear science has contributed directly to key areas of this technological revolution, including energy production and medicine, and indirectly through training of a scientific workforce. New applications derived from basic research in nuclear science continue to be developed. As a recent example, nuclear scientists are directly involved in developing techniques that are based on basic nuclear science research for scanning cargo containers for dangerous materials as they enter our ports or pass our border stations. While we cannot say what new applications will be developed in the future, we can say that there will be many based on past history.

Through applications, nuclear science provides a return on the federal investment made to support the program of basic research. Recognizing this, we welcome closer ties between basic research and the applications of our trade.

RESOURCES

NSAC was provided budget guidance by DOE and NSF in the charge letter that requested this new Long Range Plan. Specifically, the letter stated: "The projected funding for DOE is compatible with implementing the 12 GeV Upgrade of the Continuous Electron Beam Facility, and starting construction of a rare isotope beam facility that is less costly than the proposed Rare Isotope Accelerator (RIA) facility early in the next decade. At NSF the process has been put in place for developing a deep underground laboratory project and bringing this project forward in a funding decision."

Implementing the four principal recommendations of this Plan can be accomplished with a funding profile consistent with doubling the DOE's Office of Nuclear Physics budget, in actual year dollars, over the next decade together with NSF funding for DUSEL, including some of the equipment for experiments to be carried out in DUSEL. Following a staged approach, construction funds for the upgrade of CEBAF at JLab, Recommendation I, will begin in FY2009 and continue until FY2015. Work toward the construction of FRIB, Recommendation II, will begin in FY2009 with major construction funding occurring between FY2012 and FY2017. An NSF panel recently chose the Homestake mine as the site for DUSEL. Preparation of a full proposal for this new facility, which is part of Recommendation III, is the next step in making it a reality. Detector upgrades at RHIC, which are part of Recommendation IV, will proceed over several years and the accelerator modifications needed to implement beam cooling, which will significantly increase the RHIC luminosity, will be carried out between 2012 and 2015. Following this course, nuclear science in the U.S. will continue to have world-leading facilities into the future.

3. EXOTIC NUCLEI—STRUCTURE AND REACTIONS

Mirror Nuclei: Symmetry Breaking and Nuclear Dynamics

Silvia M. Lenzi

Dipartimento di Fisica dell'Università and INFN, I-35131 Padova, Italy

Abstract. Differences in excitation energy between isobaric analogue states are measured to study the validity of isospin symmetry in medium-mass nuclei as a function of the angular momentum. These differences are of the order of tens of keV and can be well reproduced by state-of-the-art shell model calculations. Several nuclear structure properties can be deduced from these data, such as the alignment of nucleons along rotational bands, the evolution of the nuclear radius and the appearance of pure single particle configurations across two main shells. In addition, the isospin breaking of the nuclear interaction is suggested by the systematic comparison with the data.

Keywords: Coulomb energy differences, shell model, gamma spectroscopy
PACS: 21.10.Sf, 21.10.-k, 21.60.Cs, 23.20.-g

INTRODUCTION

Examination of symmetries is a key question in physics as they help to interpret Nature. The appearance of symmetries is related to conserved quantities and, in quantum mechanics it gives rise to good quantum numbers. Among the variety of symmetries we encounter in nuclear physics, the isospin symmetry is one of the basic ones. When the neutron was discovered by Chatwick in 1932, it was observed that the nuclear interaction acts on it in the same way than on the proton. This charge symmetry induced Heisenberg to treat both proton and neutron as the same particle, the nucleon, characterised by a quantum number $t = 1/2$ and projection $t_z = -1/2$ and $t_z = 1/2$, respectively. This was later generalised by Wigner to the whole nucleus, where the projection of the isospin results from the contribution of the A nucleons

$$T_z = \sum_{i=1}^{A} t_{z,i} = \frac{N-Z}{2}, \tag{1}$$

and the total isospin T can get the values

$$|\frac{N-Z}{2}| \leq T \leq \frac{N+Z}{2}. \tag{2}$$

While T_z is always unique in a given nucleus, T can take different values, following eq. (2). In general (but there are many exceptions), the isospin of the ground and low-lying states is $T = |T_z| = |\frac{N-Z}{2}|$, the larger values of T lying higher in energy.

Let's look at the simplest nucleus, the deuteron. It is composed by a proton and a neutron, therefore $T_z = 0$ and $T = 0,1$. The $J = 1, T = 0$ state is the ground state while the state $J = 0, T = 1$ is not bound. ($J + T$ has to be odd as the wave function of the

CP972, *Exotic Nuclei and Nuclear/Particle Astrophysics (II)—Proceedings of the Carpathian Summer School of Physics 2007,* edited by L. Trache and S. Stoica

two nucleons is antisymmetric.) This is telling us that the nuclear force is stronger in the $T = 0$ state. A system formed by two protons or two neutrons can be coupled to $J = 0, T = 1$, but not to $J = 1, T = 0$ as it is forbidden by the Pauli principle. The three states $J = 1, T = 0$ in the proton-proton, neutron-neutron and proton-neutron pairs form a triplet of isobaric analogue states. (It is well known that such two-nucleon states are not bound.)

In general we can talk about *isobaric multiplets*, a group of nuclei with the same mass. The simplest example is a pair of mirror nuclei, which have exchanged the number of protons and neutrons. Such a pair should have identical structure in the hypothesis of the charge symmetry of the strong interaction. Another example of isobaric multiplets is the quadruplet with $A = 21$, ${}^{21}_{9}F_{12}$, ${}^{21}_{10}Na_{11}$, ${}^{21}_{11}Ne_{10}$ and ${}^{21}_{12}Mg_{9}$. In these nuclei we can identify analogue states, i.e. those states that have the same J and T and the same configuration in the isospin formalism. In this quadruplet, the mirror pair ${}^{21}_{10}Na_{11}$, ${}^{21}_{11}Ne_{10}$ has $|T_z| = 1/2$ and the pair ${}^{21}_{9}F_{12}$, ${}^{21}_{12}Mg_{9}$ has $|T_z| = 3/2$. There will be analogue states with $T = 1/2$ only in the first pair, while analogue states with $T = 3/2$ can be found in all the four nuclei.

Isobaric analogue states in multiplets with $T \geq 1$ are, however, difficult to reach experimentally and most of these multiplets are limited to low masses and spins. This is due to the fact that, while nuclei with neutron excess are easily reachable, the proton-rich members of the multiplets are farther from the stability line and the cross section for its population via fusion-evaporation or transfer reactions using stable beams and targets are very low. Recently, excited states up to spin $J \sim 10-12$ have been observed in $T = 1$ isobaric triplets thanks to important improvements in gamma-ray spectroscopic techniques. Using radioactive beams, more recently, the $J^{\pi} = 2^{+}$ $T = 2$ state in ${}^{36}Ca$ has been observed and compared to the analogue state in its mirror ${}^{36}S$ both in GSI and GANIL [1, 2]. It is thus clear why the most studied mirror pairs are those with $|N - Z| = 1$, $T_z = \pm 1/2$.

Why is it interesting to study isobaric multiplets and what can we learn from these investigations? The answer comes from the breakdown of the isospin symmetry. In fact, in a nucleus the interaction between nucleons has two terms, a nuclear term and a Coulomb term. The latter acts only between protons and therefore it is not isospin symmetric. Therefore, when we compare two mirror nuclei, for example, we will find differences between analogue states induced by the Coulomb interaction. These differences are important if we compare the absolute energy of the states as the Coulomb contribution to the bulk energy is of the order of hundreds of MeV. This can be easily verified by a simple estimate. Consider the nucleus as a sphere of radius R_C. The electrostatic energy is given by

$$E_C = \frac{3Z(Z-1)e^2}{5R_C}. \tag{3}$$

For a nucleus of mass $A = 49$, $Z = 25$ the (bulk) electrostatic energy amounts to $E \sim 120$ MeV. Taking the difference in energy with the analogue ground state in its mirror with $A = 49$, $Z = 24$ ($T = 1/2$ states), we obtain $\Delta E \sim 10$ MeV. These differences are called Coulomb Displacement Energies (CDE). If we now consider *excited* states in both nuclei and calculate the differences between the excitation energies (referred to the ground state in each partner), the differences we measure are generally of the order of tens of keV (three orders of magnitude smaller that the CDE) and in some few cases they can reach few hundreds of keV. These (small) differences have been given the name Coulomb Energy Differences (CED).

It is on these CED where we will go a little deeper to see how they can tell us many interesting features of the *nuclear* structure of medium-light nuclei. In particular, we will see how the nuclear structure changes as a function of the angular momentum. These studies have become experimentally and theoretically feasible in the last decade when large arrays of Ge detectors were constructed and shell model codes that could deal with large-dimension matrices were developed.

In the laboratory, these nuclei are populated in general using fusion-evaporation reactions and the excited states are determined by measuring the gamma-rays emitted in the decay along the yrast line. In such a reaction many channels open and detection systems with high sensitivity and resolving power are needed to get information on the weak proton-rich channels. Nuclei in the $f_{7/2}$ shell have been studied extensively from both experimental and theoretical sides, and it was the collaboration between experimentalists and theoreticians which allow to have a complete picture of the physics in this mass region.

In the following sections I will concentrate in some few examples to illustrate the different aspects of the nuclear structure that are evidenced when studying CED.

SHELL MODEL DESCRIPTION OF MED

Nucleon alignment: The mirror pair $^{50}_{26}$Fe$_{24}$- $^{50}_{24}$Cr$_{26}$

These nuclei are deformed in the ground state ($\beta \sim 0.26$) and present a rotational-like band [3, 4]. As these systems are rather small, above some value of the nuclear spin, it results energetically favourable to increase the angular momentum by aligning the spins of their valence nucleons. This causes a change in the shape of the nucleus and a discontinuity in the regular sequence of transition energies between consecutive levels. In a plot of the angular momentum vs. rotational frequency, this translates in a backbending at the aligning frequency. Two identical nucleons in the $f_{7/2}$ shell that are couplet to $J = 0$, can recouple gradually to higher values of J until they reach the maximum allowed value $J = 6$, or can abruptly align from $J = 0$ to $J = 6$. As we are dealing with nuclei near the $N = Z$ line and in the present case both protons and neutrons are even, it is not a priori easy to predict which type of nucleons will align first.

The analysis of the CED can give the answer to this question. This is due to the fact that when a pair of protons align or recouple there is, in addition to a nuclear effect (that is the same for the neutrons), a change of the Coulomb energy. It can be seen as follows: in the $J = 0$ coupled configuration, the overlap of the wavefunctions of the protons is maximum, and so is the Coulomb energy. As they are recoupled to larger values of J, the interaction decreases and in the maximum aligned $J = 6$ configuration, the Coulomb energy is minimum. As the Coulomb interaction is repulsive, this translates into a more bound state in the nucleus where the protons align, with respect to the mirror partner where a neutron pair aligns. This effect was suggested for the first time by Cameron and collaborators [5] in the case of the $T = 1/2$ $A = 49$ mirror pair.

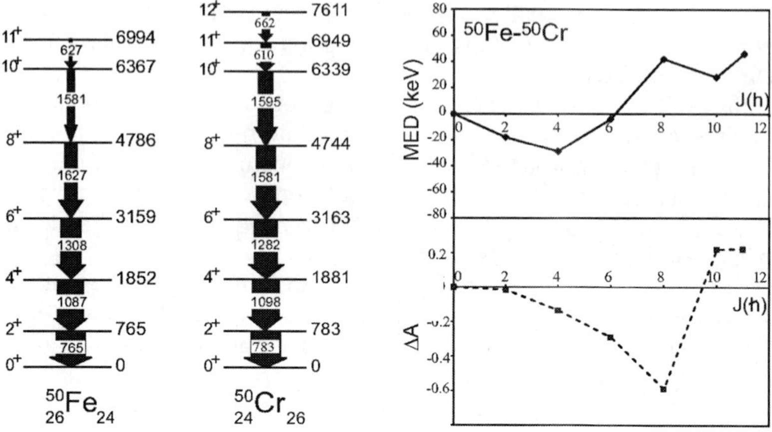

FIGURE 1. MED and alignment for the mirror pair ^{50}Fe-^{50}Cr as a function of the angular momentum [4]. (a) The experimental MED. (b) The calculated "alignment" in the shell-model framework – see text for details.

By comparing the level schemes of two mirror nuclei, the CED is obtained as

$$CED_{J,T} = E^*_{J,T,T_z} - E^*_{J,T,T_z+n}, \tag{4}$$

where n is the difference between the number of neutrons and protons, $n = |N - Z|$, and J is the spin of the state. An alignment along a rotational band will show as a jump in the CED: a positive value will indicate that a pair of protons have aligned in the $T_z + n$ ($Z_<$) partner, while a negative value will correspond to the alignment of a proton pair in the T_z ($Z_>$) one. Simultaneously, of course, a pair of neutrons will align in the corresponding mirror partners.

In the shell model framework, the alignment can be computed by *counting* the number of proton pairs in a j orbital coupled to the maximum spin ($J = 2j - 1$). In the $f_{\frac{7}{2}}$ shell, this can be performed by calculating the expectation value of the operator (that acts only on protons) $A = [(a^+ \; a^+)^{J=6} \; (a \; a)^{J=6}]^0$ for each excited state of the rotational band. Doing this for both nuclei, one can then calculate the difference $\Delta A = A(Z_>) - A(Z_<)$ for the mirror pair, as a function of the angular momentum. If the alignment of a pair of protons in the nucleus with charge $Z_>$ – and, consequently, the alignment of a pair of neutrons in the $Z_<$ – occurs first, ΔA will increase, whilst it will decrease if the opposite happens. This was introduced by Poves and Sánchez-Solano in ref. [6] for the case of $A = 51$.

In the left-hand side of Fig. 1 the level schemes of the mirror pair ^{50}Fe-^{50}Mn are reported. Only those states in ^{50}Cr relevant for the comparison are shown, as just states up to 11^+ have been identified in the proton-rich ^{50}Fe. The similarity between them is evident. In the right-hand side of Fig. 1, the experimental CED, called for the case of mirror pairs Mirror Energy Differences (MED=$E^*(^{50}\text{Fe})$-$E^*(^{50}\text{Cr})$), are shown in the upper panel. In the lower panel, instead, the result of the shell model calculation of ΔA is reported. From this figure we can deduced that a gradual alignment of protons occurs up to spin $J = 8$ in ^{50}Cr (neutrons in ^{50}Fe), followed by the sudden alignment of a pair of neutrons at $J = 10$ (protons in ^{50}Fe). It is thus clear how we can get information of the nuclear structure properties by just looking at the MED in mirror nuclei. Other features necessitate a more deep understanding of the wavefunctions and a good shell model description of the spectroscopy of the nuclei of interest is mandatory.

In this mass region, exact shell model calculations in the full fp space are feasible. For the specific case of $A = 50$, both the excitation energy and transition probabilities can be reproduced with good accuracy [3, 7]. This encourages to perform the MED with such calculations, although these energy differences are small and it is not a priori obvious that this goal could be achieved. In fact first attempts of reproducing the MED in nuclei of the $f_{7/2}$ shell failed, as some important effects were not taken into account. It was in ref. [8] where Zuker et al. gave a complete and successful description of all the CED so far measured in this mass region. In the following section the different contributions to the MED deduced from ref. [8] are described.

Monopole Coulomb contributions: The mirror pair ^{48}Mn-^{48}V

By adding the Coulomb potential to the nuclear effective interaction and performing the diagonalization of the corresponding shell model matrix in the valence space, only the multipole contribution of the Coulomb interaction (V_{CM}) is considered. Earlier tentatives of reproducing the MED as a function of the angular momentum using this approach failed as monopole contributions were not taken into account [9]. In ref. [4] a renormalization of the multipole Coulomb interactions was suggested together with the introduction of a monopole term called *radial term*. This term takes into account the changes of shape (radius) of the nucleus along the yrast sequence.

In $f_{\frac{7}{2}}$-shell nuclei, the occupation of orbits different from the $f_{\frac{7}{2}}$, important to create collectivity near the ground state, decreases along the rotational bands, due to the alignment process, producing changes of the nuclear radius. This affects the MED as valence protons in orbitals with smaller radii are *nearer* to the charged core and have *more* Coulomb energy.

Following refs. [4, 8], the monopole Coulomb contribution to the MED can be deduced by considering the Coulomb energy of a uniformly charged sphere of radius R_C (eq. (3)). When calculating the mirror energy differences for each state of spin J as a function of the angular momentum, we refer the MED(J) values to the ground state. On doing so, the monopole effect of E_C almost vanishes. A small contribution remains, however, due to the change in charge radius with the angular momentum, as discussed above. The monopole Coulomb radial contribution to the MED can be written:

$$MED^J(V_{Cr}) \quad = \quad \Delta E_C(J) - \Delta E_C(0)$$

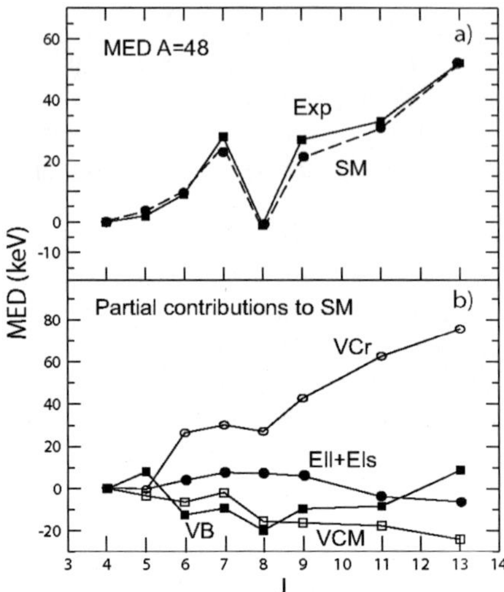

FIGURE 2. MED for the mirror pair ^{48}Mn-^{48}V as a function of the angular momentum [11]. (a) The experimental MED in comparison with shell model calculations (SM). (b) The different contributions to the shell model MED (see text for details).

$$
\begin{aligned}
&= \frac{3}{5}n(2Z-n)e^2\left(\frac{1}{R_C(J)}-\frac{1}{R_C(0)}\right) \\
&= -\frac{3}{5}n(2Z-n)e^2\frac{\Delta R(J)}{R_C^2}
\end{aligned}
\quad (5)
$$

where $\Delta R(J) = R_C(J) - R_C(0)$ and we assume, following refs. [4, 8, 10], that it is *the same in both mirror nuclei*. In the fp shell, p-orbits have larger radius than f-orbits and therefore, the Coulomb repulsion increases as the protons pass from the $p_{\frac{3}{2}}$ to the $f_{\frac{7}{2}}$ orbit. In other words, at high spin, when all nucleons are filling the $f_{\frac{7}{2}}$ shell, the monopole Coulomb contribution is larger than at low spin, where there is a significant $p_{\frac{3}{2}}$ contribution to the wavefunction.

Changes in the charge radius can be accounted for, within the shell model, by considering the evolution of the occupation numbers of the different orbits as a function of the angular momentum along the yrast bands [4]. As stated above, for nuclei in the $f_{\frac{7}{2}}$ shell, it is the relative occupation of the $p_{\frac{3}{2}}$ orbit which determines the main changes of radii. The assumption of equal radii of the mirror partners means that a calculation of the *average* of proton and neutron occupation numbers of the $p_{\frac{3}{2}}$, m_π and m_ν, respectively, is required: $\frac{m_\pi + m_\nu}{2}$. The contribution of the monopole Coulomb radial term to the MED of mirror nuclei with $|T_z| = k/2$ can be parametrised as,

$$
MED^J(V_{Cr}) = k\,\alpha_r\left(\frac{m_\pi(0)+m_\nu(0)}{2} - \frac{m_\pi(J)+m_\nu(J)}{2}\right) \quad (6)
$$

where the constant α_r can be deduced from the single-particle relative energies in mass $A = 41$, as discussed in ref. [8]. It is important to note, however, that the radial term in the calculation of the MED is not a single-particle effect and therefore it cannot be accounted for by setting different single-particle energies for protons and neutrons in the shell-model calculation.

A good example of mirror nuclei where this term gives a significant contribution, and where multipole effects almost cancel due to the symmetry between particles and holes in the $f_{7/2}$ shell for these nuclei, is the odd-odd pair $A = 48$, $T = 1$ [11]. They are described in detail in ref. [11]. Here, the MED are reported in Fig. 2a) and compared to the shell model calculations, the agreement is excellent. In Fig. 2b), the different terms entering the calculation are displayed. In particular, the radial term (V_{Cr}) gives the main contribution. The other (multipole) terms are V_{CM}, the multipole Coulomb, and V_B, the isospin symmetry breaking term, that will be discussed in the next section. Finally, monopole corrections to the single particle energies are also shown in the figure. These latter do not have a sizable contribution in the MED of $f_{7/2}$-shell nuclei but are very important in sd-shell nuclei. A detailed description can be found in the review article of ref. [12].

The isospin-breaking "nuclear" term: The mirror pair ^{54}Ni-^{54}Fe

Although the identification of the different electromagnetic terms that enter in the calculation of the MED allows to improve the description of the data, a satisfactory solution was only reached when A.J. Zuker [8] introduced the V_B contribution to the MED. This is a schematic term, deduced from the data of the $A = 42$ isobaric triplet. By considering the yrast states $J = 0,2,4,6; T = 1$ in the mirror nuclei ^{42}Ti and ^{42}Ca and assuming that they have essentially $f_{7/2}^2$ configurations, the Coulomb contribution to the MED can be calculated. In the hypothesis of isospin symmetry of the nuclear interaction, the experimental MED values and those obtained using the Coulomb interaction should be similar. This is, however, not the case for $A = 42$. The result of subtracting to the experimental MED the Coulomb contribution is of the same order of magnitude of the latter. The maximum value is obtained at $J = 2$.

This means that isospin-symmetry breaking (ISB) effects cannot be ignored in the calculation of the excitation energy differences. This effect was already noted in early studies performed by Brown and Sherr in ref. [13] but the origin of this charge-dependent interaction is still an open question. This was considered of nuclear nature in ref. [8], but contributions from the renormalisation of the Coulomb interaction could also contribute. In ref. [8], an extremely simple ansatz was proposed which consists on constructing an ISB hamiltonian in the fp shell by just taking the $f_{\frac{7}{2}}^2$ matrix element for $J = 2$ with a strength of $\sim$100 keV, as suggested from the data of $A = 42$.

$$V_B = V_{B,f_{\frac{7}{2}}}(J = 2) \tag{7}$$

where $V_{B,f_{\frac{7}{2}}}(J)$ is a matrix elements of value $\sim$100 keV. The results for the MED obtained using this schematic term together with those introduced above (monopole and

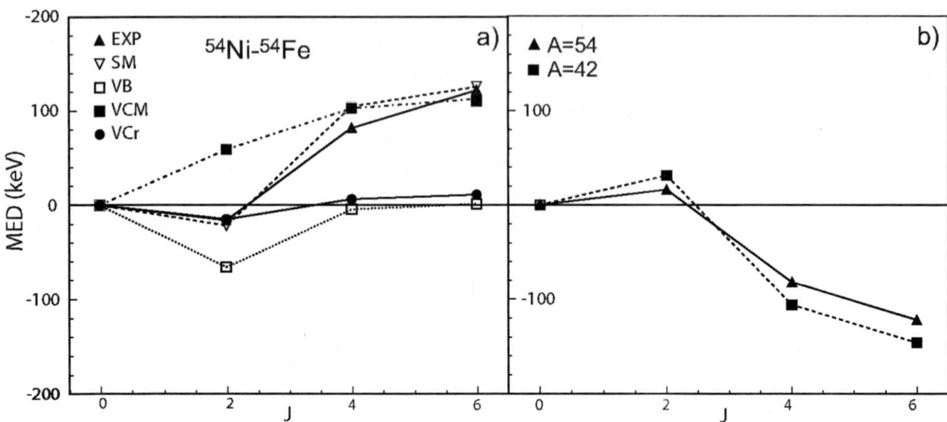

FIGURE 3. MED for the mirror pair ^{54}Ni-^{54}Fe as a function of the angular momentum [14]. (a) The experimental MED in comparison with shell model calculations (SM) and the different contributions to the latter (see text for details). (b) Comparison between the experimental MED for the cross-conjugate mirror pairs $A = 54$ and $A = 42$.

multiple Coulomb) are very successful. Using the same parametrization of eq. (7), the data on MED for all the $f_{\frac{7}{2}}$-shell isobaric multiplets measured so far can be reproduced with very good accuracy [8, 12]. The fact that the ISB is maximum at $J = 2$ and that the same schematic term is absolutely needed to reproduce the data along the whole $f_{7/2}$ shell is very significant. In Fig. 3 the MED values for the $T = 1$ mirror pair ^{54}Ni-^{54}Fe are reported. The role of the V_B term is clearly evidenced and of the same order of magnitude than the V_{CM} contribution. The contribution of V_{Cr} is small, which can be understood because these nuclei, that are near closed shell, are not deformed in the ground state. In Fig. 3b), the experimental MED for $A = 54$ are compared to those for $A = 42$. These nuclei are symmetric in the number of particles/holes in the $f_{7/2}$ shell and are therefore called cross-conjugate. (Actually, for $A = 42$ the values reported in the figure correspond to -MED to reflect this symmetry between particles and holes.) The similarity between the two MED curves is impressive. This clearly shows that the ISB term is important all across the $f_{7/2}$ shell. Moreover, analysing the wave functions for $A = 42$ and $A = 54$ it is clear that they are different: while in $A = 42$ the $f_{7/2}^2$ configuration is dominating, only 60% of the wave functions in $A = 54$ are pure $f_{7/2}^{-2}$. As the MED reflect the changes of configuration as a function of the angular momentum, the similarity of both MED, in spite of the fact that the wave functions are quite different, indicates that the configurations must be very much independent of J. In other words, the pure $f_{7/2}$ description works well because the terms that may break cross-conjugation symmetry do exist but they have J-independent effects that vanish for the MED.

CONCLUSIONS

In this work, the main features that can be put in evidence by analysing the energy differences between excited analogue states in mirror nuclei have been briefly discussed. Although not treated in here, also the energy differences among analogue states in $T = 1$ triplets (which include also the $N = Z$ member of the multiplet) can be reproduced very well with the shell model. A more complete and extensive description of both theoretical and experimental methods to study CED can be found in the review article of ref. [12]. What makes this subject fascinating is the fact that such small energy differences, caused mainly by the Coulomb potential, can reveal so many details on the *nuclear* structure that cannot be evidenced by other, more standard, methods. The bonus of the possible evidence of ISB nuclear interaction, makes it still more interesting and merits further investigation.

The extension of these studies to other mass regions such as the sd or fp shells, which have been recently addressed [12], has put in evidence the importance of another monopole term: the electromagnetic spin-orbit interaction. This changes the single particle energy of both protons and neutrons and has sizable effects on the MED whenever single-particle excitations are dominant in the wave functions.

Exploration of the isospin degree of freedom is certainly one of the key nuclear-structure goals of the present and future radioactive beam facilities. Exciting opportunities will open that will allow to test the present methods in more extreme situations.

ACKNOWLEDGMENTS

The results presented in this work have been obtained in collaboration with several colleagues to whom I am very grateful, in particular, I would like to acknowledge A.P. Zuker and M.A. Bentley.

REFERENCES

1. P. Doornenbal *et al.*, Phys.Lett. B 647, 237 (2007)
2. A. Buerger *et al.*, Acta Phys.Pol. B38, 1353 (2007)
3. S.M. Lenzi *et al.*, Phys. Rev. C **56**, 1313 (1997)
4. S.M. Lenzi *et al.*, Phys. Rev. Lett. **87**, 122501 (2001)
5. J.A. Cameron *et al.*, Phys. Rev. C **44**, 1882 (1991)
6. M.A. Bentley *et al.*, Phys. Rev. C **62**, 051303 (2000)
7. F. Brandolini *et al.*, Nucl. Phys. **A642**, 387 (1998)
8. A.P. Zuker, S.M. Lenzi, G. Martinez-Pinedo and A. Poves, Phys. Rev. Lett. **89**, 142502 (2002)
9. M.A. Bentley *et al.*, Phys. Lett. **B437**, 243 (1998)
10. J. Duflo and A.P. Zuker, Phys. Rev. C **66**, 051304(R) (2002)
11. M.A. Bentley *et al.*, Phys. Rev. Lett. **97**, 132501 (2006)
12. M.A. Bentley and S.M. Lenzi, Prog. Part. Nucl. Phys. **59**, 497 (2007)
13. B.A. Brown and R. Sherr, Nucl. Phys. **A322**, 61 (1979)
14. A. Gadea *et al.*, Phys. Rev. Lett. **97**, 152501 (2006)

Correlations in Nuclei: Configuration Interaction, Coupled-Cluster and Beyond

Mihai Horoi

Department of Physics, Central Michigan University, Mount Pleasant, Michigan 48859, USA

Abstract. Many-body correlations play a central role in the structure of nuclei. The mean-field methods, such as Skyrme Hartree-Fock and its extensions to Energy Density Functionals, induce correlations dynamically at the functional level, leaving the explicit wave functions uncorrelated. The Configuration Interaction (CI) method treats the many-body correlations exactly, but in a small configurations space. It is therefore desirable to include as many correlations as possible in larger model spaces. In this contribution I compare the results of Coupled Cluster techniques and of extrapolations methods, such as the Exponential Convergence Method, with the full CI results for nuclei around 56Ni.

Keywords: Correlations, configuration interaction, coupled-cluster, double beta decay
PACS: 21.60.Cs,31.15.Dv

THE NUCLEAR SHELL MODEL, ALSO KNOWN AS CONFIGURATION INTERACTION (CI)

The nuclear shell model is one venues of describing many-body correlations beyond the mean field. The strong character of nuclear forces requires nonperturbative many body techniques to tackle the complexity of the resulting nuclear structure. The terminology of "nuclear shell model" (NCM) has some connections with the single particle spherical shell model (SPSSM), which was so successful in describing key features of the nuclear stability and the low-lying spectra. The NSM is often viewed as a simple extension of SPSSM, and the commonality of their name is prone to confusions. In addition, it was recently shown [1] that NSM can describe in the same time the ground band and collective band(s) of double magic nuclei if one takes into consideration the interaction of a large number of configurations. Therefore, a more appropriate name for the NSM method would be the configuration interaction (CI) method, name which is used in quantum chemistry for a similar method.

The road to getting CI effective interactions from first principles will be reviewed in Section 2. This road is not complete and it's paved with ups and downs. However, the CI method with phenomenological two body effective interactions[2] was proven quite successful[3] in describing the nuclear spectroscopy all over the nuclear chart. Recent developments in CI computational software and techniques showed that adding more configurations improves the quality of the computed low-lying properties of nuclei, permitting o more complex analysis of the experimental results. The daunting charge remains to find a reliable phenomenological interaction, which can be used to extrapolate to more exotic nuclei and states. In this contribution we add few pieces to this success story.

CP972, *Exotic Nuclei and Nuclear/Particle Astrophysics (II)—Proceedings of the Carpathian Summer School of Physics 2007,* edited by L. Trache and S. Stoica

CI EFFECTIVE INTERACTIONS

The traditional nucleon-nucleon potentials used to describe the two nucleon scattering phase shifts have a hard core that was in the past the main concern for the many body techniques. Therefore, many of the approaches that were developed tried to eliminate the hard core by renormalizing the interaction to low energy model spaces. Examples of these renormalization techniques are the G-Matrix approach[4], the No-Core Shell Model effective interaction[5] obtained via Lee-Suzuki transformations, the V_{low-k} interaction[6], and the Unitary Correlator Model[7]. It turned out that the renormalization process alone does not lead to an effective interaction soft enough to be useful for nuclear calculations beyond few nucleons. Even in those cases the root mean squared deviation (*rms*) of the theoretical energies from the corresponding experimental values is of the order of 1 MeV per datum. The extension to heavier nuclei is made even more difficult by the exploding CI dimensions, which restrict the valence spaces to one or two major harmonic oscillator shell.

Therefore, one should include the effect of the missing core polarization effect in a perturbative approach[4]. Even when this improvement is included in valence spaces, such as sd or pf, the *rms* is still around 1 MeV, making the comparison with experimental data difficult. It is, however, known that phenomenologically adjusted two body effective interactions describe the data quite accurately, e.g the *rms* is about 140 keV for the USDB interaction [8]. One can blame the failure of the first principle approach to obtain reliable effective interactions on the neglect of bare three body interactions, which are known to have an important contribution[9]. Work including three body contributions is in progress. Meantime, one should further rely on phenomenologically adjusted two body interactions. The general recipe of obtaining this type of effective interactions is to start from a renormalized effective interaction with the core polarization effects included. Then, a number of levels in nuclei that can be described in the chosen valence space is selected and the matrix elements of the effective interaction (including the single particle energies) are adjusted to get a lower *rms* (see, e.g. Ref. [8] for details). The difficulty is that when has to use larger valence spaces, the CI dimensions explode, and the iterative procedure is computationally too expansive. Examples of such phenomenologically adjusted interaction are given in the next session.

COLLECTIVE BANDS IN ^{56}NI

The large-basis CI description of nuclei around ^{56}Ni is at the forefront of our microscopic understanding of nuclear properties. In the simplest model the $0f_{7/2}$ (denoted in the following by f) orbit is filled with sixteen nucleons and the next few unfilled orbits are $1p_{3/2}, 1p_{1/2}$ and $0f_{5/2}$ (denoted in by r). The historical development of configuration mixing in this region can be related to truncations in terms of the partitions $f^{n-t}r^{m+t}$, where is n maximum number of nucleons allowed in the f orbit, m is the minimum number of nucleons allowed in the r orbits ($n = 16$ and $m = 0$ for ^{56}Ni), and t is the maximum number of nucleons allowed to be excited from f to r. The earliest calculations were obtained with Hamiltonians designed for small t values. For example, a qualitative understanding of the ground states of ^{55,56,57}Ni with $t = 0$ and excited states in ^{56}Ni with $t = 1$ may be obtained [11]. However, there are many other low-lying states in ^{56}Ni that require a much more complete basis. In the high resolution ^{58}Ni(p,t)^{56}Ni experiment [10] eight 0^+ states are observed up to 10 MeV. In high-spin gamma spectroscopy a deformed band has been observed ending at a 2^+ state at 5.35 MeV [12]. In analogy with the low-lying deformed band in ^{16}O, this band in ^{56}Ni might be understood in terms of the states for four particles excited from f to r (4p-4h), and in [13] this idea was refined to 4p-4h relative to the correlated ^{56}Ni ground state. In the (p,t) experiment [10] the 0^+ state at 5.01 MeV was suggested to be the predominantly 4p-4h state, however no interpretation could be made for the lower 0^+ state at 3.95 MeV.

We focus on the complete pf-shell spectroscopy of ^{56}Ni up to about 10 MeV and how it can be understood in terms of simpler underlying configurations. It is now known that a consistent interpretation can only be obtained in a model space that goes up to very high t [14] with a Hamiltonian designed for this high-t model space. The most recent effective Hamiltonians GXPF1 [15, 16] and GXPF1A [17] are derived from a microscopic calculation by Hjorth-Jensen based on renormalized G matrix theory with the Bonn-C interaction [4], and are refined by a systematic fitting of the important linear combinations of two-body matrix elements to low-lying states in nuclei from A=47 to A=66, including some states in the ^{56}Ni, in a model space that in principle goes to t_{max}, although in practice involves some approximations [15]. Shell-model calculations that focus on the properties of the shape-coexistence of the ground state and deformed bands in ^{56}Ni has been presented previously [1, 18, 19].

Fig. 1 presents comprehensive results on all calculated levels up to 7 MeV, including those belonging to the collective band[12]: 0_3^+, 2_3^+, 4_5^+, and 6_3^+. The comparison with the experimental data shows that the state at about 4.8 MeV is not a negative parity states (as reported in Ref. [20]), but the 3_1^+ states. Other states reported by [20] seem to be misidentified. A full analysis based on the present calculation will be reported elsewhere. Here we emphasize that the coexistence of the collective states with the rest of the states is very well described. In particular, the competition in the decay of the collective 4_5^+ state to the collective 2_3^+ state, and the non-collective 2_1^+ and 4_1^+ states comes out very clearly.

The large density of low-lying 0^+ states is also one important feature of the ^{56}Ni structure. Fig. 2 compares the calculated and the experimental 0^+ levels up to 7 MeV. All the 0^+ levels are accounted for, including the collective 0_3^+ level, and the 0_4^+ level around 6 MeV, which was only identified in a $(^3He,n)$ experiment[10].

COMPARISON OF COUPLED-CLUSTER THEORY WITH CI

The CI calculations are strongly limited by the exploding dimensions. An interesting, computationally cost-efficient alternative to full CI is offered by coupled-cluster (CC) theory [21, 22], which sums entire classes of many-particle correlation effects, required

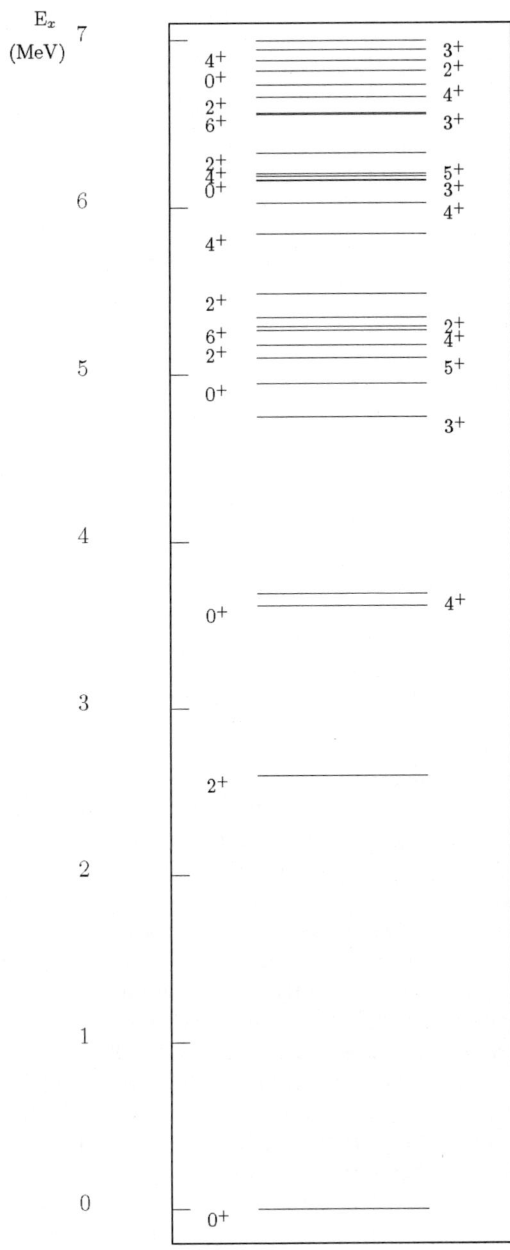

FIGURE 1. All the calculated levels in ^{56}Ni lower than 7 MeV.

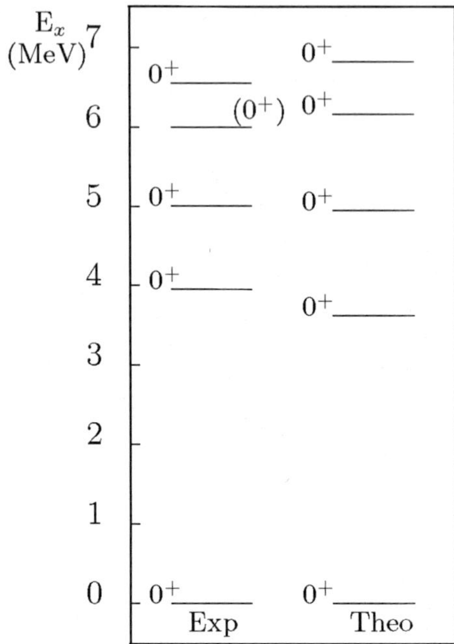

FIGURE 2. The 0^+ states in ^{56}Ni.

for an accurate description of energies and wave functions, to infinite order. The CC method started in nuclear theory [21, 23], but it has been mostly developed and applied in quantum chemistry [22, 24, 25]. It has recently been reintroduced into nuclear theory with applications to the no-core basis of light nuclei [26], but much less is known about the performance of modern CC approaches for heavier nuclei where they may become useful due to the prohibitive costs of full CI calculations. Truncated CI methods reduce these costs as well, but they are not as effective in describing particle correlations as the truncated CC models and, unlike CC methods, they are not size extensive.

It is know that the correlated CC wave functions are better described if the reference Slater determinant is a reasonable good approximation of the exact ground state (g.s.) wave function. ^{56}Ni is a double magic nucleus. With GXPF1A interaction the probability of finding the uncorrelated reference state in the exact wave function is 70%. We varied this probability by changing the gap between the lowest s.p. energies of the *f* and *r* orbits (see, e.g. Ref. [27] for more details). The results are presented in Fig. 3. When the gap shift is increased, the reference Slater determinant becomes a much better approximation to the g.s. wave function, the overlap between the reference state and the exact g.s. is closer to one, and the CC results are closer to the full CI results. In the other extreme, when the gap is decrease, the full CI states become very collective, the overlap is very small and the CC results deviate significantly from the exact results. However, in all cases, the CC results seem to be consistent with the CI results when only configuration

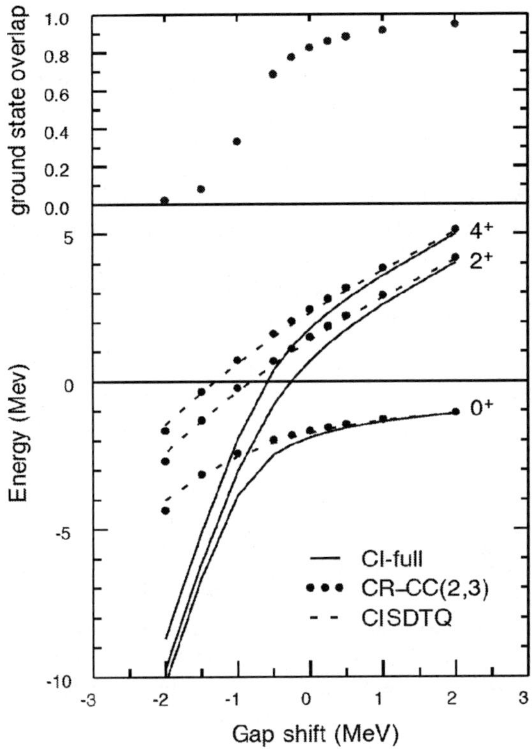

FIGURE 3. Comparison of the CI and CC calculations for the first three states in ^{56}Ni as a function of the Gap shift (see text for details).

with up to 4 particle - 4 hole excitations from the reference state are taken into account.

NEW APPROACH TO CALCULATE THE 2-NEUTRINO DOUBLE-BETA DECAY MATRIX ELEMENT

At present, the 0-neutrino double-beta ($0\nu\beta\beta$) decay is probably the most sensitive process for direct measurements of the electron neutrino mass. The calculations of the $0\nu\beta\beta$ decay probabilities are usually calibrated to the $2\nu\beta\beta$ results for which several experimental data exist. Recently[28] we calculated the $2\nu\beta\beta$ matrix element for the decay of ^{48}Ca using the CI method. Our approach was different of others[29]: we directly diagonalized 250 1^+ states in the intermediate nucleus ^{48}Sc to calculated the double Gamow-Teller sum entering the matrix element. Ref. [29] used the practical convergence of the double sum calculated with a small number of unconverged intermediate 1^+ states, which were built via Lanczos procedure starting with a Gamow-Teller door-way Lanczos vector.

Having calculated 250 fully converged 1^+ states in the intermediate nucleus, we checked the validity of the [29] method. We obtained the same results, therefore validating the approach. In addition, we found that the procedure used in [29] can be further improved by using directly the Lanczos vectors, rather than the eigenvectors. Our procedure is similar to that of Ref. [30], but we provide simple numerical techniques for the calculation of the expansion in Lanczos vectors. Our method is briefly summarized bellow.

For the $2\nu\beta\beta$ decay mode the relevant nuclear matrix element (NME) is of Gamow-Teller type, and it has the following expressions [28]:

$$M_{GT}^{2\nu}(0^+) = \sum_k \frac{\langle 0_f || \sigma\tau^- || 1_k^+ \rangle \langle 1_k^+ || \sigma\tau^- || 0_i \rangle}{E_k + E_0} . \tag{1}$$

Here E_k is the excitation energy of the 1_k^+ state of ^{48}Sc and E_0 is described in Ref. [28]. One can write the above double Gamow-Teller sum using Green's function,

$$M_{GT}^{2\nu}(0^+) = 3 \langle \sigma\tau^+ 0_f | \frac{1}{H - E_{gs} + E_0} | \sigma\tau^- 0_i \rangle . \tag{2}$$

The approach used in Ref. [29] was to study the convergence of the sum in Eq. (1) using all the (unconverged) intermediate states obtained after N Lanczos iterations starting with a door-way Gamow-Teller vector as the first Lanczos vector,

$$\sigma\tau^- |0_i\rangle = c_- |dw-\rangle = c_- |L_1^-\rangle , \tag{3}$$

or

$$\sigma\tau^+ |0_f\rangle = c_+ |dw+\rangle c_+ |L_1^+\rangle . \tag{4}$$

Ref. [30] proposed expanding the Green's function action on the initial state in terms of Lanczos eigenvectors,

$$\frac{1}{H - E_{gs} + E_0} |\sigma\tau^- 0_i\rangle \approx \frac{c_-}{H_L^N - E_{gs} + E_0} |L_1^-\rangle = c_- \left(g_1^- |L_1^-\rangle + ... + g_N^- |L_N^-\rangle \right) . \tag{5}$$

Here H_L^N is the tridiagonal Lanczos Hamiltonian obtained after N iterations. Therefore, the NME is given by

$$M_{GT}^{2\nu}(0^+) \approx 3 c_+ c_- \sum_m g_m^- \langle dw + | L_m^- \rangle . \tag{6}$$

Ref. [30] provided an analytical expression only for the g_1^- in Eq. (5). All the expansion coefficients can be numerically obtained by solving the eigenvalue problem,

$$H_L^N V = V E_L^N . \tag{7}$$

Knowing all the eigenvalues $E_L^N(1_k^+)$ and the corresponding eigenvectors, one can find all the expansion coefficients,

$$g_{\bar{m}} = \sum_{k=1}^{N} \frac{V_{1k} V_{mk}}{E_L^N(1_k^+) - E_{gs} + E_0} \quad m = 1, N \; . \tag{8}$$

Using the expansion of Eq. (6) in terms of Lanczos vectors, rather than the one in Eq. (1) in terms of the approximate (unconverged) intermediate 1^+ states has the advantage that it does not require the recovery of a large number of the full many body eigenvectors for testing the convergence of the sum. The diagonalization of the tridiagonal matrix Eq. (7) is very quick, and each new Lanczos vector provides a new term in the sum of Eq. (6) without affecting the previous ones. Overall, one can get a factor of 2-3 in the speedup of the calculations as compared with the method of Ref. [29]. We compared all three methods for ^{48}Ca and they concurred. The results will be reported elsewhere[31].

CONCLUSION AND OUTLOOK

We shown that the CI calculation with phenomenologically-adjusted effective interactions are quite successful in describing not only fragmentation of single particle strength, but also collective bands of large deformations and shape coexistence. Some of these features have 4 particle - 4 hole character, but their full description requires their entanglement with higher correlations.

The CC techniques was proven as an computationally lighter alternative to CI, quite accurate in some cases. It was proven that in all cases it accurately describes correlations up to 4p - 4h out of the reference state. Using a deformed s.p. basis may be proven as a very efficient alternative.

The status of the effective CI interaction from first principle is in full progress. However, many more studies are necessary before an effective interaction with an rms comparable to those of the phenomenologically-adjusted effective interactions is obtained.

Accurate description of the double beta decay rates depends on proper inclusion of correlations in the wave functions. Within the CI approach the sensitivity to the effective interaction is reduced due to the overall agreement of the properties of neighbors nuclides. We conclude that more effort is necessary to improve the efficiency of the CI codes and of the recipes of obtaining reliable effective interactions.

ACKNOWLEDGMENTS

The author acknowledges support from the NSF grant PHY-0555366 and DOE grant DF-FC02-07ER41457. Some of the calculations were possible due to the NSF MRI grant PHY-0619407. Support from Alexander von Humboldt Foundation was instrumental in the writing phase of this report.

REFERENCES

1. M. Horoi, B.A. Brown, T. Otsuka, M. Honma and T. Mizusaki, Phys. Rev. C **73**, 061305(R) (2006).
2. B.A. Brown and B.H. Wildenthal, Ann. Rev. of Nucl. Part. Sci. **38**, 29 (1988).
3. E. Caurier, G. Martinez-Pinedo, F. Nowacki, A. Poves and A.P. Zuker, Rev. Mod. Phys. **77**, 427 (2005).
4. M. Hjorth-Jensen, T.T.S. Kuo and E. Osnes, Phys. Rep. **261**, 125 (1995).
5. P. Navratil, G. P. Kamuntaviius1, and B. R. Barrett Phys. Rev. C **61**, 044001 (2000).
6. Scott Bogner, T. T. S. Kuo, L. Coraggio, A. Covello, and N. Itaco, Phys. Rev. C **65**, 051301R (2002).
7. R. Roth, H. Hergert, P. Papakonstantinou, T. Neff and H. Feldmeier, Phys. Rev. C **72**, 034002 (2002).
8. B. A. Brown and W. A. Richter, Phys. Rev. C **74**, 034315 (2006).
9. Petr Navratil and W. Erich Ormand, Phys. Rev. Lett. **88**, 152502 (2002); P. Navratil, V. G. Gueorguiev, J. P. Vary, W. E. Ormand, and A. Nogg, Phys. Rev. Lett. **99**, 042501 (2007).
10. H. Nann and W. Benenson, Phys. Rev. C **10**, 1880 (1974).
11. G. Kraus et al., Phys. Rev. Lett. **73**, 1773 (1994).
12. D. Rudolph, et al., Phys. Rev. Lett. **82**, 3763 (1999).
13. T. Mizusaki, T. Otsuka, M. Honma, and B.A. Brown, Phys. Rev. C **63**, 044306 (2001).
14. T. Otsuka, M. Honma and T. Mizusaki, Phys. Rev. Lett. **81**, 1588 (1998).
15. M. Honma, T. Otsuka, B.A. Brown and T. Mizusaki, Phys. Rev. C **65**, 061301(R) (2002).
16. M. Honma, T. Otsuka, B.A. Brown and T. Mizusaki, Phys. Rev. C **69**, 034335 (2004).
17. M. Honma, T. Otsuka, B.A. Brown and T. Mizusaki, Euro. Phys. Jour. A **25** Suppl. 1, 499 (2005).
18. T. Mizusaki, T. Otsuka, Y. Utsuno, M. Honma, and T. Sebe, Phys. Rev. C **59**, R1846 (1999).
19. T. Mizusaki, T. Otsuka, M. Honma, and B. A. Brown, Phys. Rev. C **63**, 044306 (2001).
20. www.nndc.bnl.gov
21. F. Coester, Nucl. Phys. **7**, 421 (1958); F. Coester and H. Kümmel, Nucl. Phys. **17**, 477 (1960).
22. J. Čıžek, J. Chem. Phys. **45**, 4256 (1966); Adv. Chem. Phys. **14**, 35 (1969).
23. H. Kümmel, K. H. Lührmann, and J. G. Zabolitzky, Phys. Rep. **36**, 1 (1978); R.F. Bishop, in Lecture Notes in Physics, Vol. 510, *Microscopic Quantum Many-Body Theories and Their Applications*, edited by J. Navarro and A. Polls (Springer, Berlin, 1998), p. 1.
24. J. Paldus and X. Li, Adv. Chem. Phys. **110**, 1 (1999.
25. P. Piecuch, K. Kowalski, I.S.O. Pimienta, and M.J. McGuire, Int. Rev. Phys. Chem. **21**, 527 (2002).
26. K. Kowalski, D.J. Dean, M. Hjorth-Jensen, T. Papenbrock, and P. Piecuch, Phys. Rev. Lett. **92**, 132501 (2004); M. Włoch, D.J. Dean, J.R. Gour, M. Hjorth-Jensen, K. Kowalski, T. Papenbrock, and P. Piecuch, Phys. Rev. Lett. **94**, 212501 (2005); J.R. Gour, P. Piecuch, M. Hjorth-Jensen, M. Włoch, and D.J. Dean, Phys. Rev. C **74**, 024310 (2006).
27. M. Horoi, J. Gour, M. Wloch, M. D. Lodriguito, B.A. Brown, and P. Piecuch, Phys. Rev. Lett. **98**, 112501 (2007).
28. M. Horoi, S. Stoica, and B.A. Brown, Phys. Rev. C **75**, 034303 (2007).
29. E. Caurier, A. Poves, and A. Zuker, Phys. Lett. B **252**, 13 (1990).
30. J. Engel, W. Haxton, and P. Vogel, Phys. Rev. C **46**, 2153(R) (1992).
31. M. Horoi, in preparation.

Elements of the *ab initio* No Core Shell Model

J. P. Vary*, P. Maris*, A. Negoita*, P. Navrátil†, V. G. Gueorguiev†, W. E. Ormand†, A. Nogga**, A. Shirokov‡ and S. Stoica§

*Department of Physics and Astronomy, Iowa State University, Ames, IA 50011, USA
†Lawrence Livermore National Laboratory, P.O. Box 808, L-414, Livermore, CA 94551, USA
**Forschungszentrum Jülich, Institut für Kernphysik (Theorie), D-52425 Jülich, Germany
‡Skobeltsyn Institute of Nuclear Physics, Moscow State University, Moscow, 119992 Russia
§Horia Hulubei National Institute for Physics and Nuclear Engineering, P.O. Box MG-6, 76900 Bucharest-Magurele, Romania

Abstract. We outline the *ab initio* no core shell model and present recent results. Nuclear properties are evaluated with two-nucleon (NN) and three-nucleon interactions (TNI) derived within effective field theory (EFT) based on chiral perturbation theory (ChPT). Fitting two available parameters of the TNI generates good descriptions of light nuclei. In another effort, referred to as the *ab exitu* approach, results are obtained with a realistic NN interaction derived by inverse scattering theory with off-shell properties tuned to fit light nuclei. Both approaches produce good results for observables sensitive to spin-orbit properties. In a third approach, we supplement the derived realistic effective interaction with phenomenological NN interaction terms with the goal to describe spectra of fp-shell nuclei in the limited basis spaces currently available.

Keywords: Light nuclear properties; Chiral effective field theory; Inverse scattering potentials, Many-body theory.
PACS: 21.60.Cs, 21.45.+v, 21.30.-x, 21.30.Fe, 27.20.+n

INTRODUCTION

Recent advances in microscopic many-body methods have opened new paths for investigating both the strong interaction itself as well as the many facets of nuclear phenomena evident in light nuclei. Once the *ab initio* no core shell model (NCSM) was introduced and shown to be reasonably convergent [1], opportunities emerged for precision testing of the properties of the strong interaction in the nuclear medium. Our focus in this presentation will be on recent and complementary efforts to determine important features of the strong interaction through the resulting properties of nuclei in the p-shell. Selected observables are especially sensitive to the three-nucleon interaction (TNI) and to the off-shell properties of the NN interaction. We also examine the degree to which phenomenological modifications of the NN interactions can improve agreement with experiment, either with phase-equivalent unitary transformations or with ad-hoc additive NN interactions.

In the first approach [2], we invoke the power of chiral perturbation theory (ChPT) [3] that provides a promising bridge with the accepted relativistic quantum field theory of the strong interactions, QCD. Beginning with the pionic or the nucleon-pion system [4] one works with systems of increasing nucleon number [5, 6, 7]. One makes use of

CP972, *Exotic Nuclei and Nuclear/Particle Astrophysics (II)—Proceedings of the Carpathian Summer School of Physics 2007,* edited by L. Trache and S. Stoica

the explicit and spontaneous breaking of chiral symmetry to systematically expand the strong interaction in terms of a characteristic small momentum of a few hundred MeV/c divided by the chiral symmetry breaking scale of about 1 GeV/c. Within the ChPT framework, the NN interactions, the TNI and also πN scattering are related to each other. Results should be trustworthy for observables dominated by momentum scales below this characteristic small momentum and thus, we expect the derived interactions to be valid for low-energy nuclear properties.

We adopt the potentials of ChPT at the orders presently available, N3LO for the NN interaction [8] and N2LO for the TNI [9] . The ChPT expansion divides the interactions into perturbative and non-perturbative elements. The latter are represented by a finite set of constants at each order of perturbation theory that are not presently calculable from QCD but can be fixed by measured properties of nuclei provided the many-body methods are sufficiently accurate. Once the non-perturbative constants are determined, the resulting Hamiltonian predicts, in principle, all other nuclear properties, including those of heavier nuclei with no residual freedom. We refer to this first effort as an application of the *ab initio* NCSM since the NN interaction is completely fixed by properties of the two-body system. Important components of TNI and higher-body interactions are also fixed by the ingredients of the NN terms. Only residual non-perturbatve chiral TNI couplings are fixed, as necessary, by the properties of nuclei beyond $A = 2$. Eventually, independent methods such as lattice QCD should fix all these parameters and complete the *ab initio* NCSM so the need for fitting would be eliminated.

In a second approach [10] the NN interaction is taken as a finite rank separable form in an oscillator representation for each partial wave. The coefficients are determined, to the extent possible, by inverse scattering techniques [11] using the available NN data. Subsequently, one investigates off-shell freedoms with phase-shift equivalent unitary transformations to tune the interaction to fit the properties of light nuclei. By fitting the ^3He and ^{16}O binding energies and the ^6Li low-lying spectra we obtain the interaction "JISP16", which represents "J-matrix Inverse Scattering Potential tuned up to ^{16}O". We achieve soft interactions with this approach that describe all the data conventionally fit by realistic NN interactions and provide good fits to light nuclear properties [10]. We display in Fig. 1 the JISP16 interaction in units of fm for the $^1S_0$ partial wave in momentum space. To obtain units of $MeV\ fm^3$ we multiply by $\frac{\hbar^2}{M_N} = 41.47\ MeV\ fm^2$. Note that this interaction is smooth over the entire range displayed.

We consider this as a phenomenological approach designed to explore regions of NN interactions that are not yet explored by other methods. We hope that the phase-shift equivalent transformation methods that prove successful in this *ab exitu* NCSM will be useful for minimizing higher-body forces in other approaches. This would facilitate efforts to access heavier nuclei within the NCSM. In many instances we find the results of the *ab initio* and *ab exitu* approaches to be similar and we cite the example of ^{10}B in the present work.

In a third approach [12, 13], we supplement the derived realistic effective NN interaction with purely phenomenological terms since the model space is so limited that we expect large corrections from neglected effective interactions such as real and effective TNI's. That is, we surrender the requirement that our effective interactions be entirely derived from the measured NN properties and the theory of many-body potentials. In-

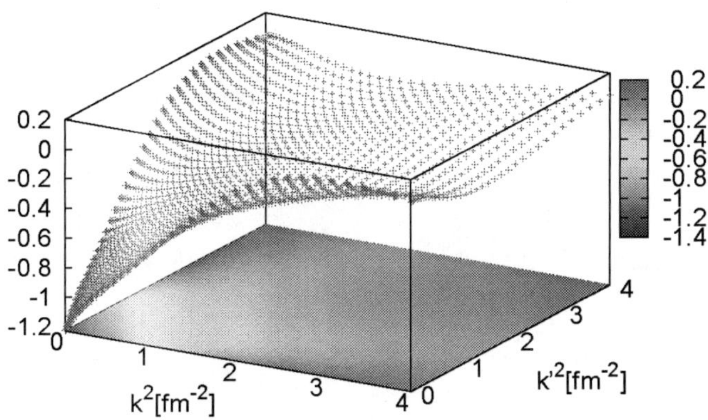

FIGURE 1. JISP16 potential for the $^1S_0$ partial wave in momentum space. The units are fm.

stead, this NCSM approach is patterned after the highly successful phenomenological shell model applied over the years to medium weight nuclei [14, 15]. However, in the NCSM all nucleons are active and we do not input any single-particle energies. That is, we still retain the pairwise interactions between all nucleons and attempt to obtain all properties, including single-particle or mean-field properties, from that pairwise interaction. In this sense, the NCSM is somewhere between the traditional shell model and the Hartree-Fock approach. We present results for ^{47}Ca that illustrate this NCSM approach.

NO CORE SHELL MODEL

The NCSM casts the diagonalization of the infinite dimensional many-body Hamiltonian matrix as a finite matrix in a harmonic oscillator (HO) basis with an equivalent "effective Hamiltonian" derived from the original Hamiltonian [1]. The finite matrix is defined by N_{max}, the maximum number of oscillator quanta shared by all nucleons above the lowest configuration. We solve for the effective Hamiltonian by approximating it as either a 3-body interaction[16] based on our chosen NN+TNI from ChPT (our *ab initio* application) or a 2-body interaction based on JISP16 (our *ab exitu* application). With these "cluster approximations", convergence is guaranteed with increasing N_{max} or with increased cluster size at fixed N_{max} [1].

The NCSM is the only approach currently available to solve the resulting many-body Schrödinger equation for mid-p-shell nuclei while preserving all symmetries when the interactions are non-local, a feature of all the interactions employed in this work.

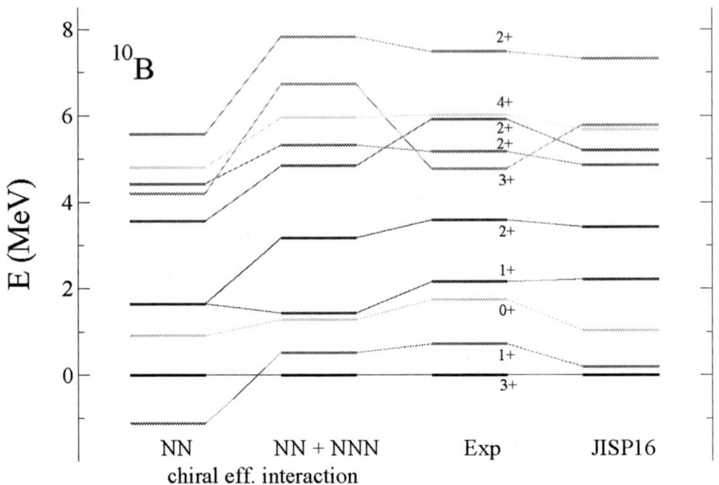

FIGURE 2. Experimental and theoretical excitation spectra of ^{10}B with respect to the lowest 3^+ state at an oscillator energy $\hbar\Omega = 14\,MeV$. The chiral effective interaction results are obtained at $N_{max} = 6$ while the JISP16 results are obtained at $N_{max} = 8$.

Ab initio NCSM with interactions from ChPT

In order to motivate the inclusion of the TNI, we begin by observing that conventional realistic NN interactions do not produce the correct ground state spin of ^{10}B: theory and experiment differ by an inversion of the two lowest levels. This inversion is illustrated in the first column of Fig. 2 for the chiral N3LO potential [8] at an oscillator energy, $\hbar\Omega = 14\ MeV$. Over the past few years, this deficiency, as well as others in mid-p-shell nuclei, such as binding energies, spectral properties and electromagnetic transition rates, have been ascribed to the need for TNI's. In terms of physics, the inadequacy of the realistic NN interactions appears likely to be insufficient spin-orbit splitting in the mean field generated by those interactions, though the mean field itself is not calculated directly. We summarize here the role of the TNI and the role of off-shell modified NN interactions in correcting this inadequacy.

We define the two non-perturbative coupling constants of the TNI, not fixed by 2-body data, as C_D (C_E), the strength of the $N-\pi-NN$ (TNI) contact term. Fig. 3 shows the trajectories of these two parameters as determined from fitting the binding energies of the $A = 3$ & 4 systems as well as the average of the two trajectories. Our approach is similar to the one used in a detailed investigation [17] of ^7Li. The ^4He results use the average of the $A = 3$ fits and the inset shows two crossing points where the ^4He binding is reproduced. Note the expanded scale. The second inset (b) depicts the corresponding rms charge radius of ^4He along with an estimate of the uncertainty in the theory arising from the folding with the proton rms charge radius.

Our results on the radii of the $A = 3$ systems are in good agreement with experiment as well. While the uncertainties in the ^3H and ^3He charge radii obscure the differences between the intersection points, the ^4He charge radius (inset (b) of Fig. 3) indicates a

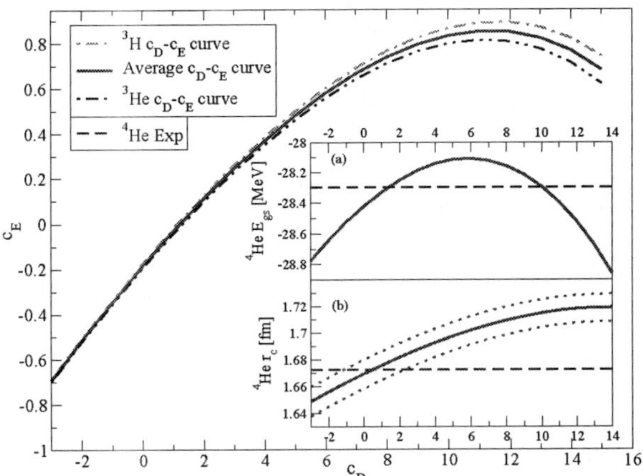

FIGURE 3. Relations between C_D and C_E for which the binding energy of ^{3}H (8.482 MeV) and ^{3}He (7.718 MeV) are reproduced. (a) ^{4}He binding energy along the averaged curve. The experimental ^{4}He binding energy (28.296 MeV) defines two points of intersection using the averaged $A = 3$ $C_D - C_E$ curve. (b) ^{4}He charge radius. Dotted lines represent the spread in r_c due to uncertainties in the proton charge radius.

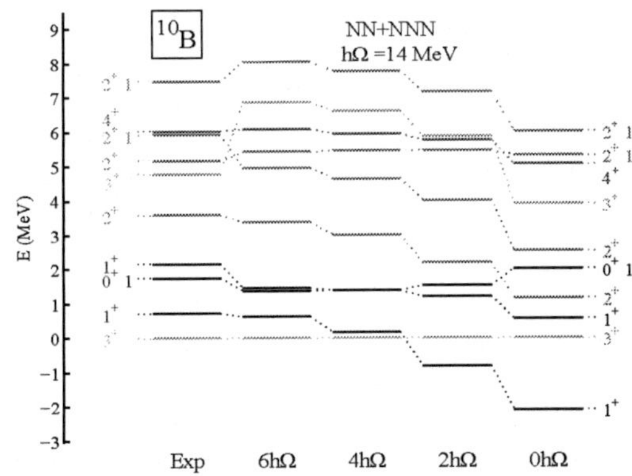

FIGURE 4. Experimental and theoretical excitation spectra of ^{10}B with respect to the lowest 3^+ state. The NCSM results are obtained with the chiral NN + TNI interaction [8, 9] at an oscillator energy, $\hbar\Omega = 14$ MeV as a function of N_{max} indicated at the bottom of each spectrum. Note the reasonable convergence as one proceeds up to $N_{max} = 6$

preference for $C_D \sim 0$ with a broad span of reasonable results around it. This led us to investigate observables in the mass 10-13 range where we find good results for $C_D = -1$, $C_E = -0.33$ [2].

An example of the improvement obtained with the TNI of ChPT for these values of C_D and C_E is shown in Fig. 2. The correct level ordering is now obtained for the lowest 5 states. In addition, the binding energy shifts towards agreement with experiment as seen below in Table 1. Refinements in our numerical methods will soon allow us to obtain the spectrum at $N_{max} = 8$ to extend the convergence trends for Fig. 4.

Ab exitu: NCSM with interaction from inverse scattering

Turning to results with the *ab exitu* NCSM we present in Fig. 2 a sample comparison of results with the *ab initio* NCSM for ^{10}B. We note that both approaches yield the correct level ordering through the first five states. Similar favorable comparisons are found in other light nuclei. It will be very helpful to have results for both approaches in larger basis spaces in order to asses more completely the distinctions between these approaches.

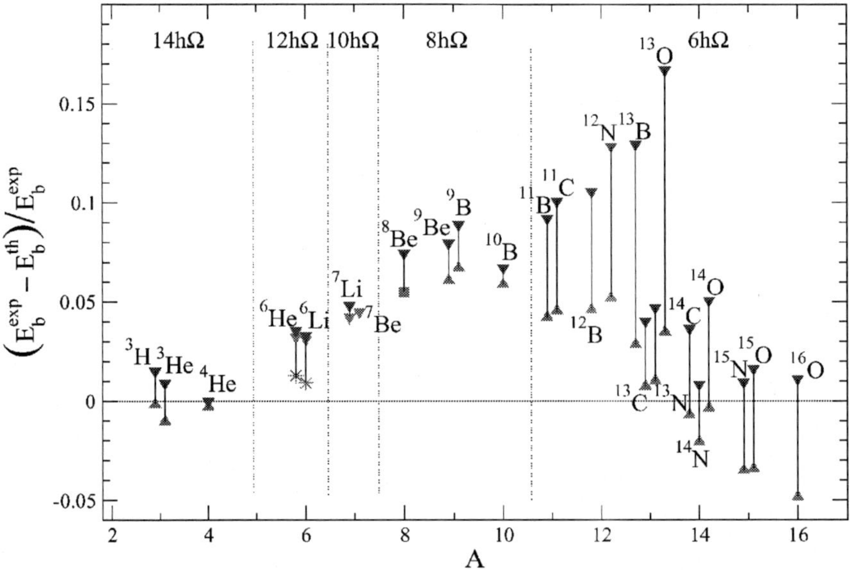

FIGURE 5. Fractional difference between theory and experiment for the binding energies of stable p-shell nuclei. The results are quoted with the specified N_{max} values and with both the bare and effective JISP16 interactions. The effective interactions are evaluated at the 2-body cluster level. The oscillator energy, $\hbar\Omega$ is taken, in each case, to be the value at which an extremum in the binding energy occurs.

For a more comprehensive view of the results with the *ab exitu* calculations, we present in Fig. 5 the binding energies of stable p-shell nuclei relative to experiment using the JISP16 interaction. Both bare and effective interaction results are presented as well as some initial extrapolations to the infinite basis limit [18]. The bare interaction results are strict upper bounds to the exact ground state energy so the bare curves will drop as the basis space is increased (direction of increased binding in the theory). The effective interaction results do not follow a variational principle. The results show a tendency to underbind nuclei in mid-p-shell and to overbind at the upper end by a few percent. Results in larger basis spaces will help clarify these trends.

Full Configuration Interaction calculations

As an alternative to deriving an effective finite many-body Hamiltonian matrix from the original Hamiltonian using the Lee-Suzuki method [1], one can solve the finite matrix defined by N_{max} for the original Hamiltonian. This has the benefit that the variational principle applies, and thus the binding energy obtained for a finite value of N_{max} is a strict upper bound for the converged result. In addition, the approach to convergence is monotonic, facilitating extrapolation techniques to infinite model-spaces. This constitutes a full Configuration Interaction (FCI) method.

In Fig. 6 we present initial results for the ground state energy of ^{12}C in an $N_{max} = 8$ basis space using the original (bare) JISP16 interaction and compare with the results obtained in smaller basis spaces. While the convergence trend is encouraging, see Fig. 7, we note that JISP16 produces a small amount of overbinding as indicated by our preliminary extropolated ground state energy and its uncertainty [18].

The convergence rate of a FCI calculation depends on the interaction: For soft interactions such as JISP16, this works quite well [18]; for other interactions convergence can be improved by a suitable renormalization of the Hamiltonian [19]. As further indication of the value of FCI, we note that other observables, such as the point proton rms radius, may be evaluated directly. Results for ^4He show excellent convergence to a result very close to experiment [18].

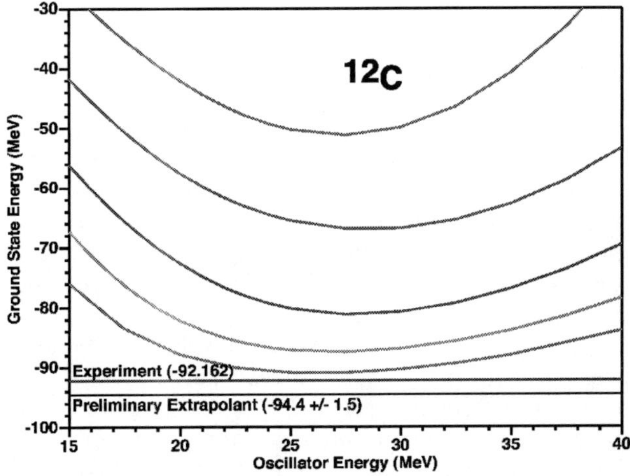

FIGURE 6. Ground state energy for ^{12}C as a function of the oscillator energy, $\hbar\Omega$, for $N_{max} = 0 - 8$ for the bare JISP16 interaction. The $N_{max} = 8$ curve is closest to experiment and each curve above it corresponds to decreases in N_{max} by two units.

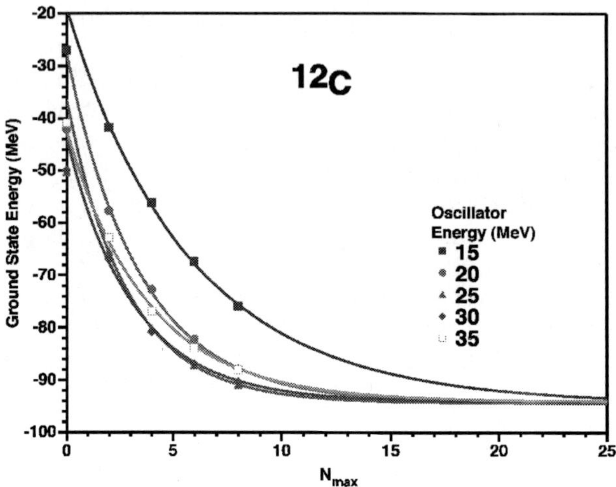

FIGURE 7. Calculated ground state energy of ^{12}C for $N_{max} = 0 - 8$ at selected values of $\hbar\Omega$. The data for $N_{max} = 2 - 8$ are fit to an exponential for each $\hbar\Omega$ plus a common constant, the asymptote. The preliminary result for the asymptote is -94.4 ± 1.5 MeV compared with the experimental ground state energy -92.16 MeV. The uncertainty is assigned using the range of the asymptotes from independent fits to the four points at each value of $\hbar\Omega$ shown here.

NCSM with interaction from theory and phenomenology

In the third approach, we work towards heavier nuclei and must restrict ourselves to basis spaces that are far from convergence as seen from the trends in light nuclei. Hence, we adopt a more phenomenological approach to the input Hamiltonian. Motivated by the success of the conventional shell model and by our successes in light nuclei, we fashion a hybrid approach that retains features of both - in particular the features that all nucleons are active and the feature that the NN interaction is adjusted to fit the properties of the heavier nuclei [12, 13].

In brief, the results are the following [12]: i) one finds that the charge dependence of the bulk binding energy of eight A=48 nuclei is reasonably described with an effective Hamiltonian derived from CD-Bonn interaction[22], while there is an overall underbinding by about 0.4 MeV/nucleon; ii) the resulting spectra are too compressed compared with experiment; iii) when isospin-dependent central NN terms plus a tensor NN interaction are added to the Hamiltonian, one achieves accurate total binding energies for eight A=48 nuclei and reasonable low-lying spectra for the three nuclei involved in double-beta decay. Only five input data were used to determine the phenomenological terms - the total binding of ^{48}Ca, ^{48}Sc, and ^{48}Ti along with the lowest positive and negative parity excitations of ^{48}Ca.

We have extended this approach to the odd-A isotopes ^{47}Ca, ^{49}Ca, ^{47}Sc and ^{47}K [13], which differ by one nucleon from ^{48}Ca. One of our goals is to test whether the same

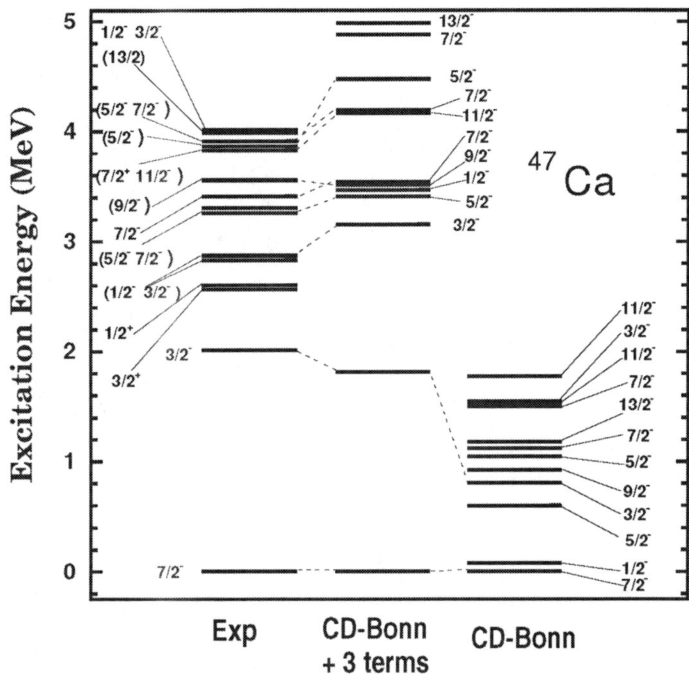

FIGURE 8. Experimental and theoretical excitation energy levels for ^{47}Ca. Both CD-Bonn and CD-Bonn+3terms results are presented.

modified effective Hamiltonian used for A=48 isotopes, is able to describe these odd-A nuclei. A particular feature of the spectroscopy of these odd nuclei is that the spin-orbit splitting gives rise to a sizable energy gap in the fp-shell between the $f_{7/2}$ and other orbitals ($p_{1/2}$, $p_{3/2}$, $f_{5/2}$) and we wanted to see if this feature can be reproduced in a NCSM where we have no input single-particle energies. The results both with and without the phenomenological terms are presented in Fig. 8.

In spite of the differences in frameworks with and without a core, we also wanted to compare our initial and modified Hamiltonian with a recent fp-shell interaction, the GXPF1, developed by Honma, Otsuka, Brown and Mizusaki [23]. We feel it is valuable to compare various fp-shell interactions in order to understand better their shortcomings and their regimes of applicability. From that comparison, notable differences are evident. For example, our diagonal two-body T=0 matrix elements are more attractive and, while our initial and modified NN Hamiltonian fp-shell matrix elements are strongly correlated, there is much less correlation with the GXPF1 matrix elements. It is worth mentioning that our interaction and Honma et al. GXPF1 interactions were also tested recently within the framework of spectral distribution theory in Ref. [24] and sizable differences were demonstrated.

TABLE 1. Properties of ^{10}B from experiment and theory. E2 transitions are in e^2 fm^4 and M1 transitions are in μ_N^2. The rms deviations of excited state energies are quoted for the lowest 9 states whose spin-parity assignments are well established and that are known to be dominated by $0\hbar\Omega$ configurations. Results are obtained in the basis spaces $N_{\max} = 6(8)$ with $\hbar\Omega = 14$ MeV for the ChPT (JISP16) interaction. In the $N3LO + TNI$ column we show selected sensitivity to changing C_D by ± 1. "N/A" indicates a result yet to be calculated. The experimental values are from Ref. [20, 21].

Property	Exp	*N3LO +TNI*	*N3LO*	*JISP16*
$\lvert E(3^+,0)\rvert$ [MeV]	64.751	64.78	56.11	59.751
r_p [fm]	2.30(12)	2.197	2.256	2.210
$Q(3_1^+,0)$ [e fm^2]	+8.472(56)	+6.327	+6.803	6.716
$\mu(3_1^+0)$ [μ_N]	+1.801	+1.837	+1.853	N/A
$E_x(3_1^-0)$ [MeV]	0.0	0.0	0.0	0.0
$E_x(1_1^-0)$ [MeV]	0.718	0.523	-1.128	0.193
$E_x(0_1\ 1)$ [MeV]	1.740	1.279	0.913	1.034
$E_x(1_2^-0)$ [MeV]	2.154	1.432	1.643	2.221
$E_x(2_1^-0)$ [MeV]	3.587	3.178	1.643	3.430
$E_x(3_2\ 0)$ [MeV]	4.774	6.729	4.193	5.792
$E_x(2_1^-1)$ [MeV]	5.164	5.315	4.419	4.861
$E_x(2_2^-0)$ [MeV]	5.92	4.835	3.555	5.201
$E_x(4_1^-0)$ [MeV]	6.025	5.960	4.790	5.685
$E_x(2_2^-1)$ [MeV]	7.478	7.823	5.565	7.320
$rms(Exp - Th)$ [MeV]	-	0.823	1.482	0.536
B(E2;$1_1^+0 \rightarrow 3_1\ 0$)	4.13(6)	3.05(62)	4.380	3.736
B(E2;$1_2^+0 \rightarrow 3_1^-0$)	1.71(0.26)	0.50(50)	0.082	0.578
B(M1;$2_1^+0 \rightarrow 3_1^+0$)	0.0015(3)	0.0000	N/A	0.0012
B(M1;$2_1^+1 \rightarrow 3_1^+0$)	0.041(4)	0.216	N/A	0.125
B(M1;$2_2^+0 \rightarrow 3_1^+0$)	0.050(12)	0.053	N/A	0.056
B(M1;$4_1^+0 \rightarrow 3_1^+0$)	0.043(7)	0.002	N/A	0.003
B(M1;$2_2^+1 \rightarrow 3_1^+0$)	-	4.020	N/A	4.148
B(GT;$3_1^+0 \rightarrow 2_1^-1$)	0.083(3)	0.07(1)	0.102	0.040
B(GT;$3_1^+0 \rightarrow 2_2^-1$)	0.95(13)	1.22(2)	1.487	1.241

CONCLUDING REMARKS

Table 1 contains selected experimental and theoretical results for ^{10}B. The binding energy and rms deviation between the experimental and theoretical excitation energies improve substantially with the inclusion of TNI. The JISP16 results lie intermediate to the N3LO and N3LO+TNI interaction results. Other observables are in reasonable accord with experiment considering that (1) we use bare electromagnetic operators, and (2) moments and transition rates are expected to be more sensitive to enlarging the basis spaces as we plan to do. The JISP16 results employ partial waves, $J \leq 4$. If we retain only $J \leq 3$ partial waves in JISP16, the excitation energies change by less than 10 keV and the binding energy decreases by 35 keV.

These results required substantial computer resources. The $N_{max} = 6$ spectrum shown in Fig. 4 and a set of additional experimental observables, takes an hour on 3500 processors on the LLNL-Thunder machine. Our largest run that is reported here, the ^{12}C with JISP16 in the $N_{max} = 8$ basis ($dimension = 6 \times 10^8$) took 2.3 hours on 15,400 processors (33,350 cpu hours) at the ORNL Jaguar facility for a single $\hbar\Omega$ value. All runs produce the lowest 15 converged eigenvectors and a suite of observables (rms radii, electromagnetic moments and transitition rates, electroweak transition rates, etc.). Development efforts underway promise to reduce the computational time by about a factor of two and to facilitate enhanced convergence as well as reaching heavier systems.

We demonstrated here that TNI's make substantial contributions to improving the spectra and other observables. In addition, phase-equivalent transformations of an interaction obtained from inverse scattering, JISP16, produces appealing fits to light nuclear properties. However, there is considerable room for further improvement in both approaches. Our leading suggestions include: (1) extend the TNI's to the order consistent with the NN interaction, N3LO; (2) extend the basis spaces to higher N_{max} values to further improve convergence; (3) examine sensitivity of TNI's to the choice of regulator; and (4) include four-nucleon interactions at a consistent order of ChPT. In addition, further exploration of the phase-shift equivalent transformations appears warranted. Finally, improving the phenomenological fits to medium mass nuclei appears within reach.

Our overall conclusion is that these results support a full program of deriving the NN interaction and its mulit-nucleon partners in the consistent approach provided by chiral effective field theory. It is straightforward, but challenging, to extend this research thrust in the directions indicated. However, the favorable results to date and the need for addressing fundamental symmetries of strongly interacting systems with enhanced predictive power firmly motivate this path. Recent developments of techniques to achieve physically-motivated larger basis spaces [25] hold great promise for attaining many of the goals we seek.

ACKNOWLEDGMENTS

This work was supported in part by the U. S. Department of Energy Grants DE-FG02-87ER40371 and DE-FC02-07ER41457. This work was also partly performed under the auspices of the U. S. Department of Energy by the University of California, Lawrence Livermore National Laboratory under contract No. W-7405-Eng-48. Support from the LDRD contract No. 04–ERD–058 and from U.S. DOE/SC/NP (Work Proposal Number SCW0498) is acknowledged. This work was also supported in part by the Russian Foundation of Basic Research.

REFERENCES

1. P. Navrátil, J. P. Vary and B. R. Barrett, *Phys. Rev. Lett.* **84**, 5728 (2000); *Phys. Rev. C* **62**, 054311 (2000).
2. P. Navrátil, V.G. Gueorguiev, J. P. Vary, W. E. Ormand and A. Nogga, *Phys. Rev. Lett* **99**, 042501 (2007).
3. S. Weinberg, *Physica* **96A**, 327 (1979); *Phys. Lett. B* **251**, 288 (1990); *Nucl. Phys.* **B363**, 3 (1991).
4. V. Bernard, *et al.*, *Int. J. Mod. Phys.* **E4**, 193 (1995).
5. C. Ordonez, L. Ray, and U. van Kolck, *Phys. Rev. Lett.* **72**, 1982 (1994); *Phys. Rev. C* **53**, 2086 (1996).

6. U. van Kolck, *Prog. Part. Nucl. Phys.* **43**, 337 (1999)
7. P. Bedaque and U. van Kolck, *Ann. Rev. Nucl. Part. Sci.* **52**, 339 (2002).
8. D. R. Entem and R. Machleidt, *Phys. Rev. C* **68**, 041001(R) (2003).
9. E. Epelbaum, W. Glöckle, and Ulf-G. Meissner, *Nucl. Phys.* **A637**, 107 (1998); **A671**, 295 (2000).
10. A. M. Shirokov, J. P. Vary, A. I. Mazur and T. A. Weber, *Phys. Letts. B* **644**, 33 (2007).
11. A. M. Shirokov, A. I. Mazur, S. A. Zaytsev, J. P. Vary and T. A. Weber, Phys. Rev. C **70**, 044005 (2004); A. M. Shirokov, J. P. Vary, A. I. Mazur, S. A. Zaytsev and T. A. Weber, *Phys. Letts. B* **621**, 96 (2005).
12. J.P. Vary, S. Popescu, S. Stoica, and P. Navrátil, nucl-th 0607041.
13. J.P.Vary, A.G. Negoita and S. Stoica, nucl-th 0612022, UCRL JRNL-226480.
14. B. A. Brown, *Progr. Part. Nucl. Phys.* **47**, 517 (2001)
15. T. Otsuka, M. Honma, T. Mizusaki, N. Shimizu and Y. Utsuno, *Prog. Part. Nucl. Phys.* **47**, 319 (2001).
16. P. Navrátil and W. E. Ormand, *Phys. Rev. Lett.*, **88**, 152502 (2002).
17. A. Nogga, P. Navrátil, B. R. Barrett and J. P. Vary, *Phys. Rev. C* **73**, 064002 (2006).
18. P. Maris, J. P. Vary and A. Shirokov, in preparation.
19. S.K. Bogner, R.J. Furnstahl, P. Maris, R.J. Perry, A. Schwenk, J.P. Vary, arXiv:0708.3754.
20. F. Ajzenberg-Selove, *Nucl. Phys.* **A490**, 1 (1988).
21. A. Ozawa, I. Tanihata, T.Kobayashi, Y. Sugihara, O. Yamakawa, K. Omata, K. Sugimoto, D. Olson, W. Christie, and H. Wieman, Nucl. Phys. **A608**, 63 (1996); H. De Vries, C. W. De Jager, and C. De Vries, *At. Data Nucl. Data Tables* **36**, 495 (1987).
22. R. Machleidt, F. Sammarruca and Y. Song, Phys. Rev. C **53**, 1483 (1996); R. Machleidt, *Phys. Rev. C* **63**, 024001 (2001).
23. M. Honma, T. Otsuka, B. A. Brown and T. Mizusaki, *Phys. Rev. C* **65**, 061301(R) (2002); *Phys. Rev. C* **69**, 034335 (2004).
24. K. D. Sviratcheva, J. P. Draayer and J. P.Vary, *Phys. Rev. C* **73**, 034324 (2006).
25. T. Dytrych, K. D. Sviratcheva, C. Bahri, J. P. Draayer and J. P. Vary, *Phys. Rev. Lett.* **98**, 162503 (2007).

Beyond Mean Field Description of the Gamow-Teller β^+ Decay of Proton-Rich Kr Isotopes

A. Petrovici

National Institute for Physics and Nuclear Engineering, R-077125 Bucharest, Romania
Institut für Theoretische Physik, Universität Tübingen, D-72076 Tübingen, Germany

Abstract. We studied the influence of shape mixing on the Gamow-Teller β^+ decay of the nucleus ^{74}Kr using the *complex* Excited Vampir variational approach for the description of the ground state of the parent nucleus and the 1^+ states in the beta window in the daughter nucleus ^{74}Br. Within the same approach the Gamow-Teller strength distributions for the decay of the ground state and the lowest-excited states in the nucleus ^{72}Kr to the 1^+ states in the nucleus ^{72}Br are obtained using a rather large model space and realistic effective interaction. The shape coexistence and mixing are consistently described for both the 0^+ and 2^+ states in the even-even parent nucleus and the 1^+ states in the odd-odd daughter nucleus. The theoretical results for the Gamow-Teller strength distributions and the accumulated strengths are compared with the available data and predictions concerning the decay of the first excited 0^+ and the yrast 2^+ in the rp-process waiting point nucleus ^{72}Kr are presented.

Keywords: shape coexistence, proton-rich nuclei, Gamow-Teller beta decay
PACS: 23.40.-s; 27.50.+e

INTRODUCTION

The proton-rich nuclei near the N=Z line in the A = 60 -90 mass region display a wide variety of interesting nuclear structure effects connected with shape-coexistence and -mixing as well as the competition between proton-neutron and like-nucleon pairing correlations. These nuclei play also an important role in nuclear astrophysics, since their weak decay determines details of the nucleosynthesis. The knowledge of the Gamow-Teller (GT) β^+ decays of the even-even N = Z nuclei near the proton drip line is of essential importance for the rapid-proton (rp) capture process. Also relevant could be the GT-strength distributions for the low-lying excited states whose thermal population may result in a significant reduction of the effective lifetime at the high temperatures of X-ray bursts [1].
Precise theoretical estimates of the beta-decay strength distribution require a reliable description of the structure of the ground state and the low-lying excited states of the parent nucleus as well as the states populated in the daughter nucleus. This is a rather complicated task, especially, if the structure of the considered nuclei is dominated by shape-coexistence and -mixing. Previous theoretical investigations [2–5] suggested that in this mass region the GT strength distributions depend sensitively on the nuclear shape. The coexistence phenomena dominating the structure and dynamics of nuclei in this mass region have been intensively investigated within the *complex* Excited Vampir varia-

CP972, *Exotic Nuclei and Nuclear/Particle Astrophysics (II)—Proceedings of the Carpathian Summer School of Physics 2007,* edited by L. Trache and S. Stoica

tional approach and many predictions that emerged have been experimentally confirmed [6–14].
The structure of ^{74}Kr was studied both theoretically and experimentally establishing that shape-coexistence and -mixing are dominating the structure of this nucleus [6–9]. Strong support for the existence of shape mixing came especially from the measurement of the electric-monopole (E0) transition from the first excited isomeric 0^+ state to the 0^+ ground state of ^{74}Kr [8]. This is in agreement with the theoretical investigations [6, 7, 9, 15], which indicated that the structure and dynamics of almost all the even-even as well as the odd-odd nuclei close to N=Z line in the A$\simeq$ 70 mass region are dominated by shape-coexistence and in some cases even -mixing, irrespective, whether low or high spin states are considered. In ^{72}Kr we predicted the existence of a low-lying isomeric 0^+ state and then it was experimentally confirmed.
The *complex* Excited Vampir variational approach uses Hartree-Fock-Bogoliubov (HFB) vacua as basic building blocks, which are only restricted by time-reversal and axial symmetry. The underlying HFB transformations are essentially *complex* and do mix proton- with neutron-states as well as states of different parity and angular momentum. The broken symmetries of these vacua (nucleon numbers, parity, total angular momentum) are restored by projection techniques and the resulting symmetry-projected configurations are then used as test wave functions in chains of successive variational calculations to determine the underlying HFB transformations as well as the configuration mixing. The HFB vacua of the above type account for arbitrary two-nucleon correlations and thus simultaneously describe like-nucleon as well as isovector and isoscalar proton-neutron pairing. Furthermore the *complex* Excited Vampir model allows the use of rather large model spaces and realistic effective interactions.
The studies which we present here represent the first atempt at a completely self-consistent calculation of the Gamow-Teller β^+ decay of the ground state of ^{74}Kr and the lowest two 0^+ states and the yrast 2^+ state in ^{72}Kr to the 1^+ states in ^{74}Br and ^{72}Br, respectively, using the *complex* Excited Vampir approach.

THE THEORETICAL FRAMEWORK

We calculated the 0^+ ground state in ^{74}Kr, the lowest 0^+ and 2^+ states in ^{72}Kr, and the lowest 1^+ states in ^{74}Br and ^{72}Br using the *complex* Excited Vampir approach. We constructed succesively in chains of independent variational calculations the Excited Vampir many-nucleon basis for each symmetry. The final solutions have been obtained diagonalizing the residual interaction between the considered Excited Vampir configurations for a given spin and parity.
As in our earlier calculations for nuclei in the $A \simeq 70$ mass region we used a ^{40}Ca core and included the $1p_{1/2}$, $1p_{3/2}$, $0f_{5/2}$, $0f_{7/2}$, $1d_{5/2}$ and $0g_{9/2}$ oscillator orbits for both protons and neutrons in the valence space. We started with an isospin symmetric basis and then introduced the Coulomb shifts for the proton single-particle levels [15]. The Coulomb-contributions to the single particle energies resulting from the ^{40}Ca core have been obtained performing spherically symmetric Hartree-Fock calculations using the Gogny-interaction D1S in a 21 major-shell basis.
The effective two-body interaction is constructed from a nuclear matter G-matrix

TABLE 1. The amount of mixing for the ground state of ^{74}Kr

monopole shifts	o-mixing	p-mixing
*ms*1	56(2)(1)(1)%	35(3)(1)(1)%
*ms*2	39(2)(1)(1)%	51(3)(1)(1)%
*ms*3	28(1)(1)%	65(3)(2)%

based on the Bonn One-Boson-Exchange potential (Bonn A). This G-matrix was modified by adding a short-range (0.707 fm) Gaussian with strength -35 MeV in the T=1 (proton-proton, neutron-neutron and proton-neutron) channnel and a Gaussian of the same range with strength -70 MeV in the T=0 channel in order to increase the pairing-correlations in these channels. In addition the isoscalar interaction was modified by monopole shifts of -500 keV for all T=0 matrix elements of the form $\langle 1p1d_{5/2};IT=0|\hat{G}|1p1d_{5/2};IT=0\rangle$, where 1p denotes either the $1p_{1/2}$ or the $1p_{3/2}$ orbit and three different sets of monopole shifts for those of the form $\langle 0g_{9/2}0f;IT=0|\hat{G}|0g_{9/2}0f;IT=0\rangle$ with $0f$ denoting either the $0f_{5/2}$ or the $0f_{7/2}$ orbit. These monopole shifts had been introduced in the earlier calculations in order to influence the onset of deformation. These earlier studies [7, 9] revealed futhermore that the monopole shifts for protons and neutrons occupying the $0g_{9/2}$ and $0f_{5/2}$ or $0f_{7/2}$ influence also the relative position of the oblate- and prolate-deformed configurations. In the present investigations the matrix elements $\langle 0g_{9/2}0f_{5/2};IT=0|\hat{G}|0g_{9/2}0f_{5/2};IT=0\rangle$ and $\langle 0g_{9/2}0f_{7/2};IT=0|\hat{G}|0g_{9/2}0f_{7/2};IT=0\rangle$ were shifted by -0.590 MeV and -0.060 MeV (*ms*1), -0.500 MeV and -0.150 MeV (*ms*2), and -0.400 MeV and -0.250 MeV (*ms*3), respectively.

RESULTS AND DISCUSSION

The Gamow-Teller β^+ Decay of ^{74}Kr

For the ground state of ^{74}Kr the up to 14 energetically lowest 0^+ Excited Vampir configurations have been included. In the daugther nucleus ^{74}Br up to 74 1^+ states have been taken into account.

In Table 1 the amount of mixing for the ground state of the ^{74}Kr nucleus is presented. The main components of the wave function for the (*ms*1)-monopole shifts are characterized in the intrinsic system by $\beta_2=-0.34$ and $\beta_2=0.33$ for the oblate and prolate configurations, respectively. It should be mentioned that the configuration contributing 3% of the total amplitude is almost spherical having $\beta_2=0.07$ in the intrinsic system. The first excited 0^+ state is obtained at an excitation energy of 737 keV. Experimentally the first excited 0^+ state occurs at 508 keV.

As in the ^{74}Kr ground state also for the 1^+ states in ^{74}Br strong mixing of differently deformed oblate and prolate configurations in the final wave functions is observed. The amount of mixing of the differently deformed oblate and prolate Excited Vampir

TABLE 2. The amount of mixing for the lowest calculated 1^+ states of ^{74}Br with significant B(GT) ($ms1$)

o − mixing / **p-mixing**
94(3)(3)%
61(35)(2)(1)%
89(3)(2)(2)(1)(1)(1)%
44(28)(19)(4)(1)(1)(1)%
97%
69(19)(5)(2)(2)%
70(7)(3)(2)(1)(1)(1)%**4(3)(2)(2)%**
9(3)% **25(24)(11)(10)(3)(2)(2)(1)(1)(1)(1)(1)(1)(1)%**
7(1)(1)% **71(8)(5)(1)(1)(1)(1)(1)%**
43(12)(9)(7)(7)(3)(3)(2)(2)(2)(1)(1)(1)(1)(1)(1)(1)%
57(3)(2)(1)(1)(1)(1)%**13(5)(4)(2)(2)(1)(1)(1)(1)%**
26(1)(1)% **36(20)(4)(3)(2)(1)(1)(1)%**
21(21)(14)(14)(6)(5)(4)(2)(2)(1)(1)(1)(1)(1)(1)%
1%**26(20)(8)(7)(6)(5)(4)(3)(2)(2)(2)(1)(1)(1)(1)(1)(1)(1)(1)(1)(1)(1)%**
2(1)(1)(1)% **36(14)(12)(6)(6)(5)(4)(2)(2)(1)(1)(1)(1)(1)(1)%**
10(2)(2)(1)(1)% **27(13)(11)(9)(3)(3)(2)(2)(2)(2)(1)(1)(1)(1)(1)(1)(1)(1)%**
50(16)(9)(5)(3)(3)(2)(2)(2)(1)(1)(1)% **2(2)%**
33(21)(12)(8)(5)(5)(3)(3)(2)(2)(1)(1)% **1(1)(1)%**
1(1)%**34(18)(13)(9)(4)(3)(2)(2)(2)(2)(1)(1)(1)(1)(1)(1)%**

configurations is illustrated in Table 2 for the lowest calculated 1^+ states with significant B(GT) using the ($ms1$)-monopole shifts. The spectroscopic quadrupole moments for the calculated 1^+ states reveal the large variety of deformations characterizing the Excited Vampir states. In order to study the influence of the oblate-prolate mixing in the ground state of the parent as well as in the structure of the 1^+ states in the daughter nucleus we slightly changed the Hamiltonian and repeated the calculations using the $ms2$ and $ms3$ sets of monopole shifts. Though the relative positions of the oblate with respect to the prolate deformed configurations are changed, the main characteristics of the Excited Vampir many-nucleon basis like, e.g., the quadrupole, octupole, and hexcadecapole moments in the intrinsic system, as well as pairing properties are almost the same as those obtained in the calculations using the $ms1$ monopole shifts. Also for the lowest calculated 1^+ states in the ^{74}Br nucleus one obtains a rather similar structure as for $ms1$. However, since the density of states is very high and even 10 1^+ states appeared within 400 keV in some regions of excitation energy, which is in agreement with the experimental data, the final mixing of the states is now changed with respect to the calculations using the monopole shifts $ms1$. Consequently the spectroscopic quadrupole moments are modified, but again a large variety of deformations is revealed.
The mixing of the states is reflected in the Gamow-Teller strength distributions which we present in Figures 1–3 for the three sets of monopole shifts together with the experimental data [16]. In the experiments at CERN/ISOLDE the Gamow-Teller strength distribution over most of the $Q_{EC} = 3.140(62)$ MeV window and the total strength were measured for states with excitation energies below 3 MeV using a total absorption spectrometer particulary dedicated to the study of very short-lived nuclei [16].

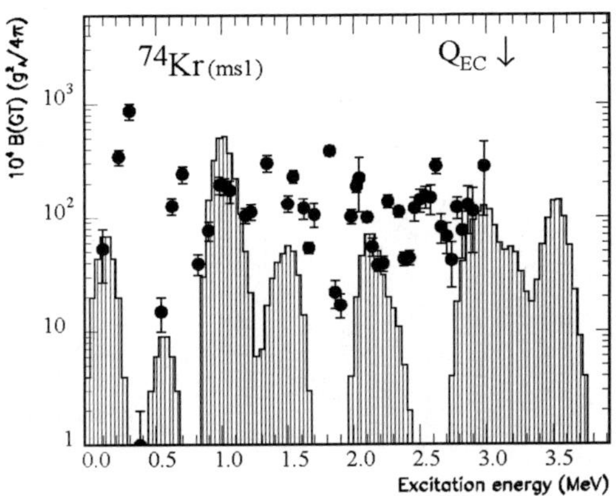

FIGURE 1. The theoretical Gamow-Teller strength distribution for ^{74}Kr obtained within the *complex* Excited Vampir approach with monopole shifts *ms*1 is compared to the experimental data [16]

To simulate the finite experimental resolution the theoretical results have been folded with Gaussian functions whose width of 60 keV was taken from the experimental resolution of the total absorption spectrometer [16]. In Figures 1–3 the folded theoretical results are the displayed as histogram with 40 keV bins. We did not include in the figures the contribution from the highest excited states since their structure is likely to be modified significantly by even higher excited configurations not included in the present many-nucleon Excited Vampir basis. To some extent this holds even for the states below 4 MeV excitation energy. Also these could still get more correlations from excitations above the lowest 1^+ states considered here. The influence of the shape mixing appears in the Gamow-Teller strength distributions which indicate maximum fragmentation for the Hamiltonian involving the *ms*3 set of monopole shifts connected with maximum contribution of prolate deformed configurations in the structure of the ground state of ^{74}Kr. The accumulated GT strenghts are similar for the three investigated cases with more contribution from the states below 1 MeV excitation energy for the *ms*3 Hamiltonian, and maximum total strength for the *ms*1 one when the ground state of ^{74}Kr is dominated by oblate deformed configurations.

The Gamow-Teller β^+ Decay of ^{72}Kr

Using the Hamiltonian corresponding to the *ms*3 set of monopole shifts we calculated the lowest 0^+ and 2^+ states of the even-even parent ^{72}Kr nucleus, and the lowest 1^+ states in the beta window in the odd-odd daughter ^{72}Br nucleus. For the description of the ground state and the first excited 0^+ state of ^{72}Kr we included in the Excited Vampir many-nucleon basis the energetically lowest 17 0^+ configurations. For the description of the yrast 2^+ state the lowest 9 2^+ configurations have been taken into account. In Table 3 the amount of mixing for the calculated states of the ^{72}Kr nucleus is presented indicating only the contributions larger than 1% of the total amplitude. In the ^{72}Br nucleus we

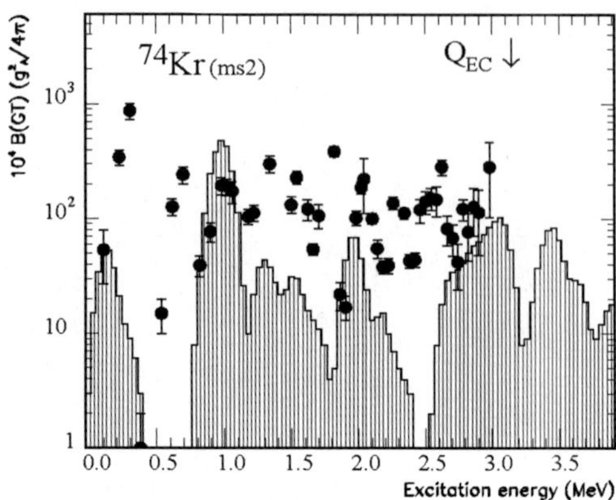

FIGURE 2. The same as in Figure 1, but for the *ms*2 monopole shifts

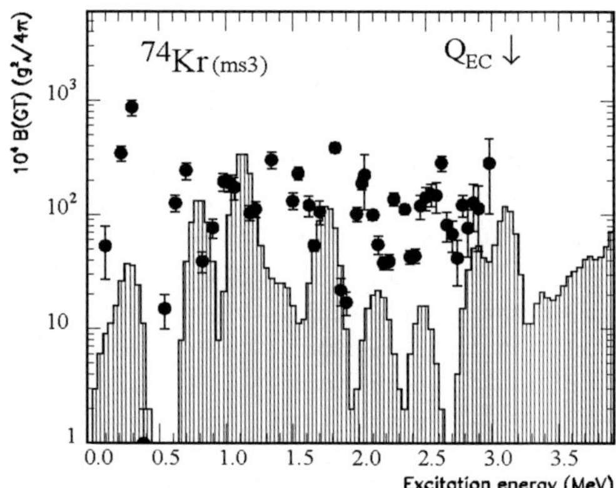

FIGURE 3. The same as in Figure 1, but for the *ms*3 monopole shifts

calculated the lowest 118 1^+ configurations in order to describe the states within the Q_{EC} window. The resulting lowest 118 1^+ states are all situated within 7.1 MeV excitation energy. As for the lowest two 0^+ states in ^{72}Kr also for the 1^+ states in ^{72}Br a variable sometimes very strong mixing of differently deformed oblate and prolate configurations in the final solutions is obtained.
In the experiments at CERN/ISOLDE the Gamow-Teller strength distribution for the ^{72}Kr β^+ decay could only be established up to 2 MeV excitation energy in ^{72}Br while the whole beta window amounts to $Q_{EC} = 5.040(375)$ MeV [17]. The theoretical results for the Gamow-Teller strength distribution for the decay of the ground state of ^{72}Kr over the whole Q_{EC} window is presented in Figure 4 together with the available data [17]. In Figures 5 and 6 we present the results for the decay of the first excited 0^+ and the yrast 2^+ state of ^{72}Kr to the calculated 1^+ states in ^{72}Br, respectively. In Figure 7 we present the accumulated Gamow-Teller strength for the decay of the ground state of

TABLE 3. The amount of mixing for the considered states of the ^{72}Kr nucleus

$I^{\pi}[\hbar]$	o-mixing	p-mixing
0_1^+	64(2)%	29(2)(1)(1)%
0_2^+	35(2)%	57(3)(1)(1)%
2_1^+	92(1)%	6%

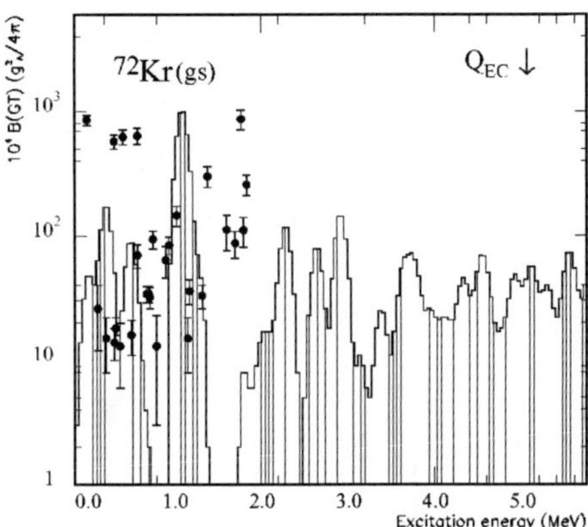

FIGURE 4. The theoretical Gamow-Teller strength distribution for the decay of the ground state of ^{72}Kr obtained within the *complex* Excited Vampir approach is compared to the experimental data [17]

^{72}Kr in 40 keV bins and compare it to the available data [17]. In Figure 8 we compare the accumulated GT strength curves for the decay of the lowest two 0^+ and the yrast 2^+ states of ^{72}Kr to the 1^+ states of ^{72}Br.
We would like to stress that in the present study (because of the large model space) no quenching was needed. It should also be noted that here for the first time strength distributions not only for the ground state but also for the decay of excited states have been calculated. Of course the theoretical results on the Gamow-Teller strength distributions could be changed, at least for the high excitation energy region, by the higher-lying configurations not included in the *complex* Excited Vampir many-nucleon basis. Also changes in the renormalization of the effective interaction could influence the oblate-prolate mixing in the structure of the wave functions of both parent and daughter states. This could perhaps also shift the 1^+ excitation energies to somewhat lower values as it seems to be suggested by experiment. In the near future systematic investigations on the decay properties of other waiting point nuclei, potential bottlenecks for the rp-process in this mass region, will be accomplished.

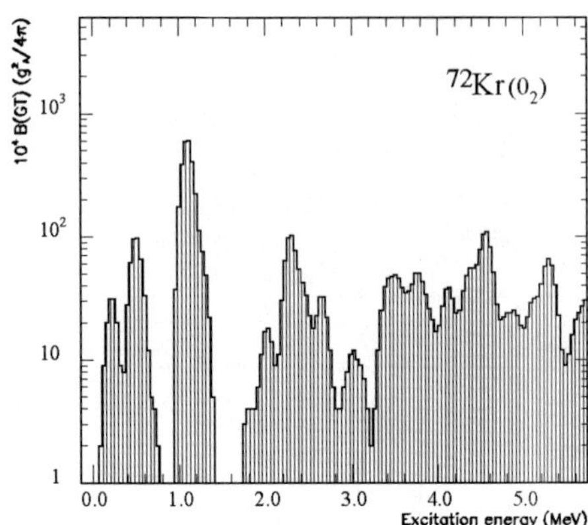

FIGURE 5. The theoretical Gamow-Teller strength distribution for the decay of the first excited 0^+ state of ^{72}Kr obtained within the *complex* Excited Vampir approach

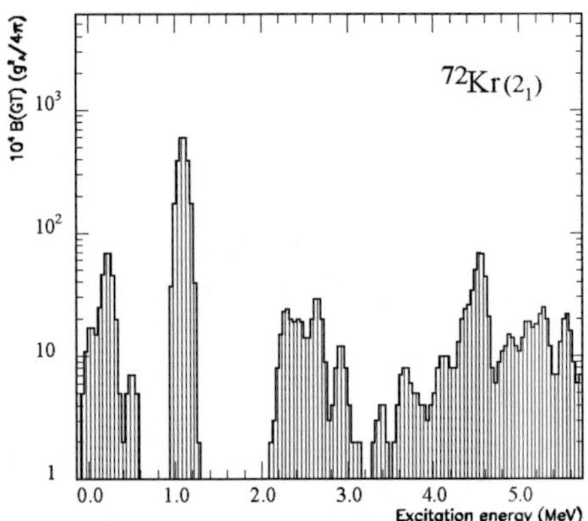

FIGURE 6. The same as in Figure 5, but for the yrast 2^+ state of ^{72}Kr

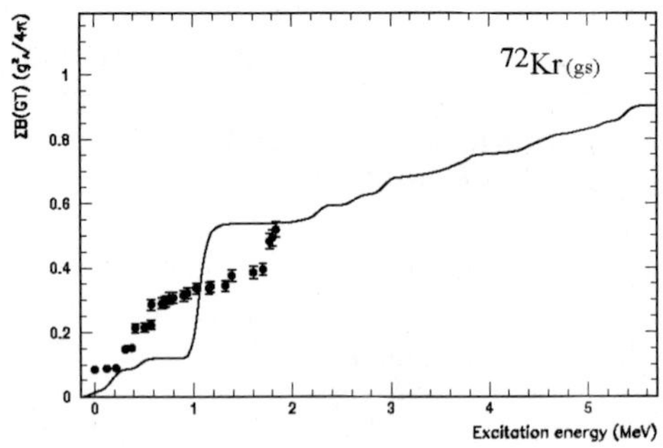

FIGURE 7. The theoretical Gamow-Teller accumulated strength for the decay of the ground state of ^{72}Kr obtained within the *complex* Excited Vampir approach is compared to the experimental data [17]

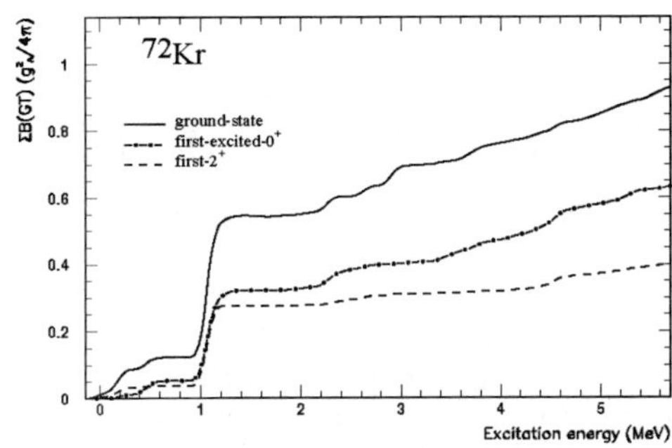

FIGURE 8. Comparison of the Gamow-Teller accumulated strength for the decay of the ground, first excited 0^+ and yrast 2^+ states of ^{72}Kr to the 1^+ states in ^{72}Br obtained within the *complex* Excited Vampir approach

ACKNOWLEDGMENTS

The results presented here have been obtained in collaboration with K.W. Schmid, O. Radu, and Amand Faessler. We wish to acknowledge the support from ANCS, Romania, under contract 54/2007 and DFG, Germany under contract 436RUM 113/20/0-1.

REFERENCES

1. H.Schatz, A. Aprahamian, J. Görres, M. Wiescher, T. Rauscher, J.F. Rembges, F.-K. Thielemann, B. Pfeiffer, P. Möller, K.-L. Kratz, H. Herndl, B.A. Brown, H. Rebel, *Phys. Rep.* **294**, 167–264 (1998).
2. I. Hamamoto, X.Z. Zhang, *Z. Phys.* **A353**, 145–148 (1995).
3. F. Frisk, I. Hamamoto, X.Z. Zhang, *Phys. Rev.* **C52**, 2468–2474 (1995).
4. P. Sarriguen, E. Moya de Guerra, A. Escuderos, *Nucl. Phys.* **A658**, 13–44 (1999).
5. P. Sarriguen, E. Moya de Guerra, A. Escuderos, *Nucl. Phys.* **A691**, 631–648 (2001).
6. A. Petrovici, K.W. Schmid, A. Faessler, *Nucl. Phys.* **A605**, 290–300 (1996).
7. A. Petrovici, K.W. Schmid, A. Faessler, J.H. Hamilton, A.V. Ramayya, *Progr. Part. Nucl. Phys.* **43**, 485–535 (1999).
8. C. Chandler, P.H. Regan, C.J. Pearson, B. Blank, A.M. Bruce, W.N. Catford, N. Curtis, S. Czajkowski, W. Gelletly, R. Grzywacz, Z. Janas, M. Lewitowicz, C. Marchand, N.A. Orr, R.D. Page, A. Petrovici, A.T. Reed, M.G. Saint-Laurent, S.M. Vincent, R. Wadsworth, D.D. Warner, J.S. Winfield, *Phys. Rev.* **C56**, R2924–R2928 (1997).
9. A. Petrovici, K.W. Schmid, A. Faessler, *Nucl. Phys.* **A665**, 333–350 (2000).
10. A. Petrovici, *Nucl. Phys.* **A704**, 144c–153c (2002).
11. A. Petrovici, K.W. Schmid, A. Faessler, *Nucl. Phys.* **A708**, 190–204 (2002).
12. A. Petrovici, K.W. Schmid, A. Faessler, *Nucl. Phys.* **A728**, 396–414 (2003).
13. A. Petrovici, K.W. Schmid, O. Radu, A. Faessler, *J. Phys. G:Nucl. Part. Phys.* **32**, 583–597 (2006).
14. A. Petrovici, K.W. Schmid, O. Radu, A. Faessler, *Eur. Phys. J.* **A28**, 19–25 (2006).
15. A. Petrovici, K.W. Schmid, O. Radu, A. Faessler, *Nucl. Phys.* **A747**, 44–52 (2005).
16. E. Poirier, F. Maréchal, Ph. Dessagne, A. Algora, M.J.G. Borge, D. Cano-Ott, J.C. Caspar, S. Courtin, J. Devin, L.M. Fraile, W. Gelletly, G. Heitz, A. Jungclaus, G. Le Scornet, Ch. Miehé, E. Nácher, B. Rubio, P. Sarriguren, J.L. Tain, O. Tengblad, C. Weber, and the ISOLDE Collaboration, *Phys. Rev.* **C69**, 034307–034307-8 (2004).
17. I. Piqueras, M.J.G. Borge, Ph. Dessagne, J. Giovinazzo, A. Huck, A. Jokinen, A. Knipper, C. Longour, G. Marguier, M. Ramdhane, V. Rauch, O. Tengblad, G. Walter, Ch. Miehé, and the ISOLDE Collaboration, *Eur. Phys. J.* **A16**, 313–329 (2003).

A Revisited View of the N=20 and N=28 Magic Numbers in the Neutron-Rich Nuclei

S. Grévy

Grand Accélérateur National d'Ions Lourds (GANIL), CEA/DSM - CNRS/IN2P3, Bd Henri Becquerel, BP 55027, F-14076 Caen Cedex 5, France

Abstract.
The shell model description of the nucleus and the magic numbers resulting from such an approach is one of the cornerstone of the nuclear structure. Nevertheless, the question of the persistence of the 'spherical' magic numbers in the very neutron-rich nuclei is highly debated. We discuss in the following experimental results obtained in the last years at GANIL concerning the N=20 and N=28 shell closures in the neutron-rich nuclei.

Keywords: shell closure, magic numbers
PACS: 21.10.-k; 23.20.Lv; 27.30.+t; 27.40.+z

WHY MAGIC NUMBERS DO EXIST ?

Different ingredients are needed in order to have magic numbers. The first one is a shell structure. This concept of shell structure in the nucleus first came from different observations made at the early beginning of the study of the nuclear structure. Already in 1935 Elsasser [1] reported about "the structure of the complex atomic nuclei" and noticed the 'regularities' in the abundance of the stable nuclei. The second ingredient is the Pauli principe. Without such a principle, all the nucleons would stay on the same orbital, the most deeply bound, and no other orbitals would be filled. Finally, we can turn to a simple 3-dimensional Fermi gas model in order to understand how magic numbers are generated. The energies of such a system are given, for the different states, as a function of the radial number n and the angular momentum ℓ. The radial number gives the number of nodes in the wave function and then an higher n corresponds to a lower wave length λ and consequently to an higher energy. In the same way, an higher angular momentum means higher centrifugal potential and then higher energy also. Since the energy of a given configuration increase both with n and ℓ, it is possible to counterbalance the effect of both to create groups of levels with similar energies, given birth to magic numbers in the shell structure. The magic numbers of the harmonic oscillator (among them 2, 8 and 20 are also magic numbers for the nuclear structure) originate only from this n and ℓ degeneracy. In order to obtain the known magic numbers of the nuclear structure above 20, other effects are important to introduce in the description and it is finally in 1949 that a complete description has been given by Maria Goeppert Mayer [2] who introduced the famous spin-orbit potential. Nevertheless, from an original picture where the magic numbers were believed to be somehow 'universal', being valid for the stable as well as for the 'exotic' nuclei, we moved to a picture in which the shell structure can change drastically with the isospin degree of freedom resulting in the collapse of some

CP972, *Exotic Nuclei and Nuclear/Particle Astrophysics (II)—Proceedings of the Carpathian Summer School of Physics 2007,* edited by L. Trache and S. Stoica

magic numbers and/or the birth of new ones. In the following, we are going to illustrate these phenomena by discussing the evolution of the N=20 and N=28 magic numbers in the neutron rich nuclei.

N=20 AND THE ISLAND OF INVERSION

Disappearance of N=20

A typical example of sudden change in the shell structure is known to occur at N=20 for the nuclei located in or close to the so-called 'island of inversion' where the intruder configurations can dominate the ground-state structure of the nuclei leading to deformed ground states. The first evidence has been found through the measurement of unexpected masses [3]. Then, the measurement of the energy of the first 2^+ state in ^{32}Mg [4] was the first direct manifestation of change in the shell structure of this nucleus. Indeed, nuclei having a magic number such as ^{40}Ca were known to have a first 2^+ state at relatively high energy characterizing the difficulty to excite particles over a large energy gap associated with a magic number. The 2^+ energy measured for ^{32}Mg - 885 keV - approximately 5 times less than expected for a magic nucleus was then interpreted as an indication of a decrease of the gap N=20 leading to the feeding of intruder orbitals. This interpretation was confirmed later by the measurement of a relatively large transition probability B(E2) [5].

In 2001 Otsuka [6] showed that the evolution of the single particle orbitals in the region of the island of inversion can be understood by considering the tensor part of the proton-neutron interaction. Indeed, by removing protons in the $d_{5/2}$ orbital, the attractive interaction on the neutron $d_{3/2}$ orbital is reduced and as a consequence the neutron $d_{3/2}$-$f_{7/2}$ gap (namely N=20) is reduced due to the migration of the neutron $d_{3/2}$ orbital. As a consequence, it was also predicted that the $s_{1/2}$-$d_{3/2}$ gap would increase giving birth to a new N=16 magic number. In such a picture, the neutron rich isotopes of oxygen are very interesting since it is experimentally proved that the last bound isotope is ^{24}O having N=16 whereas ^{28}O, despite a neutron magic number (N=20) has never been observed [7]. The particle instability of the heavier oxygen isotopes can be understood in such a scenario, the neutron $d_{3/2}$ orbital becoming unbound for the most neutron rich isotopes. Many experimental effort have been done in order to confirm this hypothesis.

Birth of N=16

In a single particle picture the position of the first $3/2^+$ state in ^{23}O (N=15) gives direct information on the size of the N=16 gap. In the same way, a large N=16 gap would reflect in a first 2^+ state located at high excitation energy in ^{24}O.

The measurement of such states has been performed at GANIL by in-beam gamma spectroscopy taking advantage of the double fragmentation method [8]. A first fragmentation in the SISSI device of a ^{36}S beam at 75MeV/u produced a cocktail beam optimized for the production of nuclei around ^{28}Na. These radioactive ions, selected by the alpha spectrometer and a Be degrader were transported at the target point of the SPEG and identified event by event by their energy loss and their time-of-fligt from the exit of the alpha spectrometer to the secondary target. The cocktail beam impinged a secondary C target (112 mg/cm^2) and the nuclei produced in the second fragmentation were transmitted through the SPEG spectrometer with a Bρ value adjusted to maximize the collection of the nuclei of interest. They were unambiguously identified through their energy loss in an Ionization Chamber and their masse over charge ratio deduced from time-of-flight and Bρ measurement. The secondary target was surrounded by the 4π gamma array 'Chateau de Cristal' consisting in 74 BaF_2 scintillators. The prompt gamma rays were selected by the time-of-flight measured between the detection of the incident nuclei and the detection of the gamma-rays in the BaF_2 detector. The time resolution allows to remove neutrons and light charged particles emitted mainly at forward angles in the fragmentation process. The photopeak efficiency and energy resolution have been measured with stationary sources and are 33% and 10% at 1 MeV, respectively. The efficiency for a moving source with a beta of 0.3 has been obtained from a Monte-Carlo simulation. The gamma spectra have been obtained by selecting, event-by-event, the incoming nuclei and the ejectiles after the secondary target. The Doppler correction has been performed with the assumption that the reaction occurs in the middle of the target. An addback procedure consisting to sum the energies of the neighbors detectors in the events with multiplicity higher than one has been applied and significantly reduces the background at low energies.

The spectra obtained for ^{23}O and ^{24}O are shown on figure 1. The flat behavior indicate that no transition is observed and consequently that these nuclei have no bound excited states. Considering the neutron separation energies which are 2.7 and 3.7 MeV in ^{23}O and ^{24}O respectively, it indicates that the N=16 gap is larger than 3.7 MeV in these isotopes.

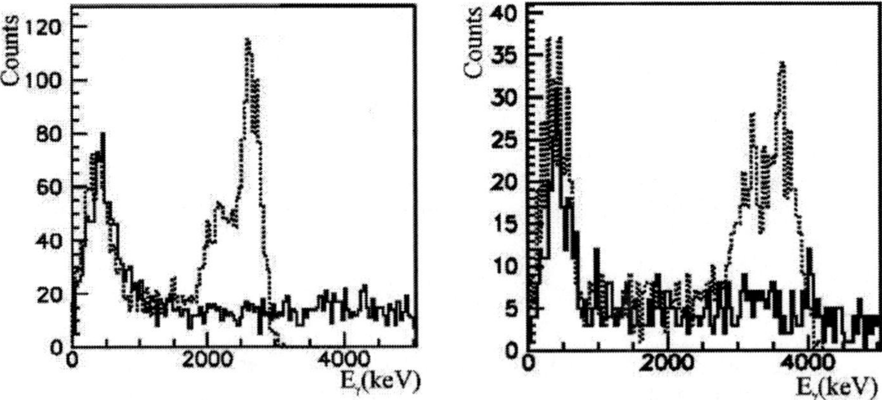

FIGURE 1. ^{23}O (left) and ^{24}O (right) Doppler corrected gamma-ray spectra and the corresponding Monte Carlo generated spectrum (dashed line) where a 2.7 MeV and 3.7 MeV simulated level deexcites in ^{23}O and ^{24}O, respectively.

The evolution of the shell structure around the island of inversion is then mainly understood (at least at the first order) as a consequence of the migration of the neutron

$d_{3/2}$ orbital due to the action of the tensor force.

N=28 : MAGICITY VERSUS DEFORMATION

Particular interest exists for the N=28 shell closure which is the lightest having the spin-orbit splitting as origin. It has been suggested for example that a reduction of the spin-orbit potential for the most neutron-rich isotopes resulting from a larger surface diffuseness could result in a complete disappearance of the N=28 shell gap [9]. More recently, Otsuka [10] highlighted the importance of the tensor force in the evolution of the shell structure around N=28.

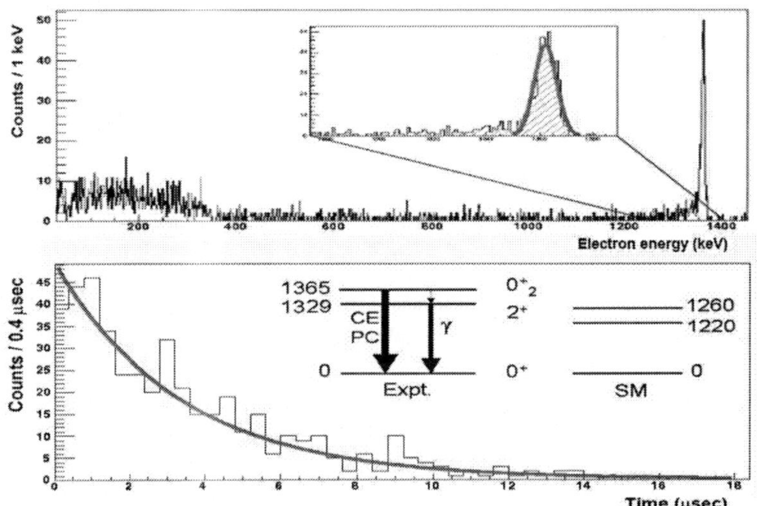

FIGURE 2. Energy (top) and decay-time (bottom) spectra of the conversion electrons corresponding to the E0 transition detected in a Si(Li) detector obtained in coincidence with ^{44}S ions. The inset on the top spectra shows a zoom on the region of the peak. The resulting level scheme is drawn on the bottom spectra with the corresponding shell model predictions.

In the last decade, evidences have been found for the presence of deformation at N=28 below ^{48}Ca, in particular in the sulfur isotopes. The halflife of ^{44}S measured at GANIL in 1993 [11] was the first indication of modification in the shell structure of the N=28 isotones below ^{48}Ca. Comparison with QRPA calculations suggested that the deformation could play an important role in this region. In 1997, Glasmacher and collaborators [12] measured the first 2^+ energy and the B(E2) of ^{44}S and interpreted these results as an indication of collectivity induced by a reduction of the N=28 shell gap. It is from more recent in-beam gamma spectroscopy experiment that it has been shown that the proton induced collectivity due to the quasi degeneracy of the $\pi s_{1/2}$ and $\pi d_{3/2}$ orbitals [13] is also essential to take into account. In other words, the reduction of the N=28 gap alone could not be strong enough to drive these nuclei towards deformation. Recently, the energy of the 0_2^+ state in ^{44}S has been measured at GANIL [14] and is found in perfect agreement with the shell model predictions. This isomer is interpreted as a direct evidence of shape coexistence in that nucleus (see fig. 2)

From a theoretical point of view, all mean field calculations [15, 9, 16, 17] have predicted deformation in this region and in particular an oblate configuration for ^{42}Si with deformation parameter β between -0.3 and -0.4. From shell model calculations, the deformation below ^{48}Ca is also well described, in particular for the sulfur isotopes [18]. The ^{40,42}S are deformed γ-soft nuclei while ^{44}S exhibits mixing configurations in the low lying states resulting in a shape coexistence, in agreement with the observations mentioned above. The shell model predictions for ^{42}Si are of particular interest since it has been shown that they are very sensitive to the detail of the interaction. Indeed, the position of the first 2^+ was first predicted at relatively high energy (2.49 MeV) [19] giving a 'magic' character to this isotope but a modified interaction based on accurate data obtain in ^{35}Si [20] leads to a new prediction at 1.49 MeV whilst the other neighboring nuclei where much less affected.

In this context, two important experiment have been performed at GANIL in order to better understand how deformation can develop below ^{48}Ca and to what extent the reduction of the N=28 shell closure is responsible of such structural changes.

- The first experiment aimed at determine the spin-orbit reduction between the Ca and the Ar isotopes by measuring the position of the neutron single particle orbitals in ^{47}Ar [21]. Within this information in both ^{49}Ca and ^{47}Ar, it is possible to experimentally determine the reduction of the N=28 shell gap when removing two protons to ^{48}Ca and to extrapolate what would be the reduction in ^{42}Si after the removal of 6 protons to ^{48}Ca.
- The second experiment aimed to determine the position of the first 2^+state in ^{42}Si [22]. Indeed, the ^{42}Si (Z=14) which is expected to have a proton closed shell $(\pi d5/2)^6$ configuration could be less sensitive to the deformation leading to a spherical 'magic' nucleus. This hypothesis is defended by Fridmann *et al* [23] who measured a significantly small two protons removal cross section ^{44}S$\rightarrow$^{42}Si. This result has been interpreted, by comparison with theoretical calculations [24], as an evidence for a large Z=14 sub-shell gap. In such an interpretation, the protons do not contribute to the deformation and the reduction of the N=28 shell closure is not strong enough to generate collectivity. On the other hand, two experimental evidences were reported previously as support of a strong deformation in Si isotopes, the β-halflife of ^{42}Si [25] and the particle stability of ^{43}Si [26].

(d,p) transfer reaction on ^{46}Ar

The (d,p) transfer reaction in inverse kinematics allows to identify the particle neutron states in the daughter nucleus by measuring the properties of the recoiling protons. Such an experiment as been performed at the SPIRAL/GANIL facility to identify the $f_{7/2}$, $p_{3/2}$, $p_{1/2}$ and $f_{5/2}$ states in ^{47}Ar. A beam of ^{46}Ar accelerated at 10.2 MeV/u with a intensity of 2.10^4 pps impinged a 0.38 mg/cm^2 thick CD_2 target. The energy and the angle of the protons were measured with the MUST detector array comprising height highly segmented double-sided Si detectors covering angles from 110 to 170 degrees. The excitation energy spectrum of ^{47}Ar was determined from the energy and angle of the detected protons. The angular distributions have been then determined for the different states and compared to distorded wave Born approximation (DWBA) calculations in order to extract information on the transferred angular momentum ℓ and the vacancy of individual orbitals. Several states with angular momentum equal to ℓ=1 (p state) and to ℓ=3 (f state) have been identified. Then, the structure of ^{47}Ar was calculated in the shell model framework using the code ANTOINE and the sdpf interaction [cau04] and compare to the experimental results to deduce the neutron single particle energies (SPE) in ^{47}Ar. The comparison between SPE energies in ^{47}Ar and ^{49}Ca is shown on the figure 3.

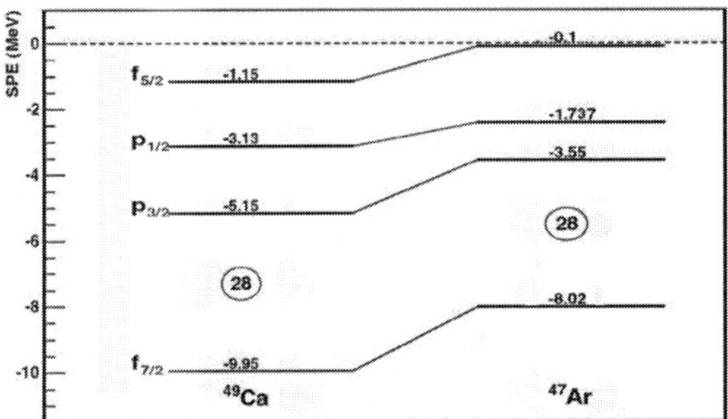

FIGURE 3. Neutron single-particle energies (SPE) of the *fp* orbitals for the $^{47}Ar_{29}$ and $^{49}Ca_{29}$ nuclei extracted from transfer reactions. The reduction of the N=28 shell gap amounts for 330 keV.

By comparing the single particle energies in ^{47}Ar and ^{49}Ca, Gaudefroy and collaborators were able to conclude that the gap N=28 is reduced by almost 330 keV when removing two protons from the ^{48}Ca [21]. As the $d_{3/2}$ and $s_{1/2}$ orbitals are quasidegenerate in ^{48}Ca, the protons are considered to be removed equiprobably from both orbitals. Then, because of the proton-neutron interaction, the removal of protons induces a differential energy change of the neutron orbitals. This interaction acts on the $f_{7/2}$ and $p_{3/2}$ orbitals to modify the size of the N=28 gap and on the $\ell\uparrow$ and $\ell\downarrow$ orbitals to reduce the spin orbit splittings. A major contribution of these changes is due to the monopole part of the proton-neutron matrix elements V^{pn}. Taken these matrix element from the nuclear data obtained on the N=21 isotones, the calculated reduction of the N=28 gap is 362 keV, in very good agreement with the experimental observation.

In beam gamma spectroscopy of ^{42}Si

Extrapolating the measured reduction of the N=28 shell gap after the removing of 2 protons up to Z=14 (removal of 6 protons), a reduction of almost 1 MeV of the N=28 gap in ^{42}Si is anticipated. Moreover, in the light of the contradictory experimental results about possible deformation in this nucleus, spectroscopic information is needed in order to better understand the structure of this nucleus.

The aim of the experiment reported here was therefore to measure the energy of the first 2^+ level in ^{42}Si and to perform spectroscopy in the neighbors nuclei by in beam γ-spectroscopy using a double fragmentation. This method has been already successfully applied to the spectroscopy of lighter nuclei around N=16 [8] as explained above. The experiment was carried out at the GANIL facility. It is important to note that the production rate obtained for the most neutron-rich ^{42}Si was around 8 atoms per day. Despite this extremely low production rate, the gamma spectroscopy of the nuclei of interest has been performed. 132 nuclei of ^{42}Si have been produced in the 2 protons removal reaction ^{9}Be(^{44}S,^{42}Si). It corresponds to a cross section of 80±20 μbarn.

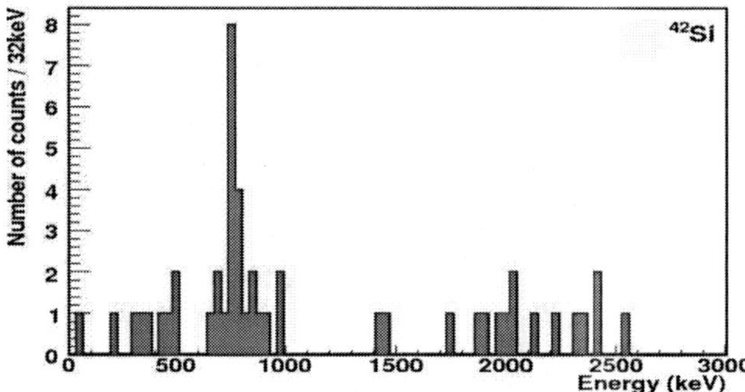

FIGURE 4. Gamma spectrum in coincidence with the detection of a ^{42}Si nucleus. The single peak is attributed to the $2^+ \to 0^+$ transtion.

The ^{42}Si spectrum shown in figure 4 exhibits a single peak at 770±19 keV with 25±5 counts between 620 and 920 keV (59 in the whole spectrum). The width is in agreement with the expected resolution of 13%. This γ-ray has been attributed to the $2^+ \to 0^+$ transition as it is the only one in the spectrum. It correspond to a feeding of the 2^+ state of 44±10 % in the 2p removal reaction ^{44}S→^{42}Si. The observed intensity is also compatible with the non-observation of this transition in [23] considering the difference in γ-efficiency between both experiments.

The energies of the 2^+ state in the Si and Ca isotopes, from N=18 to N=28, are displayed in figure 5. For Si and Ca, the 2^+ energies at N=20 increase above 3 MeV instead of 2 MeV at N=18 and less than 1.5 MeV at N=22. This sudden increase is linked to the closed shell and spherical configuration at N=20 giving a 'magic' character to these isotopes. For N=28, in contrast with the Ca, the Si develop a minimum excitation energy of the 2^+. The energy reported here - 770±19 keV - is one of the smallest 2^+ observed so far in even-even nuclei. It is also the smallest along the N=28 isotones. From this observation, it has been concluded that ^{42}Si is a collective nucleus [22] and not spherical-magic as claimed in [23]. This conclusion is similar to those given from the measurement of the short half-life of ^{42}Si [25].

Shell Model calculation using the code ANTOINE and the SDPF-NR have been performed. The interaction has been adjusted in order to reproduce the measured energy

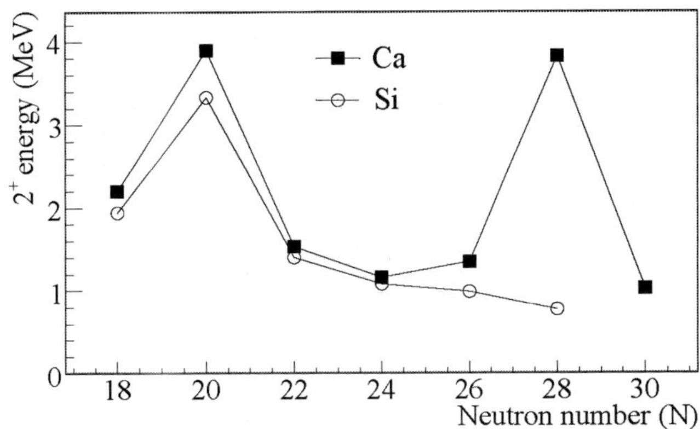

FIGURE 5. Energies of the 2^+ states measured in the Ca and Si isotopes from N=18 to N=28. The drop of the energy for the ^{42}Si is taken as a manifestation of the absence of magicity in that nucleus.

in ^{42}Si. It leads also to a better reproduction of the $5/2^+$ states the ^{37,39}P and of the 2^+ states in ^{36,38,40}Si. From these calculations, it turns out that ^{42}Si is a well deformed oblate nucleus ($\beta \sim$-0.45) with a neutron gap N=28 of 3.5 MeV and a proton gap Z=14 of 5.8 MeV. Despite these energy gaps, the $\Delta\ell$=2 excitations both for protons and neutrons seems to favor quadrupole collectivity which are not hindered by a parity change as it is the case at N=20.

CONCLUSIONS

We discussed here the evolution of the N=20 and N=28 magic numbers in the neutron rich nuclei. We have shown that the experimental work performed in the last years in gamma spectroscopy allows to understand the main evolution of the nuclear structure within the isospin degree of freedom. Whereas the N=20 shell closure is preserved up to ^{34}Si after the removal of 6 protons to the doubly magic ^{40}Ca, the removal of 2 more protons from the $d_{5/2}$ orbital drastically influence the structure of ^{32}Mg. This behavior is understood as the result of the reduction of the strong $\pi\ d_{5/2}$-$\nu d_{3/2}$ tensor interaction. By consequence, a new shell gap N=16 is created between the neutron $s_{1/2}$ and $d_{3/2}$ orbitals. The sudden change is also the result of the hindrance of quadrupole excitations above N=20 due to the parity change. The N=28 shell closure is not as robust as N=20. The combined effects of the compression of the protons and neutrons orbitals, plus the quadrupole excitations produce a rich variety of behaviors and shapes: spherical ^{48}Ca, oblate non collective ^{46}Ar, coexistence in ^{44}S and finally a oblate rotor in ^{42}Si.

ACKNOWLEDGMENTS

This work is the product of several experiment performed by a large collaboration. I thank all my colleagues and in particular Mihai Stanoiu and Laurent Gaudefroy for the authorization to discuss their results.

REFERENCES

1. W. M. Elsasser *et al.*, Annales de l'I. H. P, tome 5, n3, 223(1935).
2. M. Goeppert Mayer, *Phys. Rev* **75**, 1969 (1949).
3. C. Thibault *et al.*, *Phys. Rev. C* **12**, 644 (1975).
4. D. Guillemaud-Mueller*et al.*, *NPA* **426**, 37 (1984).
5. T. Motobayashi *et al.*, *Phys. Lett. B* **346**, 9 (1995).
6. T. Otsuka *et al.*, *Phys. Rev. Lett.* **87**, 082502 (2001).
7. O. Tarasov *et al.*, *Phys. Lett. B* **409**, 64 (1997).
8. M. Stanoiu *et al.*, *Phys. Rev. C* **69**, 034312 (2004).
9. J. A. Lalazissis *et al.*, *Phys. Rev. C* **60**, 01431 (1999).
10. T. Otsuka *et al.*, *Phys. Rev. Lett.* **95**, 232502 (2005).
11. O. Sorlin *et al.*, *Phys. Rev. C* **47**, 2941 (1993).
12. T. Glasmacher *et al.*, *Phys. Lett. B* **395**, 163 (1997).
13. D. Sohler *et al.*, *Phys. Rev. C* **66**, 054302 (2002).
14. S. Grévy *et al.*, Eur. Phys. J. A **25**, s01-111 (2005).
15. T. R. Werner *et al.*, *Nucl. Phys. A* **597**, 327 (1996).
16. S. Péru *et al.*, *Eur. Phys. J. A* **9**, 35 (2000).
17. R. Rodríguez-Guzmán *et al.*, *Phys. Rev. C* **65**, 024304 (2002).
18. E. Caurier *et al.*, *Nucl. Phys. A* **742**, 14 (2004).
19. J. Retamosa *et al.*, *Phys. Rev. C* **55**, 1266 (1997).
20. S. Nummela *et al.*, *Phys. Rev. C* **63**, 044316 (2000).
21. L. Gaudefroy *et al.*, *Phys. Rev. Lett.* **97**, 092501 (2006).
22. B. Bastin *et al.*, *Phys. Rev. Lett.* **99**, 022503 (2007).
23. J. Fridmann *et al.*, *Nature* **435**, 03619 (2005).
24. J. A. Tostevin *et al.*, *Phys. Rev. C* **70**, 064602 (2004).
25. S. Grévy *et al.*, *Phys. Lett. B* **594**, 252 (2004).
26. M. Notani *et al.*, *Phys. Lett. B* **542**, 49 (2002).

Direct Reactions with Exotic Nuclei

Angela Bonaccorso [1]

Istituto Nazionale di Fisica Nucleare, Sez. di Pisa,
Largo Pontecorvo 3, 56127 Pisa, Italy.

Abstract. This talk will review the state of art of nuclear and Coulomb breakup theories and their relation to optical models of elastic scattering of exotic projectiles. The effect of the final state interactions between the breakup particle and the core and target nuclei will be clarified and some typical numerical calculations for the relevant observables will be presented and compared to experimental data. Finally new results will be shown to demonstrate the feasibility of a novel type of experiment involving heavy projectiles far from stability on heavy targets.

Keywords: Neutron, proton breakup, projectile fragmentation, unbound and borromean nuclei.
PACS: 21.10.Jx, 24.10.-i, 25.60.Gc, 27.30.+t

INTRODUCTION

Nuclear reaction theory has experienced a great revival in the last twenty years following the large increase in quantity as well as quality of experiments with exotic beams. Exotic nuclei are located away from the stability valley and have large differences in the number of neutrons and protons. Their valence particle separation energies S_n are smaller than the average 8 MeV expected in nuclear matter. In the extreme case of halo nuclei such as several beryllium and lithium isotopes (^{11}Be,^{12}Be,^{14}Be,^{11}Li), S_n can be even less than 1 MeV. As a consequence as much as 10% of the total reaction cross section is due to just one channel: 1n or 2n breakup. Therefore, *out of necessity* (N. Orr), breakup has been the most studied reaction for very weak beams and the one for which several new models have been developed. In this short review I will first present the mechanisms which lead to breakup and the relative observables that are measured. For each of them a description will follow of the most recent advances in the models used for theoretical calculations.

CROSS SECTION

Most theoretical methods used so far to describe breakup rely on a basic approximation to describe the collision with only the three-body variables of nucleon coordinate, projectile coordinate, and target coordinate. Thus the dynamics is controlled by the three potentials describing nucleon-core, nucleon-target, and core-target interactions. Often the projectile-target relative motion is treated semiclassically by using a trajectory of the

---

[1] In collaboration with C.A. Bertulani, G. Blanchon, D.M. Brink, F. Carstoiu, A. García-Camacho, A.A. Ibraheem, J. Margueron, N.Vinh Mau.

CP972, *Exotic Nuclei and Nuclear/Particle Astrophysics (II)—Proceedings of the Carpathian Summer School of Physics 2007,* edited by L. Trache and S. Stoica

center of the projectile relative to the center of the target $\mathbf{R}(t) = \mathbf{b}_c + vt\hat{\mathbf{z}}$ with constant velocity v in the z direction and impact parameter $\mathbf{b}_c$ in the xy plane. This approximation makes the formalism applicable for incident energies above the Coulomb barrier. Along the semiclassical trajectory the amplitude for a transition from a nucleon state ψ_i bound in the projectile, to a final continuum state ψ_f, is given by [1, 2]

$$A_{fi} = \frac{1}{i\hbar}\int_{-\infty}^{\infty} dt \langle \psi_f(\mathbf{r},t)|V(\mathbf{r},\mathbf{R}(t))|\psi_i(\mathbf{r},t)\rangle, \tag{1}$$

where V is the interaction responsible for the transition which will be specified in the following. The probabilities for different processes can be represented in terms of the amplitude as $dP/d\xi = \sum |A_{fi}|^2 \delta(\xi - \xi_f)$ where ξ can be momentum, energy or any other variable for which a differential cross section is measured.

Direct one-particle re-arrangment reactions of the peripheral type in presence of strong core-target absorption can be described by an equation like [2]-[5] $d\sigma/d\xi_f = C^2S\int d\mathbf{b_c} P_{ct}(b_c) dP(b_c)/d\xi_f$, (see Eq. (2.3) of [3]) and C^2S is the spectroscopic factor for the initial single particle state. The core survival probability is defined in terms of a S-matrix function of the core-target distance of closest approach b_c. A simple parameterisation is $P_{ct}(b_c) = |S_{ct}|^2 = e^{(-\ln 2 exp[(R_s - b_c)/a])}$. It takes into account the peripheral nature of the reaction and naturally excludes the possibility of large overlaps between projectile and target. The strong absorption radius $R_s \approx 1.4(A_p^{1/3} + A_t^{1/3})$ fm is defined as the distance of closest approach for a trajectory that is 50% absorbed from the elastic channel and a=0.6 fm is a diffuseness parameter. The values of R_s thus obtained agree within a few percent with those of the Kox parameterization [6].

BREAKUP MECHANISMS

Based on the time dependent amplitude Eq.(1) and the classical projectile-target trajectory of relative motion given above, in Ref.[4] we considered the breakup of a halo nucleus like ^{11}Be consisting of a neutron bound to a ^{10}Be core in a collision with a target nucleus. The system of the halo nucleus and the target was described by Jacobi coordinates $(\mathbf{R},\mathbf{r})$ where $\mathbf{R}$ is the position of the center of mass of the halo nucleus relative to the target nucleus and $\mathbf{r}$ is the position of the neutron relative to the halo core, and the coordinate $\mathbf{R}$ is assumed to move on a classical path. The Hamiltonian of such a system is $H = T_R + T_r + V_{nc}(\mathbf{r}) + V_{nt}(\beta_2\mathbf{r} + \mathbf{R}) + V_{ct}(\mathbf{R} - \beta_1\mathbf{r})$, with β_1 and β_2 the mass ratios of neutron and core, respectively, to that of the projectile. T_R and T_r are the kinetic energy operators associated with the coordinates $\mathbf{R}$ and $\mathbf{r}$ and V_{cn} is a real potential describing the neutron-core final state interaction. V_{cn} was neglected under the hypothesis that the observables measured and calculated did not depend significantly on it. In the "Projectile Fragmentation" section we will discuss cases in which such an interaction dominates instead the measured data. The potential $V_2 = V_{nt} + V_{ct}$ describes the interaction between the projectile and the target. Both V_{nt} and V_{ct} are represented by complex optical potentials. The imaginary part of V_{nt} describes absorption of the neutron by the target to form a compound nucleus. It gives rise to the stripping part [2] of the halo breakup. The imaginary part of V_{ct} describes reactions of the halo core with the target. The potential

V_{ct} also includes the Coulomb interaction between the halo core and the target. This part of the interaction is responsible for Coulomb breakup. In fact the Coulomb force does not act directly on the neutron but it affects it only indirectly by causing the recoil of the charged core.

Since the mass ratio β_1 is small for a halo nucleus with a heavy core ($\beta_1 \approx 0.1$ for ^{11}Be) the Coulomb potential was approximated by the dipole term in Ref.[4]. Then, making an eikonal approximation for the neutron final state $\psi_f(t) = \exp(i\mathbf{k}\cdot\mathbf{r} - i\varepsilon_{\mathbf{k}}t/\hbar)\exp\left(-\frac{1}{i\hbar}\int_t^\infty V_2(\mathbf{r},t)\,dt\right)$, the amplitude became

$$A_{lm}(\mathbf{k},\mathbf{b}_c) = \frac{1}{i\hbar}\int d^3\mathbf{r}\; dt e^{-i\mathbf{k}\cdot\mathbf{r}+i\omega t} e^{\left(\frac{1}{i\hbar}\int_t^\infty V_2(\mathbf{r},\mathbf{R}(t))dt\right)} V_2(\mathbf{r},\mathbf{R}(t))\,\phi_{lm}(\mathbf{r}) \tag{2}$$

where $\omega = (\varepsilon_{\mathbf{k}} - \varepsilon_i)/\hbar$ and ϕ_{lm} is the radial part of the neutron initial wave function. $\mathbf{k}$ is the neutron-core relative momentum vector in the final state. The corresponding energy is $\varepsilon_{\mathbf{k}}$, while ε_i is the initial state binding energy. Eq. (2) is appropriate to calculate the coincidence cross section $A_p \rightarrow (A_p - 1) + n$.

The components V_{nt} and V_{ct} of V_2 are treated differently because of the long range of the Coulomb interaction. The neutron-target interaction is strong and has a short range. We assumed that the interaction time τ for this part of the interaction is very short in the sense that $\omega\tau$ is small compared with unity. On the other hand the long range Coulomb interaction between the halo core and the target is weaker and changes more slowly. The way to treat it is discussed in the next subsection.

COULOMB BREAKUP

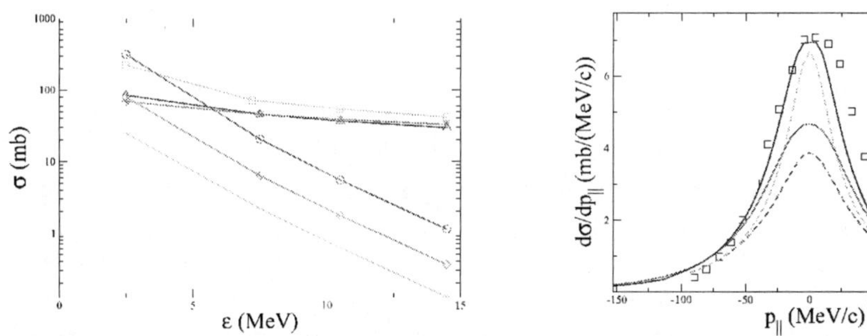

FIGURE 1. LHS: Integrated breakup cross-section for a hypothetical ^{34}Si beam against Pb at 70 A.MeV as a function of the neutron separation energy. Different initial parameters: circles (squares) are for Coulomb (nuclear) breakup from an initial s-wave; diamonds (triangles) for Coulomb (nuclear) breakup from a d-wave; pluses (stars) for Coulomb (nuclear) breakup from an initial f-wave. Nuclear breakup is the sum of diffractive and stripping contributions [2, 3]. RHS: Calculated momentum distribution of ^{7}Be fragments after proton-removal from ^{8}B against Pb at 936 MeV/A. Both dipole and full multipole results are shown for the ground state and first excited state. Calculations according to this section and Ref.[8] where more details can be found. Data are from Ref.[15].

Introducing the notation $V_c = Z_c Z_t e^2$, $V_v = Z_v Z_t e^2$ and $V_C = (Z_v + Z_c) Z_t e^2$, with $Z_v = 0, 1$ for a neutron and for a proton respectively, the Coulomb potential can be written as $V(\mathbf{r},\mathbf{R}) = \frac{V_c}{|\mathbf{R}-\beta_1\mathbf{r}|} + \frac{V_v}{|\mathbf{R}+\beta_2\mathbf{r}|} - \frac{V_C}{R}$. Here β_1 can be also the mass ratio of the proton to that of the projectile. In Ref.[7] we have shown that the Coulomb phase $\chi_{eff}(\mathbf{b}_c,\mathbf{r},k) = \int dt e^{i\omega t} V_C(\mathbf{r},t)/\hbar$. which is solvable in the dipole approximation, can similarly be calculated with the whole multipole expansion if a screening term is added and subtracted to the potential.

Aiming for an all-order formalism, in [4] it was shown that a possible way to achieve this is to use the sudden approximation, subtract the first order term, which diverges for large impact parameter, and then to add a first order term calculated in time-dependent perturbation theory. The sudden limit ($\omega \to 0$) must be therefore taken in the expression for χ_{eff}, yielding $\chi_{eff}^{sudd} = \frac{2V_C}{\hbar v}\log\frac{b_c}{R_\perp}$. Following a procedure analogous to that of [7], the Coulomb phase for the proton is shown to be [8] $\chi^p = \frac{2}{\hbar v}\left(V_c e^{i\beta_1\omega z/v}K_0(\omega b_c/v) - V_C K_0(\omega R_\perp/v) + V_v e^{-i\beta_2\omega z/v}K_0(\omega b_v/v)\right)$. Since $V_C = V_c + V_v$, the phase can be written as $\chi^p = \chi(\beta_1, V_c) + \chi(-\beta_2, V_v)$ where χ is the χ_{eff} of the previous section and $\mathbf{b}_v = \mathbf{b}_c + \mathbf{r}_\perp$ is the proton impact parameter with respect to the target. The Coulomb phase is therefore the sum of two terms: one of them describes the recoil of the core whereas the other accounts for the direct proton-target Coulomb interaction. Of course, in the case of the neutron the latter vanishes and the phase reduces to the one derived in [7]. It is easy to see that the expansion of χ_{eff} to first order in $\mathbf{r}$ yields the well known dipole approximation to the phase: which only differs from the neutron breakup case in the different constant factor, which is now $(V_c\beta_1 - V_v\beta_2)$ instead of $V_c\beta_1$ of Ref. [7]. The probability amplitude can be written as the sum of three contributions. The recoil term,

$$A^{rec} = \int d\mathbf{r} e^{-i\mathbf{k}\cdot\mathbf{r}}\phi_i(\mathbf{r})\left(e^{i\frac{2V_c}{\hbar v}\log\frac{b_c}{R_\perp}} - 1 - i\frac{2V_c}{\hbar v}\log\frac{b_c}{R_\perp} + i\chi(\beta_1, V_c)\right), \tag{3}$$

where, according to the discussions in [4, 7], the sudden limit has been used in order to include all orders in the interaction. The direct proton Coulomb interaction term A^{dir} which has the same form as Eq.(3) but with the substitution $V_c \to V_v$, $\beta_1 \to -\beta_2$ and $b_c \to b_v$. The nuclear part is given by $A^{nuc} = \int d\mathbf{r} e^{-i\mathbf{k}\cdot\mathbf{r}}\left(e^{i\chi_{nt}(b_v)} - 1\right)\phi_i(\mathbf{r})$, which is the well known eikonal form of the diffractive nuclear breakup with the nucleon-target phase $\chi_{nt}(\mathbf{b}_v) = \int dz V_{nt}(\mathbf{b}_v, z)/(v\hbar)$. Finally the expression for the differential cross-section is $\frac{d\sigma}{d\mathbf{k}} = \frac{1}{8\pi^3}\frac{mk}{\hbar^2}\int d\mathbf{b}_c|S_{ct}(b_c)|^2|A^{nuc} + A^{dir} + A^{rec}|^2$. In a number of papers higher order effects [9] and proton breakup have been discussed, among which we recall Refs.[10, 11, 12]. Some works have also addressed the problem of asymmetry in the core parallel momentum distribution after proton knockout [13, 14]. The fact that this asymmetry comes from high order terms can be directly extracted from our formalism. If the Coulomb part of the amplitude is simply expanded to first order in χ, it can be written, in terms of the one-dimensional Fourier transform in $z-$direction of the initial wave function $\tilde{\phi}_i$, as

$$\begin{aligned} A^{Cou} \simeq & \int d\mathbf{r}_\perp e^{-i\mathbf{k}_\perp\cdot\mathbf{r}_\perp}\frac{2}{\hbar v}\left(V_c K_0(\omega b_c/v)\tilde{\phi}_i(\mathbf{r}_\perp, k_z - \beta_1\omega/v) - V_C K_0(\omega R_\perp/v)\tilde{\phi}_i(\mathbf{r}_\perp, k_z)\right. \\ + & \left. V_v K_0(\omega b_v/v)\tilde{\phi}_i(\mathbf{r}_\perp, k_z + \beta_2\omega/v)\right). \end{aligned} \tag{4}$$

Thus the Coulomb breakup probability amplitude can be regarded as a coherent sum of three terms, each of which contains a shifted $z-$Fourier transform. The shifts are in opposite directions, $\beta_1\omega/v$ and $-\beta_2\omega/v$, but they are not visible directly in the calculated momentum distributions as ω depends on k itself. Moreover, the $1/v$ factor indicates that the asymmetry decreases as the beam energy increases. In the dipole approximation, however, the amplitude does not contain any asymmetry for the momentum distribution as it involves square modulii of $\tilde{\phi}_i(\mathbf{r}_\perp,k_z)$ separately. Hence we have confirmed analytically that the asymmetry in Coulomb breakup parallel momentum distributions is due to the presence of higher multipole terms, in agreement with earlier works [9, 13, 14]. However, the presence of the nuclear interaction introduces an interference that does depend on the sign of k_z and thus an additional asymmetry to that due to higher multipole terms in the Coulomb interaction.

In Fig.1LHS we show that the Coulomb break up cross section on heavy targets would be negligible compared to the total nuclear breakup cross section for heavy exotic nuclei in which the valence nucleons are expected to be in d or f orbits and with separation energies of the order of 10 MeV. For those nuclei such novel type of experiments would be therefore possible and more appropriate than the use of light targets for which the strong absorption model for the core-target interaction is not appropriate. Fig. 1RHS compares data [15] for proton breakup to calculations from our new model [8] shortly described above.

PROJECTILE FRAGMENTATION

We call *projectile fragmentation* the elastic breakup (diffraction dissociation) discussed above, when the observable studied is the neutron-core relative energy spectrum. This kind of observable has been widely measured in relation to the Coulomb breakup on heavy target. Results on light targets have also been presented [16]. These data enlighten the effect of the neutron final state interaction with the core of origin, neglected in the previous sections, while observables like the core energy or momentum distributions enlighten the effect of the neutron final state interaction with the target.

Projectile fragmentation has also been used experimentally with two neutron halo projectiles [17]-[34]. The reaction might proceed by the simultaneous emission of the two neutrons or by successive emissions [17]. The successive emission can be due to a mechanism in which one neutron is stripped by the interaction with the target, as in the one-neutron fragmentation case, while the other is left behind, for example in a resonance state, which then decays. The emission can be expected sequential if the two neutrons are not strongly correlated.

If the two neutrons are strongly correlated they will preferentially be emitted simultaneously. If the neutron which is not detected is stripped while the other suffers an elastic scattering on the target, then in both cases to first order in the interaction the neutron ends-up in a plane wave final state [2]. It can then re-interact with the core which, for example, is going to be ^{10}Be in the case of the one-neutron halo projectile ^{11}Be, while it will be ^{12}Be in the case of the projectile fragmentation of ^{14}Be, since ^{13}Be is not bound or ^{9}Li in the case of the projectile fragmentation of ^{11}Li, since in this case ^{10}Li is not bound. Experiments with a ^{14}B projectile [19] have also been performed, in which the n-^{12}Be relative energy spectra have been reconstructed by coincidence measurements. In such a nucleus the valence neutron is weakly bound, while the valence proton is strongly bound. Thus the neutron will probably be emitted in the first step and then re-scattered by the core minus one proton nucleus. The projectile-target distances at which this kind of mechanism would be relevant are probably not so large to neglect the effect of the neutron-target interaction.

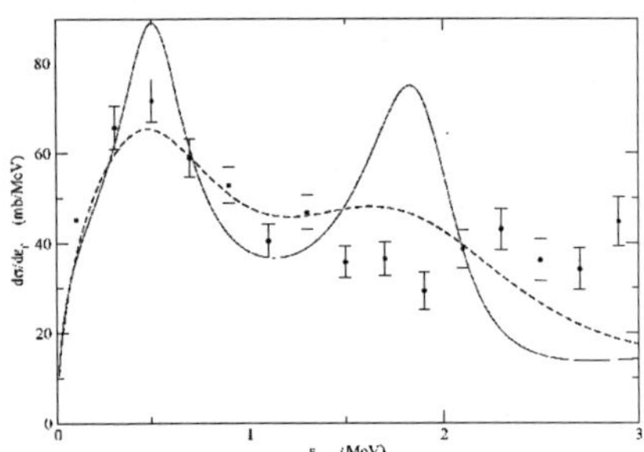

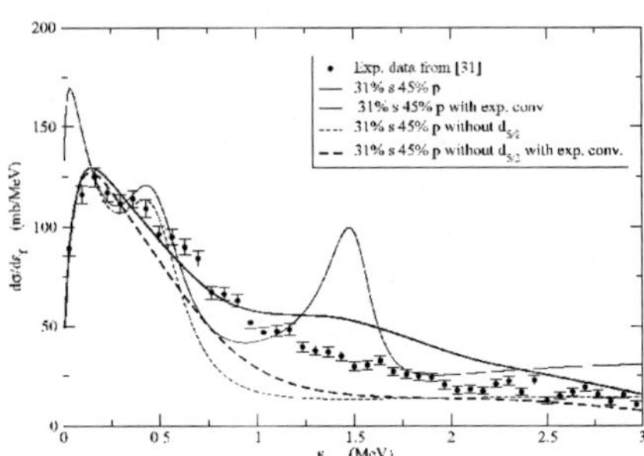

FIGURE 2. LHS: n-^{12}Be relative energy spectrum for the reaction ^{14}Be+^{12}C $\rightarrow$ n+^{12}Be+X, (ε_f is the same as $\varepsilon_{\mathbf{k}}$ of the text); the cross section is given by the sum of all transitions from the s initial state with ε_i=-1.85 MeV (solid line). Experimental points from H. Simon et al. [24] for the same reaction at 250 A.MeV. Dashed line is the folding of the calculated spectrum with the experimental resolution curve. RHS: n-^{9}Li relative energy spectrum, for the reaction ^{11}Li+^{12}C $\rightarrow$ n+^{9}Li+X at 264 A.MeV. Only the contributions from an s and p initial state with experimental spectroscopic factors [35] C^2S= 0.31 and 0.45 respectively and separation energy 0.3 MeV are included. The thin solid curve is the total calculated result. The thick solid curve curve is after convolution with the experimental resolution function. The thin dashed curve is the calculation without the d-resonance while the thick dashed curve is the same calculation after convolution. The symbols with error bars are the experimental points from Ref.[24]. Calculations are normalised to the data.

To first order the inelastic-like excitations can be described again by the time dependent perturbation amplitude Eq.(1) [1, 2]. In this section also, the potential $V(\mathbf{r},\mathbf{R}(t))$, which is the interaction responsible for the neutron transition, moves past on a constant velocity path as described in the previous sections. The radial part $\phi_i(\mathbf{r})$ of the single particle initial state wave function $\psi_i(\mathbf{r},t)$ is calculated in a potential $V_{WS}(r)$ which is fixed in space. The coordinate system and other details of the calculations can be found in Ref.[20]. In the special case of exotic nuclei the traditional approach to inelastic excitations needs to be modified. For example the final state can be eigenstate of a potential V_1 modified with respect to V_{WS} because some other particle is emitted during the reaction process as discussed in the introduction. The final state interaction might also have an imaginary part which would take into account the coupling between a continuum state and an excited core. The first order time dependent perturbation amplitude then reads $A_{fi} = \frac{1}{i\hbar v}\int_{-\infty}^{\infty} dxdydz\ \phi_f^*(x,y,z)\phi_i(x,y,z)e^{iqz}\tilde{V}(x-b_c,y,q)$, where $\tilde{V}(x-b_c,y,q) = \int_{-\infty}^{\infty} dzV(x-b_c,y,z)e^{iqz}$, and we changed variables and put $z' = z - vt$ or $t = (z-z')/v$, $q = \varepsilon_{\mathbf{k}} - \varepsilon_i/\hbar v$.

The target represented by $\tilde{V}$ perturbs the initial bound state wave function and allows the transition to the continuum by transferring some momentum to the neutron. Then it is enough to choose a simplified form of the interaction, such as a delta-function potential $V(r) = v_2\delta(x)\delta(y)\delta(z)$. The value of the strength $v_2 \equiv$ [MeV fm^3] used in the calculation is taken equal to the volume integral of the appropriate neutron-target interaction. It is clear that while in the sudden approach the initial and final state overlap is taken in the whole coordinate space, irrespective of the target and of the beam velocity, here the overlap of the initial and final wave functions depends on the core-target impact parameter. The neutron is emitted preferentially on the reaction plane and the z-component, being along the relative velocity axis is boosted by a momentum q.

Due of the strong core absorption discussed in the first part of the paper these calculations are also performed using the asymptotic form of the initial and final state wave functions. Introducing the quantization condition according to Ref.[2] the probability spectrum reads

$$\frac{dP_{in}}{d\varepsilon_k} = \frac{2}{\pi}\frac{v_2^2}{\hbar^2 v^2}C_i^2\frac{m}{\hbar^2 k}\frac{1}{2l_i+1}\Sigma_{m_i,m_f}|1-\bar{S}_{m_i,m_f}|^2|I_{m_i,m_f}|^2. \tag{5}$$

The generalization including spin is given in Appendix B of Ref.[20] and $|I_{m_i,m_f}|^2 = \left|k\frac{i}{2}i^{l_i}\int_{-\infty}^{\infty}dze^{iqz}\gamma h_{l_i}^{(1)}(i\gamma r)Y_{l_i,m_i}(\theta,0)h_{l_f}^{(-)}(kr)Y_{l_f,m_f}(\theta,0)\right|^2$. The quantity $\bar{S} = e^{2i(\delta+\nu)}$ is an off-the-energy-shell S-matrix representing the final state interaction of the neutron with the projectile core. It depends on a phase which is the sum of δ, the free particle n-core phase shift, plus ν the phase of the matrix element I. Some examples of our calculations [20] are shown in Fig. 2 and compared to recent data. See also Ref.[21].

Beside the total results which look quite convincing it is also interesting to look at some intermediate steps of our calculations. In particular we would like to compare our method to the sudden formula, as used in Ref.[18], which in our notation reads

$$\begin{aligned}\frac{d\sigma}{d\varepsilon_f} &\approx \frac{1}{k}\left(\frac{k\cos\delta-\gamma\sin\delta}{\gamma^2+k^2}\right)^2 \\ &\approx \frac{1}{k}|\sin(\delta+\beta)|^2\end{aligned} \tag{6}$$

where now $\beta = arctan(k/\gamma)$. This formalism also predicts the presence of an extra phase shift β with respect to the free particle scattering determined by δ. Similarly to our case the effect of β would be to modify the resonance-like structures. In both cases then $\bar{S}$ could be interpreted as an off-the-energy-shell S-matrix. On the other hand our additional phase ν contains an explicit dependence on the impact parameter and we use an optical model S-matrix for the description of the final state neutron-core interaction.

We show then in the LHS and RHS of Fig. (3) the calculation of an s to s transition from the fragmentation of ^{11}Li and ^{14}Be respectively. In the case of n-^{9}Li spectrum, LHS, the sudden formula and the fragmentation method give almost identical results. Thus the s-continuum state in ^{10}Li appears to be a proper virtual state with a large scattering length (-17.2 fm). In the case of ^{13}Be instead, there is no way to find agreement between the two methods, due to the fact that in order to fit the experimental spectrum one has to use a s-state with a small scattering length. The fragmentation method is more

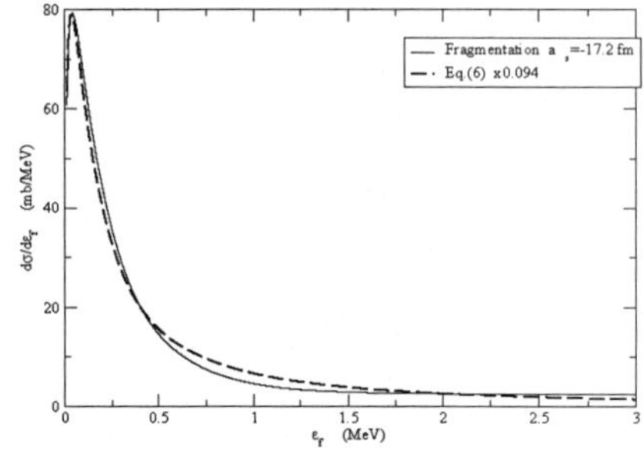

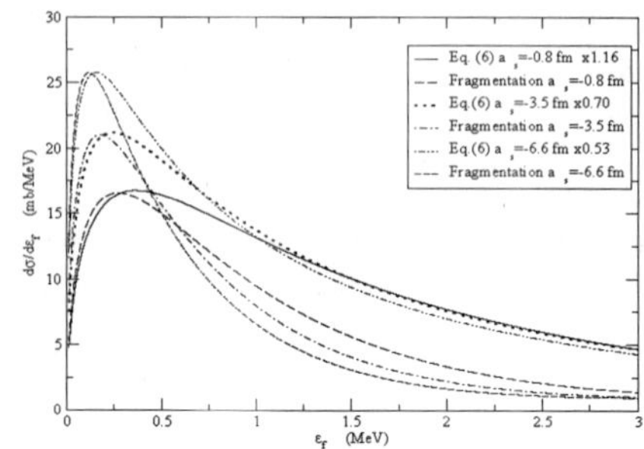

FIGURE 3. LHS: n-^{9}Li relative energy spectrum; comparison for the s to s transition of the fragmentation method result (solid curve) with that according to the sudden formula Eq. (6) (dashed curve). Both calculations use the exact phase shifts. RHS: n-^{12}Be relative energy spectrum; comparison of fragmentation method for an s to s transition with results of Eq.(6) using the same optical model phase shifts corresponding to scattering lengths as indicated. For each case we give in the legenda the normalization factor between the two calculations.

accurate since it gives to the s-state a small overall strength. The sudden formula would give a s state exhausting 80% of the total spectrum, in contradiction with the small value (-0.8 fm) of the scattering length [24].

We conclude by mentioning the importance of taking into account the nuclear and Coulomb breakup channels in the optical potentials of projectiles off stability. In Refs.[36, 37] we have shown that this can be done microscopically for the imaginary part of the optical potential. Because of the form factor behaviour such potential must be of the surface type. Its large diffuseness is due to the large decay length of the initial wave function and to the adiabaticity parameter of Coulomb breakup. Finally we refer to two books [38, 39] and recent reviews [40, 41] in which further discussion and bibliography can be found.

CONCLUSIONS AND OUTLOOK

The field of Rare Isotopes Studies is very active, growing steadily and rapidly. Some recent achievements in reaction models for neutron and proton breakup from exotic beams have been presented. From the structure point of view, in the search for the dripline position, a very important role is played by the study of nuclei unstable by neutron emission. On the other hand increasing the mass of the projectiles produced we are going to face the problem of envisaging new experiments to study them. These two are among the most important subjects which need to be adressed and further developed in the near future and for which some suggestions have been presented.

REFERENCES

1. K. Alder and A. Winther, Electromagnetic Excitation, North-Holland, 1975.
2. A. Bonaccorso and D.M. Brink, Phys. Rev. C **38** (1988) 1776; Phys. Rev. C **43** (1991) 299; Phys. Rev. C **44** (1991) 1559.
3. A. Bonaccorso, Phys. Rev. C **60** (1999) 054604.
4. J. Margueron, A. Bonaccorso and D.M. Brink, Nucl. Phys. **A 703** (2002) 105; Nucl. Phys. **A 720** (2003) 337.
5. R. A. Broglia and A. Winther, Heavy Ion Reactions, Benjamin, Reading, Mass, 1981.
6. S. Kox et al., Phys. Rev. C **35** (1987) 1678.
7. A. García-Camacho, A. Bonaccorso and D.M. Brink, Nucl. Phys. **A 776** (2006) 118.
8. A. García-Camacho, G. Blanchon, A. Bonaccorso and D.M. Brink, Phys. Rev. C **76** (2007) 014607.
9. H. Esbensen, G.F. Bertsch and C. Bertulani, Nucl. Phys. **A 581** (1995) 107.
10. J.H. Kelley et al., Phys. Rev. Lett. 77 (1996) 5020 and references therein.
11. K. Hencken, G. Bertsch and H. Esbensen, Phys. Rev. C **54** (1996) 3043.
12. A. Bonaccorso, D. M. Brink and C. A. Bertulani, Phys. Rev. C **69** (2004) 024615..
13. H. Esbensen and G.F. Bertsch, Nucl. Phys. **A 600** (1996) 37; Phys. Rev. C **59** (1999) 3240; Nucl. Phys. **A 706** (2002) 383.
14. B. Davids et al., Phys. Rev. Lett. **81** (1998) 2209; Phys. Rev. C **63** (2001) 0654806.
15. D. Cortina-Gil et al., Nucl. Phys. **A 720** (2003) 3.
16. N. Fukuda et al., Phys. Rev. C **70** (2004) 054606.
17. F.M. Marqués et al., Phys. Rev. C **64** (2001) 061301(R). N.Orr, Prog. Theor. Phys. Suppl. 146 (2003) 201.
18. G.F. Bertsch, K. Hencken and H. Esbensen, Phys. Rev. C **57** (1998) 1366.
19. J.L. Lecouey, Few Body Syst. **34** (2004) 21.
20. G. Blanchon, A. Bonaccorso, D.M. Brink, A.García-Camacho and N. Vinh Mau. Nucl. Phys. **A 784** (2007) 49.
    G. Blanchon, A. Bonaccorso, D.M. Brink, and N. Vinh Mau. Nucl. Phys. **A 791** (2007) 303.
21. G. Blanchon, A. Bonaccorso and N. Vinh Mau, Nucl. Phys. **A 739** (2004) 259.
22. M. Thoennessen et al., Phys. Rev. C **59** (1999) 111; Phys. Rev. C 60 (1999) 027303; Proceedings the Erice Int. School of Heavy-Ion Physics, 4th Course. Eds. R.A. Broglia and P.G. Hansen. World Scientific, Singapore 1998, pag.269.
23. M. Labiche et al. Phys. Rev. Lett. **86** (2001) 600.
24. H. Simon et al., Nucl. Phys. **A 734** (2004) 323 and Nucl. Phys. **A 791** (2007) 267.
25. A.A. Korsheninnikov et al., Phys. Lett. **B 343** (1995) 53.
26. G.F. Bertsch and H. Esbensen, Ann. Phys. (N.Y.) **209** (1991) 327.
27. I.J. Thompson and M.V. Zhukov, Phys. Rev. C **53** (1996) 708.
28. N. Vinh Mau and J.C. Pacheco, Nucl. Phys. **A 607** (1996) 163.
29. J.C. Pacheco and N. Vinh Mau, Phys. Rev. C **65** (2002) 044004.
30. P. Descouvemont, Phys. Lett. **B 331** (1994) 271; Phys. Rev. C **52** (1995) 704.
31. A. Adachour, D. Baye and P. Descouvemont, Phys. Lett. **B 356** (1995) 445.
32. D. Baye, Nucl. Phys. A 627 (1997) 305.
33. T. Tarutina, I.J. Thompson, J.A. Tostevin, Nucl. Phys. **A 733** (2004) 53.
34. M. Labiche, F.M. Marques, O. Sorlin and N. Vinh Mau, Phys. Rev. C **60** (1999) 027303.
35. S. N. Ershov et al., Phys. Rev. C **70** (2004) 054608.
36. A. Bonaccorso and F. Carstoiu Nucl. Phys. **A 706** (2002) 322.
37. A. A. Ibraheem, A. Bonaccorso. Nucl. Phys. **A 748** (2005) 414.
38. C.A. Bertulani, M.S. Hussein, G. Münzenberg. Physics of Radioactive Beams. Nova Science Publ. 2001.
39. Y. Suzuki, R. G. Lovas, K. Yabana, K. Varga. Structure and Reactions of Light Exotic Nuclei. Taylor and Francis Eds. 2003.
40. L.F. Canto, P.R.S. Gomes, R. Donangelo, M.S. Hussein, Phys. Rep. **424** (2006) 1.
41. B. Jonson, Phys. Rep. **389** (2004) 1.

Dissociation of Unstable Nuclear Projectiles

F. Carstoiu

IFIN-HH, P.O.Box MG6, 76900 Bucharest-Magurele, Romania

Abstract. We propose a scheme for including final state interaction effects in the Glauber model for nucleon removal reactions. This scheme can faciltate the extraction of astrophysical factors from proton removal reactions of weakly bound nuclei as an indirect method for nuclear astrophysics, as well as the study of unbound neutron rich nuclei.

Keywords: Diffraction dissociation, Coulomb dissociation
PACS: 23.20.-g, 24.10.-i, 25.60.Gc

INTRODUCTION

The investigation of fragmentation reactions using radioactive beams has lead to the recognition of narrow fragment momentum distributions (FWHM$\sim$ 50 MeV/c) as a signature of the extended valence nucleon distributions in halo nuclei [1]. Originally, these momentum distributions were assumed , in terms of the transparent limit of the Serber model [2], to be a direct mapping , via the Fourier transform, of the valence nucleon wave function. Detailed investigation of the role played by reaction mechanisms, particularly in the case of single neutron removal reactions in *psd* shell [3, 4], demonstrated that such an interpretation is not generally applicable. The fragment longitudinal momentum distributions, considered as the most cleaner probe of the valence nucleon wave function, are influenced by the core shadowing and show significant asymmetries toward low momenta. We investigate here such asymmetries by including final state interactions effects into the Glauber model for nucleon diffraction dissociation.

GLAUBER MODEL FOR NUCLEAR BREAKUP

We give here a summary of the Glauber model reaction model for breakup reactions involving weakly bound nuclei. Some details are reported in a recent paper by Sauvan et al. [3, 4] where the model has been tested on 23 reactions in p-sd shell. We shall assume that the ground state of the projectile (J^{π}) can be approximated by a superposition of configurations of the form $[I_c^{\pi_c} \otimes lj]^{J^{\pi}}$, where $I_c^{\pi_c}$ denotes the core states and nlj are quantum numbers for the single particle wave function, taken here as Woods-Saxon wave functions evaluated with the effective separation energy $S_{eff} = S_{1n} + E_x$ (E_x being the excitation energy of the core state and S_{1n} is the nucleon separation energy). We neglect coupling of core states to the final state and dynamical excitation of excited core states in the reaction. In this approximation the reaction can populate a given core state only to the extent that there is a nonzero spectroscopic factor $C^2S(c,nlj)$ in the projectile ground state. When more than one configuration contributes to a given core state, then

CP972, *Exotic Nuclei and Nuclear/Particle Astrophysics (II)—Proceedings of the Carpathian Summer School of Physics 2007,* edited by L. Trache and S. Stoica

the total cross section for one particle removal is written as an incoherent superposition of single particle cross sections,

$$\sigma_{1n} = \sum C^2 S(c,nlj)\sigma_{sp}(nlj) \ . \tag{1}$$

A similar relation holds for momentum distributions.

The projectile moves in z direction on a eikonal trajectory of impact parameter $\mathbf{b}$. In the no-recoil approximation, the core (b_c) and the valence particle (b_n) impact parameters are given by $\mathbf{b}_c \approx \mathbf{b}$ and $\mathbf{b}_n \approx \mathbf{b}+\mathbf{s}$, where $\mathbf{s}$ is the nucleon separation with respect to the core, projected onto the impact parameter plane. Let $S_c(b_c)$ and $S_n(b_n)$ be the S-matrix elements for core-target and n-target interactions in impact parameter representation. Since the energy is high, the interaction time with the target is short and the final state wave function is given by,

$$|f> = S_c(b_c)S_n(b_n)|0> \equiv S|0> \ , \tag{2}$$

where $|0>$ is the initial bound state. Note that the final state $|f>$ is a mixture of bound and continuum states. The elastic probability is given by the overlap with the ground state,

$$P_{el}(b) = |<f|0>|^2 = |<0|S|0>|^2 \ . \tag{3}$$

The scattering states are given by,

$$|s> = |f> - |0> \equiv (S-1)|0> \ . \tag{4}$$

The probability for projectile excitation is obtained by taking the overlap of the scattered wave function with the excited states of the core-nucleon system, ($|k>$)

$$P_{diff}(b) = \sum_k |<k|s>|^2 = \sum_k |<k|S-1|0>|^2 \equiv \sum_k |<k|S|0>|^2 = \sum_k |<k|f>|^2 \ . \tag{5}$$

This is known as elastic nuclear dissociation, or diffraction dissociation.

The "-1" does not contribute since $<k|0> = 0$. If the ground state is the only bound state of the system (as in e.g. ^{8}B), one can use the closure relation to eliminate the excited states,

$$\sum_k |k><k| + |0><0| = 1, \tag{6}$$

to obtain,

$$P_{diff}(b) = <0|S^2|0> - |<0|S|0>|^2 \ . \tag{7}$$

The absorption probability, which accounts for the loss of the incident flux is calculated from the norm of the wave function after the collision,

$$P_{abs}(b) = <0|0> - <f|f> \equiv <0|1-S^2|0> \tag{8}$$

In fact the general absorption operator can be decomposed as follows:

$$1-|S_c|^2|S_n|^2 = |S_c|^2(1-|S_n|^2) + |S_n|^2(1-|S_c|^2) + (1-|S_n|^2)(1-|S_c|^2) \ . \tag{9}$$

The first term corresponds to particle absorption, the second to core absorption and the third to the simultaneous core and valence particle absorption. Usually, we are interested in the first component of the total absorption, called stripping probability. Finally we note the closure relation,

$$P_{el}(b) + P_{diff}(b) + P_{abs}(b) = 1 , \tag{10}$$

which shows the reaction mechanisms accounted for by the Glauber model. The cross section for core absorption and simultaneous core and particle absorption, will be discussed elsewhere.

In order to develop a computational scheme, we neglect the coupling with core degrees of freedom and use a simple single particle wave function,

$$\psi_{JM}(\mathbf{r}) = \mathscr{R}_L(r) \sum_{M_L M_s} C^{L\ \frac{1}{2}\ J}_{M_L\ M_s M} Y_{LM_L}(\hat{\mathbf{r}}) \chi_{\frac{1}{2}M_s}(\sigma) \tag{11}$$

and introduce the density matrix,

$$\rho^J_{M'M}(\mathbf{r}) = < \psi_{JM'} | \psi_{JM} >_{spins} , \tag{12}$$

where $<>_{spins}$ means average over spin coordinates. Using the usual angular momentum algebra,

$$\rho^J_{M'M}(\mathbf{r}) = \mathscr{R}^2_L(r) \frac{\hat{L}^2 \hat{J}}{\sqrt{4\pi}} \sum_{\lambda\mu} C^{LL\lambda}_{000} W\left(JLJL, \frac{1}{2}\lambda\right) (-)^{J-\frac{1}{2}} C^{J\ \lambda\ J}_{M'\mu M} Y_{\lambda\mu}(\hat{\mathbf{r}}) , \tag{13}$$

where we use the notation $\hat{L} = \sqrt{2L+1}$, $C^{ll'\lambda}_{mm'\mu}$ are Clebsh-Gordan coefficients and W are the Wigner-6j coefficients.

The density matrix has the property,

$$Tr\rho = \frac{1}{\hat{J}^2} \sum_M \rho^J_{M'M} \delta_{M'M} = \mathscr{R}^2_L(r)/4\pi \equiv \rho_{val}(r) , \tag{14}$$

with $\mathscr{R}_L(r)$ the radial part of the single particle wave function. It is useful also to introduce the projected density,

$$\tilde{\rho}^J_{M'M}(\mathbf{s}) = \int_{-\infty}^{\infty} dz \rho^J_{M'M}(\mathbf{r}) , \qquad \tilde{\rho}_{val}(s) = \sum_M \tilde{\rho}^J_{MM}(s) . \tag{15}$$

With these preliminaries the stripping cross section is,

$$\sigma_{str} = \int d\mathbf{b} d\mathbf{s} (1 - |S_n(\mathbf{b}+\mathbf{s})|^2) |S_c(b)|^2 \tilde{\rho}_{val}(s) \equiv \int d\mathbf{s} D_{str}(s) \tilde{\rho}_{val}(s) , \tag{16}$$

with an evident definition for the distortion kernel D_{str} [5]. This kernel has a very intuitive physical interpretation: it is the convolution product of survival probability for the core and the probability for the absorption of the valence particle in complete analogy with the result of Hűfner and Nemes [6]. Eq. (16) is consistent with the spectator model for inclusive grazing reactions of Hussein and McVoy [7].

The diffraction cross section corresponding to the amplitude (7) is given by,

$$\sigma_{diff} = \int d\mathbf{b}d\mathbf{s}\ |S_c(b)|^2 |1 - S_n(\mathbf{b}+\mathbf{s})|^2\ \tilde{\rho}_{val}(s)$$
$$-\frac{1}{\hat{J}^2}\sum_{M}\int d\mathbf{b}\ |S_c(b)|^2 \left|\int d\mathbf{s}(1 - S_n(\mathbf{b}+\mathbf{s}))\tilde{\rho}^{J}_{MM}(s)\right|^2 . \tag{17}$$

The first term is dominant and to first order in valence density one can write,

$$\sigma^{(1)}_{diff} = \int d\mathbf{s}\, D_{diff}(s)\ \tilde{\rho}_{val}(s) \tag{18}$$

with $D_{diff} \simeq D_{str}$. Note that the kernels $D_{str,diff}$ account for the so called core shadowing and make the difference to the transparent limit of the Serber model.

Coulomb dissociation

We describe the Coulomb dissociation in an eikonal approximation which is formally equivalent to a first order perturbation theory. We use the long wavelength approximation for the transition operator and obtain a general formula for any electric multipolarity $(E\lambda\mu)$. However, the calculations are done only for the electric dipole and quadrupole transitions, since these give the main contribution to the cross sections.

The Coulomb excitation amplitude of a projectile in the strong field of a target can be expressed in terms of the electric multipoles matrix elements characterizing the electromagnetic decay of nuclear states. If the charge distribution of the two nuclei do not overlap during the collision, then the relative motion takes place along a classical Rutherford trajectory. At very high energies, the trajectory is well approximated by a straight line and the first order eikonal approximation should give reasonable results for momentum distributions and integrated cross sections.

Our starting point for the Coulomb amplitude is the result obtained by Bertulani and Nathan [8] in first order eikonal approximation (formally equivalent to first order Born approximation)

$$f_c(\mathbf{Q},\mathbf{q}) = i\frac{Z_t\alpha}{\beta\gamma}k\sum_{\lambda\mu} i^{\mu}\left(\frac{\omega}{c}\right)^{\lambda}\hat{\lambda}\Omega_{\mu}(Q)G_{\lambda\mu}(\beta)\mathscr{M}(E\lambda\mu) \tag{19}$$

where Z_t is target charge number, α is the fine structure constant, β is incident velocity in units of light speed, $\gamma = \left(1-\beta^2\right)^{-1/2}$, and

$$\Omega_{\mu}(Q) = \int_0^{\infty} db\, b\, J_{\mu}(Qb)K_{\mu}\left(\frac{\omega b}{\gamma v}\right)S_{pt}(b)\ , \tag{20}$$

where $Q = 2k\sin(\theta/2)$, k is center of mass bombarding momentum and θ is the scattering angle. $J_m(K_m)$ are the cylindrical (modified) Bessel function of order m. The functions $G_{\lambda\mu}(\beta)$ are the relativistic functions of Winther and Adler[9] and tabulated by Bertulani and Baur[17]. $S_{pt}(b)$ in eq. 20 is the scattering matrix for the c.m. scattering of the projectile + target and

$$E_x = \hbar\omega = \varepsilon + E_q = \varepsilon + \frac{\hbar^2 q^2}{2\mu}$$

is the excitation energy, i.e. the sum of the (absolute) binding energy and the kinetic energy of the separated clusters.

$\mathscr{M}(E\lambda\mu)$ are the matrix elements for electric transition of multipolarity $(\lambda\mu)$

$$\mathscr{M}(E\lambda\mu) = \sum_{j=1,2} Z_j \int \varphi_f^*(\mathbf{r}) r_j^\lambda \, Y_{\lambda\mu}(\hat{\mathbf{r}}_j)\, \varphi_i(\mathbf{r})\, d\mathbf{r}\,, \tag{21}$$

and $\varphi_{i(f)}$ are initial (final) state of the projectile. The two clusters are characterized by masses (m_j), charges (Z_j), and associate momenta $(\mathbf{k}_j)$. The relative motion of the outgoing clusters is described in terms of the relative momentum

$$\mathbf{q} = (m_2\mathbf{k}_1 - m_1\mathbf{k}_2)/(m_1 + m_2)\,.$$

The transition matrix elements (21) are calculated using a simple shell model wave function for the ground state

$$\varphi_i(\mathbf{r}) = \mathscr{R}_L(r) Y_{LM}(\hat{r})\,. \tag{22}$$

and a realistic wave function for continuum states

$$\varphi_f(\mathbf{r}) = \frac{4\pi}{q} \sum_{L'M'} i^{L'} e^{-i\delta_{L'}(q)} \frac{u_{L'}}{r} Y^*_{L'M'}(\hat{\mathbf{q}}) Y_{L'M'}(\hat{\mathbf{r}}) \tag{23}$$

Under the above approximations, the matrix element in (21) is given by

$$\begin{aligned} \mathscr{M}(E\lambda\mu) &= \left(Z_1 \left(\frac{m_2}{m_1+m_2}\right)^\lambda - (-)^\lambda Z_2 \left(\frac{m_1}{m_1+m_2}\right)^\lambda \right) \\ &\cdot \int d\mathbf{r}\, \varphi_f^*(\mathbf{r})\, r^\lambda\, Y_{lm}(\hat{\mathbf{r}})\, \mathscr{R}_L(r)\, Y_{LM}(\hat{\mathbf{r}}) \\ &= \frac{\sqrt{4\pi}}{q} Z_\lambda^{eff}\, \hat{\lambda}\, \hat{L} \sum_{L'M'} i^{-L'} e^{i\delta_{L'}} \hat{L'}^{-1} Y_{\lambda\nu}(\hat{q})\, C^{L\lambda L'}_{0\,0\,0}\, C^{L\,\lambda\,L'}_{M\,\mu\,M'} \\ &\cdot \int_0^\infty r^2 dr r^\lambda \frac{u_{L'}(r)}{r} \mathscr{R}_L(r)\,, \end{aligned} \tag{24}$$

with an obvious notation for the effective charge. In the following, the radial integrals appearing in (24) will be denoted by $I_{L\lambda L'}(q)$. Since the spin orientations are not specified, the differential cross section for Coulomb excitation is obtained by averaging over the magnetic projections the square of the Coulomb amplitude:

$$d^4\sigma = \hat{L}^{-1} \sum_{\mathrm{M_{spins}}} |f_{E1}(\mathbf{Q},\mathbf{q}) + f_{E2}(\mathbf{Q},\mathbf{q})|^2 \frac{d\mathbf{q}QdQ}{(2\pi)^2k^2} \tag{25}$$

Consider the following integral appearing in (25):

$$f_{\mu_1\mu_2}(\frac{\omega}{\gamma v}) = \left(\frac{\omega}{\gamma v}\right)^2 \int_0^\infty dQ\, Q\, \Omega^*_{\mu_1}(Q)\Omega_{\mu_2}(Q)\,. \tag{26}$$

The function (26) is obtained in general by numerical integration except for the diagonal term which admits a simple analytical expression. One can perform first the integration over the azimuthal angle (ϕ_q) of the relative momentum

$$d\mathbf{q} = q^2 dq \sin\theta_q d\theta_q d\phi_q\,.$$

This integration automatically selects only the diagonal terms in (26) and, using the closure relation of the Bessel functions,

$$\int_0^\infty dQ\, Q\, J_m(Qb)\, J_m(Qb') = \frac{1}{b}\delta(b-b')\,, \tag{27}$$

one obtains,

$$f_{\mu\mu} = \int_0^\infty b\, db K^2_\mu(\frac{\omega b}{\gamma v})\, |S_{pt}(b)|^2$$

After some simple calculation we obtain the following expression for the differential cross section,

$$\frac{d^2\sigma}{q^2 dq \sin\theta d\theta} = \frac{Z_t^2\alpha^2}{\sqrt{\pi}q^2\beta^2\gamma^2} \sum (-)^{L+\mu}(-)^{\frac{L_2-L_1+\lambda_1-\lambda_2}{2}}\, \hat{\lambda}_1^2\hat{\lambda}_2^2\hat{L}_1\hat{L}_2\hat{S}^{-1}\left(\frac{\omega}{c}\right)^{\lambda_1+\lambda_2} \tag{28}$$

$$\times\, e^{i\delta_{\lambda_2}-i\delta_{\lambda_1}} Z^{eff}_{\lambda_1}\, Z^{eff}_{\lambda_2}\, \overline{G}_{\lambda_1\mu}\, \overline{G}_{\lambda_2\mu}\, f_{\mu\mu}$$

$$\times\, I_{L\lambda_1L_1}(q)\, I_{L\lambda_2L_2}(q)\, C^{L\lambda_1L_1}_{0\,0\,0}\, C^{L\lambda_2L_2}_{0\,0\,0}\, C^{L_1L_2S}_{0\,0\,0}\, C^{\lambda_1\lambda_2S}_{\mu-\mu\,0} \begin{Bmatrix} S\lambda_1\lambda_2 \\ LL_2L_1 \end{Bmatrix} \overline{Y}_{S0}(\hat{\mathbf{q}})$$

The summation runs over all quantum numbers except L. The functions $\overline{G}_{\lambda\mu}$ are defined by factorizing the complex phase from the Winther -Adler relativistic functions: $G_{\lambda\mu}(\beta) = -i^{\lambda+\mu}\overline{G}_{\lambda\mu}(\beta)$. The angular dependence of the cross section is given by functions $\overline{Y}_{S0}$ which are the standard spherical harmonics defined without the phase factor $e^{im\phi}$ (variable ϕ is already integrated). The energy dependence of the reaction mechanism is controlled by functions $\overline{G}$ and $f_{\mu\mu}$. However, the magnitude of the cross section for a given multipolarity transition is mainly determined by the effective charge. If the clusters have equal charge to mass ratio, then the dipole transition has vanishing cross section in this approximation. This is readily understood from classical arguments since in this case the dipole field pushes the two clusters with the same acceleration in the same direction and does not lead to their dissociation. This is a consequence of the assumption of a well defined cluster structure of the projectile. In real cases, one finds experimentally an appreciable suppression of the E1 multipolarity, but not completely.

Finally, the angle integrated cross section is given by:

$$\frac{d\sigma}{q^2 dq} = \frac{Z_t^2 \alpha^2}{\pi \beta^2 \gamma^2 q^2} \sum_{\lambda L_1 \mu} \hat{\lambda}^2 \left(\frac{\omega}{c}\right)^{2\lambda} \left(Z_\lambda^{eff}\right)^2 \overline{G}^2_{\lambda\mu} f_{\mu\mu} \left(C^{L\,\lambda\,L_1}_{0\,0\,0}\right)^2 |I_{L\lambda L_1}|^2 \qquad (29)$$

The cross section for Coulomb dissociation is obtained from (29) by numerical integration.

MOMENTUM DISTRIBUTION FOR NUCLEAR DISSOCIATION

The differential cross section with respect to the core-valence particle relative momentum is given by:

$$\frac{d\sigma}{d\mathbf{k}} = \frac{1}{(2\pi)^3 \hat{L}_0^2} \sum_{M_0} \int d\mathbf{b} \left| \int d\mathbf{r} \Phi^\star_{\mathbf{k}}(\mathbf{r})(S-1)\Phi_{L_0 M_0}(\mathbf{r}) \right|^2 , \qquad (30)$$

where $\Phi_{L_0 M_0}(\mathbf{r}) = \mathscr{R}_{L_0}(r) Y_{L_0 M_0}(\hat{\mathbf{r}})$ is the single particle initial state and $\Phi_{\mathbf{k}}(\mathbf{r})$ are the continuum states. Since $< \mathbf{k}|L_0 M_0 >= 0$, the factor "-1" is harmless but we keep it here for convergence purposes.

It is useful to ask from the beginning what components of the scattering operator are important for nuclear diffraction. One can decompose the operator S as follows:

$$1 - S = S_c(1 - S_n) + S_n(1 - S_c) + (1 - S_c)(1 - S_n) , \qquad (31)$$

where the first term corresponds to nucleon scattering and core as a spectator, the second corresponds to core scattering and the third is the double scattering term. In principle, the first term will dominate but we keep here all terms.

We discuss first the breakup probability at fixed impact parameter b which is much simpler to calculate numerically. This is given by

$$\frac{dP(b)}{d\mathbf{k}} = \frac{1}{(2\pi)^3 \hat{L}_0^2} \sum_{M_0} \left| \int d\mathbf{r} \Phi^\star_{\mathbf{k}}(\mathbf{r})(S-1)\Phi_{L_0 M_0}(\mathbf{r}) \right|^2 . \qquad (32)$$

We dicuss a convenient numerical scheme to calculate (32). We first expand the continuum wavefunction into partial waves:

$$\Phi_{\mathbf{k}}(\mathbf{r}) = \frac{4\pi}{k} \sum_{LM} i^L e^{-i\delta_L(k)} \frac{u_L(r)}{r} Y^\star_{LM}(\hat{\mathbf{k}}) Y_{LM}(\hat{\mathbf{r}}) , \qquad (33)$$

where $u_L(r)$ are the reduced partial waves and δ_L the phase shifts. The radial wavefunction. are solutions of the equation:

$$u_L'' - \frac{L(L+1)}{r^2} u_L - \frac{2\mu}{\hbar^2} V(r) u_L + k^2 u_L = 0 . \qquad (34)$$

This is solved with boundary condition [16]

$$u_L \sim_{r\to\infty} \sqrt{\frac{2}{\pi}} \sin(kr - L\frac{\pi}{2} + \delta_L) \ . \tag{35}$$

When the valence particle is a proton one replaces (35) by

$$u_L \sim_{r\to\infty} \cos\delta_L F_L(kr) + \sin\delta_L G_L(kr) \tag{36}$$

with $F(G)$ the regular(irregular) Coulomb wave functions. One expands also the transition operator into spherical harmonics:

$$S - 1 = \sum_{\lambda\mu} \mathscr{S}_{\lambda\mu}(b,r) Y_{\lambda\mu}(\hat{r}) \ . \tag{37}$$

Inverting (37) one has

$$\mathscr{S}_{\lambda\mu}(b,r) = \int_0^{\pi} \int_0^{2\pi} \sin\theta \ \theta \ d\phi \ [S_c(b_c) S_n(b_n) - 1] Y^{\star}_{\lambda\mu}(\theta,\phi) \tag{38}$$

with $b_c = \sqrt{b^2 - 2\alpha br \sin\theta\cos\phi + \alpha^2 r^2 \sin^2\theta}$ and $b_n = \sqrt{b^2 + 2\beta br \sin\theta\cos\phi + \beta^2 r^2 \sin^2\theta}$, $\alpha = \frac{1}{A_c+1}$, $\beta = \frac{A_c}{A_c+1}$, where A_c is the core mass number. Parameters α and β account for recoil effects. Inserting (37) and (33) into (32), one obtains

$$\frac{dP(b)}{d\mathbf{k}} = \frac{1}{(2\pi)^3} (\frac{4\pi}{k})^2 \frac{1}{4\pi} \sum_{M_0} \left| \sum_{LM\lambda\mu} \frac{\hat{\lambda}}{\hat{L}} i^{-L} \ e^{i\delta_L} Y_{LM}(\hat{\mathbf{k}}) \ C^{L_0 \ \lambda \ L}_{0 \ 0 \ 0} C^{L_0 \ \lambda \ L}_{M_0 \ \mu \ M} I^{\lambda\mu}_{LL_0}(b) \right|^2 \tag{39}$$

where

$$I^{\lambda\mu}_{LL_0}(b) = \int_0^{\infty} dr \ r^2 \frac{u_L}{r} \mathscr{S}_{\lambda\mu}(b,r) R_{L_0}(r) \ . \tag{40}$$

It is convenient to introduce momentum amplitudes,

$$S_{kLML_0M_0}(b) = \sum_{\lambda\mu} \frac{\hat{\lambda}}{\hat{L}} i^{-L} \ e^{i\delta_L} C^{L_0 \ \lambda \ L}_{0 \ 0 \ 0} C^{L_0 \ \lambda \ L}_{M_0 \ \mu \ M} I^{\lambda\mu}_{LL_0} \tag{41}$$

Here momentum dependence enters through phase shifts and radial integrals.The final formula for momentum distribution is,

$$\frac{dP(b)}{d\vec{k}} = \frac{1}{(2\pi)^3} (\frac{4\pi}{k})^2 \frac{1}{4\pi} \sum_{M_0} \left| \sum_{LM} S_{kLML_0M_0}(b) \ Y_{LM}(\hat{\mathbf{k}}) \right|^2 \tag{42}$$

Energy distribution in diffraction dissociation

The energy spectrum in diffraction dissociation has never been calculated in Glauber type models. The starting point is Eq. (39), which we rewrite in the form:

$$\frac{dP(b)}{d\mathbf{k}} = \frac{1}{(2\pi)^3} (\frac{4\pi}{k})^2 \frac{1}{(4\pi)^{3/2}} \sum_{L_1 L_2 \lambda_1 \mu_1 \lambda_2 \mu_2 LM} i^{\lambda_1 - \lambda_2} e^{i\delta_{L_1} - i\delta_{L_2}} (-)^{L_0 + \mu_2} \tag{43}$$

$$\times\hat{\lambda}_1\hat{\lambda}_2\hat{L}_1\hat{L}_2\hat{L}^{-1}C^{L_0\lambda_1L_1}_{0\ 0\ 0}\,C^{L_0\lambda_2L_2}_{0\ 0\ 0}\,C^{L_1L_2L}_{0\ 0\ 0}\,C^{\lambda_1\ \lambda_2\ L}_{\mu_1-\mu_2M}$$

$$\times Y^{\star}_{LM}(\hat{k})\left\{\begin{matrix}L\lambda_1\lambda_2\\L_0L_2L_1\end{matrix}\right\}I^{\lambda_1\mu_1}_{L_1L_0}\,I^{\star\lambda_2\mu_2}_{L_2L_0}\ .$$

The energy spectrum is obtained by integrating over angles $\hat{\mathbf{k}}$:

$$\frac{dP(b)}{k^2dk}=\frac{1}{(2\pi)^3}(\frac{4\pi}{k})^2\frac{1}{4\pi}\sum_{L\lambda\mu}(C^{L_0\ \lambda\ L}_{0\ 0\ 0})^2\left|I^{\lambda\mu}_{LL_0}\right|^2\ , \tag{44}$$

and $\frac{dP(b)}{dE}\sim\frac{dP(b)}{kdk}$. The angle integrated probability is an incoherent superposition of multipoles. No interference term survives in the energy spectrum. An equivalent formulation could be done in terms of momentum amplitudes Eq.(41):

$$\frac{dP(b)}{k^2dk}=\frac{1}{(2\pi)^3}(\frac{4\pi}{k})^2\frac{1}{4\pi}\sum_{LMM_0}|S_{kLMLoM_0}(b)|^2\ . \tag{45}$$

Parallel momentum distribution

For parallel momentum distribution it is convenient to change the variable $(k_x,k_y,k_z)\rightarrow(k_\perp,\phi,k_z)$. With these new variables we have $d\vec{k}=k_\perp dk_\perp d\phi dk_z$. writing $Y_{lm}(\theta,\phi)=\bar{Y}_{lm}(\theta)e^{im\phi}$, the integration of Eq. (42) leads to

$$\frac{dP(b)}{dk_z}=\frac{1}{(2\pi)^3}(4\pi)^2(2\pi)\int_0^\infty\frac{k_\perp dk_\perp}{k^2}\sum_{MM_0}\left|\sum_{L\geq|M|}S_{kLMLoM_0}\bar{Y}_{LM}(\theta)\right|^2 \tag{46}$$

and $\cos\theta=k_z/k$.Taking into account $k^2=k_\perp^2+k_z^2$ one has $kdk=k_\perp dk_\perp$, one finally has

$$\frac{dP(b)}{dk_z}=\frac{1}{(2\pi)^3}(4\pi)^2(2\pi)\sum_{MM_0}\int_{k_z}^\infty\frac{dk}{k}\left|\sum_{L\geq|M|}S_{kLMLoM_0}\bar{Y}_{LM}(\frac{k_z}{k})\right|^2 \tag{47}$$

which holds for any value of k_z. Since $\bar{Y}_{LM}(x)\sim P_{LM}(x)$ and $P_{LM}(-x)=(-)^{L+M}P_{LM}(x)$ terms with odd $L+M$ in Eq. (47) change sign and there is an asymmetry in the parallel momentum distribution.

Tabulation

To save computing time it is necessary we prepare a table containing the multipoles $\mathscr{S}_{\lambda\mu}(b,r)$. To do this we need a bijective (one to one) function mapping $(\lambda\mu)$ onto the natural number set. The simplest one is

$$n(\lambda,\mu)=\lambda^2+\lambda+\mu+1 \tag{48}$$

For $\lambda_{max}=8$ there are 81 multipoles to be calculated.

PLANE WAVE APPROXIMATION

Neglecting final state interactions, the core and the valence particles move freely. One sets the phase shifts to zero and replace the radial wave functions by spherical Bessel functions:

$$\Phi_{\mathbf{k}}(\mathbf{r}) = e^{i\mathbf{k}\cdot\mathbf{r}} = 4\pi \sum_{LM} i^L j_L(kr) Y^{\star}_{LM}(\hat{\mathbf{k}}) Y_{LM}(\hat{\mathbf{r}}) \ . \tag{49}$$

The radial integrals are defined now by

$$\tilde{I}^{\lambda\mu}_{LL_0}(b) = \int_0^{\infty} r^2 dr\, j_L(kr) \mathscr{S}_{\lambda\mu}(b,r) R_{L_0}(r) \tag{50}$$

Momentum amplitudes become:

$$\tilde{S}_{kLML_0M_0}(b) = \sum_{\lambda\mu} \frac{\hat{\lambda}}{\hat{L}}\, i^{-L}\, C^{L_0\ \lambda\ L}_{0\ \ 0\ \ 0}\, C^{L_0\ \lambda\ L}_{M_0\ \mu\ M}\, \tilde{I}^{\lambda\mu}_{LL_0} \tag{51}$$

Expressions for all other observable are formally identical provided we replace I and S by $\tilde{I}$ and $\tilde{S}$ respectively. For example the parallel momentum distribution reads now

$$\frac{dP(b)}{dk_z} = \frac{1}{(2\pi)^3}(4\pi)^2(2\pi) \sum_{MM_0} \int_{k_z}^{\infty} dk\, k \left| \sum_{L\geq|M|} \tilde{S}_{kLML_0M_0} \bar{Y}_{LM}(\frac{k_z}{k}) \right|^2 \tag{52}$$

ACKNOWLEDGMENTS

I am indebted to Carlos Bertulani for suggestions. This work was funded by National Council of Scientific Research (Romania) under Contract ID-765/2007.

REFERENCES

1. N. A. Orr, Nucl. Phys. **A616**, 155 (1997).
2. R.Serber, Phys. Rev. **72**, 1008 (1947).
3. E. Sauvan *et al.*, Phys. Lett. **B491**, 1 (2000).
4. E. Sauvan et al., Phys. Rev. **C69**, 044603 (2004).
5. F. Negoita *et al.*, Phys. Rev. **C59**, 2082 (1999).
6. J. Hüfner, M.C. Nemes, Phys. Rev. **C23**, 2538 (1981).
7. M.S. Hussein and K.W. Mc Voy, Nucl. Phys. **A445**, 123 (1985).
8. C.A. Bertulani and A. Nathan, Nucl. Phys. **A554**,158 (1993).
9. A. Winther and K. Adler, Nucl. Phys. **A319**, 518 (1979).
10. J.P. Jeukenne, A. Lejeune and C. Mahaux, Phys. Rev. **C16**, 80 (1977).
11. A. Bonaccorso and F. Carstoiu, Phys. Rev. **C61**, 034605 (2000).
12. L. Trache *et al.*, Phys. Rev. **C61**, 024612 (2000).
13. M. Beiner and R.J. Lombard, Ann. Phys. **86**, 262 (1974).
14. J.S. Wallace, Ann. Phys.(N.Y.) **78**, 190 (1973).
15. L. Ray, Phys. Rev. **C20**, 1857 (1979).
16. K. Henken, H. Esbensen, G. Bertsch,Phys. Rev. **C54**, 3043 (1996).
17. C. A. Bertulani and G. Baur,Nucl. Phys. **A480**, 615 (1988).

Selected Aspects of Flow Phenomena in Heavy Ion Collisions

Mihai Petrovici and Amalia Pop

National Institute for Physics and Nuclear Engineering, Bucharest - Romania

Abstract. Dynamical aspects of heavy ion collisions play an essential role in the attempt to produce and study in the laboratory states of matter supposed to exist in different phases of the Universe evolution. Flow phenomena evidenced in heavy ion interactions, caused by the gradients produced during the collision, opened the possibility to access information on in-medium effects, matter equation of state and possible phase transitions. Selected aspects of flow phenomena in heavy ion collisions will be presented.

Keywords: collective flow; heavy-ion collision
PACS: 25.70.-z;25.75.-q;25.75.Ld

INTRODUCTION

In order to answer the question if one could extract information on physics specific to states of matter supposed to be characteristic for different phases of the Universe evolution after the Big Bang, using heavy ion collisions at relativistic and ultrarelativistic energies, is mandatory to have under complete control the finite size and dynamical aspects. One of the dynamical effects is the flow phenomenon on which we will focus in this presentation. By flow one understands a smooth uninterrupted movement caused by a field gradient. Understanding the flow one could aim to extract information on the initial properties of matter created in heavy ion collisions at different energies, before the expansion and extract information on in-medium effects, equation of state, phase transitions, etc.

FLOW PHENOMENA IN DISSIPATIVE HEAVY ION COLLISIONS

The first type of flow evidenced in heavy ion collisions is induced by the isospin gradient in the isospin dependent dinuclear potential. In Fig. 1(a) is represented by contour lines the dinuclear potential-energy calculated within liquid-drop model including shell corrections as a function of proton and neutron number, ^{92}Mo + ^{238}U being the injection point. The strong gradient along the isospin direction is supposed to induce a charge drift of about three units along a constant mass. The experimental two dimensional distributions in A-Z plane at half maximum for 20 MeV consecutive bins in TKEL for ^{92}Mo + ^{238}U collision at 14.7 A·MeV, presented in Fig 1(b), show that the charge mean value of the projectile-like fragments drifts by three units towards lower value, confirming the expectations based on the largest gradient in the dinuclear potential energy along the isospin degree of freedom. This was one of the largest isospin dynamics observed

CP972, *Exotic Nuclei and Nuclear/Particle Astrophysics (II)—Proceedings of the Carpathian Summer School of Physics 2007,* edited by L. Trache and S. Stoica

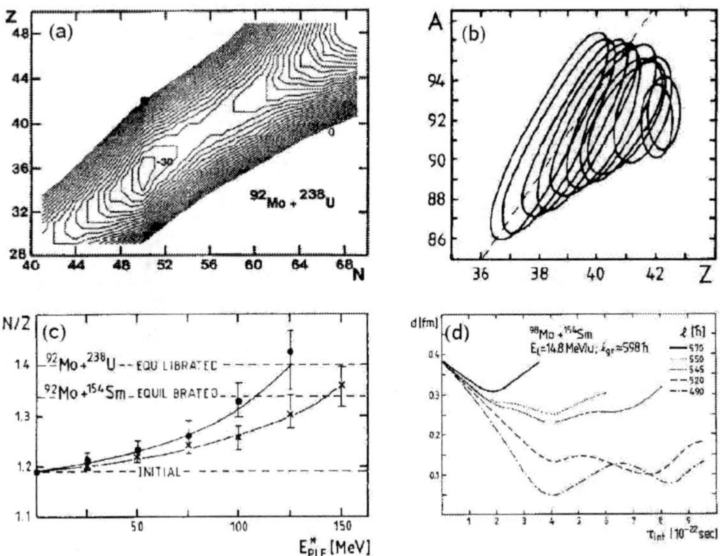

FIGURE 1. (a) A contour plot of the dinuclear potential energy surface for the $A_1 + A_2 = 330$ and $Z_1 + Z_2 = 134$, system, relative to ^{92}Mo + ^{238}U as injection point (marked by a dot); (b) Contour plots of $d^2\sigma/dA_{prim}\cdot dZ_{sec}$. (c) The N/Z ratio as a function of the excitation energy of the projectile-like fragment. (d) Time evolution of the distance between the center of mass of neutrons and protons, d of ^{98}Mo + ^{154}Sm.

in dissipative heavy ion collisions [1]. Fig.1(c) is a quantitative representation of the above expected and experimentally observed trends in terms of N/Z value as a function of excitation energy for the projectile-like fragments. The dashed lines represent the N/Z values corresponding to the minimum of the potential energy as a function of isospin. All these show that we have to do with a coherent exchange of neutrons and protons which is nothing else than a flow in the isospin degree of freedom. The collective nature of this dynamics is also confirmed by the results of time dependent Hartree-Fock (TDHF) model calculations. Fig. 1(d) shows the time dependence of the distance between the center of mass of protons and neutrons for different impact parameters (initial angular momenta). Oscillations characteristic for a not completely damped collective motion during the interaction time are visible.

ROTATIONAL-LIKE, SQUEEZE-OUT AND ELLIPTIC TYPE AZIMUTHAL DISTRIBUTIONS

What happens when increasing the incident energy? If one compares the interaction time estimated by a phenomenological model of dissipative heavy ion collisions [2] with the statistical model decay-time for light particle emission, represented in Fig.2(a) [3], for 23.7 A·MeV for example, for a total kinetic energy loss of 600 MeV, the interaction time is more than 200 fm/c longer than the emission time. As a consequence, the particles be-

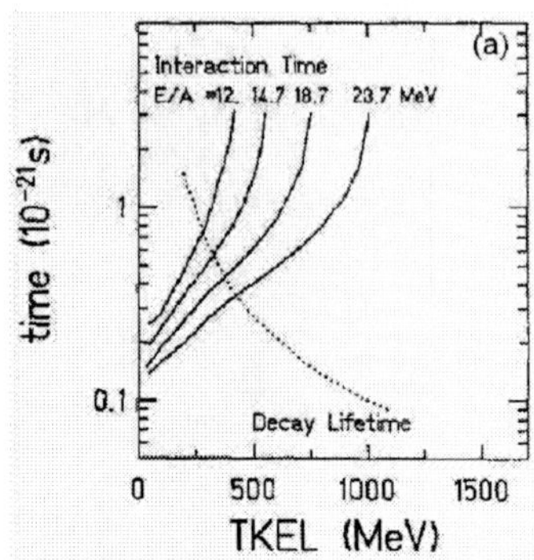

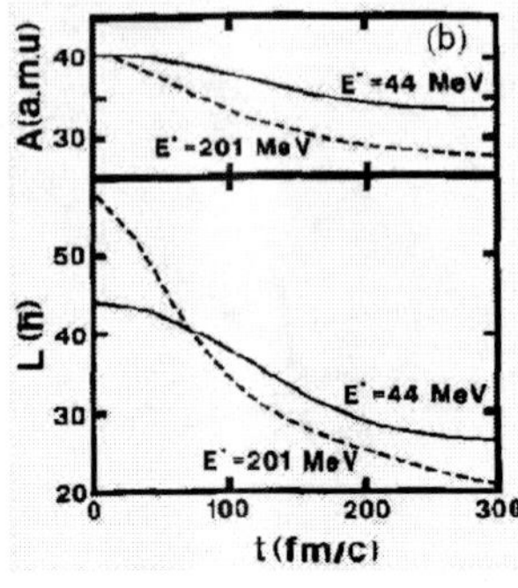

FIGURE 2. (a) Interaction time and decay lifetime as a function of total kinetic energy loss [3]; (b) Mass and angular momentum loss as a function of time for excited ^{40}Ca nucleus [4].

ing emitted during the rotation of the dinuclear system, will be preferentially focused in the reaction plane due to angular momentum conservation. This scenario is confirmed by model calculations. Fig.2(b) shows the result of a Landau-Vlasov method to solve the Boltzmann equation of n-body distribution for a ^{40}Ca nucleus at initial excitation energy and angular momentum of 44 MeV and 45 $\hbar$ and respectively, 201 MeV and 55 $\hbar$. By emitting particles, the mass and angular momentum of the parent nucleus are decreasing, later emitted particles being less aligned relative to the rotational plane. Such rotational-like pattern of the azimuthal distribution of emitted particles is expected to survive up to the incident energies where the dinuclear configuration of interacting nuclei is replaced by a participant-spectator scenario. The participating zone, compressed and excited, expands in the presence of spectator nuclear matter and a squeeze-out pattern, in the direction perpendicular to the reaction plane is expected to take place. At even higher incident energies, the Lorentz contraction of the spectators becomes important, the transit time of the spectator matter is shorter than the expansion time of the fireball and pressure gradient larger in the reaction plane could contribute to an enhanced emission of different species preferentially in the reaction plane - elliptic flow. A quantitative representation of these expected qualitative trends using the second term of a Fourier expansion in azimuth is given in Fig.3. Fig.3(a) shows a clear transition from rotational-like to squeeze-out pattern in the energy region of 100 A·MeV while in Fig. 3(b) is evidenced a transition from squeeze-out to elliptic flow at about 4 A·GeV. A detailed analysis of the energy where the transition from rotational-like to squeeze-out pattern takes place, E_{tran}, revealed a strong dependence as a function of particle species, transverse momentum ($p_t^{(0)}$) and centrality [5]. At low values of $p_t^{(0)}$, E_{tran} is decreasing with increasing mass of the reaction products. For all reaction products E_{tran} is decreasing as a function of $p_t^{(0)}$ such that at large $p_t^{(0)}$, within the error bars, all particles have the same E_{tran} value. Calculations based on a hybrid model in which a hydrodynamic expansion is coupled at the break-up moment with a statistical fragmentation [8] show that at low $p_t^{(0)}$ lighter particles are preferentially emitted earlier in the expansion process relative to the heavier ones while at large $p_t^{(0)}$ values, the maximum of the yield

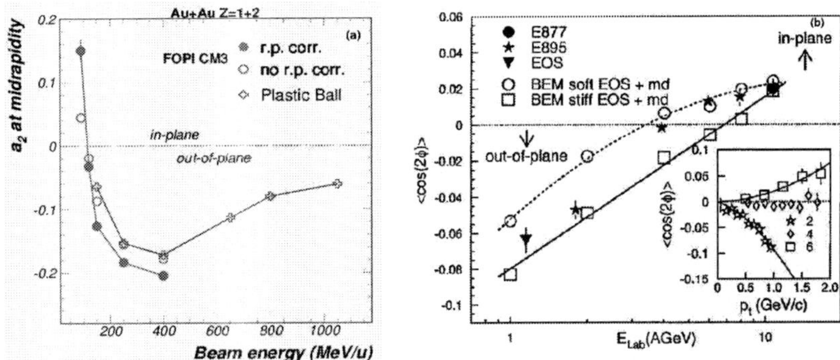

FIGURE 3. (a) The excitation function of a_2 in the low energy region measured by the FOPI Collaboration [5] compared with the Plastic Ball data [6]; (b) The excitation function of a_2 at higher energy measured by the EOS Collaboration [7].

for all species is located at the same break-up moment [9]. As a consequence, at lower $p_t^{(0)}$ values, heavier fragments being emitted later in the expansion process, will be less influenced by the rotation of the emitting source, therefore having lower E_{tran} value.

AZIMUTHALLY SYMMETRIC FLOW

Let's concentrate now a bit more on the expansion of the fireball. For highly central collisions of Au + Au, selected by imposing condition on the sum of ratios between transverse and longitudinal energies for all charged particles, the mean kinetic energy of fragments as a function of their mass was observed to have a dependence specific for a thermal motion superimposed onto a collective velocity field [8, 10]:

$$\langle E_{kin}^{cm} \rangle \approx \frac{1}{2} A \cdot m_0 < \beta_{flow}^2 > + \frac{3}{2} "T", \tag{1}$$

where A is the fragment mass number, m_0 is the nucleon rest mass, β_{flow} the expansion velocity and "T" the temperature which includes Coulomb contribution. Fitting the mean kinetic energy as a function of mass by such an expression, the flow energy per nucleon E_{coll} is obtained. The results of such an analysis at mid rapidity in two regions of polar angles as a function of incident energy for Au+Au, Xe+CsI and Ni+Ni collisions are represented in Fig.4(a). While in the forward region all the three systems show the same E_{coll} values, in transverse direction, at θ_{cm} around 90°, E_{coll} is lower, the increase as a function of incident energy is not that steep and at the same incident energy the collective expansion increases with the mass of the colliding system. Calculations based on IQMD model [11] show that the nucleons emitted at 90° suffer in the average 3.5 collisions relative to 1.7 of those emitted at smaller polar angles [12], see Fig. 4(b). This tells us that at 90° the observed effects are most probably coming from an equilibrated fireball, while at forward polar angles the effects due to transparency, Corona and uncertainty in the impact parameter selection play an important role. Comparison between experimental

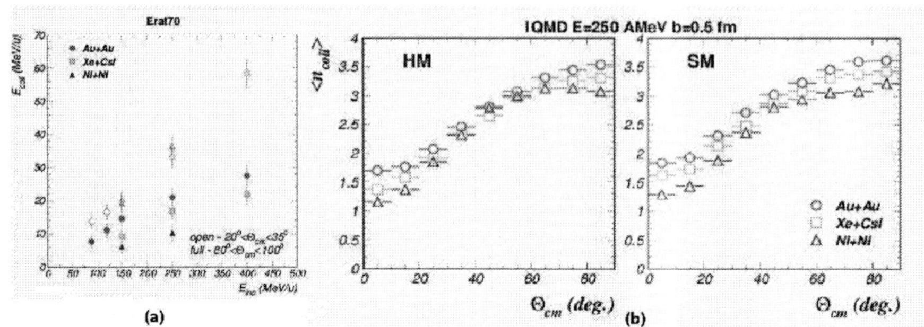

FIGURE 4. (a) Collective energy per nucleon as a function of incident energy for highly central Au + Au, Xe + CsI and Ni + Ni collisions, 70 mb geometrical cross section for $20° < \theta_{cm} < 35°$ and $80° < \theta_{cm} < 100°$; (b) The average number of collisions suffered by nucleons as a function of polar angle predicted by IQMD for hard(HM) and soft(SM) equations of state.

data and IQMD [11] and BNV [13] transport model calculations shows that both models reproduce the 90° collective flow using a soft EoS.

DETAILS ON SQUEEZE-OUT PHENOMENA

Is there anything new what we could learn studying the collective expansion of the fireball in the presence of spectator matter? In order to answer this question one has to perform similar studies for mid-central collisions [14]. Fig. 5(a) shows the azimuthal distributions of E_{coll} for Au+Au collision at 90, 120, 150, 250 and 400 A·MeV and ER4 centrality (b=2-4 fm). How much do we understand from this behaviour based on simple considerations? One could calculate for a given collision geometry the distance, respective the time with which the spectator matter has to move to let the observer at azimuth ϕ to see directly the fireball. Thus, under the hypothesis that the expansion starts at the maximum overlap, one could integrate from that moment to the break-up time the energy distribution delivered by our hybrid model [8] and obtain the mean kinetic energy for different species and collective expansion as a function of azimuth [9, 14]. Based on these simple considerations and nice qualitative agreement with the experimental trends, we could conclude that different regions of azimuth capture different periods of the central fireball expansion. If this is the case, the azimuthal distribution of E_{kin} and E_{coll} could be considered sensitive probes for the equation of state of matter at high density and temperature. With this in mind we performed transport model calculations based on BUU code [15] using momentum dependent mean-field, in-medium cross section, soft (K=210 MeV) and hard (K=380 MeV) equation of state [14]. The experimental and theoretical E^0_{coll} and ΔE_{coll} values obtained by fitting the azimuthal distributions of the flow energy with $E_{coll} = E^0_{coll} - \Delta E_{coll} \cdot \cos 2\Phi$ are represented in Fig.5(b). As one could see, the most sensitive quantity to the equation of state is the amplitude of the azimuthal oscillation of E_{coll} and the agreement with the experimental data is obtained for a soft EoS.

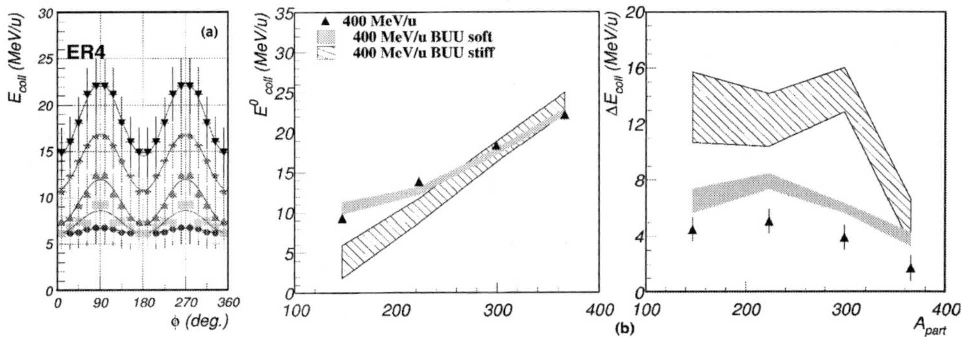

FIGURE 5. (a) The azimuthal distributions of E_{coll} for Au+Au collision at 90, 120, 150, 250 and 400 A·MeV and ER4 centrality (b=2-4 fm); (b) E^0_{coll} and ΔE_{coll}, as a function of A_{part}. The BUU results are represented by gray zones for soft EoS and dashed zones for stiff EoS, respectively.

ISOSPIN EFFECTS

One of the ingredients used in the microscopic transport models is the nucleon-nucleon cross section as a function of energy and angle. These quantities are rather well known for free space case while for in-medium case there are only theoretical estimates. Is there an observable which is most sensitive to the in-medium isospin dependent nucleon-nucleon cross section? We present in Fig.6(a) the proton azimuthal distributions for Ru + Zr and Zr + Ru collisions at 400 A·MeV, measured in the azimuthal angular range of 0°-60° and 300°-360°, symmetrized relative to 90° and 270°, respectively and relatively normalized at 90°, for CM3 centrality (b=4-6 fm) [16]. QMD model calculations [17] confirm our expectation that the squeeze-out signal for isospin asymmetric spectators is mainly sensitive to the in-medium isospin dependent nucleon-nucleon cross section. What about isospin distribution within the fireball? There are a few experimental data [18] which show a large difference in $< E_{kin} >$ of ^3H and ^3He at mid rapidity, larger than the expected difference due to Coulomb repulsion at the freeze-out moment and not reproduced by transport models. Based on the dynamics of the expansion suggested by the hybrid model mentioned above, one could imagine that due to isospin effects, a different distribution of protons relative to neutrons is established in the fireball which survives up to the break-up moment. If this is such that the outer layers of the fireball are proton rich relative to the inner zones which are neutron rich, then one expects ^3He fragments originating with higher probability from the outer regions, characterized by larger expansion velocity relative to the inner regions producing with higher probability ^3H but characterized by lower expansion [19]. If this is the case, then, based on the considerations presented in Section 4, one could expect a larger squeeze-out signal for ^3He relative to the one of ^3H. We performed such studies for Ru - Zr collisions at 400 A·MeV, summing up all 4 combinations in order to increase the statistics. The results are presented in Fig.6(b). As could be seen, ^3He shows systematically larger squeeze-out signal relative to ^3H at all centralities. This could be a strong indication that in such collisions, neutron rich matter could be obtained in the inner zones of the fireballs, such

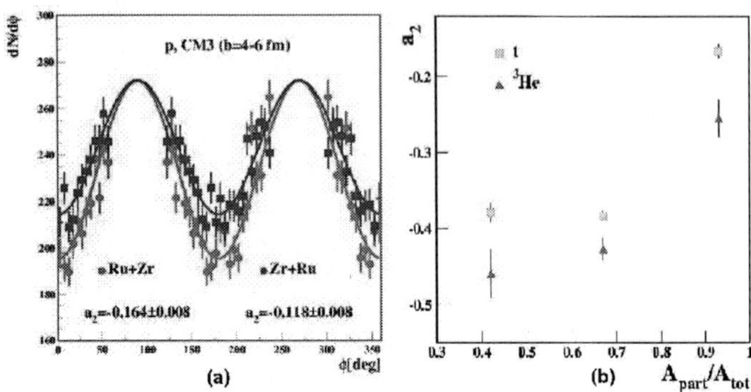

FIGURE 6. (a) Proton azimuthal distributions for Ru + Zr and Zr + Ru collisions at 400 A·MeV; (b) a_2 value for ^3H and ^3He for Ru(Zr)+Ru(Zr) at 400 A·MeV incident energy $0.8 < p_t^{(o)} < 2.0$ as a function of participating nucleons.

observable being recommended to extract information on the symmetry energy.

FLOW EFFECTS AT ULTRA-RELATIVISTIC ENERGIES

At sufficiently high energy densities matter is no longer confined to hadrons but exists in a deconfined state of quarks and gluons, the quark-gluon plasma (QGP). This new state of matter is expected to be produced in relativistic heavy ion collisions. QGP signals may manifest in the bulk properties of the collision at final state. At these energies the number of the created particles is very large and their behaviour dictates the observed flow effects. Since the particle distributions to first order only reflect the conditions in the final state of the system, signatures originating from the early stage are needed to conclude weather the system passes through a partonic phase and extract information on this phase. An observable sensitive to the early stage of the collision is the second harmonic coefficient, v_2, of the Fourier expansion of the azimuthal momentum distribution, called elliptic flow. In the coalescence scenario, which assumes elliptic flow on the constituent quark level, the data scale with the number of constituent quarks (n_q) as expected. Such scaling is observed in n_q scaled elliptic flow versus n_q scaled transverse kinetic energy (KE_T) (Fig. 7(a)) [20]. The ϕ meson exhibits the same v_2 as other particles thus indicating a partonic flow, the hadronic interaction cross section of the ϕ meson being much smaller (Fig. 7(b)) [21]. In central collisions, the final collective transverse radial flow, due to its cumulative nature, contains any collective flow generated at partonic and hadronic stages while particle yield ratios reflect mainly the statistical nature of the hadronization process. This can be demonstrated by analyzing the particle momentum spectra for common and rare particles [22]. The "blast-wave" model is used to extract the kinetic freeze-out temperature and the mean flow velocity as fit parameters. Statistical predictions of the particle ratios are used to extract the chemical freeze-out temperature. Thus the chemical freeze-out temperature is close to the Lattice QCD predicted critical temperature of $T_c \approx 170$ MeV. The data seem to indicate a sequential kinetic freeze-out

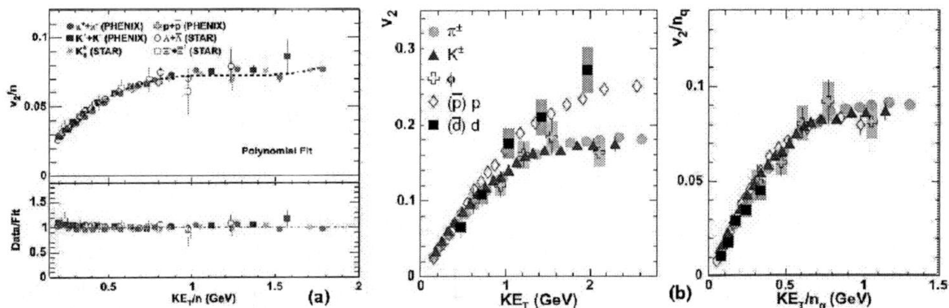

FIGURE 7. (a) v_2/n_q vs KE_T/n_q for different particle species in minimum bias Au+Au collisions; (b) v_2 vs KE_T (left) and v_2/n_q vs KE_T/n_q (right) for particle species in mid-central (20-60%) Au+Au collisions. The shaded bands indicate systematic error estimates for $(\overline{d})$d and ϕ mesons.

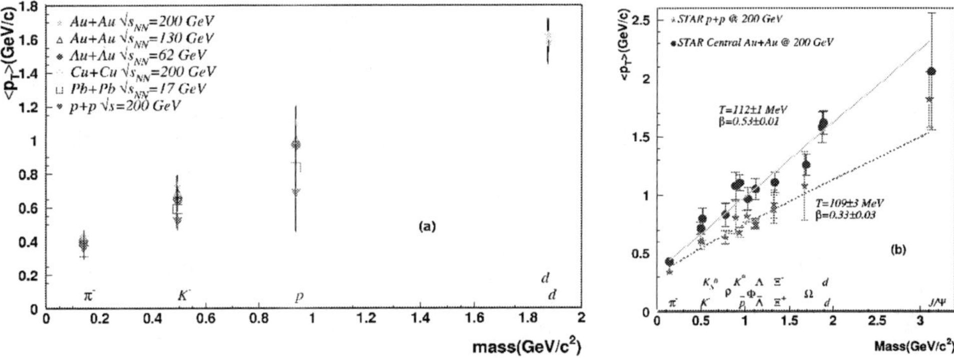

FIGURE 8. (a) Average transverse momentum as a function of particle mass; (b) $<p_T>$ at mid rapidity versus particle mass for different particles measured at STAR [23]. The lines are the fit with a formula derived from the caloric equation of state of a relativistic perfect gas boosted by a collective flow.

of particles. Thus Ω and Ξ, decoupled from the system just after chemical freeze-out possibly due to their smaller interaction cross-section with the hadronized environment, are followed by other particles(π, K, p). An useful observable for detailed study of such a scenario is the average transverse momentum ($<p_T>$) of the particle at mid rapidity. In Fig. 8(a) the average transverse momentum is seen to be an increasing function of particle mass. At the same time it is shown an increasing transverse flow strength from pp to A+A collisions and with increasing incident energy. $<p_T>$ at mid rapidity versus particle mass for different particles measured at RHIC in pp and Au+Au collisions at $\sqrt{s_{NN}}$=200 GeV is shown in Fig. 8(b) [23]. Error bars include systematic errors. The solid lines represent the fit of the data with a formula derived from the caloric equation of state of a relativistic perfect gas boosted by a collective flow. The results of the fit show a larger collective velocity in the case of the Au+Au collision compared to the pp collision. A temperature close to the critical one and a lower flow velocity is obtained by fitting separately the points corresponding to Λ, $\overline{\Lambda}$,$\Xi^{\pm}$, Ω^{-} and J/Ψ (T=172 ±7 MeV, β=0.36 ±0.02) as compared to the case of fitting only the points corresponding to π,K$^{\pm}$,K*,$\overline{p}$, d, $\overline{d}$ particles (T=104 ±4 MeV, β=0.59 ±0.02), in agreement with previous studies, suggesting that the strange baryons and J/Ψ reflect the dynamics and properties of the partonic stage of the reaction.

CONCLUSIONS AND OUTLOOK

The selected aspects of flow phenomena summarized in this lecture show that a detailed study of them is crucial for extracting information on the initial phase of matter, before expansion, its equation of state, in-medium effects and transition between different phases of matter. The low energy data evidenced that the cleanest signal on flow phenomena, associated mainly with the expansion of locally thermalized matter, can be obtained at mid rapidity, in the transverse direction relative to the collision axis and for highly central collisions. This conclusion applies also at the highest energies where the existence of flow in the deconfined phase followed by a further expansion and cooling after hadronization, seems to be confirmed by similar studies. It is a real challenge for the future experiments at LHC to use the potentiality of flow phenomena to extract new physics in an energy domain of about a factor of 30 larger than the highest one attainable at RHIC today.

REFERENCES

1. M. Petrovici et al., *Nucl. Phys. A* **477**, 277–317 (1988).
2. G. Wolschin and W. Nörenberg, *Z. Physik* **284**, 209–216 (1978).
3. R. J. Charity et al., *Z. Physik* **341**, 53–73 (1991).
4. F. Garcias and FOPI Collaboration, *Phys.Lett. B* **255**, 311–315 (1991).
5. A. Andronic and FOPI Collaboration, *Nucl.Phys. A* **679**, 765–792 (2001).
6. H. H. Gutbrod et al., *Phys.Rev.C* **42**, 640–651 (1990).
7. C. Pinkenburg and EOS Collaboration, *Phys.Rev.Lett.* **83**, 1295–1298 (1999).
8. M. Petrovici and FOPI Collaboration, *Phys. Rev. Let.* **74**, 5001–5004 (1995).
9. M. Petrovici et al., *Fizika B* **12**, 165–180 (2003).
10. S. C. Jeong and FOPI Collaboration, *Phys.Rev.Lett.* **72**, 3468–3471 (1994).
11. C. Hartnack et al., *Phys.Lett B* **336**, 131–135 (1994); *Mod.Phys.Lett. A* **9**, 1151–1157 (1994).
12. A. Andronic, *PhD Thesis* - Bucharest, 1998; M. Petrovici, Proc. Int. Symposium *Advances in Nuclear Physics*, Eds. D. Poenaru and S. Stoica, World Scientific, ISBN 9789810242763, 2000, pp. 242–249.
13. R. Bougault et al., *Nouvelles de GANIL*, no.58, 1996.
14. G. Stoicea and FOPI Collaboration, *Phys.Rev.Lett.* **92**, 072303 (2004).
15. P. Danielewicz and Q. Pan, *Phys.Rev. C* **46**, 2002–2011 (1992); P. Danielewicz, *Nucl.Phys.A* **673**, 375–410 (2000).
16. M. Petrovici et al., *GSI Sci. Rep.*, ISSN 0174-0814, 2000, p.38.
17. F. S. Zhang et al., *nucl-th/0011015*
18. G. Poggi and FOPI Coll., *Nucl.Phys. A* **586**, 755–776 (1995); M. Lisa and EOS Coll., *Phys.Rev.Lett.* **75**, 2662–2665 (1995); D. Guorio et al., *Eur.Phys.J. A* **7**, 245–253 (2000).
19. M. Petrovici et al., *GSI Sci. Rep.*, ISSN 0174-0814, 2000, p.39.
20. A. Adare and PHENIX Collaboration, *Phys. Rev. Lett.* **98**, 162301 (2007).
21. S. Afanasiev and PHENIX Collaboration, *Phys. Rev. Lett.* **99**, 052301 (2007).
22. Olga Barannikova (for the STAR Collaboration), *nucl-ex/0403014*.
23. B. I. Abelev and STAR Collaboration, *Phys. Rev. C* **75**, 064901 (2007).

Exotic Resonant States Whose Wave Functions Are Localized In The Barrier Region

N. Grama

H.Hulubei National Institute of Physics and Nuclear Engineering, Bucharest, Romania

Abstract. A unified treatment of all natural modes of the system particle+potential (bound and resonant states) is obtained in the Riemann surface approach to bound and resonant states. Exotic resonant states with properties that differ from the properties of usual resonant states are identified for various shapes of potential, provided that a potential barrier exists. In the case of a short range potential with Coulomb and centrifugal barrier it is shown that the exotic resonant states corresponding to poles in the neighborhood of certain attractors in the k-plane are the parent quasimolecular states. The properties of the parent quasimolecular states (energy, widths, deviation from the linear dependence of the energy on $l(l+1)$, doorway character, and criteria for observability) result naturally from the general properties of the exotic resonant states.

Keywords: natural modes, resonant states, quasimolecular states
PACS: 03.65.Nk, 34.50.-s, 34.80.Bm

NATURAL MODES OF THE PARTICLE+POTENTIAL SYSTEM

The natural modes of the particle+potential system are solutions $\phi_m(\vec{r})$ of the Schrödinger equation

$$\Delta\psi + \left(k^2 - V(\vec{r})\right)\psi = 0, \tag{1}$$

that are regular functions in R^3 and satisfy the radiation condition for $r \to \infty$ [1]. For short-range potentials ($lim_{r\to\infty}V(\vec{r}) \sim \Theta(r^{-\varepsilon})$, $\varepsilon > 2$) the radiation condition is [2]

$$lim_{r\to\infty}(\partial\phi_m/\partial r - ik\phi_m) = 0. \tag{2}$$

For long-range potentials ($lim_{r\to\infty}V(\vec{r}) \sim \Theta(r^{-\varepsilon})$, $\varepsilon < 2$) the radiation condition is expressed through the solution of the eikonal equation [3, 4]. By incorporating the radiation condition into the Schrödinger equation the Lippmann-Schwinger equation is obtained

$$\phi_m(\vec{r}) = \int d^3r'\, G^{(+)}(\vec{r},\vec{r'})\, V(\vec{r'})\, \phi_m(\vec{r'}), \tag{3}$$

whose solutions ϕ_m which are well behaved throughout the whole space and are outgoing to infinity define the natural modes. The natural modes may be either radiative (in which case the energy E is complex) or nonradiative (in which case E is real). We identify the radiative natural modes with resonant states and the nonradiative ones with bound states. For a central potential of strength g, i.e. $gV(r)$, by using the partial wave expansions of the wave and Green functions we obtain from Eq. 3

$$\mathscr{F}_l(g,k) = W(\phi_l(r), f_l^{(+)}(r)) = 0, \tag{4}$$

CP972, *Exotic Nuclei and Nuclear/Particle Astrophysics (II)—Proceedings of the Carpathian Summer School of Physics 2007,* edited by L. Trache and S. Stoica

where W is the Wronskian of the regular solution $\phi_l(r)$ and of the irregular solution $f_l^{(+)}(r)$ of the radial Schrödinger equation which satisfies the radiation condition, respectively. Eq. 4 defines the natural modes $k_l(g)$. Taking into account the definition of the S-matrix element S_l

$$S_l = \frac{W(\phi_l(r), f_l^{(-)}(r))}{W(\phi_l(r), f_l^{(+)}(r))} \tag{5}$$

it results that the natural modes correspond to the S-matrix poles.

RIEMANN SURFACE APPROACH TO NATURAL MODES

The pole distribution in the k-plane as a function of the potential strength g has been extensively studied for various shapes of the potential by many authors, using the "pole trajectory method": a particular path in the complex g-plane has been chosen and the corresponding trajectories of the S-matrix poles in the k-plane have been determined. In the "pole trajectory method" only a poor description of the function $k = k_l(g)$ is obtained. One can never be sure that all the poles in the k-plane have been found, or that the same pole is followed. In order to have a complete description of the S-matrix poles in the k-plane a global analysis of the function $k = k_l(g)$ is necessary. The global method for all S-matrix poles analysis involves the construction of the Riemann surface $R_g^{(l)}$ over the complex g-plane on which the pole function $k = k_l(g)$ is single valued and analytic. This implies the division of the Riemann surface into sheets and the construction of the Riemann sheets images in the k-plane. By keeping the Riemann sheets apart the single pole laying on each sheet image in the k-plane is identified. A schematic illustration of the method is given in Figure 1. In Figure 1(a) the pole function $k = k_l(g)$ defined on the complex g-plane is multiple-valued, i.e. there are many k values that correspond to a given g value. In Figure 1(b) the pole function $k = k_l(g)$ defined on the Riemann surface $R_g^{(l)}$ over the complex g-plane is single-valued. If g takes a value on a sheet $\Sigma_n^{(l)}$ of the Riemann surface $R_g^{(l)}$ then the function $k = k_l(g)$ takes only one value on the k-plane image $\Sigma_n'^{(l)}$, or in other words, there is only a single pole on each sheet image. Taking into account that one and only one branch of $k = k_l(g)$ is associated with a sheet $\Sigma_n^{(l)}$, the label n of this sheet will be used as a new quantum number for the S-matrix pole belonging to the corresponding k-plane image $\Sigma_n'^{(l)}$ and for the corresponding natural mode. The method has been used for various shapes of the potential: a rectangular well followed by a rectangular barrier (P1), a rectangular or Woods-Saxon well (P2), and a rectangular or Woods-Saxon well with Coulomb barrier (P3) [5, 6, 7].

PROPERTIES OF THE EXOTIC RESONANT STATE POLES AND STATES

For all mentioned potential barrier shapes, which also contain a centrifugal component resulted from partial wave decomposition, a new class of poles (exotic) has been identi-

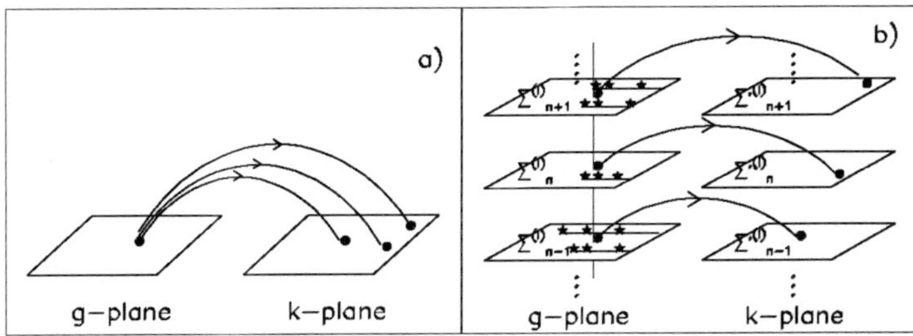

FIGURE 1. (a) The multiple-valued function $k = k_l(g)$ defined on the complex g-plane. (b) The Riemann surface $R_g^{(l)}$ over the complex g-plane of the function $k = k_l(g)$. Sheets $\Sigma_n^{(l)}$ of the Riemann surface $R_g^{(l)}$ and their k-plane images $\Sigma_n'^{(l)}$ are shown. The branch-points denoted by $\star$ and the branch-lines that join the branch-points are indicated. One can see that if g takes a value on a sheet $\Sigma_n^{(l)}$, then the function $k = k^{(l)}(g)$ takes only one value on the image of this sheet, $\Sigma_n'^{(l)}$.

fied. The exotic resonant state (ERS) poles have the following unusual properties:

i. Instead of becoming bound or virtual state poles when the strength of the potential well increases to infinity, the ERS poles remain in some bound regions of the k plane, in the neighborhood of some attractors. The number and position of these bound regions depend on the shape and height of the barrier. They occur only if the absorptive potential strength $\Im\ g > 0$ belongs to a certain window ($t_1 \leq |\Im\ g| \leq t_2$). ERS poles exist only on some Riemann sheet images, depending on the shape of the potential barrier, as illustrated in Figure 2. In Figure 2a) and 2b) the sheets and the aggregate of the sheet images for potential P1 with $l = 0$ are given; in Figure 2c) and 2d) the sheets and the aggregate of the sheet images for potential P2 with $l = 2$ are given; in Figure 2e) and 2f) the sheets and the aggregate of the sheet images for potential P3 with $l = 0$ are given. One can see that for potential P1 there are exotic poles only for strong absorptive potentials. There are an infinite number of Riemann sheet images on which there are situated ERS poles. On each Riemann sheet image there is only one bound region for the ERS pole. In the case of potential P2 there are ERS poles on a finite number of Riemann sheet images, the number of these sheet images increasing as the orbital angular momentum l increases. The ERS poles occur for either weak or strong absorptive potentials. In the case of potential P3 there are an infinite number of Riemann sheet images where ERS poles are situated, and these ERS poles occur for either strong or weak absorption.

ii. ERS poles go to some attractors (which will be denoted k_z) when $g \rightarrow \infty$. The way the ERS pole approaches the attractor in the case P3 with the Coulomb parameter c=9.5 and $l = 0$ is shown in Figure 3.

iii. The wave functions of the resonant states corresponding to poles situated in the neighborhood of the attractors are almost completely confined to the region of the barrier. The localization of the wave function in the case of potential P1 is illustrated in Figure 4.

The case of two channel model with square potentials has also been studied [8] and it was shown that besides the branch-points, which originate from the branch-points for the uncoupled channels there are new branch-points, which are intrinsic to the coupling.

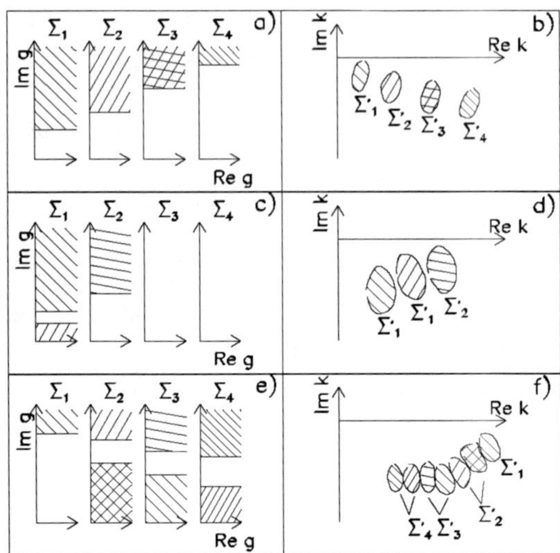

FIGURE 2. The first four sheets Σ_n, $n = 1, 2, 3$, and 4 and the aggregates of their k-plane images Σ_n', for the three mentioned shapes of the potential.

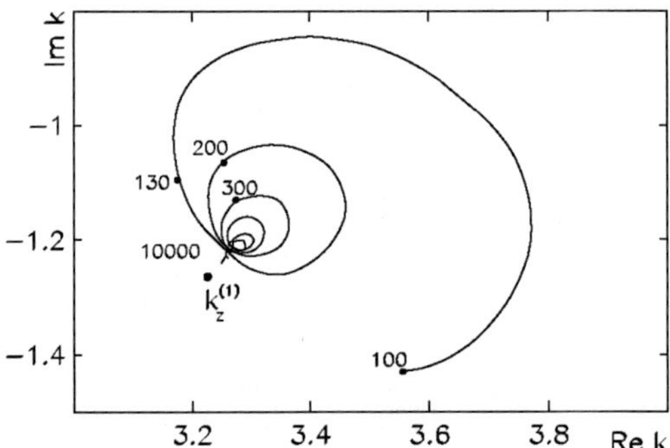

FIGURE 3. The spiral curve along which the exotic pole moves to the attractor on the k-plane image Σ_1' for the potential made of a square well with a Coulomb barrier with c=9.5 in the case $l = 0$ and g varying as $g = \lambda(10+i)$. The numbers beside the curve give the values of the parameter λ. The attractor is marked by •

The Riemann surface approach provided a new insight into the nature of the resonant states for two coupled channels. The distinct origin and the relationship of the Feshbach and Newton-Fonda resonant states have been clarified.

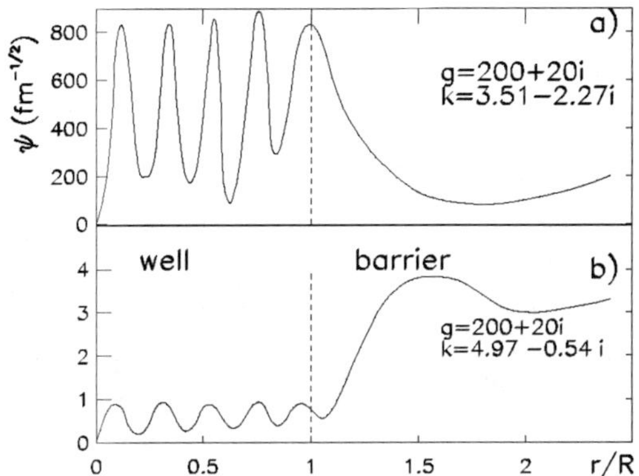

FIGURE 4. The moduli of the wave functions of an usual resonant state (a) and of an ERS (b) for the potential P1 with equal radii for well and barrier. The strength of the potential well depth g and the corresponding poles in the sheet images Σ_1' (for the ERS pole) and Σ_5' (for the usual resonant state pole) are given.

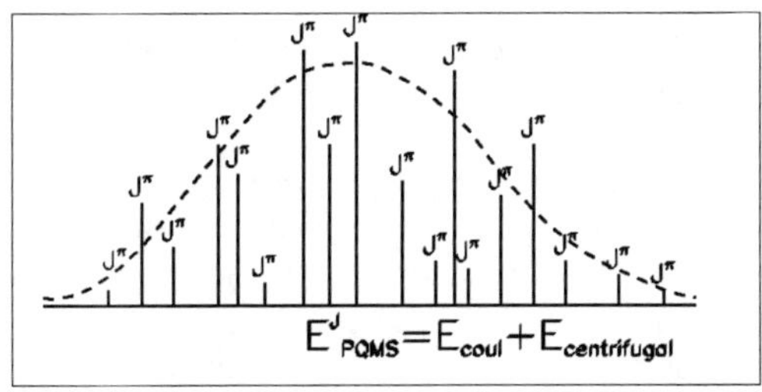

FIGURE 5. A di-nuclear PQMS with a width of several MeV (dashed line) is fragmented into a number of narrow QMS (having widths of order of hundreds of keV) with the same J^π.

PARENT DI-NUCLEAR QUASIMOLECULAR STATES: A PARTICULAR CASE OF ERS

Di-nuclear quasimolecular states (QMS) represent the most important cluster phenomenon in nuclear physics. They are nuclear resonant states (RS) excited in heavy-ion reactions, having special properties: *a)* QMS are highly excited RS in nuclei above the threshold E_{th} of two-body decay into the cluster constituents; *b)* The probability of QMS decay into cluster constituents is large, while the probabilities for decay into other channels are small; *c)* QMS have good spins and parities (J^π); *d)* QMS excitation energies are situated around the top of the total barrier $(E^J_{QMS} \simeq E_{coul} + E_{centrifugal})$, in a region of high-level density. Their widths Γ_{QMS} are of the order of hundred of keV; *e)* For a given J there are several QMS. For the same J the centroids of QMS energies form a straight line in the plane E_{ex} vs $J(J+1)$, appropriate to a rotational band.

As suggested by Feshbach [9] the narrow QMS with the same J^π are fragments of a di-nuclear parent quasimolecular state (PQMS) of a given J^π, as illustrated in Figure 5.

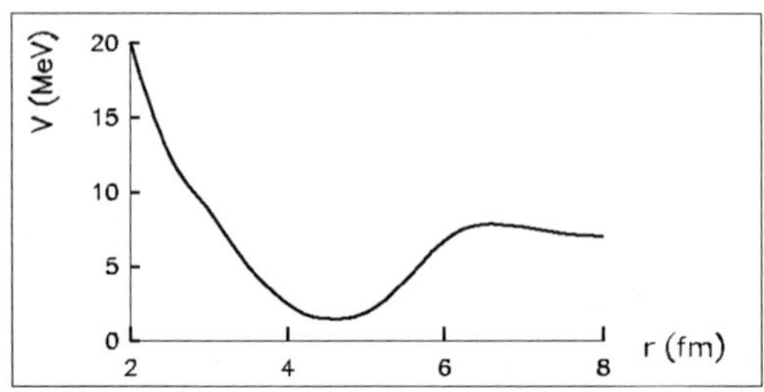

FIGURE 6. The potential for the $^{12}C+^{12}C$ scattering in the s state.

In spite of the great deal of research work done in the field of QMS there are main open questions still left :
a) Stability against the dissolution into the complex neighboring compound nuclear states: Because of the very high level densities at the excitation energies of QMS it is expected that QMS be totally damped, and consequently unobservable in the experiment. The fact that QMS have been observed in a region of high-level density suggests that they belong to a new class of states of the nuclear system, that fulfill extraordinary conditions which prevent them from spreading out. The nature of the structural difference between the QMS and the underlying compound nuclear states should be explained.
b) The QMS wave function is localized in the barrier region: It is a great puzzle to understand how a short range nuclear attraction can be operative in order to generate a RS having the wave function localized outside the potential well, in the region of the barrier. For example, in the ^{12}C+^{12}C system the distance of the closest approach at $E_{cm} = 5.6$ MeV (the lowest energy at which resonances have been observed) is 9.3 fm, while the ^{12}C radius is 3 fm. This means that the surfaces of the two nuclei will be separated by 3.3 fm in the closest proximity [10]. How could a short range attractive potential of about 4 fm (see Figure 6) be operative in this configuration, leading to the formation of a RS? Some attempts have been done to construct an attractive long range "effective" potential that could operate at the distance of the closest approach [10, 11], but this potential contradicts the usual potential shape. The large diffuseness used in [10, 11] decreases the Coulomb barrier so that the states are located above the barrier and consequently they are echoes and not RS.

The present work answers the above questions. By using an usual potential we are looking for the possibility to generate a class of ERS that could be candidates for PQMS, i.e. RS which have the wave function localized in the region of the barrier, particularly stable with respect to the dissolution into the neighboring compound nucleus resonant states. In order to do this we have to identify all classes of RS and to study their properties. The existence of different types of RS must be reflected in the existence of different types of S-matrix poles. Consequently we use the global method to identify simultaneously all the S-matrix poles, based on the Riemann surface approach to natural modes.

Taking into account the properties of the ERS given above we identify the PQRS observed in heavy ions collisions as ERS corresponding to poles located at the first attractors $k_z^{(1)}(c,l)$ for the potential P3:

1. They both form rotational bands of simple configuration states at high energies in nuclei, near the top of the total barrier. The ERS corresponding to poles at the first attractor $k_z^{(1)}$ form rotational bands. An approximation of the energy and width of a ERS corresponding to a pole in the neighborhood of $k_z^{(1)}$ has been obtained from the asymptotic expression of $k_z^{(1)}$ for large values of the Coulomb parameter $c = Z_1 Z_2 e^2 M/\hbar^2$.

$$E \simeq \frac{\hbar^2}{2MR^2}\left\{k_B^2 - \alpha_1 X - \frac{\alpha_1^2}{2}\frac{X^2}{k_B^2} - \frac{k_B^2}{X}\right\}, \tag{6}$$

$$\Gamma \simeq \frac{\hbar^2}{2MR^2}\left\{2\alpha_2 X - \alpha_1\alpha_2\frac{X^2}{k_B^2} - \alpha_1\alpha_2\frac{k_B^2}{X}\right\}, \tag{7}$$

where $k_B = [2c + l(l+1)]^{1/2}$, $X = [k_B^2 + l(l+1)]^{2/3}$ and $\alpha_1 = Re\ \alpha$ and $\alpha_2 = Im\ \alpha$ with $\alpha = (9\pi/8)^{2/3} exp(i\pi/3)$. The first term in Eq. 6 indicates a rotational character of the energy spectrum, while the other terms in Eq. 6 give corrections to the linear dependence of E on $l(l+1)$ and at the same time they show that the energy levels are situated below the barrier top $\hbar^2 k_B^2/2MR^2$. In the first approximation the width of the resonant state that corresponds to a pole situated in the neighborhood of $k_z^{(1)}(c,l)$ increases as $[c + l(l+1)]^{2/3}$ with respect to c and l. In the first approximation the slope of the tangent to the curve Γ versus $l(l+1)$ decreases as l increases, i.e. the widths tend to have a slower variation with respect to l for large l values. For lighter systems there is a stronger dependence of the widths on the orbital angular momentum. A good agreement was obtained in the comparison of the theoretical energies given by Eq. 6 with the experimental energies of the quasimolecular states reported in literature, from ^{12}C+^{12}C up to ^{28}Si+^{28}Si by using the same prescription for the radius $R = r_0(A_1^{1/3} + A_2^{1/3})$ with r_0=1.3 fm for all nuclear systems. The comparison for the systems ^{12}C+^{12}C, ^{12}C+^{16}O, and ^{16}O+^{16}O is shown in Figure 7.

2. The experimentally observed deviation from the linear dependence of E on $l(l+1)$ in the light systems, such as ^{12}C+^{12}C, ^{12}C+^{16}O and ^{16}O+^{16}O, results from the dependence of the slope of the tangent to the curve E vs $l(l+1)$ with respect to c and l. As one can see, for light systems (small c) there is a rather strong variation of the slope with respect to l, especially for low l values. The calculated energies in Figure 7 are obtained with the same radius prescription, both below and above the Coulomb barrier. Other authors [12, 13] have been forced to use two rotational bands for light heavy-ions systems, separated by the Coulomb energy E_{Coul}. In other words, these light systems (e.g. ^{12}C+^{12}C, ^{12}C+^{16}O and ^{16}O+^{16}O) were considered as dinuclear systems of two separated spheres below the Coulomb barrier and of two partially overlapping spheres above the Coulomb barrier.

3. The small amplitude of the wave function of a ERS inside the well leads to a small overlap with the adjacent resonant states of the continuum that are mostly confined to the compound nucleus radius, which is smaller than the potential well radius. This small overlap confers stability to such a state against the dissolution into the complex neighboring compound nuclear states. The ERS are a new kind of doorway states, whose stability is a consequence of the localization of the wave function, rather than of a symmetry.

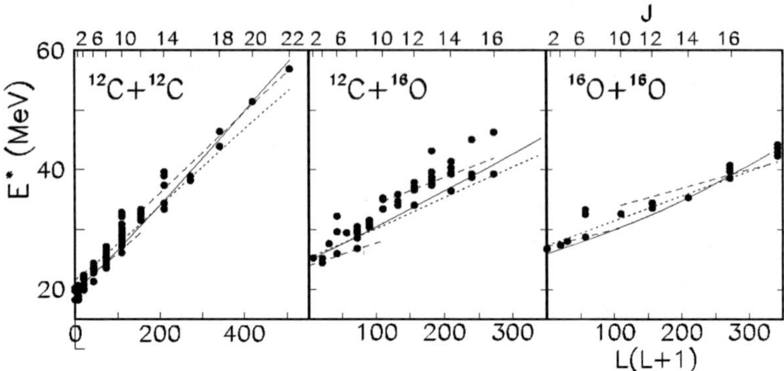

FIGURE 7. Comparison of the experimental excitation energies of PQMS with the ERS energies given by Eq. 6 (continuous curves). The dot line is the grazing line. The dashed lines are the E vs $l(l+1)$ lines for different configurations (two separate spheres below the Coulomb barrier and two partially overlapping spheres above the Coulomb barrier).

4. A good agreement was found between the calculated widths of the ERS given by Eq. 7 and the experimental widths of the parent quasimolecular states obtained by the reconstruction from the the spread fragments [14], [15].

5. Eq. 7 provides information on the observability of the quasimolecular resonant states. For the light systems the width increases rapidly as l increases so that it becomes very large for a relatively low l value and the corresponding resonance in the cross section is not observed. For heavier systems the widths of the resonant states increase slowly as l increases and consequently the states can be observed even at very large l values. For example, in the ^{12}C+^{12}C system the highest l-value of the observed quasimolecular resonance is l=18, while in the ^{28}Si+^{28}Si quasimolecular resonances with l=42 could be observed [16].

6. Systematic heavy ions scattering studies [17] have shown that an enhanced gross structure both in the angular distribution and excitation functions for light nucleus-nucleus systems is produced by a weakly absorbing four parameter optical potential of Woods-Saxon form with an energy dependent imaginary part of the potential

$$W = A + BE_{cm}. \tag{8}$$

For a given system we can calculate the weak absorption thresholds q_l for which the ERS occur as a function of l. A linear dependence $q_l = a + bl(l+1)$ has been noticed. Taking into account Eq. 8 one obtains the energy dependence of the thresholds q_l

$$q_l = (a - b\alpha/\beta) + bE/\beta. \tag{9}$$

Eq. 9 has to be compared to Eq. 8. The gross structure in the angular distribution and excitation function has been reproduced [17] in the case of $^{12}C+^{12}C$ system by a Woods-Saxon potential with the parameters V=14 MeV, R=6.13 fm, d=0.49 fm and an energy dependent imaginary part $W = 0.4 + 0.14E_{cm}$. In Figure 8 by hatching the weak absorption window for the occurrence of ERS in the $^{12}C+^{12}C$ system is shown. One can see that the optical model potential satisfies the constraints imposed to the complex potential strength g in order to generate ERS. This is a new proof that the quasimolecular states are ERS. The absorption that explains the scattering data belongs to the absorption

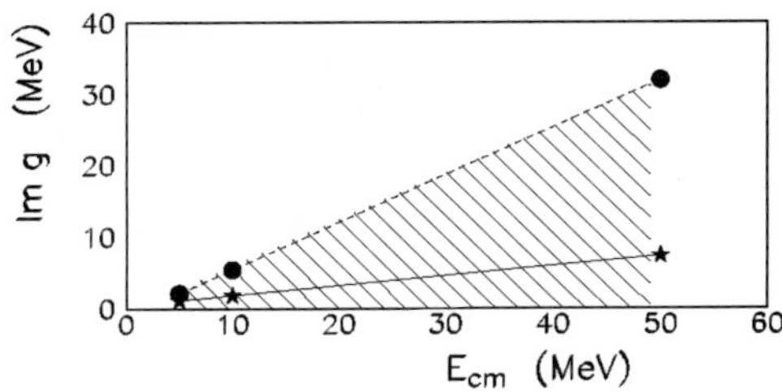

FIGURE 8. The figure shows that the imaginary part of the potential which reproduces the $^{12}C+^{12}C$ data belongs to the absorption window which allows the occurrence of the ERS (indicated by hatching). By $\star$ the imaginary part of the potential (8) is marked. By $\bullet$ the curve that gives the energy dependence of the thresholds of the absorption window is given.

window that ensures the occurrence of the ERS.

REFERENCES

1. D. N. Pattanayak and E. Wolf, *Phys. Rev. D*, **13**, 2287 (1976).
2. A. Sommerfeld, *Partial Differential Equations in Physics*, Academic Press Inc., 1949, p. 189.
3. Y. Saito, *Pacific Journal of Mathematics* **126**, 331 (1987).
4. G. Barles, *Commun. in Partial differential equations* **12**, 263–283 (1987).
5. C. Grama, N. Grama and I. Zamfirescu, *Ann. Phys.(N.Y.)*, **218**, 346 (1992).
6. C. Grama, N. Grama and I. Zamfirescu, *Ann. Phys.(N.Y.)*, **232**, 243 (1994).
7. C. Grama, N. Grama and I. Zamfirescu, *Phys. Rev. A* 61 (2000) 032716.
8. C. Grama, N. Grama and I. Zamfirescu, *J. Phys. B At. Mol. Opt. Phys.* **40**, 1035 (2007).
9. H. Feshbach, J. Phys. (Paris), **C5**, Suppl 11, 177 (1976).
10. L. Satpathy, in *Prog. in Particle and Nuclear Physics*, **29**, Pergamon Press, 1992, p.327.
11. L. Satpathy, P. Sarangy, A. Faessler,
12. D. A. Bromley, in *Proc. 4th Int. Conf. on Clustering Aspects of Nuclear Structure and Nuclear Reactions*, eds. J. S. Lilley, M. A. Nagarajan, Reidel, Dordrecht, 1985, p. 1.
13. H. S. Khosha, S. S. Malik, R. K. Gupta, Nucl. Phys. A, **513**, 115 (1990).
14. R. R. Betts, Nucl. Phys. A, **447**, 257c (1985).
15. H. Fröhlich et al., in *Lecture Notes in Physics* **156**, ed. Eberhardt K.A., (1982) p. 79.
16. R. R. Betts, S. B. Di Cenzo, J. F. Petersen, Phys. Lett. B, **100**, 117 (1981).
17. R. H. Siemssen, in *Nuclear Molecular Phenomena, Proc. Int. Conf. on Resonances in Heavy Ions Reactions, Hvar, Yugoslavia*, eds. N. Cindro, 1977, p.79.

4. EXOTIC MATTER AND EXOTIC DECAYS

Superallowed Nuclear Beta Decay: A Window on the Weak Interaction

J.C. Hardy

Cyclotron Institute, Texas A&M University, College Station, TX, 77843, USA

Abstract.
Measurements on superallowed $0^+ \to 0^+$ nuclear beta transitions currently provide the most demanding test of the Conserved Vector Current (CVC) hypothesis and the most precise value for the up-down element, V_{ud}, of the Cabibbo-Kobayashi-Maskawa (CKM) matrix. Both are sensitive probes for physics beyond the Standard Model. Analysis of the experimental results depends on small radiative and isospin-symmetry-breaking corrections whose uncertainties now dominate those from experiment. Recent experiments have been focusing on tests of these corrections with a view to reducing their uncertainties. An overview is presented together with a description of measurements at Texas A&M on the superallowed decay of ^{34}Ar.

PACS: 23.40.Bw, 12.15.Hh

OVERVIEW

Beta decay between nuclear analog states of spin-parity, $J^\pi = 0^+$, and isospin, $T = 1$, has a unique simplicity: it is a pure vector transition and is nearly independent of the nuclear structure of the parent and daughter states. The measured strength of such a transition – expressed as an "ft value" – can then be related directly to the vector coupling constant, G_V with the intervention of only a few small ($\sim$1%) calculated terms to account for radiative and nuclear-structure-dependent effects. Once a reliable value has been determined for G_V, it is only a short step to obtain from it the value for V_{ud}, the up-down mixing element of the Cabibbo-Kobayashi-Maskawa (CKM) matrix; and only another short step to the most demanding available test of the unitarity of that matrix, one of the basic precepts of the electroweak standard model.

In dealing with these decays, which are referred to as superallowed, it is convenient to combine some of the small correction terms with the measured ft-value and define a "corrected" $\mathscr{F}t$-value. Thus, we write [1]

$$\mathscr{F}t \equiv ft(1+\delta_R')(1+\delta_{NS}-\delta_C) = \frac{K}{2G_V^2(1+\Delta_R^V)}, \tag{1}$$

where $K/(\hbar c)^6 = 2\pi^3\hbar\ln 2/(m_e c^2)^5 = 8120.278(4)\times 10^{-10}\,\mathrm{GeV}^{-4}\mathrm{s}$; δ_C is the isospin-symmetry-breaking correction and Δ_R^V is the transition-independent part of the radiative correction. The terms δ_R' and δ_{NS} comprise the transition-dependent part of the radiative correction, the former being a function only of the electron's energy and the Z of the daughter nucleus, while the latter, like δ_C, depends in its evaluation on the details of nuclear structure. From this equation, it can be seen that a measurement of any one of these superallowed transitions establishes an individual value for G_V; moreover, if the

CP972, *Exotic Nuclei and Nuclear/Particle Astrophysics (II)—Proceedings of the Carpathian Summer School of Physics 2007,* edited by L. Trache and S. Stoica

Conserved Vector Current (CVC) assertion is correct that G_V is not renormalized in the nuclear medium, all such values – and all the $\mathscr{F}t$-values themselves – should be identical within uncertainties, regardless of the specific nuclei involved.

This assertion of CVC can be tested and a value for G_V obtained with a precision considerably better than 0.1% if experiment can meet the challenge, since the four small corrections terms only contribute to the overall uncertainty at the 0.03% level. As it turns out, experiment has exceeded that goal, leaving theory currently as the dominant contributor to the uncertainty.

The ft-value that characterizes any β-transition depends on three measured quantities: the total transition energy, Q_{EC}, the half-life, $t_{1/2}$, of the parent state and the branching ratio, R, for the particular transition of interest. The Q_{EC}-value is required to determine the statistical rate function, f, while the half-life and branching ratio combine to yield the partial half-life, t. In 2005, a survey of world data on superallowed $0^+ \rightarrow 0^+$ beta decays appeared [1], in which all previously published measurements were included, even those that were based on outdated calibrations if enough information was provided that they could be corrected to modern standards. In all, more than 125 independent measurements of comparable precision, spanning four decades, made the cut. A total of nine transitions yielded ft values with 0.1% precision or better, and three more with precision between 0.1 and 0.4%.

From those results a number of important conclusions were drawn. First, the $\mathscr{F}t$ values for all twelve transitions, covering the range from A=10 to A=74, formed a consistent set, which confirmed the constancy of G_V to 1.3 parts in 10^4, the tightest limit ever set. Second, the survey results set a limit on any possible contribution from scalar currents. The presence of a scalar current – induced or fundamental – would manifest itself as a Z-dependence in the $\mathscr{F}t$ values that would be most evident at low Z. There was no hint of any such curvature and a limit could be set on the scalar relative to the vector current of $|C_S/C_V| \leq 0.0013$, again the tightest limit ever set. Finally, with the test of CVC passed, it was possible to use the average value of G_V to obtain the up-down element of the CKM matrix via the relation $V_{ud} = G_V/G_F$, where G_F is the well known [2] weak-interaction constant for purely leptonic muon decay.

Since the 2005 survey closed, experiments in this field have intensified, with 10 additional measurements already published [3, 4, 5, 6, 7, 8, 9, 10, 11, 12]; in fact, a thirteenth precisely measured transition has been added to the set. We have now included those new results in an interim update, and present the resultant ft and $\mathscr{F}t$ values in Figure 1. A comparison of the left and right panels in the figure demonstrates that the calculated corrections act very well to remove the considerable scatter that is evident in the former but is missing in the latter. Since δ_R' has essentially the same value for all transitions other than those from ^{10}C and ^{14}O, the removal of the scatter must be attributed to the calculated nuclear-structure-dependent corrections, δ_{NS} and δ_C, that have been applied in the derivation of $\mathscr{F}t$. The calculation of these correction terms [13] was based on nuclear-structure models that were solidly grounded in a wide variety of independent nuclear measurements, none involving superallowed β decay. Thus the consistency of the $\mathscr{F}t$ values is a powerful validation of those calculated corrections. All thirteen $\mathscr{F}t$ values form a statistically consistent set – with a normalized chi-square of 0.9 – over a wide range of nuclear masses, from A=10 to A=74, a conclusion

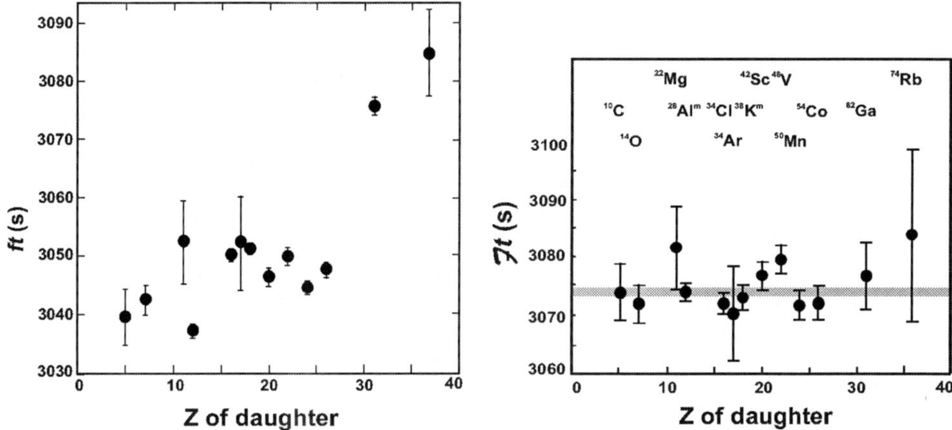

FIGURE 1. Results from the 2005 survey [1] updated with more recent published results [3, 4, 5, 6, 7, 9, 10, 11]. The uncorrected ft values for the thirteen best known superallowed decays (left) are compared with the same results after corrections have been applied (right). The grey band in the right-hand panel is the average $\mathscr{F}t$ value, including its uncertainty.

that is entirely consistent with CVC expectation. The average of all thirteen cases is $\overline{\mathscr{F}t}$ = 3073.9(8)s.

From the $\mathscr{F}t$-value data in Fig. 1 and the recently improved calculation of Δ_R^V [14] we obtain the result,

$$V_{ud} = 0.97378(26). \qquad \text{[nuclear]} \tag{2}$$

This number can be compared with our previous value, quoted in the 2005 survey, V_{ud} = 0.97380(40). The new result is completely consistent with the earlier one, but it has a considerably reduced uncertainty – due mostly to the improvement in the calculated "inner" radiative correction, Δ_R^V, but also to the new experimental results, which in many cases have pushed to even higher precision than was possible before. With our new value for V_{ud} and the other elements of the top row of the CKM matrix taken from the most recent Particle Data Group review [2], we obtain for the unitarity sum,

$$|V_{ud}|^2 + |V_{us}|^2 + |V_{ub}|^2 = 0.9992(11), \tag{3}$$

which is consistent with the matrix being unitary.

The nuclear result for V_{ud} in Eq. 2 can be compared with results [2] obtained solely from neutron decay – 0.97460(190) – and from pion decay – 0.97490(260). These values are consistent with the nuclear result but their uncertainties are much larger. Evidently, for now the value of V_{ud} is determined entirely by the nuclear measurements, and its precision is limited only by the experimental uncertainties in the nuclear measurements and the theoretical uncertainties attached to their applied correction terms. Since the uncertainty in the unitarity sum, Eq. 3, comes roughly equally from the V_{ud} and V_{us} terms, it too can be significantly improved if the nuclear uncertainties can be reduced. There is thus considerable motivation to continue the search for improvements in both the theory and experiments that focus on superallowed beta decay.

The uncertainty given for V_{ud} in Eq. 2 is dominated by the uncertainty in the theoretical correction terms. In fact, experiment contributes a mere 0.00007 to the total uncertainty of 0.00026. This does not mean, however, that experiment has no future role to play until the theory can be improved. To the contrary, carefully chosen experiments can test the theory and guide it towards improvements. The approach is best illustrated by Fig. 1 and the observation that the calculated corrections are validated by their success in removing the ft-value scatter. Improvements in experimental precision would test the calculations' effectiveness even more severely, as would new examples of $0^+ \rightarrow 0^+$ transitions, specifically selected for having *larger* calculated corrections.

Although the first option, improving the precision on already well known transitions, might have seemed less appealing than the option to measure new cases, this approach has actually produced some of the most striking results in the past two years. Historically, the Q_{EC} value of a superallowed transition has been a very challenging property to measure with sufficient precision. Since f depends on the decay energy to the fifth power, these Q values have to be determined to about $\pm 0.01\%$, or ± 500 eV for a 5-MeV β decay. Up to the time of the 2005 survey, the only measurements to meet these standards were of reaction Q-values, typically obtained from a (p,n) or (^{3}He,t) reaction in which the β-decay daughter nucleus was used as the target. The situation has completely changed since then. For the first time, with on-line Penning traps it has now become possible to measure parent and daughter masses individually and obtain the difference between them to a precision of a few hundred eV in light nuclei [3, 6, 7, 8]. This has been, and still is, a very active research area, which almost from the start proved its value by revealing an anomaly in the results, visible in the $\mathscr{F}t$ values in Fig. 1 at A = 42 and 46. This has already inspired improvements to the nuclear-structure-dependent corrections [15] and will undoubtedly impact the value of V_{ud}.

The second experimental option to test the theoretical correction terms is to study new examples of superallowed transitions, particularly those with large calculated corrections. The reasoning is that if the ft values measured for cases with large calculated corrections also turn into corrected $\mathscr{F}t$ values that are consistent with the others, then this must verify the calculations' reliability for the existing cases, which have smaller corrections. In fact, the cases of ^{22}Mg, ^{34}Ar, ^{62}Ga and ^{74}Rb were all chosen for this reason. It can be seen from Fig. 1 that all are characterized by large correction terms. However it can also be seen that their error bars are still rather too large to seriously constrain the calculations. The reason for the large error bars in the cases of ^{22}Mg and ^{34}Ar is quite different from the reason for ^{62}Ga and ^{74}Rb. That difference is important to appreciate.

Most of the well-known superallowed emitters, like ^{26}Al^m and ^{34}Cl, are odd-odd $T_z = 0$ nuclei that decay to even-even $T_z = +1$ daughters. For these cases, there are no 1^+ states in the daughter that are energetically available to β decay, so the superallowed transition is the only branch that occurs with appreciable intensity: its branching ratio is essentially 100% with negligible uncertainty. In contrast, ^{22}Mg and ^{34}Ar are even-even $T_z = -1$ nuclei that decay to odd-odd $T_z = 0$ daughters, in which low-lying 1^+ states do occur. For such cases, the superallowed transition is only one of several strong β-decay branches, so its branching ratio must be measured, and the uncertainty on that measurement feeds directly into the uncertainty on its ft value. Branching-ratio measurements with 0.1% precision are extremely challenging. Even so, in the next section I will give an example of a measurement at Texas A&M that takes on this challenge and promises ultimately to reduce the uncertainty on the ^{34}Ar ft value to the 0.1% level.

The cases of ^{62}Ga and ^{74}Rb are quite different. They are $T_z = 0$ nuclei feeding even-even daughters, like the historically well-known cases. However, these being heavier nuclei, some 1^+ states do appear within the β-decay energy window, and weak but important competing transitions can occur. This does introduce some complications, but recent experiments have demonstrated that these can be brought under control [11]. The real difficulty lies with the nuclear models that must be used to calculate the nuclear-structure-dependent corrections. While the p, sd and $f\frac{7}{2}$ shells have been reasonably well characterized by shell-model effective interactions, for nuclei in the region above A=60 experimental information is, as yet, relatively sparce, and there are no model calculations that can be regarded as reliable. Consequently, the calculated correction terms that depend on nuclear structure carry very large uncertainties for these nuclei [13, 15]. Unfortunately what this means is that, although the ft-value measurements for these nuclei can provide a valuable check on the overall trend of the correction-term calculations, no matter how precise the measurements are, they cannot help to reduce the uncertainties on V_{ud} until we have a better grasp of the nuclear structure in this region of nuclei. This is undoubtedly an important task for radioactive-beam facilities in future.

^{34}AR DECAY MEASUREMENTS

Precise half-life and branching-ratio measurements need high-purity radioactive beams, a requirement that is even more important for cases like ^{34}Ar where the daughter nucleus, ^{34}Cl, is itself radioactive. We achieved this goal by using a production reaction with inverse kinematics, ^{1}H (^{35}Cl, 2n) ^{34}Ar, and selecting the desired reaction product with the Momentum Achromatic Recoil Separator (MARS). A primary beam of 30-AMeV ^{35}Cl from the Texas A&M superconducting cyclotron impinged on a liquid-nitrogen-cooled hydrogen gas target operated at 1.6-atm. The resultant 26-AMeV beam of ^{34}Ar separated by MARS exited the vacuum chamber through a thin Kapton window and then passed through a 0.3-mm-thick plastic scintillator and a series of Al degraders, which were adjusted to ensure the implantation of the ^{34}Ar nuclei at the center of a 76-μm-thick aluminized Mylar tape, part of our fast tape-transport system. With an ^{34}Ar beam intensity of about 3×10^4 particles/s, we collected a radioactive sample, >99.8% pure, for typically 1s, then turned off the beam and transported the sample in 180 ms to a shielded counting location where data were collected for a preset length of time, after which the cycle was repeated. This sequence was repeated until sufficient statistics had been collected. The experimental arrangement is shown in Figure 2.

What detection equipment was placed at the counting location depended on whether a branching-ratio or a half-life was being measured. The figure shows the arrangement used for the former, in which a spectrum of γ-rays was recorded for those events in the HPGe detector that were observed to be in coincidence with positrons in the thin plastic β detector, and the β-decay branching ratios were determined from the corresponding

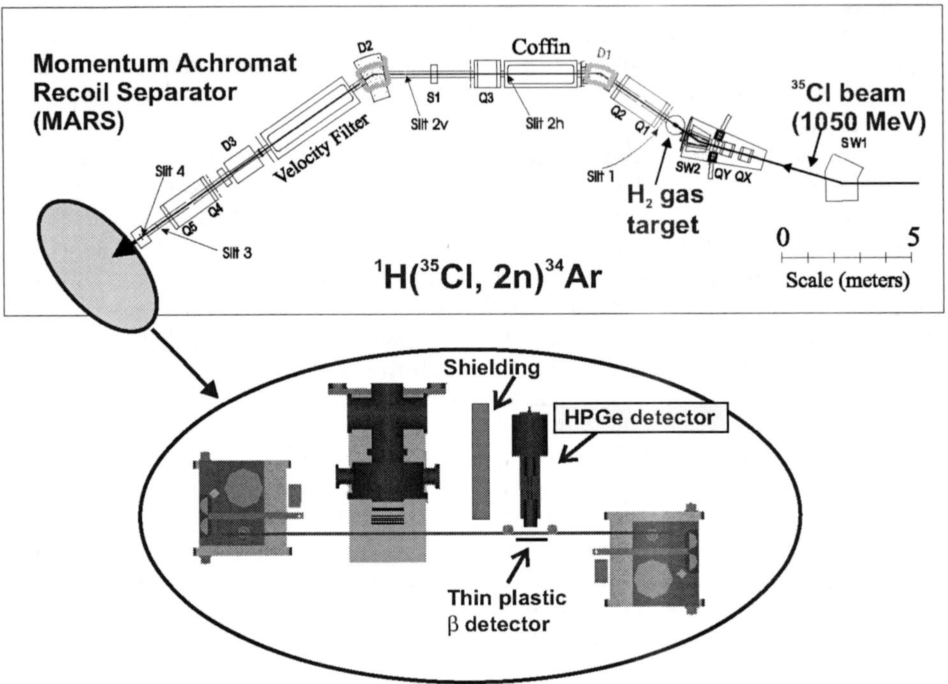

FIGURE 2. Experimental arrangement used for the ^{34}Ar branching-ratio measurement. For the half-life measurements, the HPGe detector and thin plastic β detector were replaced by a 4π gas proportional counter; the tape passed through the center of that detector.

intensities of the β-delayed γ-ray peaks. For the latter, a 4π gas proportional counter split into two halves was used, with the tape passing between the halves and the collected sample being stopped exactly at the center for each measurement period.

The spectrum of recorded β-delayed γ rays is shown in Figure 3, where no impurity peaks can be detected. The β-branching ratio, R_i, for a particular transition populating state i, which decays by emitting γ ray, γ_i, is determined via the equation

$$R_i = \frac{N_{\beta\gamma_i}}{N_\beta \varepsilon_{\gamma_i}} k, \tag{4}$$

where $N_{\beta\gamma_i}$ is the total number of β-γ coincidences measured in the γ_i peak, N_β is the total number of β singles, ε_{γ_i} is the detector efficiency for γ ray, γ_i, and k is a small correction factor (*i.e.* $k \sim 1$) that, among other things, takes into account the differences in the β-detector efficiency for the different transitions participating in ^{34}Ar decay. (See Ref. [16] for a detailed description of the factors that contribute to k.) This relation highlights the importance of a precise absolute efficiency calibration for the γ-ray detector and a reasonable knowledge of relative efficiencies in the beta detector. Our HPGe detector's absolute efficiency is accurately known (to $\pm 0.2\%$ for 50-1400 keV γ rays and $\pm 0.4\%$ up to 3500keV) from source measurements and Monte

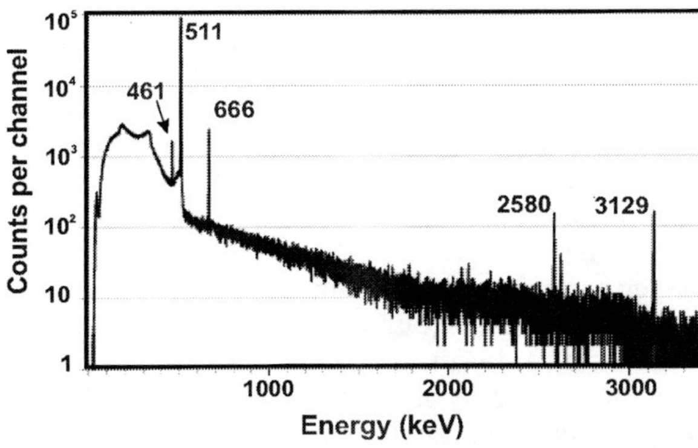

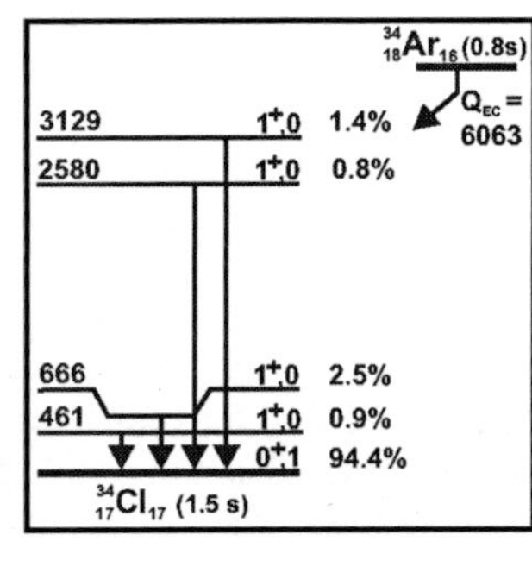

FIGURE 3. Spectrum of γ rays observed in coincidence with positrons from collected ^{34}Ar sources. Also shown is the β-decay scheme for ^{34}Ar.

Carlo calculations [17]. The relative efficiency as a function of β energy in the plastic scintillator was determined by Monte Carlo calculations and checked by comparison with measurements on conversion-electron sources.

Preliminarily we find the sum of the branching ratios for all transitions populating excited states in ^{34}Cl to be 5.64(8)%. These branches are all Gamow-Teller in character, and it is the ground-state transition that is the superallowed one. Subtracting the sum of excited-state transitions from 100%, we obtain a branching ratio of 94.36(8)% for the superallowed transition. Although the precision on the measured sum of branching ratios is only 1.4% – largely determined by counting statistics – the subtraction from 100% leads to a precision of 0.08% for the superallowed branching ratio, quite good enough to yield a very useful uncertainty on the ft value.

For the half-life measurement, the signals from the gas detector were amplified and sent to a discriminator, whose output was split and processed in parallel by two separate systems, each with a different pre-set dominant non-extendable dead-time. The two data streams allowed us to test that our dead-time-corrected result was independent of the actual dead time of the circuit. In addition to this check, the overall measurement was also subdivided into many separate runs, differing from one another only in their detection parameters: dominant dead-time, detector bias and discriminator threshold. No systematic dependence of the half life on any of these parameters could be detected. The details of the measurement, from which we obtained a half-life for ^{34}Ar of 843.8(4) ms, have already been published [10].

With the preliminary value quoted here for the superallowed branching ratio of ^{34}Ar, the $\mathcal{F}t$ value for this decay becomes 3073.1(37)s, which agrees very closely with the average $\overline{\mathcal{F}t}$ value for all superallowed transitions, and now has a precision (0.12%) that is comparable to the best-known cases. This result stringently confirms the nuclear-structure-dependent correction terms within the sd shell, where many of those best-known cases occur.

CONCLUSIONS

Measurements of superallowed $0^+ \rightarrow 0^+$ nuclear beta decay have achieved remarkable precision and are playing a important role in delimiting the electroweak standard model. To date, CVC has been verified to 0.013%, the scalar current limited to at most 0.13% of the vector current and CKM unitarity demonstrated to within 0.11%. Current activity in the field is focused on improving this precision even further, with special emphasis on reducing, through experiment, the uncertainties currently attached to the theoretical correction terms that must be applied to the experimental results before fundamental weak interaction parameters can be extracted from them. Recent measurements have in some cases confirmed the correction terms and in others have led the way towards refinements. There seems little doubt that the value of V_{ud} can still be improved by nuclear measurements.

ACKNOWLEDGMENTS

All of the work reported here was done in collaboration with others: on the theoretical side with I.S. Towner; on the experimental side with V.E. Iacob, V. Golovko, N. Nica, L. Trache, R.E. Tribble, C. Gagliardi, H.I. Park, J. Goodwin and others. The work was supported financially by the U.S. Department of Energy under Grant No. DE-FG03-93ER40773 and by the Robert A. Welch Foundation under Grant No. A-1397.

REFERENCES

1. J.C. Hardy and I.S. Towner, Phys. Rev. C **71**, 055501 (2005); Phys. Rev. Lett. **94**, 092502 (2005).
2. W.-M. Yao *et al.*, Journal of Physics G **33**, 1 (2006).
3. G. Savard, F. Buchinger, J.A. Clark, J.E. Crawford, S. Gulick, J.C. Hardy, A.A. Hecht, J.K.P. Lee, A.F. Levand, N.D. Scielzo, H. Sharma, K.S. Sharma, I. Tanihata, A.C.C. Villari, and Y. Wang, Phys. Rev. Lett. **95**, 102501 (2005).
4. I.S. Towner and J.C. Hardy, Phys. Rev. C **72**, 055501 (2005).
5. B. Hyland, D. Melconian, G.C. Ball, J.R. Leslie, C.E. Svensson, P. Bricault, E. Cunningham, M. Dombsky, G.F. Grinyer, G. Hackman, K. Koopmans, F. Sarazin, M.A. Schumaker, H.C. Scraggs, M.B. Smith and P.M. Walker, J. Phys. G: Nucl. Part. Phys. **31**, S1885 (2005).
6. T. Eronen, V. Elomaa, U. Hager, J. Hakala, A. Jokinen, A. Kankainen, I. Moore, H. Penttiliä, S. Rahaman, S. Rinta-Antilla, A. Saastamoinen, T. Sonoda, J. Äystö, A. Bey, B. Blank, G. Canchel, C Dossat, J. Giovinazzo, I Matea, N. Adimi, Phys. Lett. B **636**, 191 (2006).
7. T. Eronen, V. Elomaa, U. Hager, J. Hakala, A. Jokinen, A. Kankainen, I. Moore, H. Penttilä, S. Rahaman, J. Rissanen, A. Saastamoinen, T. Sonoda, J. Äystö, J.C. Hardy, and V.S. Kolhinen, Phys. Rev. Lett. **97**, 232501 (2006).
8. G. Bollen, D. Davies, M. Facina, J. Huikari, E. Kwan, P.A. Lofy, D.J. Morrissey, A. Prinke, R. Ringle, J. Savory, P. Schury, S. Schwarz, C. Sumithrarachchi, T. Sun, L. Weissman, Phys. Rev. Lett. **96**, 152501 (2006).
9. P.H. Barker and A.P. Byrne, Phys. Rev. C **73**, 064306 (2006).
10. V.E. Iacob, J.C. Hardy, J.F. Brinkley, C.A. Gagliardi, V.E. Mayes, N. Nica, M. Sanchez-Vega, G. Tabacaru, L. Trache, R.E. Tribble, Phys. Rev. C **74**, 055502 (2006).
11. B. Hyland, C.E. Svensson, G.C. Ball, J.R. Leslie, T. Achtzehn, D. Albers, C. Andreoiu, P. Bricault, R. Churchman, D. Cross, M. Dombsky, P. Finlay, P.E. Garrett, C. Geppert, G.F. Grinyer, G. Hackman, V. Hanemaayer, J. Lassen, J.P. Lavoie, D. Melconian, A.C. Morton, C.J. Pearson, M.R. Pearson, A.A.

Phillips, M.A. Schumaker, M.B. Smith, I.S. Towner, J.J. Valiente-Dobón, K.Wendt, and E.F. Zganjar, Phys. Rev. Lett. **97**, 102501 (2006).
12. J.T. Burke, P.A. Vetter, S.J. Freedman, B.K. Fujikawa and W.T. Winter Phys. Rev. C **74**, 025501 (2006).
13. I.S. Towner and J.C. Hardy, Phys. Rev. C **66**, 035501 (2002).
14. W.J. Marciano and A. Sirlin, Phys. Rev. Lett. **96**, 032002 (2006).
15. I.S. Towner and J.C. Hardy, Phys. Rev. C, to be published.
16. V.E. Iacob, J.C. Hardy, C.A. Gagliardi, J. Goodwin, N. Nica, H.I. Park, G. Tabacaru, L. Trache, R.E. Tribble, Y. Zhai and I.S. Towner, Phys. Rev. C **74**, 015501 (2006).
17. J.C. Hardy, V.E. Iacob, M. Sanchez-Vega, R.T. Effinger, P. Lipnik, V.E. Mayes, D.K. Willis and R.G. Helmer, Appl. Radiat. Isot. **56**, 65 (2002); R.G. Helmer, J.C. Hardy, V.E. Iacob, M. Sanchez-Vega, R.G. Neilson and J. Nelson, Nucl. Instr. Meth. Phys. Res. A **511**, 360 (2003); and R.G. Helmer, N. Nica, J.C. Hardy and V.E. Iacob, Appl. Radiat. Isot. **60**, 173 (2004).

Nuclear Matrix Elements for $0\nu\beta\beta$ Decay: Recent Advances

J. Suhonen and M. Kortelainen

Department of Physics, University of Jyväskylä, P.O. Box 35, FIN-40351 Jyväskylä, Finland

Abstract. The neutrinoless double beta ($0\nu\beta\beta$) decay of atomic nuclei plays a key role in the search for massive Majorana neutrinos and their mass scale. To extract the necessary information from the measured data the nuclear-structure effects have to be accounted for. The nuclear matrix elements (NME's) for the $0\nu\beta\beta$ decays crystallize these effects, and the calculation of these matrix elements has become of paramount importance in the community of neutrino physics. In this article we discuss the NME's for the light-neutrino exchange mechanism, computed by using the proton-neutron quasiparticle random-phase approximation (pnQRPA). Recent developments in this field relate to the handling of the nucleon-nucleon short-range correlations in the NME's and independent experimental probes of the wave functions relevant for the NME's.

Keywords: Neutrinoless double beta decay, Nuclear matrix elements, Short-range correlations, Unitary correlation operator method, beta decays, nuclear muon capture
PACS: 14.60.Hc,21.30.Fe,21.60.Cs,27.50.+e,27.60.+j

INTRODUCTION

In the present review we assume that the neutrinoless double beta ($0\nu\beta\beta$) decay proceeds via massive Majorana neutrinos (Majorana antineutrino equals Majorana neutrino) and thus is a lepton-number violating process. This means that the $0\nu\beta\beta$ decay cannot proceed in the standard model of electroweak interactions where only massless Dirac neutrinos (Dirac antineutrino is different from Dirac neutrino) are allowed. Different scenarios going beyond the standard model have been proposed to account for the $0\nu\beta\beta$ decay: grand unification theories (GUT's), supersymmetric extensions of the standard model, leptoquarks, etc. (see, e.g. [1, 2, 3, 4].

The double beta decay measurements are performed under massive amounts of rock either in deep underground caverns or inside mountains. This is needed since otherwise the cosmic rays interfere strongly with the detection of the two electrons emitted in the double beta decay. Since the half-lives of these decays are extremely long, large amounts of the decaying material are needed and the background reduction becomes essential. Experimental aspects of $0\nu\beta\beta$ decay have been recently reviewed in [5]. At present, two experiments are running: the NEMO 3 [6] at Modane underground laboratory and CUORICINO [7] at Gran Sasso. These experiments are in a position to possibly verify the alleged discovery of the $0\nu\beta\beta$ decay by the Heidelberg group [8, 9]. Several important future experiments are under R&D planning and construction [5, 10]. Most of these large-scale experiments bear a price tag of the order of 100 million dollars or euros.

In double beta decay experiments the decay half-life is measured by detecting the sum energy of the two emitted electrons. There are two types of double beta decay: in

CP972, *Exotic Nuclei and Nuclear/Particle Astrophysics (II)—Proceedings of the Carpathian Summer School of Physics 2007,* edited by L. Trache and S. Stoica

the $0\nu\beta\beta$ decay the two electrons are the only final-state leptons since the antineutrino, emitted in the first beta-decay vertex, is absorbed in the second one where an inverse beta decay takes place. In order this to be possible one needs the neutrino to be a massive Majorana particle, not occurring in the standard model of electroweak interactions. The neutrino propagates as a virtual particle thus creating a 'neutrino exchange' potential, close to the Coulomb type $1/r$ radial dependence. The two-neutrino double beta ($2\nu\beta\beta$) decay is a second-order weak process within the standard model and it looks much like two ordinary beta decays where the two created antineutrinos are emitted as final-state leptons. The different numbers of the final-state leptons make the shapes of the sum-energy spectra appear different for the two modes of double beta decay. The spectrum is just one peak for the $0\nu\beta\beta$ mode and continuous (somewhat like in the ordinary beta decay) for the $2\nu\beta\beta$ mode. The $2\nu\beta\beta$ mode has been measured in the laboratory for nine nuclear systems [11], and with increasing sensitivity of the experiments this decay mode has become a disturbing background for the measurement of the neutrinoless mode.

The measured $0\nu\beta\beta$-decay half-life is connected with the effective neutrino mass $\langle m_\nu \rangle$ through the relation

$$\left[t_{1/2}^{(0\nu)}(0_{\mathrm{i}}^+ \rightarrow 0_{\mathrm{f}}^+)\right]^{-1} = G_1^{(0\nu)} \left(\frac{\langle m_\nu \rangle}{m_{\mathrm{e}}}\right)^2 \left(M^{(0\nu)}\right)^2 , \tag{1}$$

where $G_1^{(0\nu)}$ is the leptonic phase-space factor which can be calculated in a straightforward way from the Q value of the decay [12]. The nuclear matrix element (N.M.E.) $M^{(0\nu)}$ will be discussed later in more detail. The effective neutrino mass is defined through

$$\langle m_\nu \rangle = \sum_{j=1}^{3} \lambda_j^{\mathrm{CP}} m_j |U_{\mathrm{e}j}|^2 , \tag{2}$$

where λ_j^{CP} are the Majorana CP phases and $U_{\mathrm{e}j}$ is a component of the neutrino mixing matrix. The mixing matrix has been measured to relatively high accuracy by the various neutrino oscillation experiments (see, e.g. [13, 14]). The CP phases are still unknown. Unlike the oscillation experiments, the $0\nu\beta\beta$ decay experiments can access the Majorana nature of the neutrino, its absolute mass scale, and possibly the Majorana CP phases [2, 12, 15, 16, 17].

For the running and planned future double-beta experiments the computed values of the involved N.M.E. $M^{(0\nu)}$ have become an important issue [14, 16, 17, 18]. They are needed when one wants to extract quantitative neutrino properties from the measured data. Nuclear models of different type have been deviced to compute the NME's of the $0\nu\beta\beta$ decay [12]. There are essentially two competing families of nuclear models on the market, the nuclear shell model [19, 20] and the proton-neutron quasiparticle random-phase approximation (pnQRPA) [12, 21, 22]. In the pnQRPA calculations both modes of the double beta decay are treated as two-step processes: the initial even-even ground state 'decays' to the virtual states of the adjacent odd-odd nucleus, which subsequently decay to the ground (or excited) state of the final nucleus. The complete set of virtual states is usually referred to as the intermediate states of double beta decay. For the $2\nu\beta\beta$

mode only the 'allowed' transitions to the intermediate 1^+ states bear relevance [23]. For the $0\nu\beta\beta$ mode the complete set of virtual states is added by expanding the neutrino-exchange potential in multipoles. Now, in addition to the allowed transitions, also the 'forbidden' transitions to other multipole states than 1^+ are important. This stems from the large average momentum of the exchanged virtual neutrino.

The pnQRPA is constructed to describe the energy levels of odd-odd nuclei and their beta decays to the neighboring even-even nuclei [24]. Also its derivative, renormalized pnQRPA [25], has been used [26, 27] to compute double-beta matrix elements. The pnQRPA (and the renormalized pnQRPA) calculations can be fine tuned by using the so-called particle-particle strength parameter, g_{pp}, that controls the magnitude of the proton-neutron two-body interaction for the 1^+ intermediate states in double beta decay [21, 22]. There are two ways to fix the value of this parameter, either by using the data on $2\nu\beta\beta$ decays [27] or the data on single beta decays [28, 29]. The interference between the different virtual transitions through the intermediate states of angular momentum and parity J^π play an important role in both modes of the double beta decay. Independent probes of these transitions are thus called for. Such measurements start to be feasible at the moment, and this will be discussed later in this review. The technically important aspects in the calculations of $0\nu\beta\beta$ NME's, namely the inclusion of *short-range correlations, higher-order terms of nucleonic weak current*, and the *nucleon's finite size*, are taken up in the following section.

SHORT-RANGE CORRELATIONS AND THE NEUTRINOLESS DOUBLE BETA DECAY

In the mass mode of the $0\nu\beta\beta$ decay a light virtual Majorana neutrino is exchanged by the two decaying neutrons of the initial nucleus. The average exchanged momentum is large, of the order of 100 MeV, so that the two neutrons tend to overlap. To prevent this a Jastrow type of correlation function [30] has been traditionally introduced into the $0\nu\beta\beta$ calculations [27, 31]. The Jastrow procedure is a very rudimentary way to introduce the *short-range correlations* into many-nucleon systems. A more sophisticated microscopic approach for the short-range correlations is the unitary correlation operator method (UCOM) [32]. In the UCOM one obtains the correlated many-particle state from the uncorrelated one by a unitary transformation and thus the norm of the correlated state is conserved and no amplitude is lost in the relative wave function. In [33, 34, 35] it was demonstrated that the Jastrow procedure leads to the excessive reduction of 25% – 40% in the magnitudes of the $0\nu\beta\beta$ nuclear matrix elements. At the same time the UCOM reduces the magnitudes of the matrix elements only by 4% – 16%.

Here we present the results of nuclear-structure calculations for the $0\nu\beta\beta$ ground-state-to-ground-state decays of ^{76}Ge, ^{82}Se, ^{96}Zr, ^{100}Mo, ^{116}Cd, ^{128}Te, ^{130}Te, and ^{136}Xe. To the pnQRPA calculated 'bare' matrix element we add corrections emerging from the *higher-order terms of nucleonic weak current* and the *nucleon's finite size* using the recipes of [27, 36]. Contributions from these corrections have been thoroughly analyzed in [34, 35]. For the short-range correlations we use both the Jastrow and the UCOM methods. Details about the used model space and single-particle energies are given in [34, 35].

We have used the Bonn-A G-matrix as the two-body interaction and we have renormalized it in the standard way [37, 38]. To obtain physical values of the proton-neutron particle-particle interaction strength, $g_{\rm pp}$, of the pnQRPA, we fitted by it the experimental rates of the $2\nu\beta\beta$ decay. The fitting included the experimental errors and the uncertainty $1.0 \leq g_{\rm A} \leq 1.25$ in the value of the axial-vector coupling constant $g_{\rm A}$. For ^{136}Xe only the experimental lower limit of the $2\nu\beta\beta$ half-life is available. This lower limit gives the value $g_{\rm pp} = 0.74$ with $g_{\rm A} = 1.0$. The upper limit for $g_{\rm pp}$ we obtain from the point where the calculated $2\nu\beta\beta$ half-life goes to infinity, i.e. the corresponding N.M.E. goes to zero. This procedure gives $g_{\rm pp} = 0.83$ with $g_{\rm A} = 1.25$. The extracted values of $g_{\rm pp}$ and the corresponding values of $g_{\rm A}$ are listed in Table 1.

We also used the available β^- decay data to fix the value of $g_{\rm pp}$. Unfortunately, these data cover only the nuclei ^{100}Mo, ^{116}Cd, and ^{128}Te. The lowest value of $g_{\rm pp}$, compatible with the β^- decay data, is also included in Table 1. For ^{100}Mo the interval determined by the β^- decay is contained in that of the $2\nu\beta\beta$ decay. From Table 1 one notices that for ^{116}Cd and ^{128}Te the β^- determined interval is in conflict with the one determined through the $2\nu\beta^-\beta^-$ data. This discrepancy is especially striking for ^{116}Cd, as was already pointed out in [28]. The reason for this discrepancy is not clear but an interesting observation relates to the single-state dominance [39], i.e. to the situation where the lowest state of the intermediate odd-odd nucleus is a 1^+ state and it exhausts practically all of the $2\nu\beta\beta$-decay amplitude. As clearly shown in Table 1 of [39] the ^{116}Cd decay obeys very closely the single-state dominance hypothesis whereas the ^{128}Te decay obeys it slightly less and the ^{100}Mo decay the least.

In Table 1 we also give the values of the nuclear matrix element of (1), calculated either by including the Jastrow or the UCOM short-range correlations on top of the other mentioned corrections. In Table 1 we also include the corresponding half-lives $t_{1/2}^{(0\nu)}$, expressed in units of $\mathrm{yr}/(\langle m_\nu\rangle[\mathrm{eV}])^2$. As seen, the UCOM matrix elements are considerably larger in magnitude than those computed by using the traditional Jastrow approach. Such big difference between the Jastrow and the UCOM results reflects on the estimated neutrino-mass sensitivities of the presently running and the future double-beta-decay experiments. Detailed analyses of the Jastrow and UCOM contributions in terms of the multipole decompositions of $M^{(0\nu)}$ are performed in [33, 34, 35].

Fig. 1 presents the half-lives of Table 1 as bars with the lower and upper half-life limits. In the cases of ^{116}Cd and ^{128}Te the isolated short horizontal lines correspond to the lower limit set by the beta-decay data. The scale of the vertical axis is logarithmic. It is evident that the sensitivities to the neutrino mass do not vary drastically from nucleus to nucleus. Notable exception is ^{128}Te, standing out by its small available phase space.

INDEPENDENT PROBES OF THE $0\nu\beta\beta$ MATRIX ELEMENTS

In this section we discuss some independent probes of the virtual transitions constituting the $0\nu\beta\beta$ N.M.E's. We first make a list of the possible probes and then discuss some of these probes in more detail. At least the following probes have been suggested:

- *Ordinary beta-decay transitions* from the intermediate states of double beta decay. This probe will be more extensively addressed below.

TABLE 1. Calculated $0\nu\beta\beta$ nuclear matrix elements, the used $g_{\rm pp}$ and $g_{\rm A}$ values and the resulting half-lives. The Jastrow (J) or UCOM (U) plus the other corrections are included. The half-lives $t_{1/2}^{(0\nu)}$ are expressed in units of $\mathrm{yr}/(\langle m_\nu\rangle[\mathrm{eV}])^2$.

Nucleus	$g_{\rm pp}$	$g_{\rm A}$	$M^{(0\nu)}$(J)	$t_{1/2}^{(0\nu)}$(J)	$M^{(0\nu)}$(U)	$t_{1/2}^{(0\nu)}$(U)
^{76}Ge	1.02	1.00	−5.077	4.0×10^{24}	−6.555	2.4×10^{24}
	1.06	1.25	−4.029	2.6×10^{24}	−5.355	1.4×10^{24}
^{82}Se	0.96	1.00	−3.535	1.9×10^{24}	−4.597	1.1×10^{24}
	1.00	1.25	−2.771	1.2×10^{24}	−3.722	6.9×10^{23}
^{96}Zr	1.06	1.00	−3.131	1.2×10^{24}	−4.319	6.1×10^{23}
	1.11	1.25	−2.065	1.1×10^{24}	−3.117	4.7×10^{23}
^{100}Mo	1.07	1.00	−3.526	1.2×10^{24}	−4.849	6.2×10^{23}
	1.09	1.25	−2.737	7.9×10^{23}	−3.931	3.8×10^{23}
^{116}Cd	0.82 (β^- decay)	1.25	−3.981	3.5×10^{23}	−4.928	2.3×10^{23}
	0.97	1.00	−3.681	1.0×10^{24}	−4.682	6.3×10^{23}
	1.01	1.25	−3.034	6.1×10^{23}	−3.935	3.6×10^{23}
^{128}Te	0.86 (β^- decay)	1.25	−4.068	9.5×10^{24}	−5.509	5.2×10^{24}
	0.89	1.00	−4.279	2.1×10^{25}	−5.841	1.1×10^{25}
	0.92	1.25	−3.383	1.4×10^{25}	−4.790	6.9×10^{24}
^{130}Te	0.84	1.00	−4.061	9.5×10^{23}	−5.442	5.3×10^{23}
	0.90	1.25	−2.993	7.0×10^{23}	−4.221	3.5×10^{23}
^{136}Xe	0.74	1.00	−2.864	1.8×10^{24}	−3.719	1.1×10^{24}
	0.83	1.25	−2.053	1.4×10^{24}	−2.802	7.6×10^{23}

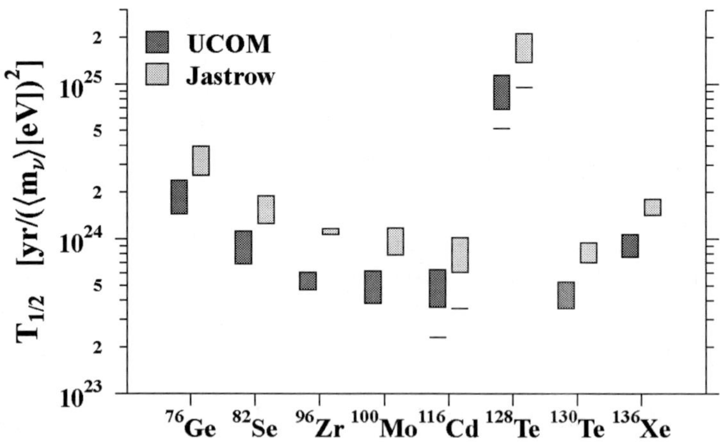

FIGURE 1. The half-lives of Table 1 presented as bars in logarithmic scale. The horizontal black lines are the lower limit extracted from the β^- data.

- *Charge exchange reactions.* This is a very powerful method and it can probe directly the $2\nu\beta\beta$ N.M.E's. However, for the $0\nu\beta\beta$ N.M.E's the method does not produce the virtual transition amplitudes directly due to different radial dependence of the involved excitation operators. In this case only a strength function for states of a given multipolarity is provided. The mostly used reactions are the (p,n) and (^{3}He,t) to the β^- direction, and (n,p) and (d,^{2}He) to the β^+ direction. Aspects of these reactions have been recently reviewed in [40]. Contributions to the $2\nu\beta\beta$ N.M.E's have been already extracted from measurements performed at the KVI, Groningen and RCNP, Japan. The studied N.M.E's relate to the $2\nu\beta\beta$ decays of ^{48}Ca [41], ^{116}Cd [42, 43], and ^{64}Zn, ^{76}Ge, and ^{96}Zr [44]. For studies of $0\nu\beta\beta$ N.M.E's some cases, like the pf-shell nuclei, could offer a good testing ground to see the importance of the first 2^- excitation in the intermediate nucleus [29].
- *Neutrino-nucleus charged-current scattering.* This proposition to test the $0\nu\beta\beta$ N.M.E.'s was put forward in [45]. At the moment it seems that this method is realizable only if beta beams or neutrino factories will produce intense enough neutrino fluxes in the far future.
- The $0\nu\beta\beta$ N.M.E.'s can be probed starting from the $2\nu\beta\beta$ N.M.E.'s, as discussed in the previous section. This idea was put forward recently in [46].
- *Ordinary muon capture* by nuclei. This probe will be discussed in detail below.

Beta decays as probes of virtual transitions

The most natural probes of double beta decays are the measurements of single beta decays from the intermediate odd-odd nucleus of the double beta decay to the adjacent even-even nuclei. Extensive theoretical work on this probe has been done along the years [38, 47, 48, 49, 50].

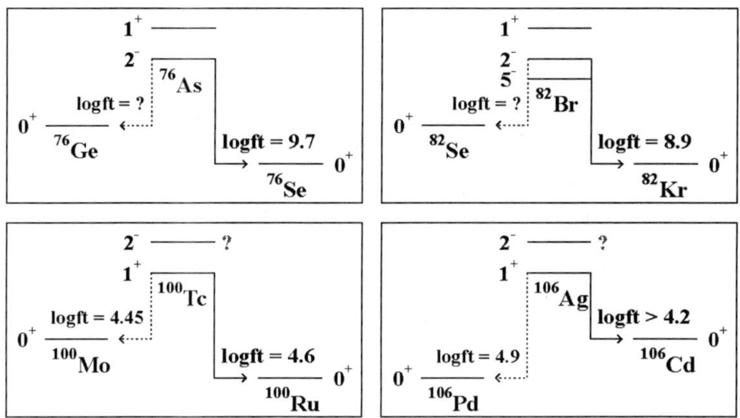

FIGURE 2. Available experimental data for the beta decays of medium-mass odd-odd nuclei involved in double beta decays.

Typical examples of the available data are shown in Fig. 2 for double-beta chains involving medium-mass nuclei. As seen from the figure the available data on relevant beta decays are rather scarse and more beta-decay measurements are called for [51]. Particularly, in case of a possible single-state dominance [39], beta decays are a powerful tool to experimentally determine the $2\nu\beta\beta$-decay rate [52, 53]. Recently a big program has been launched in the TITAN ion trap facility at TRIUMF to measure electron-capture branching ratios in ^{76}As, ^{82m}Br, ^{100}Tc, ^{110}Ag, ^{114}In, ^{116}In, and ^{128}I [54].

Nuclear muon capture as a probe of virtual transitions

Recently it was proposed [55] to use the ordinary muon capture (OMC) as a probe of the virtual transitions to the J^π states of the intermediate odd-odd nucleus. Due to the large mass of the captured muon (roughly 100 MeV) the OMC can lead to highly excitated states of the final nucleus. In particular, the OMC can access, in principle, the energy region of the Gamow–Teller giant resonance. However, the large muon mass also leads to a more complicated theoretical treatment of this process as compared to the beta decay. In particular, the induced terms of the nucleonic weak current are activated [56]. Of the various terms of the induced current, access to the magnitude of the induced pseudoscalar contribution has been problematic [57]. In contrast to the beta decay, the large mass of the muon allows high transition rates also for the so-called forbidden transitions. This can be used to analyze the structure of all J^π states in the intermediate nuclei, in particular the 2^- states which are expected to be very relevant in the $0\nu\beta\beta$ decay [29].

It seems that the OMC is a powerful probing tool for light nuclei [58]. According to the calculations the leading contributions to the $2\nu\beta\beta$-decay matrix elements stem from those intermediate 1^+ states which are strongly fed also in the OMC. In this way the OMC is able to probe wave functions of those states which are the most relevant for the double beta decay. After the data analyses of the PSI experiments the calculations can be tested in the near future.

The study [58] on the relation of the OMC with the $2\nu\beta\beta$ decay can be extended to comparison of the OMC with the $0\nu\beta\beta$ decay. Let us take the case of $A = 48$ as an example. In Fig. 3 the relevant multipole contributions $J^\pi = 1^+, 2^+, 3^+, 4^+$ to the Gamow–Teller part $M_{\rm GT}^{(0\nu)}$ of the matrix element $M^{(0\nu)}$ of (1) (see, e.g. [51]) are contrasted against strong muon-capture transitions $^{48}\mathrm{Ti} \to {}^{48}\mathrm{Sc}(J^\pi)$. From the figure one can conclude that the OMC feeds mainly those intermediate states which are relevant building blocks of $M_{\rm GT}^{(0\nu)}$. It is then obvious that, like in the $2\nu\beta\beta$ case, the OMC is a powerful probe of the wave functions of the intermediate states for the $0\nu\beta\beta$ decays, as well.

ACKNOWLEDGMENTS

This work has been partially supported by the Academy of Finland under the Finnish Centre of Excellence Programme 2006-2011 (Nuclear and Accelerator Based Pro-

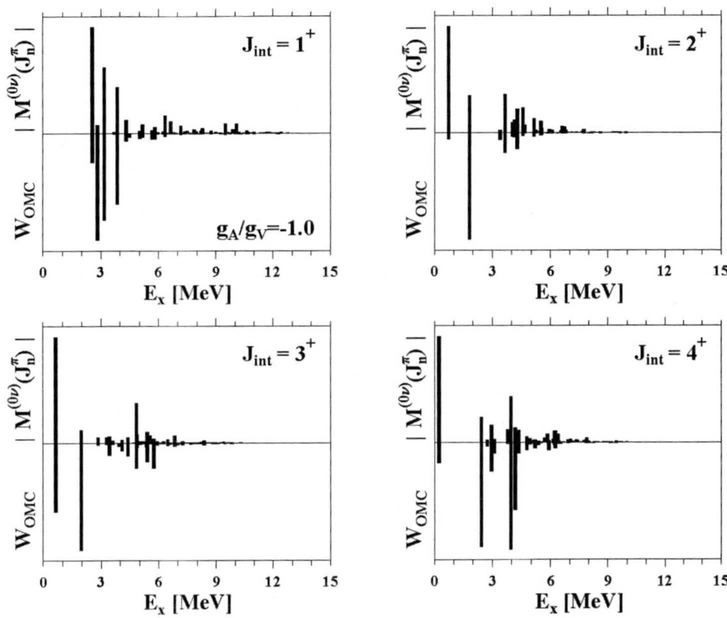

FIGURE 3. Contributions of the J^π states in ^{48}Sc to the matrix element $M^{(0\nu)}_{GT}$ (upper part of each panel, in arbitrary units) for the $0\nu\beta\beta$ decay of ^{48}Ca, and the OMC rates from ^{48}Ti to the same J^π states (lower part of each panel, in arbitrary units). The horizontal axes give the excitation energy in ^{48}Sc.

gramme at JYFL). We thank also the EU ILIAS project under the contract RII3-CT-2004-506222.

REFERENCES

1. J.D. Vergados, *Phys. Reports* **133**, 1 (1986).
2. A. Faessler, and F. Šimkovic, *J. Phys.* **G24**, 2139 (1998).
3. H.V. Klapdor-Kleingrothaus, *Springer Tracts in Modern Physics* **163**, pp. 69-104, Springer, Berlin, 2000.
4. J.D. Vergados, *Phys. Reports* **361**, 1 (2002).
5. F.T. Avignone III, S.R. Elliott, and J. Engel, arXiv:0708.1033, and to be published in *Rev. Mod. Phys.* (2007).
6. R. Arnold *et al.*, *Phys. Rev. Lett.* **95**, 182302 (2005).
7. C. Arnaboldi *et al.*, *Phys. Rev. Lett.* **95**, 142501 (2005).
8. H.V. Klapdor-Kleingrothaus *et al.*, *Mod. Phys. Lett.* **A16**, 2409 (2001).
9. H.V. Klapdor-Kleingrothaus *et al.*, *Phys. Lett.* **B586**, 198 (2004).
10. A.S. Barabash, *Phys. Atom. Nucl.* **67**, 458 (2004).
11. A.S. Barabash, *Czech. J. Phys.* **56**, 437 (2006).
12. J. Suhonen, and O. Civitarese, *Phys. Reports* **300**, 123 (1998).
13. S. Fukuda *et al.*, *Phys. Rev. Lett.* **86**, 5651 (2001) ; Q.R. Ahmad *et al.*, *Phys. Rev. Lett.* **89**, 011301 (2002) ; H. Hagiwara *et al.*, *Phys. Rev.* **D66**, 010001 (2002) ; K. Eguchi *et al.*, *Phys. Rev. Lett.* **90**, 021802 (2003) ; M.H. Ahn *et al.*, Phys. Rev. Lett. **90**, 041801 (2003) ; S.N Ahmed *et al.*, Phys. Rev.

Lett. **92**, 181301 (2004).
14. S.M. Bilenky, *J. Phys.* **G32**, R127 (2006).
15. M. Doi, T. Kotani, and E. Takasugi, *Prog. Theor. Phys. Suppl.* **83**, 1 (1985).
16. S.R. Elliott, and P. Vogel, *Annu. Rev. Nucl. Part. Sci.* **52**, 115 (2002).
17. S.R. Elliott, and J. Engel, *J. Phys.* **G30**, R183 (2004).
18. V.M. Gehman, and S.R. Elliott, *J. Phys.* **G34**, 667 (2007).
19. E. Caurier, F. Nowacki, A. Poves, and J. Retamosa, *Nucl. Phys.* **A654**, 973c (1999).
20. E. Caurier *et al.*, *Rev. Mod. Phys.* **77**, 427 (2005).
21. P. Vogel, and M.R. Zirnbauer, *Phys. Rev. Lett.* **57**, 3148 (1986).
22. O. Civitarese, A. Faessler, and T. Tomoda, *Phys. Lett.* **B194**, 11 (1987).
23. O. Civitarese, and J. Suhonen, *Nucl. Phys.* **A607**, 152 (1996).
24. J. Suhonen, *From Nucleons to Nucleus: Concepts of Microscopic Nuclear Theory*, Springer, Berlin, 2007.
25. J. Toivanen, and J. Suhonen, *Phys. Rev. Lett.* **75**, 410 (1995).
26. J. Toivanen, and J. Suhonen, *Phys. Rev.* **C55**, 2314 (1997).
27. V.A. Rodin, A. Faessler, F. Šimkovic, and P. Vogel, *Nucl. Phys.* **A766**, 107 (2006) ; Erratum in arXiv:0706.4304v1 [nucl-th], to appear in *Nucl. Phys.* **A**.
28. J. Suhonen, *Phys. Lett.* **B607**, 87 (2005).
29. O. Civitarese, and J. Suhonen, *Phys. Lett.* **B626**, 80 (2005), *ibid. Nucl. Phys.* **A761**, 313 (2005).
30. G.A. Miller and J.E. Spencer, *Ann. Phys.* **100**, 562 (1976).
31. T. Tomoda, *Rep. Prog. Phys.* **54**, 53 (1991).
32. H. Feldmeier, T. Neff, R. Roth, and J. Schnack, *Nucl. Phys.* **A632**, 61 (1998).
33. M. Kortelainen, O. Civitarese, J. Suhonen, and J. Toivanen, *Phys. Lett.* **B647** (2007) 128.
34. M. Kortelainen, and J. Suhonen, *Phys. Rev.* **C75**, 051303(R) (2007).
35. M. Kortelainen, and J. Suhonen, *Phys. Rev.* **C76**, 024315 (2007).
36. F. Šimkovic, G. Pantis, J.D. Vergados, and A. Faessler, *Phys. Rev.* **C60**, 055502 (1999).
37. J. Suhonen, T. Taigel, and A. Faessler, *Nucl. Phys.* **A486**, 91 (1988).
38. J. Suhonen, *Nucl. Phys.* **A563**, 205 (1993), *ibid.* **A700**, 649 (2002).
39. O. Civitarese, and J. Suhonen, *Phys. Rev.* **C58**, 1535 (1998) ; *Nucl. Phys.* **A653**, 321 (1999).
40. D. Frekers, *Prog. Part. Nucl. Phys.* **57**, 217 (2006).
41. S. Rakers *et al.*, *Phys. Rev.* **C70**, 054302 (2004).
42. S. Rakers *et al.*, *Phys. Rev.* **C71**, 054313 (2005).
43. M. Sasano *et al.*, *Nucl. Phys.* **A788**, 76c (2007).
44. E.-W. Grewe, and D. Frekers, *Prog. Part. Nucl. Phys.* **57**, 260 (2006).
45. C. Volpe, *J. Phys.* **G31**, 903 (2005).
46. V.A. Rodin, A. Faessler, F. Šimkovic, and P. Vogel, *Phys. Rev.* **C68**, 044302 (2003).
47. M. Aunola and J. Suhonen, *Nucl. Phys.* **A602**, 133 (1996).
48. A. Griffiths and P. Vogel, *Phys. Rev.* **C46**, 181 (1992).
49. J. Suhonen and O. Civitarese, *Phys. Rev.* **C49**, 3055 (1994).
50. J. Suhonen and O. Civitarese, *Nucl. Phys.* **A584**, 449 (1995).
51. J. Suhonen, *Nucl. Phys.* **A752**, 53c (2005).
52. A. García *et al.*, *Phys. Rev.* **C47**, 2910 (1993).
53. M. Bhattacharya *et al.*, *Phys. Rev.* **C58**, 1247 (1998).
54. D. Frekers, J. Dilling, and I. Tanihata, *Can. J. Phys.* **85**, 57 (2007).
55. M. Kortelainen, and J. Suhonen, *Europhys. Lett.* **58**, 666 (2002), *ibid. Nucl. Phys.* **A713**, 501 (2003).
56. T. Siiskonen, J. Suhonen, V.A. Kuz'min, and T.V. Tetereva, *Nucl. Phys.* **A635**, 446 (1998), *ibid. Nucl. Phys.* **A651**, 435 (1999).
57. T. Siiskonen, J. Suhonen, and M. Hjorth-Jensen, *Phys. Rev.* **C59**, R1839 (1998) ; *J. Phys.* **G25**, L55 (1999).
58. M. Kortelainen, and J. Suhonen, *J. Phys.* **G30**, 2003 (2004).

Properties of Finite Nuclei and the Nuclear Matter Equation of State

Shalom Shlomo

Cyclotron Institute, Texas A&M University, College Station, TX 77843, USA

Abstract. We first review the success of the Hartree-Fock (HF) theory with Skyrme type effective nucleon-nucleon interaction in describing the ground states properties of finite nuclei and of nuclear matter (NM). Next we review the current status of the HF based random phase approximation (RPA) as applied to the study of properties of giant resonances. We then discuss, in particular, the properties of compression modes and the extracted value of the nuclear matter incompressibility coefficient which is needed to extend our knowledge of the NM equation of state.

Keywords: properties of nuclei and nuclear matter, giant resonances, nuclear compressibility, Hartree-Fock and random phase approximation
PACS: 21.60Jz,24.30Cz,21.65+f

INTRODUCTION

Nuclear matter (NM) properties, such as binding energy per nucleon, neutron and proton densities, incompressibility coefficient and symmetry energy coefficient, are important physical quantities needed for determining the nuclear matter equation of state (EOS) of NM. The EOS is an important ingredient in the study of properties of nuclei, supernova collapse, neutron star and heavy-ion collisions. The saturation point of the EOS for symmetric NM (SNM) is well determined from ground state properties of nuclei, such as binding energies and central densities, by extrapolation to infinite NM. To extend our knowledge of the EOS beyond the saturation point of SNM, accurate value of the incompressibility coefficient, K, which is directly related to the curvature of the EOS, is needed. An accurate value of the symmetry energy coefficient, J, is needed for the EOS of asymmetric nuclear matter.

In the following we first review the Hartree-Fock (HF) method and its success in providing a good description of ground state properties of finite nuclei and present a modern effective nucleon-nucleon (NN) interaction of Skyrme type with enhanced predictive power. We then present the HF based random phase approximation (RPA) method for providing a microscopic description of properties of nuclear giant resonances (GR) and discuss the experimental method for extracting the strength distributions of GR from excitation cross sections. Finally we present some results of our calculations and review the current status of K as determined from the strength distributions of compression modes in nuclei, mainly the isoscalar giant monopole resonance (ISGMR) and the isoscalar giant dipole resonance (ISGDR).

CP972, *Exotic Nuclei and Nuclear/Particle Astrophysics (II)—Proceedings of the Carpathian Summer School of Physics 2007,* edited by L. Trache and S. Stoica

SKYRME HARTREE-FOCK APPROXIMATION

The nucleus is a finite many-body system of strongly interacting nucleons which exhibits fascinating phenomena of single particle motion and correlated motions of a few and of many nucleons [1]. Since it is not possible to solve the nuclear many body Schroedinger equation, one resorts to approximate methods. In particular, in the HF approach one assumes an independent particle motion of the nucleons which is govern by a mean field, approximating the interactions of each nucleon with all the other nucleons. In the HF approximation, the antisymmetric nuclear many body wave function is given as, $\Phi = det[\phi_\alpha]$, and the HF equations for the single particle wave functions, ϕ_α, are derived by minimizing the energy of the nucleus $E = \langle\Phi|H|\Phi\rangle$. One obtains the equations

$$\begin{aligned} &- \frac{\hbar^2}{2m}\triangle\phi_{\alpha_i}(\mathbf{r}) + \sum_j^A \int \phi^*_{\alpha_j}(\mathbf{r}')V(\mathbf{r},\mathbf{r}')\phi_{\alpha_i}(\mathbf{r})\phi_{\alpha_j}(\mathbf{r}')d\mathbf{r}' \\ &- \sum_j^A \int \phi^*_{\alpha_j}(\mathbf{r}')V(\mathbf{r},\mathbf{r}')\phi_{\alpha_i}(\mathbf{r}')\phi_{\alpha_j}(\mathbf{r})d\mathbf{r}' = \varepsilon_{\alpha_i}\phi_{\alpha_i}(\mathbf{r}). \end{aligned} \tag{1}$$

These coupled equations, which are solved by iteration, can be simplified significantly [2] by employing a zero range interaction such as the Skyrme type interaction,

$$\begin{aligned} V_{12} &= t_0(1+x_0P^\sigma_{12})\,\delta(\mathbf{r}_1-\mathbf{r}_2) + \frac{1}{2}t_1(1+x_1P^\sigma_{12}) \times \left[\overleftarrow{k}^2_{12}\delta(\mathbf{r}_1-\mathbf{r}_2) + \delta(\mathbf{r}_1-\mathbf{r}_2)\overrightarrow{k}^2_{12}\right] \\ &+ t_2(1+x_2P^\sigma_{12})\,\overleftarrow{k}_{12}\delta(\mathbf{r}_1-\mathbf{r}_2)\overrightarrow{k}_{12} + \frac{1}{6}t_3(1+x_3P^\sigma_{12})\,\rho^\alpha\left(\frac{\mathbf{r}_1+\mathbf{r}_2}{2}\right)\delta(\mathbf{r}_1-\mathbf{r}_2) \\ &+ iW_0\overleftarrow{k}_{12}\delta(\mathbf{r}_1-\mathbf{r}_2)(\overrightarrow{\sigma_1}+\overrightarrow{\sigma_2}) \times \overrightarrow{k}_{12}, \end{aligned} \tag{2}$$

where t_i, x_i, α and W_0 are the parameters of the interaction, P^σ_{12} is the spin exchange operator, $\overrightarrow{\sigma}_i$ is the Pauli spin operator, $\overrightarrow{k}_{12} = -i(\overrightarrow{\nabla}_1 - \overrightarrow{\nabla}_2)/2$, and $\overleftarrow{k}_{12} = -i(\overleftarrow{\nabla}_1 - \overleftarrow{\nabla}_2)/2$. Here, the right and left arrows indicate that the momentum operators act on the right and on the left, respectively.

Since the early work of Brink and Vautherin [2], which first demonstrated the success of the Skyrme HF approach in reproducing bulk properties of nuclei, continuous efforts have been made to readjust the parameters of the Skyrme interaction to better reproduce experimental data. Most of the parameters of the Skyrme interactions which are available in the literature were obtained by fitting the HF results to experimental data on bulk properties of a few stable closed-shell nuclei. Recently [3], we have implemented the simulated annealing method (SAM) to fit the values of the parameters of the Skyrme interaction by searching for the global minimum in the hypersurface of the χ^2 function. We have fitted the values of the Skyrme parameters to an extensive set of experimental data (see Table 1) together with a few additional constraints, such as the Landau stability conditions for NM and a positive slope $L = 3\rho dS/d\rho$ of the symmetry energy density $S(\rho)$ for densities up to $3\rho_0$, a condition imposed by the neutron star models. The parameters of the KDE0 Skyrme interaction obtained in our fit [3], are given in Table 2 together with the parameter sets of the SK255 [4] and SGII [5] interactions. The

TABLE 1. Selected experimental data for the binding energy B, charge rms radius r_{ch}, rms radii of valence neutron orbits r_v, spin-orbit splitting S-O, breathing-mode constrained energy E_{con}, and critical density ρ_{cr} used in the fit to determine the parameters of the Skyrme interaction.

Properties	Nuclei
B	^{16,24}O, ^{34}Si, ^{40,48}Ca, ^{48,56,68,78}Ni, ^{88}Sr, ^{90}Zr, ^{100,132}Sn, ^{208}Pb
r_{ch}	^{16}O, ^{40,48}Ca, ^{56}Ni, ^{88}Sr, ^{90}Zr, ^{208}Pb
$r_v(\nu 1d_{5/2})$	^{17}O
$r_v(\nu 1f_{7/2})$	^{41}Ca
S-O	$2p$ orbits in ^{56}Ni
E_{con}	^{90}Zr, ^{116}Sn, ^{144}Sm, ^{208}Pb
ρ_{cr}	Nuclear matter

TABLE 2. The values of the Skyrme parameters for the KDE0 [3] interaction were obtained by minimizing the χ^2. For the sake of comparison we have also listed the values of the parameters for the SK255 [4] and SGII [5] interactions.

Parameter	KDE0	SK255	SGII
t_0(MeV·fm^3)	-2526.5110	-1689.35	-2645
t_1 (MeV·fm^5)	430.9418	389.30	340
t_2 (MeV·fm^5)	-398.3775	-126.07	-41.9
t_3(MeV·fm$^{3(1+\alpha)}$)	14235.5193	10989.60	15595
x_0	0.7583	-0.1461	0.09
x_1	-0.3087	0.1160	-0.0588
x_2	-0.9495	0.0012	1.425
x_3	1.1445	-0.7449	0.06044
W_0 (MeV·fm^5)	128.9649	95.39	105
α	0.1676	0.3563	0.1667

selection of the extensive experimental data, together with the constraints which were include in the fit, ensures that the KDE0 interaction can be used for the study of the bulk properties of nuclei at and away from the stability line, as well as the properties of neutron stars.

SELF-CONSISTENT HF-RPA APPROACH

In the microscopic and self-consistent Hartree-Fock (HF)-based random-phase-approximation (RPA) approach one starts by adopting a specific effective nucleon-nucleon interaction, V_{12}, such as the Skyrme type interaction, with parameters determined by a fit to experimental data on bulk properties of a wide range of nuclei, such as nuclear masses, charge root mean square (rms) radii, etc. The properties of the giant resonances are then calculated within the self-consistent HF-RPA approach by solving the RPA equations using the particle-hole (p-h) interaction V_{ph} which corresponds to V_{12}, i.e, employing all the components of the V_{12} interaction. Various numerical methods have been adopted in the literature to solve the RPA equations. In particular, in the Green's function approach [6] one evaluates the RPA Green's function G,

$$G = G_0(1 + V_{ph}G_0)^{-1} , \tag{3}$$

where G_0 is the free p-h Green's function.

In general the transition operator (for nuclei) has the form

$$\hat{F} = \sum \hat{f}(i), \qquad \hat{f} = f(r)\,[Y_L \times S^{\sigma}]^J\, T^{\tau}. \tag{4}$$

Here $J = (1,2,...)$, $L = (0,1,2,...)$, $\sigma = (0,1)$ and $\tau = (0,1)$ are the total angular momentum, orbital angular momentum, spin and isospin transfered to the ground state in the excitation. Note that $T^{\tau} = 1$ (isoscalar) or $\vec{\tau}$ (isovector) and $S^{\sigma} = 1$ (electric) or $\vec{\sigma}$ (magnetic). The strength function $S(E)$ associated with the scattering operator $\hat{F}$ is given by

$$S(E) = \sum_n \left|\langle 0 \mid \hat{F} \mid n\rangle\right|^2 \delta(E_n - E_0 - E) = \frac{1}{\pi} Im\left[Tr(\hat{f}G\hat{f})\right], \tag{5}$$

where the sum is over the excited states $\mid n\rangle$ with eigenenergies E_n of the nucleus, and the transition density $\hat{\rho}_t = \rho_t(r)\,[Y_L \times S^{\sigma}]^J\, T^{\tau}$ is obtained from,

$$\hat{\rho}_t(E) = \frac{\Delta E}{\sqrt{S(E)\Delta E}}\frac{1}{\pi} Im[\hat{f}G]. \tag{6}$$

Note that $\hat{\rho}_t$ as defined in Eq. (6) is associated with transition strength in the excitation energy region of $E \pm \Delta E/2$.

For the ISGMR we adopt the scattering operator $r^2 Y_{00}$. Considering the ISGDR, we first point out that the scattering operator $\hat{f}_1 = \vec{r}$ leads to a coherent spurious state which is associated with the center of mass motion. The spurious state transition density has the form $\rho_{ss} \sim d\rho_0/dr$. In a fully self-consistent HF-RPA calculation the spurious state $(T = 0, L = 1)$ appears at zero energy and no spurious state mixing (SSM) with the ISGDR takes place. However, in some numerical implementation of the HF-RPA theory, self-consistency is violated, leading to SSM in the ISGDR. We have shown recently (see Ref. [7] and references therein), that in order to correct for the effects of SSM on $S(E)$ and ρ_t, the scattering operator $\hat{f}$ for the ISGDR should be replaced by

$$\hat{f}_\eta = \hat{f} - \eta \hat{f}_1, \qquad \eta = \frac{\langle \hat{f}\rho_{ss}\rangle}{\langle \hat{f}_1\rho_{ss}\rangle}. \tag{7}$$

Note that for the scattering operator $\hat{f} = r^3 Y_1$ one has $\eta = 5\langle r^2\rangle/3$.

The steps involved in the relativistic mean field based RPA (RRPA) calculations are analogous to those for the non-relativistic HF-RPA approach [8]. The nucleon-nucleon interaction are generated through the exchange of various mesons. The Lagrangian parameters are usually obtained [9], as in the case of non-relativistic mean field calculations, by a fitting procedure to some bulk properties of a set of nuclei.

DWBA APPROACH FOR EXCITATION CROSS-SECTION

In the distorted wave Born approximation (DWBA) the cross-section for $\alpha-$nucleus scattering, $\alpha + A \to \alpha + A^*$, is given by,

$$\frac{d\sigma^{DWBA}}{d\Omega} = \left(\frac{\mu}{2\pi\hbar^2}\right)^2 \frac{k_f}{k_i} |T_{fi}|^2, \tag{8}$$

where μ is the reduced mass and k_i and k_f are the initial and final linear momenta of the α-nucleus relative motion, respectively. The transition matrix element T_{fi} is given by,

$$T_{fi} = \langle \chi_f^{(-)} \Psi_f | V | \chi_i^{(+)} \Psi_i \rangle, \tag{9}$$

where V is the α-nucleon interaction, Ψ_i and Ψ_f are the initial and final states of the nucleus and $\chi_i^{(+)}$ and $\chi_f^{(-)}$ are the corresponding distorted wave functions of the $\alpha-$nucleus relative motion, respectively. To evaluate T_{fi}, Eq. (9), one may adopt one of the following approaches:

(i) In the approach which is usually employed by experimentalists one first carries out the integrals in (9) over the coordinates of the nucleons, obtaining the transition potential

$$\delta U \sim \int \Psi_f^* V \Psi_i. \tag{10}$$

Then the excitation cross section is calculated using a certain DWBA code with δU and the optical potential $U(r)$ as input. Note that in the deformed collective model [1], δU is obtained from U.

(ii) In the approach which is usually employed in theoretical calculations one first integrate over the relative $\alpha-$nucleus coordinates to obtain the scattering operator,

$$\mathscr{O} \sim \int \chi_f^{(-)*} V \chi_i^{(+)}, \tag{11}$$

and then calculate the matrix element $< \Psi_f | \mathscr{O} | \Psi_i >$ within a theoretical model for Ψ_i and Ψ_f. We emphasize that it is quite common in theoretical work to adopt for (11) the very simple scattering operator of Eq. (4) and evaluate the strength function $S(E)$ instead of the excitation cross section.

The folding model (FM)-DWBA approach to the evaluation of U and δU provides a direct link to the description of $\alpha-$nucleus scattering based on microscopic HF-RPA results. Within the FM-DWBA approach, the optical potential $U(r)$ is given by:

$$U(r) = \int d\mathbf{r}' V(|\mathbf{r}-\mathbf{r}'|, \rho_0(r'))\rho_0(r') \tag{12}$$

where $V(|\mathbf{r}-\mathbf{r}'|, \rho_0(r'))$ is the $\alpha-$nucleon interaction, which is generally complex and density dependent, and $\rho_0(r')$ is the ground state HF density of a spherical target nucleus. In our calculations, both the real and imaginary parts of the $\alpha-$nucleon interaction were chosen to have Gaussian forms with density dependence [7]:

$$V(|\mathbf{r}-\mathbf{r}'|, \rho_0(r')) = -V(1+\beta_V \rho_0^{2/3}(r'))e^{-\frac{|\mathbf{r}-\mathbf{r}'|^2}{\alpha_V}} - iW(1+\beta_W \rho_0^{2/3}(r'))e^{-\frac{|\mathbf{r}-\mathbf{r}'|^2}{\alpha_W}}. \tag{13}$$

The parameters V, β_V, α_V, W, β_W and α_W in Eq. (13) were determined by a fit to the elastic scattering data. The radial form $\delta U_L(r,E)$ of the transition potential, for a state with the multipolarity L and excitation energy E, is obtained from:

$$\delta U(r,E) = \int d\mathbf{r}' \delta\rho_L(\mathbf{r}',E) \left[V(|\mathbf{r}-\mathbf{r}'|,\rho_0(r')) + \rho_0(r') \frac{\partial V(|\mathbf{r}-\mathbf{r}'|,\rho_0(r'))}{\partial\rho_0(r')} \right], \quad (14)$$

where $\delta\rho_L(\mathbf{r}',E)$ is the transition density for the considered state.

We point out that within the "microscopic" FM-DWBA approach to the $\alpha-$nucleus scattering, both ρ_0 and ρ_L, which enter Eqs. (12) and (14), respectively, are obtained from the self-consistent HF-RPA calculations. Within the "macroscopic" FM-DWBA approach, employed by experimentalists, one adopts collective transition densities, ρ_{coll}, which are assumed to have energy-independent radial shapes and are obtained from the ground state density using a collective model. Therefore, for a proper comparison between experimental and theoretical results for $S(E)$, one should adopt the "microscopic" FM approach in the DWBA calculations of $\sigma(E)$.

RESULTS

During the last two decades it was claimed that (see for example [10]): (i) The non-relativistic HF-RPA calculations for the centroid energy E_0 of the ISGMR, with Skyrme interaction predict a value of $K = 210 - 220$ MeV, whereas the Gogny interaction predicts $K = 230 - 240$ MeV (ii) Considering the early ISGDR data for the centroid energy E_1, the non-relativistic HF-RPA calculations predict a value of $K = 160 - 180$ MeV. (iii) Recent relativistic RPA (RRPA) calculations for E_0 of the ISGMR yield a value of $K = 250 - 270$ MeV. Since, to a good approximation, E_0 and E_1 are $\propto \sqrt{K}$, the discrepancies of about 20% in the value of K obtained from the E_0 and E_1 data and from the relativistic and non-relativistic models, are quite significant in view of the accuracy of $2 - 3\%$ in the experimental data for E_0 [11]. In the following we present modern results of highly accurate calculations of strength functions, energy and and excitation cross-sections of giant resonances and provide simple explanations of these discrepancies between theory and experiment.

Fully self-consistent HF-RPA calculations

Although the HF based RPA approach is conceptually well understood, in actual calculations one makes compromises for reasons of simplicity or numerical expense. One can hardly avoid limitations in the numerical basis and the RPA particle-hole configuration space. Furthermore, most of the presently available HF-RPA calculations omit the painful to evaluate pieces of the RPA residual interaction, namely its spin-orbit and/or Coulomb parts. Unfortunately, apart from some fully self-consistent calculations, most earlier HF-RPA calculations are contaminated by self consistency violations (SCV).

To demonstrate the importance of carrying out fully self-consistent calculations, we present in Figure 1 recent results [12] for $S(E)$ of isoscalar giant resonances in ^{208}Pb

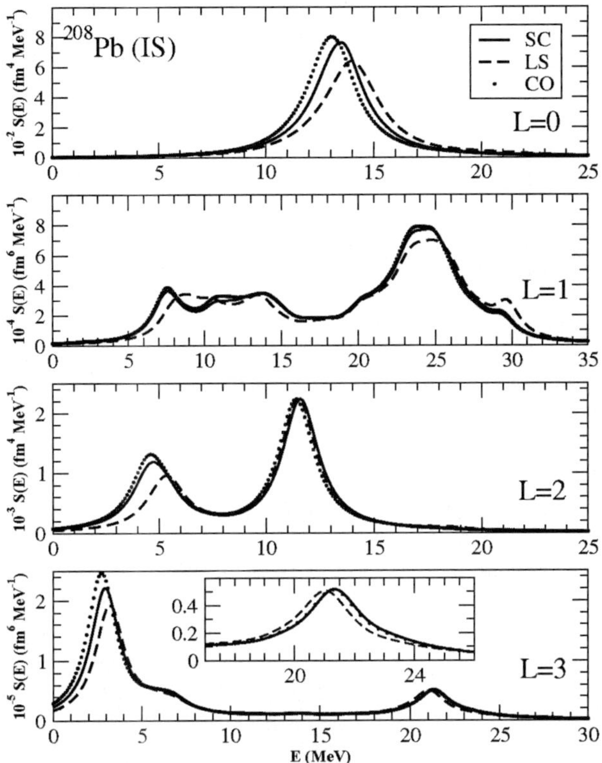

FIGURE 1. HF-RPA results for isoscalar strength functions of ^{208}Pb for $L = 0-3$ multi-polarities are displayed. SC (full line) corresponds to the fully self-consistent calculation where LS (dashed line) and CO (open circle) represent the calculations without the ph spin-orbit and Coulomb interactions in the RPA, respectively. A magnified giant resonance is shown in the inset (lowest panel) for $L = 3$. The Skyrme interaction SGII [5] was used.

with multipolarities $L = 0-3$, see also Ref. [13]. The interaction SGII [5] was used. It is seen from the Figure 1 that the effects of violation of self-consistency due to the neglect of the particle-hole (p-h) spin-orbit or Coulomb interactions in the RPA calculations are most significant for the ISGMR. For example, due to the neglect of the p-h spin-orbit interaction the shift in the centroid energy E_0 of the ISGMR in ^{208}Pb is about 0.8 MeV, which is 3 times larger than the experimental uncertainty. The upward shift of E_0 obtained with non-self-consistent calculations with Skyrme interactions explains the discrepancy in the value of K obtained with Skyrme interactions (210 MeV) and with Gogny interactions (230 MeV).

Relativistic and non-relativistic models

For a meaningful comparison between the predictions of the relativistic and the non-relativistic models for the value of K, as deduced from data on the ISGMR, we have recently generated [4] new parameter sets for the Skyrme interaction by a least square fitting procedure using exactly the same experimental data for the bulk properties of nuclei considered in Ref. [9] for determining the NL3 parameterization of an effective Lagrangian used in the relativistic mean field (RMF) models. The center of mass correction to the total binding energy, finite size effects of the protons and Coulomb energy were treated in a way similar to that employed in determining the NL3 interaction in Ref. [9]. Further, it was required in the fitting procedure that the values of K, symmetry energy coefficient (J) and the charge rms radius of the ^{208}Pb nucleus be very close to 271.76 MeV, 37.4 MeV and 5.50 fm, respectively, as obtained with the NL3 interaction. The parameter set SK255 having $K = 255$ MeV with $J = 37$ MeV was also generated, see details in Ref. [4].

In Table 3, we present the results of fully self-consistent HF based RPA calculations for the ISGMR centroid energy (E_0) obtained using the Skyrme interactions SGII [5], KDE0 [3] and the new parameter set SK255 [4], and compare them with the RMF based RPA results [14] for the NL3 interaction [9]. It is seen from Table 3 that the new parameter set SK255 of the Skyrme interaction yields for the ISGMR centroid energies E_0 values which are quite close to the RRPA results with the NL3 interaction. Moreover, for the SK255 parameter set, one finds a good agreement with experimental data for

TABLE 3. Fully self-consistent HF-RPA results for ISGMR centroid energy $E0$ (in MeV) obtained using the Skyrme interactions SK255 [4], SGII [5] and KDE0 [3], and compared with the RRPA results [14] with NL3 interaction [9]. The excitation energy range of the integration of the strength function is $\omega_1 - \omega_2$. Also given are the corresponding values of the incompressibility, K, and the symmetry energy, J, coefficients. The experimental data are from Ref. [11].

Nucleus	$\omega_1 - \omega_2$	Expt.	NL3	SK255	SGII	KDE0
^{90}Zr	0-60		18.7	18.90	17.89	18.03
	10-35	17.81± 0.30		18.85	17.87	17.98
^{116}Sn	0-60		17.1	17.31	16.36	16.58
	10-35	15.85± 0.20		17.33	16.38	16.61
^{144}Sm	0-60		16.1	16.21	15.26	15.46
	10-35	15.40± 0.40		16.19	15.22	15.44
^{208}Pb	0-60		14.2	14.34	13.57	13.79
	10-35	13.96± 0.20		14.38	13.58	13.84
K (MeV)			272	255	215	229
J (MeV)			37.4	37.4	26.8	33.0

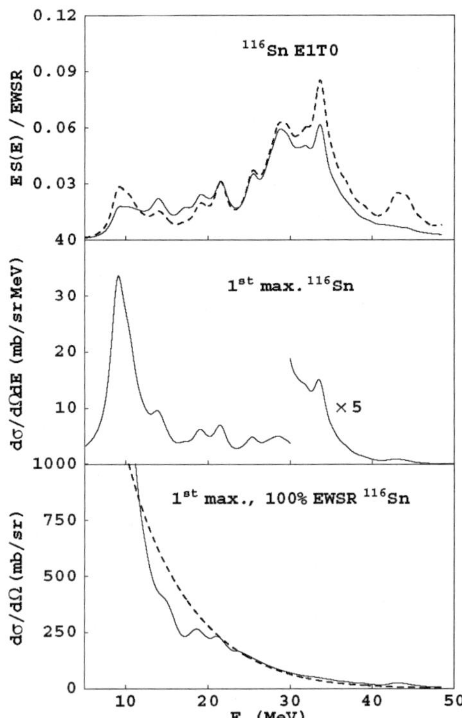

FIGURE 2. Reconstruction of the ISGDR EWSR in ^{116}Sn from the inelastic $\alpha-$particle cross sections. The middle panel: maximum double differential cross section obtained from ρ_t (RPA). The lower panel: maximum cross section obtained with ρ_{coll} (dashed line) and ρ_t (solid line) normalized to 100% of the EWSR. Upper panel: The solid and dashed lines are the ratios of the middle panel curve with the solid and dashed lines of the lower panel, respectively (taken from Ref. [7].

E_0 for all the nuclei considered, provided, the corresponding excitation energy ranges used in determining E_0 are the same as those used in obtaining the experimental data. We have thus resolved the discrepancy in the extracted value of K within relativistic and non-relativistic models [4] and conclude that the difference in the extracted values of K is mainly due to the differences in the values of the symmetry energy coefficient J and its slope L associated with these models (see also Ref. [8]).

Nuclear compressibility from ISGMR and ISGDR

In Figure 2, we present the results [7] of microscopic calculations of the excitation cross section of the ISGDR in ^{116}Sn by 240 MeV α-particle scattering within the FM-DWBA, using the HF results for the ground state density, ρ_0, and the HF based RPA results for the transition densities, ρ_t. We note that the use of the collective model transition density, ρ_{coll}, increases the energy weighted sum rule (EWSR) by about 15% and may shift the centroid energy by only a few percent. However, it is important to note that the calculated maximum cross section for the ISGDR drops below the experimental sensitivity of about 3 mb/sr/MeV for excitation energy above 30 MeV. Thus the high excitation energy region which contains about 20% of the EWSR is missing from the experimental data. This missing strength leads to a reduction of about 3.0 MeV in the ISGDR centroid energy and thus explains the discrepancy between theory and experiment concerning the value of K extracted by using the ISGDR data. More recent experimental work [11] seems to confirm this prediction.

CONCLUSIONS

We have reviewed the success of the HF-RPA theory in describing ground properties of nuclei and presented: (i) a modern Skyrme type NN interaction (KDE0) with an enhanced predictive power for properties of nuclei, (ii) results of fully self-consistent HF based RPA calculations of the centroid energies E_0 of the ISGMR using various Skyrme type effective interactions and compared them with the RRPA results using the NL3 interaction, (iii) results of microscopic calculations of α-nucleus excitation cross sections $\sigma(E)$ for the ISGDR, using the FM-DWBA. The results of our calculations led to the resolution of the long standing discrepancies in the value of K, as obtained from the ISGMR and ISGDR data within relativistic and non-relativistic mean-field based RPA. We thus conclude (see also the review in Ref. [15]) that $K = 240 \pm 20$ MeV. The uncertainty of 20 MeV in K is mainly due to the present uncertainty in the value of J.

Acknowledgments This work was supported in part by the US National Science Foundation under Grant No. PHY-0355200 and the US Department of Energy under Grant No. DOE-FG03-93ER40773.

REFERENCES

1. A. Bohr, and B. R. Mottelson, *Nuclear Structure* (Benjamin, London, 1975), Vol. II.
2. D. Vautherin and D. M. Brink, Phys. Rev. C **5**, 626 (1972).
3. B. K. Agrawal, S. Shlomo and V. Kim Au, Phys. Rev. C **72**, (2005) 014310.
4. B. K. Agrawal, S. Shlomo and V. Kim Au, Phys. Rev. C **68**, 031304(R) (2003).
5. Nguyen Van Giai and H. Sagawa, Phys. Lett. B **106**, 379 (1981).
6. S. Shlomo, and G. F. Bertsch, Nucl. Phys. **A243**, 507 (1975).
7. S. Shlomo and A. I. Sanzhur, Phys. Rev. C **65**, 044310 (2002).
8. J. Piekarewicz, Phys. Rev. C **66** 034305 (2002).
9. G. A. Lalazissis, J. Konig, and P. Ring, Phys. Rev. C **55**, 540 (1997).
10. Nguyen Van Giai, P. F. Bortignon, G. Colo, Zhongyu Ma and M. Quaglia, Nucl. Phys. **A687**, 44c (2001).
11. D. H. Youngblood *et al*, Phys. Rev. C **69**, 034315 (2004); Phys. Rev. C **69**, 054312 (2004).
12. T. Sil, S. Shlomo, B. K. Agrawal, and P.-G. Reinhard, Phys. Rev. C **73**, 034316 (2006).
13. B. K. Agrawal and S. Shlomo, Phys. Rev. C **70**, 014308 (2004).
14. T. Niksic, D. Vretenar and P. Ring, Phys. Rev. C **66**, 064302 (2002).
15. S. Shlomo, V. M. Kolomietz and G. Colo Eur. Phys. J. A **30**, 23 (2006).

Probing Nuclear Structure by Cold Emission Processes

D.S. Delion

Horia Hulubei National Institute of Physics and Nuclear Engineering, 407 Atomistilor, POB MG-6, Bucharest-Măgurele, Romania

Abstract. Cold emission processes (one and two-proton emission, alpha-decay, heavy cluster emission and cold binary or ternary fission) are presently among important tools to investigate the structure of rare nuclei far from the stability line. We analyze the coupling between collective excitations of the emitted fragments and the relative motion, in terms of the coupled channels technique. It turns out that partial decay widths to excited states of emitted fragments are very sensitive to the nuclear structure details.

Keywords: α-decay, Cluster decay, Cold fission, Half-life, Resonant states, Barrier penetration, Spectroscopic factor, Coupled channels method
PACS: 21.10.Jx, 21.10.Tg, 24.10.Eq, 25.70.Ef

INTRODUCTION

Nuclear physics was born with the discovery of the natural α, β and γ radioactivities of Uranium by H. Becquerel in 1896. Physicists realized that α-decay (emission of ^{4}He nuclei) involves a new type of short range force called strong interaction, while β-decay (electron emission) is connected with the so-called weak interaction.

In 1980, long time after the discovery of the α-decay, based on the large binding energy of ^{208}Pb, it was predicted the emission from the ground state (cold emission) of heavier clusters like ^{14}C, ^{24}Ne and ^{28}Mg [1]. This new kind of natural radioactivity was soon experimentally found [2].

The first discovery of proton spontaneous emission was made in 1970 [3] and today are known about 50 proton emitters. A very exotic decay, namely two proton emission, is supposed to exist for some nuclei, where this process is energetically allowed. It was predicted in 1960 [4], but only relative recently it was detected [5].

Nuclear fission accompanied the emission of neutrons was discovered in 1939 by O. Hahn and F. Strassmann, but only in 1962 some fission experiments reported binary fission fragments with very large kinetic energy [6]. In this case the excitation energy of fragments shows that they are close to their ground states. Therefore this process was called cold (neutronless) fission. The first direct observation of the cold binary fission of ^{252}Cf was reported at Oak Ridge National Laboratory [7, 8].

In this review we introduce the main tools to describe emission processes from deformed nuclei, namely Gamow resonances and the coupled channels method. We suppose that the dynamics of the decaying systems is fully described by a phenomenological potential. We analyze the fine structure in alpha decay and double fine structure in cold fission.

CP972, *Exotic Nuclei and Nuclear/Particle Astrophysics (II)—Proceedings of the Carpathian Summer School of Physics 2007,* edited by L. Trache and S. Stoica

EMISSION THEORY FOR DEFORMED NUCLEI

A binary decay process is allowed as soon as the final configuration is energetically more favorable than the initial one, i.e. the so-called Q-value of the process is positive. The shortest measurable half-life is at present $T = 10^{-12}$ sec which is much longer than the characteristic nuclear time $T_N \approx 10^{-21}$ *sec*. Thus, the decay process may be considered stationary, but with a complex energy

$$E_r = E^{(0)} - \frac{i}{2}\Gamma , \qquad (2.1)$$

as first proposed Gamow in his famous paper explaining α-decay as a quantum penetration process [9]. Thus, the dynamics of the decaying system is described by the stationary Schrödinger equation, with the following Hamiltonian

$$\mathbf{H} = -\frac{\hbar^2}{2\mu}\Delta_r^2 + \mathbf{H}_1(\mathbf{x}_1) + \mathbf{H}_2(\mathbf{x}_2) + V(\mathbf{x}_1, \mathbf{x}_2, \mathbf{r}) , \qquad (2.2)$$

where $\mathbf{x}_k$ are internal coordinates of fragments and $\mathbf{r}$ the distance between them. We denoted by V the inter-fragment potential and by $\mathbf{H}_k$ the Hamiltonians describing the internal motion of the emitted fragments with eigenvalues E_k. The external solution can be written as a superposition of different outgoing channels $c \equiv (J_1 J_2 J j)$, describing all possible splittings

$$\Psi_{J_i M_i}(\mathbf{x}_1, \mathbf{x}_2, \mathbf{r}) = \sum_c \frac{f_c(r)}{r} \mathscr{Y}^{(c)}_{J_i M_i}(\mathbf{x}_1, \mathbf{x}_2, \omega) , \qquad (2.3)$$

where, with $\omega \equiv (\phi, \theta)$, we introduced the core-angular harmonics depending on fragment wave functions and relative angular harmonic as follows

$$\mathscr{Y}^{(J_1 J_2 J j)}_{J_i M_i}(\mathbf{x}_1, \mathbf{x}_2, \omega) \equiv \left\{ [\Phi_{J_1}(\mathbf{x}_1) \otimes \Phi_{J_2}(\mathbf{x}_2)]_J \otimes \mathscr{Y}_j(\omega) \right\}_{J_i M_i} . \qquad (2.4)$$

These harmonics, as well as their components, are mutually orthonormal. We split the potential into a spherical and a deformed component, i.e. $V = V_0 + V_d$. By projecting out a given channel c and taking into account the orthonormality condition one obtains the coupled channels system of equations for the radial wave functions as follows

$$\frac{d^2 f_c(r)}{dr^2} = \left\{ \frac{l_c(l_c+1)}{r^2} + \frac{2\mu}{\hbar^2}[V_0(r) - E_c] \right\} f_c(r) + \frac{2\mu}{\hbar^2} \sum_{c'} V_d^{(cc')}(r) f_{c'}(r) , \qquad (2.5)$$

where the channel energy is defined as $E_c = E - E_1 - E_2$. The emitted fragments interact via short range nuclear forces. Therefore at large distances only the spherical Coulomb interaction becomes active and the system (2.5) becomes diagonal, depending on the reduced radius $\rho_c = \kappa_c r$ and the Coulomb parameter (twice the Somerfeld parameter)

$$\chi_c = 2\frac{Z_1 Z_2 e^2}{\hbar v_c} , \qquad (2.6)$$

written in terms of the asymptotic velocity and momentum for a given channel "c". The outgoing solutions are the standard Coulomb-Hankel waves [10] $H_l^{(+)}(\chi,\rho) = G_l(\chi,\rho)+iF_l(\chi,\rho)$, where $G_l(\chi,\rho)/F_l(\chi,\rho)$ is the irregular/regular Coulomb function.

We will present the numerical integration procedure of the system (2.5). We determine two sets of N linear independent vector functions. One is regular in origin and the other is diagonal and outgoing at large distances, i.e.

$$\begin{aligned} \mathscr{R}_{ca}(r) &\to_{r\to 0} \delta_{ca} r^{l_c+1} \\ \mathscr{H}_{ca}^{(+)}(r) &\to_{r\to\infty} \delta_{ca} H_{l_c}^{(+)}(\kappa_c r) . \end{aligned} \tag{2.7}$$

The matrix $\mathscr{H}$ is found by backward integration starting from large distances. The c-th component of the internal/external solutions can be written as a superposition of these fundamental solutions

$$\begin{aligned} f_c^{(int)}(r) &= \sum_a \mathscr{R}_{ca}(r) M_a \\ f_c^{(ext)}(r) &= \sum_a \mathscr{H}_{ca}^{(+)}(r) N_a \to_{r\to\infty} N_c H_{l_c}^{(+)}(\kappa_c r) . \end{aligned} \tag{2.8}$$

The matching constants are found by using the continuity of internal and external functions and their derivatives. In order to find a non-trivial solution the following secular equation has to be fulfilled

$$det \begin{bmatrix} \mathscr{R}(R) & \mathscr{H}^{(+)}(R) \\ \frac{d}{dr}\mathscr{R}(R) & \frac{d}{dr}\mathscr{H}^{(+)}(R) \end{bmatrix} = 0 . \tag{2.9}$$

The solutions provide complex energies (2.1) which are the S-matrix poles. They define the so-called Gamow resonances. By using the continuity equation and a wave function normalized to unity inside some large sphere the decay width becomes proportional to the total flux through the surface of this sphere, i.e.

$$\Gamma = \sum_c \hbar v_c \lim_{r\to\infty} \oint r^2 d\omega \int d\mathbf{x_1} \int d\mathbf{x_2} |\Psi_c|^2 = \sum_c \hbar v_c |N_c|^2 \equiv \sum_c \Gamma_c . \tag{2.10}$$

By using (2.8) one obtains the factorization of the partial decay width

$$\Gamma_c = 2P_{l_c}(R)\,|\gamma_c(R)|^2 , \tag{2.11}$$

where the penetrability has the standard form, i.e.

$$P_{l_c}(E_c,R) \equiv \frac{\kappa_c R}{F_{l_c}^2(\chi_c,\kappa_c R)+G_{l_c}^2(\chi_c,\kappa_c R)} = \frac{\kappa_c R}{\left|H_{l_c}^{(+)}(\chi_c,\kappa_c R)\right|^2} . \tag{2.12}$$

The reduced decay width is given by

$$\gamma_c(R) = \sqrt{\frac{\hbar^2}{2\mu R}} s_c(R) , \quad s_c(R) = \sum_{c'} H_{l_c}^{(+)}(\chi_c,\kappa_c R) \left[\mathscr{H}_{cc'}^{(+)}(R)\right]^{-1} f_{c'}^{(int)}(R) . \tag{2.13}$$

The most general method to estimate the interaction between two massive fragments is given by the double folding procedure between nuclear densities. This procedure is described in many text-books and review papers, e.g. [11] and consists in the following double integral

$$V(\mathbf{x_1},\mathbf{x_2},\mathbf{r}) = \int d\mathbf{r}_1 \int d\mathbf{r}_2 \rho_1(\mathbf{x_1},\mathbf{r}_1)\rho_2(\mathbf{x_2},\mathbf{r_2})v(\mathbf{r}+\mathbf{r}_1-\mathbf{r}_2) \,, \quad (2.14)$$

where $\mathbf{r}_k$ is the radius giving the nucleon position inside the k-th nucleus, with the density ρ_k and v is the nucleon-nucleon force. This integral can be expanded in terms of fragment Euler angles ω_k as follows

$$V(\omega_1,\omega_2,\mathbf{r}) = \sum_{\lambda_1\lambda_2\lambda_3} \frac{4\pi}{\hat{\lambda}_1\hat{\lambda}_2} V_{\lambda_1\lambda_2\lambda_3}(r)\mathscr{Y}_0^{(\lambda_1\lambda_2\lambda_3)}(\omega_1,\omega_2,\omega) \quad (2.15)$$

where the corresponding core-angular harmonics is a particular case of (2.4)

$$\mathscr{Y}_0^{(\lambda_1\lambda_2\lambda_3)}(\omega_1,\omega_2,\omega) = \left\{\left[Y_{\lambda_1}(\omega_1)\otimes Y_{\lambda_2}(\omega_2)\right]_{\lambda_3}\otimes Y_{\lambda_3}(\omega)\right\}_0 \,. \quad (2.16)$$

One of the most popular two-body interaction, used to describe heavy ion scattering is given by a superposition of three Yukawa potentials, simulating the exchange of different mesons, called also M3Y interaction [12]. The above described double folding procedure supposes that the two fragments are already born. But the ratio between the computed half-life, by using the double folding potential (2.14), and the experimental value is less than unity and it is called spectroscopic factor $S = T_{th}/T_{exp}$. This is due to the fact that in reality the emitted fragments do not exist during the decay process, but they are born with certain probability.

FINE STRUCTURE IN α-DECAY

We will first investigate the so-called fine structure in α-decays to J^+ states in even-even nuclei. At present there are a lot of high precision data on α-decay fine structure to rotational levels, see e.g. [13]. A clear difference between vibrational-transitional and rotational regions is given in figure 1 (a), where we plotted the logarithm of the intensity ratio

$$I_c = log_{10}\frac{\Gamma_0}{\Gamma_c} \,. \quad (3.1)$$

where the channel is defined by $c = J$, versus the energy E_2. All values of the rotational region with $E_2 \leq 0.1\ MeV$ are concentrated around $I \approx 0.5$, while for the other nuclei one has an increasing trend, given by a linear ansatz (solid line) $I_{exp} = 4.361\ E_2 + 0.924\ ,\ \ \sigma = 0.774$. In figure 1 (b) we plotted the same quantity, but versus the neutron number.

To analyze rotational nuclei we used the coupled channels formalism by considering $J_i = 0$, $\Phi_{J_2=0}(\mathbf{x}_2) = 1$ in Eq. (2.4) and in the expansion of the potential (2.15) $\lambda_1 =$

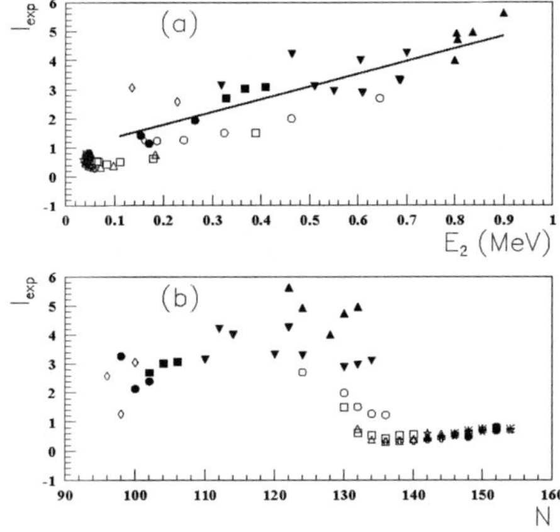

FIGURE 1. (a) The experimental values of the logarithm of the intensity ratio (3.1) versus the energy E_2. Various symbols denote different isotopic chains. The straight line fits the values of vibrational-transitional emitters. (b) The experimental values of the logarithm of the intensity ratio (3.1) versus the neutron number.

λ_3, $\lambda_2 = 0$. As usually in our computations we use the M3Y nucleon-nucleon [12] plus Coulomb force.

We described the density of the daughter nucleus by an axially deformed Woods-Saxon shape, by using quadrupole and hexadecapole deformation parameters of the nuclear surface. The α-particle density is given by a Gaussian distribution. Concerning the α-nucleus interaction we multiplied the double folding integral by a strength parameter v_a, in order to simulate the spectroscopic factor multiplying the wave function, given by the initial not-corrected potential, i.e.

$$\overline{V}_\lambda(r) = v_a V_\lambda(r) \, . \tag{3.2}$$

These multipoles are plotted in figure 2.

We used the standard value of the diffusivity $a = 0.5 \ fm$, the α-particle size parameter $b = 1.19 \ fm$ [14] and the deformations β_2, β_4 given by the systematics of Ref. [15]. To this potential we added a repulsive core, depending upon one independent parameter

$$\begin{aligned} \overline{V}_0(r) &= v_a V_0(r) \, , \quad r > r_m \\ &= c(r - r_0)^2 - v_0 \, , \quad r \le r_m \, . \end{aligned} \tag{3.3}$$

It simulates the Pauli principle and adjusts the energy of the system to the experimental Q-value. This interaction is given in figure 2 by two pocket-like curves. The curves labelled by (1) and (2) give the same Q-value. We showed in Ref. [16] that the total half-life and the partial decay widths do not depend upon the shape of this repulsive potential.

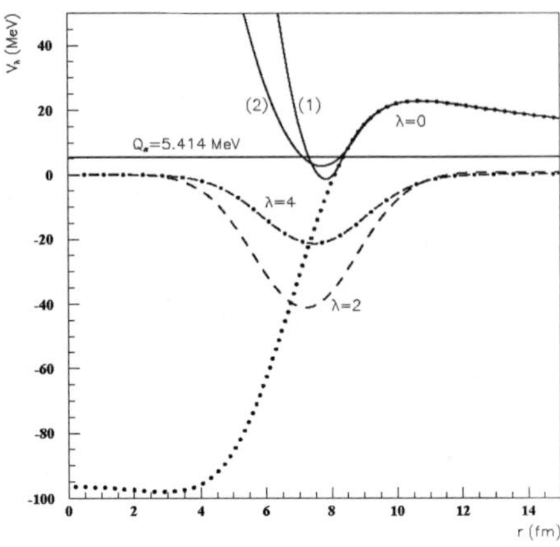

FIGURE 2. The radial components of the renormalized α-nucleus potential (3.2) for λ=0 (dots), λ=2 (dashes) and λ=4 (dot-dashes). The solid pocket-like curves (1) and (2) are the monopole parts of the interaction (3.3), giving the same Q-value. Their parameters are (1) $c = 90.117$ $(MeV\ fm^{-2})$, $Q_\alpha + v_0 = 10.272$ (MeV) and (2) $c = 30.296$ $(MeV\ fm^{-2})$, $Q_\alpha + v_0 = -3.816$ (MeV). The horizontal line denotes the Q-value. The decay process is $^{232}Pu \rightarrow ^{228}U + \alpha$.

We stress the fact that only three parameters, namely v_a, v_0, c, are independent, because the radii r_0, r_m are determined by using the matching conditions connecting internal repulsion with double folding surface potential.

It turns out that, in order to satisfy the relation between the half-life and Q-value, the quenching strength should be of the form $v_a = 0.668 - 0.004\ (A - 208)$. We have also checked that an alternative rule for the quenching factor depending upon the charge number, i.e $v_a = 0.7 - 0.0125\ (Z - 82)$, gives very close results. The value of the repulsive strength was taken $c = 100\ MeV\ fm^{-2}$. As mentioned, we adjusted the energy of the first resonant state in the pocket-like potential by using the repulsive depth v_0. The computed half-lives practically reproduce the experimental values.

In Figure 3 (a) are given the quadrupole (squares) and hexadecapole deformations (triangles), while in (b) the experimental values of the ratios I_2, I_4 (squares) versus the number of the decay process, given by the first column of the Table I in ref. [16]. We also give by open circles the results of our calculation for I_2 and I_4. One sees that we obtained a good agreement with experimental data for I_2. Concerning I_4 a very good agreement was achieved only for the $Z = 90$ isotope chain. For the last Z=96 and Z=98 chains the agreement is within a half-order of magnitude, while the central peak, around the Z=94 chain, is not reproduced. The difference between experiment and theory in this region is about 1.5 orders of magnitude.

The main trend of the experimental and computed I_2 values in Figure 3 (b) is clearly correlated with the quadrupole deformation in (a). The same is true, but mainly for the computed I_4 values, which are correlated with β_4 in (a).

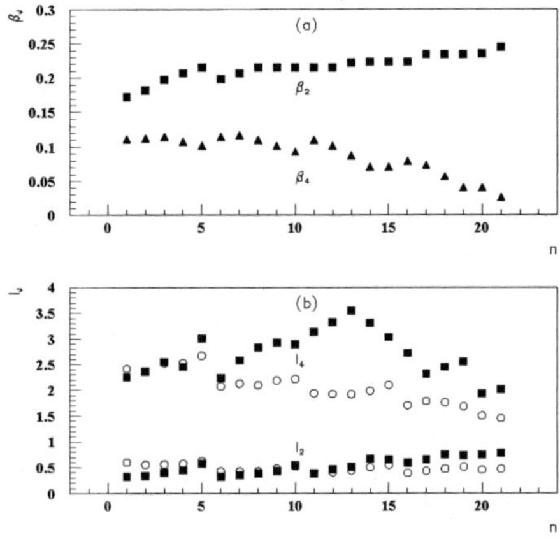

FIGURE 3. (a) Deformation parameters β_2 (squares) and β_4 (triangles) versus the decay number given by the first column in Table I of ref. [16]. (b) Experimental ratios I_2, I_4 defined by Eq. (3.1) (squares) and the corresponding computed values (open circles). The attractive quenching parameter is given by the rule $v_a = 0.668 - 0.004\,(A - 208)$. The corresponding data in Table I of ref. [16] are labelled by the index (a).

DOUBLE FINE STRUCTURE IN COLD FISSION

In the last years an intense experimental activity was performed in order to investigate cold (neutron-less) binary and ternary fission process of ^{252}Cf [7, 8, 17, 18]. It involved modern facilities, as the Gammasphere and Eurogam, which were able to identify this rare process using the triple γ-rays coincidence technique.

The yields of rotational states were extracted from the intensities of γ-rays emitted in coincidence during the deexcitation of fragments for ^{104}Mo-^{148}Ba and ^{106}Mo-^{146}Ba [17]. It was shown that the cold fission population is centered around the low-lying 2^+ and 4^+ states and the states higher than 6^+ are practically not populated.

We expect an important dependence of the wave function upon the neutron diffusivity. Therefore the double fine structure seen in emission fragments is an important tool to probe the densities of neutron rich fragments [19]. Thus, the nuclear densities are parametrized in the intrinsic system of coordinates by deformed Fermi distributions as follows

$$\begin{aligned} \rho^{(k)}(\mathbf{r}_k) &= \frac{\rho_0^{(k)}}{1+exp\left[\left(r_k - c^{(k)}\right)/a^{(k)}\right]}\,, \quad k=1,2\,, \\ c^{(k)} &= c_0^{(k)}\left[1+\sum_{\lambda\geq 2}\beta_\lambda^{(k)}Y_{\lambda 0}(\omega_k')\right]\,, \end{aligned} \tag{4.1}$$

where the nuclear radius is given by

$$c_0^{(k)} = (r_0 + w^{(k)})A^{1/3} \;, \quad r_0 = (d - 1/d)A^{-1/3} \;. \tag{4.2}$$

The most favorable fissioning configuration is the pole-to-pole ($p-p$) one, with $\omega_1 = \omega_2 = \omega$, where the Coulomb barrier has the lowest possible value. In our calculation we considered as a spherical component the pole-to-pole interaction $V_{p-p}(R)$ until the intersection with the true spherical part $V_0(R)$.

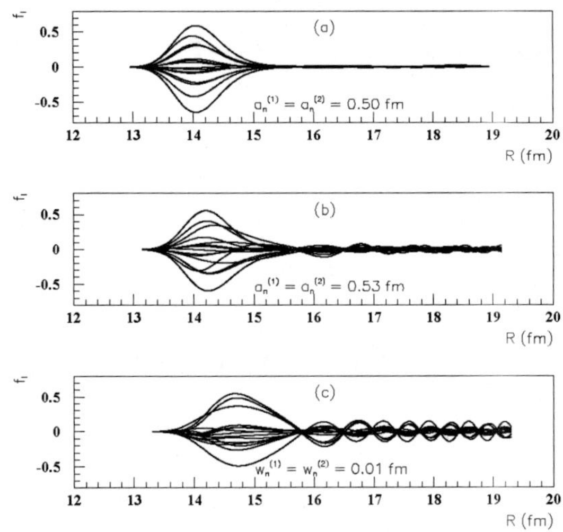

FIGURE 4. The radial wave function components $f_l(R)$ for neutron density parameters in Eq. (4.1) $a_n = 0.50\ fm$, $w_n = 0$ (a), $a_n = 0.53\ fm$, $w_n = 0$ (b) and $w_n = 0.01\ fm$, $a_n = 0.5\ fm$ (c). The fission process is $^{252}\mathrm{Cf} \rightarrow {}^{104}\mathrm{Mo} + {}^{148}\mathrm{Ba}$.

We analyzed the cold fission (neutron-less) process, i.e. by supposing ground-state deformations of final fragments from the systematics given in Ref. [15]. The corresponding values are $\beta_2^{(1)} = 0.349, \beta_4^{(1)} = 0.030$ for ^{104}Mo and $\beta_2^{(2)} = 0.236, \beta_4^{(2)} = 0.131$ for ^{148}Ba. The energy of the resonant state is given by the Q-value, i.e. $E = 214.67\ MeV$. We adjust it by using the strength of a repulsive core.

For proton densities we considered a standard diffusivity $a_p^{(k)} = 0.5\ fm$ and radial parameter $w_p^{(k)} = 0$. The only free parameters, according to (4.1) remain $a_n^{(k)}$ and $w_n^{(k)}$. First of all we studied the influence of the neutron diffusivity on the decay widths by considering $w_n^{(k)} = 0$. Our calculation showed that partial decay widths remain practically unchanged for a constant value of the sum $a_n^{(1)} + a_n^{(2)}$. We will present our results for a common value $a_n \equiv a_n^{(1)} = a_n^{(2)}$.

It turns out that a small increase of this parameter from $a_n = 0.50\ fm$ up to $a_n = 0.53\ fm$ changes the behavior of radial components beyond the external turning point.

The radial probabilities in this region are proportional to the partial decay widths. The 15 radial components of the wave function $f_l(R)$, $l \rightarrow (l, I_1, I_2)$ are plotted in Figure 4 (a) for $a_n = 0.50\ fm$ and in 4 (b) for $a_n = 0.53\ fm$. One can see an increasing of the oscillatory tail in the case (b) with respect to (a).

Then we investigated the influence of the radial parameter $w_n = w_n^{(1)} = w_n^{(2)}$ by keeping a constant diffusivity $a_n = 0.5\ fm$. The results remain again qualitatively unchanged for a constant sum. We changed this parameter from 0 up to $0.01\ fm$. This corresponds to an increasing of the neutron radius by $0.05\ fm \approx 0.01\ A^{1/3}$. In Figure 4 (c) we give the radial components for $w_n = 0.01\ fm$. The oscillatory tails are even larger in this case.

CONCLUSIONS

Particle, cluster emission and fission from deformed nuclei is a field of research that only recently has been pursued with great vigor, because the emission processes allow us to study exotic nuclei close to the proton/neutron drip lines, or superheavy nuclei. It turns out that the wave function of the relative motion between emitted fragments is very sensitive to the nuclear surface shape. The conclusion of this presentation is that emission processes indeed offer an unique tools to investigate the structure of rare nuclei, which due to their short lifetime are otherwise very difficult to explore.

ACKNOWLEDGMENTS

This paper has been supported by the Contract PN 06 35 01 01/2006 and IDEI-14/2007.

REFERENCES

1. Săndulescu A, Poenaru D N and Greiner W 1980 Fiz. Elem. Chastits At Yadra **11** 1334.
2. Rose H J and Jones G A 1984 Nature (London) **307** 245.
3. K.P. Jackson, et. al., Phys. Lett. B**33**, 281 (1970).
4. V.I. Goldansky, Nucl. Phys. **19**, 482 (1960).
5. R.A. Kryger, *et. al.* Phys. Rev. Lett **74**, 860 (1995).
6. Milton J C D and Fraser J S 1962 Can. J. Phys. **40** 1626.
7. Hamilton J H *et al.* 1994 J. Phys. G**20** L85.
8. G.M. Ter-Akopian, et al., Phys. Rev. Lett. **73** (1994) 1477.
9. G. Gamow, Z. Phys. **51**, 204 (1928).
10. M. Abramowitz and I. A. Stegun, eds., *Handbook of Mathematical Functions*, (Dover Publications Inc., New York, 1983).
11. F. Carstoiu and R.J. Lombard, Ann. Phys. (N.Y.) **217** (1992) 279.
12. G. Bertsch, J. Borysowicz, H. McManus, and W.G. Love, Nucl. Phys. A **284** (1977) 399.
13. Y.A. Akovali, Nucl. Data Sheets **84**, 1 (1998).
14. Dao T. Khoa, Phys. Rev. C **63**, 034007 (2001).
15. P. Möller, R.J. Nix, W.D. Myers, and W. Swiatecki, At. Data Nucl. Data Tables **66**, 131 (1995).
16. D.S. Delion, S. Peltonen, and J. Suhonen, Phys. Rev. C **73**, 014315 (2006).
17. A. Săndulescu, A. Florescu, F. Carstoiu, W. Greiner, J.H. Hamilton, A.V. Ramayya, and B.R.S. Babu, Phys. Rev. C **54** (1996) 258.
18. J.H. Hamilton, et. al., Yad. Fiz. **65** (2002) 677; Phys. At. Nucl. **65** (2002) 695.
19. D.S. Delion, A. Sandulescu, and W. Greiner, Phys. Rev. C **68**, 041303(R) (2003).

Studies with Exotic Nuclei: Two Proton Radioactivity

C. Borcea[a], B. Blank[a], G. Canchel[a], C.E. Demonchy[a], J. Giovinazzo[a], L. Hay[a], J. Huikari[a], S. Leblanc[a], I. Matea[a], J.-L. Pedroza[a], J. Pibernat[a], L. Serani[a], F. de Oliveira Santos[b], S. Grevy[b], L. Perrot[b], C. Stodel[b], J.-C. Thomas[b], C. Dossat[c]

[a]*Centre d'études nucléaires de Bordeaux-Gradignan, Université Bordeaux 1 - UMR 5797 CNRS/IN2P3, Chemin du Solarium, BP 120, F-33175 Gradignan Cedex, France*

[b]*Grand Accélérateur National d'Ions Lourds, CEA/DSM - CNRS/IN2P3, Bvd Henri Becquerel, BP 55027, F-14076 CAEN Cedex 5, France*

[b]*DAPNIA, CEA Saclay, F-91191 Gif-sur-Yvette Cedex, France*

Abstract. In the present paper, we present measurements that led to the discovery of two-proton radioactivity. After the first observation of this decay mode for ^{45}Fe, new measurements evidenced this decay mode also for ^{54}Zn and most likely ^{48}Ni. A new detector based on the time-projection chamber principle allowed now to visualize the two protons directly.

Keywords: two-proton radioactivity, time-projection chamber
PACS: 23.50.+z, 23.90.+w, 21.10.-k, 27.40.+z

INTRODUCTION

The study of nuclei ever further away from the valley of stability became customary nowadays, especially with the technological developments allowing obtaining and using secondary beams of radioactive species. The nuclei at the limits of stability are under scrutiny and their study revealed a number of phenomena which are typical for these regions of the chart of nuclides. Perhaps the most frequently mentioned are: the reconsideration of magic numbers and their associated properties, the halo nuclei and the new types of radioactivity.

The discovery of the radioactive decay was of crucial importance for the further development of the nuclear physics and a perfect illustration of a typical quantum phenomenon: the tunneling through a barrier. After alpha, beta and gamma decays, around 1940 followed the discovery of spontaneous fission decay. It took another 40 years until the next discoveries around 1980: one proton radioactivity [1] and cluster radioactivity [21]. The last on the list was two proton radioactivity discovered in the 21st century. Though predicted by Goldanskii and others in the 1960 [3], its discovery was conditioned by the before mentioned technological progress in producing and identifying very exotic nuclei close to the limits of stability.

CP972, *Exotic Nuclei and Nuclear/Particle Astrophysics (II)—Proceedings of the Carpathian Summer School of Physics 2007,* edited by L. Trache and S. Stoica

TWO-PROTON RADIOACTIVITY: A PROTON DRIP LINE PHENOMENON

Nuclei far from stability undergo beta decay and as a result come closer to the stability line. Neutron rich nuclei decay by β^- decay while the proton rich ones decay by β^+ decay. Far from the stability valley, the available energy for these decays, (the Q_β value) is high. This is due to the fact that the gain in stability following the decay is more important far from stability than close to stability. In terms of binding energies, a "cut" through the chart of nuclides along a line of isobars shows (fig. 1) a parabolic shape with steep increase far from stability.

The high Q value of the decay allows populating highly excited states in the daughter nucleus. Sometimes, the excitation energy exceeds the proton/neutron binding energy in the daughter nucleus and the particle emission after beta decay becomes possible. In such a case a so called β-delayed particle emission takes place. This kind of decay is a powerful spectroscopic tool and has been used extensively to obtain information on nuclei far from stability, difficult to achieve otherwise.

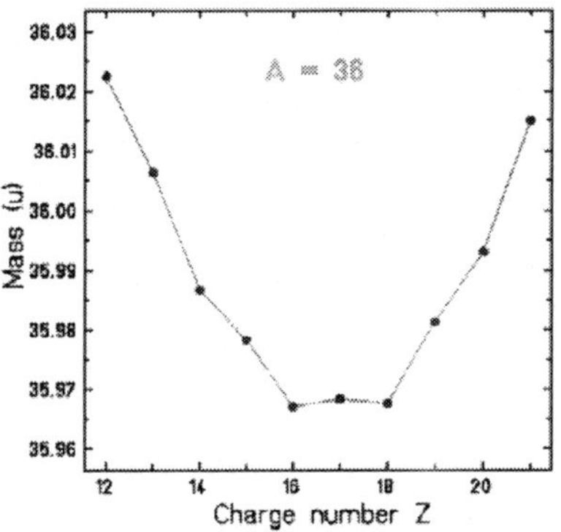

FIGURE 1: Mass excess of A=36 isobars as a function of Z. On the valley of stability (Z=16 to 18) the mass is low as it incorporates more binding. Remark the steep increase far from stability.

For the proton rich nuclei, the last orbital contains protons which are merely bound. The Coulomb potential creates a barrier that may keep the nucleus bound against proton decay for some time even if the protons of the last orbital are placed above the Fermi level (see fig.2). Depending on the energy balance of a particular situation, this barrier may be tunneled with a sizable probability. The emission thus observed of one or two protons represents a new kind of radioactivity. Moreover, sometimes the daughter nucleus is also proton unstable and its decay can be very fast. What is then observed experimentally is a sequence of two proton emissions leading to a daughter nucleus with a Z diminished by two units (see diagram in fig.3).

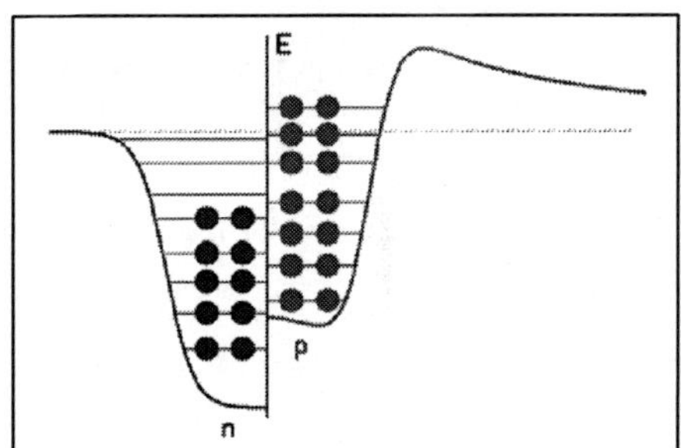

FIGURE 2: Schematic description of the orbital filling for protons and neutrons in the case of a proton rich nucleus. Due to the Coulomb potential, protons may occupy orbitals above the Fermi level.

However, this is not always the case. If the Z-1 nucleus is more bound than the Z parent but the decay to the Z-2 daughter is still energetically possible, then a pure two proton radioactivity occurs. Both processes are sketched in fig.3.

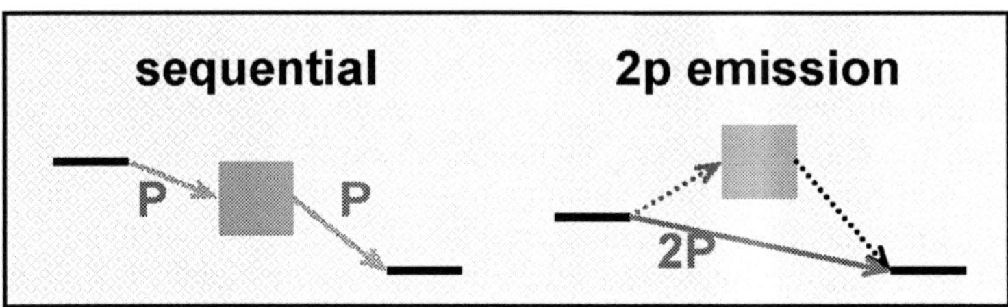

Figure 3: The differences between a sequential emission of two protons and a true 2p emission: the neighboring Z+1 nucleus is more bound than the decaying nucleus in the case of true 2p emission.

A practical illustration of the second process is provided by the case of ^{45}Fe in fig. 4: the Z-1 nucleus, ^{44}Mn is more bound by about 130 keV and the width of its ground state is of only 1 eV. Therefore the transition from ^{45}Fe to ^{44}Mn is forbidden while the decay by two proton emission toward ^{43}Cr is allowed with a Q value of 1.15 MeV. One should keep in mind however that this decay mode is in competition with the β^+ decay of ^{45}Fe to the isobaric analogue and other states in ^{45}Mn followed by proton emission, as shown in fig. 5. The experimental discrimination of the two modes is also explained by the figure: while the 2p emission shows up in a e.g. silicon detector as a sharp line, the β-p process leads to a line overlapped onto a continuum of energies due to the simultaneous detection of the proton and the (variable) energy deposit of β particles.

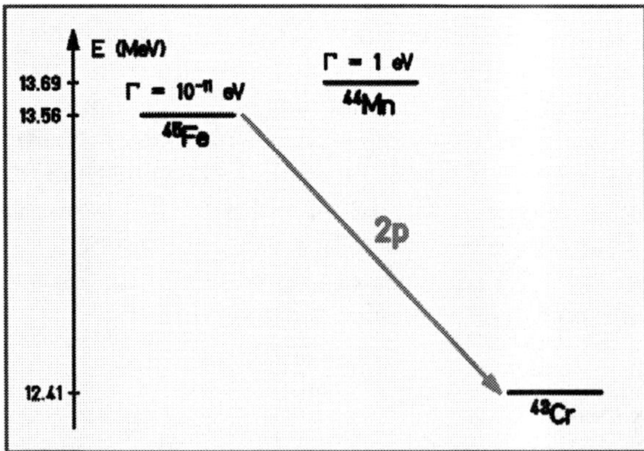

FIGURE 4: The disposal of ground state energies of ^{45}Fe, ^{44}Mn and ^{43}Cr showing the inaccessibility of the state in ^{44}Mn that makes ^{45}Fe a true 2p emitter.

Finding good candidates for 2p decay is not easy as a number of conditions must be fulfilled. The actual problem with these nuclei very difficult to produce is that their masses are not measured or, when measured, they have large error bars. Therefore, one should relay on model predictions where, however, the spread of the values for the mass is fairly large.

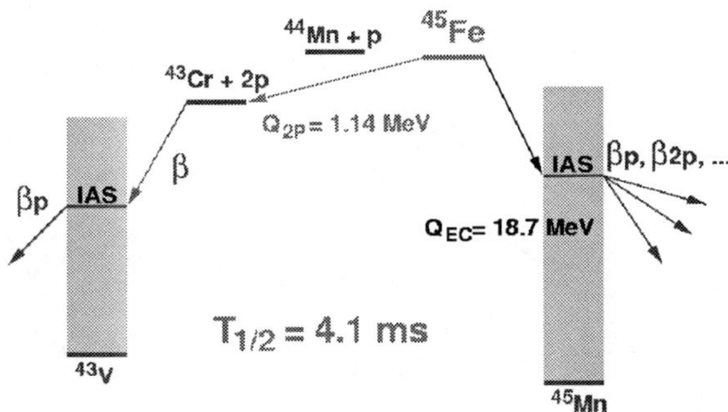

FIGURE 5: Decay scheme for ^{45}Fe: left side via 2p emission and sub sequential β decay, right side β decay followed by one or more proton emission.

When calculating the tunneling time, this spread translates into many orders of magnitude of half-life uncertainty. Anyway, even if known precisely, yet the Q value should not be too high, otherwise the lifetime of the decay will be too short, but also not too low, in which case the lifetime is too long and the decay goes overwhelmingly via the competing β decay channel. In the above calculations for the lifetime, the usual tunneling formalism is used for the two protons taken as a single entity. However, inside the barrier, the two protons are paired. This pairing energy places them at a different energy level: the system is more bound and as an effect, the decay Q value decreases. The barrier to tunnel is wider and the lifetime increases. Moreover, as a result of the tunneling, the two outgoing protons are no longer paired. So the object at the beginning and at the end of the process is not the same, which means a change in the barrier itself. Another change is due to the decrease of the charge defining the height of the barrier by two units. These processes are not accounted for in the classical treatment of the tunneling in which the potential well and the associated barrier do not change during the process, as do not change the nature and intrinsic energy of the tunneling particle. Though the corrections associated with the true process of 2p emission could be small, they may still translate in important changes in the lifetime.

A number of theoretical approaches have been developed to describe the two proton radioactivity. This is quite a challenge because they have to explain not only the lifetime but also the mechanism itself, including the way the decay energy is shared between the three products in the exit channel: daughter nucleus and the two protons. The main three approaches present in the literature are: an R-matrix approach with the two protons in a resonant state [4], a shell model embedded in the continuum approach [5], and a three body approach using Jacobi transformations [6]. The performance of these approaches can be judged by the comparison with the experimental results for ^{45}Fe, ^{48}Ni and ^{54}Zn concerning the lifetime.

Two-proton radioactivity was first observed in experiments at GANIL [7] and GSI [8] in the decay of ^{45}Fe and later for ^{54}Zn [9] and possibly ^{48}Ni [10]. The experiments will be described below.

PRODUCTION AND DECAY OF 2P RADIAOCTIVE NUCLEI

In the following, the beam preparation at GANIL will be described. The method itself is in its principle the same for other places (e.g. GSI, MSU).

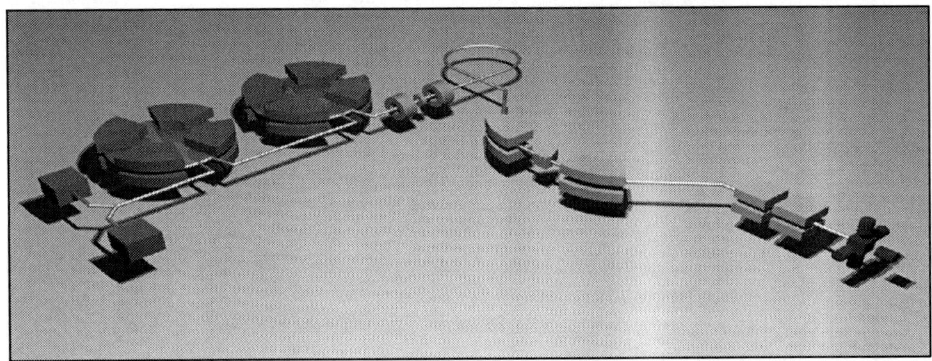

FIGURE 6: Overlay of the secondary beam preparation and transport at GANIL. Order of elements in sequence from left to right: injector cyclotrons, coupled cyclotrons, SISSI device, alpha spectrometer, LISE spectrometer, Wien filter, experimental setup.

The two coupled cyclotrons of GANIL delivered a beam of ^{58}Ni with an intensity of 3-4μA and energy of 75 MeV/A (see fig. 6). This beam was sent on a natural Ni target with a thickness of 200mg/cm^2 placed inside the SISSI device. SISSI is made of two superconducting solenoids: the first one strongly focuses the beam on the target and the second is tuned to refocus into the beam line the fragmentation products of interest. Further, this beam composed of yet many species is brought via a transport line to the LISE achromatic spectrometer followed by a Wien filter for further purification. A 50μm Be degrader is placed in between the two dipoles of LISE. In the final focal point of the spectrometer is placed the detection-identification setup.

The detection-identification setup consists of four silicon detectors to which is added a time pickup detector based on micro-channel plates. Additional timing is provided by the cyclotron radiofrequency (RF). Of the four silicon detectors, the first two are for energy loss. The spectrometer is tuned such that the nuclei of interest, in particular ^{45}Fe are stopped in the third silicon detector which is a double sided strip detector (DSSSD) providing not only the energy but also x and y coordinate of the particle. The last detector is a thick one, serving as a veto for light particles that have a longer range than the studied ones. This fairly compact detection system is surrounded by four large volume Ge detectors disposed in cross in a close geometry. The DSSSD is very important in order to unambiguously correlate the implant of the particle of interest and its decay, be it proton, 2p, beta or gamma.

The tuning of SISSI and the LISE spectrometer are of utmost importance in order to find the best conditions to produce select and transport to the detecting device the (very) rare species under study with the maximum efficiency. This operation is continuously assisted by the simulation program LISE and confronted with the results obtained for more abundant species as a reliability check. The identification is made by energy loss and time of flight of the implanted nuclei. A high degree of redundancy is introduced by the use of more detectors and two different time references. Only the particles satisfying all criteria are accepted. Two independent and time correlated acquisition systems are used for implanted ions and for detection of their radioactivity. Fig. 7 shows an identification matrix obtained for the observation of ^{48}Ni at GANIL in 2004 [10]. The four implantation events could be correlated with 4 decay events. One of them is compatible with a 2p emission for both the estimated energy and the decay time. It is worth mentioning that the production cross section for ^{48}Ni estimated from the integral current and the observed events is of only 0.04pb.

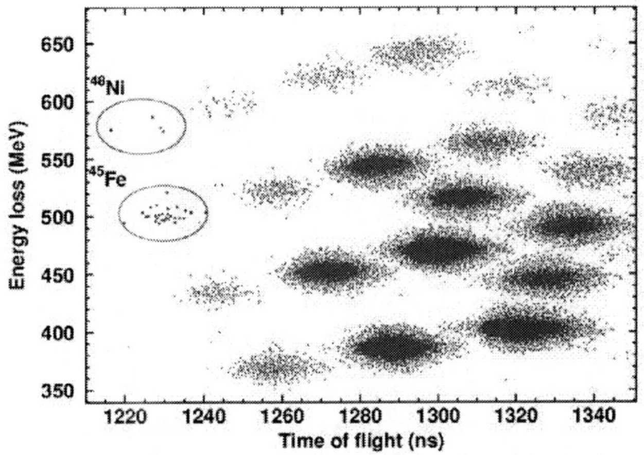

FIGURE 7: Identification matrix (energy loss versus time of flight) from the 2004 experiment at GANIL showing the 2p emission candidates, ^{45}Fe and ^{48}Ni.

RESULTS: ^{45}FE AND ^{54}ZN

Besides the result for ^{48}Ni already mentioned experimental efforts during 2000-2005 lead to a clear evidence for the 2p decay of ^{45}Fe and of ^{54}Zn. These results are presented together with the main measured quantities (decay energy, decay time, branching ratio) in fig. 8. This information is very valuable and put already severe constraints on the theories aiming to describe 2p radioactivity. Some of the previous evaluations have been already re-examined. However, for clarifying the mechanism of this new decay type it is necessary to measure the complete kinematics of the decay events. This can not be done with the above described detection and therefore constituted the main motivation for creating a new tool, a time projection chamber (TPC) that will be described below.

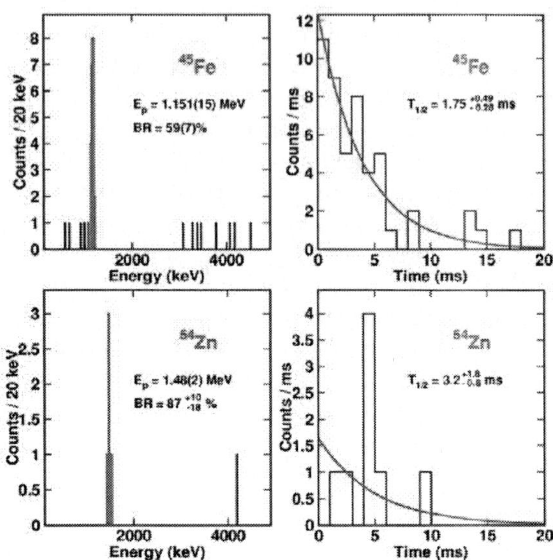

FIGURE 8: Top panels: Results for ^{45}Fe. Left: Energy spectra of observed proton decays. In red are the 2p emissions, in black the emission after beta decay. Right: Half-life of ^{45}Fe as determined with all decay events. Lower panels: Results for ^{54}Zn showing energy (left) and time (right) spectra with the same convention.

TPC DESCRIPTION AND FIRST EXPERIMENTAL RESULTS

The ions of interest are injected and stopped in a parallelepiped shaped chamber filled up with P10 gas (90% Ar – 10% CH_4) at atmospheric pressure. Once stopped, in a few milliseconds they may undergo 2p decay. Upper and lower planes are respectively cathode and anode electrodes, placed at a distance of 15 cm from each other. The electrons from the ionization created by this three-arm event (incoming ion and the two emitted protons) will start drifting in the electric field toward the anode. The lower electrode consists of a double sided micro-strip detector developed at CERN. It is made up of a grid of micro-strips of copper onto a thin support (200 μm of capton) for the upper part and a grid of similar micro-strips on the lower face, disposed perpendicularly with respect to the upper ones. The width and pitch of the grids are of 200 μm. Everything is placed onto an epoxy frame. There are 768 strips on each side, covering about 15 cm (see fig. 9). The electrons are in the vicinity of the anode multiplied by four consecutive layers of gas electron multipliers (GEM). This electron flux arrives on the anode and creates the projection of the event on the anode plane. The upper and lower strips will share the electrons roughly equally and the created signal is transmitted to a preamplifier attached to each strip.

The drifting time is a measure of the height above the anode plane at which the ionization was produced. Measuring the arrival time of the electrons at the anode with respect to the arrival of the first electrons one can reconstruct the z coordinate.

A sketch of the TPS is shown in Fig.10 together with the main features of the detector. The associated integrated electronics is made out of chips (VAT/TAT by

IDEAS-Norway) with 32 energy and time channels each. The time to amplitude converter covers a window of 10 μs and the data acquisition is realized in standard PXI-VME.

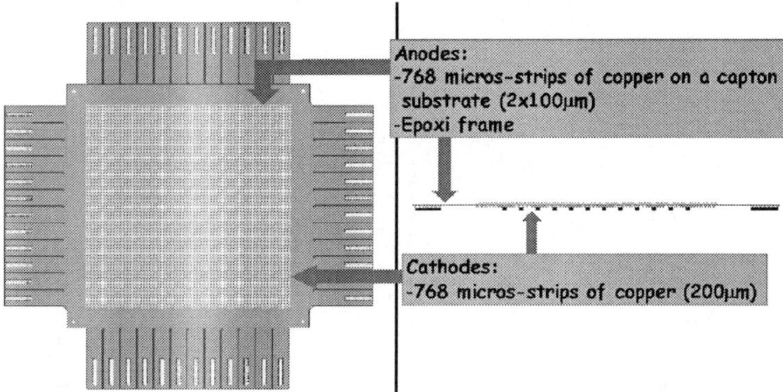

FIGURE 9: Schematic view of the double-sided micro strip gas detector. Some details are also given.

After tests, the detector was used in an experiment at GANIL in September 2006 [11]. Though not all channels worked perfectly, the trajectories could be successfully reconstructed. Fig. 11 shows one event of 2p decay for ^{45}Fe and one event of β-2p decay of ^{43}Cr. The upper panels indicate the x and y position of the stopped ion, while the lower panels indicate the x and y coordinates of the two protons.

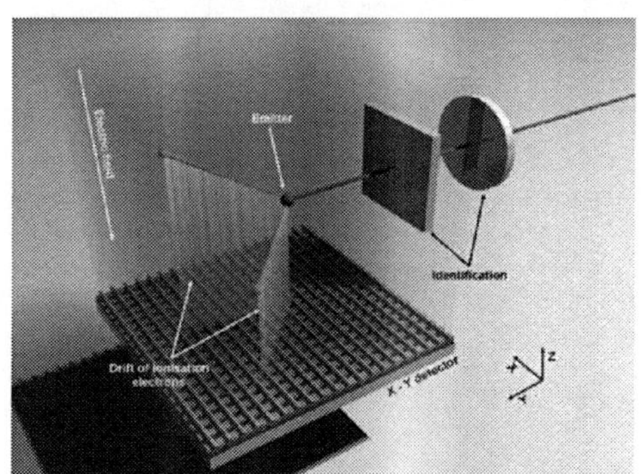

FIGURE 10: Schematic representation of the TPC. The beam enters from the right, traverses two identification silicon detectors and is stopped in the active volume of the TPC. There the radioactive decay happens and the particles are view by the 2D detector on the bottom of the TPC

The statistics accumulated during the experiment was low; therefore more efforts are needed in order to obtain enough information about the correlation between the two protons. On the other hand, the experiment proved that the detector works at the foreseen parameters and is an excellent and unique tool for this kind of studies.

SUMMARY

We have presented 2p radioactivity studies carried out over the last few year at the LISE3 separator of GANIL. These experiments, together with an experiment performed at the FRS of GSI, lead to the discovery of 2p radioactivity for ^{45}Fe and ^{54}Zn, and possibly also of ^{48}Ni. By means of a time-projection chamber, the presence of two protons in this decay could be demonstrated clearly. Further studies will try to determine the degree of angular and energy correlations between the two protons.

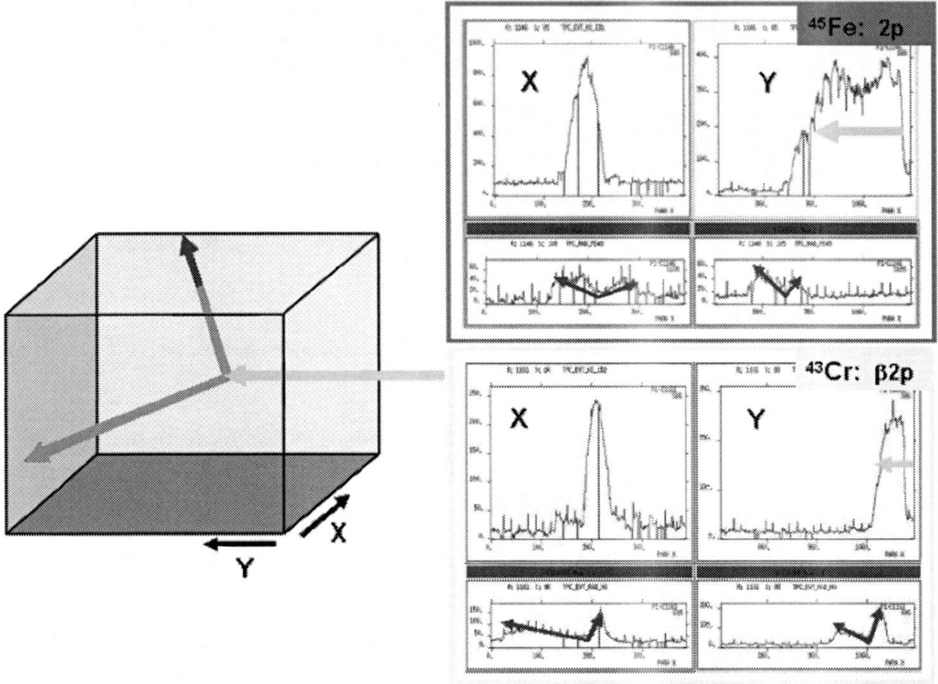

FIGURE 11: First results obtained with the TPC for ^{45}Fe 2p decay (upper right panel) and ^{43}Cr β-2p decay (lower right panel). The upper part of the left figures shows the results for x and y coordinates of the implanted ions. The green arrow indicates the direction of the incoming ions into the gas volume. Lower parts show the results for the two proton decay events. The blue arrows indicate the directions as in the schematically shown TPC volume on the left.

REFERENCES

1. S. Hofmann et al., Z. Phys. A305 (1982) 111
2. H.J. Rose and G.A. Jones, Nature 307 (1984) 245
3. V. Goldanskii, Nucl. Phys. 19 (1960) 482
4. B.A. Brown, F.C. Barker, and D.J. Millner, Phys. Rev. C65 (2002) 051309
5. J. Rotureau, J. Okolowicz, M. Ploszajczak, Nucl. Phys. A767 (2006) 13
6. L.V. Griogorenko et al., Phys. Rev. C64 (2001) 054002
7. J. Giovinazzo et al., Phys. Rev. Lett. 89 (2002) 102501
8. M. Pfützner et al., Eur. Phys. J. A14 (2002) 279
9. B. Blank et al., Phys. Rev. Lett. 94 (2005) 232501
10. C. Dossat et al., Phys. Rev. C72 (2005) 054315
11. J. Giovinazzo et al., Phys. Rev. Lett. 99 (2007) 102501

Shell Corrections Stabilizing Superheavy Nuclei and Semi-spheroidal Atomic Clusters

Dorin N Poenaru

Horia Hulubei National Institute of Physics and Nuclear Engineering (IFIN-HH), P O Box MG-6, RO 077125 Bucharest-Magurele, Romania

Abstract. The macroscopic-microscopic method is used to illustrate the shell effect stabilizing superheavy nuclei and to study the stability of semi-spheroidal clusters deposited on planar surfaces. The alpha decay of superheavy nuclei is calculated using three models: the analytical superasymmetric fission model; the universal curve, and the semiempirical formula taking into account the shell effects. Analytical relationships are obtained for the energy levels of the new semi-spheroidal harmonic oscillator (SSHO) single-particle model and for the surface and curvature energies of the semi-spheroidal clusters. The maximum degeneracy of the SSHO is reached at a super-deformed prolate shape for which the minimum of the liquid drop model energy is also attained.

Keywords: macroscopic-microscopic method, shell and pairing corrections, superheavy nuclei, deposited atomic clusters on surfaces
PACS: 24.75.+1, 21.60.Ev, 21.10.Pc, 36.40.Qv, 31.50.-x

INTRODUCTION

In 1939, based on the macroscopic liquid drop model (LDM) [1, 2], Lise Meitner understood that very heavy nuclei (with $A \geq 100$) would never exist because there would be no potential barrier against spontaneous fission. The expectations changed in the sixties (see [3] and the refs. therein) when nuclear scientists were able to take properly into account not only the collective motion of nucleons (LDM) but also their individual characteristics, described within a quantum single-particle model like the two center shell model. When a shell and pairing correction energy, according to the macroscopic-microscopic (M-M) approach [4], is added to the LDM deformation energy, a potential barrier shows up stabilizing superheavy nuclei. The double shell closures have been initially $Z = 114$, $N = 184, 196$. Other islands were thought to be centered on $Z = 164$, $N = 318$. Presently we should also add the possible spherical magic numbers $N = 172$, $Z = 120, 126$ and the deformed ones $N = 152, 162$ and $Z = 108$.

The shell effect is so important, that in fact the majority of proton-rich superheavy nuclei produced until now are mainly decaying by α emission instead of fission [5, 6, 7]. We can estimate the α-decay lifetimes by using the methods based on the fission approach (see [8] and the refs. therein) we developed to predict and study heavy particle radioactivities [9, 10] (analytical superasymmetric fission (ASAF) model, universal curves (UNIV)), and the semi-empirical formula taking into account the shell effects (semFIS). The experimental data for 142 transuranium α emitters including superheavies are compared with our calculations [11, 12, 13].

The M-M method may be also adapted to study the stability of atomic clusters. We developed a new single-particle shell model [14]. The semi-spheroid is in the same

CP972, *Exotic Nuclei and Nuclear/Particle Astrophysics (II)—Proceedings of the Carpathian Summer School of Physics 2007,* edited by L. Trache and S. Stoica

time an equipotential surface of this model and the basic shape of the corresponding LDM [15]. In all previous studies the maximum degeneracy of the quantum states was reached for a spherical shape, explaining the high stability of the doubly magic nuclei or of the metal clusters with spherical closed shells. We obtained a striking result: the maximum degeneracy of the semi-spheroidal harmonic oscillator (SSHO) occurs at a superdeformed prolate shape (semiaxes ratio a/c=1/2). In the same time the superdeformed prolate semi-spheroid is the equilibrium shape within LDM. Some of the shapes of atomic clusters deposited on planar surfaces, observed with a scanning probe microscope, are very similar to a semi-spheroid [16, 17].

SUPERHEAVY NUCLEI

Yukawa-plus-exponential Model Macroscopic Energy

In a fission process $^{A}Z \rightarrow {}^{A_1}Z_1 + {}^{A_2}Z_2$ the phenomenological energy E_{Y+EM} is calculated within Yukawa-plus-exponential model (Y+EM) [18] by taking into account the difference between charge and mass asymmetry [19]. The deformation energy is defined as

$$E_{Y+EM} = (E_Y - E_Y^0) + (E_c - E_c^0) = E_Y^0[B_Y - 1 + 2X(B_c - 1)] \tag{1}$$

where $E_Y^0 = a_2 A^{2/3}\{1 - 3x^2 + (1 + 1/x)[2 + 3x(1 + x)]\exp(-2/x)\}$ and $E_c^0 = a_c Z^2 A^{-1/3}$ are energies corresponding to spherical shape and $a_2 = a_s(1 - \kappa I^2)$, $I = (N - Z)/A$, $x = a/R_0$, $R_0 = r_0 A^{1/3}$. The relative Yukawa and Coulomb energies B_Y and B_c are functions of the nuclear shape. For a binary fragmentation with charge densities ρ_{1e} and ρ_{2e}, one has [19] a relative energy

$$B_Y = \frac{E_Y}{E_Y^0} = \frac{a_{21}}{a_{20}} B_{Y1} + \frac{\sqrt{a_{21}a_{22}}}{a_{20}} B_{Y12} + \frac{a_{22}}{a_{20}} B_{Y2} \tag{2}$$

with axially-symmetric shape-dependent terms expressed by triple integrals [8].

In a similar way the Coulomb relative energy is given by

$$B_c = \frac{E_c}{E_c^0} = \left(\frac{\rho_{1e}}{\rho_{0e}}\right)^2 B_{c1} + \frac{\rho_{1e}\rho_{2e}}{\rho_{0e}^2} B_{c12} + \left(\frac{\rho_{2e}}{\rho_{0e}}\right)^2 B_{c2} \tag{3}$$

with detailed expressions given elsewhere [8]. The integrals are computed numerically by Gauss-Legendre quadratures.

Microscopic Shell and Pairing Corrections

We would like to outline the calculations of the shell [4] and pairing [20] corrections $\delta E = \delta U + \delta P$ leading to the total deformation energy

$$E_{def}(R, \eta) = E_{Y+EM}(R, \eta) + \delta E(R, \eta) \tag{4}$$

More details are given in [21]. By choosing two intersected spheres for nuclear shape parametrization one can take the separation distance between fragment centers, R, as a deformation parameter. Initially, for a parent nucleus $R_i = R_0 - R_2$. At the touching point $R_t = R_1 + R_2$. The mass asymmetry $\eta = (A_1 - A_2)/A$.

The two-center shell model [22] gives at every pair of coordinates (R, η) the sequence of doubly degenerate discrete energy levels $\varepsilon_i = E_i/\hbar\omega_0^0$ in units of $\hbar\omega_0^0 = 41A^{-1/3}$, arranged in order of increasing energy. The smoothed-level distribution density is obtained by averaging the actual distribution over a finite energy interval $\Gamma = \gamma\hbar\omega_0^0$, with $\gamma \simeq 1$,

$$\tilde{g}(\varepsilon) = \left\{ \sum_{i=1}^{n_m} [2.1875 + y_i(y_i(1.75 - y_i/6) - 4.375)] e^{-y_i} \right\} (1.77245385\gamma)^{-1} \quad (5)$$

where $y = x^2 = [(\varepsilon - \varepsilon_i)/\gamma]^2$. The summation is performed up to the level n_m fulfilling the condition $|x_i| \geq 3$. The Fermi energy, $\tilde{\lambda}$, and the total energy of this distribution are given by

$$N_p = 2\int_{-\infty}^{\tilde{\lambda}} \tilde{g}(\varepsilon) d\varepsilon \; ; \; \tilde{u} = \tilde{U}/\hbar\omega_0^0 = 2\int_{-\infty}^{\tilde{\lambda}} \tilde{g}(\varepsilon)\varepsilon d\varepsilon \quad (6)$$

with $N_p = Z$ for proton levels and $N_p = A - Z$ for neutron levels, leading to a non-linear equation in $\tilde{\lambda}$, solved numerically. In units of $\hbar\omega_0^0$ the shell corrections are calculated for each pair (R, η):

$$\delta u(n, R, \eta) = \sum_{i=1}^{n} 2\varepsilon_i(R, \eta) - \tilde{u}(n, R, \eta) \quad (7)$$

$n = N_p/2$ particles. Then $\delta u = \delta u_p + \delta u_n$.

Similarly, for pairing corrections we take the doubly degenerate levels $\{\varepsilon_i\}$ in units of $\hbar\omega_0^0$. $Z/2$ levels are occupied with n levels below and n' above Fermi energy contributing to pairing, $n = n' = \Omega\tilde{g}_s/2$. The cutoff energy, $\Omega \simeq 1 \gg \tilde{\Delta} = 12/\sqrt{A}\hbar\omega_0^0$. The gap Δ and the Fermi energy λ are solutions of the BCS system [23, 20] of two equations:

$$0 = \sum_{k_i}^{k_f} \frac{\varepsilon_k - \lambda}{\sqrt{(\varepsilon_k - \lambda)^2 + \Delta^2}} \; ; \; \frac{2}{G} = \sum_{k_i}^{k_f} \frac{1}{\sqrt{(\varepsilon_k - \lambda)^2 + \Delta^2}} \quad (8)$$

where $k_i = Z/2 - n + 1$, $k_f = Z/2 + n'$, and $2/G \simeq 2\tilde{g}(\tilde{\lambda}) \ln\left(2\Omega/(\tilde{\Delta})\right)$.

As a consequence of the pairing correlation, the levels below the Fermi energy are only partially filled, while those above the Fermi energy are partially empty. The occupation probability by a quasiparticle (v_k^2) or a hole (u_k^2) is given by $v_k^2 = [1 - (\varepsilon_k - \lambda)/E_k]/2$; $u_k^2 = 1 - v_k^2$. The quasiparticle energy is expressed as $E_v = \sqrt{(\varepsilon_v - \lambda)^2 + \Delta^2}$.

The pairing correction $\delta p = p - \tilde{p}$, represents the difference between the pairing correlation energies for the discrete level and for the continuous level distributions

$$p = \sum_{k=k_i}^{k_f} 2v_k^2\varepsilon_k - 2\sum_{k=k_i}^{Z/2} \varepsilon_k - \frac{\Delta^2}{G} \; ; \; \tilde{p} = -(\tilde{g}\tilde{\Delta}^2)/2 = -(\tilde{g}_s\tilde{\Delta}^2)/4 \quad (9)$$

Compared to shell correction, the pairing correction is out of phase and smaller leading for R = constant to a smoother total curve $\delta e(\eta) = \delta u(\eta) + \delta p(\eta)$ where $\delta p = \delta p_p + \delta p_n$. The PES versus the normalized separation distance $\xi = (R - R_i)/(R_t - R_i)$

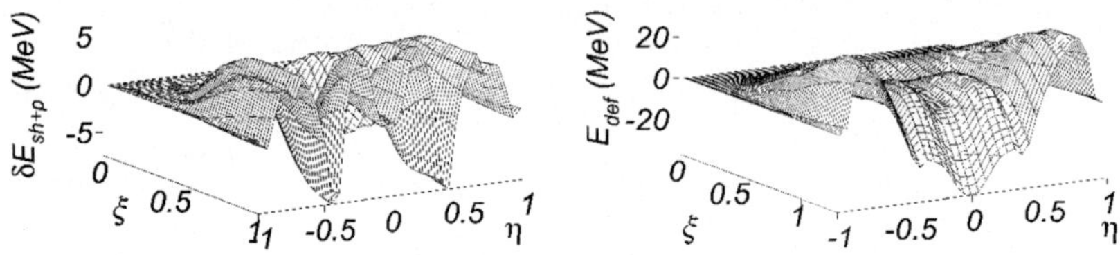

FIGURE 1. Potential energy surfaces of $^{284}112$ vs. separation distance $\xi = (R - R_i)/(R_t - R_i)$ and mass asymmetry $\eta = (A_1 - A_2)/(A_1 + A_2)$. Shell + Pairing corrections (left), and total deformation energy (right). The very deep valley around $\eta = 0.465$ shown at the left-hand side corresponds to the doubly magic numbers of the heavy fragment ^{208}Pb with $N_1 = 126$, $Z_1 = 82$.

and the mass asymmetry η are plotted in Fig. 1. Two valleys present in the shell plus pairing correction PES (left) at a large value of mass asymmetry are not seen on the total deformation energy (right) because they are added to the Businaro-Gallone mountain of the Y+EM energy. The very deep valley around $\eta = 0.465$ shown at the left-hand side corresponds to the doubly magic numbers of the heavy fragment ^{208}Pb with $N_1 = 126$, $Z_1 = 82$. The shell and pairing effects are essential for stabilizing the properties of a superheavy nucleus.

Alpha-decay Half-lives

Fission theory, and in particular the ASAF model, was very successful in predicting half-lives against heavy particle radioactivities [24, 25] and reproducing lifetimes for α-decay [26]. All experiments on cluster radioactivities performed until now confirmed the predicted half-lives. From the beginning our fission models ASAF and the numerical (NuSAF) superasymmetric fission model, as well as the universal curve $\log T = f(\log P)$ [27] have been first tested on α decay. There are many semiempirical relationships [26, 28] allowing to estimate the disintegration period if the kinetic energy of the emitted particle $E_\alpha = QA_d/A$ is known. Q is the released energy and A_d, A are the mass numbers of the daughter and parent nuclei. The description of data in the neighbourhood of the magic proton and neutron numbers, where the errors of the other relationships are large, was improved by deriving a new formula (SemFIS) based on the fission theory of α-decay [29]. In the region of superheavy nuclei the Viola-Seaborg (SemV-S) [30] formula (SemV-S) is the most frequently used.

We rely on the experimental values of atomic masses [31], released energies and halflives [25, 32, 33, 34, 35, 36]. The deviations of calculated values of half-lives against α-decay from the experimental ones for 136 even-even nuclei are presented in the Refs. [11, 12, 13]. Three kinds of fission models (ASAF, UNIV, and SemFIS) are compared with the SemV-S. An overall estimation of the accuracy, gives the standard rms deviation of $\log T$ values σ. For the 163 data points we obtained: 0.390 for ASAF; 0.325 for UNIV; 0.192 for SemFIS, and 0.556 for SemV-S.

SEMI-SPHEROIDAL ATOMIC CLUSTER

By analyzing some shapes of cluster deposited on a surface obtained by using scanning probe microscopy [16, 17], we expect that a semi-spheroid with the z axis of cylindrical symmetry oriented perpendicularly on the surface plane may be a good starting point to study some of the possible shapes encountered in practice if the interaction energy between the deposited cluster and the substrate may be neglected. The oblate and prolate semi-spheroid is in the same time an equipotential surface of the single-particle shell model and the basic shape of the corresponding liquid drop model (LDM).

The electronic shell structure in monovalent free-electron metal clusters [37] has shown a strong analogy with the single-particle states of atomic nuclei, despite gross differences in the physical forces binding the two systems. Several theories and computation techniques from nuclear physics could be adapted to atomic clusters [38, 39]. We should mention in particular the M-M method [4, 40, 41] different from that used in the present work.

Liquid Drop Model

The cylindrical symmetry of the shapes we consider allows to use the dimensionless coordinates (ρ, z), the length scale being the radius of a semi-sphere with the same volume, $R_s = 2^{1/3}R_0 = 2^{1/3}r_sN^{1/3}$, in which r_s is the Wigner-Seitz radius (2.117 Å for Na [42, 43]) and $\rho = \rho(z)$ is the surface equation given by

$$\rho^2 = \begin{cases} (a/c)^2(c^2 - z^2) & z \geq 0 \\ 0 & z < 0 \end{cases} \tag{10}$$

where a is the minor (major) semiaxis for prolate (oblate) semi-spheroid and c is the major (minor) semiaxis for prolate (oblate) semi-spheroid. Volume conservation leads to $a^2c = 1$. The (quadropolar) deformation parameter, δ [44], and the eccentricity, e, are defined by

$$a = \left(\frac{2-\delta}{2+\delta}\right)^{1/3} \; ; \; c = \left(\frac{2+\delta}{2-\delta}\right)^{2/3} \; ; \; e^2 = \begin{cases} 1 - a^2/c^2 & prolate \; (a < c) \\ a^2/c^2 - 1 & oblate \; \; (a > c) \end{cases} \tag{11}$$

The deformation energy with respect to semi-spherical shape of a neutral cluster is given by

$$E - E^{s0} = (E_s - E_s^{s0}) + (E_c - E_c^{s0}) = E_s^{s0}\left(B_{surf}^s - 1\right) + E_c^{s0}\left(B_{curv}^s - 1\right) \tag{12}$$

where for a semi-sphere one has

$$E_s^{s0} = 3\pi R_s^2\sigma = 3\pi 4^{1/3}R_0^2\sigma = \frac{3}{4^{2/3}}E_s^0 \; ; \; E_c^{s0} = 2\pi R_s\gamma_c = 2\pi 2^{1/3}R_0\gamma_c = \frac{E_{curv}^0}{4^{1/3}} \tag{13}$$

with E_s^0 and E_{curv}^0 the surface and curvature energy of a sphere with the same volume.

The liquid drop part (volume, surface, and curvature terms) of the binding energy of Na semi-spherical clusters are given, in eV, by:

$$E_{sN} = -2.252N + \frac{3}{4^{2/3}} 0.541 N^{2/3} + \frac{1}{4^{1/3}} 0.154 N^{1/3} \tag{14}$$

When $a > c$ (oblate semi-spheroid), $e^2 = a^2/c^2 - 1$, we obtain for the shape dependent terms:

$$B_{surf}^s = \frac{a}{3}\left[2a + \frac{c}{e}\ln\left(e + \frac{a}{c}\right)\right] \; ; \; B_{curv}^s = \frac{c}{2} + \frac{1}{2ce^2}\left(\frac{e}{c}\,\mathrm{arctan}\,e\right) = \frac{c}{2} + \frac{a^2}{2ce}\,\mathrm{arctan}\,e \tag{15}$$

When $c > a$ (prolate semi-spheroid), $c^2 - a^2 = e^2c^2$, we get

$$B_{surf}^s = \frac{a}{3}\left(2a + \frac{c}{e}\arcsin e\right) \; ; \; B_{curv}^s = \frac{c}{2} + \frac{a^2}{4ce}\ln\left|\frac{1+e}{1-e}\right| \tag{16}$$

For a semi-spherical shape $B_{surf}^s = B_{curv}^s = 1$. Typical variations of the deformation energy, E_{LD}, with the shape coordonate δ are shown in figure 2, for semi-spheroidal sodium clusters. The minimum corresponds to $\delta = 2/3$, that is a superdeformed ($a/c = 1/2$) prolate shape.

Semi-spheroidal Harmonic Oscillator (SSHO)

The threedimensional SSHO was developed [14] by using just the half of the parabolic potential along the axis of symmetry $V_z(z)$ without taking into account an additional term proportional to $(\mathbf{l}^2 - \langle \mathbf{l}^2 \rangle_n)$ which was considered [44] in order to explain the experimental magic numbers larger than 40. In this way we obtain analytical solutions for the electronic levels. The potential along the symmetry axis, $V_z(z)$, has a wall of an infinitely large height at $z = 0$, and concerns only positive values of z, implying opacity of the surface. As a consequence, only negative parity states along the z axis (with integer quantum number $n_z = 1, 3, 5, \ldots n$) are allowed. In units of $\hbar\omega_0$ the variation of eigenvalues with deformation, $\varepsilon(\delta) = E(\delta)/(\hbar\omega_0)$, is given by

$$\varepsilon = \frac{2}{(2-\delta)^{1/3}(2+\delta)^{2/3}}\left[n + \frac{3}{2} + \delta\left(n_\perp - \frac{n}{2} + \frac{1}{4}\right)\right] \tag{17}$$

Each level may be labeled by two quantum numbers, nonnegative integers, $n, n_\perp$. The eq. (17) is the same as for the spheroidal harmonic oscillator, but for semi-spheroidal shapes one should only allow the values of n and $n_\perp$ for which $n_z = n - n_\perp \geq 1$ are odd numbers.

For the semi-spherical shape ($\delta = 0$) the magic numbers are 2, 6, 14, 26, 44, 68, 100, 140, ... For $\delta = -0.4$ ($a/c = 3/2$) the magic numbers are: 2, 6, 12, 22, 36, 54, 78, 108, 144, ... They are: 2, 8, 18, 20, 34, 38, 50, 58, 64, 80, 92, 100, ... for $\delta = 0.4$ ($a/c = 2/3$) and 2, 8, 10, 14, 22, 26, 46, 54, 66, 84, 96, 114, 138, 156, ... for the hyperdeformed prolate shape $\delta = 1$ ($a/c = 1/3$). The largest degeneracy is obtained for a superdeformed prolate shape with $\delta = 2/3$ ($a/c = 1/2$) with the magic numbers 2, 8, 20, 40, 70, 112, 168, ... identical to those of the spherical harmonic oscillator. The single-particle shell gap of the SSHO in eV is given by $\hbar\omega_0(N) = \frac{13.72}{r_s R_s}$ where r_s and R_0 are expressed in Å.

Shell and Pairing Corrections. Total Deformation Energy

The deformation energy in the framework of the M-M method is given by macroscopic (LDM) + (shell + pairing) corrections

$$E_{def} = E_{LDM} + \delta E = E_{LDM} + \delta U + \delta P \tag{18}$$

There is a difference between the calculations of $\tilde{\Delta}$ for nuclear and atomic level schemes. For a sodium atomic cluster $\tilde{\Delta} = 1.3416/\sqrt{N}\hbar\omega_0$ unlike $\tilde{\Delta} = 12/\sqrt{A}\hbar\omega_0^0$ in the nuclear case. The mass number A is replaced by the number of atoms N for a monovalent metalic cluster. As a result of taking into account the pairing corrections, δP, the total shell

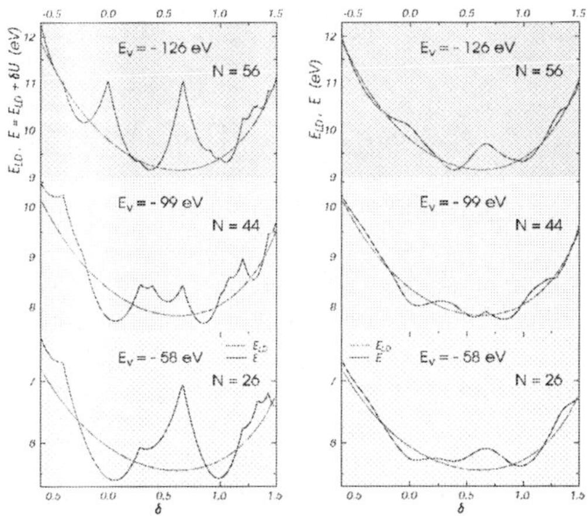

FIGURE 2. Liquid drop model deformation energy, E_{LD}, the total deformation energy including shell corrections, $E = E_{LD} + \delta U$ (left-hand side), for SSHO energy levels vs. δ. On the right-hand side the pairing corrections are included, $E = E_{LD} + \delta U + \delta P$. Three different numbers of Na atoms ar considered: $N = 26$, $N = 44$, and $N = 56$. The volume energy $E_v = E_v^0$ was not included in E_{LD}.

and pairing corrections, $\delta E = \delta U + \delta P$, are smoother compared to δU (compare the right-hand side of the figure 2 with the left-hand side one). The figure 2 refers to two semi-spheroidal Na clusters with semi-spherical magic numbers, $N = 26$ and $N = 44$, and a mid-shell one, $N = 56$. In fact $N = 26$ is also magic number at $\delta = 1$, $N = 44$ is very close to the magic number $N = 46$ at $\delta = 1$, and $N = 56$ is magic number at $\delta = -0.8/3$ and not far from $N = 58$ which is magic number at $\delta = 0.4$ and $N = 54$ — magic number at $\delta = 1$. In this way one can explain the position of minima of $E_{LD} + \delta U$ in the left-hand side part of the figure 2. The ground state deformation of Na_{26} is $\delta_{gs} = 0.97$, and for tw isomeric states it has $\delta_{is1} = 0.375$ and $\delta_{is2} = 0.06$. Similarly

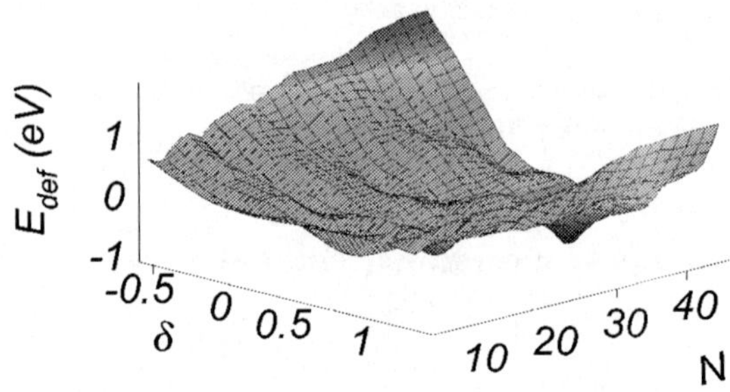

FIGURE 3. Potential energy surface $E_{def} = E_{LD} - E^{0s}_{LD} + \delta E$ versus the deformation δ and the number of atoms N for semi-spheroidal atomic clusters of sodium. E^{0s}_{LD} is the liquid drop energy of a semi-spherical shape.

for Na_{44}, $\delta_{gs} = 0.83$, $\delta_{is1} = 0.565$, and $\delta_{is2} = 0.07$. For Na_{56} we found $\delta_{gs} = 0.39$, and $\delta_{is} = 0.99$. All deformations are different from the LDM most stable value of $\delta = 0.63$ (prolate superdeformed); they are higher (hyperdeformed) in case of relatively small clusters with $N = 26, 44$.

The smoothing effect of the pairing corrections, δP, which is out of phase and smaller in amplitude compared to the shell corrections, δU, can be seen from the example given at the bottom of figure 2.

In figure 3 one can see a potential energy surface (PES) representing the deformation energy $E_{def} = E_{LD} - E^{0s}_{LD} + \delta E$ versus the deformation δ and the number of atoms N for semi-spheroidal atomic clusters of sodium. E^{0s}_{LD} is the liquid drop energy of a semi-spherical shape. From this figure it is clear that the deepest minimum corresponds to the super-deformed prolate semi-spheroid with $\delta = 2/3$ and $N = 40$. Also, one can see that the majority of semi-spheroidal clusters of Na with $N \leq 46$ will have the ground-state deformation around $\delta = 2/3$. Consequently, if some super-deformed semi-spheroidal shapes will be observed in an experiment of cluster deposition on a planar surface, the fact may be interpreted as an evidence for a negligible interaction energy between the deposited cluster and the surface material.

ACKNOWLEDGMENTS

This work was partly supported by Deutsche Forschungsgemeinschaft, Bonn, the network of excellence EXCELL of the European Commission, and by Ministry of Education and Research, Bucharest within the "IDEI" programme.

REFERENCES

1. L. J. W. S. Rayleigh, *Phil. Mag.* **14**, 184–186 (1878).
2. N. Bohr, *Nature* **137**, 344, 351 (1936).

3. W. Greiner, *Phys. Atom. Nucl.* **66**, 1009–1014 (2003).
4. V. M. Strutinsky, *Nuclear Physics, A* **95**, 420 (1967).
5. S. Hofmann, and G. Münzenberg, *Review of Modern Physics* **72**, 733–767 (2000).
6. Y. T. Oganessian, *J. Phys. G: Nucl. Part. Phys.* **34**, R165–R242 (2007).
7. K. Morita et al., *J. Phys. Soc. Jpn.* **76**, 043201 (2007).
8. D. N. Poenaru, editor, *Nuclear Decay Modes*, Institute of Physics Publishing, Bristol, UK, 1996.
9. A. Săndulescu, D. N. Poenaru, and W. Greiner, *Sov. J. Part. Nucl.* **11**, 528–541 (1980).
10. W. Greiner, and D. N. Poenaru, "Radioactivity," in *Encyclopedia of Condensed Matter Physics, Vol. 5*, edited by F. Bassani et al., Elsevier, Oxford, 2005, pp. 106–116.
11. D. N. Poenaru, I. H. Plonski, R. A. Gherghescu, and W. Greiner, *J. Phys. G: Nucl. Part. Phys.* **32**, 1223–1239 (2006).
12. D. N. Poenaru, R. A. Gherghescu, and N. Carjan, *Europhysics Letters (EPL)* **77**, 62001 (2007).
13. D. N. Poenaru, I. H. Plonski, and W.Greiner, *Phys. Rev., C* **74**, 014312 (2006).
14. D. N. Poenaru, R. A. Gherghescu, A. V. Solov'yov, and W. Greiner, E-print: arXiv:0704.0847v1 [physics.atm-clus], 2007.
15. D. N. Poenaru, R. A. Gherghescu, A. V. Solov'yov, and W. Greiner, *Europhysics Letters (EPL)* **79**, 63001 (2007).
16. K. Seeger, and R. E. Palmer, *Appl. Phys. Lett.* **74**, 1627–1629 (1999).
17. B. Bonanni, and S. Cannistraro, *J. Nanotechnology Online* **1**, 1–14 (November 2005), doi: 10.2240/azojono0105.
18. H. J. Krappe, J. R. Nix, and A. J. Sierk, *Phys. Rev. C* **20**, 992–1013 (1979).
19. D. N. Poenaru, M. Ivaşcu, and D. Mazilu, *Computer Physics Communications* **19**, 205–214 (1980).
20. M. Brack, J. Damgaard, A. S. Jensen, H. C. Pauli, V. M. Strutinsky, and C. Y. Wong, *Rev. Mod. Phys.* **44**, 320–405 (1972).
21. D. N. Poenaru, and W. Greiner, chap. 6 in [8], pp. 275–336.
22. R. A. Gherghescu, *Phys. Rev.* **C 67**, 014309 (2003).
23. J. M. Eisenberg, and W. Greiner, *Microscopic Theory of the Nucleus*, Nuclear Theory Vol. III, 2nd printing, North-Holland, Amsterdam, 1976.
24. D. N. Poenaru, Y. Nagame, R. A. Gherghescu, and W. Greiner, *Phys. Rev. C* **65**, 054308 (2002).
25. D. N. Poenaru, D. Schnabel, W. Greiner, D. Mazilu, and R. Gherghescu, *Atomic Data Nucl. Data Tables* **48**, 231–327 (1991).
26. D. N. Poenaru, E. Hourany, and W. Greiner, chap. 4 in [8], pp. 204–236.
27. D. N. Poenaru, and W. Greiner, *Physica Scripta* **44**, 427–429 (1991).
28. A. Parkhomenko, and A. Sobiczewski, *Acta Phys. Polonica B* **36**, 3095–3108 (2005).
29. D. N. Poenaru, and M. Ivaşcu, *Journal de Physique (Paris)* **44**, 791–796 (1983).
30. V. E. Viola Jr., and G. T. Seaborg, *J. Inorg. Nucl. Chem.* **28**, 741 (1966).
31. G. Audi, A. H. Wapstra, and C. Thibault, *Nucl. Phys., A* **729**, 337–676 (2003).
32. A. Rytz, *Atomic Data Nucl. Data Tables* **47**, 205–239 (1991).
33. G. Audi, O. Bersillon, J. Blachot, and A. H. Wapstra, *Nucl. Phys., A* **729**, 3–128 (2003).
34. M. Gupta, and T. W. Burrows, *Nuclear Data Sheets* **106**, 251–365 (2005).
35. G. Shanmugam, S. Sudhakar, and S. Niranjani, *Phys. Rev.* **C 72**, 034310 (2005).
36. P. R. Chowdhury, C. Samanta, and D. N. Basu, *Phys. Rev.* **C 73**, 014612 (2006).
37. W. D. Knight et al., *Phys. Rev. Lett.* **52**, 2141–2143 (1984).
38. R. Schmidt, et al., editors, *Nuclear Physics Concepts in the Study of Atomic Cluster Physics*, vol. 404, Lecture Notes in Physics, Springer, Berlin, 1992.
39. J. P. Connerade, A. V. Solov'yov, and W. Greiner, *Europhysics News* **33**, 200 (2002).
40. H. Koizumi, S. Sugano, and Y. Ishii, *Z. Phys. D* **28**, 223–234 (1993).
41. C. Yannouleas, and U. Landman, *Phys. Rev. B* **48**, 8376–8387 (1993).
42. M. Brack, *Phys. Rev. B* **39**, 3533–3542 (1989).
43. C. Yannouleas, and U. Landman, *Phys. Rev. B* **51**, 1902–1917 (1995).
44. K. L. Clemenger, *Phys. Rev. B* **32**, 1359–1362 (1985).

5. SUBNUCLEONIC DEGREES OF FREEDOM

pp Collisions at RHIC

C. A. Gagliardi

Cyclotron Institute, Texas A&M University, College Station, TX, 77843, USA)

Abstract. RHIC, the world's first polarized hadron collider, has opened a new window to explore the spin structure of the nucleon. Collisions with longitudinally polarized beams are providing information about the polarization of the gluons in the proton and will provide information about the polarization of the anti-quarks. Collisions with transversely polarized beams may be yielding hints of parton orbital motion. Measurements with unpolarized beams are exploring the range of applicability of perturbative QCD and determining the non-perturbative inputs that are needed for precise QCD calculations. A brief overview is given of results to date and planned future directions.

Keywords: Gluon polarization; Sivers effect; Collins effect; inclusive cross sections; fragmentation functions.
PACS: 13.88.+e, 13.85.Ni, 13.87.Ce, 12.38.Qk.

INTRODUCTION

RHIC, the Relativistic Heavy-Ion Collider, is a very flexible accelerator facility. It is capable of colliding heavy ion beams at center-of-mass energies of up to 200 GeV/nucleon-nucleon pair or proton beams at center-of-mass energies of up to 500 GeV. RHIC can study collisions of identical nuclei (*e.g.*, p+p, Cu+Cu, Au+Au) or different nuclei (*e.g.*, d+Au). Furthermore, the proton beams are polarized, making RHIC the world's first polarized hadron collider.

RHIC was built with three goals in mind: to search for and study the quark-gluon plasma, to explore the partonic structure of the proton, and to determine the partonic structure of nuclei. This talk provides a very brief overview of the RHIC effort to explore the partonic structure of the proton. The investigations of the quark-gluon plasma are the subject of other talks at this meeting [1,2].

The two RHIC beams cross at six interaction regions where collisions can be studied. Initially, experiments were mounted at four of those six points. Two large experiments, PHENIX and STAR, are general purpose facilities which provide complementary abilities to measure many features of p+p and heavy ion collisions. In contrast, the two small experiments, BRAHMS and PHOBOS, were designed with specific limited physics goals in mind. The small experiments have now completed their programs and been decommissioned. BRAHMS, PHENIX, and STAR have participated in the p+p part of the program that is discussed here.

CP972, *Exotic Nuclei and Nuclear/Particle Astrophysics (II)—Proceedings of the Carpathian Summer School of Physics 2007,* edited by L. Trache and S. Stoica

UNPOLARIZED PROTON COLLISIONS

Unpolarized p+p collisions at RHIC provide "baseline" physics. For the heavy-ion program, they establish an essential baseline for comparison in order to identify what phenomena in heavy-ion collisions are novel. For the p+p program, they establish the applicability of perturbative QCD (pQCD) to interpret results from polarized p+p collisions and constrain the non-perturbative inputs that are required to perform pQCD calculations.

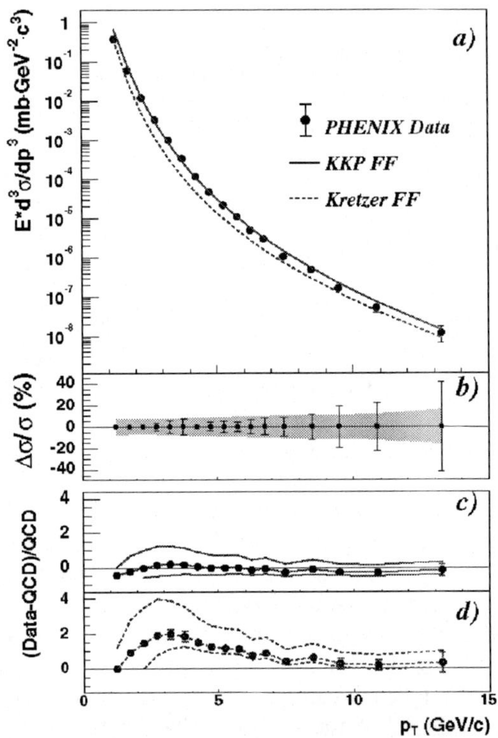

FIGURE 1. Measurement of the inclusive π^0 production cross section in mid-rapidity p+p collisions at 200 GeV. Panels (c) and (d) compare the results to pQCD predictions based on the KKP and Kretzer fragmentation functions, respectively. This figure is taken from Ref. [3].

Particle production in p+p collisions at RHIC has been studied over a broad range of transverse momenta (p_T) and pseudo-rapidity (η). Measurements have now been reported for pions, kaons, protons, lambdas, direct photons, and jets. For example, an early PHENIX result [3] demonstrated that the cross section for mid-rapidity π^0 production agrees with pQCD predictions over eight orders of magnitude, as shown in Fig. 1. Recently, PHENIX has extended this measurement out to $p_T \sim 20$ GeV/c, and the good agreement continues [4]. This overall agreement is a general feature of all similar comparisons that have been performed to date. But it is nonetheless important to note from the lower panels of Fig. 1 that quantitative agreement seems to require a particular choice of fragmentation functions, in this case the KKP fragmentation functions [5].

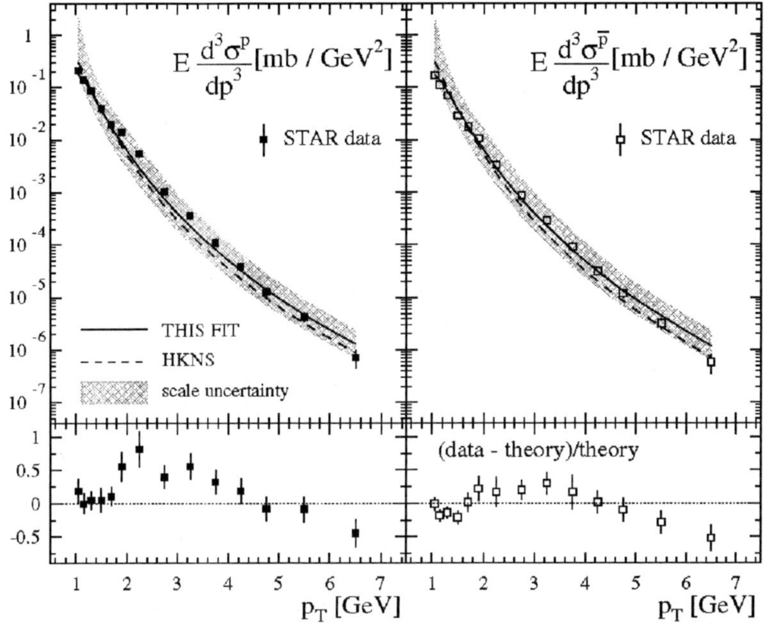

FIGURE 2. A recent fit to mid-rapidity inclusive proton (left) and anti-proton (right) cross section measurements in p+p collisions at 200 GeV. The fit was performed while deriving a new set of global proton and charged hadron fragmentation fuctions. This figure is taken from Ref. [6].

Inclusive hadron calculations in pQCD require non-perturbative inputs that must be taken from experiment in the form of parton distribution functions and fragmentation functions. Over the years, several global analyses have determined parton distribution functions for the proton with high precision. In contrast, to date fragmentation functions are only determined with modest precision. Thus, one needs to take care that agreement as illustrated in Fig. 1 is not fortuitous. In fact, much of the new high-p_T inclusive hadron data coming from RHIC are sensitive to fragmentation functions in regions that are not well constrained by previous measurements. For example, Fig. 2 illustrates a recent fit [6] to STAR mid-rapidity proton and anti-proton production data that was performed as part of a new global analysis of proton fragmentation functions. Calculations based on previous fragmentation functions either underpredict the observed yield by up to an order of magnitude [5] or overpredict it by a factor of two [7]. Those previous fragmentation functions were derived from very similar input data sets that did not include the recent RHIC inclusive hadron results. This indicates that the behavior of the previous fits in this region were not driven by data, but rather by different phenomenological assumptions regarding the overall fragmentation function shapes.

Therefore, it is very important to perform pQCD tests that are insensitive to the choice of fragmentation functions. Two such tests have been performed to date at

RHIC. PHENIX has measured the direct photon yield [8], and STAR has investigated the inclusive jet production cross section [9]. Both are in good agreement with pQCD expectations. The latter comparison is shown in Fig. 3.

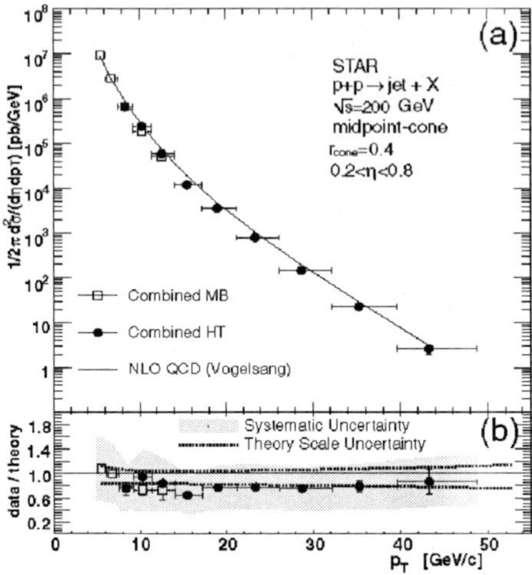

FIGURE 3. A comparison of the mid-rapidity inclusive jet cross section in 200 GeV p+p collisions to next-to-leading-order pQCD calculations. This figure is taken from Ref. [9].

LONGITUDINALLY POLARIZED PROTON COLLISIONS

Polarized deep-inelastic lepton scattering experiments have demonstrated that only ~25% of the spin of the proton arises from the spins of its quarks. This surprising result has come to be known as the "spin crisis". The rest of the spin of the proton must arise from the spins of the gluons or from orbital angular momentum contributions. The polarized deep-inelastic scattering measurements only provide weak, indirect limits on the polarization of the gluons. In contrast, a large fraction of the high-p_T cross section in p+p collisions arises at the parton level from quark+gluon (*qg*) and gluon+gluon (*gg*) scattering. Thus, polarized p+p collisions at RHIC provide excellent sensitivity to gluon polarization, and determining the gluon polarization in the proton is a major goal of the RHIC spin program.

Over the past several years, PHENIX and STAR have focused on measurements of the longitudinal double-spin asymmetry A_{LL} for a variety of inclusive channels. These processes have sizable cross sections and, thus, provide good sensitivity during an era when the beam luminosities and polarizations have been increasing. To date, the most sensitive measurements have been inclusive π^0 asymmetries from PHENIX [4] and inclusive jet asymmetries from STAR [10]. The most recent results that have been submitted for publication involve data that were recorded during 2005 and are shown in Figs. 4 and 5. In both cases, the results are compared to next-to-leading-order pQCD calculations that are based on fits to the polarized deep-inelastic scattering data which make four different assumptions regarding the gluon polarization. The experimental data preclude the maximum scenario and disfavor the minimum scenario. The "standard" scenario, which is derived from the best-fit to the polarized inclusive deep-inelastic scattering data, and the unpolarized gluon scenario are both allowed by the 2005 data. More recent preliminary results from the two collaborations that are based on analyses of 2006 data now clearly preclude the minimum scenario, and disfavor the standard scenario as well. It seems that the either the gluon polarization is small or negative, or the momentum-dependence of the gluon polarization is more complex than has been assumed by the global analyses that have been performed to date.

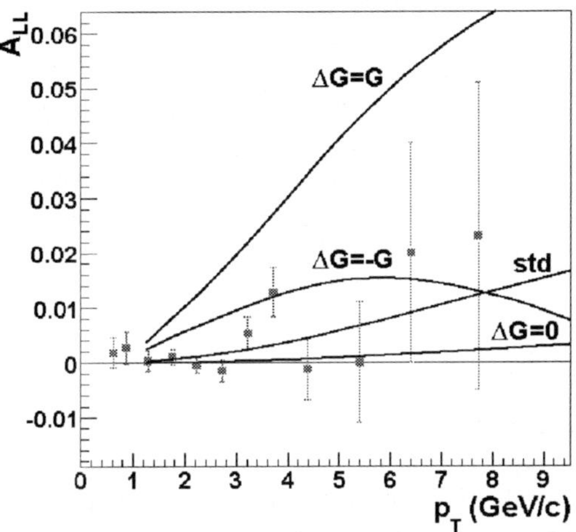

FIGURE 4. PHENIX measurement of the longitudinal double-spin asymmetry A_{LL} for inclusive π^0 production in mid-rapidity 200 GeV p+p collisions, based on data that were recorded in 2005. The curves show pQCD predictions based on four different fits to polarized deep-inelastic scattering measurements. This figure is taken from Ref. [4].

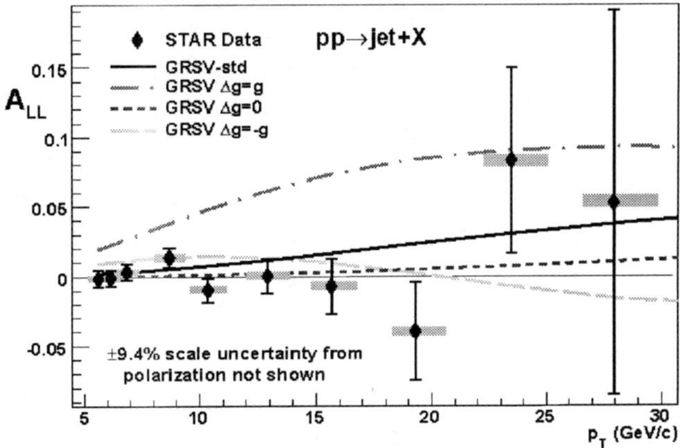

FIGURE 5. A STAR measurement of the longitudinal double-spin asymmetry A_{LL} for inclusive jet production in mid-rapidity 200 GeV p+p collisions, based on data that were recorded in 2005. The curves show pQCD predictions based on four different fits to polarized deep-inelastic scattering measurements. This figure is taken from Ref. [10].

TRANSVERSELY POLARIZED PROTON COLLISIONS

In naïve pQCD, transverse single-spin asymmetries are expected to be very small. However, the Fermilab E704 Collaboration observed very large transverse single-spin asymmetries, A_N, for inclusive pion production at large Feynman-x, x_F [11], and STAR has demonstrated that these large single-spin asymmetries persist to RHIC energies [12]. Today, these asymmetries are believed to arise from a parton transverse motion preference in a transversely polarized proton, the "Sivers effect", or from a parton transverse spin preference in a transversely polarized proton combined with a transverse asymmetry in fragmentation, the "Collins effect". Furthermore, the HERMES Collaboration has shown that both the Sivers and Collins effects are non-zero in semi-inclusive deep-inelastic scattering [13].

It is very important to unravel the dynamics that are producing the large single-spin asymmetries in the forward direction at RHIC. On the one hand, if they arise from the Sivers effect, they provide an indirect way to explore parton orbital motion. On the other hand, if they arise from the Collins effect, they provide a way to measure quark transversity in the proton, a fundamental property that is closely related to quark polarization, but not equivalent. It is even possible that both effects are contributing simultaneously.

Recently, BRAHMS has reported very large asymmetries for charged pions at energies of 200 and 62 GeV [14] that are similar to the previous results from Fermilab. BRAHMS has also extended the measurements to include charged kaons, protons, and anti-protons. Surprisingly, the K^+, K^-, and anti-proton asymmetries all demonstrate the same sign as the π^+ asymmetries, and opposite to π^-. In contrast, the observed proton asymmetry at forward rapidity is very close to zero.

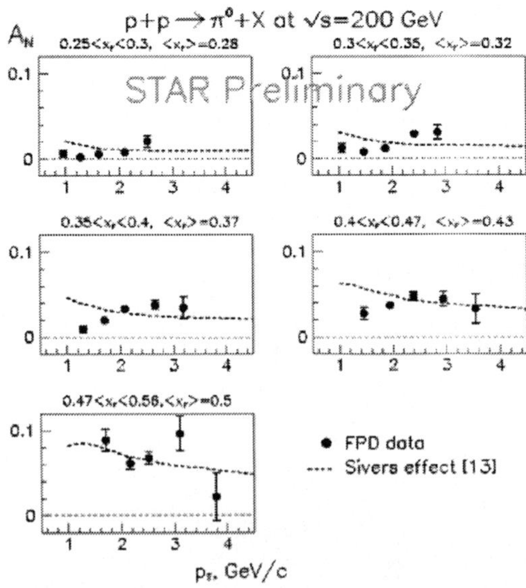

FIGURE 6. The first separation of the x_F and p_T dependences of A_N for inclusive π^0 production at forward rapidity. This figure is taken from Ref. [15].

All the previous transverse single-spin asymmetry measurements have been presented as a function of either x_F or p_T, while summing over the other variable. In order to gain additional insight into the single-spin dynamics, STAR has performed the first determination of A_N as a function of p_T for narrow bins in x_F [15]. The expectation was that A_N would be proportional to $1/p_T$. The results shown in Fig. 6 demonstrate that the p_T dependence is substantially more complex than had been anticipated.

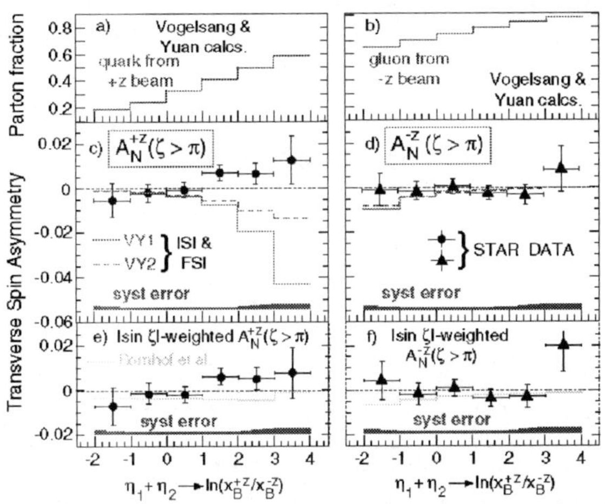

FIGURE 7. Measurement of the tranverse spin dependence of mid-rapidity di-jet acoplanarity. The asymmetries in the right-hand panels are primarily sensitive to the gluon Sivers function. Those in the left-hand panels are sensitive to the gluon Sivers function for negative $\eta_1 + \eta_2$, and the quark Sivers function for $\eta_1 + \eta_2 > 2$. This figure is taken from Ref. [16].

Forward rapidity A_N measurements are primarily sensitive to valence quark effects in the proton. Gluons can also participate in the Sivers effect (but not the Collins effect). To explore this possibility, STAR has measured the acoplanarity of back-to-back di-jets as a function of transverse spin [16]. The observed asymmetries, shown in Fig. 7, are an order of magnitude smaller than the Sivers effect seen in semi-inclusive deep-inelastic scattering by HERMES. The small quark Sivers results may arise from a detailed cancellation of *u* vs. *d* quark effects and initial vs. final state effects in p+p collisions. Meanwhile, it appears that the gluon Sivers effect is small.

CONCLUSION

During the next few years, the RHIC longitudinal spin program will continue to investigate gluon polarization. Upcoming measurements will focus on more sensitive, but also more rare, probes. Di-jet production will provide a way to reconstruct the underlying parton kinematics in leading order, thus allowing the first direct measurement of the momentum dependence of the gluon polarization. Direct photon production is dominated by quark-gluon Compton scattering, which provides high sensitivity to gluon polarization without the complications of mixing *qg* and *gg* scattering effects. Finally, p + p → γ + jet + X will combine the high sensitivity and selectivity of direct photon production with the di-jet ability to reconstruct partonic kinematics. Thus, it has often been called the "golden channel" to measure gluon polarization. In parallel, enhanced transverse spin measurements are planned that will extend the forward single-spin asymmetry studies to much higher p_T. They will also provide an unambiguous separation of the Sivers and Collins effects for event kinematics where large single-spin asymmetries have been observed to date.

In the further future, the RHIC spin program will study the longitudinal single-spin asymmetry A_L for $W^{\pm}$ production in order to measure the anti-quark polarization in the proton.

ACKNOWLEDGMENTS

This work was supported in part by the U.S. Department of Energy.

REFERENCES

1. R. Rapp, these proceedings.
2. S. Mioduszewski, these proceedings.
3. S.S. Adler *et al.* (PHENIX Collaboration), *Phys. Rev. Lett.* **91**, 241803 (2003).
4. A. Adare *et al.* (PHENIX Collaboration), arXiv:0704.3599.
5. B.A. Kniehl, G. Kramer, and B. Potter, *Nucl. Phys.* **B582**, 514 (2000).
6. D. de Florian, R. Sassot, and M. Stratmann, *Phys. Rev.* D **76**, 074033 (2007).
7. S. Albino, B.A. Kniehl, and G. Kramer, *Nucl. Phys.* **B725**, 181 (200); **B734**, 50 (2006).
8. S.S. Adler *et al.* (PHENIX Collaboration), *Phys. Rev. Lett.* **98**, 012002 (2007).
9. B.I. Abelev *et al.* (STAR Collaboration), *Phys. Rev. Lett.* **97**, 252001 (2006).
10. B.I. Abelev *et al.* (STAR Collaboration), arXiv:0710.2048.
11. D.L. Adams *et al.*, *Phys. Lett.* **B261**, 201 (1991); **B264**, 462 (1991).
12. J. Adams *et al.* (STAR Collaboration), *Phys. Rev. Lett.* **92**, 171801 (2004).
13. A. Airapetian *et al.* (HERMES Collaboration), *Phys. Rev. Lett.* **94**, 012002 (2005).
14. J.H. Lee and F. Videbaek (for the BRAHMS Collaboration), Proc. XV Int. Work. Deep-Inelast. Scatt. (DIS2007), in press; http://www4.rcf.bnl.gov/brahms/WWW/publications.html#conference.
15. L. Nogach (for the STAR Collaboration), arXiv:hep-ex/0612030.
16. B.I. Abelev *et al.* (STAR Collaboration), *Phys. Rev. Lett.* **99**, 142003 (2007).

An Introduction to High p_T Physics at RHIC

Saskia Mioduszewski

Cyclotron Institute, MS #3366
Texas A&M University
College Station, TX 77843-3366

Abstract. We introduce the physics being studied at the Relativistic Heavy Ion Collider (RHIC), with particular attention to high transverse momentum (p_T) phenomena. An overview of results and conclusions from the high p_T program at RHIC is presented.

Keywords: RHIC, Jet quenching, High p_T, Relativistic heavy-ion collisions.
PACS: 25.75.-q

INTRODUCTION

A property of Quantum-Chromo-Dynamics (QCD), the theory of the strong interaction, is confinement. One of the main goals in relativistic heavy-ion collisions is to "melt" nuclear matter at extreme temperatures and densities into a deconfined state of freely interacting quarks and gluons, called the Quark-Gluon Plasma (QGP). According to lattice QCD calculations [1], such a phase transition is predicted at critical temperature $T_c \sim 170$-190 MeV [2] or critical energy density $\varepsilon_c \sim 0.7$ GeV/fm^3 (approximately 5 times the density of normal nuclear matter). RHIC was proposed in 1983 to help achieve this goal. The Relativistic Heavy Ion Collider (RHIC) at Brookhaven National Laboratory has provided high-energy heavy-ion collisions since the year 2000. In the meantime, much has been learned about the matter created in these collisions. We have observed large collectivity in the particles emitted from the heavy-ion collisions, which can only be achieved by early thermalization [3]. In fact, the matter produced is not only liquid-like but nearly an ideal fluid [4], described well by hydrodynamics with no viscosity. We have also observed a very large, factor of 5, suppression of hadrons at large transverse momenta ($p_T > 2$ GeV/c) [5-7]. This can be described by calculations of parton energy loss in the matter created [8-11] and has led to the conclusion that the matter created at RHIC is very dense, more than 100 times the density of normal nuclear matter [12].

CP972, *Exotic Nuclei and Nuclear/Particle Astrophysics (II)—Proceedings of the Carpathian Summer School of Physics 2007,* edited by L. Trache and S. Stoica

COMPARING AU+AU COLLISIONS TO P+P COLLISIONS

In order to interpret the measurements from Au+Au collisions, the more "simple" p+p collisions are used as a baseline measurement. A comparison to p+p collisions can reveal new physics in the Au+Au collision that is due to the matter that is created rather than the larger Au+Au system being a mere superposition of individual nucleon+nucleon (p+p, p+n, and n+n) collisions. To make the comparison between the two collision systems, a scaling of the expected number of individual nucleon+nucleon collisions ("binary collisions" N_{binary}) is calculated, based on the centrality of the Au+Au collision. For a given impact parameter, a Glauber model can be used to calculate the mean number of binary collisions [13]. Since the impact parameter is not experimentally known, the Au+Au collisions are divided into event centrality classes based on observables that are correlated to the centrality of the collision [14]. In order to quantify the differences from p+p collisions, the nuclear modification factor,

$$\begin{aligned} R_{AA}(p_T) &= \frac{\text{Yield per A + A collision}}{N_{binary} \times (\text{Yield per p + p collision})} \\ &= \frac{d^2N^{A+A}/dp_T\, d\eta}{(N_{binary})(d^2\sigma^{p+p}/dp_T\, d\eta)/\sigma^{p+p}_{inelastic}} \\ &= \frac{d^2N^{A+A}/dp_T\, d\eta}{(T_{AA})(d^2\sigma^{p+p}/dp_T\, d\eta)} \end{aligned} \quad (1)$$

(where T_{AA} is the nuclear thickness function) is calculated from the yields measured in Au+Au collisions, for a chosen centrality class, and p+p collisions as a function of transverse momentum (p_T).

HIGH P_T RESULTS FROM RHIC

In order to understand the complicated dynamics of a heavy-ion collision, one needs a probe of the produced matter. Hard scatterings, or collisions with a large momentum transfer between partons, provide a particularly valuable probe because they occur early in the collision, leaving the hard-scattered partons sensitive to the properties of the collision medium. A hard scattering occurs on a timescale of ~$1/p_T$; which, for $p_T = 2$ GeV, is approximately 0.1 fm/c. This is smaller than the time it takes for the system to equilibrate. Therefore, the hard-scattered partons experience the entire space-time of the system and serve as a probe of the produced medium. It has been predicted that these partons will reinteract in a deconfined medium of free color charges, losing much of their energy [15]. Since hard-scattered partons eventually fragment into hadrons, the products are two cones of correlated (in angle) hadrons, called jets. The predicted parton energy loss would result in a suppression of the high transverse momentum tail of the measured hadron spectrum, where the hadrons are likely to be the leading particles of jets, and is known as "jet quenching."

Due to the high multiplicity environment, jets cannot be easily observed in a heavy-ion collision. When interpreting the modification of the particle spectra at high p_T as an effect of the medium on the hard-scattered partons, one assumes that the hadrons measured at high p_T emanate from jets. One of the first results from RHIC showed a suppression in the hadron yields at large transverse momentum with respect to the binary-scaled yields measured in p+p collisions. Figure 1 shows the nuclear modification factor (as defined in Eq. 1) measured for neutral pions in the most central Au+Au collisions and d+Au collisions [7,16]. This Au+Au data can be described by theoretical calculations of parton energy loss in the matter created in Au+Au collisions [8-11]. From these theoretical frameworks, we have learned that the gluon density dN_g/dy must be approximately 1000 [8], and the energy density of the matter created in the most central collisions must be approximately 15 GeV/fm^3 to account for the large suppression observed in the data [12].

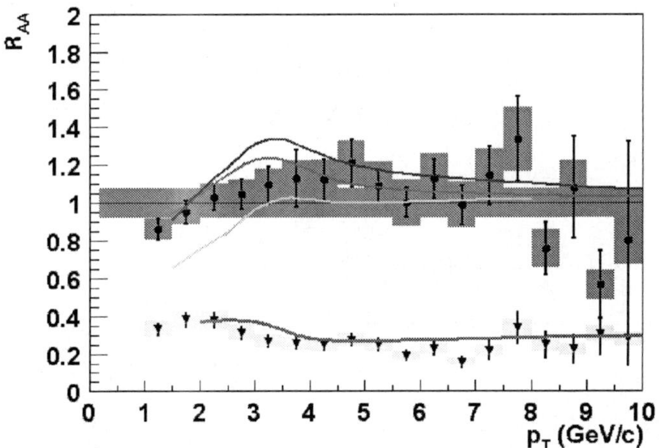

FIGURE 1. Nuclear modification factor of neutral pions measured in the most central (0-10% centrality) Au+Au collisions (black triangles) [7] and in d+Au collisions (black circles) [16]. Lines are theoretical predictions [17] both for the cold nuclear matter, d+Au, and for parton energy loss in the medium created in central Au+Au collisions.

Since the d+Au system does not produce a hot, dense medium, the d+Au data shows only the effect of cold nuclear matter on the hadron yields. As expected from the theoretical calculations that attribute the suppression of high-p_T hadron yields in the Au+Au data to parton energy loss in the dense matter, the measured yields in d+Au collisions are not suppressed at high p_T.

A method to statistically detect di-jets (back-to-back jets) in a high-multiplicity environment is via two-particle angular correlation measurements. In the measured correlation function, the "near-side" jet can be seen as a correlation around $\Delta\phi=0°$, and the "away-side" around $\Delta\phi=180°$. To extract the jet signal from the correlation distributions, other sources of correlation, such as flow and resonance decays, and the combinatorial background need to be disentangled from the correlations due to jets. In such a measurement, the STAR experiment found that the away-side jet disappears in central Au+Au collisions [18] for trigger particles with p_T>4 GeV/c and associated particles in the p_T range of 2-4 GeV/c. This result is shown in Fig. 2. When triggering on a high-p_T particle, the near-side correlation looks very similar to the correlation due to jets measured in p+p collisions (black histogram); but there is no jet correlation seen on the away-side in central Au+Au events, contrary to what is measured in p+p collisions. An interpretation of this result is that the hadrons measured at high p_T in central Au+Au collisions come from hard scatterings near the surface of the system. This means that those jets that we measure (trigger on) were able to escape the medium; while the corresponding jet must traverse the dense medium where it could interact, lose much its energy, and become lost in the low-p_T soft part of the spectrum.

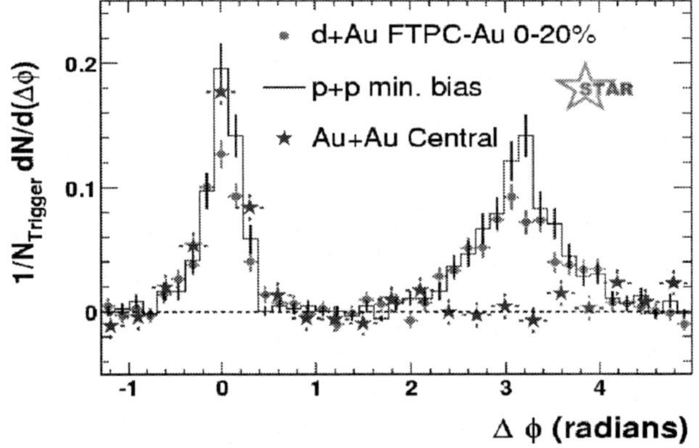

FIGURE 2. Yield of associated particles per trigger particle as a function of $\Delta\varphi$ (azimuthal angle of associated particle relative to trigger particle). The trigger particle has $p_T > 4$ GeV/c, and the associated particles have $2 < p_T < 4$ GeV/c. Uncorrelated "pedestal" and the v_2 modulated flow have been subtracted.

With increasingly large data sets as the RHIC program progressed, the measurement of such correlations was pushed to higher p_T, where the backgrounds from (correlated) collective motion and from soft uncorrelated particles become smaller. At sufficiently high trigger p_T, the away-side jet reappears, although suppressed in strength [19]. Figure 3 shows the evolution of the correlation function with increasing p_T(assoc) for trigger particle with $8<p_T<15$ GeV/c. The background is not subtracted in these plots, but becomes less significant with increasing p_T of the associated particles. A peak on the away-side, especially for the highest p_T selections, is clearly visible even without any background subtraction. This suggests that some of the jets with p_T exceeding 6 GeV/c, or so, indeed escape the dense medium. On the other hand, a comparison to the d+Au data reveals that, although the away-side yield in central Au+Au is still suppressed at these large values of p_T, the width of the away-side peak does not seem to be modified in central Au+Au collisions. This may be due to a bias in which only tangentially produced jets (in which both near-side and away-side are near the surface) are observed, while those produced within the medium are again completely absorbed.

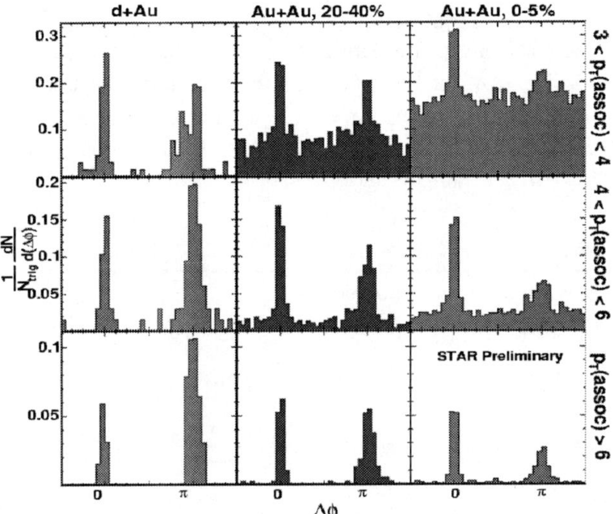

FIGURE 3. Yield of associated particles per trigger particle as a function of $\Delta\varphi$ (azimuthal angle of associated particle relative to trigger particle). The trigger particle has $8 < p_T < 15$ GeV/c, and the associated particles in three ranges indicated on the right side of the figure. No background has been subtracted in these plots.

Some theoretical calculations, which include the energy loss of hard-scattered partons within the bulk matter that is itself evolving (e.g. hydrodynamically), estimate the origin of the observed jets and conclude that the surface bias in these di-jet measurements is not so extreme [20].

THE FUTURE OF PHOTON-HADRON CORRELATIONS

The correlation measurement of a direct photon with hadrons (γ-jet) has been considered the "golden probe" of the dense matter created in heavy-ion collisions [21]. Since the photon does not interact strongly with the medium, it provides a clean measure of the energy of the jet produced together with the photon in the initial hard scattering. Any modification to the jet on the away-side of the trigger photon is a direct handle on the modification due to the medium. Jet yields from the correlation measurement, extracted as a function of associated-particle p_T, produce a (modified) fragmentation function, to be compared with the fragmentation function measured in p+p collisions. This measurement, however, requires a large amount of statistics. The dominant source of photons is hadronic decays, making it difficult to extract the *direct* photon-hadron correlation signal. The advantage in Au+Au collisions is the suppression of hadrons at high p_T, increasing the ratio of signal photons to background photons with increasing centrality. According to measurements of direct photons in Au+Au collisions, the ratio of direct photons to background photons is 1:1 at p_T of approximately 7 GeV/c in the most central Au+Au collisions [22] (almost an order of magnitude better signal to background than in p+p collisions at the same p_T). Figure 4 shows the nuclear modification factor for direct photons and π^0, integrated over p_T>6 GeV/c as a function of centrality in Au+Au collisions. The π^0 yields gradually become more suppressed as the centrality increases, reaching a factor of 5 in the most central collisions, while direct photons show no suppression.

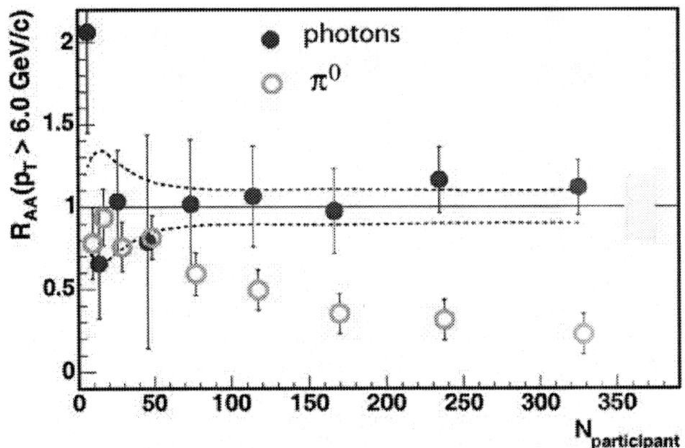

FIGURE 4. Nuclear modification factor for direct photon and π^0 yields integrated over $p_T > 6$ GeV/c as a function of centrality, shown as the number of participants N_{part} in Au+Au collisions [22].

With the latest RHIC run of Au+Au collisions, the correlation measurement of direct photons to hadrons is now feasible with sufficient statistics for a significant result of the "modified fragmentation function" in Au+Au collisions, with more than one million events containing a trigger particle with energy exceeding 7.5 GeV recorded in the STAR experiment.

CONCLUSIONS

The goal of colliding heavy ions at high energies is to detect and study the properties of a state of matter made up of deconfined quarks and gluons (QGP). Jets from hard scatterings serve as probes of the matter that is created in these collisions, as modifications to the jets result from the effect of the dense medium on partons from "hard scatterings" that occur early in a Au+Au collision. Hadrons at high transverse momentum exhibit significant suppression in central Au+Au collisions relative to binary-scaled p+p collisions, which requires a very dense medium (~100 x normal nuclear matter). Angular correlation measurements between hadrons also show a strong effect of the medium on jet yields as a function of trigger and associated p_T. However, in hadron-hadron correlation measurements, the hadrons that originate from di-jets both suffer the effect of the medium and thus make it difficult to quantify the effect as a function of the original jet energy. Direct photons, on the other hand, are not significantly affected by the strongly interacting medium and thus provide a measure of the original energy of the parton, produced together with the photon in the hard scattering. The ultimate goal of measuring γ-jet correlations, and thus measuring the effect of the medium as a function of the true (unmodified) jet energy, will be accessible with the data from the latest RHIC run (Run-7).

ACKNOWLEDGMENTS

We acknowledge support from the Office of Nuclear Physics in the D.O.E. Office of Science and the Alfred P. Sloan Foundation.

REFERENCES

1. F. Karsch, Lect. Notes Phys. **583**, 209 (2002).
2. M. Cheng et al, Phys. Rev. **D74**:054507, 2006.
3. U.W. Heinz and P.F. Kolb, Nucl. Phys. **A702**, 269-280 (2002).
4. E.V. Shuryak, Prog. Part. Nucl. Phys. **53**, 273-303 (2004).
5. K. Adcox et al, Phys. Rev. Lett. **88**: 022301 (2002).
6. C. Adler et al, Phys. Rev. Lett. **89**: 202301 (2002).
7. S.S. Adler et al, Phys. Rev. Lett. **91**: 072301 (2003).
8. I. Vitev and M. Gyulassy, Phys. Rev. Lett. **89**:252301 (2002).
9. P. Levai et al., Nucl. Phys. **A698**, 631 (2002).
10. X.-N. Wang, Phys. Lett. **B595**, 165-170 (2004); X.-N. Wang, Phys. Lett. **B579**, 299-308 (2004).
11. C.A. Salgado and U.A. Wiedemann, Phys. Rev. **D68**:014008, 2003.
12. I. Vitev, J. Phys. **G30**, S791-S800 (2004).
13. R. J. Glauber and G. Matthiae, Nucl. Phys. **B21**, 135 (1970).
14. M. Miller et al, arXiv:nucl-ex/0701025v1.
15. M. Gyulassy and M. Plümer, Phys. Lett. **B243**, 432 (1990); X.N. Wang and M. Gyulassy, Phys. Rev. Lett. **68**, 1480 (1992); R. Baier et al., Phys. Lett. **B345**, 277 (1995).

16. S.S. Adler et al, Phys. Rev. Lett. **91**, 072303 (2003).
17. I. Vitev and M. Gyulassy, Nucl. Phys. **A715**, 779-782 (2003); I. Vitev, Phys. Lett **B562**, 36-44 (2003).
18. C. Adler et al, Phys. Rev. Lett. **90**:082302, 2003.
19. J. Adams et al, Phys. Rev. Lett. **97**:162301, 2006.
20. T. Renk and K.J. Eskola, Phys. Rev. **C75**:054910, 2007.
21. X.-N. Wang et al., Phys. Rev. Lett. **77**, 231-234 (1996).
22. S.S. Adler et al, Phys. Rev. Lett. **94**:232301, 2005.

Phase Transitions with Subnucleonic Degrees of Freedom

R. Rapp

Cyclotron Institute and Physics Department, Texas A&M University, College Station, Texas 77843-3366, U.S.A.

Abstract. We first give a short and elementary overview of standard-model phase transitions (electroweak and strong) with emphasis on how they generated the visible mass in the early universe. Subsequently, we focus on transitions associated with the strong interaction, in particular the mechanisms and observable consequences of spontaneous chiral symmetry breaking and its restoration. Recent advances in the interpretation of dilepton spectra in relativistic heavy-ion collisions, which provide the unique opportunity to study QCD phase changes in the laboratory, are discussed.

Keywords: Quark-Gluon Plasma, Early Universe, Chiral Symmetry Restoration, Ultrarelativistic Heavy-Ion Collisions, Dileptons
PACS: 12.38.Mh, 24.10.Pa, 24.85.+p, 25.75.-q

INTRODUCTION: MASS GENERATION AFTER THE BIG BANG

The evolution of the (very) early universe is characterized by a fascinating sequence of phase transitions which are fundamental to the understanding of how the universe acquired its present day appearance. These transitions probably occurred close to (local) thermal equilibrium and involve varying degrees of freedom (particles and composites) and their interactions, giving rise to several fundamental phenomena many of which are still subject of frontier research efforts in modern particle and nuclear physics. Within the scope of the standard model of particle physics, the electroweak and strong phase transitions are of special significance. Both involve gauge interactions based on a non-abelian group structure implying multiply valued charges of the underlying matter constituents (fermions), as well as force carriers (vector bosons) which themselves carry charge. This is quite different from the well-known (abelian) gauge theory of electromagnetism, Quantum Electro-Dynamics (QED), which is based on a single charge (and its anti-charge) and a neutral force carrier, the photon. However, even QED, as the underlying microscopic theory of condensed matter physics, produces a rich matter structure of different phases and materials with a large variety of magnetic and electric properties. Needless to say that the phase diagram of a nonabelian theory can be expected to be equally rich, even with qualitatively new phenomena.

Returning to the evolution of the universe, Fig. 1 summarizes the most important snapshots from a nuclear physics (or nuclear synthesis) point of view. Starting point at a few microseconds after the Big Bang is the Quark-Gluon Plasma (QGP), a state of matter where quarks and gluons roam (and interact) freely and are not yet confined into protons and neutrons (or other hadrons such as π, K, ρ, etc.). The gauge theory of strong interactions, Quantum Chromo-Dynamics (QCD), predicts an associated phase transition

CP972, *Exotic Nuclei and Nuclear/Particle Astrophysics (II)—Proceedings of the Carpathian Summer School of Physics 2007,* edited by L. Trache and S. Stoica

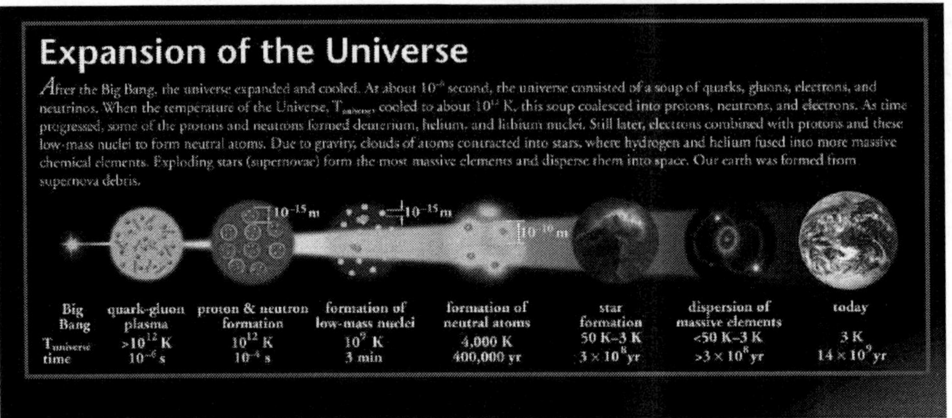

FIGURE 1. Pictorial representation of the nuclear physics aspects of the evolution of the universe from the big bang until today. The QCD phase transition is indicated to the very left, occurring at about $10^{-5}s$ after the Big Bang at a temperature of $\sim 2\times10^{12}K \simeq 200$ MeV. Even before that, at $\sim 10^{-12}s$ and a temperature of $\sim 2\times10^{15}K \simeq 200$ GeV, the electroweak transition is believed to have taken place. Chart taken from Ref. [1].

(or at least a rapid cross over) when heating ordinary nuclear matter to temperatures of $T_c \simeq 200$ MeV. At even earlier times, $\sim 10^{-12}s$, one believes the electroweak (EW) phase transition to have occurred at temperatures' $T_c \simeq 100-200$ GeV (again, a cross over is also possible).

A fundamental aspect shared by the EW and QCD phase transitions is their intimate connection to generating the visible mass of the universe. In both cases, the mass generation phenomenon is related to spontaneous symmetry breaking mechanisms (similar to a ferromagnet, where below the critical (Curie) temperature, a nonzero magnetization develops breaking the rotational symmetry of the system). Below the respective critical temperatures, vacuum condensates form, which in the EW is due to the Higgs-boson field, while in QCD it is a scalar quark-antiquark composite. The vacuum of the Standard Model is therefore a very "dense" state, and particles "ploughing" through it (i.e., interacting with it) acquire a mass! The discovery of the Higgs boson is a central objective of the upcoming experiments at the Large Hadron Collider (LHC), but the direct study of EW *matter* is out of question. A particularly exciting aspect of the QCD phase transition is that there is a realistic possibility to observe it in laboratory experiments, by colliding heavy atomic nuclei at high energies.

In the remainder of this paper, we will give an introduction to theoretical aspects of hot and dense QCD matter, with emphasis on the problem of mass generation. We will review some of the exciting observations that have been made in ultrarelativistic heavy-ion collisions, and why we think that (and possibly what) they teach us about the QGP. In the last part we will focus on a particular observable – dilepton spectra – which is arguably the most promising probe of mass generation at the QCD phase transition.

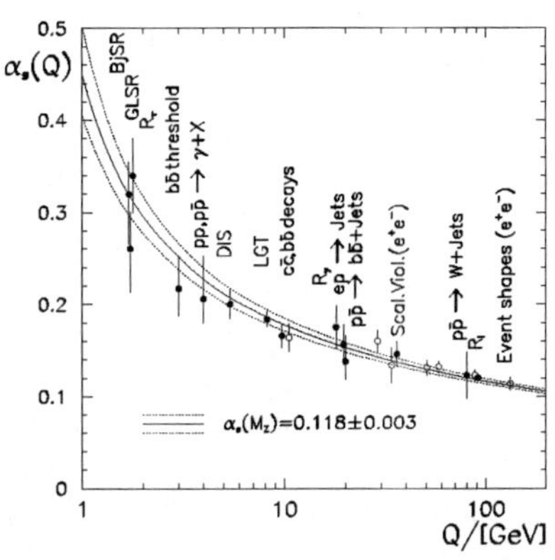

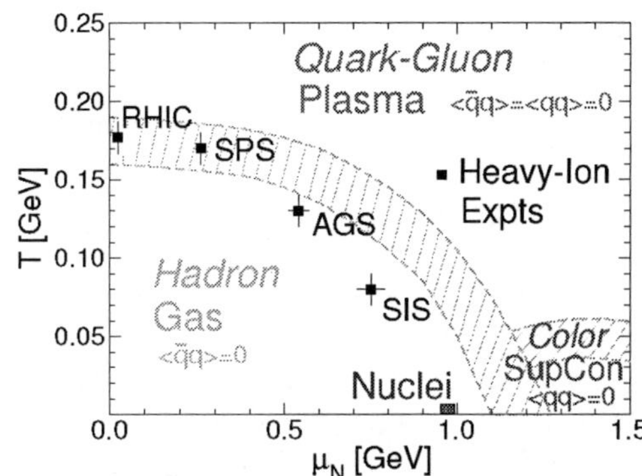

FIGURE 2. Left panel: "running" coupling constant of QCD as a function of momentum transfer, Q, as measured in various experiments (figure taken from Ref. [2]). Right panel: QCD phase diagram in the plane of nucleon (or baryon) chemical potential ($\mu_{N(B)}$), and temperature (T). The hatched regions schematically separate the HG at low T and μ_N, the QGP at high T and colorsuperconducting phases at high μ_N. Lattice QCD computations indicate that the finite-T transition at μ_N=0 is most likely a cross over, while theoretical models suggest that the finite-μ_N transition at T=0 is presumably of first order (connecting phases with different nonvanishing order parameters), implying the existence of a second order critical point in between. The "experimental" points represent thermal model fits to measured hadron multiplicities in heavy-ion collisions at various collision energies [9].

BUILDING BLOCKS OF MATTER AND THE VACUUM

In the standard model of particle physics (SM), the condensation of the Higgs field, ϕ, induces a spontaneous breaking of the $SU(2)\times U(1)$ gauge symmetry which gives rise to the (bare) masses of the fermions (6 quarks, 6 leptons and antiparticles) according to (cf., e.g., Ref. [3] for a pedagogical introduction)

$$m_f \sim G_f \, \langle 0|\phi|0\rangle \ , \tag{1}$$

where $\langle 0|\phi|0\rangle$ denotes the vacuum expectation value and G_f are the Higgs-fermion coupling constants (which are parameters of the SM). In addition, the originally massless gauge bosons (as well as the Higgs particle itself) acquire masses via the coupling to the condensate (except for the photon corresponding to an unbroken $U(1)$). Since EW symmetry breaking pertains to a local (gauge) symmetry, the (would-be) Goldstone bosons (associated with degenerate vacua in group space) are "eaten up" by the (new) longitudinal components of the massive vector bosons $W^{\pm}$ and Z. The stable matter around us, however, is only made of up (u) and down (d) quarks, as well as electrons, whose EW masses are in the range $m_{e,d,u} \simeq 0.5-10$ MeV/c^2. This is dramatically smaller than the proton mass, $M_p = 939$ MeV/c^2, the lightest bound state of 3 quarks and elementary building block of atomic nuclei. Furthermore, no isolated quarks have ever been observed. Clearly, the origin of these phenomena has to be sought for in strong interaction.

The theory of strong interactions is determined by the QCD Lagrangian [3],

$$\mathscr{L}_{\rm QCD} = \bar{q}\,(i\,\partial\!\!\!/ + g\,A\!\!\!/ - m_q)\,q - \frac{1}{4}G_{\mu\nu}G^{\mu\nu}\ , \tag{2}$$

formulated in terms of quark (q) and gluon fields ($A^\mu = A^\mu_a \lambda^a/2$, where λ^a are the 3×3 Gell-Mann matrices which generate the $SU(3)$ gauge group). g (or $\alpha_s = g^2/4\pi$) denotes the dimensionless gauge coupling, and m_q are the (bare) quark masses generated in the EW sector, while the gluons are massless. The field-strength tensor, $G_{\mu\nu} = \partial_\mu A_\nu - \partial_\nu A_\mu + ig[A_\mu, A_\nu]$, includes a commutator term induced by the matrix structure of the $SU(3)$ gauge group giving rise to gluon self-interactions. The famous "asymptotic freedom" property of QCD follows from the renormalization of the strong coupling resulting in a logarithmic dependence of the strong coupling on the momentum transfer, Q, of a given interaction process: $\alpha_s(Q^2) = 12\pi/[(33-2N_f)\log(Q^2/\Lambda^2_{\rm QCD})]$, ($N_f$: number of active quark flavors), cf. left panel of Fig. 2. On the one hand, this implies that the strong interactions are reduced in strength at high Q^2, which is the regime where perturbation theory applies and QCD is well-tested experimentally. On the other hand, it implies that around the scale $\Lambda_{\rm QCD} \simeq 200$ MeV (which is the only scale intrinsic to QCD), the interactions become truly strong and non-perturbative phenomena take over. Most notably, the degrees of freedom change (quarks and gluons are confined into hadrons) and the "constituent quarks" within hadrons acquire a typical mass of $m_q^* \simeq 350$ MeV, about 2 orders of magnitude larger than their "bare mass"! Similar to the EW case, the mechanism responsible for the constituent-quark mass is the formation of a condensate. The latter is induced by a strong attraction in the scalar-isoscalar quark-antiquark channel, $\bar{q}q$, whose subsequent condensation leads to new ground state ("vacuum") of the theory. This mechanism requires a minimal interaction strength above which the new ground state is energetically favored over the trivial vacuum – consistent with the strong coupling regime of the QCD coupling at low Q^2. Upon summing over the light-quark flavors (u, d and s (strange)), the scalar-pair density amounts to about

$$\langle 0|\bar{q}q|0\rangle = \langle 0|\bar{q}_L q_R + \bar{q}_R q_L|0\rangle \simeq 5\ {\rm fm}^{-3} \tag{3}$$

(strictly speaking, the $\bar{q}q$ condensate is negative indicating the gain in the free energy of the vacuum; we ignore this minus sign in our discussion here).

An important consequence of the scalar quark condensate is its symmetry breaking pattern. In addition to (unbroken) local gauge invariance, the QCD Lagrangian, Eq. (2), features several global (i.e., space-time independent) symmetries. In particular, in the limit of vanishing bare masses for u and d quarks ($m_{u,d} \to 0$), $\mathscr{L}_{\rm QCD}$ is invariant under simultaneous isospin and chiral rotations, $e^{i(\vec{\alpha}_V + \gamma_5\vec{\alpha}_A)\cdot\vec{\tau}} \in SU_V(2) \otimes SU_A(2)$, corresponding to a vector times axialvector representation ($\alpha_{V,A}$ are the 3-dimensional rotation angles). This symmetry group can also be decomposed into a chiral (left-/right-handed) quark basis of two isospin groups, $SU_R(2) \otimes SU_L(2)$. This form reflects the fact that massless quarks cannot change their handedness, i.e., propagate as either left- or right-handed modes, even in the presence of gluon exchange. The condensate maximally breaks the axialvector part of the symmetry, as indicated by the mixing of left- and right-handed modes in the first equality of Eq. (3). Thus, not surprisingly, the condensate formation is

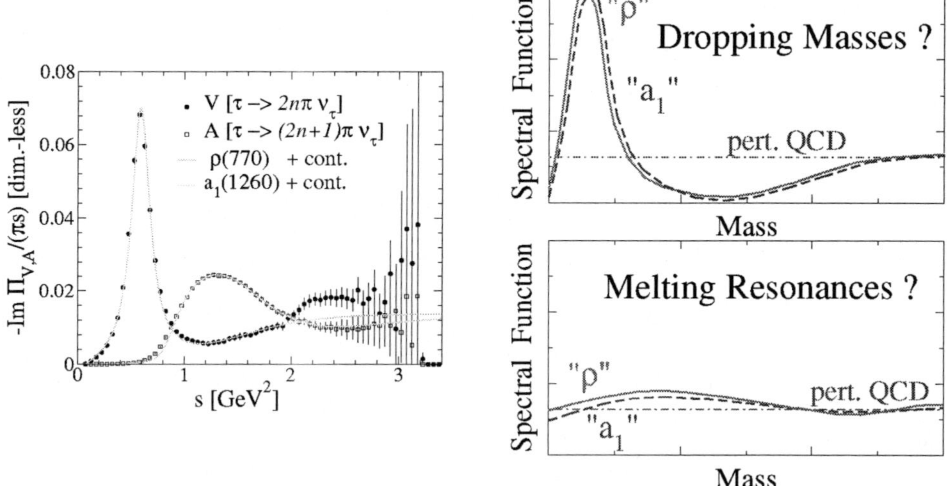

FIGURE 3. Left panel: vector-isovector (ρ channel) and axialvector-isovector (a_1 channel) spectral functions in the vacuum as a function of invariant mass, $s = M^2$, as measured in hadronic τ decays [10]. Right panel: possible scenarios for chiral symmetry restoration in the vector-axiavector channels.

associated with dynamical mass generation,

$$m_q^* \propto G\langle 0|\bar{q}q|0\rangle \ , \tag{4}$$

where G denotes a nonperturbative coupling constant (induced by, e.g., instanton configurations [4]). Both the constituent quark mass and the scalar condensate are not experimental observables, but there are strong evidences in the measured hadron spectrum to support the notion of the "Spontaneous Breaking of Chiral Symmetry" (SBCS): (i) The breaking of a global symmetry entails the emergence of massless Goldstone bosons which are identified with the pions; their mass is "anomalously" small compared to all other hadrons, and its finiteness is believed to be a consequence of the small bare masses of u- and d-quarks, as quantified by the Gell-Mann-Oakes-Renner (GOR) relation, $m_\pi^2 f_\pi^2 \simeq m_q\langle 0|\bar{q}q|0\rangle$; also note that the pions come in a (almost) degenerate isospin triplet as expected from the unbroken group-space directions corresponding to the $SU_V(2)$ (part of the small, isospin-violating mass splitting between neutral and charged pion masses is associated with the *difference* in the bare u- and d-quark masses). The pion decay constant, $f_\pi \simeq 93$ MeV, is interpreted as the "pole strength" of the Goldstone modes (and the GOR relation nicely identifies it on the same level as the $\bar{q}q$-condensate as an order parameter of SBCS); (ii) Hadronic states which transform into each other under chiral transformations (e.g., $\pi(140)$-$\sigma(400\text{-}1200)$, $\rho(770)$-$a_1(1260)$, $N(940)$-$N^*(1535)$, etc.) are called chiral multiplets (or chiral partners). If chiral symmetry were manifest in the QCD vacuum, these multiplets should be degenerate in mass; rather, in nature chiral partners are split by a large margin of typically $\Delta M \simeq 500$ MeV. One of the best examples of SBCS are the vector (V) and axialvector (A) spectral functions, $\text{Im}\Pi_{V,A}$, which have been precisely measured in hadronic τ decays into an even and odd number of pions, respectively, cf. right panel of Fig. 3. The V and A spectral functions are essentially proportional to the spectral functions of ρ and a_1 mesons, $\mathrm{Im}D_{\rho,a_1}$, up to masses of $M \simeq 1-1.5$ GeV, merging into a perturbative continuum beyond that (reconfirming that SBCS is a low-energy phenomenon). In fact, the integrated difference of V and A spectral functions can be related to the pion decay constant [5],

$$f_\pi^2 = -\int \frac{ds}{\pi s} \left[\mathrm{Im}\Pi_V(s) - \mathrm{Im}\Pi_A(s)\right] , \tag{5}$$

which exemplifies the intimate connection between chiral order parameters and hadronic spectroscopy.

A central question is whether the condensate structure of the SM can be tested by heating and/or exciting the vacuum with hot and dense matter, as in the early universe. Naively, one would expect the condensates to "melt" at temperatures corresponding to the characteristic scales of their formation, i.e., $T_{\mathrm{EW}} \simeq M_{W,H} \simeq 100-200$ GeV and $T_{\mathrm{QCD}} \simeq \Lambda_{\mathrm{QCD}}$. This has indeed been observed in numerical lattice computations of the finite-T partition function [6]. In addition, in the QCD case, chiral symmetry restoration has been found to coincide with the deconfinement of quarks and gluons, which has not been understood yet. The critical energy density to induce this transition is roughly $\varepsilon_c \simeq 1$ GeV/fm^{-3} (about 5-10 times that of normal nuclear matter in atomic nuclei), at a critical temperature of $T_c = 160-190$ MeV. A semi-schematic characterization of the QCD phase diagram in terms of its condensate structure is illustrated in the right panel of Fig. 2. It turns out that at high baryon chemical potential but moderate temperatures an additional condensate develops, i.e., BCS-type diquark pairing (so-called colorsuperconductivity) [7, 8], which results in a rather rich phase structure in this regime which we will not discuss any further here. Also note that chiral symmetry restoration, i.e., the vanishing of the quark-antiquark condensate, must be accompanied by a dramatic reshaping of the hadronic spectrum toward and above T_c, in the form of degenerate chiral partners. In principle, there are many ways how this can be realized, two of which are depicted in the right panel of Fig. 3: the hadronic masses could drop together to zero (which would provide an intuitive connection to the vanishing of the constituent quark mass), or the hadronic spectral functions could become very broad, essentially "melting" the resonance structures (which could provide an intuitive merging to the perturbative quark-antiquark continuum). A particularly exciting aspect of the QCD phase diagram is that (unlike the EW one) it can be studied experimentally, by colliding atomic nuclei at high energies. In the following we will focus on the study of QCD matter using heavy-ion collisions.

THE LITTLE BANG IN HEAVY-ION COLLISIONS

When colliding heavy nuclei at ultrarelativistic energies ($\sqrt{s} \gg M_N$), several hundreds to thousands of particles (hadrons) are produced, e.g., at the CERN Super Proton Synchrotron (SPS, $\sqrt{s} = 17$ GeV) and at the Brookhaven Relativistic Heavy-Ion Collider (RHIC, $\sqrt{s} = 20-200$ GeV). The first question to be asked is whether these particles re-interact, or better yet, equilibrate, to justify the notion of *matter*. The answer is a definite *yes*: momentum spectra of pions π, kaons (K), protons (p), Lambdas (Λ), etc., look quite different in heavy-ion collisions compared to p-p collisions at the same energy. Systematic analysis reveals that the final-state hadron spectra represent a "thermal freeze-out" (i.e., the moment when the (short-range) strong interactions cease and the hadrons start streaming freely to the detector). The thermal freezeout can be quantified by a temperature of $T_{\rm fo} \simeq 100-120$ MeV and a collective expansion velocity which can reach more than half the speed of light – a truly explosive expansion! The next piece of evidence is the chemical composition of the fireball: it turns out that the ratios of produced hadrons represent a "chemical freezeout" at around $T_{\rm ch} \simeq 160-180$ MeV [9], which is tantalizing close to the expected critical temperature, T_c. A significantly earlier "chemical" than "thermal" decoupling is expected since *inelastic* hadronic reaction cross section, e.g., $\pi\pi \to K\bar{K}$, (which are underlying chemical equilibration), are known to be factors of 10-100 smaller than *elastic* ones (underlying thermal equilibration), such as $\pi\pi \to \pi\pi$ or $\pi N \to \pi N$ which are dominated by strong resonances such as ρ and Δ. Another valuable observable is the so-called "elliptic flow" of hadrons, $v_2^h(p_T)$, which develops in non-central nuclear collisions: here, the nuclear overlap (or reaction) zone is of an almond-like shape; if the system thermalizes rapidly, the pressure gradients across the "short" axis of the almond will be larger than along its "long" axis, driving an elliptic asymmetry in the final momentum spectra. This asymmetry turns out to rather sensitive to the early thermalization of the matter, since any initial free streaming will dilute the effect. At RHIC energies, the measured v_2 of hadrons is large, consistent with hydrodynamic simulations if a thermalization time of $\tau_0 \leq 1$ fm/c is assumed [11, 12]. The hydrodynamic predictions are consistent with the measured hadron spectra up to momenta of $p_T \simeq 2-3$ GeV, comprising more than 95% of all produced hadrons. Combining the initial nuclear system size with the number of produced hadrons, one can estimate the initial energy density at thermalization, resulting in $\varepsilon_0 \simeq 20$ GeV/fm^3 at RHIC, about ~20 times larger than the critical one and thus deep inside the Quark-Gluon Plasma (QGP)! Another rather dramatic phenomenon observed at RHIC is the suppression of hadrons with high transverse momentum, $p_T \geq 5$ GeV. Compared to baseline spectra in p-p collisions, a 5-fold suppression of the pion spectra has been observed. The only viable explanation of this effect is an energy loss of fast moving partons (quarks or gluons) when traversing the QGP, the so-called jet quenching [13, 14]. The estimated initial energy densities in these calculations are quite consistent with what is extracted from hydrodynamic calculations for describing the elliptic flow. The interactions needed to explain both the elliptic flow and jet quenching are so strong that the picture of a weakly interacting QGP has been invalidated, leading to the notion of a strongly coupled QGP (sQGP). The microscopic properties of the relevant interactions in the sQGP are under intense investigation. In addition, no direct signals of the QCD phase transition itself could be identified thus far. This will be the topic of the next section, leading us back to the question of mass generation.

DILEPTONS AND CHIRAL RESTORATION

A particularly valuable probe of the hot/dense reaction zone in heavy-ion collisions are electromagnetic observables, i.e., photons and dileptons. Once produced, they decouple from the interacting fireball and traverse it without significant rescattering. Especially

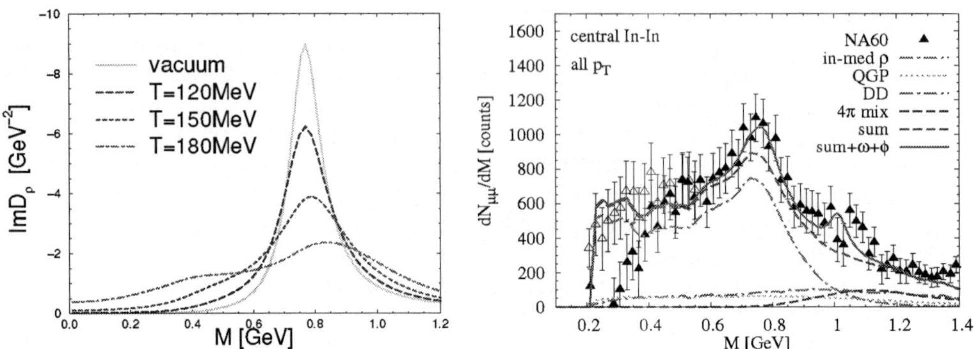

FIGURE 4. Left panel: ρ-meson spectral functions in the vacuum and in hadronic matter at increasing temperature and baryon density [20]. Right panel: theoretical calculations for dimuon invariant-mass spectra based on the spectral functions in the left panel [21], compared to data from central In-In collisions at CERN-SPS energies (final-state hadron decays have been subtracted) [16].

the exclusive dilepton decays of vector mesons, $\rho, \omega, \phi \to e^+e^-, \mu^+\mu^-$, are of prime interest, since the invariant mass of the lepton pair, $M^2 = (p_{e^-} + p_{e^+})^2$ (where $p_{e^\pm}$ are the electron and positron 4-momenta), directly reflects the mass of the parent vector meson at the moment of decay. The study of in-medium modifications of vector mesons and corresponding signals in dilepton spectra in heavy-ion collisions have therefore been subject of intense research over the last 20 years. On the experimental side, substantial excess radiation in heavy-ion collisions over sources expected from *p*-*p* collisions (or from electromagnetic final-state decays of produced hadrons) has been consistently observed [15, 16, 17]. The dielectron invariant mass spectra measured by the CERES/NA45 collaboration in Pb-Au collisions [15] furthermore indicate substantial medium modifications of the ρ-meson spectral function (its contribution to thermal dilepton radiation dominates over those from ω and ϕ decays by a large factor, ~10). However, the somewhat limited statistics of the CERES data did not allow for definite conclusions. On the theoretical side, an enormous activity (see Ref. [18] for a review) has been triggered by the suggestion that the ρ mass should drop in the medium as a direct consequence of the reduction of the quark condensate (and thus of chiral symmetry restoration) [19]. Alternatively, hadronic many-body calculations of the ρ meson propagator,

$$D_\rho(M,q;\mu_B,T) = \left[M^2 - m_\rho^2 - \Sigma_{\rho\pi\pi} - \Sigma_{\rho B} - \Sigma_{\rho M}\right]^{-1}, \quad (6)$$

have been performed, where one evaluates in-medium selfenergy contributions from interactions of the ρ and its pion cloud with the surrounding baryons (*B*) and mesons (*M*) of the heat bath. The rather generic result of these calculations for the ρ spectral function (i.e., imaginary part of the propagator) is a strong increase of its width with little shift of its mass [20], cf. left panel of Fig. 4. In fact, when extrapolated to the expected phase boundary, the ρ resonance essentially melts. It is important to note that the medium effects induced by baryons play the prevalent role.

To confront the theoretical spectral functions with experimental dilepton spectra, the pertinent thermal dilepton emission rates have to be integrated over the space-time evolution of the heavy-ion collision system under consideration [18]. Such an evolution can be modeled, e.g., by hydrodynamic simulations or simplified expanding fireballs which are constrained by the observed hadron spectra. In the right panel of Fig. 4 the theoretical predictions [21] following from the "melting" ρ-meson spectral function are compared to dimuon spectra in central In(158 AGeV)-In collisions as recently measured by the NA60 collaboration at the CERN-SPS [16]. For the first time, the true experimental excess radiation from the nuclear fireball could be separated from the final-state hadron decays (e.g., $\eta \to \gamma\mu^+\mu^-$, $\omega \to \pi^0\mu^+\mu^-, \mu^+\mu^-$ and $\phi \to \mu^+\mu^-$). A fair agreement between data and theory is found; one also sees that at masses above $M \simeq 1$ GeV continuum radiation from both the hadronic and QGP phases becomes relevant, indicative for thermal radiation from matter close to T_c [21].

SUMMARY AND CONCLUSIONS

Our present understanding of the evolution of the early universe implies several dramatic phase changes. On the GeV/TeV scale, which is the realm of the standard model, these are associated with the electroweak and strong interactions, both of which are intimately related with the problem of mass generation. The latter is believed to be triggered by the formation of Higgs and quark-antiquark condensates, respectively. In fact, the major portion of the mass of stable (visible) matter is created during the QCD transition, at a critical temperature of about 200 MeV, a few microseconds after the big bang. Over the past 30 years, the investigation of such matter has become possible in laboratory experiments, by colliding heavy nuclei at high energies. Recent experiments at RHIC have provided strong evidence that a thermalized medium with energy densities well above the critical one has been produced, quite reminiscent to the primordial Big Bang matter. Direct signals of the phase transition itself, i.e., quark deconfinement and chiral symmetry restoration, have been elusive thus far. The most promising observable in this regard are dilepton invariant-mass spectra, with which one hopes to probe the fate of the hadronic (ρ-meson) mass. Recent accurate measurements of dilepton spectra at the CERN-SPS have confirmed theoretical calculations which predict a "melting" of the ρ resonance when approaching the critical temperature. Further theoretical studies are needed to establish a direct relation to chiral symmetry restoration, by evaluating the in-medium properties of the chiral partner of the vector channel (ρ), i.e., the axialvector (a_1) spectral function. Several other exciting questions in this field are being addressed, e.g., the understanding of the properties of the sQGP, direct signals of deconfinement, or the discovery of the putative chiral critical point in the phase diagram. Future experiments at RHIC, LHC and FAIR, together with sustained theoretical efforts, will provide deeper insights and most likely raise new questions.

ACKNOWLEDGMENTS

I am grateful to the organizers of this school for the kind invitation and a well organized meeting. I thank H. van Hees for collaboration on some of the presented topics, and for a careful reading of the ms. This work is supported by a U.S. National Science Foundation CAREER Award under contract no. PHY-0449489.

REFERENCES

1. Taken from: http://www.lbl.gov/abc/wallchart/
2. M. Schmelling, arXiv:hep-ex/9701002.
3. F. Halzen and and A.D. Martin, *Quarks and Leptons*, John Wiley & Sons (New York, 1984).
4. T. Schäfer and E. V. Shuryak, Rev. Mod. Phys. **70**, 323 (1998)
5. S. Weinberg, Phys. Rev. Lett. **18**, 507 (1967).
6. F. Karsch, arXiv:0711.0661 [hep-lat].
7. R. Rapp, T. Schäfer, E.V. Shuryak and M. Velkovsky, Phys. Rev. Lett. **81**, 53 (1998).
8. M. Alford, K. Rajagopal and F. Wilczek, Phys. Lett. **B422**, 247 (1998).
9. P. Braun-Munzinger, K. Redlich and J. Stachel, arXiv:nucl-th/0304013.
10. R. Barate *et al.* [ALEPH Collaboration], Eur. Phys. J. **C4**, 409 (1998).
11. D. Teaney, J. Lauret and E. V. Shuryak, Phys. Rev. Lett. **86**, 4783 (2001)
12. P. F. Kolb and U. W. Heinz, arXiv:nucl-th/0305084.
13. M. Gyulassy, I. Vitev, X. N. Wang and B. W. Zhang, arXiv:nucl-th/0302077.
14. S. Mioduszewski, these proceedings.
15. G. Agakichiev *et al.* [CERES Collaboration], Eur. Phys. J. C **41**, 475 (2005).
16. R. Arnaldi *et al.* [NA60 Collaboration], Phys. Rev. Lett. **96** (2006) 162302.
17. S. Afanasiev *et al.* [PHENIX Collaboration], arXiv:0706.3034 [nucl-ex].
18. R. Rapp and J. Wambach, Adv. Nucl. Phys. **25**, 1 (2000).
19. G.E. Brown and M. Rho, Phys. Rev.Lett. **66**, 2722 (1991).
20. R. Rapp and J. Wambach, Eur. Phys. J. **A6**, 415 (1999).
21. H. van Hees and R. Rapp, Phys. Rev. Lett. **97**, 102301 (2006)

The Physics of Deep-Inelastic Scattering at HERA

Cristinel Diaconu

Centre de Physique des Particules de Marseille and
Deutsches Elektronen Synchrotron, Notkestrasse 85, 22607 Hamburg, Germany

Abstract. In this paper an introduction to the physics of deep-inelastic scattering is given together with an account of some of the most recent results on the proton structure obtained in electron– and positron–proton collisions at the HERA collider.

Keywords: Deep inelastic scattering, proton structure, HERA collider , QCD.
PACS: 13.60.-r, 14.20.-c, 12.38.-t

INTRODUCTION

The investigation of the matter structure using particle collisions started in early XX century with Geiger, Mardsen and Rutherford discovery of the atomic nucleus using α-particle scattering on a gold foil. The principle of the measurement is related to the spatial resolution obtained using high energy particles. The 'resolving power' can be expressed as $\delta =\sim 200\ MeV/Q\ [10^{-15}\ \mathrm{m}]$ and is related to the uncertainity principle: the higher the transfer momentum (denoted by Q), the smaller the details that can be "flashed" and imprinted in the distribution of the scattered (point-like) particle. The search for further substructure levels continued with the scattering of leptons on light nuclei (H,D) in order to investigate the structure of protons and neutrons, the main components of the nuclear matter. The finite size of the proton was established in elastic scattering and its components, the quarks and the gluons, were discovered using electrons of higher and higher energies, which were ultimately able to break the nucleon in the so called deep inelastics scattering (DIS). In this paper, an introduction [1] to the physics of DIS is given using as examples recent measurements of the proton structure performed at HERA electron-proton collider[1].

THE OBSERVABLES OF LEPTON–HADRON SCATTERING

The lepton–hadron scattering is described in the framework shown in figure 1. The scattering can occur via the exchange of γ or Z bosons (neutral currents NC) or via W bosons (charged currents CC). In the later case, a neutrino is expected in the final state. The incoming electron (with a four-momentum k) scatters off the proton (P) to a

[1] The paper is based on an introductory lecture presented at the "Carpathian Summer School of Physics on Exotic Nuclei and Nuclear/Particle Astrophysics" , August 20-31, 2007, Sinaia, Romania.

CP972, *Exotic Nuclei and Nuclear/Particle Astrophysics (II)—Proceedings of the Carpathian Summer School of Physics 2007,* edited by L. Trache and S. Stoica

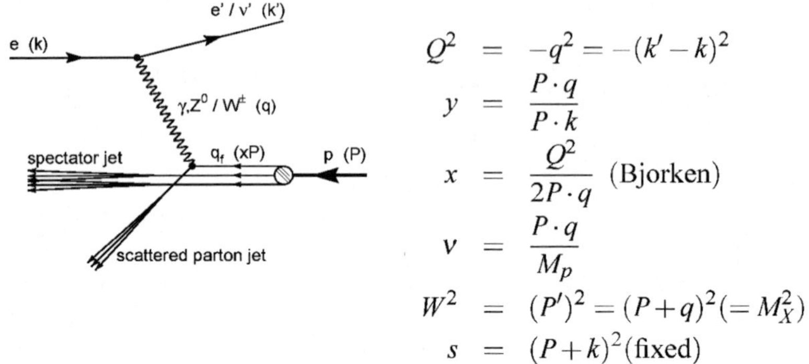

FIGURE 1. Lepton–hadron scattering: an exchange of a boson in the t–channel.

final state electron with four–momentum k' via a virtual photon γ^* or a weak boson with a virtuality Q^2 in the t-channel. The Bjorken variable x is associated with the fraction of the momentum of the proton carried by the struck parton. The total centre-of-mass energy is given by $\sqrt{s}$ and the energy of the $\gamma^* p$ system is given by W, which is equivalent to the total mass of the hadronic system in the final state M_X. In the case of elastic scattering $M_X = M_p$ and from M_X expression follows that $Q^2 = 2Pq$ and $x = 1$ (the whole proton interacts). Only two variables are independent, since the reaction is completely defined by the scatering angle and by the electron-parton centre-of-mass energy. The variable ν has a simple meaning in the proton rest frame, as the energy lost by the electron during the scattering $\nu = E_e - E'_e$, while y represents the fractional energy loss $y = \frac{E_e - E'_e}{E_e}$. Q^2 can be expressed as a function of the electron energy and scattered angle $Q^2 = 4E_eE'_e \sin^2\frac{\theta}{2}$. From these relations, it is obvious that the DIS kinematics can be calculated from the measurement of the scattered electron only. The measurement of the hadrons in the final state, if available, can be exploited as an extra constraint or used in case of CC reactions, where the outgoing neutrino is not measured.

THE HERA PROJECT

The idea for a large electron–proton collider to mark a new step in the studies for proton structure, beyond the fixed target experiments, was promoted already in seventies [3]. The HERA collider project started in 1985 and produced the first electron–proton collisions in 1992. It is composed of two accelerators designed to store and collide counter rotating electrons (e^-) or positrons (e^+) with an energy of 27.5 GeV and protons with an energy of 920 GeV. The operations came to an end in june 2007 and the final analyses using the collected data are in progress at present.

HERA ring hosted two collider mode detectors H1 and ZEUS. They were build as hermetic (4π) multi-purpose detectors equipped with internal trackers able to measure charged particle momenta and calorimeters completing the measurement of the energy flow. Two other experiments used $e^\pm$ or p beams for fixed target studies: HERMES, dedicated to the study of polarised $e^\pm p(N)$ collisions and (until 2003) HERA-B dedicated to the study of beauty production in hadronic collisions. Since 2003, the $e^{\pm}$ beam were longitudinally polarised in collision mode with an average polarisation of $P_{e^{\pm}} = 30-40\%$. An integrated luminosity of 300^{-1} (200 pb^{-1}) has been collected in $e^{+}p$ ($e^{-}p$) by each the two collider mode experiments, H1 and ZEUS.

DEEP-INELASTIC SCATTERING MEASUREMENTS AT HERA

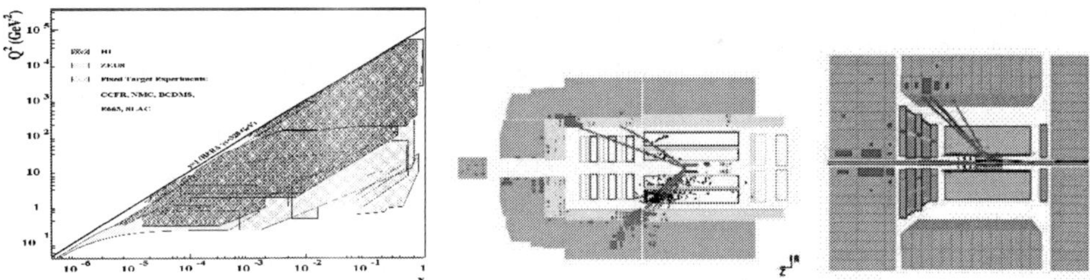

FIGURE 2. The kinematic plane accessible at HERA compared to former fixed target experiments (left) and event displays of a neutral current scattering event measured by H1 (center) and charged current scattering event measured by ZEUS (right).

The H1 and ZEUS experiments measured both neutral current (NC) and charged current (CC) processes. The kinematic (x, Q^2) plane accessible at HERA is shown in figure 2 (left) with a Q^2 domain up to 50000 GeV^2 and x down to 10^{-5}. The NC events contain a prominent electron and a jet of particles measured in the calorimeter, while in CC events only the jet is visible since the outgoing neutrino in not detected. Examples of such events are shown in figure 2.

Since a large domain in x and Q^2 is accessed, the NC cross section is sensitive to weak force effects. The proton structure, as revealed by the photon and Z^0 boson in DIS, can be incorporated into the so-called generalised structure functions. The cross section is parameterised as following:

$$\frac{\mathrm{d}^2\sigma^{\pm}_{\mathrm{NC}}}{\mathrm{d}x\mathrm{d}Q^2} = \frac{2\pi\alpha^2}{xQ^4}(Y_+\tilde{F}_2 \mp Y_- x\tilde{F}_3 - y^2\tilde{F}_L) \ , \tag{1}$$

The generalised structure functions $\tilde{F}_2$ and $x\tilde{F}_3$ can be further decomposed as [4]

$$\tilde{F}_2 \equiv \ F_2 \ - v_e \frac{\kappa Q^2}{(Q^2+M_Z^2)} F_2^{\gamma Z} + (v_e^2 + a_e^2)\left(\frac{\kappa Q^2}{Q^2+M_Z^2}\right)^2 F_2^Z \, , \tag{2}$$

$$x\tilde{F}_3 \equiv \ - a_e \frac{\kappa Q^2}{(Q^2+M_Z^2)} xF_3^{\gamma Z} + (2v_e a_e)\left(\frac{\kappa Q^2}{Q^2+M_Z^2}\right)^2 xF_3^Z \, , \tag{3}$$

with $\kappa^{-1} = 4\frac{M_W^2}{M_Z^2}(1-\frac{M_W^2}{M_Z^2})$ in the on-mass-shell scheme [5]. The quantities v_e and a_e are the vector and axial-vector weak couplings of the electron or positron to the Z^0 [5]. The electromagnetic structure function F_2 originates from photon exchange only and

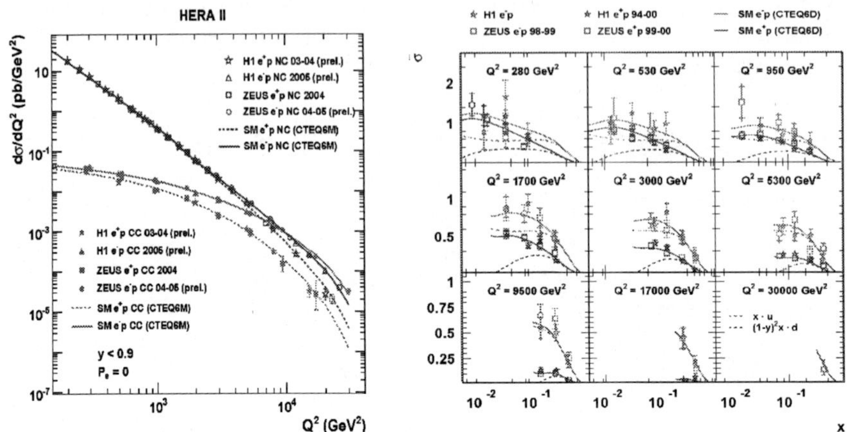

FIGURE 3. Left: the charged current and neutral current cross section as a function of Q^2 measured in $e^{\pm}p$ collisions at HERA. Right: the charged current reduced cross section $\tilde{\sigma}_{CC}$ as a function of x for various Q^2 values measured in electron– and positron–proton collisions.

dominates over the vast majority of the measured phase space. The functions F_2^Z and xF_3^Z are the contributions to $\tilde{F}_2$ and $x\tilde{F}_3$ from Z^0 exchange and the functions $F_2^{\gamma Z}$ and $xF_3^{\gamma Z}$ are the contributions from γZ interference. These contributions are significant only at high Q^2. The structure functions provide however direct information on the proton components. Within the so-called quark-parton model (QPM), the proton is composed of spin one half partons, called quarks $q = u, d$ with fractionary charges e_q. This model predicts the structure functions as combinations of quark densities $q(x)$. For instance $F_2 = \sum_q e_q q(x)$ and is independent of Q^2 (Bjorken scaling). This model is improved within the Quantum Chromodynamics (QCD) predicting quark interactions via gluons, the careers of the strong force, and leading to F_2 dependence on Q^2 (scaling violation).

The charged current (CC) interactions, $e^{\pm}p \to \overset{(-)}{\nu}_e X$, are mediated by the exchange of a W boson in the t channel. The cross section is parameterised as:

$$\frac{d^2\sigma^{CC}(e^{\pm}p)}{dxdQ^2} = \frac{G_F^2}{2\pi x}\left[\frac{M_W^2}{M_W^2+Q^2}\right]^2 \tilde{\sigma}_{CC}^{\pm}(x,Q^2), \tag{4}$$

with $\tilde{\sigma}_{CC}^{\pm}(x,Q^2) = \frac{1}{2}\left[Y_+ W_2^{\pm}(x,Q^2) \mp Y_- xW_3^{\pm}(x,Q^2) - y^2 W_L^{\pm}(x,Q^2)\right]$. $\tilde{\sigma}$ is the reduced cross section, G_F is the Fermi constant, M_W, the mass of the W boson, and W_2, xW_3 and W_L, CC structure functions. In the quark parton model, the structure functions $W_2^{\pm}$ and $xW_3^{\pm}$ may be interpreted as lepton charge dependent sums and differences of quark and anti-quark distributions: $W_2^+ = x(\overline{U}+D)$, $xW_3^+ = x(D-\overline{U})$, $W_2^- = x(U+\overline{D})$, $xW_3^- = x(U-\overline{D})$, whereas $W_L^{\pm} = 0$. The terms xU, xD, $x\overline{U}$ and $x\overline{D}$ are defined as the sum of up-type, of down-type and of their anti-quark-type distributions.

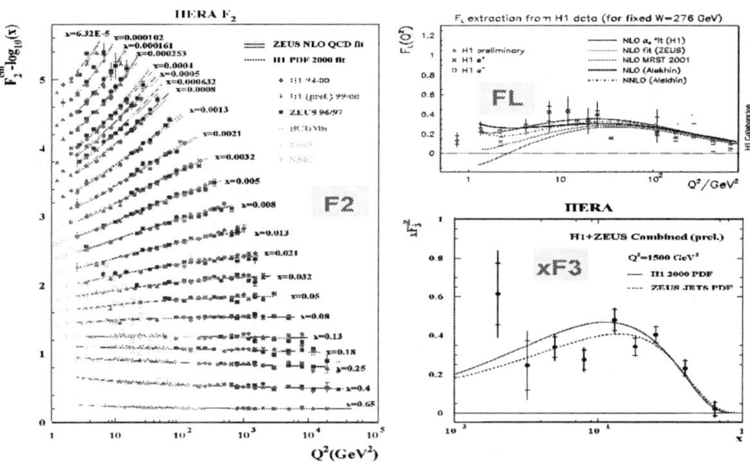

FIGURE 4. The determination of the structure functions F_2, F_L and xF_3 from HERA measurements.

The differential NC and CC cross sections as a function of Q^2 are shown in figure 3 (left) for $e^{\pm}p$ collisions. At low Q^2 the NC cross section, driven by the electromagnetic interaction, is two orders of magnitude larger than the CC cross section which correspond to a pure weak interaction. At large $Q^2 \sim M^2_{W,Z}$ the two cross sections are similar, which can be interpreted as a hint for electroweak unification. The largest Q^2 measurement corresponds to a resolution $\delta \simeq 10^{-18}$ m, i.e. 1/1000 the proton size. The agreement between the measurement and the prediction based on improved parton model constitutes a spectacular confirmation of QCD and suggests that no evidence for quark substructure is observed at present.

The double differential reduced cross section $\tilde{\sigma}_{CC}(x,Q^2)$ is shown in figure 3 (right). The CC processes are sensitive to individual quark flavours, especially visible at large Q^2: the e^+p collisions probe the d quark distribution, while e^-p are more sensitive to the u distribution. This is a very useful feature of the CC processes compared to the NC, where the quark flavour separation is weaker.

STRUCTURE FUNCTIONS MEASUREMENTS: F_2, F_L AND xF_3

The NC cross section is dominated over a large domain by the F_2 contributions, defined in equation 1. The measurement of the NC cross section at HERA can therefore be translated into an F_2 measurement, which is shown in figure 4 together with the previous measurements performed at fixed target experiments. One can observe the Bjorken scaling in the region at high x, but obvious scaling violation at lower x. This can be understood in terms of DGLAP equations [2] as a contribution driven by the gluon $\partial F_2(x,Q^2)/\partial \ln(Q^2) \approx (10\,\alpha_s(Q^2)/27\pi)xg(x,Q^2)$. while at high x the quark pdf's are the major contributors to the evolution in DGLAP equations.

From the F_2 measurements at fixed Q^2 one can observe a steep increase of F_2 towards

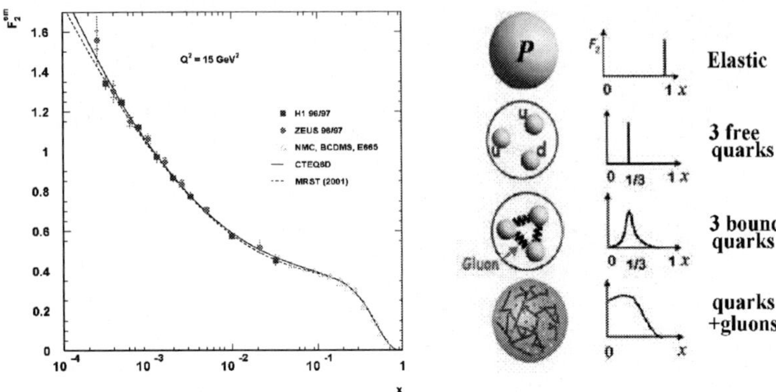

FIGURE 5. Left: The measurement of F_2 as a function of x for $Q^2 = 15$ GeV. Right: a sketch of the correspondence between F_2 shape as a function of x and the quark-parton model.

low x, as shown in figure 5 (left). The region at low x is populated by quarks which have undergone a hard or multiple gluon radiation and carry a low fraction of the proton momentum at the time of the interaction. The observation of such large fluctuations to very high parton density is driven by the uncertainty principle, which requires that the interaction time be very short and therefore at high Q^2. In this regime, it is expected that the structure function grows at low x and shrinks at large x, confirmed by the experimental observation. The rise of the structure functions at low x is one of the most surprising observations at HERA. It is predicted in the double leading log limit of QCD [6]. It can be intuitively understood in terms of gluon driven parton production at low x, as depicted in figure 5 (right).

The longitudinal structure function F_L is usually a small correction, only visible at large y. The F_L measurement from the cross section has to proceed in such a way that F_2 contribution is separated. Indirect methods assume some parameterisation of F_2 to extract F_L. Using this method, an F_L determination can be performed and is shown in figure 4 at fixed W (the $\gamma^* p$ centre-of-mass energy). In the naive QPM the longitudinal structure function $F_L = F_2 - 2xF_1 \equiv 0$ and therefore F_L contains by definition the deviations from the Callan-Gross relation. It can be shown that F_L is directly related to the gluon density in the proton [7, 8] $xg(x) = 1.8[\frac{3\pi}{2\alpha_s} F_L(0.4x) - F_2(0.8x] \simeq \frac{8.3}{\alpha_s} F_L$ meaning that at low x, to a good approximation F_L is a direct measure for the gluon distribution.

A direct measurement of F_L can be performed if the cross section $\sigma \sim F_2(x, Q^2) + f(y)\, F_L(x, Q^2)$ is measured at fixed x and Q^2 but variable y. This can only be performed if the centre-of-mass energy is varied, for instance by reducing the proton beam energy. Eliminating $F2(x, Q2)$, F_L can be directly measured with reduced uncertainties from the difference of cross sections: $F_L \sim C(y) * (\sigma(E_p^1) - \sigma(E_p^2))$. Measurements of DIS at HERA at lower proton energies of 460 GeV and 575 GeV has been performed at the end of the run in 2007 in order to perform the first direct measurement of F_L in the low x and high Q^2 regime.

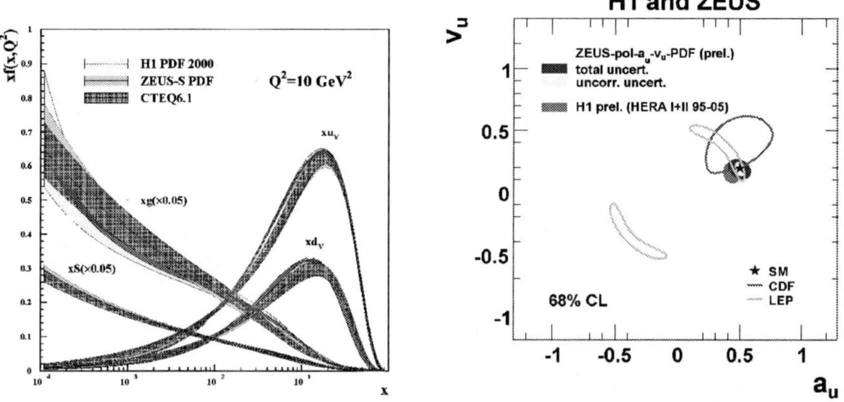

FIGURE 6. Left: The parton distribution functions extracted from HERA data. Right: Axial and vector couplings of the u–quark measured from the combined electroweak–QCD fit at HERA and compared with measurements from LEP (using light quarks production at Z pôle $e^+e^- \to q\bar{q}$) and Tevatron (from Drell-Yan electron pair production $q\bar{q} \to e^+e^-$).

The structure function $x\tilde{F}_3$ can be obtained from the cross section difference between electron and positron unpolarised data $x\tilde{F}_3 = \frac{Y_+}{2Y_-}\left[\tilde{\sigma}^-(x,Q^2) - \tilde{\sigma}^+(x,Q^2)\right]$ The dominant contribution to xF_3 arises from the γZ interference. In leading order QCD the interference structure function $xF_3^{\gamma Z}$ can be written as $xF_3^{\gamma Z} = 2x[e_u a_u(U - \overline{U}) + e_d a_d(D - \overline{D})]$, with $U = u + c$ and $D = d + s$ thus provides information about the light quark axial vector couplings (a_u, a_d) and the sign of the electric quark charges (e_u, e_d). The averaged $xF_3^{\gamma Z}$, determined by H1 and ZEUS for a Q^2 value of 1500 GeV^2, is shown in figure 4.

PARTON DISTRIBUTION FUNCTIONS AND ELECTROWEAK EFFECTS

The NC and CC cross section measurements are used in a global fit in order to extract the parton distribution functions (pdf's) [9, 10]. The shapes for the quarks $q(x,Q_0^2)$ and gluon $g(x,Q_0^2)$ distributions are parametrised as a function of x at a given scale Q_0^2 and evolved using DGLAP equations [2] to each (x,Q^2) point where the cross section has been measured. The theoretical cross section can therefore be accurately calculated as a function of the pdf's parameters. A χ^2 is then built using the measurements and the predictions for all measurements points and minimised to extract the non-perturbative pdf's parameters. Since the number of parameters (typically 10) is much lower than the number of measurements (several hundred) the fit also consitutes a very powerful test of QCD. The structure functions from the fit are compared with data in figure 4. The parton distribution functions are extracted using the decomposition of the structure function described above. As an example, the pdf's obtained for $Q^2 = 10$ GeV^2 are shown in figure 6. The valence distributions peak at $1/3$ as expected from simple

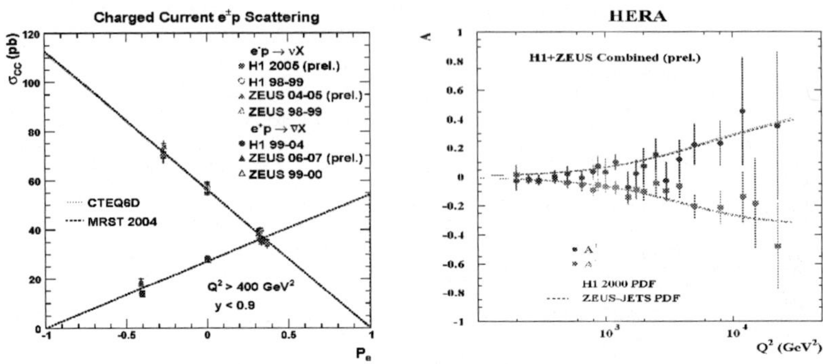

FIGURE 7. Left: The dependence of the charged current cross section on the electron or positron beam polarisation at HERA. Right: The polarisation asymmetry of the NC cross section at HERA.

counting with u_V twice as large as d_V. Gluon distribution is enhanced at low x. The knowledge of the proton structure deduced from inclusive CC/NC measurements can be used to calculate the cross section of exclusive processes leading to a specific final state FS as a convolution of the parton level cross section with pdf's, for instance: $\sigma_{ep\to FS} = \sigma_{\mathrm{eq->FS}} \otimes q(x,Q^2)$. This factorisation can also be used to calculate the cross section of processes produced in proton–proton collisions using the pdf's measured in DIS.

Recently, a new approach has been adopted by the H1 and ZEUS collaborations[11], performing a combined QCD–electroweak fit. The strategy is to leave free in the fit the EW parameters together with the parameterisation of the parton distribution functions. Due to the t-channel electron-quark scattering via Z^0 bosons, the DIS cross sections at high Q^2 are sensitive to light quark axial (a_q) and vector (v_q) coupling to the Z^0. This dependence includes linear terms with significant weight in the cross section which allow to determine not only the value but also the sign of the couplings. The measurements of the u–quark couplings obtained at HERA, LEP and Tevatron are shown in figure 6.

$e^{\pm}p$ COLLISION WITH A POLARISED LEPTON BEAM

The polarisation of the electron beam at HERA II allows a test of the parity non-conservation effects typical of the electroweak sector. The most prominent effect is predicted in the CC process, for which the cross section depends linearly on the $e^{\pm}$–beam polarisation: $\sigma^{\mathrm{e^{\pm}p}}(\mathrm{P}) = (1\pm\mathrm{P})\sigma^{\mathrm{e^{\pm}p}}_{\mathrm{P=0}}$. The results[13] obtained for the first time in $e^{\pm}p$ collisions are shown in figure 7. The expected linear dependence is confirmed and provides supporting evidence for the V-A structure of charged currents in the Standard Model.

Due to parity violating couplings of the Z boson, the $e^{\pm}$ beam polarisation effects can also be measured in NC processes at high Q^2. The charge dependent longitudinal polarisation asymmetries of the neutral current cross sections, defined as

$$A^{\pm} = \frac{2}{P_R - P_L} \cdot \frac{\sigma^{\pm}(P_R) - \sigma^{\pm}(P_L)}{\sigma^{\pm}(P_R) + \sigma^{\pm}(P_L)} \simeq \mp k a_e \frac{F_2^{\gamma Z}}{F_2}, \tag{5}$$

measure to a very good approximation the structure function ratio. These asymmetries are proportional to combinations $a_e v_q$ and thus provide a direct measure of parity violation. In the Standard Model A^+ is expected to be positive and about equal to $-A^-$. At large x the asymmetries measure the d/u ratio of the valence quark distributions according to $A^{\pm} \simeq \pm k \frac{1+d_v/u_v}{4+d_v/u_v}$. The measurement from ZEUS and H1 [14], shown in figure 7, are in agreement with the theoretical predictions.

OUTLOOK

The study of deep-inelastic scattering is a fundamental branch of high energy physics. The structure of matter was last time resolved to new components, the quarks, in the first break-up of the proton at SLAC in 1968. Since then, the Standard Model of particle physics has become a well established theory with the last quark, the top, discovered in 1994. The knowledge of the structure of the baryonic matter, dominating the visible universe, has made huge progress in the last decades, thanks to an impressive effort to unravel the nucleon structure in fixed target experiments and at the HERA *ep* collider. The knowledge acquired at HERA is invaluable also for the physics of the Large Hadron Collider, foreseen to start *pp* collisions at 14 TeV in 2008. By enabling even more ambitious DIS experiments [15] beyond HERA, the matter structure investigations may gain a new momentum.

REFERENCES

1. More details about DIS physics can be found in the following monographies: R.K. Ellis, W.J. Sterling and B.R. Webber, "QCD and Collider Physics", Cambridge, ISBN 521 58189 3, (1996);
   R. Devenish and A. Cooper-Sarkar, "Deep Inelastic Scattering", Oxford University Press, ISBN13: 978-0-19-850671-3, (2005) ;
2. V.N. Gribov and L.N. Lipatov, Sov. J. Nucl. Phys. 15 (1972) 438, 675; L.N. Lipatov, Sov. J. Nucl. Phys. 20 (1975) 94; Yu. L. Dokshitzer, Sov. Phys. JETP 46 (1977) 641; G. Altarelli and G. Parisi, Nucl. Phys. B126 (1977) 298.
3. A. Febel, H. Gerke, M. Tigner, H. Wiedemann and B. H. Wiik, IEEE Trans. Nucl. Sci. **20** (1973) 782.
4. M. Klein and T. Riemann, Z. Phys. C **24** (1984) 151.
5. S. Eidelman *et al.* [Particle Data Group Collaboration], Phys. Lett. **B592** (2004) 1.
6. A. De Rújula *et al.*, Phys. Rev. D**10** (1974) 1649; R. D. Ball and S. Forte, Phys. Lett. B **336** (1994) 77 [arXiv:hep-ph/9406385]; R. D. Ball and S. Forte, Phys. Lett. B **335** (1994) 77 [arXiv:hep-ph/9405320].
7. G. Altarelli and G. Martinelli, Phys. Lett. B **76** (1978) 89.
8. A. M. Cooper-Sarkar *et al.*, Z. Phys. C **39** (1988) 281.
9. C. Adloff *et al.* [H1 Collaboration], Eur. Phys. J. C **30**, 1 (2003) [hep-ex/0304003].
10. S. Chekanov *et al.* [ZEUS Collaboration], Phys. Rev. D **67**, 012007 (2003) [hep-ex/0208023].
11. H1 and ZEUS Collaboration, contribution to EPS Conference, Manchester, UK, july 2007 [H1prelim-07-041, ZEUS-prel-06-003].
12. D. Acosta *et al.* [CDF Collaboration], Phys. Rev. D **71**, 052002 (2005) [hep-ex/0411059].

6. NUCLEAR PHYSICS FOR ASTROPHYSICS

^{7}Be in Stars and in the Laboratory

Michael Hass and Vivek Kumar

Department of Particle Physics, Weizmann Institute of Science, Rehovot, Israel

Abstract. We discuss results and future plans for low-energy reactions that play an important role in current nuclear astrophysics research and that happen to concentrate around the region of $A = 7$. The ^{7}Be(p, γ)^{8}B and the ^{3}He(^{4}He, γ)^{7}Be reactions are crucial for understanding the solar-neutrino oscillations phenomenon and the latter one plays a central role in the issue of cosmic ^{7}Li abundance and Big-Bang Nucleosynthesis. The electron-capture (EC) decay rate of ^{7}Be in metallic Cu host and the β^{-}-decay rate of ^{198}Au in the host alloy Al-Au have been measured *simultaneously* at several temperatures, ranging from 0.350 K to 293 K. The resulting null temperature dependence is discussed in terms of the inadequacy of the often-used Debye-Hückel model for such measurements.

Keywords: Cross sections of solar fusion reactions; Neutrino oscillations; Big-bang nucleosynthesis; Host and temperature dependence of the EC half-life.
PACS: 25.40.Lw, 26.35.+c, 26.65.+t, 27.20.+n

SOLAR FUSION REACTIONS, SOLAR NEUTRINOS AND BIG-BANG NUCLEOSYNTHESIS

The Cross Section of the ^{7}Be(P, γ)^{8}B Reaction

Our planet is bombarded every second with a large number of charge-less, mass-less neutrinos, originating in the nuclear fusion reactions that power the energy production in the Sun. A major, long-lasting research effort, focusing on the apparent shortfall of detected solar neutrinos as compared to theoretical predictions, culminated recently with the results of the Sudbury Neutrino Observatory (SNO) experiment [1], affirming the notion of neutrino oscillations (and hence neutrino mass). The SNO experiment, as well as the preceding Homestake [2] and Super-Kamiokande [3] experiments are sensitive only to a small fraction of the solar neutrino spectrum, to the high-energy neutrinos that are emitted as a result of the fusion of protons with the nucleus ^{7}Be and the subsequent beta decay of the ^{8}B product nucleus. The cross section of this reaction has been measured in the laboratory several times. However, mostly due to difficulties with the preparation and homogeneity of the radioactive ^{7}Be target, large discrepancies still persist in the extracted cross section values. The present experiment uses in a novel way a 2 mm diameter target of the ^{7}Be radioactive nuclei (with a half-life of 53 days), prepared by direct implantation at the ISOLDE (CERN) laboratory and brought to the Van de Graaff accelerator of the Weizmann Institute, Israel, for the measurement of the reaction. The results of these experiments have been published in detail in several previous papers [4–8]. Fig. 1 presents a brief summary of the results.

With the present determination of the p+^{7}Be cross section at solar energies to an accuracy of better than 4%, this important nuclear physics quantity ceases to be the largest source of error in the Standard Solar Model [9] estimates of the ^{8}B neutrino flux.

CP972, *Exotic Nuclei and Nuclear/Particle Astrophysics (II)—Proceedings of the Carpathian Summer School of Physics 2007,* edited by L. Trache and S. Stoica

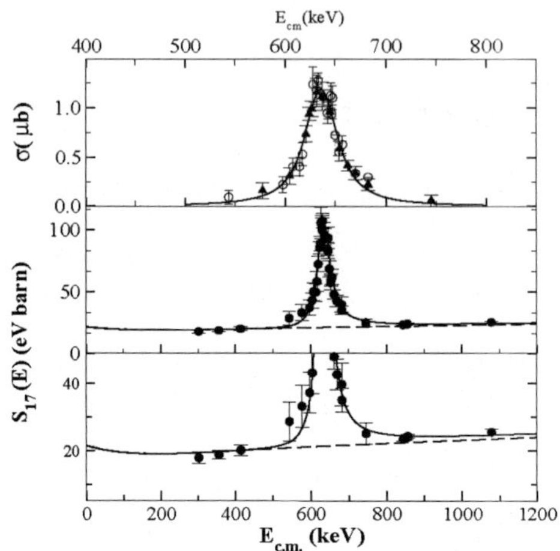

FIGURE 1. The astrophysical S_{17} factor (the cross section of the reaction, with the penetrability through the Coulomb barrier factored out) from the present measurement. The extrapolation to $E = 0$ yields $S_{17}(0) = 21.5(7)$ eV b. Depicted in the figure are the regions of the M1 resonance and those above and below it.

The Cross Section of the ^{3}He(^{4}He, γ)^{7}Be Reaction

The ^{3}He(^{4}He,γ)^{7}Be reaction is another major source of uncertainty in determining the solar neutrino flux that is proportional approximately [9] to the astrophysical S factor of this reaction, $S_{34}(0)^{0.8}$. A compilation of previous data, together with the latest measurement by the authors, is presented by Hilgemeier *et al.* in Ref.[10], from which a significant scatter in $S_{34}(0)$ can be seen, resulting in a considerable uncertainty on the adopted $S_{34}(0)$. It is therefore highly important to provide a more precise value than the presently recommended value of $S_{34}(0)$=0.51±0.02 keV b [10]. We have recently initiated a new precision measurement of this cross section at energies around $E_{c.m.}$ = 500 - 950 keV to serve as a normalization to the $S_{34}(E)$ curve, using a ^{3}He beam and a ^{4}He gas cell and determining the cross section by measuring the ensuing ^{7}Be activity. Since the energy dependence among various theoretical approaches and experimental determinations agrees quite well [11, 12], the normalized curves will then be compared with S(E) values from our ongoing experiment to obtain $S_{34}(0)$. We note here also our earlier attempt towards a new determination of $S_{34}(0)$ by using the accelerator mass spectrometry technique [13].

A schematic diagram of our experimental setup is shown in Fig. 2. The ^{3}He beam from the 3 MV Van de Graaff accelerator at the Weizmann Institute of Science enters the ^{4}He gas cell through a nickel (Ni) window of either 0.5 or 1 μm and is raster-scanned in order to avoid excessive localized heating of the Ni window. The beam direction is defined by an upstream slit at 2 meters from the center of the cell and two Ta collimators of 3 mm, one at the entrance and the other at the exit of the chamber. An aperture operated at −400 V, placed before the Ni window, serves as a secondary electron suppressor. The gas cell is insulated from the beam line and the entire chamber, including the Cu stopper that is

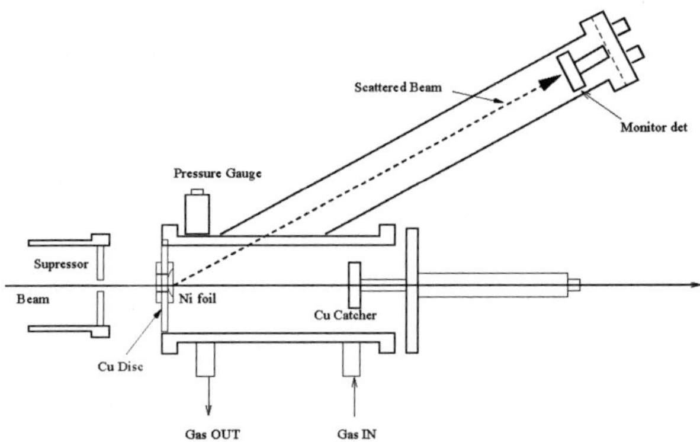

FIGURE 2. A schematic drawing of the chamber. The various items described above, such as the gas foil, electron suppression, the Si detector and the Cu stopper, are depicted.

TABLE 1. Results from the present experiment at $E_{c.m.} = 952$ keV. The cross section and S-factor values are quoted in arbitrary units for both current integration (CI) and elastic scattering (Rut.) beam-particle normalizations as described in the text. The total number N(^{7}Be, $t = 0$) of ^{7}Be nuclei is normalized to correspond to the end of the implantation.

$E_{c.m.}$	**N(^{7}Be, t=0)**	σ_{CI} **[arb. units]**	σ_{Rut} **[arb. units]**	**S (E) [arb. units]**	S_{Rut}**(E) [arb. units]**
952	6.64 10^6 (3.5%)	1000	1000	0.400	0.400
952	8.61 10^6 (3.0%)	911	893	0.364	0.357
952	6.89 10^6 (4.1%)	1024	1089	0.409	0.434
952	5.86 10^6 (4.5%)	952	1012	0.380	0.405
951	10.4 10^6 (3.6%)	1006	940	0.402	0.377

in electric contact with the chamber, serves as Faraday cup to determine the number of impinging ^{3}He particles. ^{7}Be nuclei produced by fusion of ^{3}He on ^{4}He in the cell move forward in the laboratory system and are implanted in the Cu catcher at depths of few microns. The catcher is kept at distances of 10.3 or 13.9 cm from the Ni foil and the ^{4}He gas pressure inside the cell is accordingly adjusted to obtain $\sim$120 μg/cm^2 thickness of gas. This corresponds to an energy width of the target of $\sim$200 keV at $E_{cm} = 1.1$ MeV. The gas pressure was monitored and maintained at a constant pressure. In addition, we monitor on-line the elastically scattered ^{3}He from the Ni window, using a Si surface barrier detector placed at 44.68° with respect to the beam direction. This provides a crosscheck of the beam current measurement.

The ^{7}Be nucleus decays by a 478 keV γ-ray transition from the first excited state in ^{7}Li with a lifetime of $T_{1/2} = 53.29 \pm 0.07$ d and a branching ratio of $10.45 \pm 0.04\%$ [see, e.g., Refs. 7, 8, and references therein]. The number of ^{7}Be nuclei produced in the reaction is obtained by measuring the 478 keV γ-activity at the NRC-Soreq laboratory, using a large-volume Ge detector, placed in a specially shielded environment for low count rate

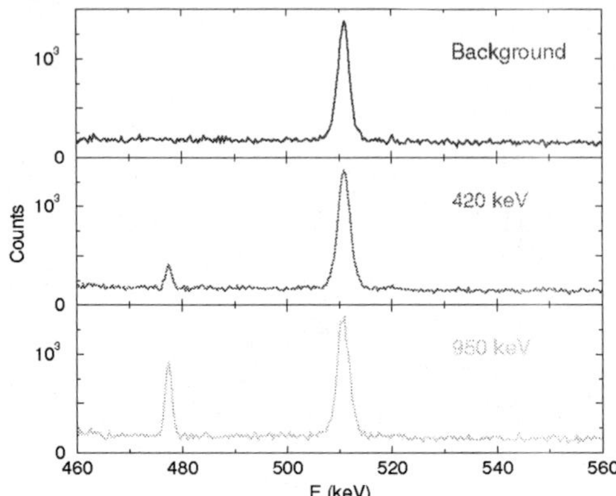

FIGURE 3. A typical spectrum from the decay of 6×10^6 atoms of ^{7}Be, corresponding to one of the measurements presented in Table I. The 478 keV γ-peak can be seen prominently with very small background.

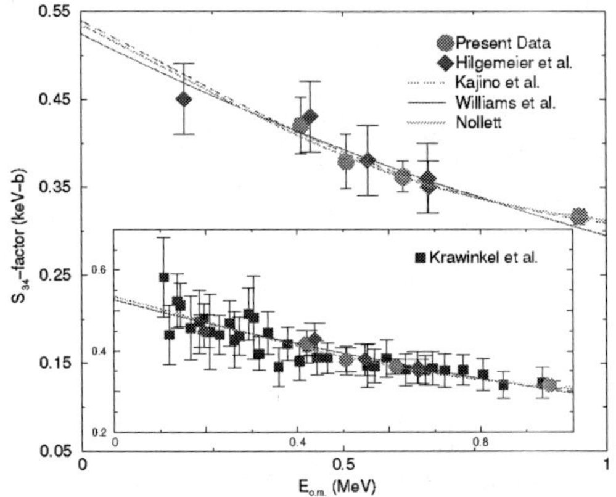

FIGURE 4. Present data together with that of previous results and representative theoretical fits, yielding $S_{34}(0) = 0.533(20)(7)$ keV b.

measurements. A similar setup was used previously by our group to determine with high accuracy the number of ^{7}Be atoms in our implanted ^{7}Be target for the determination of $S_{17}(0)$ [6, 7]. A typical γ-ray spectrum is shown in Fig. 3. The results from our recent experiment [14] are shown in Table I and in Fig. 4.

The value of the S factor we obtain is $S_{34}(0) = 0.53(2)(1)$ keV b, in excellent agreement with earlier compilation but with a much reduced error. Our present result is also in excellent agreement with a recent experiments of the Luna collaboration [15, 16]. Recently, a slightly higher value of $S_{34}(0)$ [17] has been obtained. We note the timeliness of the precise determination of $S_{34}(0)$ in view of the recent commencement of data taking of the BOREXINO experiment [18].

HOST AND TEMPERATURE DEPENDENCE OF THE ^{7}BE HALF LIFE

The decay rate of radioactive nuclei that undergo orbital Electron Capture (EC) depends on the properties of the atomic electron cloud around the nucleus. Hence, EC may exhibit varying decay rates if the nucleus is implanted into host materials with different properties of their corresponding electron clouds. The first suggestion of this effect in ^{7}Be, which is the lightest nucleus that decays by EC, and reports of experiments trying to investigate this phenomenon, have been presented by Segré *et al.* [19–21]. This effect has been qualitatively attributed in the past to the influence of the electron affinities of neighboring host atoms [22]. The electron density of the ^{7}Be atom in a high-electron affinity material such as gold is decreased via the interaction of its 2s electrons with the host atoms, resulting in a lower decay rate (longer half-life). Recently, the life-time modification has been suggested to stem from differences of the Coulomb screening potential [23] between conductors and insulators.

Several experimental and theoretical investigations were conducted during recent years to study the host material effect on the decay rate of ^{7}Be [22, 24–28], with a somewhat confusing scattering of experimental results. It was found that the half-life of ^{7}Be encapsulated in a fullerene C_{60} cage and ^{7}Be in Be metal is 52.68(5) and 53.12(5) days, respectively, amounting to a difference of 0.83(13)% [27]. A smaller effect of $\approx$0.2% was measured for the half-life 53.64(22) d of ^{7}Be in C_{60} and 53.60(19) d of ^{7}Be in Au [28]. A recent theoretical evaluation shows that short and long half-lives 52.927(56) d and 53.520(50) d were measured for ^{7}Be in Al_2O_3 and ^{7}Be in (average of BeO, BeF_2 and $Be(C_5H_5)_2$), respectively [22]. These results yield the magnitude of the effect to be as large as 1.1%. Another experimental investigation has shown that the half-life increases by 0.38% from ^{7}Be in graphite, 53.107(22) d, to ^{7}Be in Au, 53.311(42) [25], while a very recent investigation [29] has seen no effect to within 0.4%. The great interest in this phenomenon for ^{7}Be arises also from the need to explore the possible contribution of the half-life of ^{7}Be to the measurement of the cross section of the two fusion reactions, ^{7}Be(p, γ)^{8}B and ^{3}He(^{4}He, γ)^{7}Be, that play an important role in determining the solar neutrino flux [6, 14].

The present work has been undertaken in order to probe this phenomenon yet again in an experimental approach that takes full advantage of the experience gained in measuring implanted ^{7}Be activity in a controlled and precise manner for cross section determinations of the solar fusion reactions mentioned above [6, 14]. As a demonstration of the quality of the γ-activity measurement, we cite the results of [6, 7] where two independent determinations of the *absolute* activity of ^{7}Be, at the Soreq laboratory and at Texas A&M University, were in excellent agreement to within 0.7%. The same setup has also been used for determining the ^{7}Be activity ensuing from the ^{3}He(^{4}He, γ)^{7}Be reaction [14]. We report the measurement of the half-life of ^{7}Be implanted in four host materials: copper, aluminum, aluminum oxide (sapphire - Al_2O_3) and PVC (polyvinyl chloride - $[C_2H_3Cl]_n$).

The primary source of ^{7}Be for implantation was a graphite target, from the Paul Scherrer Institute (PSI), used routinely for the production of π mesons [30]. Many spallation products are accumulated in the target, including ^{7}Be. Graphite material from the PSI meson production target was placed in an ion-source canister and was brought to ISOLDE (CERN); ^{7}Be was extracted at ISOLDE by selective ionization using a resonance laser ion source. Direct implantation of ^{7}Be at 60 keV in the host material was subsequently followed. A detailed description of the extraction and implantation of ^{7}Be at ISOLDE is provided in detail in Refs. [7, 31]. This procedure facilitated a precision measurement of the cross section of the reaction ^{7}Be(p, γ)^{8}B. The implantation spot was defined by a 2 mm collimator positioned at close proximity to the target for the Cu sample and a 5 mm collimator for the other samples. This small change in the ensuing counting geometry has been well investigated for the measurement of ^{7}Be activity [14] and does not affect the results in any significant manner. The implantation process provided full control of the spot composition (^{7}Be; ^{7}Li) as well as a radial and depth profiles. For earlier implantations, at a density of ^{7}Be in Cu far exceeding that of the present experiment, the spot was found to be robust and the ^{7}Be inventory in the spot was stable [5, 7], excluding naturally radioactive decay. The copper, aluminum and PVC host material targets consisted of disks of 12 mm diameter and 1.5 mm thickness, while the sapphire target was a square of 10.2 mm × 10.2 mm. The median implantation depth of ^{7}Be into these materials has been estimated using the SRIM code [32] and found to be 12, 24, 470 and 37 μm, respectively, i.e. all implantation depths were well below the surface.

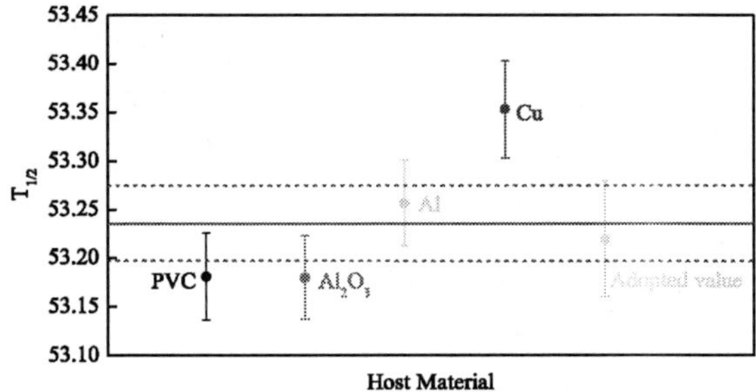

FIGURE 5. The half life of ^{7}Be in 4 host materials. The solid line represents the weighted average and the broken lines correspond to a $\pm 1\sigma$ interval. Also shown is the adopted value in the literature [33].

The results of the four host materials, Al and Cu (conductors) and PVC and Al_2O_3 are presented in Fig. 5. We found a null (or very small) change for conductors versus insulators at room temperature [34]. Even though the statistical test of the present data supports a null effect within $\pm$ 1σ.

Recently, we have measured *simultaneous* the temperature dependence of the EC decay rate of ^{7}Be in metallic Cu host and of the β^--decay rate of ^{198}Au in the host alloy Al-Au, ranging from 0.350 K to 293 K. For this we have used the "double-source assembly" experiment based on the 2 mm diameter ^{7}Be source (as explained earlier). Adjacent to the 2 mm spot of ^{7}Be activity on the Cu disc, a ^{198}Au ($T_{1/2}$ = 2.6956(3) d [35]) source was attached. This radionuclide disintegrates by 100% β^- emission with a 95.58(12)% branch of a 411.8 keV γ-ray [36], very close in energy to the γ line of ^{7}Be.

A 0.51 mm thick Al-Au (0.135% of Au) alloy wire of mass 13.7 mg was irradiated in the nuclear reactor at the Soreq Nuclear Research Centre, Israel, to produce ^{198}Au by neutron activation - ^{197}Au(n,γ)^{198}Au. The 20 min irradiation in the pneumatic transfer tube facility ("Rabbit") of the reactor induced an activity of approximately 2.8×10^4 Bq at the end of irradiation, 1-2 days prior to the commencement of the γ measurements. The active Al-Au wire was glued to the copper disk, adjacent to the ^{7}Be spot, by a heat-conducting, low-temperature adhesive, thus assuring identical thermal as well as geometrical (thermally affected expansion and contraction) properties for both ^{7}Be and ^{198}Au radionuclide.

We have used two experimental arrangements for the series of temperature dependence measurements, one for a $T = 293$ K to $T = 12.5$ K comparison ("experiment 1") and one for a $T = 4$ K to $T = 0.350$ K comparison ("experiment 2"). Schematics of the set-up for "experiment 1" and "experiment 2" are shown in Fig. 6(a) and in Fig. 6(b), respectively.

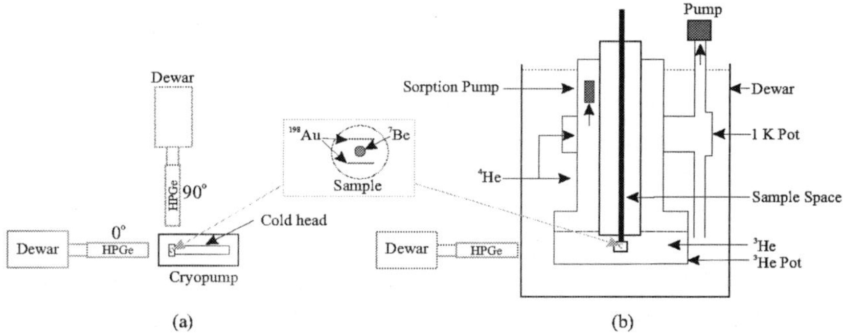

FIGURE 6. (a) Schematics of the set-up for the measurements at 12.5 K and 293 K ("experiment 1") at the 0° and 90° geometry assemblies; (b) Schematics of the set-up for the measurement at 0.350 K and 4K ("experiment 2") at the 90° geometry assembly.

The count rates of ^{7}Be and ^{198}Au were calculated for every measured spectrum and corrected for decay using the published half-lives 53.353 and 2.6956 d. In this correction, the two reference times for the 0° and 90° measurements were the middle times of the first room-temperature runs in each of these measurements. The weighted averages of the ratios of count rates at different temperatures are presented in Fig. 7.

The individual normalized count rates of ^{7}Be and ^{198}Au at the 0° position for 293 K were: 2.8462(44) and 236.05(6). For 12.5 K the normalized rates were: 2.8121(29) and 232.59(20), respectively. These results demonstrate that at this position, the count rates of both radionuclides are lower by about (1.20 ± 0.18) to (1.46 ± 0.09)% at the low-temperature (12.5 K) compared to the room temperature (293 K). The corresponding count rates of ^{7}Be and ^{198}Au at the 90° position for 293 K were 0.8030(22) and 94.53(24) and for 12.5 K these were 0.8092(29) and 95.57(44), respectively. This indicates an opposite finding at the 90° position, but with much smaller precisions. A statistical test supports a null-effect, which is anticipated because the thermal contraction induces a negligible change in the distance between the detector and the sources. We emphasize again that these effects cancel out in very good approximation by the ^{7}Be/^{198}Au ratios.

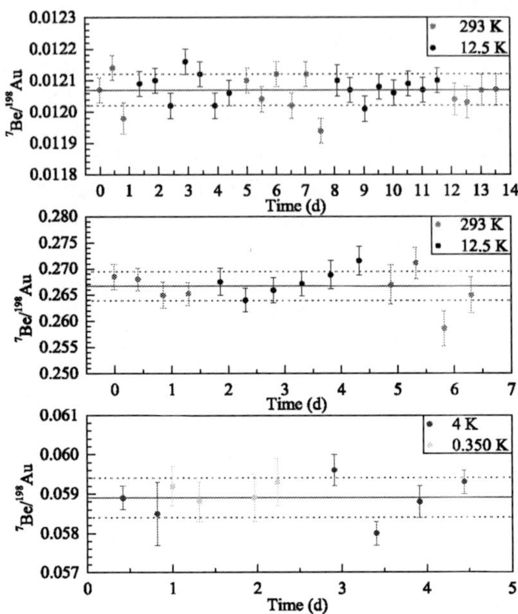

FIGURE 7. Top: Ratio of rates (^{7}Be/^{198}Au) at 0° for the temperature at 12.5 K and 293 K; Middle: Ratio of rates (^{7}Be/^{198}Au) at 90° for the temperature at 12.5 K and 293 K; Bottom: Ratio of rates (^{7}Be/^{198}Au) at 90° for the temperature at 4 K and 0.350 K. The solid line represents the weighted average and the broken lines correspond to a $\pm 1\sigma$ interval.

In summary, the half-life of ^{198}Au in the alloy Al-Au (0.135% of Au) was measured and a null-effect of (0.02 ± 0.11)% was found at 12.5 K relative to 293 K. The count rate of ^{7}Be in copper was measured at these temperatures relative to that of ^{198}Au. The normalized ratio of ratios (ratio ^{7}Be/^{198}Au at 12.5 K to ratio ^{7}Be/^{198}Au at 293 K), at the 0° and 90° are 1.0015(16) and 1.0032(58), which are equal within the experimental uncertainty, yielding a null temperature effect of (0.15 ± 0.16)%, i.e., the half-life of ^{7}Be at 12.5 K is shorter by 0.15 ± 0.16 than the value of 53.353 d as was determined at 293 K [34]. The same ratios of ^{7}Be to ^{198}Au yields 0.0590(3) and 0.0589(3) at 0.350 K and 4 K, respectively. The ratio of ratios is therefore 1.0028(71), again a null temperature dependence of (0.28 ± 0.71)% for these two temperatures.

We note that the Debye-H*ü*ckel screening model, used in previous publications to explain the apparent temperature dependence, is not applicable for a solid (strong-coupled plasma) but rather for a weak-coupled plasma. Furthermore, even within the Debye-H*ü*ckel screening model itself, the explicit temperature dependence of U_D - the Coulomb energy of the Debye cloud - of the form $1/\sqrt{T}$, is valid only for temperatures much higher than the Fermi temperature, E_F, being of the order of thousands of degrees for Cu [37–39]. Below the Fermi temperature (certainly for 293 K and lower), there should be no temperature dependence. The present results for ^{7}Be and ^{198}Au validate the theoretical picture above and do not support earlier experimental observations.

ACKNOWLEDGMENTS

I would like to thank my close co-workers at the Weizmann Institute and all my other colleagues for a fruitful collaboration. We acknowledge the support of the Israel Science Foundation (ISF).

REFERENCES

1. Q. R. Ahmad *et al.* (SNO collaboration), *Phys. Rev. Lett.* **87**, 071301 (2001); **89**, 011301 (2002).
2. B.T. Cleveland *et al.*, *Astrophys. J.* **496**, 505 (1998).
3. S. Fukuda *et al.*, *Phys. Rev. Lett.* **86**, 5651 (2001).
4. L. Weissman *et al.*, *Nucl. Phys. A***630**, 678 (1998).
5. M. Hass *et al.*, *Phys. Lett. B* **462**, 237 (1999).
6. L.T. Baby *et al.*, *Phys. Rev Lett.* **90**, 022501 (2003).
7. L.T. Baby *et al.*, *Phys. Rev. C* **67**, 065805 (2003).
8. L.T. Baby *et al.*, *Nucl. Phys. A* **718**, 487c (2003).
9. See, e.g., J.N. Bahcall *et al.*, *Astrophys. J.* **555**, 990 (2001), and references therein.
10. M. Hilgemeier *et al.*, *Z. Phys. A* **329**, 243 (1988).
11. T.A. Tombrello *et al.*, *Phys. Rev.* **131**, 2582 (1963).
12. R.D. Williams *et al.*, *Phys. Rev. C* **23**, 2773 (1981).
13. C. Bordeanu *et al.*, Prog. Part. Nucl. Phys. **46**, 97 (2001).
14. B.S. Nara Singh *et al.*, *Phys. Rev. Lett.* **93**, 262503 (2004).
15. D. Bemmerer *et al.*, *Phys. Rev. Lett.* **97**, 122502 (2006).
16. F. Confortola *et al.*, *Phys. Rev. C* **75**, 065803 (2007).
17. T.A.D. Brown *et al.*, *Phys. Rev. C* **76**, 055801 (2007).
18. C. Arpesella *et al.* (Borexino Collaboration), http://arxiv.org/abs/0708.2251v2.
19. E. Segré, *Phys. Rev.* **71**, 274 (1947).
20. E. Segré and C. E. Wiegand, *Phys. Rev.* **75**, 39 (1949).
21. R.F. Leininger *et al.*, *Phys. Rev.* **76**, 897 (1949).
22. P. Das and A. Ray, *Phys. Rev. C* **71**, 025801 (2005).
23. K.U. Kettner *et al.*, *J. Phys. G* **32**, 489 (2006); and references therein.
24. A. Ray *et al.*, *Phys. Lett. B* **455**, 69 (1999).
25. E.B. Norman *et al.*, *Phys. Lett. B* **519**, 15 (2001).
26. A. Ray *et al.*, *Phys. Lett. B* **531**, 187 (2002).
27. T. Ohtsuki *et al.*, *Phys. Rev. Lett.* **93**, 112501 (2004).
28. A. Ray *et al.*, *Phys. Rev. C* **73**, 034323 (2006).
29. B.N. Limata *et al.*, *Eur. Phys. J. A Direct* **27**, 193 (2006).
30. G. Heidenreich *et al.*, *AIP Conf. Proc.* **642**, 122 (2002).
31. U. Köster *et al.*, *Nuc. Inst. Meth. B* **204**, 343 (2003).
32. SRIM package from www.srim.org.
33. D.R. Tilley *et al.*, *Nucl. Phys. A* **708**, 3 (2002).
34. Y. Nir-El *et al.*, *Phys. Rev. C* **75**, 012801(R) (2007).
35. J.K. Tuli, *Nuclear Wallet Cards*, 7th ed. (2005), www.nndc.bnl.gov.
36. R.B. Firestone and V.S. Shirley (Eds.), *Table of Isotopes*, 8th ed., John Wiley & Sons, New York, (1996).
37. N.W. Ashcroft and N.D. Mermin, *Solid State Physics*, Thomson Learning Inc., (1976).
38. K. Czerski *et al.*, *Eur. Phys J. A Direct* **27**, 83 (2006).
39. S. Mehedinteanu, *Acta Physica Polonica B* **38**, 3287 (2007).

The Coulomb Dissociation of ^{8}B: A New Tool in Nuclear Astrophysics [1]

Moshe Gai

Laboratory for Nuclear Science at Avery Point, University of Connecticut, 1084 Shennecossett Rd, Groton, CT 06340-6097.
and
Department of Physics, WNSL Rm 102, Yale University, PO Box 208124, 272 Whitney Avenue, New Haven, CT 06520-8124.

e-mail: moshe.gai@yale.edu, URL: http://www.phys.uconn.edu

Abstract.
The GSI1, GSI2 (as well as the RIKEN2 and the corrected GSI2) measurements of the Coulomb Dissociation (CD) of ^8B are in good agreement with the most recent Direct Capture (DC) $^7Be(p,\gamma)^8B$ reaction measurement performed at Weizmann and in agreement with the Seattle result. Yet it was claimed that the CD and DC results are sufficiently different and need to be reconciled. We show that these statements arise from a misunderstanding (as well as misrepresentation) of CD experiments. We recall a similar strong statement questioning the validity of the CD method due to an invoked large E2 component that was also shown to arise from a misunderstanding of the CD method. In spite of the good agreement between DC and CD data the slope of the astrophysical cross section factor (S_{17}) can not be extracted with high accuracy due to discrepancies among the most recent DC data as well as a discrepancies among the three reports of the GSI CD data. The slope is directly related to the d-wave component that dominates at higher energies. This d-wave component must be subtracted from measured data to extrapolate to zero energy. Hence the uncertainty of the measured slope leads to an additional downward uncertainty ($^{+0.0}_{-3.0}$ eV-b) of the extrapolated zero energy cross section factor, $S_{17}(0)$. Such an uncertainty is also consistent with the smaller value of $S_{17}(0)$ extracted using the ANC method. This uncertainty must be alleviated by future experiments to allow a precise determination of $S_{17}(0)$, a goal that so far has not be achieved in spite of strong statement(s) that appeared in the literature.

Keywords: Coulomb Dissociation, Direct Capture, Astrophysical Cross Section Factor, Solar Neutrinos.
PACS: 26.30.+k, 21.10.-k, 26.50.+k, 25.40.Lw

INTRODUCTION

The Coulomb Dissociation (CD) method was developed in the pioneering work of Baur, Bertulani and Rebel [1] and has been applied to the case of the CD of ^8B [2, 3, 4, 5] from which the cross section of the $^7Be(p,\gamma)^8B$ reaction was extracted. This cross section is essential for calculating the ^8B solar neutrino flux. The CD data were analyzed with a remarkable success using only first order Coulomb interaction that includes only E1 contribution. An early attempt (even before the RIKEN data were published)

[1] Work Supported by USDOE grant No. DE-FG02-94ER40870

CP972, *Exotic Nuclei and Nuclear/Particle Astrophysics (II)—Proceedings of the Carpathian Summer School of Physics 2007,* edited by L. Trache and S. Stoica

to refute this analysis by introducing a non-negligible E2 contribution [6] was shown [7] to arise from a neglect of the angular acceptance of the RIKEN1 detector and a misunderstanding of the CD method. Indeed the CD of ^8B turned out to be a testing ground of the very method of CD. Later claims by the MSU group for evidence [8] of non-negligible E2 contribution in **inclusive measurement** of an asymmetry, were disputed in a recent **exclusive measurement** of a similar asymmetry by the GSI2 collaboration [5].

In contrast, Esbensen, Bertsch and Snover [9] recently claimed that higher order terms and an E2 contribution are an important correction to the RIKEN2 data [3]. It is claimed that "S_{17} values extracted from CD data have a significant steeper slope as a function of E_{rel}, the relative energy of the proton and the ^7Be fragment, than the direct result". However they find a substantial correction only to the RIKEN2 CD data and claim that this correction(s) yield a slope of the RIKEN2 data in better agreement with Direct Capture (DC) data. In addition it is stated [9] that "the zero-energy extrapolated $S_{17}(0)$ values inferred from CD measurements are, on the average 10% lower than the mean of modern direct measurements". The statements on significant disagreement between CD and DC data are based on the re-analyses of CD data by the Seattle group [10]. In this paper we demonstrate that an agreement exists between CD and DC data and the statements of the Seattle group [10] are based on misunderstanding (as well as misrepresentation) of CD data.

In spite of the general agreement between CD and DC data, still the the slope of astrophysical cross section factor measured between 300 - 1,500 keV can not be extracted with high accuracy. This hampers our ability to determine the d-wave contribution that dominates the cross section of the $^7Be(p,\gamma)^8B$ reaction at higher energies and must be subtracted for extrapolating the s-wave to zero energy. Lack of accurate knowledge of the d-wave contribution to data (even if measured with high accuracy), precludes accurate extrapolation to zero energies. We show that this leads to additional downward uncertainty ($^{+0.0}_{-3.0}$ eV-b) of the extrapolated $S_{17}(0)$. We doubt the strong statement that $S_{17}(0)$ was measured with high accuracy (see for example [10]).

THE SLOPE OF S_{17} ABOVE 300 KEV

Early on it was recognized that s-wave capture alone yields an s-factor with a negative slope. This is due to the Coulomb distortion of the s-wave at very low distances. The observation of a positive slope of S_{17} measured at energies above 300 keV was recognized as due to the d-wave contribution. It was also recognized that the d-wave contribution is very large at measured energies and in fact it dominates around 1.0 MeV. The d-wave contribution must be subtracted from data measured at higher energies in order to allow for an accurate extrapolation of the s-wave to zero energy (where the d-wave contribution is very small, of the order of 6%). The (large) contribution of the d-wave at energies above 300 keV leads to a linear dependence of S_{17} on energy (with a positive slope of the order of +7.0 eV-b/MeV). An accurate extrapolation of S_{17} must rely on an accurate knowledge of the d-wave contribution or the slope at energies above 300 keV.

In Fig. 1 we show the slope parameter (S' = dS/dE) extracted from both DC and CD

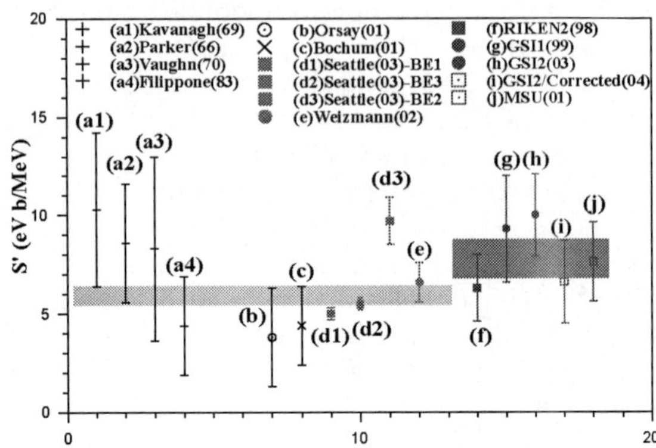

FIGURE 1. The measured slopes (S' = dS/dE) of world data measured between 300 and 1500 keV, as discussed in the text. The range of "average values" is indicated and discussed in the text.

data in the energy range of 300 - 1500 keV. We refer the reader to [11] for detailes on data used to extract the slope shown in Fig. 1. We conclude from Fig. 1 that the slope parameter can not be extracted from DC data [10, 13, 14, 15, 16, 17, 18, 19] with high accuracy as claimed. The DC data are not sufficiently consistent to support this strong statement [10]; for example there is not a single data point measured by the Bochum group [14] that agrees with that measured by the Seattle group [10], where we observe that some of the individual data points disagree by as much as five sigma. The disagreement of the three slopes measured by the Seattle group and the disagreement with the Weizmann slope are most disturbing. In the same time the dispersion among slopes measured in CD is also of concern. However, it is clear that the over all agreement between CD and DC data (1.7 sigma) is better than the agreement among specific DC data. We do not support the strong claim of substantial disagreement between slopes measured in DC and CD [10].

The lack of evidence for substantial difference between CD and DC results leads to doubt on the very need to reconcile these data [12]. Furthermore, in Fig. 2 we show the slope obtained by EBS after their attempt to reconcile the slope of CD with the slope of DC data. Clearly the original slope of the RIKEN2 data obtained using only first order E1 interactions is in considerably better agreement with DC data than the so called reconciled slope obtaine by EBS.

$S_{17}(0)$ EXTRACTED FROM CD DATA

In Fig. 20 of the Seattle paper [10] they show extracted $S_{17}(0)$ from CD using the extrapolation procedure of Descouvemont and Baye [20], and based on this analysis it is stated [9] that "the zero-energy extrapolated $S_{17}(0)$ values inferred from CD measurements are, on the average 10% lower than the mean of modern direct measurements". The extracted $S_{17}(0)$ shown in Fig. 20 [10] are only from data measured at energies below 425 keV and the majority of CD data points that were measured above 425 keV

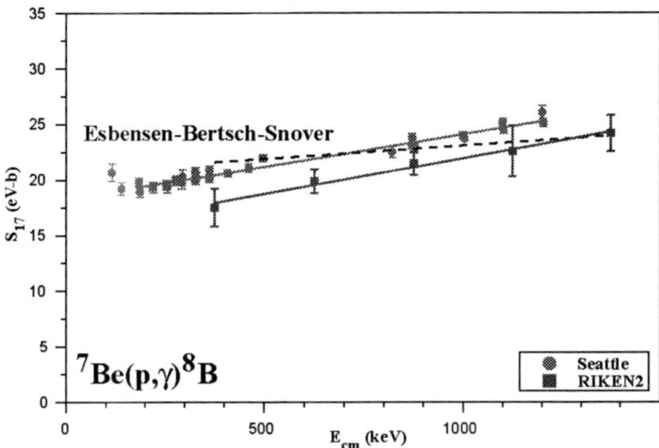

FIGURE 2. Extracted S_{17} from the RIKEN2 CD data [3] using first order electric dipole interaction as shown in [5], compared to the DC capture data published by the Seattle group [10] and the so called reconciled slope calculated by EBS [9]. The shown RIKEN2 data include systematic uncertainties (equal or slightly smaller) as published [3].

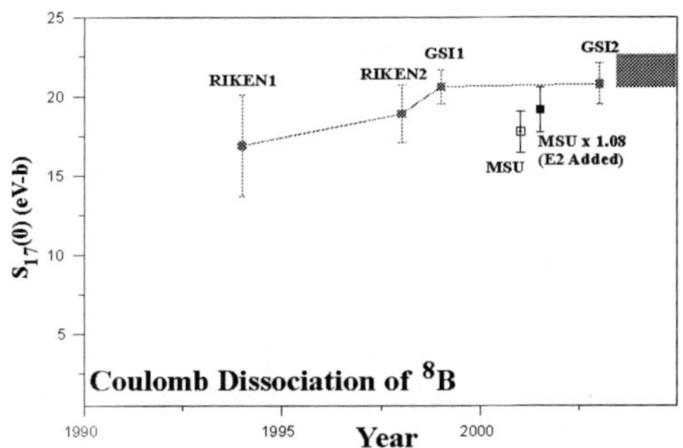

FIGURE 3. Measured $S_{17}(0)$ as originally published by the authors who performed the CD experiments. These analyses include all measured data points [2, 3, 4, 5, 8] using the extrapolation procedure of Descouvemont and Baye [20]. We also plot the MSU data as published as well as with the E2 correction ($\approx 8\%$) [8] added back to the quoted $S_{17}(0)$, as discussed in the text. The range of $S_{17}(0)$ results from the measurements of DC by the Seattle [10] and Weizmann groups [15] is indicated.

were excluded in Fig. 20 [10].

This arbitrary exclusion of (CD) data above 425 keV has no physical justification (especially in view of the fact that the contribution of the 632 keV resonance is negligible in CD). For example as shown by Descouvemont [21] the theoretical error increases to approximately 5% at 500 keV and in fact it is slightly decreased up to approximately 1.0 MeV, and there is no theoretical justification for including data up to 450 keV but excluding data between 500 keV and 1.0 MeV.

Thus when excluding the CD data above 425 keV, the Seattle group excluded the data that were measured with the best accuracy and with smallest systematical uncertainty. If in fact one insists on such an analysis of CD data, one must estimate the systematic

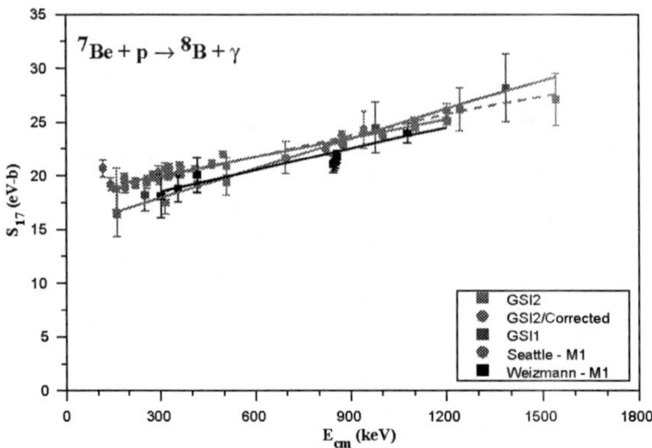

FIGURE 4. A comparison of the most recent DC data with the GSI1 and GSI2 results.

uncertainty due to this selection of data. This has not been done in the Seattle re-analyses of CD data [10].

Instead we rely here on the original analyses of the authors that published the CD data. In Fig. 3 we show the $S_{17}(0)$ factors extracted by the original authors who performed the CD experiments. These results include all measured data points up to 1.5 MeV, and are analyzed with the same extrapolation procedure of Descouvemont and Baye [20].

We note that the (four) CD results are consistent within the quoted error bars, but they show a systematic trend of an increased $S_{17}(0)$ (to approximately 20.7 eV-b), while the error bars are reduced. We obtain a $1/\sigma$ weighted average of $S_{17}(0) = 20.0 \pm 0.7$ with $\chi^2 = 0.5$, which is in excellent agreement with the measurement of the Weizmann group [15] and in agreement with the measurement of the Seattle group [10].

EXTRAPOLATING $S_{17}(0)$ FROM WORLD DATA

The current situation with our knowledge of S_{17} and the extrapolated $S_{17}(0)$ is still not satisfactory. The main culprit are major disagreements among DC data. It is clear for example that the systematic disagreements between the Orsay-Bochum [13, 14] and the Weizmann-Seattle [10, 15] results must be resolved before these data are included in a so called "world average". In Fig. 4 we compare the most recent Seattle-Weizmann data (with M1 contribution subtracted) with the GSI1 and GSI2 (as well as corrected GSI2) results. While the data appear in agreement we still observe a systematic disagreement between all measured slopes. The DC data of the Seattle and the Weizmann groups have different slopes as do the GSI1, GSI2 and corrected GSI2 data. The slope above 300 keV is directly related to the d-wave contribution that dominates at measured laboratory energies. The d-wave component must be subtracted from measured data to extrapolate to solar burning energies. This disagreement does not allow for an accurate (better than 5% accuracy) extrapolation of $S_{17}(0)$ and must be resolved by future experiments. A reasonable systematic error of +0.0 -3.0 eV-b due to extrapolation seems to be required by current data. Such an additional downward systematical error is also consistent with the results of data analysis using the ANC method that yields smaller values of $S_{17}(0)$ [22]. The results of the ANC method do not rely on a theoretical extrapolation as in the case of the DC and CD data that are measured at higher energies.

REFERENCES

1. G. Baur, C.A. Bertulani, and H. Rebel; Nucl. Phys. **A458**(1986)188.
2. T. Motobayashi *et al.*; Phys. Rev. Lett. **73**(1994)2680.
3. T. Kikuchi *et al.*; Phys. Lett. **B391**(1997)261, ibid E. Phys. J. **A3**(1998)213.
4. N. Iwasa *et al.*; Phys. Rev. Lett. **83**(1999)2910.
5. F. Schumann *et al.*; Phys. Rev. Lett. **90**(2003)232501, ibid Phys. Rev. **C73**(2006)015806.
6. K. Langanke and T.D. Shoppa; Phys. Rev. C52(1995)1709.
7. M. Gai and C.A. Bertulani; Phys. Rev. **C52**(1995)1706.
8. B.S. Davids *et al.*; Phy. Rev. **63**(2001)065806.
9. H. Esbensen, G.F. Bertsch, and K. Snover; Phys. Rev. Lett. **94**(2005)042502.
10. A.R. Junghans *et al.*; Phys. Rev. **C68**(2003)065803.
11. M. Gai; Phys. Rev. **C74**(2006)025810.
12. M. Gai; Phys. Rev. Lett. **96**(2006)159201.
13. F. Hammache *et al.*; Phys. Rev. Lett. **86**(2001)3985.
14. F. Strieder *et al.*; Nucl. Phys. **A696**(2001)219.
15. L.T. Baby Phys. Rev. **C67**(2003)065805, ER **C69**(2004)019902(E).
16. B.W. Filippone *et al.*; Phys. Rev. **C28**(1983)2222.
17. F.J. Vaughn *et al.*, Phys. Rev. **C2**(1970)1657.
18. P.D. Parker, Phys. Rev. **150**(1966)851.
19. R.W. Kavanagh, T.A. Tombrello, T.A. Mosher, and D.R. Goosman, Bull. Amer. Phys. Soc., **14**(1969)1209.
20. P. Descouvemont and D. Baye; Nucl. Phys. **A567**(1994)341.
21. P. Descouvemont; Phys. Rev. **C70**(2004)065802.
22. L. Trache *et al.*; Phys. Rev. C **R67**(2003)062801.

Reaction Rates for H-burning in Stars from Experiments with Radioactive Nuclear Beams

Livius Trache

Cyclotron Institute, Texas A&M University, College Station, TX 77843-3366

Abstract. In this lecture some of the indirect methods for nuclear astrophysics that can be applied to extract reaction rates for H-burning in stars are treated. Instead of reviewing all methods, I chose for exemplification four different experiments of our Texas A&M group, each involving a different method, but all related to the study of the proton rich nucleus ^{23}Al. They involve the use of radioactive nuclear beams. I will describe the use of one-nucleon transfer reactions (the ANC method), breakup reactions at intermediate energies, β-decay and β-delayed proton-decay to extract nuclear structure information that is then used for nuclear astrophysics. I will show the specificities of each method, talk about their complementarities and redundancies.

Keywords: nuclear astrophysics; indirect methods; H-burning; transfer reactions; breakup; β-decay; β -delayed proton-decay.
PACS: 26.30.+k, 25.60-t, 25.70.Hi, 23.40.-s, 23.50.+z

FOREWORD

Today's sessions are called "Nuclear Physics for Astrophysics", a term coined by professor Rolfs when "Nuclear Astrophysics" became too broad, encompassing all phenomena in cosmos involving nuclei and nuclear processes, a much broader realm indeed. In the present lecture I will stick very closely to what the name of the section means and talk not about stellar reactions, but about nuclear physics experiments used to determine those nuclear reaction rates in stellar environments, reactions for H-burning in particular. What connects them is the fact that they are all indirect methods and use radioactive nuclear beams. The emphasis will be to present the methods, rather than to communicate results (values for reaction rates).

As organizer, I had the advantage to know from the outset that I could count on the excellent introduction to the field made by prof. Rolfs in the first lecture this morning (*Editor's note: unfortunately missing from the present Proceedings*), including his briefing about its main notions, tools and challenges. I also knew I could count on the very good examples contained in the lectures of the two speakers following. I will use these advantages and go directly into the subject! The reader in need for a good introduction could go to his excellent book [1] or to my more modest papers in the proceedings of last year's Carpathian school in Predeal [2].

CP972, *Exotic Nuclei and Nuclear/Particle Astrophysics (II)—Proceedings of the Carpathian Summer School of Physics 2007,* edited by L. Trache and S. Stoica

INDIRECT METHODS IN NUCLEAR ASTROPHYSICS

It was said and shown earlier in the day that an important challenge of direct nuclear astrophysics measurements stems from the very low cross sections involved when reactions between charged particles are involved, due to Coulomb repulsion. This is the case with radiative proton capture reactions occurring in H-burning. Another important challenge in the study of stellar reactions is that in many important cases the targets involved are unstable nuclei. These two reasons combined lead to the use of indirect methods.

A number of indirect methods have been used for nuclear astrophysics, with some of them tailored specifically for this purpose. The list, however, is not very long: Coulomb dissociation [3,4], transfer reactions (the ANC method) [5], breakup at intermediate energies [6,7], the Trojan horse method [8], as well as other spectroscopic methods, in particular the location and study of resonances. Some can be applied to reactions important in H-burning.

Our MARS group at texas A&M University was involved in a nuclear astrophysics program for over a decade now. Our studies had two main characteristics: (a) were done using indirect methods and (b) involved experiments with both radioactive nuclear beams (RNB) and with stable beams. I select here four different ones that are related to one specific nuclear astrophysics problem. Space requirements of this volume will make me be more schematic and focused in print than in the actual presentation.

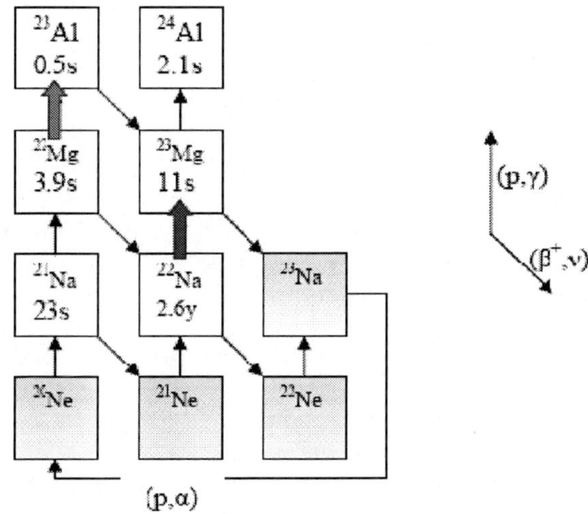

FIGURE 1. The reaction flow around the NeNa cycle in H-burninng in ONe novae. The two radiative proton capture reactions on ^{22}Mg and ^{22}Na treated here shown by the thick arrows.

The Nuclear Astrophysics Motivation

Space-based gamma-ray telescopes have had for some time the ability to detect gamma rays of cosmic origin, proving that nucleosynthesis is an ongoing process in our own Galaxy. Gamma rays following the β-decay of long-lived ^{26}Al, ^{44}Ti, ^{56}Ni, etc., have been detected. It was also predicted that ^{22}Na ($t_{1/2}$=2.6 y) is going to be produced in the thermonuclear runaway of the so-called ONe (Oxygen-Neon) novae [1,9]. The novae models using the currently available nuclear reaction rates led to the conclusion that enough ^{22}Na is produced and survives as part of the reaction chain in the hot NeNa cycle ^{20}Ne(p,γ)^{21}Na(p,γ)^{22}Mg(βγ)^{22}Na for the gamma-ray line E_γ=1275 keV from its decay to be observed at the sensitivity level of the current satellite based telescopes. However, the line was not observed and the question arising is why? The nuclear data came under scrutiny. Two reactions were proposed as mechanisms for the depletion of ^{22}Na or its precursor ^{22}Mg: ^{22}Na(p,γ)^{23}Mg [10] and ^{22}Mg(p, γ)^{23}Al [11]. These radiative proton capture reactions are marked with thick arrows in Figure 1 which shows the reaction flow in this mass region in ONe novae. In the following, we study these two reactions, or better said, we use indirect methods to obtain nuclear structure information from which we extract stellar reaction rates.

Transfer reactions – the ANC method

We have two contributions with two examples on this subject at this conference [12,13]. The Asymptotic Normalization Coefficient (ANC) method is the indirect method we introduced more than a decade ago to determine astrophysical S-factors for the non-resonant component of radiative proton capture at low energies (tens or hundreds of keV) from one-proton transfer reactions involving complex nuclei at laboratory energies (10 MeV/u) [5]. Later we have also introduced the idea of using neutron-transfer reactions with stable beams and charge symmetry to determine proton ANC for cases where RNB experiments are impossible or very difficult. Here, instead of using the proton transfer reaction (^{22}Mg,^{23}Al) to measure the g.s. ANC for ^{23}Al, we have studied its mirror nucleus ^{23}Ne using the mirror neutron-transfer reaction ^{13}C(^{22}Ne,^{23}Ne)^{12}C and charge symmetry. This experiment and its results are described in the contribution [12], I refer the reader there.

Breakup reactions at intermediate energies: the breakup of ^{23}Al

The previous experiment gives information about the g.s. of ^{23}Al provided it is the analog of the g.s. of ^{23}Ne. However, the ground state spin and parity for ^{23}Al was uncertain (before our experiments), with assignments that include $1/2^+$, $3/2^+$ and $5/2^+$. The mirror nucleus ^{23}Ne has $J^\pi=5/2^+$ for its ground state and it was natural to assume the same spin for ^{23}Al. However, recently it was claimed [14,15] that proton-rich ^{23}Al is a halo nucleus and that this can only be explained if the last proton is in the $2s_{1/2}$ orbital, not in $1d_{5/2}$ (level inversion), i.e. $J^\pi=1/2^+$ for ^{23}Al g.s. Using $1/2^+$ instead of $5/2^+$, we calculated [7] the astrophysical S-factor and stellar reaction rate for ^{22}Mg(p,γ)^{23}Al and found an increase of 30-50 times over the current estimate for the temperature range T_9=0.1-0.3. This would result in a significant depletion of ^{22}Mg before it β decays into ^{22}Na and, if confirmed, could explain the non-observation of the 1.275 MeV γ-ray. It was important for both nuclear structure and for nuclear astrophysics to determine unambiguously the spin and parity of the ground state of ^{23}Al. We proposed to do it in two ways. One was by measuring the core momentum distribution from ^{23}Al breakup at intermediate energies. The other way was to use β-decay measurements.

We proposed an experiment that could determine the spin and parity, the ANC and the configuration mixing in the ground state of ^{23}Al from the measurement of the momentum distribution of the core after the proton breakup on a light target. We calculated that the momentum distribution is about 2 times narrower for a $2s_{1/2}$ orbital than for a $1d_{5/2}$ orbital. The method to use breakup reaction for nuclear astrophysics was proposed earlier by our group [6], and this particular case was described in Ref. 7. At the end of 2004 the experiment E491 entitled "**^{23}Al proton breakup at intermediate energies and astrophysical consequences**" proposed by a group of scientists from TAMU, IFIN-HH Bucharest, LPC Caen, France and a number of British universities, was accepted by the PAC at GANIL Caen. In September-October 2006 the experiment was carried out using a special experimental setup. The Ge clover detectors of EXOGAM were arranged in a new configuration at the target position of the magnetic spectrograph SPEG. Also 12 NaI detectors were included in the system. The setup was meant for two experiments measuring coincidences between gamma-rays detected at the target position and the core remnants after breakup, measured at the focal plane of SPEG. The two experiments were aimed at two different classes of nuclei: our experiment E491 was doing it for a number of radioactive **proton-rich nuclei around ^{23}Al**, the companion experiment E452 concentrated on **neutron-rich nuclei** around ^{23}O. They were the first experiments of this class to use high resolution Ge detectors for the detection of gamma-rays from the core excitations.

The primary beam 95 A MeV ^{32}S was delivered on a 450 mg/cm^2 carbon target in SISSI. SISSI was tuned for a ^{23}Al secondary beam at magnetic rigidity Bρ=1.95 Tm. That gave about 15 kHz of secondary beam on the second target (C at 180 mg/cm^2), a cocktail made mostly of ^{24}Si, ^{23}Al, ^{22}Mg and ^{21}Na, with ^{23}Al @ 50 MeV/u being about 1.4% of total.

With SPEG tuned at 1.75 Tm (later 1.76 Tm) we could see that we measured the momentum distributions for the cores of the beam nuclei after the removal of one proton: ^{23}Al, ^{22}Mg, ^{21}Na and ^{20}Ne. We have inclusive momentum distributions and in coincidence with gammas. The data analysis is still in progress. The conclusion we have from the analysis at this point is that the method works and that there is a large amount of configuration mixing in the g.s. of ^{23}Al, a finding which supports the small value of the spectroscopic factor $S(1d_{5/2})\approx 0.4$ found in the TAMU experiment for its mirror. We also found, for the first time in such measurements, that beside one-proton-removal we have an important charge-exchange component (for example the reaction ^{12}C(^{22}Mg,^{22}Na)^{12}N, see Fig. 2). Another conclusion is that we will be able to determine ANCs for a number of other nuclei present in the beam cocktail used, ANCs to be used to determine astrophysical S-factors and rates relevant for reactions in H-burning, all impossible to determine from any other measurements. This is the main advantage of breakup reactions at intermediate energies: it can be used for cocktail beams of low quality.

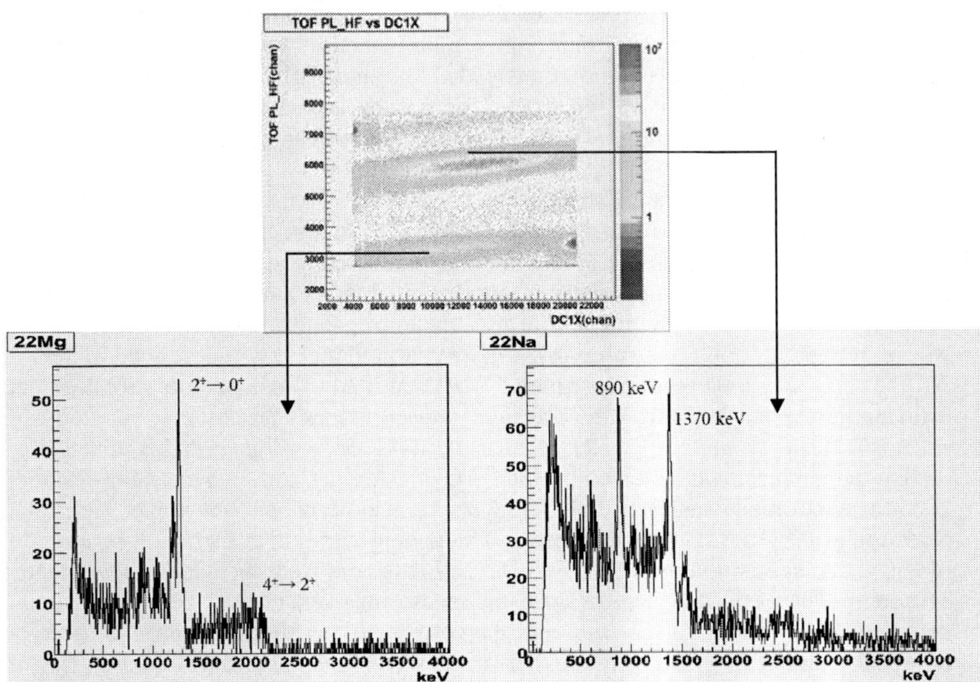

FIGURE 2. Illustration of the superposition of the two reaction mechanisms discussed in text. Top part shows for the same mass A=22 separated in the focal plane of SPEG split into ^{22}Mg and ^{22}Na. Bottom parts show the gamma-rays in coincidence with each one, separately.

β-decay of ^{23}Al

In 2005 and 2006, we had experiments at TAMU to produce and study the β-decay of ^{23}Al. We produced ^{23}Al using a 48 MeV/nucleon ^{24}Mg beam from the K500 cyclotron via the ^{24}Mg(p, 2n)^{23}Al reaction on a hydrogen gas cryogenic target cell cooled by LN_2. With MARS recoil separator a ^{23}Al beam of about 4000 pps at the extraction slits was separated in the focal plane. This was the first time pure and intense ^{23}Al samples were ever produced and separated. Its β-decay was further studied using a fast tape transport system [16]. Because the ranges of impurities in the beam (5%) are different from that of ^{23}Al, a pure ^{23}Al sample stops in the tape. In our measurement, we collected ^{23}Al for 1 second. Then we shifted the RF phase of the cyclotron to stop the primary beam, moved the ^{23}Al sample in 0.175 s with the tape transport system to a counting station consisting of a HPGe γ detector and a β detector. β and β-γ coincidence were measured for 3.2 seconds. These cycles were precisely clock controlled and were repeated continuously. Time-tagged coincidence data were stored event by event in the computer. This experimental setup is a typical one for measuring β-γ coincidences [16], except that the HPGe detector was closer than usual (at 4.9 cm). We have measured the γ-energy range 0-9 MeV. We also separated pure samples of ^{24}Al by tuning MARS for this product, and did a similar β–γ measurement. We use its known gamma-rays for energy and efficiency calibration in the range E_γ=4-

9 MeV, a range not covered by previous efficiency measurements. This was very important, as we use a method of detailed balance of the β flux to obtain absolute branching ratios and absolute *logft* values for transitions to final states in ^{23}Mg. This method was used here for the first time for a nucleus with a complex decay scheme. We have also measured the lifetime more precisely by a new technique, tagging with known gamma-lines from ^{23}Al decay. Part of the data were already published [17] and are part of a thesis [18].

From the measured β singles and β-γ coincidence decay spectrum (Fig. 3) we have established the ^{23}Al β-decay scheme and the absolute β-branching ratios. From the decay scheme we clearly determined that ^{23}Al ground state spin and parity is $J^{\pi}=5/2^+$. We also measured more precisely the lifetime $t_{1/2}=0.447(4)$ sec. A spectacular cleaning of the background of gamma-ray spectrum at large energies was obtained with the BGO shield, as shown in Figure 3. We have determined the log*ft* values for decays to 15 final states in ^{23}Mg. Among others, we found a doublet with small *ft*-values at 7803(2) keV and 7787(2) keV. We have determined for the first time the *ft*-value for the transition to the state at 7803 keV: *ft*=2042(120) s, a fact which identifies it as the Isobar Analog State (IAS) of ^{23}Al ground state. Both states are resonances contributing to the depletion reaction ^{22}Na(p,γ)^{23}Mg in novae. For the latter resonance at E_{res}=207(3) keV, combining our data with data from 3 other experiments, we find its resonance strength to be $\omega\gamma$=2.6(9) meV, making it the dominant contribution in the reaction rate at the temperatures of explosive H burning in ONe novae. To further improve our knowledge about this reaction rate we determined that a re-measurement of the β-delayed proton and gamma decay of these ^{23}Mg states is needed.

Parts of the results were already published [1], determining that the ^{22}Mg(p,γ)^{23}Al reaction is not able to explain the non-observation of the 1.275 MeV gamma-ray from the decay of long-lived ^{22}Na from space telescopes.

Study of β-delayed p-decay of proton-rich nuclei

Efforts were made for a long time to determine the reaction rate for the proton capture reaction ^{22}Na(p,γ)^{23}Mg, the most credible candidate for depleting ^{22}Na out of the NeNa cycle [11]. Data have shown early on that resonant capture plays the overwhelming role. States in ^{23}Mg above the proton separation energy are resonances in this reaction and can play roles in the rate of this reaction at astrophysical energies. The problem received much attention related to the anomalous Ne isotopic ratio found in some meteorites (Ne-E anomaly) [19,20] and to the breakout of the NeNa cycle cited above. Confusion still exists about the precise position of resonances and their strength even after a few direct measurements [11,21] and many spectroscopic studies [22-25], leading to uncertainties for the reaction rate in stellar environments that may be yet of a few orders of magnitude [26,25]. Some of these states are populated in the decay of ^{23}Al, including the isobaric analog state of its ground state which can be identified by its preferential population.

Before our experiments, β-decay of ^{23}Al could only be studied using the small p-decay branches following the β-decay to levels in ^{23}Mg above the proton threshold at S_p=7580 keV. Two different experiments [24,27] concur in their results at proton energies above E_p=500 keV, but differ at lower energies, in particular in the region E_p=200 keV where the important resonances lay, and in particular in the decay of the IAS. Also the protons to gamma branching ratios were only poorly determined in this nucleus. All of the above led us to seek the study of β-delayed proton decay of this nucleus using our well separated, high energy ^{23}Al beam. We came up with a novel technique that we have proved to work well and which we intend to extend to other nuclei in the same region.

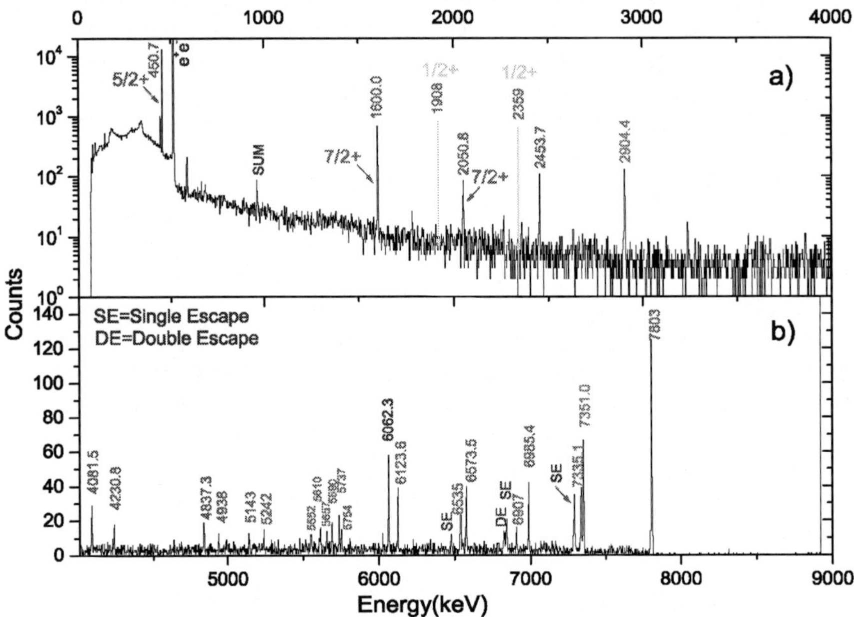

FIGURE 3. ^{23}Al β-γ coincidence spectrum taken with a BGO Compton shield around the HpGe detector. From Ref. [18].

To measure protons at such low energies, the windows of the detectors are always a problem. We decided to avoid it by implanting the source in the detector. This can be done given the large kinetic energy of our sources produced in inverse kinematics. A setup consisting of a thin Si strip detector (p-detector) and a thick Si detector (β-detector) was designed and realized (Figure 4). This telescope is at 45° to the beam axis to allow for good gamma-ray detection with the HpGe detector situated at 90° outside the small chamber. A variable energy degrader consisting of an Al foil (0.8 mm thick in our case) on a computer controlled rotating feed-through was used to stop the source nuclei in the middle of the thin p-detector. We pulsed the beam from the cyclotron, implanting the source nuclei (for 1 sec.), then switch the beam off and, after a very short delay (10 msec) and measured for β–p and β–γ coincidences simultaneously (1 sec. again). Si detectors are also sensitive to positrons, and the total signal in the implantation detector is the sum of the proton and beta contributions. The latter is a continuum, which shifts the proton peak and produces an asymmetry on the high energy side of it. It also produces a large background at lower energies. To minimize these effects, the p-detector must be as thin as possible, and the size of the pixel must be small.

The setup was used in a very recent experiment (May 2007). Colleagues from the University of Jyvaskyla, Finland, the University of Edinburgh, UK and the University of York, UK, have joined the experiment done in our laboratory. The measurements used the telescope in two different modes:

a) – the implantation control mode, working as ΔE-E HI telescope, used to control the implantation of the desired source in the first detector.

b) – the decay mode, in which the gain was adjusted to accommodate the detection of low energy protons (up to 2.5 MeV in the p-detector) and betas (5-6 MeV in the β-detector) in coincidence.

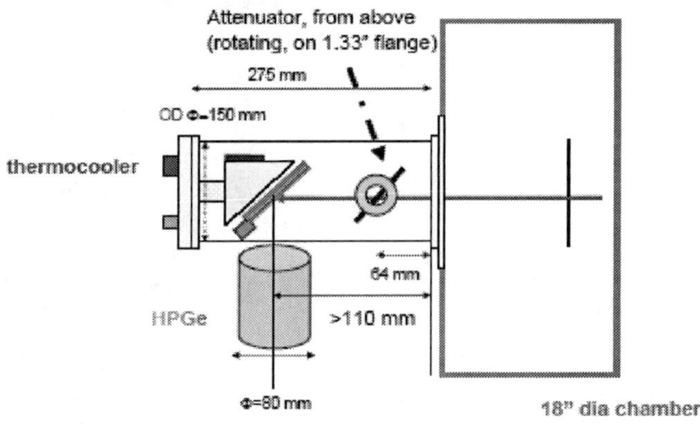

FIGURE 4. The experimental setup used to measure β-delayed p-decay from implanted sources.

We have measured with the setup above, the following:

- implantation control in the "HI telescope mode" for ^{24}Al, ^{23}Al, ^{22}Mg, ^{20}Na, and ^{21}Mg
- ^{23}Al β-delayed p-decay with ^{23}Al implanted in the p-detector. β-p and β-γ coincident spectra were measured here to identify the proton peaks and get the proton/gamma branching ratios.
- β-background with ^{22}Mg implanted in the p-detector.
- p-detector calibration with: ^{21}Mg (known βp emitter) and ^{20}Na (known $\beta\alpha$ emitter) implanted in the p-detector.

The method proved to be very selective: with a careful tuning we could implant the desired isotope in the p-detector and measure efficiently its β-delayed proton or alpha decay. As an example, the calibration spectrum in the bottom right of Fig. 5 was acquired in about 8 hours with ^{21}Mg rates below 1 pps. The final data analysis is in progress.

Figures from TAMU Run0507

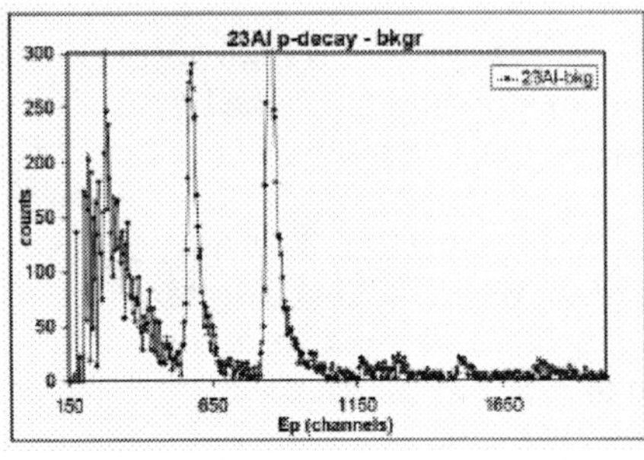

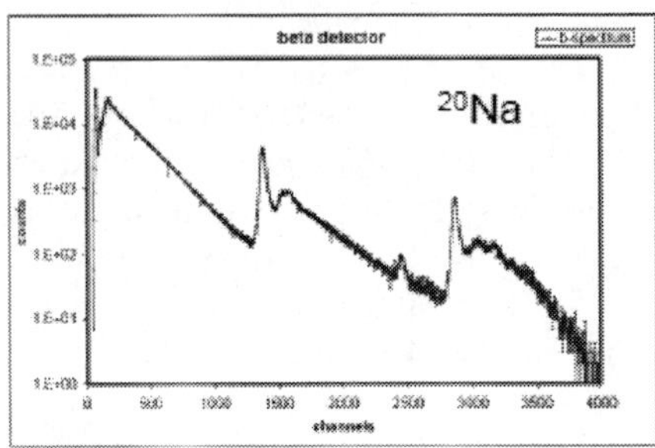

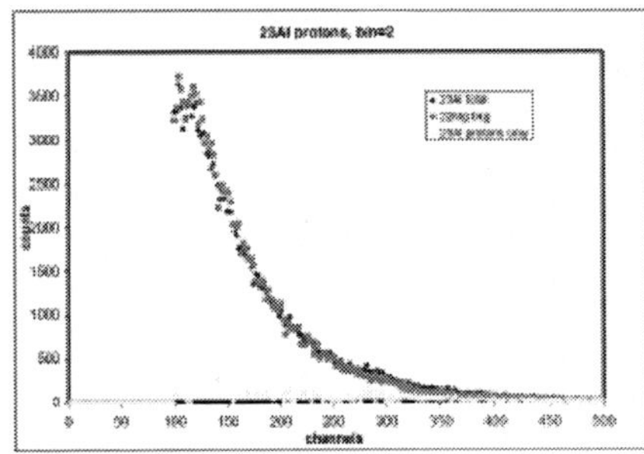

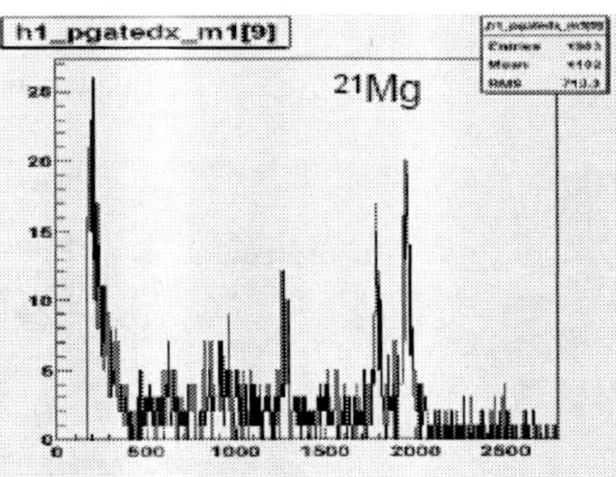

FIGURE 5. Spectra from the study of β-delayed p-decay of ^{23}Al. Preliminary results only! Top left: a proton spectrum with β background subtracted. The proton peaks at 550 and 824 keV are very well separated. Bottom left: comparison of lower energy part of the spectra in p-detector from the decay of ^{23}Al (β and p-decay) and of ^{22}Mg (β only). Right: calibration spectra. Top: of the thick β-detector using implanted ^{20}Na (β and α), bottom: of the thin p-detector using implanted ^{21}Mg.

We intend to continue our studies to nuclei with similar properties and similar importance in explosive H-burning in stars: ^{27}P and ^{31}Cl using the same methods.

CONCLUSIONS

I have illustrated four indirect methods used in nuclear astrophysics with radioactive nuclear beams. I have grouped them by a single nuclear astrophysics motivation, even though they were spanning a broad range of nuclear physics experiments and tools: (a) one-nucleon transfer reactions at about 10 MeV/u, (b) breakup at intermediate energies, (c) β-decay and (d) β-delayed proton-decay of proton rich nuclei. The use of RNBs and indirect methods allows us to extend our reach to regions of unstable nuclei of crucial importance for our understanding of cosmic nucleosynthesis.

ACKNOWLEDGMENTS

This paper is based on work done along the years with my colleagues from the MARS group at the Cyclotron Institute, Texas A&M University: RE Tribble, CA Gagliardi, JC Hardy, AM Mukhamedzhanov, VE Iacob, N. Nica, G. Tabacaru, A. Banu and many students. F. Carstoiu of IFIN-HH Bucharest very beneficially collaborated with us over many years. For the GANIL E491 experiment, work was done with the participation of several people from GANIL and the groups of N. Orr of LPC, Universite de Caen, of M. Chartier at the University of Liverpool, of W. Catford at the University of Surrey and our colleagues F. Negoita and F. Rotaru from IFIN-HH Bucharest. In the p-decay experiment J. Aysto and A. Saastamoinen (Jyvaskyla), P. Woods and T. Davinson (Edinburgh), D. Jenkins and M. Bentley (York) contributed. I thank them all. The original articles or planned publications are cited throughout this paper.

This work was supported by the U.S. Department of Energy under Grant No. DE-FG03-93ER40773.

REFERENCES

1. C. Rolfs and W.S. Rodney, *Cauldrons in the Cosmos*, University of Chicago Press, 1988.
2. L. Trache, in *"Collective Motion and Phase transitions in Nuclear Systems" Proceedings of the Predeal Summer School of Physics 2006*, edited by A.A. Raduta et al., World Scientific Publ. Co, Singapore, 2007, pp. 141-156; ibidem *Rom. J. Phys.* **58**, 823-845 (2007).
3. G. Baur, CA Bertulani and H. Rebel, *Nucl. Phys.* **A458**, 188 (1986).
4. G. Baur and H. Rebel, *Ann. Rev. Nucl. Sci.* **46**, 321 (1996).
5. see e.g. A.M. Mukhamedzanov, C.A. Gagliardi and RE Tribble, *Phys. Rev. C* **63**, 024612 (2001)and contributions by TA Abdullah and A. Banu in these Proceedings.
6. L. Trache, F. Carstoiu, C.A. Gagliardi and RE Tribble, *Phys. Rev. Lett.* **87**, 271102 (2001).
7. L. Trache, F. Carstoiu, C.A. Gagliardi and RE Tribble, *Eur. Phys. J. A* **27**, s01, 237 (2006).
8. G.R. Pizzone et al., these Proceedings, and references therein.
9. J. Jose, A. Coc, M. Hernanz, *Astroph. J.* **520**, 347(1999); **560,** 897 (2001).
10. M. Wiescher et al., *Nucl. Phys.* **A484**, 90 (1988).
11. See e.g., F. Stegmuller et al., *Nucl. Phys.* **A601**, 168 (1996) and references therein.
12. T. Al-Abdullah, thesis, Texas A&M University, 2007, and these Proceedings.
13. A. Banu et al., these Proceedings and to be published.
14. X.Z. Cai et al., *Phys. Rev. C* **65**, 024610 (2002).
15. Z. Hu-Yong et al., *Chin. Phys. Lett.* **19**, 1599 (2002); **20**, 1234 (2003).
16. J. C. Hardy et al., *Phys. Rev. Lett.* **91**, 082501 (2003).
17. V.E. Iacob et al, *Phys. Rev. C* **74**, 045810 (2006).
18. Y. Zhai, thesis, Texas A&M University, 2007, and to be published.
19. D.C. Black, *Geoch. Cosmoch. Acta* **36**, 347 (1972).
20. M. Arnould and W. Beelen, *Astron. Astroph.* **33**, 215 (1974); M. Arnould and H. Norgaard, *Astron. Astroph.* **64**, 195 (1978).
21. S. Seuthe et al., *Nucl. Phys.* **A514**, 471 (1990).
22. S. Schmidt et al. *Nucl. Phys.* **A591**, 227 (1995).
23. S. Kubono, *Nucl. Phys.* **A588**, 305c (1995).
24. R. J. Tighe et al., *Phys. Rev. C* **52**, R2298 (1995).
25. D.G. Jenkins et al., *Phys. Rev. Lett.* **92**, 031101 (2004).
26. C. Angulo et al., *Nucl. Phys.* **A656**, 3 (1999).
27. K. Perajarvi et al., *Phys. Lett.* **B492**, 1 (2000).

Coulomb Dissociation of Proton-Rich sd-shell Nuclei for Astrophysical Interest and Their Implications

Ken-ichiro YONEDA[a]

[a]*RIKEN Nishina Center, 2-1 Hirosawa, Wako, Saitama 351-0198, Japan*

Abstract. Two astrophysical reactions, ^{22}Mg(p,γ)^{23}Al and ^{26}Si(p,γ)^{27}P, were investigated by using the Coulomb breakup of proton-rich isotopes ^{23}Al and ^{27}P. The radiative widths determined from the cross sections suggest that the β^+ decay of ^{22}Mg is favored over the ^{22}Mg(p,γ)^{23}Al reaction, while ^{26}Si(p,γ)^{27}P reaction is faster than the β^+ decay of ^{26}Si around the peak temperature of heavy novae. Two-neutron knockout reactions of the proton-rich sd-shell isotopes ^{26}Si, ^{30}S, and ^{34}Ar were also studied. The measured branching ratios are well reproduced by a theoretical calculation which takes into account nucleon-nucleon correlation in a nucleus.

INTRODUCTION

Since the dawn of radioactive isotope beam physics, technologies of heavy ion acceleration and radioactive isotope production have made drastic progress, yielding broader range of neutron-rich and proton-rich isotopes accessible for experimental studies. Such studies have evinced nuclear structures which appear only for isotopes with large neutron or proton excess, such as exotic density distribution which is referred to as neutron halo or skin [1]. Changes of the shell structure were also found experimentally by the disappearance of neutron magic numbers N = 8 and 20 [2] as well as by the neutron number N = 16 which behaves like a magic number in the vicinity of the neutron drip line [3].

Progress of radioactive isotope beams has also opened opportunities for experimental studies of nuclear astrophysics. It is suggested that nucleosynthesis processes occur in the hot and dense stellar environment such as supernovae and X-ray bursts [4]. In such environment, nuclear reactions of short-lived radioactive isotopes contribute significantly [5]. In the proton-rich sd-shell region, the so-called rp-process occurs, and heavier elements are synthesized through the sequence of proton capture reactions and β^+ decays [6]. Since the main reaction flow is determined by their competition, accumulation of experimental nuclear data of this region is severely desired. While the measurements of capture cross sections have been difficult with the limited statistics of radioactive isotopes, recent development of the radioactive isotope beam facilities reduced the statistical difficulty and measurements are gradually turning feasible even for the isotopes far from stability.

CP972, *Exotic Nuclei and Nuclear/Particle Astrophysics (II)—Proceedings of the Carpathian Summer School of Physics 2007,* edited by L. Trache and S. Stoica

In this contribution, two types of experimental studies using radioactive isotope beams are presented. The first is the Coulomb breakup reactions of ^{23}Al [7] and ^{27}P [8]. This study is relevant to stellar reactions, ^{22}Mg(p,γ)^{23}Al and ^{26}Si(p,γ)^{27}P, which are key reactions in clarifying the rp-process path. As a related topic, the two-neutron knockout reactions from the proton-rich sd-shell isotopes ^{26}Si, ^{30}S, and ^{34}Ar [9] are shown. The measured cross sections are compared with a theoretical calculation which takes into account nucleon-nucleon correlation in a nucleus [10]. The quenching of the shell occupancies and its separation energy dependence are also discussed.

COULOMB BREAKUP OF ^{23}AL AND ^{27}P

Experimental studies of the stellar reactions ^{22}Mg(p,γ)^{23}Al and ^{26}Si(p,γ)^{27}P were performed by using the Coulomb breakup of radioactive isotopes ^{23}Al and ^{27}P. Among the rp-process branches, these two reactions are of particular importance in connection with the galactic γ emitters [11]. ^{22}Na, which is a daughter isotope of the β^+ decay of ^{22}Mg, is a galactic γ emitter, and the rate of ^{22}Mg(p,γ)^{23}Al reaction is one of the components needed to interpret the amount of galactic γ rays. ^{26}Al is also a galactic γ emitter [12], and the interpretation of its galactic γ rays requires the rate of the ^{26}Si(p,γ)^{27}P reaction.

The present study aims at determining the radiative widths Γ_γ from the cross section of the Coulomb breakup reactions. Generally, cross sections of low-energy proton capture are too small to measure for radioactive isotopes which are subject to quite limited statistics compared to stable isotopes. The Coulomb breakup is equivalent to the absorption of the virtual photons followed by the proton emission, and is thus the inverse process of the (p,γ) reaction of astrophysical interest. Experimental cross sections are far larger than the direct capture reaction, mainly owing to the large number of virtual photons at intermediate energy heavy ion collision [13]. Furthermore, the resonant capture rate is proportional to $\Gamma_\gamma\Gamma_p/(\Gamma_\gamma+\Gamma_p)$ where Γ_p denotes a proton decay width, and is approximated to Γ_γ when $\Gamma_p >>\Gamma_\gamma$, and the Coulomb excitation cross section is related to Γ_γ. Therefore, the Coulomb breakup measurement is sensitive to the resonant capture reaction rates.

Experiment

The experiment was performed at the radioactive isotope beam facility RIPS [14] in RIKEN Nishina Center. The radioactive isotope beams of ^{23}Al and ^{27}P were produced by the projectile fragmentation reactions of ^{28}Si at 135 MeV/nucleon and ^{36}Ar at 115 MeV/nucleon, respectively, incident on a beryllium target. The rates of the ^{23}Al (^{27}P) beams were 7 x 10^3 (2.8 x 10^3) cps, and the purities were 7% (1%). The beam particles were identified on an event-by-event basis by RIPS, and were transported to 87 mg/cm^2 ^{208}Pb (125 mg/cm^2 ^{nat}Pb) targets at the final focus of RIPS. The energies of the ^{23}Al (^{27}P) beams were 50 MeV/nucleon (57 MeV/nucleon).

The products of the breakup reaction, ^{22}Mg (^{26}Si) and a proton, were detected by a strip silicon telescope and a plastic scintillator hodoscope located 51 cm and 2.78 m downstream of the target, respectively. Figure 1 shows the schematic view of the setup.

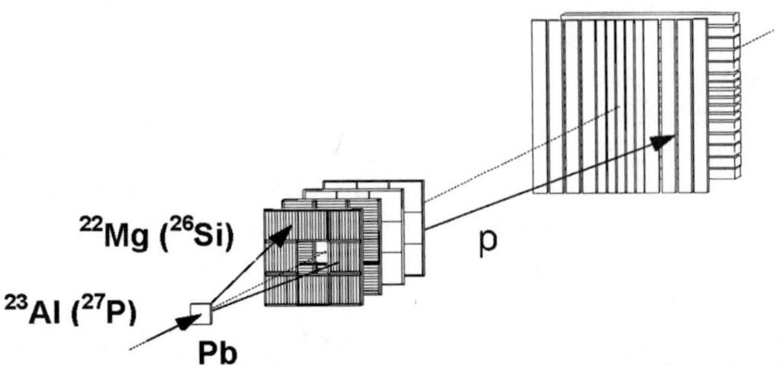

Figure 1. Schematic view of the experimental setup. The radioactive isotope ^{23}Al(^{27}P) from radioactive isotope separator RIPS bombarded a lead target. Breakup products ^{22}Mg (^{26}Si) were detected by a strip silicon telescope. A plastic hodoscope detected protons emitted in the breakup reactions.

The strip silicon telescope consisted of four layers of detectors of 0.5 mm thickness. Each layer was of eight silicon detectors with 50 x 50 mm^2 effective area. The eight detectors in each layer formed a 3 x 3 matrix with a hole in the center, which allowed most of the unreacted beam particles to pass through the telescope without touching the detector sensing area so that the count rate of the detector was below 2 x 10^3 cps with the total beam rate of 1 x 10^5 cps. The first two layers had 5 mm wide strip electrodes, which provided position information in the transverse and vertical directions. The detectors in the third and fourth layers were of single electrode.

The ^{22}Mg (^{26}Si) isotopes produced in the breakup reaction were stopped in the fourth layer, providing ΔE-E information used for particle identification. The protons punched through the silicon telescope and stopped in the plastic scintillator hodoscope which has 1 x 1 m^2 active area consisting of a 5mm thick ΔE plane and a 60 mm thick E plane. The protons were identified by the ΔE, E, and TOF information.

The energies of ^{22}Mg (^{26}Si) were measured by the energy deposit in the strip silicon telescope, while the proton energy was determined through the time-of-flight information between the target and plastic hodoscope. The emission angles of both ^{22}Mg (^{26}Si) and proton were determined from the position information in the silicon telescope. The momentum vectors of ^{22}Mg (^{26}Si) and proton were thus measured, which were used to construct the relative energy spectrum.

A NaI(Tl) scintillator array [15] with sixty-eight detectors was placed around the target (not shown in the Fig. 1). This detector was used to measure γ rays from the excited states in ^{22}Mg (^{26}Si) so that we could obtain relative energy spectrum of events which directly connected to the ground state on ^{22}Mg (^{26}Si) by subtracting the coincidence component corresponding to the decay via excited states in ^{22}Mg (^{26}Si).

Result and discussion

$^{22}Mg(p,\gamma)^{23}Al$

In Fig. 2 (Left), the relative energy spectrum obtained for proton and ^{22}Mg in the breakup reaction of $^{208}Pb(^{23}Al,p^{22}Mg)^{208}Pb$ is shown [7]. The spectrum is after

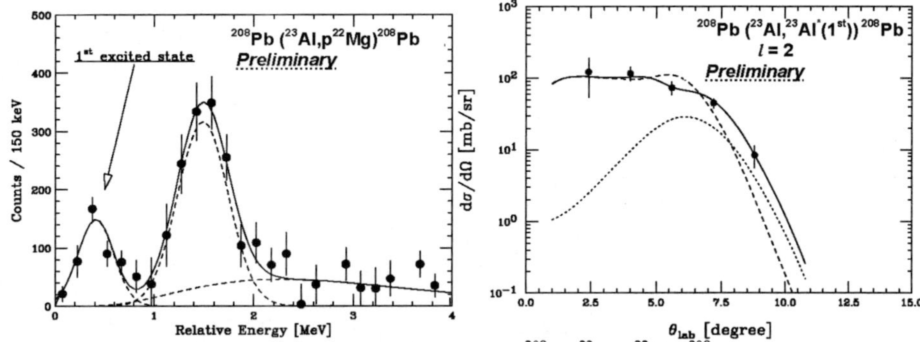

Figure 2. (Left) Relative energy spectrum obtained for $^{208}Pb(^{23}Al,p^{22}Mg)^{208}Pb$. The solid curve represents the result of fitting with two Gaussian functions and a non-resonant component. The dashed curves show each component separately. (Right) Differential cross section for ^{23}Al excitation to the first excited state. The dashed and dotted curves represent the theoretical calculations assuming the l=2 transition for the Coulomb and nuclear parts, respectively. The solid curve corresponds to their coherent sum with the assumption $\beta_C = \beta_N$.

subtraction of the relative energy distribution of the decay through the excited states in ^{22}Mg. In the spectrum, a peak at around 400 keV is seen, which corresponds to the first excited state at 405 keV [16]. Another peak at around 1500 keV may be constituted by two resonance states at 1450 keV [17] and 1773 keV [16], and the energy is too high to have significance in the stellar reactions. The Coulomb dissociation yield was obtained by fitting the relative energy spectrum with three components, two Gaussians and a distribution corresponding to the non-resonant proton capture process.

The differential cross sections obtained for the excitation to the 405 keV resonant state in ^{23}Al is seen in Fig. 2 (Right). The angle is defined as that of the center-of-mass of ^{22}Mg and proton with respect to the direction of the incident ^{23}Al. The theoretical angular distributions by the coupled channel code ECIS79 [18] are also shown in the figure. The differential cross sections are well reproduced by the calculation assuming $l = 2$, therefore confirming the E2 excitation from the $5/2^+$ ground state to the $1/2^+$ first excited state [19]. Here, equal Coulomb and nuclear deformation parameters $\beta_C = \beta_N$ are assumed. This fit preliminarily gives Γ_γ of 7.0 (1.3) x 10^{-7} eV. The quoted error includes uncertainties of choice of the optical potential and also ambiguities from possible difference of the deformation parameters β_C and β_N [20].

The Γ_γ value measured in this work is to be incorporated into the reaction network calculations of nucleosynthesis. The comparison between the direct capture rate by the theoretical calculation [16] and resonant capture reaction through the $1/2^+$ resonant state determined in this work suggested that both are comparable in the temperature region of about 0.3 – 0.7 GK, while the direct capture is dominant in the other temperature region. The shell model calculation used in the network calculation gives a consistent value of Γ_γ, and the network calculation claims that β^+ decay is dominant in the novae environment. The result thus suggests that the $^{22}Mg(p,\gamma)^{23}Al$ reaction cannot influence the amount of the galactic γ emitter ^{22}Na.

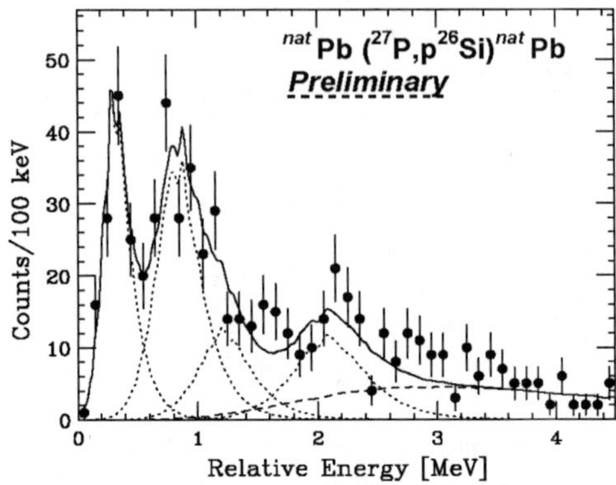

Figure 3. Relative energy spectrum obtained for $^{nat}Pb(^{27}P,p^{26}Si)^{nat}Pb$. The solid curve represents the result of fitting with four resonant components and a non-resonant component. The dashed curves show each component separately.

$^{26}Si(p,\gamma)^{27}P$

Figure 3 shows the relative energy spectrum obtained for proton and ^{26}Si in the breakup reaction of $^{nat}Pb(^{27}P,p^{26}Si)\,^{nat}Pb$ [8]. In the case of ^{27}P breakup, the number of γ rays observed in coincidence was negligible small. The spectrum was fit with four resonant state components and a continuum part which corresponds to the non-resonant proton capture. The lower two are known states, but the other two have not been reported yet. The third one is particularly small and the peak is not clearly visible in the spectrum, but the existence is likely since both the shell model [21] and level sequence in the mirror nucleus ^{27}Mg [22] show three low-lying excited states in ^{27}P at similar energies.

In the case of ^{27}P breakup, the yield is not enough to analyze the differential cross sections. Hence the cross sections were deduced from the peak yield in the relative energy spectrum given by the fitting. The spins and parities of the ground, first, second, and third excited states are assumed to be $1/2^+$, $3/2^+$, $5/2^+$, and $5/2^+$, respectively. These spins and parities are given consistently from the shell model calculation [21] and the known states in the mirror nucleus ^{27}Mg [22]. The energy of the fourth resonant state is too high and its astrophysical contribution is negligible.

The excitation to the first excited state is of M1/E2 multipolarities, and the present measurement is sensitive to E2 [23]. Hence, the M1/E2 mixing ratio is needed to deduce Γ_γ of this resonant state. We estimated the ratio from the measured mixing ratio of the mirror nucleus ^{27}Mg [24], and the ratio of the mixing ratios of ^{27}P and ^{27}Mg are calculated by a shell model [21]. This procedure exhibited the best reproduction of the experimental mixing ratios for the pairs of the mirror nuclei in the sd-shell region. Γ_γ of the first excited state is thus extracted preliminarily to be 5.0(2.9) x 10^{-3} eV. For the second and third excited states, the excitation is pure E2 and the preliminary Γ_γ values are 2.7(0.8) x 10^{-4} eV and 3.4(0.9) x 10^{-4} eV.

As for the resonant capture rate compared to the direct capture, the resonant capture via the first excited state is faster compared to the calculated direct capture rate [16] over a broad range of temperature of the novae environment. Compared to the β^+ decay rate, the capture reaction becomes faster around the peak temperature of heavy novae. Since the experimental Γ_γ's obtained in this experiment are consistent with the shell model calculations which have been used in the network calculation of nucleosynthesis, the result of this experiment is to be interpreted as validation of the network calculations at this site of the reaction chain.

TWO-NEUTRON KNOCKOUT FROM ^{26}SI, ^{30}S, AND ^{34}AR

Direct knockout reactions from fast heavy-ion beams have been used for nuclear structure studies, and one-nucleon knockout has been established as a versatile tool of nuclear structure study even for the isotopes far from stability [25]. In the scheme of one nucleon knockout, a nucleon is removed from a fast heavy ion typically at about 100 MeV/nucleon. Since the momentum of the reaction residue reflects the momentum which the nucleon carries before it is knocked out, the momentum analysis of the reaction residue provides information of the orbital angular momentum of the knocked-out nucleon. Coincidence measurement of de-excitation γ rays of the reaction residue makes this method more powerful, since it extracts a single l component and facilitates disentanglement of the l components mixed in the original momentum distribution. This method is also in good harmony with the theoretical calculation, i.e. the reaction is well described by a few-body reaction theory in eikonal approximation [26]. From the ratio of the measured cross section to the calculated one, we can discuss the spectroscopic factor and the shell occupancies of the projectile isotope.

Concerning the two-nucleon knockout, it was shown that two proton removal from neutron-rich isotopes proceed as a direct reaction [27], and two neutron removal from proton rich isotopes are expected to proceed as a direct process as well. This is likely since the most probable indirect path is one neutron knockout followed by one neutron evaporation, while the final state of the one neutron removal favors proton emission because of the lower proton energy threshold.

Recently, a sophisticated theoretical approach of the direct two-nucleon removal process was issued [10], in which the cross sections are calculated with inputs of the two-nucleon amplitudes which are to be good indices of the nucleon-nucleon correlation in a nucleus. This is intriguing since the two-nucleon removal may provide a probe of nucleon-nucleon correlation in a nucleus via this theoretical calculation. Nevertheless, only limited experimental data which can be used to test this calculation existed. This work aimed at providing the first case of two-neutron knockout which is to be compared with this theory. The reactions in the sd shell, ^{9}Be(^{34}Ar,^{32}Ar), ^{9}Be(^{30}S,^{28}S), and ^{9}Be(^{26}Si,^{24}Si), were chosen, since the shell model is relatively matured and small ambiguity from nuclear structure was expected.

Experiment

The experiment was performed using the A1900 fragment separator [28] and S800 spectrograph [29] in the Coupled Cyclotron Facility of the National Superconducting Cyclotron Laboratory at Michigan State University. The radioactive isotopes, ^{26}Si, ^{30}S, and ^{34}Ar, were produced by the projectile-fragmentation reaction of a 150

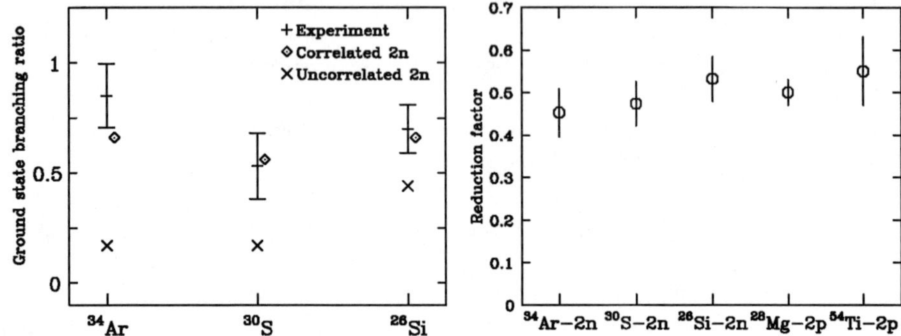

Figure 4. (Left) Ground state branching ratios in the two-neutron removal from ^{26}Si, ^{30}Si and ^{34}Ar. The results of the calculation which takes into account the nucleon-neutron correlation as well as without correlation effect are also plotted. (Right) Reduction factor of the two-nucleon removal cross sections.

MeV/nucleon ^{36}Ar primary beam on a beryllium target. The isotopes impinged on a 376 mg/cm^2 beryllium secondary target at the target position of the S800 spectrograph. The mid-target energy was about 110 MeV/nucleon. The incoming particles were tagged by the time-of-flight measurement in the A1900 separator, and the reaction residues were identified by using the S800 spectrograph. The target was surrounded by the germanium detector array SeGA [30], an array of 17 32-fold segmented high-purity germanium detectors to measure γ rays from the reaction residues. Partial cross sections of populating individual states were extracted from the yield of the γ rays.

Result and Discussion

As a result of the experiment [9], the inclusive cross sections were obtained from the yield of reaction residues to be 1.01(10) mb, 0.73(8) mb, and 0.48(6) mb for ^{26}Si, ^{30}S, and ^{34}Ar, respectively. Two γ transitions in the reaction residue ^{24}Si and a transition in ^{28}S and ^{32}Ar were observed. The γ transition in ^{28}S was newly found in this measurement (E_γ = 1512(8) keV). From the yield of the γ peaks, the partial cross sections were obtained to be 0.15(4) mb and 0.14(4) mb for the first excited 2$^+$ state and second excited state of ^{24}Si, 0.34(8) mb for the first 2$^+$ state of ^{28}S, and 0.07(4) mb for the first 2$^+$ state of ^{32}Ar.

In the left panel of Fig 4, the measured ground state branching ratios, which are given by the ratio of the partial cross section of populating the ground state to the inclusive one, are compared to the theoretical ones. The theoretical calculations were made by taking into account the neutron-neutron correlation in a nucleus (diamonds), while calculated values without correlation effect (crosses) are also shown. In the calculation two-nucleon amplitudes given by a shell model based on the USD interaction [31] are used. The calculations without correlation significantly underestimate the experimental values. By taking into account the two-neutron correlation, removal of two-neutrons pair coupled to 0$^+$ is enhanced. The calculated values are in good agreement with the experimental ratios.

It has been believed that the measured cross sections of single nucleon knockout are always smaller than the theoretical ones, and the ratio, so-called reduction factor, is almost constant at about 0.6 [25]. But recently, it was reported that this ratio has a dependence of separation energy of the removed nucleon [32]. In order to see the ratio for the case of two nucleon removal, the ratios are plotted for the five available cases including the result of this experiment [10]. One can see that the ratios are almost 0.5 for all the cases. Note that the two-nucleon separation energies are almost the same at about 30 MeV for all these cases. This result hence may imply that the reduction factor has a relationship with the separation energy also for the two-nucleon removal. More experimental results with larger variety of separation energies are desired for further investigation of the separation energy dependence of the reduction factor.

SUMMARY AND FUTURE OUTLOOK

Experimental studies of the radiative capture reactions ^{22}Mg(p,γ)^{23}Al and ^{26}Si(p,γ)^{27}P were performed by using the Coulomb dissociation of radioactive isotope beams of ^{23}Al and ^{27}P on a lead target. The radiative widths Γ_γ's of the low-lying resonant states in ^{23}Al and ^{27}P were determined. Γ_γ of the first excited state in ^{23}Al suggests that the resonant capture rate is comparable with the direct one in a temperature range of 0.3 – 0.7 GK, and ^{22}Mg β^+ decay is favored over the ^{22}Mg(p,γ)^{23}Al reaction in the novae condition. As for the case of ^{27}P, the capture reaction via the first excited state is suggested to be faster than the direct capture. This gives rise to the dominance of the proton capture reaction compared to the β^+ decay of ^{26}Si around the peak temperature of heavy novae. Since the Γ_γ's measured in the present work are consistent with the theoretical ones which have been used in the network calculation of nucleosynthesis, the results suggest verification of the theoretical predictions so far made concerning the resonant capture at these sites of the reaction chain.

Two-neutron knockout reactions of the proton-rich nuclei ^{26}Si, ^{30}S, and ^{34}Ar were measured. The inclusive cross sections as well as partial ones feeding to the individual bound states in the reaction residues were extracted. The obtained branching ratios are in good harmony with a theoretical calculation which takes into account nucleon-nucleon correlation in a nucleus. The reduction factors of the cross sections, which are given by the ratio of the experimental cross sections to the theoretical ones, are almost constant for the two-nucleon removal reactions with similar separation energies, which is consistent with the case of single nucleon removal reactions.

In RIKEN Nishina Center, the next generation radioactive isotope beam facility project called RI Beam Factory has just started [33]. In this project, the rates of radioactive isotope beams become drastically higher, which afford enormous opportunities of the reaction studies. The Coulomb breakup of proton-rich isotopes for astrophysical interest is one of the major subjects, and we expect to extend the experimental field to the heavier region up to ^{100}Sn. This study requires not only high rate radioactive isotope beams but also a spectrometer which has a large momentum byte covering heavy ions and proton at the same time. For this purpose, we plan to construct a large-gap superconducting spectrometer magnet, called SAMURAI (Superconducting Analyzer for MUlti-particle with RAdioactive Isotope beams). This magnet is used for multipurpose, not only for proton and heavy ion coincidence experiments but also neutron and heavy ion coincidence, missing mass spectroscopy, d-p scattering for three-nucleon force studies, and so on. The technical design is now being made and we hope to be ready to start construction as soon as possible.

ACKNOWLEDGMENTS

The study of Coulomb breakup was supported in part by a Grant-in Aid for Scientific Research (No. 15204025) from the Ministry of Education, Culture, Sports, Science, and Technology in Japan. The work of two-neutron removal reactions was supported by the U.S. National Science Foundation under Grants No. PHY-0110253 and No. PHY-0244453, and the United Kingdom Engineering and Physical Sciences Research Council (EPSRC) Grant No. EP/D003628.

REFERENCES

1. I. Tanihata, J. Phys. G 22, 157 (1996).
2. e.g. N. Aoi et al., Phys. Rev. C 66, 014301 (2002); H. Iwasaki et al., Phys. Lett. B 491, 8 (2000); H. Iwasaki et al., Phys. Lett. B 481, 7(2000); T. Motobayashi et al., Phys. Lett. B 346, 9 (1995).
3. A. Ozawa et al., Phys. Rev. Lett. 84 5493 (2000).
4. R. Wallace and S.E. Woosley, Astrophys. J. Suppl. 45, 389 (1981); A.E. Champagne and M. Wiescher, Annu. Rev. Nucl. Part. Sci. 42 (1992)39.
5. e.g. C.E. Rolfs and W.S. Rodney, *"Cauldrons in the Cosmos"* (The University of Chicago Press).
6. F. –K. Tielemann et al., Phys. Rev. C 63, 5 (2001).
7. T. Gomi et al., Nucl. Phys. A734, E77 (2004).
8. Y. Togano et al., Eur. Phys.J. A 27, Suppl. 1, 233 (2006).
9. K. Yoneda et al., Phys. Rev. C 74, 021303 (2006).
10. J.A. Tostevin and B.A. Brown, Phys. Rev. C 74, 064604 (2006).
11. S. Starrfield, et al., Astrophys. J. Suppl. 127, 485 (2000).
12. R. Diehl et al., Astron. Astrophys. Suppl. Ser. 97, 181 (1993).
13. G. Bauer, C.A. Bertulani, H. Rebel, Nucl. Phys. A458, 188 (1986).
14. T. Kubo et al., Nucl. Instr. and Meth. B70, 309 (1992).
15. T. Motobayashi et al., Phys. Lett. B209, 9 (1995).
16. J.A. Caggiano et al., Phys. Rev. C 64, 025802 (2001).
17. B. Blank et al., Z. Phys. A357, 247 (1997).
18. J. Raynal, Coupled channel code ECIS79, unpublished.
19. A. Ozawa et al., Phys. Rev. C 74, 021301 (2006).
20. E. Khan et al., Nucl. Phys. A694, 103 (2001).
21. B.A. Brown et al., Phys. Rev. C 74 (2006) 034315.
22. P.M. Endt, Nucl. Phys. A 521 (1990) 1.
23. K. Langanke et al., Phys. Rev. C 49 (1994) 1771.
24. M.J.A. de Voigt et al., Nucl. Phys. A186 (1972) 365.
25. P.G. Hansen and J.A. Tostevin, Annu. Rev. Nucl. Part. Sci. 53, 219 (2003).
26. J.A. Tostevin, J. Phys. G 25, 735 (1999).
27. D. Bazin et al., Phys. Rev. Lett. 91, 012501 (2003).
28. D. J. Morrissey et al., Nucl. Instrum. Methods Phys. Res. B 204, 90 (2003).
29. D. Bazin et al., Nucl. Instrum. Methods Phys. Res. B 204, 629 (2003); J. Yurkon et al., Nucl. Instrum. Methods Phys. Res. A 422, 291 (1999).
30. W. F. Mueller et al., Nucl. Instrum. Methods A 466, 492 (2001).
31. B.A. Brown and B.H. Wildenthal, Annu, Rev. Nucl. Part. Sci. 38, 29 (1998).
32. A. Gade et al., Phys. Rev. Lett. 93, 042501 (2004); J. Lee et al., Phys. Rev. C 73, 044608 (2006).
33. Y. Yano, Nucl. Instrum. Methods Phys. Res. B 261, 1009 (2007).

Underground Nuclear Astrophysics at LUNA

Matthias Junker

Laboratori Nazionali Del Gran Sasso

Abstract. Nuclear cross sections play a key role in understanding stellar evolution and elemental synthesis. Also in the field of astroparticle physics precise knowledge on thermonuclear cross sections is needed to extract the particle properties from the experimental data. While it is desirable to directly measure the relevant cross sections in the energy range of interest for the specific stellar environment this proves to be difficult, if not impossible, due to the effect of the Coulomb barrier, which causes an exponential drop of the cross sections at stellar energies. Consequently direct measurements are hampered by low counting rates and background caused by cosmic rays and environmental radioactivity. In addition background induced by the beam or the target itself can disturb the measurements.

In this contribution I will discuss some of the reactions studied by LUNA in the past years to illustrate important aspects underground nuclear astrophysics.

Keywords: Underground accelerator, direct measurements, stellar nuclear synthesis
PACS: 01.10.Fv, 25.40.Lw, 26.20.Cd, 29.30.-h

INTRODUCTION

Stars generate energy and synthesize chemical elements in thermonuclear reactions [1]. In a star non–resonant reactions induced by charged particles take place in a narrow energy window called the Gamow peak, locate far below the Coulomb barrier. At these energies the cross section $\sigma(E)$ of a charged particle induced reaction drops steeply with decreasing energy due to the Coulomb barrier in the entrance channel [1]:

$$\sigma = S(E)e^{-2\pi\eta} \tag{1}$$

where $S(E)$ is the astrophysical S-factor, and η is the Sommerfeld parameter with $2\pi\eta = 31.29 Z_1 Z_2 (\mu/E)^{1/2}$. Here Z_1 and Z_2 are the charge numbers of projectile and target nucleus, respectively, μ is the reduced mass (in amu units), and E is the center of mass energy[1] in keV units.

Generally, $\sigma(E)$ has a very low value at the Gamow peak. This prevents a direct measurement in laboratories at the earth's surface as the signal to background ratio is too small due to cosmic ray interactions with detectors. Hence, cross sections are usually measured at high energies and expressed as astrophysical S-factor (Eq. (1)). The S-factor is then used to extrapolate the data to the Gamow peak region. Although $S(E)$ varies only slowly with energy for the direct capture process, resonances and resonance tails may hinder the extrapolation resulting in large uncertainties [1]. Therefore, the primary

[1] In this contribution, E denotes the energy in the center of mass system, and E_{beam} beam is the projectile energy in the laboratory system.

CP972, *Exotic Nuclei and Nuclear/Particle Astrophysics (II)—Proceedings of the Carpathian Summer School of Physics 2007,* edited by L. Trache and S. Stoica

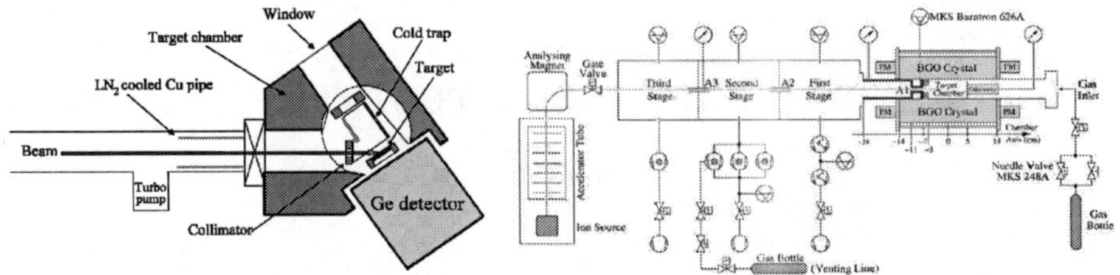

FIGURE 1. Schematic view of the solid state target setup (left) and the gas target (right) used to measure the S-factor of the reaction $^{14}N(p,\gamma)^{15}O$.

goal of experimental nuclear astrophysics remains to measure cross sections at energies inside the Gamow peak, or at least to approach it as closely as possible. In this case the data taken must contain as much nuclear information as possible (e.g. branching ratios, angular distributions).

THE LUNA UNDERGROUND ACCELERATOR FACILITY

To fully exploit this approach the LUNA[2] collaboration employs a high current accelerator located deep underground in the italian "Laboratori Nazionali del Gran Sasso" (LNGS)[3] protected from cosmic rays by 1400m of rock equivalent to 3800 m of water. This shielding suppresses the flux of cosmic ray induced muons by six orders of magnitude [2] and the neutron flux by three orders of magnitude [3].

The 400 kV electrostatic accelerator constructed by High Voltage Engineering Europe (Netherlands) is embedded in a tank filled with a gas mixture of N_2/CO_2 at 20 bar. The high voltage is generated by an Inline-Cockcroft-Walton power supply located inside the tank and capable to handle 1 mA at 400 kV. The radio-frequency ion source mounted directly on the accelerator tube provides ion beams of 1 mA hydrogen (75% H^+) and 200μA ^{4}He over a continuous operating time of about 40 days. The absolute energy of the beam is known to an accuracy of 0.3 keV and the energy spread and the long-term energy stability were observed to be 100 eV and 5 eV/h, respectively [5].The ion beam can be injected in two parallel beam lines equipped with a windowless gas target (Line 1)and a solid state target system (Line 2).

$^{14}N(P,\gamma)^{15}O$ — THE BOTTLENECK OF THE CNO CYCLE.

In stars with temperatures above $20*10^6$ K, hydrogen burning is dominated by the CNO cycle. Its rate is determined by the slowest process, the $^{14}N(p,\gamma)^{15}O$ reaction. The capture cross-section needs to be known down to $E = 30$ keV corrisponding to the

[2] **L**aboratory for **U**underground **N**uclear **A**astrophysics

[3] `http://www.lngs.infn.it`

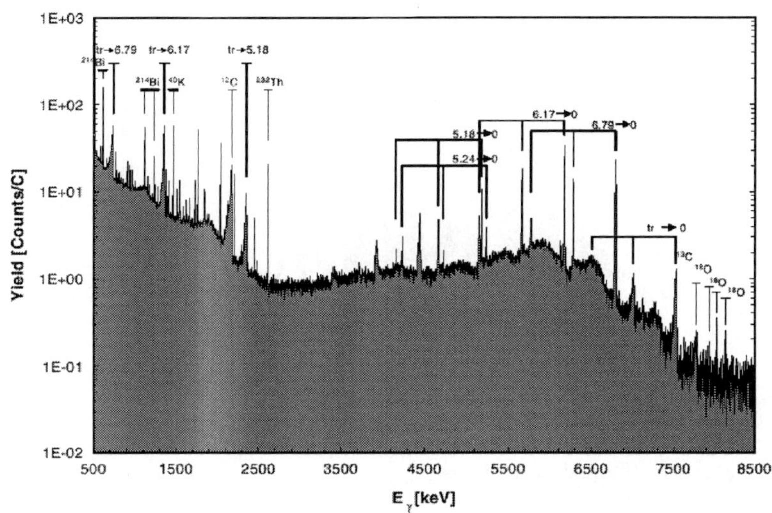

FIGURE 2. Spectra of the reaction $^{14}N(p,\gamma)^{15}O$ taken with a HPGe detector at E_{Beam}=250keV accumulating a charge of 87 Cb on the targets [13].

Gamow peak in core H-burning stars. The cross section plays a key role in determination the age of Globular Clusters, which pose a lower limit for the age of the universe [6] and in the context of solar neutrino spectroscopy [8][9][10].

At low energies the cross section of the $^{14}N(p,\gamma)^{15}O$ reaction is dominated by a narrow resonance $E = 259$ keV and by the direct capture process. In addition previous publications [11] indicate a strong influence of a sub-threshhold state at -507 keV. In any case the reaction proceeds through several levels in the ^{15}O compound nucleus. To determine the astrophysical S-factor at stellar energies LUNA adopted two different approaches, which are illustrated in the following.

Measurements with solid target and High Purity Germanium Detectors

As a first approach solid targets of TiN have been bombarded with an intens proton beam and the produced γ-rays have been detected by High Purity Germanium (HPGe) Detectors [12][13]. Typical spectra obtained are shown in FIGURE 2 and 3. Due to the low cosmic background present in the underground site and the high resolution of the detector used the spectra reveal the full decay scheme of the reaction. Even the shapes of the peaks related to primary γ-rays can be evaluated. At $E_{beam} = 140$ keV all secondary γ-rays and the ground state transition can still be identified clearly. The beam induced background can be tagged easily.

There are however some limitions in the approach to the measurement: The efficiency of the HPGe detector and the total solid angle covered are small. In addition, due to

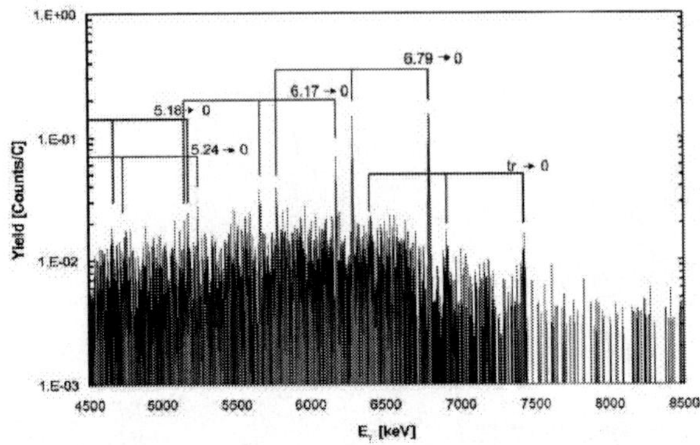

FIGURE 3. Spectrum of the reaction $^{14}N(p,\gamma)^{15}O$ taken with a HPGe Detector at E_{Beam}=140keV accumulating a charge of 210 Cb on the targets. The spectrum is dominated by the environmental background for E_γ <2.6MeV (not shown)[13].

the particular branching scheme of the reaction, the peaks are distributed over the full spectrum. While this is not a problem at "high" beam energies, at low energies the peak to background ratio decreases as can be seen in FIGURE 3. In particular below $E_\gamma = 2.614$ MeV environmental background environmental background competes with the γ-rays from the reactions as the underground location at LNGS only slightly reduces the natural radioactivity caused e.g. by ^{40}K and the RaTh decay chains. Shielding this radioactivity not only involves the use of massive lead and cupper shielding but imposes a carefull selection of all the materials close to the detector [7]. Such as system has been developed by LUNA for the $^3He(\alpha,\gamma)^7Be$ cross section measurement (see below).

Furthermore, due to the close geometry of the setup, the cascade structure of the capture process and the small branching to the ground state the "summing-in" effect (i.e. the cascade γ-rays giving a contribution to the ground state full-energy peak) and the "summing-out" effect (i.e. a full-energy detection of one cascade γ-ray and any concurrent interaction of the other member of the cascade) have to be taken into account.

For these reasons data could be taken and evaluated not lower than E = 119 keV and an extrapolation based on the R-Matrix formalism has been used to determine the S-factor in the relevant Gamow peak. In order to improve its reliability high energy data taken by [11] and corrected for the summing effect have also been included. The resulting $S_{tot}(0)$= 1.61±0.08 keV b [13] is a factor of two lower than the extrapolation published in [11].

Finally it can be concluded from this analysis that the data above the 259 keV resonance are of crucial importance for a reliable extrapolation. This finding emphasizes that as long as extrapolations are needed a low energy experiment alone cannot solve the problem at low temperatures and detailed analysis of the nuclear structure is required.

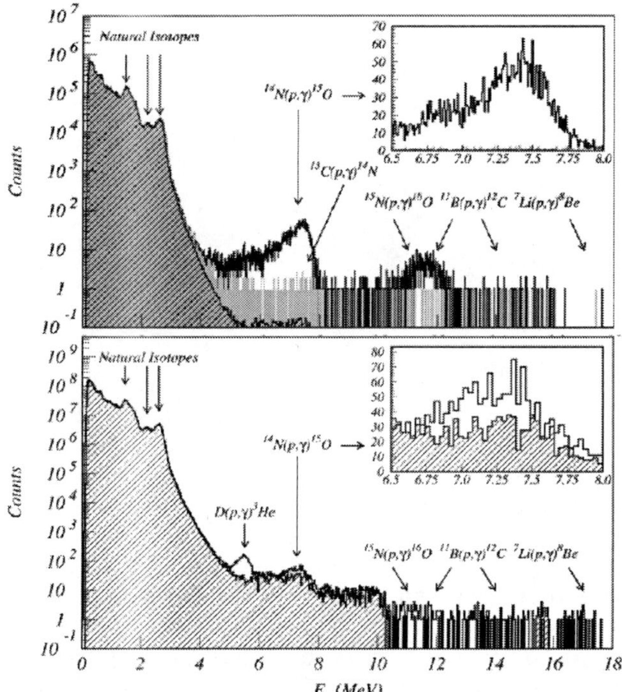

FIGURE 4. **Top**: $^{14}N(p,\gamma)^{15}O$ γ-ray spectrum at E_{beam}=141.1 keV with 1 mbar nitrogen gas (solid line). Accumulated charge 19.9 C. Background run at same beam energy, normalized to equal lifetime with the nitrogen run, with 1 mbar helium in the target (gray line), accumulated charge 20.8 C. The laboratory background without beam, normalized to equal lifetime with the nitrogen run is also shown (shaded area). In the inset the $^{14}N(p,\gamma)^{15}O$ region of interest is expanded. **Bottom**: γ-ray spectrum at E_{beam} = 80.9 keV (E_{eff}= 70 keV, solid line), running time 49.12 days, accumulated charge is 928 C. The most relevant beam induced background reactions are indicated. The laboratory background is shaded. In the inset, the region of interest (ROI) for our analysis is enlarged: $^{14}N(p,\gamma)^{15}O$ signal (solid line), and laboratory background (shaded area).

Measurements with gas target and BGO summing detector

An alternative approach persued in the framework of LUNA to determine the S-factor at low energies consists in the use of a summing detector in combination with a windowless gas target system [14]. In brief the accelerator delivered a proton beam to a differentially pumped windowless gas target located inside a 4π BGO summing crystal (FIGURE 1). Because of the high absolute detection efficiency and due to the near 4π geometry of the BGO detector, γ-rays emitted in a cascade are summed with high probability into a peak at $E_{\gamma} = Q + E$ ($Q = 7.297$ MeV is the Q-value for the $^{14}N(p,\gamma)^{15}O$ reaction) in the spectrum (FIGURE 4). The strong drop of cross section and the resulting degradation of the signal to background ratio can be clearly observed.

In the course of the experiment it turned out to be difficult to identify the origin of

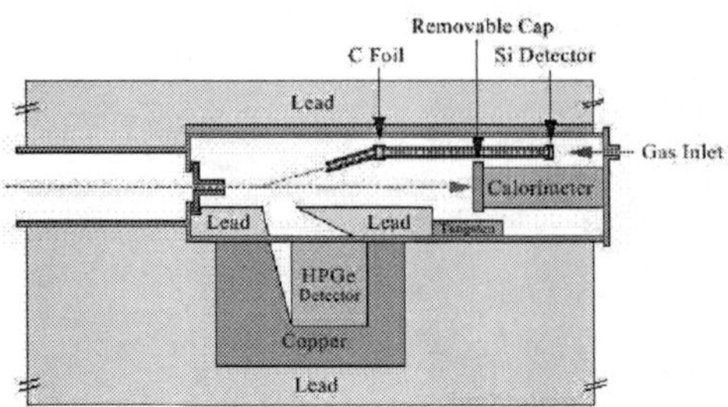

FIGURE 5. Detector setup used for the ^{3}He(α,γ)^{7}Be experiment.

the beam induced background without the use of high resolution detectors. The intrinsic radioactivity of the detector, which extents up to 10 MeV, and the low resolution of BGO detectors contributes to the bad signal to background ratio (FIGURE 4) and pose considerable limitations on the measurements. Details of the background studies can be found in [15].

The data taken with the BGO detector agree perfectly with the data from the HPGe setup in the overlapping energy range. The lower limit of the measurement is $E = 70$ keV corresponding to $T_6 = 60$. Thus for several astrophysical scenarios the uncertainties caused by extrapolations could be eliminated completely. The solar Gamow peak, which has its maximum at $E = 21$ keV, could not be covered by the measurements. For this important scenario extrapolation is still needed. Missing any kind of spectroscopic information, the data of the BGO experiment are not suited for a reliable extrapolation, which must be based on the HPGe detector experiment described previously.

^{3}HE(α,γ)^{7}BE — A KEY REACTION FOR SOLAR NEUTRINO PHYSICS

The nuclear physics input from the ^{3}He(α,γ)^{7}Be cross section is a major uncertainty in the fluxes of ^{7}Be and ^{8}B neutrinos from the sun predicted by solar models and in the ^{7}Li abundance obtained in big-bang nucleosynthesis calculations. The ^{3}He(α,γ)^{7}Be (Q-value: 1.586 MeV) reaction leads to the emission of prompt γ-rays, and the final ^{7}Be nucleus decays with a half-life of 53.22 days, emitting a 478 keV γ-ray in 10.44% of the cases.

The cross section can be measured by detecting either the induced ^{7}Be activity (activation method) or the prompt γ-rays from the reaction (prompt-γ method). Previous activation studies cover the energy range $E = 420$ to 2000 keV. Prompt-γ measurements cover $E =$ 107-2500 keV, although with limited precision at low energies. Previous activation results are, on the average, 13% higher than prompt-γ data. This discrepancy is the origin of the large uncertainty quoted on the reaction cross section. The LUNA collaboration performed a new experiment using simultaneously prompt-γ and activation method with the same experimental setup using the 400 kV accelerator at LNGS [16][17][18].

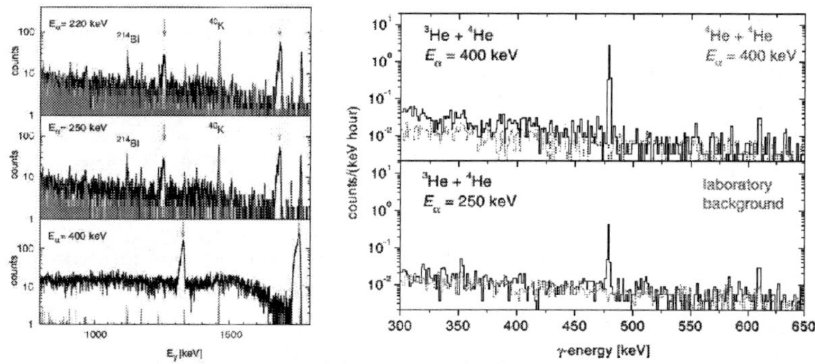

FIGURE 6. **Left**: Spectrum of the reaction ^{3}He(α,γ)^{7}Be taken at E_α= 220, 250, and 400 keV compared with natural laboratory background (in gray) normalized to beam-measurement live times (respectively: 31.2 d, 21.3 d, and 4.8 d). Arrows indicate the primary transition peaks to the first excited state and to the ground state. **right**: Offline γ-counting spectra. Solid black line: ^{3}He gas bombarded at E_α = 400 keV (top panel) and 250 keV (bottom panel), respectively. Dotted red line, top panel: ^{4}He gas bombarded at E_α = 400 keV. Dotted red line, bottom panel: laboratory background.

In the preparation of the experiment particular attention has to be given to a good shielding of the detectors used as both, prompt and β delayed γ-rays, have energies lower than 2.6 MeV. Heavy shielding and a careful choice of all materials close to the detectors are mandatory to obtain low energy data points. While a heavy shielding above ground gives rise to secondary particles causing additional background in the detector and limiting thus the effect of the shielding this effect is negligible for underground nuclear astrophysics (while it is still relevant for experiments on neutrinoless double beta decay or dark matter).

Three couples of cross section values have been measured (prompt-γ and activation) at E_α = 220, 250, and 400 keV. A sketch of the interaction chamber is given in FIGURE 5. The α-beam enters the ^{3}He extended windowless gas target through a 7 mm diameter collimator and is stopped on a detachable copper disk that serves as the primary catcher

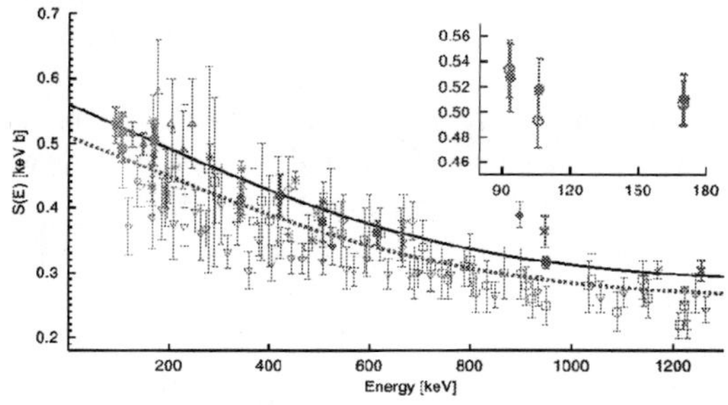

FIGURE 7. Overview of all available S-factor values for the ^{3}He(α,γ)^{7}Be reaction. Filled and open circles: LUNA work. In the inset a zoom of prompt-γ (filled circles) and activation (open circles) data obtained by LUNA.. For literature references see [18]

for the produced ^{7}Be and as the hot side of a calorimeter. The latter measures the beam intensity (about 250 μA). Prompt γ-rays are counted with a 135% ultra-low-background HPGe detector shielded with 5 cm of OFHC copper and 25 cm of lead. The detector and the shield are enclosed in a sealed plastic box flushed with dry N_2 to reduce ^{222}Rn background. Due to the underground environment, where cosmic muons are strongly reduced, the shielding's suppression factor is five orders of magnitude for γ-rays below 2 MeV.

During the runs in which prompt γ-rays are detected, the ^{7}Be nuclei produced inside the gas target get implanted into the removable calorimeter cap. To obtain the activation data after each run, the detachable disk is dismounted from the calorimeter and moved to the LNGS underground low-activity counting facility [19] where heavily shielded HPGe Detectors with a low intrinsic radioactivity are available. Spectra of both (prompt-γ and activation) measurements are shown in FIGURE 6.

The resulting S-factor is displayed in FIGURE 7. It can be seen that the difference between data taken with the prompt-γ method and the activation method is smaller than $\pm 1.4\%$. The low energy data minimize the uncertainty upon extrapolation. However, a refined measurement of the slope of the S-factor in a wide energy range would be useful to confirm theoretical calculations thus reducing the uncertainty on the extrapolated $S(0)$.

^{25}MG(P,γ)^{26}AL

The reaction ^{25}Mg(p,γ)^{26}Al (Q=6.3 MeV) is the slowest reaction of the Mg–Al cycle [1]. The β^{+} decay of $^{26}Al_{gs}$ to the exited state of ^{26}Mg gives rise to a 1.8 MeV γ-ray, one of the most important lines for γ-astronomy [20]. The aim of the experiment is to measure the strength of the resonances of particular relevance directly. It must be noted that in case of a strong resonance at low energies the Gamow peak is determined by the area of the resonance itsself. The resonances at 190 and at 89 keV in ^{25}Mg(p,γ)^{26}Al are examples of such a situations and also the resonance at 125 keV may be of interest, provided it is sufficiently strong. The existence of all resonances is confirmed by indirect studies while in particular the strength the resonances at 125 and 89 keV resonance have been determined only indirectly [20] as, due to cosmic ray background, measurements of these important resonances have proven to be very difficult or even impossible in a laboratory above ground.

Indirect studies of the reaction reveal a complicated decay scheme of all resonances resulting in complex spectra when studied with a HPGe detector. Due to the high resonance strength $\omega\gamma$ of the 304 keV resonance $8.6 * 10^{12}$ reactions occur when using the LUNA 400kV accelerator in connection with enriched ^{25}Mg targets. The decay structure can thus be identified in spite of the low detection efficiency (see FIGURE 8) . Due to there low $\omega\gamma$ this does not hold for the 125 and 89 keV resonance which result in 74 reactions per day.

As measurement of this cross section can basically be focused on those three energies at which the resonances are expected and as the determination of $\omega\gamma$ contains all the information needed for nuclear synthesis calculations, extrapolations are not necessary and consequently there is no need for collecting spectroscopic information to support

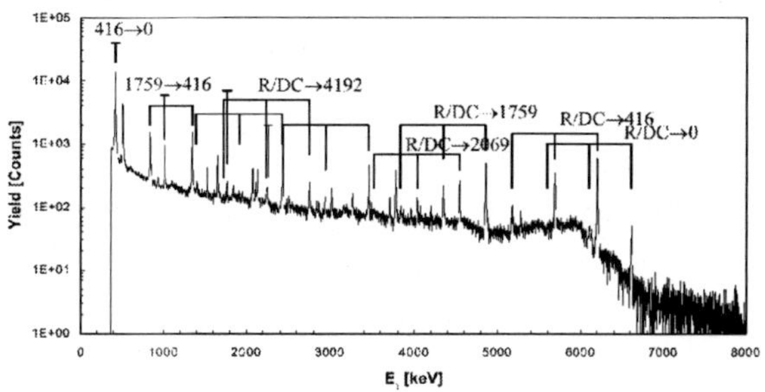

FIGURE 8. Spectrum of the $^{25}Mg(p,\gamma)^{26}Al$ on the resonance at $E = 304$ keV.

them. Thus the ideal technique to study the low energy resonance is the use of the BGO summing detector (see above) with a solid state target system. Details of the setup can be found in [21]. Due to the Q-value the summing peak is expected between 6.4 and 6.7 MeV depending on the resonance studied. Although in this area background is still present (see FIGURE 3) it turns out that it does not hamper the measurement even at the 89 keV resonance. However care must be taken to avoid beam induced background from the reactions $^{11}B(p,\gamma)^{12}C$ and $^{7}Li(p,\gamma)^{8}Be$ as both reactions have lower Coulomb barriers and higher Q-values than the reactions studied. In the course of the experiment several targets had to be discarted as they showed a contamination of one or both of these elements which covered to signal of the reaction.

The data taking for this resonance is proceeding at the moment going on.

REFERENCES

1. C. Rolfs, W.S. Rodney, Cauldrons in the Cosmos, University of Chicago Press, 1988
2. S.P. Ahlen et al., Phys. Lett. B 249 (1990), 149
3. P. Belli et al., Nuovo Cimento A 101 (1989), 959
4. U. Greife et al., Nucl. Instrum. Methods Phys. Res. A 350 (1994), 327
5. A. Formicola et al., Nucl. Instrum. Methods Phys. Res. A 507 (2003), 609
6. G. Imbriani et al., A&A 420 (2004), 625
7. G. Heuser, Ann. Rev. Part. Sci. 45 (1995), 543
8. H. Back et al., Phys Rev C 74 (2006), 045805
9. J.N. Bahcall, M.H. Pinsonneault, Phys. Rev. Lett. 92 (2004). 121301
10. S. Degl'Innocenti et al., Phys. Lett. B 590 (2004), 13
11. U. Schröder et al., Nucl. Phys. A 467 (1987), 240
12. A. Formicola et al., Phys. Lett. B, 591 (2004), 61
13. G. Imbriani et al., Eur. Phys. J. A 25 (2005), 455
14. D. Bemmerer et al., Nucl. Phys. A 779 (2006), 297
15. D. Bemmerer et al., Eur. Phys. J. A 24 (2005), 313
16. D.Bemmerer et al., Phys. Rev lett. RL 97 (2006), 122502
17. Gy. Gyürky et al., Phys. Rev. C 75 (2007), 035805
18. F. Confortola et al., Phys. Rev. C 75 (2007), 065803
19. C. Arpesella, Appl. Radiat. Isotopes 47 (1996), 991
20. LUNA Collaboration, Contrib. to LNGS Anual Report 2005, available at `www.lngs.infn.it`
21. LUNA Collaboration, Contrib. to LNGS Anual Report 2006, available at `www.lngs.infn.it`

Spectroscopy of Proton-rich Nuclei for Explosive Hydrogen Burning Reactions

Philip J Woods

School of Physics, University of Edinburgh, EH9 3JZ, UK

Abstract. Gamma-ray spectroscopic techniques developed for studies of exotic nuclei are described for the application to precise energy measurements and unambiguous state assignments of key astrophysical resonances in explosive hydrogen burning scenarios, such as novae explosions.

Keywords: Nuclear Astrophysics, Gamma Spectroscopy.
PACS: 23.20.Lv,27.30.+t:

INTRODUCTION

New generation telescopes are yielding detailed observations on explosive stellar events, providing challenges to stellar modeling of these processes. Key information comes from measurements of light curves, ejecta, cosmic gamma-ray emission, and meteoritic abundance ratios. In Novae and X-ray bursters the basic astrophysical model involves accretion of Hydrogen and Helium material from the less evolved companion star onto the compact surface of degenerate matter on a white dwarf or neutron star, respectively. Under such hot, dense, astrophysical conditions the reaction rates and nucleosynthetic flow are strongly influenced by reactions involving proton-rich radioactive nuclei, and critically depend on the properties of a few key resonances/states sitting in the region of the Gamow window.

For many years researchers have used reactions involving light nuclei, particularly transfer reactions, to obtain indirect information on these resonances. But these studies have suffered from poor resolution, and ambiguous DWBA fits and assignments.

The possibility of performing direct measurements of explosive astrophysical hydrogen and helium burning reactions has been a key, often the key, motivating factor for the development of ISOL based radioactive beam facilities. However, although a number of impressive direct measurements have been made with radioactive beams, the impact has as yet been relatively limited due to the unavailability of the right beam species, with the appropriate intensity. In responding to this situation, our collaboration has sought to utilize the state-the-art gamma-ray and separator facilities available at the Argonne National Laboratory to access and measure key astrophysical resonances in proton-rich nuclei. This approach is described below.

CP972, *Exotic Nuclei and Nuclear/Particle Astrophysics (II)—Proceedings of the Carpathian Summer School of Physics 2007,* edited by L. Trache and S. Stoica

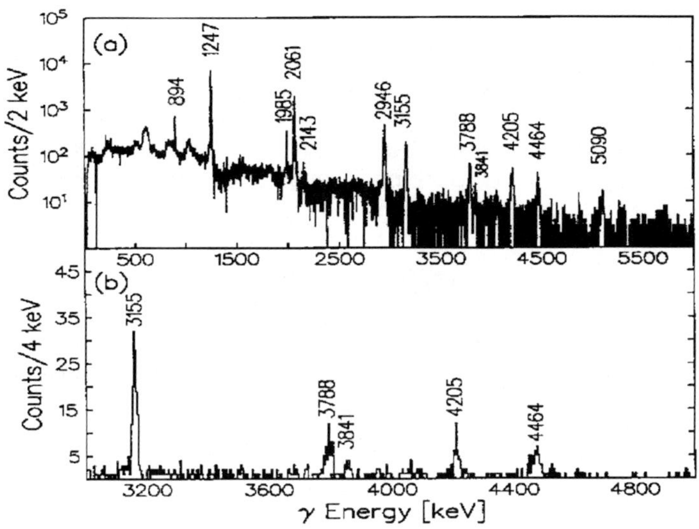

FIG. 1. (a) Gamma-ray spectrum measured in coincidence with ^{22}Mg residues. (b) High-energy γ rays detected in coincidence with the $2_1^+ \rightarrow 0_1^+$, 1247 keV transition in ^{22}Mg.

GAMMA SPECTROSCOPIC MEASUREMENTS

The proton rich nuclei are produced in heavy ion fusion evaporation reactions, for example ^{22}Mg was produced via 2 neutron evaporation by bombarding a ^{12}C target with a 15 pnA beam of 52 MeV ^{12}C ions [1]. Prompt gamma rays were detected using the highly efficient GAMMASPHERE array of Ge detectors [2] in coincidence with A =22 charge state 8^+ recoils detected at the focal plane of the Argonne Fragment Mass Analyzer (FMA) [3]. ^{22}Mg recoils were separated from the other A=22 isobars, including the well known mirror nucleus ^{22}Ne, using energy loss information from the ionization chamber situated at the focal plane. Figure 1a shows a gamma ray spectrum which is gated on ^{22}Mg recoils and Figure 1b shows gammas coincident with the 2^+ to ground state transition. Using such information it is possible to construct complete level schemes of nuclei below the particle threshold, and to observe low lying states above the particle threshold. In this case the key astrophysical resonance at an excitation energy, 5711keV, and a resonance energy of 206 keV, was identified through the coincidence gate of Fig1b. This was a particularly important result as it resolved discrepancies between astrophysical resonance energy measurements taken with the DRAGON separtor[4]of ^{21}Na(p,γ)^{22}Mg reaction relevant for production of the cosmic gamma rays from the β-decay of ^{22}Na in novae explosions, and previous high precision mass measurements relevant for the CVC hypothesis in the Standard Model [1].

This experiment, and others carried out at the FMA, illustrated that contrary to expectations, heavy ion fusion reactions in this mass region populate states of all spins significantly, albeit that high spin states are proportionately more strongly populated than low spin states. This means that for the first time full unambiguous level schemes have been constructed for a range of nuclei in this region for comparison with shell model calculations, and for determining Coulomb energy differences in the mirror nuclei, often observed in the same experiment [1,5,6].

CONCLUSION

It has been demonstrated that important precision measurements on key astrophysical resonances for explosive hydrogen burning scenarios can be performed by the innovative use of modern gamma-spectroscopic techniques, something not foreseen when such devices were originally designed. These measurements are highly complementary to direct measurements with radioactive beams, and previous studies using light ion reactions.

ACKNOWLEDGEMENTS

This programme of work has involved many people over a period of time beginning around 2000 and has grown significantly. It is not practical to acknowledge all the people involved here but I would in particular like to acknowledge the key roles in this programme of work of Dr Darek Seweryniak from Argonne National Laboratory, Dr David Jenkins from the University of York, and very importantly Gavin Lotay from Edinburgh who will be the first student for whom this will form their PhD work.

REFRENCES

1. D Seweryniak et al. Phys. Rev. Lett. **94,** 032501 (2005)
2. RVF Janssens and FS Stephens, Nucl. Phys. News **6,** 9 (1996)
3. CN Davids et al., Nucl. Instr. Meth. B**70**, 358 (1992).
4. S Bishop et al., Phys. Rev. Lett. **90,** 162501 (2003)
5 D Seweryniak et al., Phys. Lett. B**590,** 170 (2004)
6 D Seweryniak et al., Phys. Rev. C**75,** 062801(R) (2007)

The S_{E1} factor of the $^{12}C(\alpha,\gamma)^{16}O$ Reaction and the ^{16}N β-delayed α Decay

Xiaodong Tang[a], K. E. Rehm[b], I. Ahmad[b], C. R. Brune[c], A. Champagne[d], J. P. Greene[b], A. A. Hecht[b], D. Henderson[b], R.V. F. Janssens[b], C. L. Jiang[b], L. Jisonna[e], D. Kahl[b], E. F. Moore[b], M. Notani[b], R. C. Pardo[b], N. Patel[b], M. Paul[f], G. Savard[b], J. P. Schiffer[b], R. E. Segel[e], S. Sinha[b], B. Shumard[b], and A. H. Wuosmaa[g]

[a]University of Notre Dame, Notre Dame, IN 46530
[b]Physics Division, Argonne National Laboratory, Argonne, Illinois 60439, USA
[c]Ohio University, Athens, Ohio 45701, USA
[d]University of North Carolina, Chapel Hill, North Carolina 27599, USA
[e]Northwestern University, Evanston, Illinois 60208, USA
[f]Hebrew University, Jerusalem, Israel
[g]Western Michigan University, Kalamazoo, Michigan 49008, USA

Abstract. The $^{12}C(\alpha,\gamma)^{16}O$ reaction is of great importance for the understanding the nuclear synthesis and explosion mechanism of the supernova. The cross section at the relevant stellar burning energies, around 300 keV, is dominated by the sub-threshold resonances. The ^{16}N β-delayed α decay data provide the most sensitive constrain on this sub-threshold resonance. Using twin-ionization chambers, we have measured the α particles and ^{12}C recoil nuclei in coincidence. The spectrum has been fitted together with the published $^{12}C(\alpha,\gamma)^{16}O$ data and the elastic scattering phase shift data using R-matrix. The S_{E1} factor at 300 keV has been determined as 73±21 keVb.

Keywords: the $^{12}C(\alpha,\gamma)^{16}O$ reaction, ^{16}N β-delayed α decay, S_{E1} factor
PACS: 21.10Pc, 23.60.+e, 25.55.-e, 26.20+f

THE $^{12}C(\alpha,\gamma)^{16}O$ REACTION

The stellar reaction rate of the $^{12}C(\alpha,\gamma)^{16}O$ is one of the most important unsettled rates left in nuclear astrophysics [1,2,3]. The final C/O ratio at the end of helium burning, which has great influence on the subsequent heavy ion burning processes and supernova explosion, is limited by the large uncertainty associated with the $^{12}C(\alpha,\gamma)^{16}O$ reaction rate. There are two main α-capture modes for the $^{12}C(\alpha,\gamma)^{16}O$ reaction. One is E1 capture with contribution from the 1^- resonance at E_{cm}=2.418 MeV (E_x=9.585 MeV) and the sub-threshold 1^- state at E_{cm}=-45 keV (E_x=7.117 MeV). The other is E2 capture including contributions from direct capture and the tail from a sub-threshold 2^+ state at E_{cm}=-245 keV (E_x=6.917 MeV). Contributions from cascade transitions are expected to be small. At the relevant stellar burning energies, around 300 keV, the cross sections are at the order of 10^{-17} b which are dominated by the sub-threshold resonances and are beyond the present experimental capabilities. Limited by extremely low yield and background, the lowest energy reached by direct measurement is about 890 keV in center of mass where the cross section is about 10^{-11} b [4-9]. We have to rely on R-matrix theory to extrapolate the data down by 6 orders of magnitude. The values quoted in the recent NACRE compilation [10] are 79±21

CP972, *Exotic Nuclei and Nuclear/Particle Astrophysics (II)—Proceedings of the Carpathian Summer School of Physics 2007,* edited by L. Trache and S. Stoica

keVb for S_{E1} and 120±60 keVb for S_{E2}, respectively. In this paper S_{E1} and S_{E2} refer to the S-factors at E_{cm}=300 keV.

Three sets of experimental data, the $^{12}C(\alpha,\gamma)^{16}O$, the $^{12}C(\alpha,\alpha)^{12}C$ and the ^{16}N β-delayed α decay, are used in R-matrix fitting to achieve the best determination of the S_{E1} factor around E_{cm}=300 keV [11]. The cross sections around 300 keV are dominated by the sub-threshold 1^- state. The excitation energy and the radiative width of the sub-threshold 1^- state are well determined by gamma ray spectroscopy. The ^{16}N data provides the tightest constrain on the reduced α width for the sub-threshold 1^- state and therefore serves as the most sensitive constrain on the S_{E1} factors around 300 keV.

In the ^{16}N β-delayed α decay energy spectrum, there are two peaks. The big peak locating at E_{cm}~2.4 MeV is from the higher lying 1^- state (E_x=9.585 MeV). The small satellite peak is explained as the interference effect between the sub-threshold 1^- state (E_x=7.117 MeV) and the higher lying 1^- state (E_x=9.585 MeV). The reduced α width is sensitive to the peak height ratio between these two peaks [12]. The total branching ratio to the α decay channel is about 10^{-5} while the branching ratio to the small peak is a factor of 500 smaller. The tiny branching ratio of the ^{16}N decay complicates the α-decay experiments in two ways: It requires a strong and pure source of ^{16}N, and the high β background severely masks the low energy part of the α spectrum in the vicinity of the small satellite peak from which S_{E1} is obtained. This peak has been studied by several groups [11,13,14], all of them using thin Si-surface barrier detectors, which are sensitive to β particles, have dead layers and are prone to deterioration during measurements lasting several weeks.

A NEW ^{16}N DECAY EXPERIMENT

Because of the importance of the $^{12}C(\alpha,\gamma)^{16}O$ reaction, we have re-measured the ^{16}N decay using a different approach. To reduce the sensitivity to electrons, we developed an array of high-acceptance ionization chambers of minimal thickness [15], to be used for the coincident detection of ^{12}C and α particles following the ^{16}N decay. The main goal of the experiment were (1) to re-measure the height of the small satellite α peak, using a different experimental setup, subject to different systematic uncertainties, (2) to extend the measurements towards lower energies and, (3) to resolve the discrepancies, observed in earlier experiments, in the shape of the low energy part of the α spectrum.

A schematic of the experimental setup is given in Fig. 1. A 61-MeV ^{16}N ($T_{1/2}$=7.1 sec) beam is slowed down in a gas-filled attenuation cell and stopped in a 17 μg/cm^2 (nominally 10 μg/cm^2) carbon foil (foil a) mounted on a rotating wheel (diameter 18.7 cm) located in the main part of the detection chamber. The cell and the main chamber are filled with P10 counting gas at a pressure of 150 Torr. After an irradiation of 15 s, the foil (a) is rotated (in 60 ms) by 120 degree, placing it between a pair of twin-ionization chamber (pair I) for the coincident detection of ^{12}C-α pairs. The carbon foil is counted in this pair of ionization chambers (IC) for 15 sec. During this time, a second foil (b), positioned on the wheel 120 degree relative to the first foil, is irradiated. At the end of the second radiation, foil (b) is rotated to be between a second pair of ionization chambers (pair II) and counted, while foil (a) is again irradiated with ^{16}N. Detector pair I is meanwhile counting a non-irradiated foil to establish backgrounds. The same foil, which is never irradiated, is also used for measuring background in detector pair II.

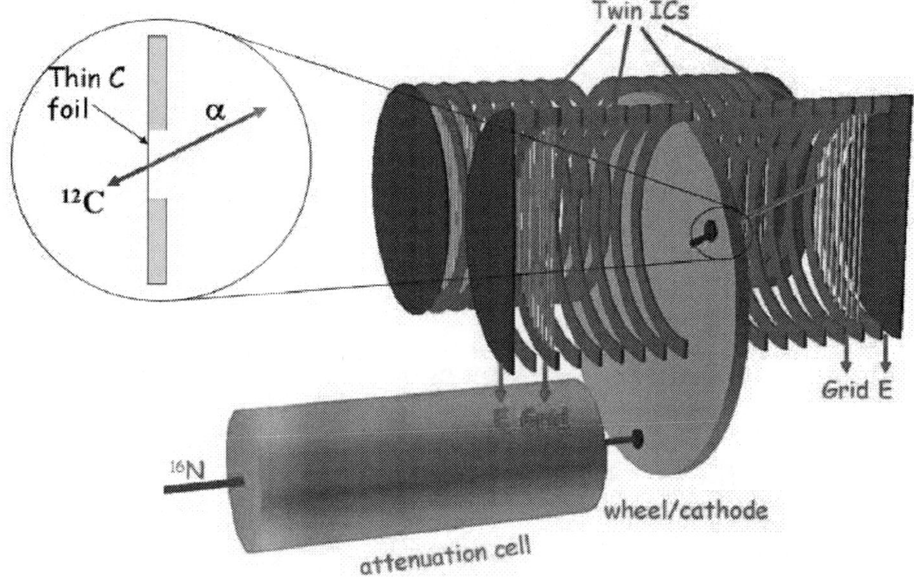

FIGURE 1. Schematic of the experimental setup used for a measurement of the β-delayed α decay of ^{16}N.

The ionization chambers are the Bragg type twin ionization chambers to measure the ^{12}C-α coincidence from the ^{16}N decay. The α's from the ^{16}N decay range from 0.6 MeV to 2.5 MeV. The energy calibration of the ionization was carried out with the α-^{7}Li and α-t coincidence obtained from the ^{10}B(n,α)^{7}Li and ^{6}Li(n,α)t, respectively, using moderated neutrons from a Pu-Be neutron source. These two reactions provided α particles of 1.472, 1.776, and 2.056 MeV energies that bracket the energy range of interest.

The natural background from a single detector is dominated by ^{210}Po α decay from the solder used in the Frisch grid with a rate about 170 counts/hour. To test the sensitivity of the ionization chambers to β particles, a ^{22}Na β^+ source with an activity of about 10^5 decays/s was mounted on the target wheel. The strength of this source is similar to the actual β particle rate from ^{16}N in the experiment. The only change observed in the spectrum was a 25% increase in the background in single detector at very low energies (E~100 keV). With the coincidence, these backgrounds from solder and β particles from ^{16}N are negligible.

The ^{16}N was produced via an In-Flight technique, by bombarding a deuterium-filled cryogenic gas cell with an 82-MeV primary ^{15}N beam from the ATLAS facility. No ^{17,18}N ions can be produced in this process. All beam contaminants are identified as stable isotopes which do not interfere the decay measurement. With 6×10^{11} /s ^{15}N incident on the deuterium filled production cell, the ^{16}N beam intensity measured at the foil was about 3×10^{6} /s. By tuning the pressure of the attenuation cell, the optimal ^{16}N collection efficient at the foil was measured as 5%.

The measured coincident energy spectra from the ^{16}N catcher foil and background foil are shown in Fig. 2a and 2b respectively. The energy ratio between α and ^{12}C from ^{16}N decay should be 3 if we ignore the tiny ^{16}O recoil energy from the β decay. The α particles from the decay have an energy range from 0.45 MeV up to 2.4 MeV and the corresponding ^{12}C energy range from 0.15 MeV up to 0.8 MeV. Due to the significant ^{12}C energy loss in the foil, the measured energy ratio between the α and ^{12}C is larger than 3. Those are considered as the good events shown symmetrically in the energy spectrum. However, some particles, either α or ^{12}C, can be caught by the target frame and only deposit their partial energy in the gas. Those events end as the two blobs close to the y-axis with downstream energy less than 300 keV. Those events can be eliminated by using the angle information from the twin-ionization chambers. In this paper, we only discuss the events where the α particle is emitted downstream and the full ^{12}C energy is recorded, represented by the events in the lower part of Fig. 2a.

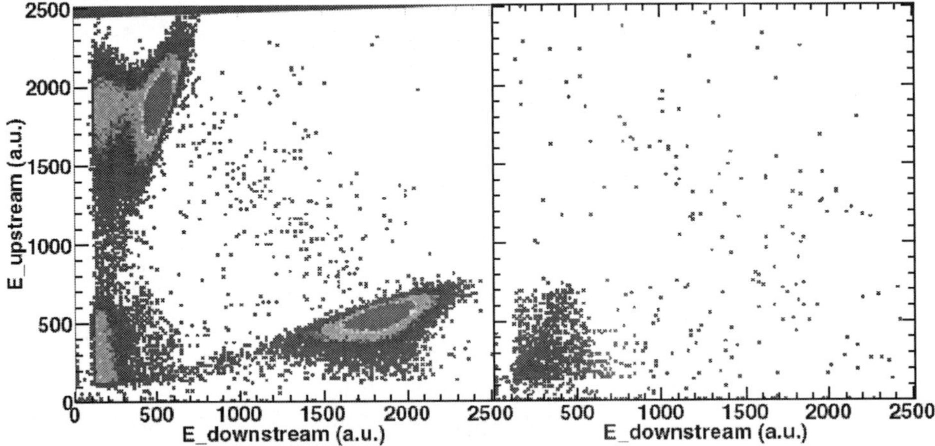

FIGURE 2. (Left) coincident energy spectrum, measured with one of the twin-ionization detectors for a foil with implanted ^{16}N particles. (Right) same as above, but for a non-irradiated foil.

Selecting events with upstream energy below channel 800, and requiring the correct pulse height ratio of α/^{12}C events, results in the α spectrum given in Fig. 3(a). The additional requirement is that all particles are in the correct Frisch-grid anode region (see [16] for details) of the upstream (^{12}C) and downstream (α) ionization chambers. While the pressure of 150 Torr (equivalent to a Si detector thickness of 7 μm) minimizes the interference from the low-energy α particles stopped in the target frame (giving rise to the group next to the origin in Fig. 2), it also leads to a distortion of the α spectrum at channel numbers above ~2000 (corresponding to energies above 1.9 MeV) due to incomplete stopping of higher energy α particles. To verify this, some of the measurements were performed at a pressure of 195 Torr [Fig. 3(b)]. At this pressure, particles up to 2.4 MeV are stopped in the ionization chamber. The solid line in Fig. 3(b) represents the normalized spectrum obtained at 150 Torr and illustrates the effects of incomplete stopping. These effects were also reproduced in Monte Carlo simulations. Correcting the 150 Torr data at energies above an α energy of 1.9 MeV, we obtain a summed spectrum shown in Fig. 4 containing 2.2×10^5 events. Due to the insensitivity of ionization chambers to the copious β's, the low-energy satellite peak in the β-delayed α spectrum of ^{16}N can now be followed, practically background-free and without correcting for pileup events from electrons or contributions from $^{17,18}N$ decays, down to a c.m.-energy of ~600 keV. The solid line is the result of an R-matrix fit, which is discussed below.

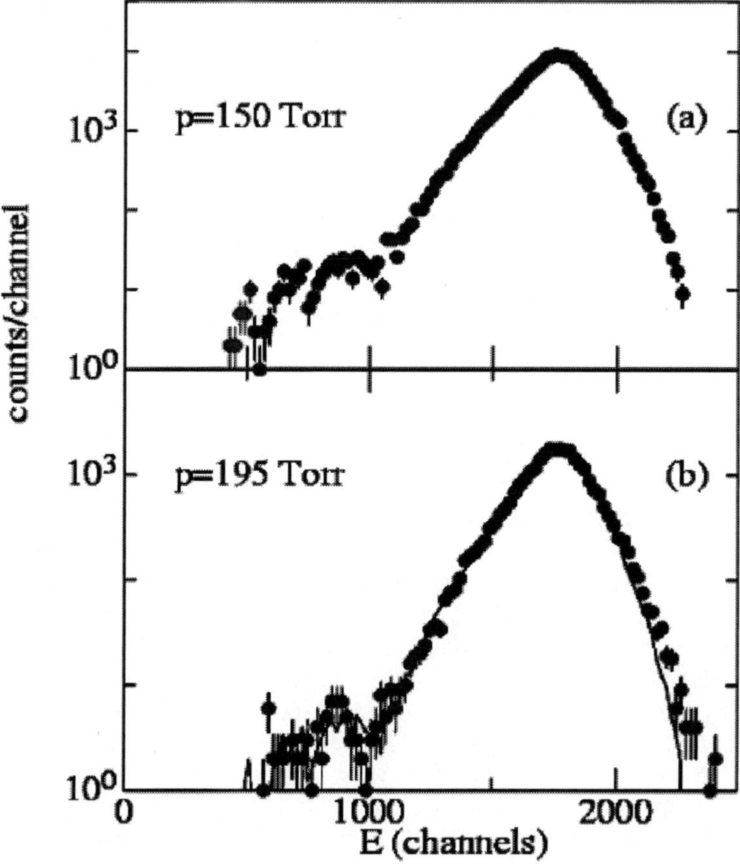

FIGURE 3. (a) The α spectrum measured in one of the twin ionization chambers filled with 150 Torr of isobutane. (b) Same as (a), but at a pressure of 195 Torr. The solid line is the spectrum obtained at 150 Torr, scaled in height.

Our final data are shown together with some previous data in Figure. 4. By comparing with TRIUMF data [11], the big peak in our spectrum is broader and in better agreement with the Mainz data [17]. The satellite peak, which originates from the interference between the subthreshold 1^- state and the higher lying 1^- state in ^{16}O, is still in agreement with earlier measurements.

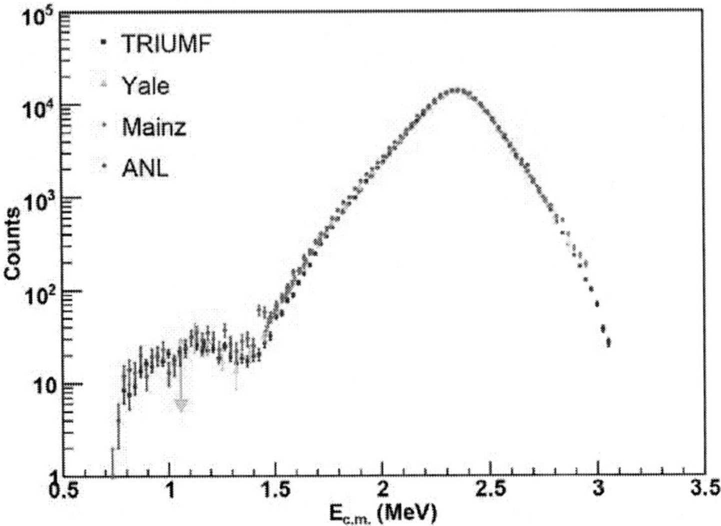

FIGURE 4. Our final ^{16}N spectrum is shown together with the previous data sets from TRIUMF, Yale and Mainz.

R-MATRIX ANALYSIS

Extracting the S_{E1} factor from the data is usually done using the R-matrix formalism through a least-square fit to data from the β-delayed α decay of ^{16}N, in combination with experimental results from the capture reaction $^{12}C(\alpha,\gamma)^{16}O$ performed at higher energies, as well as with phase shift parameters obtained from the elastic scattering of $^{12}C(\alpha,\alpha)^{12}C$. For this, we have used the R-matrix program from [11]. To be comparable with the TRIUMF result, we first fit our ^{16}N data together with the $^{12}C(\alpha,\gamma)^{16}O$ and $^{12}C(\alpha,\alpha)^{12}C$ data as used in [11]. For the E1 radiative capture data (p-wave), we considered the subthreshold 1^- state (E_x=7.12 MeV), the high lying 1^- state (E_x=9.585 MeV) and a background 1^- state for truncating all the other higher lying 1^- states. However, if we only considered p-wave components, there is a minimum in the ^{16}N spectrum at ~1.2 MeV in the center of mass. Clearly, the experimental data could not be explained only by the p-wave component. To solve this problem [11], we introduced an additional f-wave component with contribution from the three 3^- states (E_x=6.13 MeV, 11.601 MeV and a background state). For the channel radius parameter a=6.5 fm, the best fit is determined as $S_{E1} = 73 \pm 17(\text{stat})\ \text{keVb}$.

We also investigated the best value of the channel radius parameter a. Figure. 5(a) shows the dependence of the total χ^2 on S_{E1} factor and a. Unlike the TRIUMF data, the best fit to our data occurs at a=5.5 fm and $S_{E1} = 75 \pm 18(stat)$ keVb. For a=5.5 fm, the dependence of the χ^2s, the total χ^2 as well as the individual χ^2 components from different data set, on the S_{E1} factor are shown in Figure. 5(b). It clearly indicates that the ^{16}N data provides the most sensitive constrain to the S_{E1} factor.

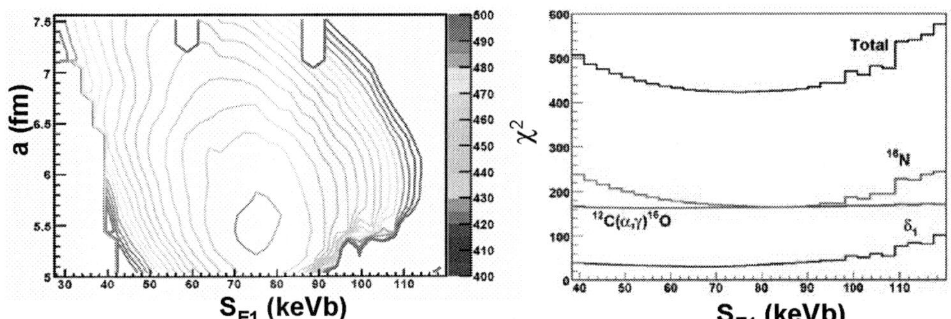

FIGURE 5. (a) Contour plot of χ^2 dependence on the channel radius a and S_{E1} factor. The S_{E1} is 73 keVb when a=6.5 fm. The best S_{E1} is determined as 75 keV when a=5.5 fm. (b) For a=5.5 fm, the total χ^2 as well as the individual χ^2 components from different data sets dependence on S_{E1} factor.

In the TRIUMF's systematic uncertainty budget, there are three major sources. The largest term is the energy calibration of their Si detector for the decay measurement which contributes a ±10 keVb systematic uncertainty. Our detectors are calibrated with three α lines from slow neutron induced (n,α) reactions on ^{10}B and ^{6}Li. This term was reduced down to ±5 keVb. The second largest term is the relative β decay branching ratio between the subthreshold 1^- state (E_x=7.117 MeV) and the higher lying 1^- state (E_x=9.585 MeV). For this, we have measured the branching ratio to subthreshold 1^- state [16]. The result is 10% higher than the NNDC compilation with an uncertainty about 2%. With this new branching ratio, the best S_{E1} factor decreases to 72 keV (a=5.5 fm) and this systematic uncertainty term is reduced to ±3 keVb. The third largest term in the TRIUMF's systematic uncertainty budget is the ^{17}N contamination. This term disappears automatically in our experiment since our ^{16}N production is ^{17,18}N free. The uncertainty of the radiative width of the subthreshold 1- state superimposes additional ±2 keVb.

Four sets of independent ^{12}C(α,γ)^{16}O measurements used in the fitting display clear systematic differences which introduce ±4 keVb systematic uncertainty. The ±10% uncertainty in the overall normalization of the ^{12}C(α,γ)^{16}O cross section data contributes an additional ±2 keVb systematic uncertainty. So, by summing all the contribution, the total systematic uncertainty of our result is ±8(sys) keVb. The two systematic uncertainties arising from the direct measurement can be avoided by using the recent published Stuggart (α,γ) data. The best S_{E1} factor with the Stuggart data, our ^{16}N data and the new branching ratio is 73±18(stat) keVb. This combination provides a better systematic uncertainty of ±7(sys) keVb. By including the statistical uncertainty, the SE1 factor is determined as 73±21 keVb.

Two partial wave components, p-wave and f-wave, are considered to fit the ^{16}N β-delayed α decay spectrum. There is a correlation between these two components. The deviation between TRIUMF and ANL spectrum at lower energies leads to a different extraction of the f-wave components as shown in Figure. 6. The correlation between p-wave and f-wave component hinders the precise extraction of the important p-wave components. To resolve this issue, another independent ^{16}N experiment will be helpful.

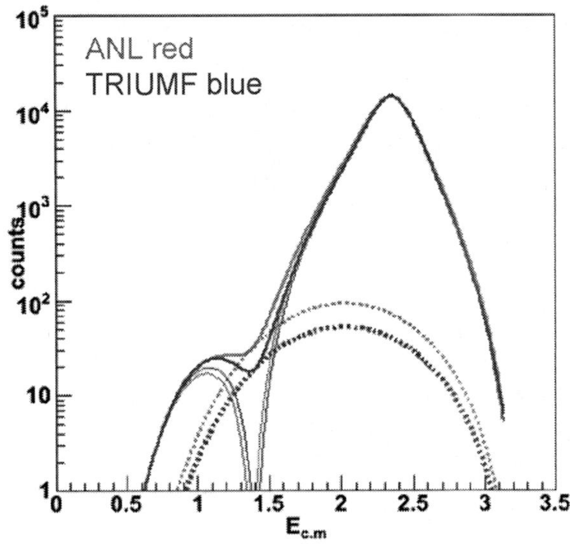

FIGURE 6. The R-matrix fitting curves for ANL (red) and TRIUF (blue) data. To be comparable, they are normalized to the same amplitude. The p-wave components are shown as thin solid line and the f-wave components as dotted lines.

THE CONCLUSION

We have performed a new measurement of the β-delayed α decay of ^{16}N using a setup that is insensitive to the high electron background. The low-energy part of the spectrum was found to be in better agreement with earlier data from the MAINZ [17] and Yale [13,14] than with the data from the TRIUMF group [11]. The satellite peak, which originates from the interference between the subthreshold 1^- state and the higher lying 1^- state in ^{16}O, was clearly resolved. Its height is in agreement with earlier measurements. Despite the significant deviation between TRIUMF and our data, the best fitting S_{E1} factors agree each other.

The desired accuracy of the S factor of $^{12}C(\alpha,\gamma)^{16}O$ reaction at 300 keV should be better than 20% [18]. However, despite of more than 40 years of effort, the goal has not been reached. With the additional constrain from ^{16}N on the α width of the subthreshold 1^- state, the S_{E1} factor still has an uncertainty about 30%. The situation the S_{E2} factor is even more difficult because it depends on the direct measurements and elastic scattering data which are less sensitive to the subthreshold resonance. To reach the final goal, we need improve the quality of the experimental data as well as the theory.

ACKNOWLEDGMENTS

We would like to thank the operations group of the ATLAS accelerator for providing the ^{16}N beam for the experiment. We also would like to thank Nick Sciezlo for calculating the Fermi function. This work was supported by the US Department of Energy, Office of Nuclear Physics, under Contract No. DE-AC02-06CH11357 and by the NSF Grant No. PHY-02-16783 (Joint Institute for Nuclear Astrophysics).

REFERENCES

1. T. A. Weaver and S. Woosley, *Phys. Rep.* **227**, 65 (1993).
2. L. R. Buchmann and C. A. Barnes, *Nucl. Phys.* **777**, 254 (2006).
3. W. Fowler, *Rev. Mod. Phys.* **56**, 149 (1984).
4. A. Redder *et al., Nucl. Phys. A* **462**, 385 (1987).
5. J. M. L. Ouellet *et al., Phys. Rev. Lett.* **69**, 1896 (1992).
6. R. Kunz *et al., Phys. Rev. Lett.* **86**, 3244 (2001).
7. R. Kunz *et al., Astrophys. J.* **567**, 643 (2002).
8. M. Assuncao *et al., Phys. Rev. C* **73**, 055801 (2006).
9. R. Plag, *Ph.D. thesis*, FZ Karlsruhe [FZKA Report No. 7099, 2005].
10. C. Angulo *et al., Nucl. Phys. A* **656**, 3 (1999).
11. R. Azuma et al., *Phys. Rev. C* **50**, 1194 (1994); L. Buchmann et al., *Phys. Rev. Lett.* **70**, 726 (1993).
12. D.Baye and P.Descouvemont, *Nucl. Phys. A* **481**, 445 (1988)
13. Z. Zhao et al., *Phys. Rev. Lett.* **70**, 2066 (1993).
14. R. H. France et al., *Nucl. Phys. A* **621**, 165c (1997).
15. C. Budtz-Jørgensen *et al., Nucl. Instrum. Methods Phys. Res., Sect. A* **258**, 209 (1987).
16. K. E. Rehm *et al.* (to be published).
17. K. Neubeck, H. Schober, and H. Waffler, *Phys. Rev. C* **10**, 320 (1974).
18. *Opportunity in Nuclear Science*, A Long Range Plan for the Next Decade, The DOE/NSF Nuclear Science Advisory Committee (2002)

Trojan Horse Method: Recent Results

R.G.Pizzone and C.Spitaleri

Dipartimento di Metodologie Fisiche e Chimiche per l'Ingegneria, Universitá di Catania and Laboratori Nazionali del Sud - INFN , Catania

Abstract. Owing the presence of the Coulomb barrier at astrophysically relevant kinetic energies, it is very difficult, or sometimes impossible to measure astrophysical reaction rates in laboratory. This is why different indirect techniques are being used along with direct measurements. The THM is unique indirect technique allowing one measure astrophysical rearrangement reactions down to astrophysical relevant energies. The basic principle and a review of the main application of the Trojan Horse Method are presented. The applications aiming at the extraction of the bare $S_b(E)$ astrophysical factor and electron screening potentials U_e for several two body processes are discussed.

Keywords: indirect methods, cross section measurements, nuclear astrophysics
PACS: 24.50.+g,26.20.Np

INTRODUCTION

Nuclear fusion reactions, that take place in the hot interiors of remote and long-vanished stars over billions of years, are the origin in the universe of nearly all the chemical elements and their isotopes [BBFH, Fowler 1984, Rolfs & Rodney, 1988]. The detailed understanding of the origin of the chemical elements and their isotopes has combined astrophysics and nuclear physics, and forms what is called nuclear astrophysics. In turn, nuclear reactions are the heart of nuclear astrophysics: they influence sensitively the nucleosynthesis of the elements in the earliest stages of the universe and in all the objects formed thereafter, and control the associated energy generation (by processes called nuclear fusion or nuclear burning), neutrino luminosity, and evolution of stars. A good knowledge of the rates of these fusion reactions is essential for understanding this broad picture [Strieder et al:, 2001].

LIMIT OF THE TWO-BODY MEASUREMENTS

In a stellar plasma the constituent nuclei are usually in thermal equilibrium at some local temperature T. Occasionally they collide with other nuclei, whereby two different nuclei can emerge from collision A+x→ c+C. The cross section $\sigma(E)$ of nuclear fusion reaction A(x,c)C is of course governed by the laws of quantum mechanics where, in most cases, the Coulomb and centrifugal barriers arising from nuclear charges and angular momenta in the entrance channel of the reaction strongly inhibit the penetration of one nucleus into another. This barrier penetration leads a steep energy dependence of the cross section. It is the challenge to the experimentalist to make precise $\sigma(E)$ measurements over a wide range of energies, as our fragmented knowledge of nuclear physics prevents us from

CP972, *Exotic Nuclei and Nuclear/Particle Astrophysics (II)—Proceedings of the Carpathian Summer School of Physics 2007,* edited by L. Trache and S. Stoica

predicting σ(E) on purely theoretical grounds. For these reasons bare nucleus cross section measurements $\sigma_{pl}(E)$(for stellar plasma) of the (p,α) reaction at the Gamow energy (E_G) should be known with an accuracy better than 10% [Rolfs & Rodney, 1988, Rolfs et al., 2001] because of their crucial role in understanding the first phases of the Universe history and the subsequent stellar evolution. Unfortunately the presence of Coulomb barrier, in the reactions with charge particles, is a limit, often insuperable, to perform measurements of the cross sections at ultralow energies. Indeed the Coulomb barrier of height E_C in charged-particle induced reactions E_C cause an exponential decrease of the cross section $\sigma_b(E)$ at $E > E_C$, $\sigma_b(E) \sim exp(-2\pi\eta)$, leading to a low-energy limit of direct $\sigma_b(E)$ measurements, which is typically much larger than E_G. Owing to the strong Coulomb suppression, the behavior of the cross section at E_G is usually extrapolated from the higher energies by using the definition of the smoother astrophysical factor S(E):

$$S_b(E) = E\sigma_b(E)exp(2\pi\eta) \tag{1}$$

where $exp(2\pi\eta)$ is the inverse of the Gamow factor, which removes the dominant energy dependence of $\sigma(E)$ due to the barrier penetrability.
Although the S_b(E)-factor allows for an easier extrapolation, large uncertainties to $\sigma_b(E_G)$ may be introduced due to for instance the presence of unexpected resonances, or high energy tails of sub-threshold resonances. In order to avoid the extrapolation procedure, a number of experimental solutions were proposed in direct measurements for enhancing the signal-to-noise ratio at E_G.
In recent years the availability of hight-current low-energy accelerators, such as that at the underground Laboratories together with improved target and detection techniques have allowed us to perform $\sigma_b(E)$ measurements in some cases down to E_G or at least close to E_G [Bonetti et al., 1999]. Then in principle no $\sigma_b(E)$ extrapolation would be needed anymore for these reactions. However, the measurements in laboratory at ultralow energies suffer from the complication due to the effects of electron screening [Assembaum et al., 1987, Strieder et al., 2001]. This leads to an exponential increase of the laboratory measured cross section $\sigma_s(E)$ [or equivalently of the astrophysical factor $S_s(E)$] with decreasing energy relative to the case of bare nuclei. This can be described by an enhancement factor defined by the relation

$$f_{lab}(E) = \sigma_s(E)/\sigma_b(E) = exp(\pi\eta U_e/E) \tag{2}$$

In this equation U_e is the electron screening potential in the laboratory which is different from the U_{pl} present in the stellar environment. Clearly, a good understanding of U_e is needed in order to calculate σ_b from the experimental data σ_s using equation (2). The effective cross section $\sigma_{pl}(E)$ in the stellar plasma, is connected to the bare nucleus cross section $\sigma_b(E)$ and to the the stellar electron screening enhancement factor f_{pl} by the relation

$$\sigma_{pl}(E) = \sigma_b(E)f_{pl}(E) = \sigma_b(E)\cdot exp(\pi\eta U_{pl}/E) \tag{3}$$

with U_{pl} is the plasma potential energy, η the Sommerfeld parameter.
If $\sigma_b(E)$ is measured at the ultralow energies E_G and U_{pl} is estimated within the framework of the Debye-Hückel theory, it is possible estimate from eq.(3) the effective

cross section $\sigma_{pl}(E)$ in the stellar plasma. In turn, the understanding of U_e may help to better understand U_{pl}, needed to calculate σ_{pl}.
Then, although it is possible to measure cross sections in the Gamow energy range, the bare nucleus cross section σ_b is extracted by extrapolating the direct data behavior at higher energies where negligible electron screening contribution is expected. In order to decrease uncertainties in the case of charged particle induced reactions a rather striking conclusion could be achieved: to avoid extrapolations, experimental techniques were improved. After improving measurements (at very low energies), electron screening effects were discovered. Finally to extract from direct (shielded) measurements the bare astrophysical S_b(E)-factor, extrapolation were performed at higher energy. In any case the extrapolation procedure is necessary and in consequence we find again the uncertainties problem in direct measurements.

THE TROJAN HORSE METHOD

Thus, alternative methods for determining bare nucleus cross sections of astrophysical interest are needed. In this context a number of indirect methods, e.g. the Coulomb dissociation (CD) [Baur 1994, Baur 1996], the Asymptotic Normalization Coefficient method (ANC)[Tang et al. 2003, Azhari et al. 2001, Mukhamedzhanov et al 2001, Mukhamedzhanov et al 1997, Mukhamedzhanov et al 1999] and the Trojan-horse method (THM) were developed (Table I). The THM has already been applied several times to reactions connected with fundamental astrophysical problems [Pizzone et al., 2003, Pizzone et al. 2005] . Some of them make use of direct reaction mechanisms, such as transfer processes (stripping and pick-up) and quasi-free reactions (knock-out reactions).
In particular, the THM is a powerful tool which selects the quasi-free (QF) contribution of an appropriate three-body reaction performed at energies well above the Coulomb barrier to extract a charged particle two-body cross section at In the framework of the extrapolation problems linked to the presence of electron screening effects a number of experimental measurements were carried out in order to measure the bare nucleus cross section in reactions of astrophysical interest (Table I).

The idea of the THM [Baur 1986] is to extract the cross section of an astrophysically relevant two-body reaction

$$A+x \rightarrow c+C \tag{4}$$

at low energies from a suitable chosen three-body quasifree reaction

$$A+a \rightarrow c+C+S \tag{5}$$

This is done with the help of direct theory assuming that the nucleus a has a strong $x \oplus S$ cluster structure. In many applications [Spitaleri et al. 2001, Spitaleri et al. 2004, Lattuada et al. 2001], this assumption is trivially fulfilled e.g. a = deuteron, x = proton, S= neutron. This three-body reaction can be described by a Pseudo-Feynmam diagram, where only the first term of the Feynman series is retained The upper pole describes the virtual break-up of the target nucleus a into the clusters x and S; S is then considered to be spectator to the $A+x \rightarrow c+C$ reaction which takes place in the lower pole.

This description, is called Impulse Approximation (IA) [Chew 1950]. In Plane Wave Impulse Approximation (PWIA) the cross section of the three body reaction can be factorized into two terms corresponding to the two poles and it is given by [Neudatchin 1965, Jacob 1966]:

$$\frac{d^3\sigma}{dE_c d\Omega_c d\Omega_C} \propto KF \left(\frac{d\sigma}{d\Omega_{cm}}\right)^{off} \cdot |\Phi(\vec{p}_s)|^2 \tag{6}$$

where:

- $[(d\sigma/d\Omega)_{cm}]^{off}$ is the off-energy-shell differential cross section for the two body A(x,c)C reaction at the center of mass energy E_{cm} given in post collision prescription by:

$$E_{cm} = E_{c-C} - Q_{2b} \tag{7}$$

  where Q_{2b} is the two body Q-value of the $A+x \to c+C$ reaction and E_{c-C} is the relative energy between the outgoing particles c and C;
- KF is a kinematical factor containing the final state phase-space factor and it is a function of the masses, momenta and angles of the outgoing particles:

$$KF = \frac{\mu_{Aa} m_c}{(2\pi)^5 \hbar^7} \frac{p_C p_c^3}{p_{Aa}} \left[\left(\frac{\vec{p}_{Bx}}{\mu_{Bx}} - \frac{\vec{p}_{Cc}}{m_c}\right) \cdot \frac{\vec{p}_c}{p_c}\right]^{-1} \tag{8}$$

- $\Phi(\vec{p}_s)$ is the Fourier transform of the radial wave function $\chi(\vec{r})$ for the x-S inter-cluster motion, usually described in terms of Hänkel, Eckart and Hultheń functions depending on the x-S system properties.

In the experimental works [Table 2] the validity conditions of the IA appear to be fulfilled. We stress that one cannot extract the absolute value of the two-body cross section. However, the absolute value can be extracted through normalization to the direct data available at energies above and/or below the Coulomb barrier. Thanks to this, since we select the region of low momentum p_s for the spectator ($p_s \leq 40$ MeV/c) the PWIA approach can be used for further analysis of the experimental results. If $|\Phi(\vec{p}_s)|^2$ is known and KF is calculated, it is possible to derive $[(d\sigma/d\Omega)_{cm}]^{exp}$ from a measurement of $d^3\sigma/dE_c d\Omega_c d\Omega_C$ by using Eq. 6.

$$\left(\frac{d\sigma}{d\Omega}\right) \propto \left[\frac{d^3\sigma}{dE_c d\Omega_c d\Omega_C}\right] \cdot [KF|\Phi(\vec{p}_s)|^2]^{-1} \tag{9}$$

The application of the quasi-free mechanism to the study of reactions at astrophysical energies, was an extension of previous researches on the quasi-free mechanism at very low energies[Spitaleri 1990] In particular it is an extension, of the indirect excitation function measurements for the two body cross section of the $^7Li(p,\alpha)\alpha$ and $^6Li(p,\alpha)^3He$ reactions [Zadro et al. 1989, Calvi et al. 1990] obtained at low energies. This phenomenological approach was supported from the theory of the THM proposed by Baur ([Baur 1986]), whose basic idea is to extract an $A+x \to C+c$ two-body reaction cross section, at ultralow energies from a suitable $A+a \to C+c+s$ quasi-free three-body reaction. Under appropriate kinematical conditions, the threee-body reaction

is considered as the decay of the "Trojan Horse" a into the cluster x and s and the interaction of A with x inside the nuclear field , whereby the nucleus s can be considered as a spectator to the reaction. The Table I shows the "Trojan Horse" nuclei used in recent experiments.
If the bombarding energy E_A is chosen high enough to overcome the Coulomb barrier in the entrance channel of the three-body reaction, both Coulomb barrier and electron screening effects are negligible. In the original paper by [Baur 1986] it was proposed that the initial velocity of the projectile A is compensated for by the Fermi-motion of particle x. In this framework, a momentum of the order of hundreds MeV/c, could be needed. However, in the case of a nuclei with a predominant l=0 inter-cluster motion, such momenta populate the tail of the momentum distribution for particle x, making very critical the separation from eventual background reaction mechanisms, like sequential decays feeding the same exit channel. Moreover, as already mentioned, the tail of the calculated momentum distribution entering Eq. 6 changes depending on the theoretical approach applied, thus a very sophisticated treatment might be required in order to get the relevant two-body cross-section.
We stress that in our have approach the initial projectile velocity is compensated for by the binding energy of particle x inside a. Thus the two-body reaction can be induced at very low (even vanishing) $A-x$ relative energy . Moreover the role of the cutoff in the momentum distribution consists in fixing the accessible astrophysical energy region, as given by:

$$E_{qf} = E_{Ax} - B_{xs} \tag{10}$$

where E_{Ax} is the beam energy in the center-of-mass of the two-body A-x system, B_{xs} represents the binding energy for the x-b system and M_{xs} describes their inter-cluster motion within the chosen cutoff in momentum. In this way it is possible to extract the two-body cross section from Eq.(9) after inserting the appropriate penetration function G_l in order to account for the penetrability effects affecting the direct data below the Coulomb barrier [Cherubini et al 1996, Spitaleri et al. 1999]. The complete formula is given by:

$$\left(\frac{d\sigma}{d\Omega}\right) \propto \left[\frac{d^3\sigma}{dE_c d\Omega_c d\Omega_C}\right] \cdot [KF|\Phi(\vec{p}_s)|^2]^{-1} \cdot G_l \tag{11}$$

As shown above, since in the experimental works the IA validity conditions are fulfilled, the PWIA was applied for the extractions of the two-body cross-section. In this approximation the differential two-body cross-section of Eq.9 is expressed by:

$$\left(\frac{d\sigma}{d\Omega}\right) = \sum_l C_l P_l \left(\frac{d\sigma_l}{d\Omega}\right) \tag{12}$$

As already mentioned, the THM data are not affected by electron screening effects. Therefore, once the behavior of the absolute bare $S_b(E)$ factor from the two-body cross-section is extracted, a model-independent estimate of the screening potential U_e can be obtained from comparison with the direct screened S(E)-factor.
Then, after normalization of the indirect data, the comparison between the two data sets can be performed down to the low energy region.

TABLE 1. Nuclei with cluster structure which can have been used as Trojan Horse nuclei and their principal properties

	TH -nucleus	binding energy MeV	Clusters	Intercluster motion
1	d	2.225	$p-n$	l=0
2	t	6.257	$n-d$	l=0
3	^{3}He	5.494	$p-d$	l=0
4	^{6}Li	1.475	α - d	l=0
6	^{7}Li	2.468	α - t	l=1
7	^{7}Be	1.587	α- ^{3}He	l=1
8	^{9}Be	2.467	α- ^{5}He	l=0

CONCLUSIONS

An experimental program has already been undertaken to study p-capture reactions on $^{6,7}Li$, main responsible for their destruction [Pizzone et al. 2005, Pizzone et al., 2003]. The extracted S(E) factors as well as electron screening potentials, extracted as stated above are shown in table III. Recently, $^3He(d,p)^4He$ and $^{11}B(p,\alpha)^8Be$ were also investigated by selecting the quasi-free contribution to the $^3He(^6Li,\alpha p)^4He$ and $^{11}B(^2H,\alpha^8Be)n$ three-body processes. Their importance is indeed strongly related to the cosmology as well as to stellar structure and evolution. The bare nucleus S(E)-factor for the $^3He(d,p)^4He$ and $^{11}B(p,\alpha)^8Be$ are reported in table 3 as well. A review of the results concerning the AGB nucleosynthesis is presented in the contribution by La Cognata et al. in this volume.

The present paper presents the basic features of the THM and a review of recent applications to several reactions of importance in astrophysics.In particular these results show the possibility of extracting the bare nucleus two-body cross section via THM.However, a lot remains to do in the future to achieve reliable information for many key reactions and processes. New theoretical developments are strongly needed especially for the study of electron screening effects in fusion reactions in order to meet progress in the application field (fusion reactors).

TABLE 2. Two-body reactions studied via Trojan Horse Method

	Direct reaction	Indirect reaction	E_{inc} (MeV)	Q (MeV)	q_t (MeV/c)	TH_{nucl}	ref
[1]	$^{7}Li(p,\alpha)^{4}He$	$^{7}Li(d,\alpha\ \alpha)n$	19,22	15.122	391	$d = (p\oplus n)$	[Lattuada et al. 2001]
[2]	$^{6}Li(d,\alpha)^{4}He$	$^{6}Li(^{6}Li,\alpha\ \alpha)^{4}He$	5	22.372	136	$^{6}Li = (\alpha\oplus d)$	[Spitaleri et al. 2001]
[2]	$^{6}Li(d,\alpha)^{4}He$	$^{6}Li(^{6}Li,\alpha\ \alpha)^{4}He$	5	22.372	136	$^{6}Li = (\alpha\oplus d)$	[Spitaleri et al. 2001]
[3]	$^{6}Li(p,\alpha)^{3}He$	$^{6}Li(d,\alpha\ ^{3}He)n$	14, 25	1.795	263	$d = (p\oplus n)$	[Tumino et al., 2002]
[4]	$^{11}B(p,\alpha)^{8}Be$	$^{11}B(d,^{8}Be\ \alpha)n$	27	6.36	370	$d = (p\oplus n)$	[Spitaleri et al. 2004]
[5]	$^{10}B(p,\alpha)^{7}Be$	$^{10}B(d,^{7}Be\ \alpha)n$	27	-1.079	360	$d = (p\oplus n)$	[Lamia et al., 2007]
[6]	$^{9}Be(p,\alpha)^{6}Li$	$^{9}Be(d,^{6}Li\ \alpha)n$	22	-0.099	360	$d = (p\oplus n)$	[Romano et al., 2006]
[7]	$^{2}H(^{3}He,p)^{4}He$	$^{6}Li(^{3}He,p\ \alpha)^{4}He$	5, 6	16.88	83-95	$^{3}He = (p\oplus n)$	[La Cognata et al., 2005]
[8]	$^{2}H(d,p)^{3}H$	$^{2}H(^{6}Li,t\ p)^{4}He$	14	2.59	112	$^{6}Li = (\alpha\oplus d)$	[Rinollo et al., 2005]
[9]	$^{12}C(\alpha,\alpha)^{12}C$	$^{12}C(^{6}Li,\alpha\ ^{12}C)^{2}H$	20	1.47	130	$^{6}Li = (\alpha\oplus d)$	[Spitaleri et al. 2000]
[10]	$^{15}N(p,\alpha)^{12}C$	$^{15}N(p,\alpha^{12}C)n$	60	2.74	645	$d = (p\oplus n)$	[La Cognata et al.(2006)]
[11]	$^{18}O(p,\alpha)^{15}N$	$^{18}O(p,\alpha^{15}N)n$	54	1.76	668	$d = (p\oplus n)$	[La Cognata et al., 2007]

REFERENCES

[Assembaum et al., 1987] H.J Assenbaum, K.Langanke, C.Rolfs, Z.Phys.,**327** 461 (1987)
[Azhari et al. 2001] A.Azhari, V.Burjan, F.Carstoiu, C.A.Gagliardi, V.Kroha, A.M.Mukhamedzhanov, F.M.Nunes, X.Tang, L.Trache, R.E.Tribble, Phys.Rev., **C63** , 055803 (2001)
[Baur 1986] G. Baur, Phys. Lett. B **178**, 135 (1986)
[Baur 1994] G. Baur and H. Rebel, J. Phys. G **20**, 1 (1994) and references therein
[Baur 1996] G. Baur and H. Rebel, Annu. Rev. Nucl. Part. Sci. **46**, 321 (1996)
[Bonetti et al., 1999] R.Bonetti et al., Phys. Rev. Lett., **82**, 5205 (1999)

TABLE 3. Two-body reactions results in recent experiments as regards the astrophysical S(E)factor and the electron screening potential

	Reaction	$U_e^{Adiab.}$ (eV)	U_e^{Dir} (eV)	U_e^{THM} (eV)	$S(0)^{Dir}$ (MeVb)	$S(0)^{THM}$ (MeVb)
[1]	$^7\mathrm{Li}+\mathrm{p}\rightarrow\alpha+\alpha$	186	300±160	330± 40	0.059	0.055±0.003
[2]	$^6\mathrm{Li}+\mathrm{d}\rightarrow\alpha+\alpha$	186	330±120	340±50	17.4	16.9±0.5
[3]	$^6\mathrm{Li}+\mathrm{p}\rightarrow^3He+\alpha$	186	440±150	450±100	2.86	3.0±0.3
[4]	$^{11}B+p\rightarrow^8Be+\alpha$	304	430±80		2.1	
[6]	$\mathrm{d}+^3He\rightarrow p+\alpha$	115	219±7	180±40	6.51	6.08±1.42

[BBFH] E.M.Burbige,G.R.Burbige,W.A.Fowler,F.Hoyle, Rev.Mod.Phys.**29**, 547 (1957)

[Calvi et al. 1990] G. Calvi, M. Lattuada, D. Miljanić, F. Riggi, C. Spitaleri, M. Zadro,Phys. Rev. C **41**, 1848 (1990)

[Cherubini et al 1996] S. Cherubini, V. N. Kondratyev, M. Lattuada, C. Spitaleri, Dj. Miljanić, M. Zadro, G. Baur, Ap. J. **457**, 855 (1996).

[Chew 1950] G.F. Chew, Phys.Rev. **80**, 196(1950)

[Fowler 1984] W.A.Fowler, Rev. Mod. Phys.**56**, 149 (1984)

[Jacob 1966] G.Jacob, Th. A. Maris, Rev. Mod. Phys. **38** , 121 (1966)

[La Cognata et al., 2005] M.La Cognata, C.Spitaleri, A.Tumino, S.Typel, S.Cherubini, L.Lamia, A.Musumarra, R:G.Pizzone, A.Rinollo, C.Rolfs, S.Romano, D.Schiurmann, F.Streider, Phys. Rev. C **72** 065802(2005)

[La Cognata et al., 2005] M.La Cognata, C.Spitaleri, A.Tumino, S.Typel, S.Cherubini, L.Lamia, A.Musumarra, R:G.Pizzone, A.Rinollo, C.Rolfs, S.Romano, D.Schiurmann, F.Streider, Phys. Rev. C **72** 065802(2005)

[La Cognata et al.(2006)] M.La Cognata, S.Romano, C.Spitaleri, R.Tribble, L.Trache, S.Cherubini, Fu Changbo, L.Lamia, A.Mukhamedzhanov, R.G.Pizzone, A.Tumino, Eur. Phys.Journ. A, **27**, 249 (2006).

[La Cognata et al., 2007] M.La Cognata,C.Spitaleri,R.E.Trbble, T.Al-bdullah, A.Banu,S.Cherubini, V.Crucillà, C.Fu,V.Goldberg, M.Gulino, L.Lamia,A.Mukhamedzhanov, R.G.Pizzone, S.M.R. puglia, G.G.Rapisarda, S.Romano, M.L.Sergi, G.Tabacaru, L.Trache, S.Tudisco, A.Tumino, S.Typel, Y.Zhai, J.Phys.G: Nucl. Part.Phys. **34**, 1(2007).

[Lamia et al., 2007] L.Lamia, S.Romano, N.Carlin, S.Cherubini, V.Crucillá M.De Moura, M.Gamiero Munhoz, M.Gulino, R.Liguori Neto, M.La Cognata, F.Mudó,R.G,Pizzone, S.M.R.Puglia,A.Suaide, E.Szanto, A.Szanto de Toledo, M.L.Sergi, C.Spitaleri,S.Tudisco, A.Tumino, Nucl. Phys. A.**787** 309 (2007)

[Lattuada et al. 2001] M. Lattuada, R.G. Pizzone, S. Typel, P. Figuera, Dj. Miljanić, A. Musumarra, M.G. Pellegriti, C. Rolfs, C. Spitaleri, H.H. Wolter, Ap.J. **562**, 1076 (2001)

[Mukhamedzhanov et al 1997] A.M.Mukhamedzhanov, A.M.Clark, C.A.Gagliardi,Y. W. Lui, L.Trache,R.E.Tribble,H.M.Xu,V.Cejpek ,V.Burjan, V.Kroha, F.Castiou, Phys. Rev. C **56**, 1302 (1997)

[Mukhamedzhanov et al 1999] A.M.Mukhamedzhanov, R.E.Tribble Phys. Rev. C**59**, 3418(1999)

[Mukhamedzhanov et al 2001] A.M.Mukhamedzhanov, C.A.Gagliardi, R.E.Tribble Phys. Rev. C**63**, art.n.024612 (2001)

[Neudatchin 1965] U.G.Neudatchin, Y.F.Smirnov, At. Energy Rev., **3** , 157 (1965)

[Pizzone et al., 2003] R.G. Pizzone, C. Spitaleri, M. Lattuada, S. Cherubini, A. Musumarra, M.G. Pellegriti, S. Romano, A. Tumino, V. Castellani, S. Degl'Innocenti, A. Imperio, Astron. & Astroph. 398, (2003) 423-427;

[Pizzone et al. 2005] R.G. Pizzone, A. Tumino, S. Degl'Innocenti, C. Spitaleri, S. Cherubini, A. Musumarra, S. Romano, Astronomy & Astrophysics 438, 779Ű784 (2005)

[Rinollo et al., 2005] A.Rinollo, S.Romano, C.Spitaleri, C.Conomo, S.Cherubini,A Del Zoppo,

P.Figuera, M.La Cognata, L.Lamia, A.Musumarra,M.G.Pellegriti, R.G.Pizzone, c.Rolfs, D.Schurmann, F.Streider,S.Tudisco, A.Tumino, Nucl. Phys. A **758**146c (2005)

[Rolfs & Rodney, 1988] C.Rolfs, W.S. Rodney, Cauldrons in the cosmos, University of Chicago Press, Chicago (1988)

[Rolfs et al., 2001] C.Rolfs, Prog. Part. Nucl. Phys., **153** , 23 (2001)

[Romano et al., 2006] S.Romano, L.Lamia, C.Spitaleri, C.Li, S.Cherubini, M.Gulino, M.La Cognata, R.G.Pizzone, A.Tumino, Eur. Phys. J. A**27**, 221 (2006)

[Spitaleri 1990] C. Spitaleri, Problems of Fundamental Modern Physics II, World Scientific,p. 21- 35, 1990 and references therein

[Spitaleri et al. 1999] C. Spitaleri, M. Aliotta, S. Cherubini, M. Lattuada, Dj. Miljanić, S. Romano, N. Soić, M. Zadro, R. A. Zappalà, Phys. Rev. C **60**, 055802 (1999)

[Spitaleri et al. 2000] C. Spitaleri, M. Aliotta, P. Figuera, M.Lattuada,R. G. Pizzone, S. Romano, A. Tumino, C. Rolfs, L. Gialanella, F. Strieder, S. Cherubini, A. Musumarra, D. Miljanić, S. Typel and H. H. Wolter, Eur. Phys. J **A 7**, 181 (2000)

[Spitaleri et al. 2001] C. Spitaleri, S. Typel, R.G. Pizzone, M.Aliotta, S. Blagus, M. Bogovac, S. Cherubini, P. Figuera, M. Lattuada, M. Milin, Dj. Miljanić, A. Musumarra, M.G. Pellegriti, D. Rendić, C. Rolfs, S. Romano, N. Soić, A. Tumino, H.H. Wolter, M. Zadro, Phys. Rev. C **63**, 005801 (2001)

[Spitaleri et al. 2004] Spitaleri, A. Di Pietro, P. Figuera, M. Lattuada, A.Musumarra, M.G. Pellegriti, R.G. Pizzone S. Romano, C. Rolfs, S. Tudisco, S. Typel, Phys. Rev.C . **69** 055806 (2004)

[Strieder et al., 2001] F. Streider, C.Rolfs, C.Spitaleri,P.Corvisiero, Naturwissenschaften **88**, 461, (2001)

[Tang et al. 2003] X.D.Tang, A.Azhari, C.A.Gagliardi, A.M.Mukhamedzhanov, F.Pirlepeson, L.Trache, R.E.Tribble, V.Burjan, V.Kroha, F.Castiou, Phys. Rev. C**67**, art.n.015804 (2003)

[Tumino et al., 2002] A. Tumino, C. Spitaleri, A. Di Pietro, P. Figuera, M. Lattuada, A.Musumarra, M.G. Pellegriti, R.G. Pizzone S. Romano, C. Rolfs, S. Tudisco, S. Typel, Nucl.Phys.**A** (2003)

[Zadro et al. 1989] M. Zadro, Dj. Miljanić, C. Spitaleri, G. Calvi, M. Lattuada, F. Riggi, Phys. Rev. C **40**, 181 (1989)

Recent Astrophysical Experiments on Cyclotron of NPI at Řež in Collaboration with Texas A&M University

V. Kroha[a], V Burjan[a], Z. Hons[a], J. Mrázek[a], Š. Piskoř[a], C.A. Gagliardi[b], A.M. Mukhamedzhanov[b], L. Trache[b], R.E. Trible[b] G. Pizzone[c], S. Romano[c], C. Spitaleri[c]

[a]*Nuclear Physics Institute CAS p.r.i., Řež, Czech Republic*
[b]*Cyclotron Institute Texas A&M University, USA*
[c]*INFN Laboratori Nazionali del Sud, Catania, Italy*

Abstract. Survey of the most important results of astrophysical research on the ^{3}He cyclotron beam in NPI Řež is presented. ANCs for some proton capture reactions of the CNO stellar cycle are discussed. Results of the last experiments on the isotopic gas target of ^{15}N are also mentioned.

Keywords: Proton capture reaction, Asymptotic normalization coefficient, R-matrix approach, Nuclear astrophysics.
PACS: 25.70.Hi, 25.55.Hp, 26.30.-k

Cooperation between Nuclear Physics Institute of Czech Academy of Sciences p.r.i. (NPI) at Řež and Cyclotron Institute of Texas A&M University (TAMU) is devoted to indirect measurements of astrophysically important reaction rates. A new indirect technique [1], known as an Asymptotic Normalization Coefficient method (ANC), was developed especially for the direct proton capture reactions which proceed in the stellar environments. This method, for the first time tested at NPI [2], is widely used in many laboratories over the world.

The first application of ANC method concerned the synthesis of ^{8}B which is the main source of high energy solar neutrinos. Measurements were realized on superconducting cyclotron of TAMU using different heavy ion beams and the complementary measurements on ^{3}He beam in NPI. . Resulting value of astrophysical S-factor is $S(0) = 17.8 \pm 2.8$ eVb [3,4] and references therein A low value of corresponding astrophysical S-factor was confirmed, in agreement with many other results.

In Fig.1 a result of analysis of the next reaction studied in both institutions is presented. Proton capture $^{9}Be(p,\gamma)^{10}B$ was investigated in NPI using $(^{3}He,d)$ reaction on ^{9}Be and in proton transfer process $^{9}Be(^{10}B,^{9}Be)^{10}B$ in TAMU. Reaction $^{9}Be(p,\gamma)^{10}B$ plays an important role in stellar nucleosynthesis of light p-shell elements. The energy behavior of S-factor has not been well understood, especially, the direct part of radiative capture was quite controversial. Reaction was analyzed in R-matrix approach

CP972, *Exotic Nuclei and Nuclear/Particle Astrophysics (II)—Proceedings of the Carpathian Summer School of Physics 2007,* edited by L. Trache and S. Stoica

and the direct component was calculated using ANCs extracted from the transfer reactions. Details of analysis are published in [5,6].

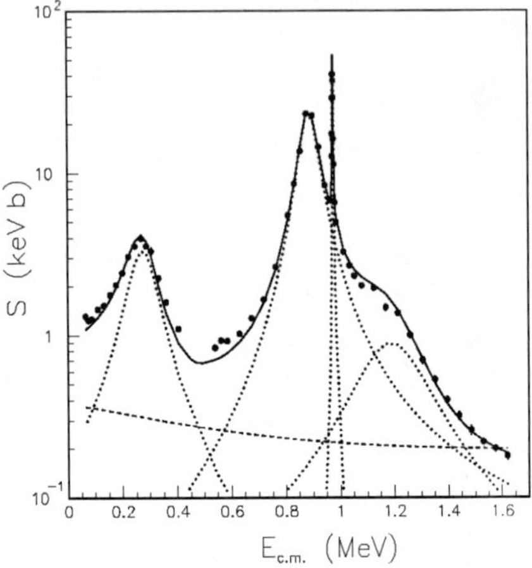

FIGURE 1. S-factor for $^{9}Be(p,\gamma)^{10}B$ capture. The dotted lines correspond to resonant components, dotted line to direct term and full line is told s-factor. The experimental points are from [7].

Reaction $^{14}N(p,\gamma)^{15}O$ is one of the most important reactions in the CNO cycle. As the slowest reaction in the cycle, it defines the rate of energy production and, hence, the lifetime of the stars. The astrophysical *S*-factor for this reaction had been measured by several different groups but their results of *S(0)* differed by about a factor of 2.

In [8] we report a new determination of the ANC for the subthreshold state at ε = -544 keV, which corresponds to 6.79 MeV level in ^{15}O. The direct capture through this state dominates the reaction rate at stellar energies. To calculate the astrophysical *S*-factor we used the *R*-matrix approach. For reliable extraction of ANCs from angular distributions of $^{14}N(^{3}He,d)^{15}O$ reaction, the analysis has been done using DWBA and CCBA calculation with the FRESCO code [9]. Result of analysis for the subthreshold state is presented in Fig. 2. Value of the total *S*-factor, extrapolated to zero energy, *S(0)* = 1.70 ± 0.22 keVb, is in very good agreement with measurements of other authors.

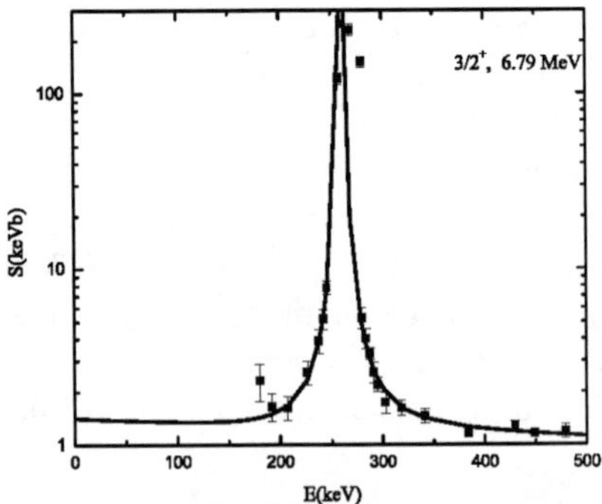

FIGURE 2. S-factor for transfer to the subthreshold state for $^{14}N(p,\gamma)^{15}O$ capture reaction. The experimental points are from [10,11].

Reaction $^{13}C(p,\gamma)^{14}N$ belongs also to the important reactions in the CNO cycle. It precedes the slowest reaction $^{14}N(p,\gamma)^{15}O$ mentioned above. In addition, $^{13}C(p,\gamma)^{14}N$ radiative capture rate is important for nucleosynthesis via the *s* process because it can deplete the seed nuclei required for the neutron generator reaction $^{13}C(\alpha,n)^{16}O$ in asymptotic giant branch stars with solar metalicity.

The most accurate measurements of the *S*-factor for this proton capture reaction were carried out by King at al. for energies E>100keV [12]. In the analysis the nonresonant amplitudes were parametrized in terms of spectroscopic factors which were treated as free parameters. To know the total capture rate an accurate determination of the nonresonant contribution is very important. We reanalyzed the experimental data of [12] within the framework of the *R*-matrix approach. The nonresonant amplitudes were parametrized in terms of the asymptotic normalization coefficients extracted from the angular distribution of $^{13}C(^{3}He,d)^{14}N$ and $^{13}C(^{14}N,^{13}C)^{14}N$ reactions. The *R*-matrix method correctly accounts for the interference between the nonresonant and resonant amplitudes. Results of our analysis are in excellent agreement with those obtained by King et al. despite the differences in interpretation. The *S*-factor that we obtained, $S(0) = 7.6 \pm 1.1$ keVb, is significantly higher than value given previously and reduces the amount of ^{13}C available for the $^{13}C(\alpha,n)^{16}O$ reaction as a neutron source to fuel heavy element synthesis through the *s* process. The energy behavior of the total *S(0)* factor is presented in Fig. 3. The dotted line is for the nonresonant part, dashed line for resonant capture and the solid line correspond to the best fit which includes both terms and their interference. The experimental detail and analysis of result are published in [13,14].

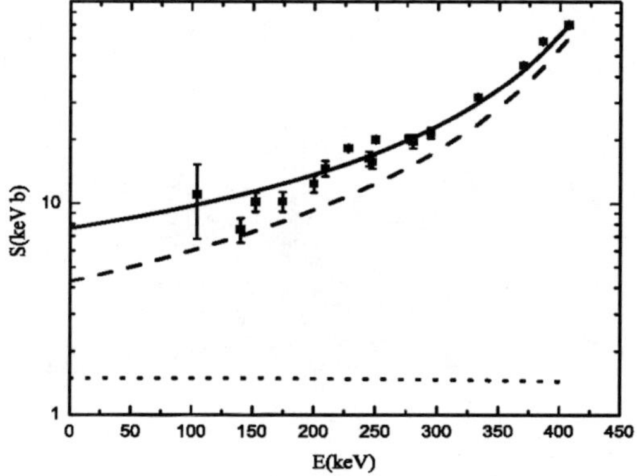

FIGURE 3. S-factor for $^{13}C(p,\gamma)^{14}N$. Dotted line is for the direct term, dashed line for resonant component and full line is total S-factor.

The next important proton capture reaction studied on NPI cyclotron was $^{20}Ne(p,\gamma)^{21}Na$. This reaction plays an important role in hydrogen burning in stars with masses larger than that of the sun. The $^{20}Ne(p,\gamma)^{21}Na$ reaction, as the first in *NeNa* cycle, determines the rate of ^{21}Na stellar production. The nuclei ^{21}Na, ^{21}Ne, ^{22}Na, ^{22}Ne and ^{23}Na are gradually created during *Ne-Na* burning. ^{21}Ne nuclei that are produced as result of the $^{21}Na(\beta^{+}\nu)^{21}Ne$ decay serve as seeds for the neutron generator reaction $^{21}Ne(\alpha,n)^{24}Mg$.

The $^{20}Ne(p,\gamma)^{21}Na$ reaction rate at stellar energies is dominated by capture to the ground state through the tail of a subthreshold resonance state at excitation energy of 2425 KeV in ^{21}Na. Both resonant and direct capture contribute while direct captures to other bound states are negligible. The overall normalization of direct part to the subthreshold state is given by asymptotic normalization coefficient which simultaneously determines the proton partial width of this subthreshold resonance. To date, the experimental data for this reaction have been measured [15] at proton energies of $E_p \geq 400$ keV, substantially above the Gamow window. To determine ANCs, the $^{20}Ne(^{3}He,d)^{21}Na$ proton transfer reaction has been measured at incident energy of 25.83 MeV. The experiment was carried out with ^{3}He beam from isochronous cyclotron at NPI. Used gas cell contained high-purity isotopic ^{20}Ne (99.99 %). The working pressure was kept at 195 mbar and monitored by gas control on-line system. The body of the target gas cell is shown in Fig.4. Reaction products were measured by pair of ΔE-E telescopes which were equipped with a pair of collimating slits for definition of solid angles. The effective target thickness at different reaction angles was determined by a Monte Carlo simulation.

FIGURE 4. Target gas cell and telescops used in experiments with isotopic gases.

Corresponding ANCs for calculation of *S*-factor for proton radiation capture were extracted from the angular distributions of deuterons in $^{20}Ne(^{3}He,d)^{21}Na$ transfer reaction by DWBA analysis. The experimental data [15] were analyzed in frame of *R*-matrix approach in which the normalization of the external nonresonant capture amplitude is governed by the ANC of the final bound state. Fig.5. presents results of our analysis. In the region of the experimental points the direct capture through the subthreshold state dominates the *S*-factor. The capture to the ground state through the subthreshold resonance dominates at lowest energies. The dotted line in Fig.5 corresponds to the direct capture to the subthreshold state, dashed one to capture to the ground state through the subthreshold resonance. The solid-line is our total astrophysical factor. Our $S(0)$ value is $S(0)$ = 5900 6 ± 1200 keVb. Description of experimental equipment and analysis of data is published in [16].

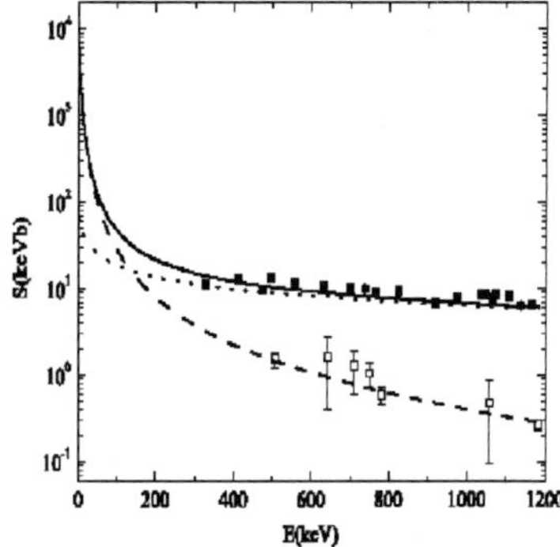

FIGURE 5. S-factor for $^{20}Ne(p,\gamma)^{21}Na$. Dotted line correspond to direct capture to the subthreshold state, dashed curse is capture to the ground state through the subthreshold resonance; the full curve is our total S-factor. Experimental points are from [15]

One of the last reaction of astrophysical importance is $^{15}N(p,\gamma)^{16}O$ proton capture. This reaction provides a leak from the CN cycle to CNO bi-cycle and the CNO tri-cycle. The loss of catalytic material via $^{15}N(p,\gamma)^{16}O$ reaction generates a new sequence of reactions $^{16}O(p,\gamma)^{17}F(e^+ \nu_e)^{17}O(p,\alpha)^{14}N$ which restores the material of the CN cycle or generates another path $^{16}O(p,\gamma)^{17}F(e^+\nu_e)$ $^{17}O(p,\gamma)^{18}F(e^+\nu_e)$ $^{18}O(p,\gamma)^{19}F(p,\gamma)^{20}N$...... that leads to formation of the heavier elements. So far the reaction experimental data were measured directly for $E_p \geq 155$ keV [17] and earlier [18] for $E_p \geq 220$ keV. At very low stellar energies the $^{15}N(p,\gamma)^{16}O$ reaction proceeds by the resonant capture to the ground state through the first two interfering 1^- s wave resonances at $E = 312$ a 964 keV [17]. Il has been also found that the interference with the direct capture has to be taken into account. To fit the experimental data in [17], the contribution from direct capture term was varied arbitrarily. To calculate more accurately the contribution from the nonresonant captures we determined the proton ANCs for the ground and seven excited state of ^{16}O by measuring the angular distributions of the peripheral $^{15}N(^3He,d)^{16}O$ proton transfer reaction at an energy of 25.83 MeV of the ^3He beam from cyclotron U-120M of NPI. The corresponding ANCs were obtained from comparison of the experimental differential cross sections with the DWBA calculations. The astrophysical S-factor for $^{15}N(p,\gamma)^{16}O$ capture was calculated in the R-matrix approach using these ANCs and proton and α resonance widths after fitting the available data from $^{13}N(p,\alpha)^{12}C$ reaction. The experiment was performed in the same geometry as in the case of $^{20}Ne(^3He,d)^{21}N$ nucleon. The target gas chamber contained the high purity ^{15}N isotope. A typical deuteron spectrum is shown in Fig. 6. The comparison of deuteron spectra from ^{15}N gas and air in insert shows that ^{15}N gas does not contain any admixture of ^{14}N.

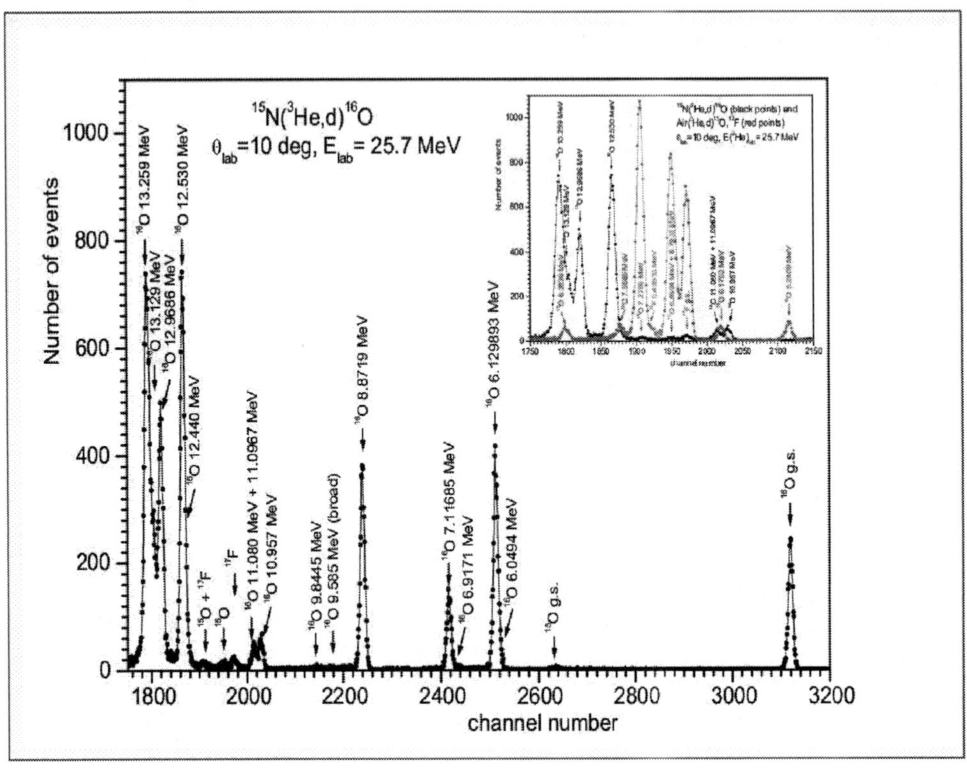

FIGURE 6. Spectrum of deuterons from $^{15}N(^{3}He,d)^{16}O$ reaction. Red spectrum in the insert corresponds to reaction (^{3}He,d) on air, see text.

For fit to the $^{15}N(p,\gamma)^{16}O$ data it was necessary extract the radiation width for both resonances by normalization of the calculated astrophysical factor to the data [17] at resonance peaks. Results of this are shown in Fig.7. Our astrophysical factor $S(0)$ = 38.0 ± 6.0 keVb and is lower then $S(0)$ = 64.0 ± 6.0 keVb in [17]. The disagreement between our fit and the fit in [17] corresponds to the contribution of the nonresonant capture to the ground state which was overestimated in by an order of magnitude. Probable reason is double counting of the internal nonresonant capture in [17].

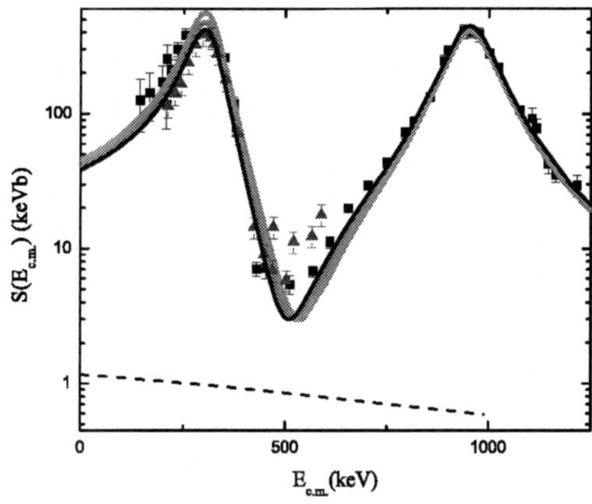

FIGURE 7. Fit of experimental data for $^{15}N(p,\gamma)^{16}O$ capture. Dashed line corresponds to direct term. Experimental points (black) is from [17], blue ones from [18].

Our obtained astrophysical factor *S(0)* for $^{15}N(p,\gamma)^{16}O$ allows us to conclude that every 1480 cycles of the main CN cycle (rather than every 880 cycles determined in [17] and every 2200 cycles from [18]) the CN catalyst is lost due to this reaction. Results of analysis are prepared to submitting for publication.

REFERENCES

1. H. M. Xu et al., *PRL*, 73(2005)2027.
2. C. A. Gagliardi et al., *PRC*, 59(1999)1949.
3. A. Azhari et al., *PRL*, 12(1999)396.
4. C. Tabacaru et al., *PRC*, 73(2006)025808.
5. A. M. Mukhamedzhanov et al., *PRC*, 56(1997)1302.
6. A. Sattarov et al., *PRC*, 60(1999)035801.
7. D. Zahnow et al., *Nucl. Phys. A*, 589(1995)95.
8. A. M. Mukhamedzhanov et al., *PRC*, 67(2003)065804.
9. I. J. Thompson, *Comput. Phys. Rep.*, 7(1988)168.
10. U. Schröder et al., *Nucl. Phys A*, 467(1987)240.
11. C. Angelo and P. Descouvemont, *Nucl. Phys. A*, 690(2001)755.
12. J. D. King et al., *Nucl. Phys. A*, 567(1994)354.
13. A. M. Mukhamedzhanov et al., *Phys. Rev. C*, 66(2002)027602.
14. A. M. Mukhamedzhanov et al., *Nucl. Phys A*, 725(2003)279.
15. C. Rolfs et al., *Nucl. Phys A*, 241(1975)460
16. A. M. Mukhamedzhanov et al., *Phys. Rev. C*, 73(2006)035806.
17. C. Rolfs and W. S. Rodney, *Nucl. Phys A*, 235(1974)450.
18. D. F. Hebbard, *Nucl. Phys*, 15(1960)289

7. NUCLEOSYNTHESIS

Big Bang Nucleosynthesis in the New Cosmology

Brian D. Fields

Center for Theoretical Astrophysics, Depts. of Astronomy and Physics, Univ. of Illinois, Urbana IL 61801 USA

Abstract. Big bang nucleosynthesis (BBN) describes the production of the lightest elements in the first minutes of cosmic time. We review the physics of cosmological element production, and the observations of the primordial element abundances. The comparison between theory and observation has heretofore provided our earliest probe of the universe, and given the best measure of the cosmic baryon content. However, BBN has now taken a new role in cosmology, in light of new precision measurements of the cosmic microwave background (CMB). Recent CMB anisotropy data yield a wealth of cosmological parameters; in particular, the baryon-to-photon ratio $\eta = n_B/n_\gamma$ is measured to high precision. The confrontation between the BBN and CMB "baryometers" poses a new and stringent test of the standard cosmology; the status of this test are discussed. Moreover, it is now possible to recast the role of BBN by using the CMB to fix the baryon density and even some light element abundances. This strategy sharpens BBN into a more powerful probe of early universe physics, and of galactic nucleosynthesis processes. The impact of the CMB results on particle physics beyond the Standard Model, and on non-standard cosmology, are illustrated. Prospects for improvement of these bounds via additional astronomical observations and nuclear experiments are discussed, as is the lingering "lithium problem."

Keywords: Origin, formation, and abundances of the elements, Big Bang nucleosynthesis
PACS: 98.80.Ft, 26.35.+c

INTRODUCTION

Big-bang nucleosynthesis (BBN) offers the deepest reliable probe of the early universe, being based on well-understood Standard Model physics. Predictions of the abundances of the light elements, D, ^{3}He, ^{4}He, and ^{7}Li, synthesized at the end of the "first three minutes" are in good overall agreement with the primordial abundances inferred from observational data, thus validating the standard hot big-bang cosmology (see [1]). This is particularly impressive given that these abundances span nine orders of magnitude — from ^{4}He/H ~ 0.08 down to ^{7}Li/H $\sim 10^{-10}$ (ratios by number). Thus BBN provides powerful constraints on possible deviations from the standard cosmology, and on new physics beyond the Standard Model [2].

We are presently entering an age of a "new cosmology," in which BBN plays a changing but crucial role. The basic world model of a hot big bang has been resoundingly confirmed by a wealth of new observations with unprecedented precision. The fundamental parameters of this world model, including the abundances of the matter and energy component of the cosmos, are now known to a few percent.

The advent of "precision cosmology" is largely spurred by measurements of the cosmic microwave background radiation (CMB). The CMB and BBN are intimately connected. This link traces back to the work of the Gamow, Alpher, and Herman, [3, 4], who determined the thermodynamic conditions needed for nucleosynthesis in the early

CP972, *Exotic Nuclei and Nuclear/Particle Astrophysics (II)—Proceedings of the Carpathian Summer School of Physics 2007,* edited by L. Trache and S. Stoica

universe, and used this to extrapolate a present radiation temperature of order $T \sim 5$ K. This historic BBN–CMB connection has recently been deepened with the advent of high precision measurements of the CMB anisotropy [5]. The measurements have led to determinations of cosmological parameters with an unprecedented accuracy. These include the baryon density, which is the sole parameter in standard BBN.

With the cosmic baryon density measured independently via BBN and the CMB, several new analyses are possible. (1) First and foremost, one can pit BBN against the CMB, asking whether the results of the independent BBN and CMB "baryometers" agree. This comparison marks a fundamental and non-trivial test of the hot big bang. We will see, in fact, that cosmology passes this test, but then other strategies are possible when one *combines* BBN and the CMB, taking advantage of the precision of the CMB bayon density determination. (2) For standard BBN, a CMB-based baryon density fixes all parameters. Then BBN simply makes definite predictions for the abundances of the light elements which can then be contrasted with their observational determinations [6]. The results shed new light on the nucleosynthesis of light elements *after* the big bang.

Finally, (3) one can relax the assumptions of standard BBN (e.g., an early universe populated only by known Standard Model particles and interactions). With the baryon density fixed by the CMB, for the first time, all light elements are available to probe early universe physics. This is perhaps the most exciting new possibility for BBN, and highlights the other aspect of the "new cosmology," namely that our view of the constituents, dynamics, and history of the cosmos has drastically changed of late. The universe appears to be accelerating today, due to a dark energy whose nature is unknown, but may provide clues to such far-reaching realms as quantum gravity. How and whether the dark energy is connected with another cosmic period of acceleration–inflation–remains to be clearly elucidated. Also, the need for and abundance of dark matter is ever more firmly established, but its nature and any connection with dark energy (or baryons!) is unclear. All of these open questions almost surely will find their answers in the early universe. And in this context it is crucial to appreciate that BBN provides our earliest reliable probe.

Thus the new cosmology presents both challenges and opportunities for BBN in general and for nuclear physics in particular. These will be the focus of this review, which draws very heavily from [7]. For other recent reviews see [1].

BIG-BANG NUCLEOSYNTHESIS THEORY

The theory of BBN consists of following the microphysics of weak and nuclear reactions in the cosmological context of an expanding, cooling universe. In fact, the essential aspects of BBN can be understood in terms of the competition between the cosmic expansion rate. and particle reaction rates. The expansion rate $H = \dot{a}/a$, with $a(t)$ the cosmic scale factor, is given by the usual Friedmann equation, applied to the early universe:

$$H^2 = \frac{8\pi G}{3}\rho \quad (1)$$

In the early universe, the energy density was dominated by relativistic species ("radiation")

$$\rho_{\rm rad} = \frac{\pi^2}{30}(2+\frac{7}{2}+\frac{7}{4}N_\nu)T^4 \equiv \frac{\pi^2}{30}g_* T^4 \tag{2}$$

which consists of photons, electrons and positrons, and N_ν neutrino flavors. Thus time and temperature scale as $t \sim 1/H \sim 1/T^2$, and in standard BBN (i.e., with $N_\nu = 3$), we roughly have $t/1\,\text{sec} \simeq (2.4/g_*^{1/2})(1\text{ MeV}/T)^2$.

The synthesis of the light elements is sensitive to physical conditions in the early radiation-dominated era at temperatures $T \lesssim 1$ MeV, corresponding to an age $t \gtrsim 1$ s. At these high temperatures, weak interactions rates were rapid compared to the expansion rate: $\Gamma_{\rm weak} \gg H$, and thus the weak interactions were in thermal equilibrium. In particular, the processes $n+e^+ \leftrightarrow p+\bar{\nu}_e$ and $n+\nu_e \leftrightarrow p+e^-$ drive the number ratio of neutrons to protons to $n/p = e^{-Q/T}$, where $Q = 1.293$ MeV is the neutron-proton mass difference. As the temperature dropped, the neutron-proton inter-conversion rates, $\Gamma_{np} \sim G_F^2 T^5$, fell faster than the Hubble expansion rate, $H \sim \sqrt{g_* G_N}\, T^2$. This resulted in breaking of chemical equilibrium ("freeze-out") at $T_{\rm fr} \sim (g_* G_N/G_F^4)^{1/6} \simeq 0.8$ MeV. The neutron fraction at this time, $n/p = e^{-Q/T_{fr}} \simeq 1/6$ is thus sensitive to every known physical interaction, since Q is determined by both strong and electromagnetic interactions while $T_{\rm fr}$ depends on the weak as well as gravitational interactions. Moreover the sensitivity to the Hubble expansion rate affords a probe of e.g. the number of relativistic neutrino species [8]. After freeze-out the neutrons were free to β-decay so the neutron fraction dropped to $\simeq 1/7$ by the time nuclear reactions began. A useful semi-analytic description of freeze-out has been given [9, 10].

The nucleosynthesis chain begins with the formation of deuterium in the process $p(n,\gamma)$D. However, the number density of photons is huge relative to the baryon density, i.e., the baryon-to-photon ratio $\eta = n_B/n_\gamma \sim 10^{-9}$; Indeed, η is the sole parameter in the standard BBN model. Because of the large number of photons per baryon, photo-dissociation delays production of deuterium (and other complex nuclei) well after T drops below the binding energy of deuterium, $B_{\rm D} = 2.23$ MeV. When the quantity $\eta^{-1}\exp(-B_{\rm D}/T) \sim 1$, the rate for deuterium destruction (D + $\gamma \to p+n$) finally falls below the deuterium production rate and the nuclear chain begins at a temperature $T \sim 0.1$ MeV. Only 2-body reactions are important because the density has become rather low by this time.

In addition to the $p(n,\gamma)$D reaction, there are 10 other major strong–i.e., nuclear–reactions reactions (along with the neutron lifetime) leading to the production of the light elements:

D(D, p)T ^{3}He(n, p)T   D(D, n)^{3}He   D(p,γ)^{3}He

Followed by the reactions producing ^{4}He:

^{3}He(D, p)^{4}He   T(D, n)^{4}He

The gap at $A = 5$ is overcome and the production of mass $A = 7$ is regulated by:

^{3}He(^{4}He,γ)^{7}Be   T(^{4}He,γ)^{7}Li ^{7}Be(n,p)^{7}Li ^{7}Li(p,^{4}He)^{4}He where ^{7}Be will decay by electron capture but only long after BBN when the universe has cooled to $T \sim$ eV.

Because ^{4}He is the most stable light element, nearly all neutrons that survive until the When nuclear reactins begins end up in this nuclide. Heavier nuclei do not form in any significant quantity both because of the absence of stable nuclei with mass number $A = 5$ or 8, and the large Coulomb barriers for reactions involving $A \geq 7$. Hence the primordial mass fraction of ^{4}He, conventionally referred to as Y_p, can be estimated by the simple counting argument $Y_p = 2(n/p)(1+n/p)^{-1} \simeq 0.25$. ^{4}He shows little sensitivity to nuclear cross section uncertainties, which are however important in determining the other "left-over" abundances: D and ^{3}He at the level of a few times 10^{-5} by number relative to H, and ^{7}Li/H at the level of about 10^{-10}. These values can be understood in terms of approximate analytic arguments [11, 10]. The experimental parameter most important in determining Y_p is the neutron lifetime, τ_n, which normalizes (the inverse of) Γ_{np}. The experimental uncertainty in τ_n used to be a source of concern but has recently been reduced substantially. The Particle Data Group [12] world average is $\tau_n = 885.7 \pm 0.8$ s.

FROM NUCLEAR DATA TO PRIMORDIAL ABUNDANCE PREDICTIONS

The homogeneous nature of BBN, and the relatively small number of key reactions, makes the abundance evolution one of the most computationally simple in all of nuclear astrophysics. This relative simplicity presents us with the opportunity to calculate the abundances with a precision and statistical rigor that also is unique to nuclear astrophysics.

The major uncertainties come from the thermonuclear reaction rates. As noted above, there are 11 key strong rates (as well as the neutron lifetime) which dominate the uncertainty budget [13, 14, 15]. In contrast to the situation for much of stellar nucleosynthesis, BBN occurs at high enough temperatures (strong rates freeze out at $T \sim 0.1$ MeV) that laboratory data exist at and even below the relevant energies, so that no extrapolation is needed; this again places BBN in a unique position.

The goal of BBN theory is to derive primordial abundances which are as precise as the nuclear data allow, and to quantify the uncertainties in the predictions. This process takes several steps, going from the nuclear data to the final abundances. (1) First, the relevant nuclear data must be cataloged and evaluated. The data deemed reliable are used to infer both the best-fit cross section (in fact, the astrophysical S-factor), and to obtain an estimate of its uncertainty δS, both as functions of energy. (2) The cross section data are then averaged over a thermal Maxwell-Boltzmann distribution to obtain thermonuclear rates; similarly, one must propagate the cross section uncertainties to obtain thermonuclear errors. (3) Finally, the thermonuclear rates and uncertainties are placed in the BBN code. Monte Carlo techniques [13, 14] are used to determine the best-fit abundances, and their uncertainties, at each η.

Recently the input nuclear data have been carefully reassessed [15, 16, 17], leading to improved precision in the abundance predictions. In addition, polynomial fits to the predicted abundances and the error correlation matrix have been given [18, 17].

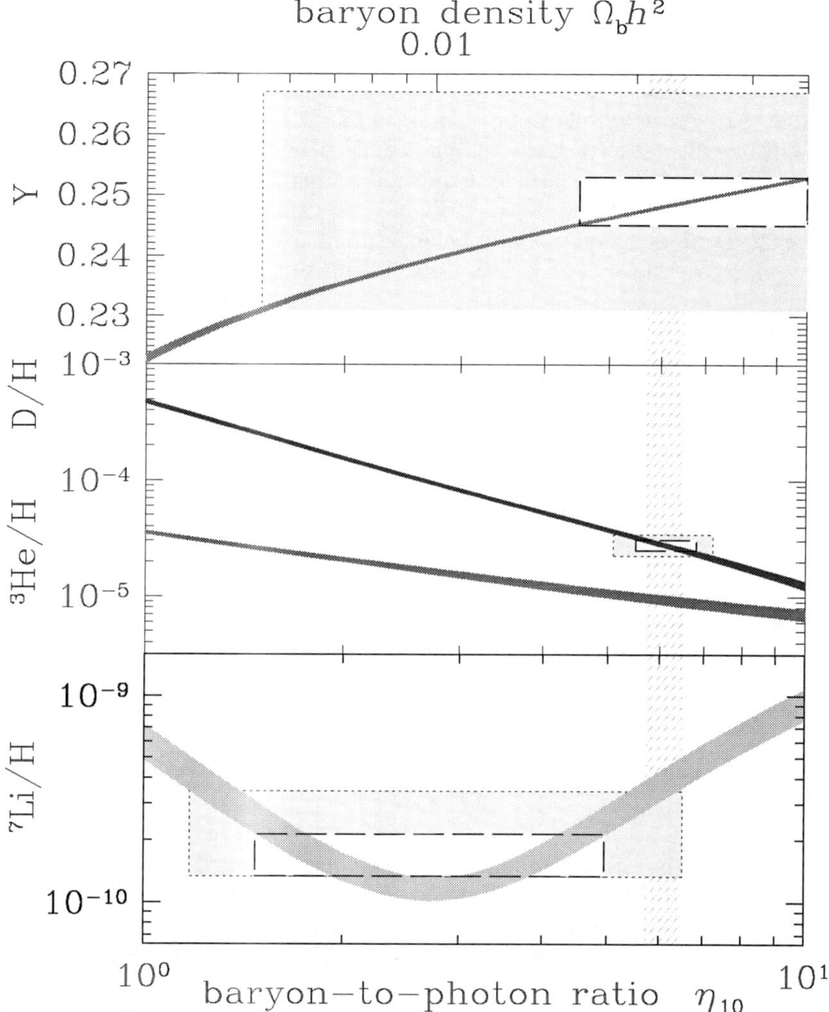

FIGURE 1. Primordial light element abundances as predicted by BBN and WMAP, shown as a function of baryon-to-photon ratio $\eta_{10} \equiv 10^{10}\eta$ or equivalently, baryon density parameter $\Omega_B h^2$. Curve widths are $2-\sigma$ theoretical uncertainties found by propagation of nuclear cross section errors [19]. For each element the rectangles give observational constraints, with statistical errors shown in smaller dotted boxes, and sytematic errors shown in the larger filled boxes. For each element, observational assessments of primordial abundance determines an allowed range in the ordinate; this sets the height of the boxes. Using the theory curves, these observational ranges correspond to a range in η which sets the widths of the curves. Since each element determines an allowed range in the one free parameter η, the problem is overconstrained. The horizontal overlap of the boxes thus indicates concordance, which represents a consistency check for BBN and a determination of the η from light element abundances alone. The WMAP results appear as the narrow vertical band. This concordance between the independently-determined BBN and CMB baryon densities marks a crucial and non-trivial test of the hot big bang cosmology.

The elemental abundances are shown in Fig. 1 as a function of η_{10} [13]. The plot is for standard BBN, which is based on Standard Model physics alone, where $N_\nu = 3$. (The implications of BBN for physics beyond the Standard Model will be considered briefly below, §5). As noted above, in the standard case, the only free parameter is the density of baryons (strictly speaking, nucleons), which sets the rates of the strong reactions. The baryon density is usually expressed normalized to the blackbody photon density as $\eta \equiv n_B/n_\gamma$. Because standard BBN is a one-parameter theory, any abundance measurement determines η, while additional measurements overconstrain the theory and thereby provide a consistency check. BBN has thus historically been the premier means of determining the cosmic baryon density.

Recently, however, a new "baryometer" has emerged in the form of the CMB. The release of the WMAP results on the anisotropy spectrum of the CMB were a landmark event for all of cosmology, but particularly for BBN. The three-year WMAP data implies

$$\eta_{10}^{\rm wmap} = 6.11 \pm 0.19 \tag{3}$$

The promise of CMB precision measurements of the baryon density suggests a new approach in which the CMB baryon density becomes an input to BBN. Thus, within the context of the Standard Model (i.e., with $N_\nu = 3$), BBN becomes a zero-parameter theory, and the light element predictions are completely determined to within the uncertainties in $\eta_{\rm CMB}$ and the BBN theoretical errors. Comparison with light element observations then can be used to restate the test of BBN–CMB consistency, or to turn the problem around and test the astrophysics of post-BBN light element evolution [19].

LIGHT ELEMENT OBSERVATIONS AND COMPARISON WITH THEORY

BBN theory predicts the universal abundances of D, ^{3}He, ^{4}He, and ^{7}Li, which are essentially determined by $t \sim 180$ s. Abundances are however observed at much later epochs, after stellar nucleosynthesis has commenced. The ejected remains of this stellar processing can alter the light element abundances from their primordial values, but also produce heavy elements such as C, N, O, and Fe ("metals"). Thus one seeks astrophysical sites with low metal abundances, in order to measure light element abundances which are closer to primordial. For all of the light elements, systematic errors are an important and often dominant limitation to the precision of the primordial abundances. Due to space considerations, we will only quote the main observational results here; for a fuller account see [7] and references therein.

In recent years, high-resolution spectra have revealed the presence of D in high-redshift, low-metallicity quasar absorption systems (QAS), via its isotope-shifted Lyman-α absorption. These are the first measurements of light element abundances at cosmological distances. The six most precise observations of deuterium [20, 21, 22, 23, 24] in QAS give D/H = $(2.84 \pm 0.26) \times 10^{-5}$, where the error includes a correction [7] for the dispersion among the measurements. The steep decrease of D/H with η makes it a sensitive probe of the baryon density. Using the WMAP value for the baryon density (3) the primordial D/H abundance is predicted to be [6]:

$(D/H)_p = 2.75^{+0.24}_{-0.19} \times 10^{-5}$. As one can see from Fig. 1, this is in excellent agreement with the QAS observations.

We observe ^{4}He in clouds of ionized hydrogen (HII regions), the most metal-poor of which are in dwarf galaxies [25, 27, 26]. Conservatively, it would be difficult at this time to exclude any value of Y_p inside the range 0.232 – 0.258 [27]. As seen in Fig. 1, this range is excellent agreement with WMAP+BBN predictions Note also that the large uncertainty ascribed to this value indicates that the while ^{4}He is certainly consistent with the WMAP determination of the baryon density, it does not provide for a highly discriminatory test of the theory at this time.

The systems best suited for Li observations are metal-poor stars in the spheroid (Pop II) of our Galaxy [28, 29, 30]. These imply a primordial value [31] $\mathrm{Li/H}|_p = (1.23 \pm 0.06) \times 10^{-10}$. The ^{7}Li abundance based on the WMAP baryon density is predicted to be [6] $^7\mathrm{Li/H} = 3.82^{+0.73}_{-0.60} \times 10^{-10}$ The clear contradiction between these results is known as the "lithium problem" and is an increasingly pressing issue in BBN. Various sources of systematic error have been identified and scrutinized. Stellar evolution effects (i.e., Li depletion) must occur at some level [32, 33], but recent data [29] suggest this effect is too small. Nuclear uncertainties also are potentially important [35], but the work of [36] on the dominant mass-7 production channel $^3\mathrm{He}(\alpha,\gamma)^7\mathrm{Be}$ argues that a "nuclear fix" to the ^{7}Li BBN problem is unlikely. Perhaps the least well-constrained and thus most important systematic error is due to stellar temperature scale determinations; a recent study [34] suggests that these could play a large role in the lithium problem.

Finally, we turn to ^{3}He. Here, the only observations available are in the solar system and (high-metallicity) HII regions in our Galaxy [37, 38]. This makes inference of the primordial abundance difficult Consequently, it is no longer appropriate to use ^{3}He as a cosmological probe [39]; instead, one might hope to turn the problem around and constrain stellar astrophysics using the predicted primordial ^{3}He abundance [40].

The overall concordance as evidenced in Fig. 1 is encouraging and will be sharpened with further WMAP data, and subsequently from *Planck Surveyor* measurements. However note that ^{7}Li is inconsistent with the CMB (as it is with D and ^{4}He) given the error budgets we have quoted. The question then becomes more pressing as to whether this mismatch comes from systematic errors in the observed abundance predictions, or whether there might be new physics at work.

BEYOND THE STANDARD MODEL

Given the simple physics underlying BBN, it is remarkable that it still provides the most effective test for the cosmological viability of ideas concerning physics beyond the Standard Model. Limits on particle physics beyond the Standard Model come mainly from the observational bounds on the ^{4}He abundance. As discussed earlier, the neutron-to-proton ratio is fixed by its equilibrium value at the freeze-out of the weak interaction rates at a temperature $T_f \sim 1$ MeV modulo the occasional free neutron decay. Furthermore, freeze-out is determined by the competition between the weak interaction rates and the expansion rate of the universe

$$G_F{}^2 T_f{}^5 \sim \Gamma_{\rm weak}(T_f) = H(T_f) \sim \sqrt{G_N g_*} T_f{}^2 \tag{4}$$

In the Standard Model, the number of relativistic particle species at 1 MeV is $g_* = 5.5 + \frac{7}{4}N_\nu$, where 5.5 accounts for photons and $e^\pm$, and N_ν is the number of (massless) neutrino flavors. The presence of additional neutrino flavors (or any other relativistic species) at the time of nucleosynthesis increases the overall energy density of the universe and hence the expansion rate leading to a larger value of T_f, (n/p), and ultimately Y_p. The dependence of the light element abundances on N_ν is signficant enough that the all light elements can be combined with the WMAP η to find [19]

$$N_\nu \lesssim 4.5 \tag{5}$$

Because of the form of Eq. (4) it is clear that just as one can place limits [8] on N_ν, any changes in the weak or gravitational coupling constants can be similarly constrained (for a discussion see ref. [41]). The limits on N_ν can be translated into limits on other types of particles or particle masses that would affect the expansion rate of the universe during nucleosynthesis. For a full review see [42].

SUMMARY

As cosmology moves into a new and high precision era, the utility of BBN is shifting from a test of the basic cosmological world model to one which is a sensitive probe of astrophysical processes which relate to the light element abundances as well as a sensitive probe of physics beyond the Standard Model. In the case of ^{4}He and D/H, within the large uncertainties in the observational data, the concordance is excellent. ^{7}Li remains problematic though several possible (if not perfect) astrophysical solutions are available. Finally, as the data on the light element abundances improves, BBN will sharpen its ability to probe physics beyond the standard model. As such it remains the only available tool for which to examine directly the very early universe.

ACKNOWLEDGMENTS

It is a pleasure to thank Livius Trache and the rest of the organizers for including me in an outstanding school. I am gratetul to acknowledge my longtime BBN collaborators, particularly Keith Olive and Richard Cyburt.

REFERENCES

1. T. P. Walker, G. Steigman, D. N. Schramm, K. A. Olive and K. Kang, Ap.J. **376** (1991) 51; K. A. Olive, G. Steigman, and T. P. Walker, Phys. Rep. **333** (2000) 389
2. S. Sarkar, RPP **59** (1996) 1493; R. A. Malaney and G. J. Mathews, Phys. Rept. **229** (1993) 145.
3. G. Gamow, Nature, **162**, 680 (1948)
4. R.A. Alpher and R. Herman, Nature, **162**, 774 (1948)

5. D. N. Spergel *et al.* [WMAP Collaboration], Astrophys. J. Suppl. **170**, 377 (2007) D. N. Spergel *et al.*, Astrophys. J. Suppl. **148** (2003) 175
6. R. H. Cyburt, B. D. Fields and K. A. Olive, Phys. Lett. B **567** (2003) 227
7. B. D. Fields and K. A. Olive, Nuclear Physics A, **777** (2006) 208; B. D. Fields, European Physical Journal A, **27** (2006) 3; B. D. Fields and S. Sarkar, "Big-Bang Nucleosynthesis" mini-review, in W. M. Yao *et al.* [Particle Data Group], J. Phys. G **33**, 1 (2006).
8. G. Steigman, D.N. Schramm, and J. Gunn, Phys. Lett. **B66**, 202 (1977).
9. J. Bernstein, L. S. Brown and G. Feinberg, Rev. Mod. Phys. **61** (1989) 25.
10. V. Mukhanov, arXiv:astro-ph/0303073.
11. R. Esmailzadeh, G. D. Starkman and S. Dimopoulos, Ap.J., **378** (1991) 504.
12. S. Eidelman et al . Phys. Lett. B (2004) 1.
13. R. H. Cyburt, B. D. Fields and K. A. Olive, New Astron. **6** (2001) 215
14. L.M. Krauss and P. Romanelli, Ap.J. **358**, 47 (1990); M. Smith, L. Kawano, and R.A. Malaney, Ap.J. Supp. **85**, 219 (1993); N. Hata, R.J. Scherrer, G. Steigman, D. Thomas, and T.P. Walker, Ap.J. **458**, 637 (1996); A. Coc, E. Vangioni-Flam, M. Casse and M. Rabiet, Phys. Rev. D **65**, 043510 (2002).
15. K. M. Nollett and S. Burles, Phys. Rev. D **61**, 123505 (2000)
16. C. Angulo et al. (NACRE Collaboration), Nuc. Phys. A **656** (1999) 3.
17. R. H. Cyburt, Phys. Rev. D **70** (2004) 023505
18. G. Fiorentini, E. Lisi, S. Sarkar and F. L. Villante, Phys. Rev. D **58**, 063506 (1998) S. Burles, K. M. Nollett and M. S. Turner, Astrophys. J. **552**, L1 (2001)
19. R. H. Cyburt, B. D. Fields and K. A. Olive, Astropart. Phys. **17** (2002) 87
20. J. M. O'Meara, S. Burles, J. X. Prochaska, G. E. Prochter, R. A. Bernstein and K. M. Burgess, Astrophys. J. **649**, L61 (2006)
21. S. Burles and D. Tytler, *Ap.J.* **499**, 699 (1998); *Ap.J.* **507**, 732 (1998).
22. J. M. O'Meara, D. Tytler, D. Kirkman, N. Suzuki, J. X. Prochaska, D. Lubin and A. M. Wolfe, Astrophys. J. **552**, 718 (2001)
23. D. Kirkman, D. Tytler, N. Suzuki, J. M. O'Meara and D. Lubin, Ap. J. Supp. **149** (2003) 1
24. M. Pettini and D. V. Bowen, Astrophys. J. **560**, 41 (2001)
25. Y.I. Izotov, T.X. Thuan, and V.A. Lipovetsky, Ap.J. **435**, 647 (1994); Ap.J.S. **108**, 1 (1997); Y.I. Izotov, and T.X. Thuan, Ap.J. **500**, 188 (1998).
26. K. A. Olive and E. D. Skillman, New Astron. **6** (2001) 119.
27. K. A. Olive and E. D. Skillman, Ap.J. (in press), arXiv:astro-ph/0405588.
28. F. Spite, and M. Spite, A.A. **115**, 357 (1982); P. Molaro, F. Primas, and P. Bonifacio, A.A. **295** , L47 (1995); P. Bonifacio and P. Molaro, MNRAS **285**, 847 (1997).
29. S.G. Ryan, J.E. Norris, and T.C. Beers, Ap.J. **523**, 654 (1999).
30. B.D.Fields and K.A. Olive, New Astronomy, **4**, 255 (1999); E. Vangioni-Flam, M. Cassé, R. Cayrel, J. Audouze, M. Spite, and F. Spite, New Astronomy, **4**, 245 (1999).
31. S.G. Ryan, T.C. Beers, K.A. Olive, B.D. Fields, and J.E. Norris, *Ap.J. Lett.* **530**, L57 (2000).
32. C.P. Deliyannis, P. Demarque, and S.D. Kawaler, Ap. J. Supp. ,**73** (1990) 21.
33. S. Vauclair,and C. Charbonnel, Ap. J. **502** (1998) 372; M. H. Pinsonneault, T. P. Walker, G. Steigman and V. K. Narayanan, Ap. J. **527** (1998) 180 M. H. Pinsonneault, G. Steigman, T. P. Walker, and V. K. Narayanan, Ap. J. **574** (2002) 398
34. P. Bonifacio, et al., Astron. Astrophys., **390** (2002) 91.
35. A. Coc, E. Vangioni-Flam, P. Descouvemont, A. Adahchour and C. Angulo, Ap. J. , **600** (2004) 544
36. R. H. Cyburt, B. D. Fields and K. A. Olive, Phys. Rev. D (in press) 2004,
37. D.S. Balser, T.M. Bania, R.T. Rood, and T.L. Wilson, Ap.J. **510** (1999) 759.
38. K. A. Olive, D. N. Schramm, S. T. Scully and J. W. Truran, Astrophys. J. **479** (1997) 752
39. T. M. Bania, R. T. Rood and D. S. Balser, *Nature* **415** (2002) 54.
40. E. Vangioni-Flam, K. A. Olive, B. D. Fields and M. Casse, Astrophys. J. **585** (2003) 611
41. E. W. Kolb, M. J. Perry and T. P. Walker, Phys. Rev. D **33**, 869 (1986); B.A. Campbell and K.A. Olive, Phys. Lett. **B345** (1995) 429; L. Bergstrom, S. Iguri, and H. Rubinstein, Phys. Rev. **D60** (1999) 045005; C. M. Mueller, G. Schaefer and C. Wetterich, arXiv:astro-ph/0405373.
42. R. H. Cyburt, B. D. Fields, K. A. Olive and E. Skillman, Astropart. Phys. **23**, 313 (2005)

The Astrophysical r-Process 50 Years after B^2FH

K.-L. Kratz*, K. Farouqi†, L.I. Mashonkina** and B. Pfeiffer*

*Max-Planck-Institut für Chemie, Otto-Hahn-Institut, D-55128 Mainz, Germany
†Department of Astronomy and Astrophysics, University of Chicago, Chicago, IL 60637, USA
**Institute of Astronomy, Russian Academy of Science, RU-119017 Moscow, Russia

Abstract. Since the historical papers by Burbidge et al. and Cameron 50 years ago, it is generally accepted that half of the chemical elements above Fe are formed in explosive stellar scenarios by a rapid neutron-capture process (the classical "r-process"). Already from their essential ideas, it became clear that a correct modelling of this nucleosynthesis process requires both, the knowledge of various nuclear properties very far from stability and a detailed description of the astrophysical environments. However, it took about three decades, until in 1986 the first experimental nuclear-physics data on the neutron-magic r-isotopes ^{80}Zn and ^{130}Cd could be obtained, which act as key "waiting points" in the respective A≃80 and 130 peaks of the Solar-System (SS) r-abundances ($N_{r,\odot}$). Since then, using steadily improved nuclear data, we have optimized our r-process calculations to reproduce the present observables of the isotopic $N_{r,\odot}$ "residuals", as well as the more recent elemental abundances in ultra-metal-poor, r-process-enriched halo stars. Concerning the latter observations, we support the basic idea about two different types of r-processes. Based on our many years' experience with the site-independent "waiting-point approach", we recently have extended our studies to fully dynamical network calculations for the most likely astrophysical r-process scenario, i.e. the high-entropy wind (HEW) of core-collapse type II supernovae (SN II). Again, an excellent reproduction of all observables for the "main" r-process has been achieved. However, a major difference is the nucleosynthesis origin of the lighter heavy elements in the $29 \leq Z \leq 45$ mass region. Here, the HEW model predicts – instead of a "weak" neutron-capture r-process component – a primary rapid charged-particle process. This may explain the recent observations of a non-correlation of these elements with the heavier "main" r-process elements.

Keywords: r-process nucleosynthesis, low metallicity stars, core collapse supernovae
PACS: 20.40.Lw, 24.10.-i, 26.30.+k, 97.10.Cv, 99.60.Bw

INTRODUCTION

Historically, the field of nuclear astrophysics has always been concerned with the interpretation of *"observables"* such as specific signatures in luminosity, or chemical abundance distributions (see, e.g. [1, 2, 3, 4], and the description of the originating nucleosynthesis processes (see, e.g. [5, 6, 7, 8]). Steadily improved observational instrumentations and spectroscopic detection methods, nowadays allow detailed analyses of such features in nature and in the ejecta of various stellar events. In the case of the rapid neutron-capture process (i.e. the *"r-process"*), the most important observables are (i) the distribution of the Solar-System (SS) isotopic abundance "residuals" ($N_\odot - N_{s,\odot} \simeq N_{r,\odot}$; [9, 10]), (ii) so-called "FUN" isotopic abundances anomalies in meteorites (see, e.g. [11]), and (iii) the recent elemental abundances in ultra-metal-poor (UMP) halo stars and globular clusters (see, e.g. [4, 12, 13, 14]). Moreover, considerable progress in experimental and theoretical knowledge of nuclear-physics properties far from β-stability now offers increasingly meaningful comparisons between predicted and observed neutron-capture

CP972, *Exotic Nuclei and Nuclear/Particle Astrophysics (II)—Proceedings of the Carpathian Summer School of Physics 2007,* edited by L. Trache and S. Stoica

abundances, which yield valuable information about the required astrophysical conditions for the r-process (see, e.g. [15, 16, 14]). With the "best" nuclear-data input available, we have performed a series of neutron-capture nucleosynthesis calculations for (i) the classical, site-independent "waiting-point" approximation (see, e.g. [15, 17, 18, 19]), and (ii) the site-specific high-entropy-wind (HEW) scenario of core-collapse SN II (see, e.g. [20, 21, 22, 23, 24, 25, 26, 27, 28, 29, 30]). These computations are compared with the $N_{r,\odot}$ pattern and the recent observations of elemental abundances in UMP halo stars (see, e.g. [12, 4, 18, 31, 16, 14, 32, 33]).

ASTROPHYSICAL MODELS

The observations of stellar spectra beyond Ba in UMP halo stars are consistent with the $N_{r,\odot}$ pattern of the classical "main" r-process (see, e.g. [31, 34, 14, 35]), where the relative abundances among individual elements are apparently constant over the whole Galactic evolution. This suggests that all stellar events produce the same abundance distribution for these heavy elements. However, such a unique picture is not obtained for the lighter elements of the classical "weak" r-process below the $A \simeq 130$ peak. Although the abundance *ratios* seem to be quite robust, recent halo-star observations in the Cu to Pd region indicate a star-to-star scatter in the *yields* by more than an order of magnitude relative to the "main" r-process element Eu (see, e.g. [12, 14, 32, 33]). This indicates that the production of the lighter elements is not strictly correlated with the formation of the trans-Ba elements, and may even come from a nucleosynthesis mode which is different from a classical rapid neutron-capture process (see, e.g. [34, 35, 32]).
In any case, a specific astrophysical site will still have (smoothly) varying conditions in different ejected mass zones. This translates into a superposition of r-process components, where for the classical r-process [5, 15, 17] a component is defined by a combination of several parameters: a neutron number density (n_n) and a temperature ($T_9=10^9$ K) – defining the corresponding neutron separation energy (S_n) of an r-process path – and a process duration time (τ). More physically for an adiabatic expansion, the above parameters can be replaced by an entropy (S), a proton/nucleon ratio ($Y_e=\langle Z/A \rangle$) and an expansion time scale (V_{exp}) [24, 26, 30, 32].

The $(n,\gamma)\rightleftharpoons(\gamma,n)$ "Waiting-Point" Assumption

Making use of the first experimental information on nuclear properties of r-process isotopes, in particular on the key N=50 and 82 "waiting-point" nuclides ^{80}Zn and ^{130}Cd ([36]), with our FK^{2}L paper in 1993 [15] we have started a series of parameter studies within the classical r-process model. With an initial abundance in the Fe group as r-process "seed" and using constant n_n and T_9 as well as the assumption of an instantaneous freezeout, as mentioned above this approach is based on a continuous superposition of *weighted* r-process components with varying flow paths related to contour lines of S_n values in the range of 4 to 1 MeV. Such a procedure is, in principle, site-independent and is essentially meant to provide in a first step a good reproduction of the three $N_{r,\odot}$ peaks at $A \simeq 80$, 130 and 195, and then in a second step also a good fit to the total r-abundance distribution. The weights $\omega(n_n)$ and process durations $\tau(n_n)$ of the individual r-process components have a behavior similar to powers of n_n, corresponding to a linear relation in $\log(n_n)$ [15, 12, 14]. This approach (although primarily only a fit and not a realistic stellar site calculation) also has the advantage that such a continuous dependence on nuclear-physics and astrophysics conditions has embedded some relation to a specific stellar site, as for example the SN II "neutrino-wind" scenario. Therefore, one expects some predictive power also for mass regions that cannot be fitted explicitly, as e.g. the trans-Pb region of the Th, U cosmochronometers (see, e.g. [12, 16, 14]). One of the major nuclear-structure conclusions in this context was the quest for *"shell quenching"* far from stability [15, 37, 38, 39] in order to avoid severe abundance deficiencies close to the 2nd and 3rd $N_{r,\odot}$ peaks, and to restrict the necessary r-process time scale ($\tau \simeq 3.5$ s) and neutron-density range ($10^{20} \leq n_n \leq 10^{28}$ at T_9=1.35) for a robust classical r-process (see, e.g. [17, 18, 14]).

The SN II "High-Entropy-Wind" Scenario

In a realistic astrophysical scenario, the required high neutron excess of 10 to 150 neutrons per seed nucleus to form the whole r-process distribution up to the Th, U region is only possible for high matter densities and low Y_e ($0.1 \leq Y_e \leq 0.3$) in neutron stars under β-equilibrium. The other option is the α-rich freeze-out in the so-called "high-entropy-wind" (HEW) scenario of core-collapse SNII with moderate Y_e ($0.4 \leq Y_e \leq 0.49$) and low matter density (see, e.g. [20, 21, 22, 23, 24, 25, 26, 27, 28, 29, 30, 32]). After the freeze-out of charged-particle reactions (at $T_9 \simeq 3$) in matter that expands from high temperatures but low densities, about 90% is locked into α-particles, which leaves a large neutron to seed ratio (Y_n/Y_{seed}) for the few existing neutron-rich r-process "seed" nuclei in the A$\simeq$90 mass region. In this scenario, Y_n/Y_{seed} is a function of and permits for high S rapid neutron captures which can form the heaviest r-process nuclei. Still in a parameterized way, within our Mainz – Basel collaboration, since the mid-1990's we have performed a series of large-scale dynamical r-process network calculations with temperature-dependent astrophysical parameters and a full modeling of the non-equilibrium freezeout phase (see, e.g. [24, 26, 30, 32]).
Most of the major conclusions for the "main" r-process drawn from our ealier classical "waiting-point" calculations were in principle confirmed, in particular the importance of the predicted *"shell quenching"* at N=82 for the overall r-process matter flow. The recent experimental confirmation of the reduction of the N=82 shell gap below ^{132}Sn [39, 19, 40] results in a speeding up of the r-process through the major A$\simeq$130 bottle neck, leading to process duration of $\leq$500 ms with a considerably lower entropy range ($110 \leq S \leq 300$ for Y_e=0.45) compared to the initial approaches a decade ago (see, e.g. [20, 21]). However, within the HEW model a major difference occurs concerning the nucleosynthetic origin of the lighter heavy elements in the $29 \leq Z \leq 45$ region. Instead of a clasical "weak" neutron-capture r-process, for low entropies ($1 \leq S \leq 110$) the freeze-out of a rapid charged-particle process occurs. The possibility of several nucleosynthesis processes adding up to the $N_{r,\odot}$ abundances has already been discussed earlier. On the one hand, Travaglio et al. [34] have discussed the possibility of an additional

"LEPP" (Light Element Primary Process) neutron-capture component, whereas Qian and Wasserburg also propose an independent charged-particle process for the lighter elements [35]. Only recently, increasing observational evidence in the LEPP region has become available, which confirm our HEW model predictions. In particular our calculations are able to explain why the production of the lighter elements is not tightly correlated with the simultaneous formation of the heavier, "main" r-process elements beyond Ba [32].

EXPERIMENTAL INFORMATION ON R-PROCESS NUCLEI

As has been discussed in detail elsewhere (see, e.g. [17, 41, 18]), there are in principle two types of nuclear-structure information that may help to solve still existing r-process problems. First, there is the quite obvious need for nuclear masses and gross β-decay properties (half-life $T_{1/2}$ and β-delayed neutron emission probability P_n) of those isotopes lying **in** the r-process boulevard. Second, also the evolution of shell structure from β-stability towards the neutron dripline must be understood. This requires detailed spectroscopic studies, in particular for the key neutron-magic "waiting-point" nuclei, as has, for example, been performed for N=50 ^{80}Zn [42] and N=82 ^{130}Cd [39].

Gross β-Decay Properties

Theoretically, the two gross β-decay parameters $T_{1/2}$ and P_n are interrelated via their usual definition in terms of the so-called β-strength function (S_β; see, e.g. [43, 44, 45]). Over the years, numerous theoretical approaches have been applied to model β-decay (for a recent review, see, e.g. [46]). These approaches range from phenomenological, algebraic expressions up to large-scale microscopic models. Some of these methods emphasize global applicability, whereas others seek self-consistency or the comprehensive inclusion of nuclear correlations. Unfortunately, even to date none of these models incorporates all important ingredients. Moreover, most models still have strong limitations, e.g. they are restricted to Gamow-Teller (GT) transitions, to spherical shapes, to even-even nuclei, or have too small shell-model spaces (for a detailed discussion, see, e.g. [40, 19]). Therefore, these models usually describe more or less successfully only small mass regions, if their locally fine-tuned parameters are kept fixed. As a consequence, for the time being they are unsuitable as a basis for large-scale r-process calculations.
There is so far only one consistent (macroscopic-) microscopic approach with global applicability of predicting a variety of nuclear properties which are needed for detailed dynamical r-process calculations: namely the macroscopic-microscopic Finite Range Droplet mass model (FRDM [44]) combined with the Quasi-Particle Random Phase Approximation (QRPA) of Möller et al. [44]. Originally restricted to GT-transitions, the latest version of this model also includes first-forbidden (ff) corrections (FRDM + QRPA(GT+ff)), as well as a spreading of the single-particle GT-transitions [45] according to experimental findings (see, e.g. [47, 48, 49, 43]). This QRPA approach shows an average error in predicting $T_{1/2} \leq 1$ s of about 3; accordingly, the respective error for P_n-values is 3.5 (see Figs. 4 and 5 in [45]). In order to check the (maximum possible)

effect of the $T_{1/2}$ uncertainties on r-process abundances calculated within the classical "waiting-point" approach, we have changed **all** $T_{1/2}$ values by a factor 3 in both directions, while keeping all other nuclear-physics quantities and astrophysics parameters constant. As shown, for example, in the lower part of Fig. 4 of [19], the deviations in the respective theoretical r-abundance distributions relative to the "standard" $N_{r,calc}$ pattern (see upper part of Fig. 4 of [19], and shown in the left part of Fig. 1 of this paper) clearly indicate significant differences in the speed of the overall r-process matter flow introduced by the different $T_{1/2}$ sets. Fortunately, today for most of the exotic isotopes around N=82 that dominate the bottle-neck behavior of the matter flow in the $A \simeq 130$ $N_{r,\odot}$ region, experimental data are available (see, e.g. Table 1 in [51]).
Nevertheless, desirable further progress in the above approaches certainly will ask for more than just the reproduction of gross β-decay properties, but rather for an understanding of the detailed nuclear structure, at least for a number of selected "key" r-process isotopes. As already mentioned above, a recent, important example in this context has been the identification of several a priori unexpected features in the decay of N=82 ^{130}Cd, i.e. the high energy of 2.12 MeV of the 1^+ 2QP ($\nu g_{7/2} \otimes \pi g_{9/2}$) state with its log(ft)=4.1, the high Q_β value of 8.34 MeV, and the non-negligible ff β-feeding of about 11.5% to low-lying states in the daughter ^{130}In [39, 51]. On the one hand, these experimental data have helped to finally "understand" the underlying nuclear structure of the gross β-decay quantities $T_{1/2}$ and P_n, first measured in 1986 [52]. On the other hand, these properties have turned out to be of particular astrophysical importance since ^{130}Cd is – for both, the classical r-process model as well as for the HEW model – the most important "waiting-point" isotope for the **"main"** r-process.

Nuclear Masses

A substantial number of theoretical approaches to calculate nuclear masses exist nowadays for various applications. Similar to the situation for the gross β-decay properties, the sophistication of these models is largely different and should therefore be understood before they are used as nuclear input. The key question in the present case – i.e. the calculation of r-process observables with theoretical S_n values – is: Are the masses calculated for unknown nuclei reliable, or do the model predictions diverge outside the experimental data set, to which their parameters were adjusted? This question is best addressed, if published theoretical masses are compared to experimental data obtained **after** the publication of the calculations (see, e.g. [44, 53, 54]).
It has been claimed repeatedly that the recent more sophisticated HFB mean-field mass models (see, e.g. [55, 56, 57]), that go beyond gross-theory and macroscopic-microscopic approaches, would by virtue of their added microscopic complexity be physically more reliable and have better predictive power for unknown nuclei. However, recent tests [53, 58] have shown that this expectation has so far not been fulfilled. Moreover, despite the admirable efforts of the Montreal–Brussels group, there seem to be fundamental problems with the HFB models. As is discussed in detail in the topical review of Rikovska Stone [58], several parameters are clearly not treated in a *"selfconsistent"* and *"fully microscopic"* way, as claimed by the authors. Furthermore, the macroscopic-

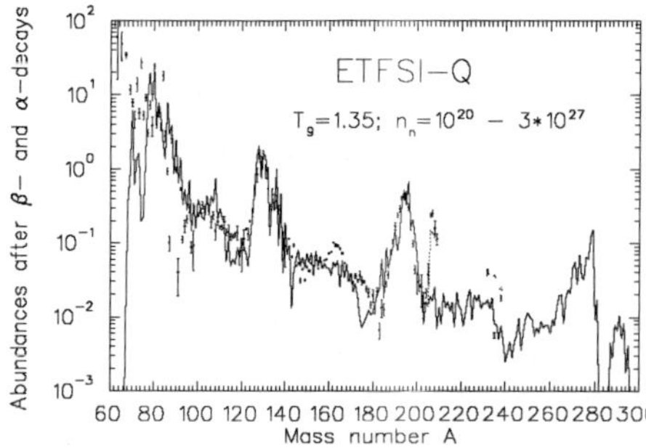

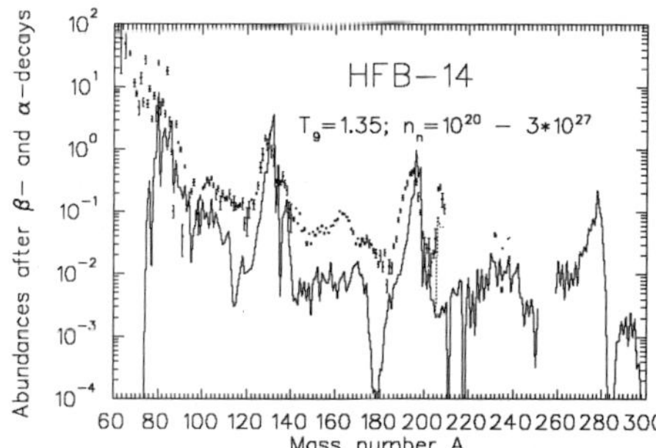

FIGURE 1. Comparison of isotopic r-abundance curves (solid line) calculated within the "waiting-point" model to the $N_{r,\odot}$ distribution (dots) using the "quenched" ETFSI-Q [50] and the new HFB-14 [61] mass models. Apart from the S_n-values, all other nuclear and astrophysics parameters, including the weighting of the n_n-components, were kept constant. For further discussion, see text.

microscopic FRDM (with its parameters fitted to the masses known already in 1992), still is doing a better job with its rms value of 601 keV for the combined experimental data from the 2003 mass evaluation [59] and the available "new" mass measurements up to 2007 [60] than even the latest HFB-14 mass model with a σ_{rms}=729 keV [61].

COMPARISON OF THEORETICAL ABUNDANCES TO R-PROCESS OBSERVABLES

Taking advantage of the site-independent "waiting-point" approach, during the last decade we have tested in great detail (i) the influence of different nuclear-physics input parameters on various r-process observables, and (ii) under which stellar conditions a successful r-process has to run (see, e.g. [15, 17, 18, 16, 51, 40, 19, 14]). More recently, we then have compared the results from the classical equilibrium model with the new, full dynamical network calculations for specific stellar site of the so-called high-entropy wind (HEW) of core-collapse SN II (see, e.g. [24, 26, 30]).

As an example for the "waiting-point" model, in Fig. 1, comparisons of calculated r-abundance distributions ($N_{r,calc}$) with the isotopic $N_{r,\odot}$ pattern for two different mass models are shown, i.e. ETFSI-Q [50] and the recent HFB-14 [61]. In both cases, experimentally known masses from [59] have been superimposed. It is quite evident, that the use of the supposed "microscopic" HFB model does not improve the overall fits. On the contrary, compared to the ETFSI-Q fit the (local) deviations become more pronounced. For the site-independent "waiting-point" approach, where the astrophysical quantities are in principle free parameters, these deficiencies may be avoided by chosing different n_n, T_9 and $\tau(n_n)$ components with specifically selected component weightings or with permitted random relations between the respective parameters [62].

In contrast, for the HEW scenario, where for any given expansion speed the other astrophysical parameters S, Y_e and Y_n/Y_{seed} as well as the (model-inherent) weighting of the respective entropy (or n_n) components are correlated, the above deficiencies with

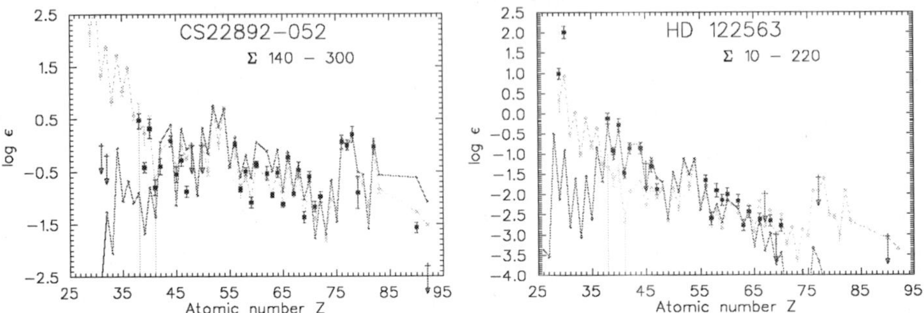

FIGURE 2. Comparisons of elemental abundances calculated within the HEW model (solid line) with the observed abundances of two selected UMP halo stars (squares) and the solar values (dashed curve). For a discussion of the respective fit criteria, see text.

the use of the HFB mass formulae can no longer be "repaired". Therefore, within the HEW approach we restrict all calculations to the overall "best" mass model, i.e. the ETFSI version with shell quenching [50]. As for example shown in Fig. 2 of [26], already a superposition of three selected S components scaled to ^{130}Te results in a remarkable agreement of the $N_{r,\odot}$ distribution of the "main" r-process. And, as depicted in Fig. 2 of [32], also the conversion from isotopic to elemental r-abundances yields good agreement with the SS r-process "residuals" beyond Ag (Z=47) over the whole Z-range via the 2nd and 3rd peaks up to Pb and Bi. However, for the full entropy range of 10<S<300 (full line with triangles in Fig. 2 of [32]), pronounced *"overabundances"* between Sr and Pd relative to the $N_{r,\odot}$pattern were predicted.

This observation is of particular interest, because some quite recently observed "r-process poor" UMP halo stars (e.g. HD 122563 [33]), indeed, seem to confirm these overabundances in the $28 \leq Z \leq 36$ region. As an example, a "best" overall fit to the measured elemental abundances of HD 122563 is shown in the right part of Fig. 2. In order to obtain this level of agreement, the full low-S ($S \geq 10$) charged-particle component of our HEW model had to be used; but on the high-S side, for the neutron-capture component the maximum entropy had to be limited to $S \simeq 220$. This cutting out the highest S components $220 \leq S \leq 300$ – actually necessary to reproduce the region of the 3rd peak – indicates that the "r-process poor" host star HD 122563 has been polluted by a SN which failed to make a complete "main" r-process.

In contrast to the above signatures, several of the earlier observed "r-process rich" halo stars (e.g. CS 22892-052 [12]) had shown pronounced *"underabundances"* in the LEPP region. The respective "best" fit to the observed abundances of CS 22892-052 is depicted in the left part of Fig. 2. In this case, the full high-S range ($140 \leq S \leq 300$) had to be used to reproduce the solar-like "main" r-process pattern between Ba and Pb; but the low-S range ($10 \leq S \leq 139$) had to be cut out to simulate the measured underabundance pattern in the LEPP region. One can speculate that, contrary to HD 122563 for which presumably the full amount of low-S material has been ejected, in the case of CS 22892-052 a significant part of those low-S inner mass zones of the high-entropy bubble may not have been ejected but fallen back to the nascent neutron star.

Recent spectroscopic data of LEPP elements in UMP halo stars (see, e.g. [12, 13, 33, 34, 35]) have shown that their abundances are not tightly correlated with the almost pure "main" r-process element Eu. As suggested independently by Travaglio et al. [34] and Qian and Wasserburg [35], this may indicate a nucleosynthetic origin different from that of the heavier neutron-capture elements. As has been discussed in some detail in [32], and is also shown exemplarily in Fig. 2, this general conclusion is confirmed and quantified by our HEW model.

In order to obtain a deeper insight into the possible production modes of the LEPP elements, in a first step we have systematically checked their relative abundances as a function of S-ranges. As is shown e.g. in Fig. 1 and Table 1 of [32], in the HEW scenario one can in principle distinguish between four specific nucleosynthesis components. In the lowest S-range ($1 \leq S \leq 50$), a "normal" α-rich freezeout occurs which directly produces stable or close to stable isotopes of the lightest LEPP elements in the region Fe to Y. Also in the next higher S-range ($50 \leq S \leq 110$) there are not yet sufficient free neutrons to start a real neutron-capture process. However, under these S-conditions at freezeout the seed composition is already shifted from the stability line to the neutron-rich side, including well-known β-delayed neutron (βdn) precursors in the $80 \leq A \leq 100$ mass region. The number density of these delayed neutrons is low, but exceeds that of the free neutrons. As a consequence, the possible βdn-recapture during freezeout can mimic a very-weak s-like LEPP mode, qualitatively supporting the earlier suggestions of Travaglio et al. [34]. However, since the initial nucleosynthesis mode is a charged-particle process, one can expect that also the elements which are predominantly formed within this S-range should not be tightly correlated with Eu. In the third S-range ($110 \leq S \leq 150$), our HEW model predicts the density of free neutrons to become high enough ($1 \leq Y_n/Y_{seed} \leq 10$) to start a *"weak"* r-process from Mo (Z=42) up to the onset of the left wing of the $A \simeq 130$ $N_{r,\odot}$ peak. The HEW neutron-capture production of these elements may, therefore be already more closely correlated with the "main" r-process component. Finally, for the highest S-range ($150 \leq A \leq 300$), the HEW model predicts a very robust *"main"* r-process.

For a better understanding of this latter neutron-capture component, which synthesizes the heavy r-isotopes up to the actinides, it is worth to have a closer look how this process operates. With $S \geq 150$, its neutron-capture flow starts from a neutron-rich seed composition in the $100 \leq A \leq 120$ mass region and produces within about 150 ms the full abundances of the N=82 progenitor nuclides from ^{125}Tc up to ^{130}Cd at the top of the $A \simeq 130$ peak. For $S \geq 190$, the matter flow succeeds to overpass the N=82 bottle-neck. The r-process now speeds up again to form the whole rare-earth element region within a rather small entropy range of $220 \leq S \leq 250$, before the flow is again slowed down at the N=126 shell closure with the build up of the $A \simeq 195$ 3rd r-peak. After a duration of about 330 ms, for entropies of $S \geq 270$ the r-process overcomes the N=126 bottle neck, again speeds up and now forms the Th, U actinide region up to the next shell closure at N=184. In this way, a total process duration of only $\tau \leq 500$ ms is needed to run a complete "main" r-process in the HEW scenario.

After this more general part, in a second step, we now can confront our HEW model predictions to the recent spectroscopic observations from the UMP halo stars, in particular as possible diagnostics of the different nucleosynthesis modes and their correlations in the LEPP region. Since so far the largest data set on abundance observations exists for

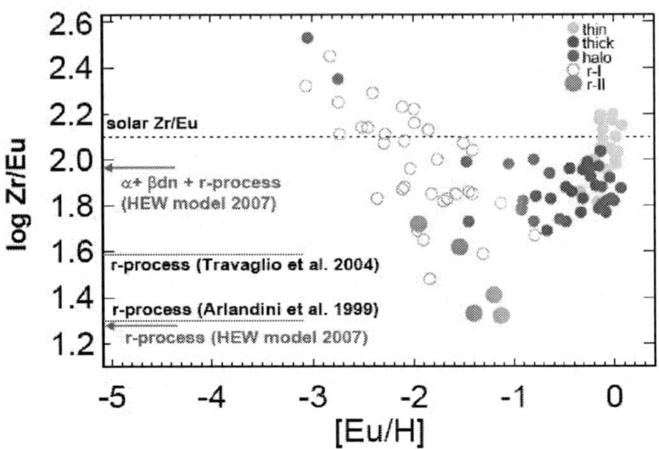

FIGURE 3. Elemental abundance ratios of log(Zr/Eu) as a function of r-process "enrichment" in units of [Eu/H] for stars in different regions of our Galaxy. The observational data [65, 66] are compared to the abundance prediction of the classical SS r-process "residual" of Zr by Arlandini et al. [10], to the r-process plus LEPP abundance estimate of Travaglio et al. [34], and to our SN II HEW nucleosynthesis components [26, 30, 32]. For discussion, see text.

Sr, Y and Zr – covering the full range between "r-process poor" stars (like HD 122563) up to "r-process rich" stars (like CS 22892-052) – in the present paper we will concentrate on these three elements. The first important result is that their abundance ratios, i.e. Sr/Y, Sr/Zr and Y/Zr, seem to be independent of both, r-process *"enrichment"* **[Eu/H]** in the range -3.5 to -0.5 (for the square-bracket notation used by astronomers, see, e.g. [12, 14]) and *"metallicity"* **[Fe/H]** in the range -3.3 to -0.9. The mean values of the observed abundance ratios (Sr/Y=6.17$\pm$1.06, Sr/Zr=1.12$\pm$0.54) and Y/Zr=0.18$\pm$0.04) are very well reproduced by the low-S charged-particle component of our HEW model (Sr/Y=6.80, Sr/Zr=1.34 and Y/Zr=0.20). However, their comparison to the respective model predictions for the "main" r-process (Sr/Y=9.74, Sr/Zr=0.042 and Y/Zr=0.004) clearly indicates that this nucleosynthesis mode can be excluded. When extending the present narrow $38<Z<40$ element range to the available data of the whole LEPP region ($29\leq Z\leq 47$), also a classical "weak" neutron-capture process [31, 63, 64] seems unlikely.

Again following astronomers' conventions to display log(X/Eu) vs. r-process enrichment [Eu/H], as an example we show in Fig. 3 such a plot for Zr. In this figure, observational data [65, 66] for the thin and thick Galatic disk, the halo and r-enriched stars are compared to the total SS abundance ratio of log(Zr/Eu)=2.1 and the predictions by (i) Arlandini et al. [10], (ii) Travaglio et al. [34], and (iii) our HEW model results. Several conclusions can be drawn from this figure. First, the highest Zr/Eu values – beyond the SS ratio – are seen for low r-process enrichments of [Eu/H]$\leq$-2.5. Second, our HEW model prediction for a pure neutron-capture process agrees with the classical "r-residual" ratio of Zr/Eu [10]. The respective low log(Zr/Eu)$\simeq$1.3 agrees with the observed ratios of the 5 r-process-rich (r-II) stars within their given errors. Third, the value of log(Zr/Eu)$\simeq$1.6, given by Travaglio et al. when adding the neutron-capture LEPP component, seems to agree with the two other r-II stars and a minor fraction of the r-I stars for a moderate r-process enrichment of [Eu/H]$\simeq$-2.2. Forth, our HEW model (with charged-particle plus neutron-capture components) predicts a considerably higher total value of log(Zr/Eu)$\simeq$1.95, reproducing the majority of the Halo and thick-disc observational data. In summary, for the nucleosynthetic origin of Zr we see a somewhat inconclusive picture, indicating e.g. a large uncertainty for its s-process contribution between about 20% and 85% of the SS value.

SUMMARY AND OUTLOOK

In summary, we have shown that progress in both experimental and theoretical nuclear-physics properties has been essential for the recent advances in the understanding of r-process nucleosynthesis. Based on our many years' experience with the classical "waiting-point" model, we have extended our systematic studies to more realistic, full dynamic network calculations within the HEW scenario of core-collapse SN II. From the combined results, we draw the following conclusions for the operation of a full r-process, as defined historically in [5, 6]:

(i) The $26 \leq Z \leq 47$ LEPP region is formed in a primary charged-particle process, not by a "weak" neutron-capture r-process component. (ii) The robustness of the "main" neutron-capture r-process originates mainly from the nuclear-structure properties at the N=82 shell closure. The resulting A$\simeq$130 peak acts as the major bottle neck in the total r-process matter flow up to the heaviest elements. (iii) The flood of very recent parameter studies for more exotic SN scenarios which request extreme nuclear and / or astrophysical conditions – such as rather unlikely low Y_e values ($0.07 \leq Y_e \leq 0.39$, e.g. in [67, 68]), unusually low freezeout temperatures (see, e.g. [69], very high expansion velocities requesting extremely short process time scales (see, e.g. [70]), strong effects of neutrino or active-sterile neutrino reactions combined with dominant fission recycling (see, e.g. [68]) – in order to obtain a steady-state-like r-process robustness, are not convincing as long as they only reproduce selected observational features, like the mere existence of the r-abundance peaks, but fail to provide a consistent picture for the **whole** SS and UMP halo-star r-process abundance distributions.

What remains to be done is to couple our r-process network caluclations to full hydrodynamical SN II calculations (see, e.g. [71, 72]). This will be the next step towards a final answer to the question of the *"origin of the heavy elements between Fe and Th, U"*, which has been considered number three of todays *The 11 Greatest Unanswered Questions in Physics* [73].

REFERENCES

1. H.E. Suess and H.C. Urey, *Rev. Mod. Phys.*, **28**, 53 (1956).
2. W.H.G. Lewin et al., *Space Sci. Rev.*, **62**, 233 (1993).
3. K. Lodders, *Ap. J.*, **591**, 1220 (2003).
4. C. Sneden and J.J. Cowan, *Science*, **299**, 70 (2003).
5. E.M. Burbidge et al., *Rev. Mod. Phys.*, **29**, 547 (1957).
6. A.G.W. Cameron, *Pub. Astron. Soc. Pac.*, **6**, 201 (1957).

7. J.J. Cowan et al., *Phys. Rep.*, **208**, 267(1991).
8. G. Wallerstein et al., *Rev. Mod. Phys.*, **69**, 995 (1997).
9. F. Käppeler et al., *Rep. Prog. Phys.*, **52**, 31 (1989).
10. C. Arlandini et al., *Ap. J.*, **525**, 886 (1999).
11. K.-L.Kratz et al., *Mem. S. A. It.*, **72**, 453 (2001).
12. J.J. Cowan et al., *Ap. J.*, **521**, 194 (1999).
13. J.J. Cowan and C. Sneden, *Nature*, **440**, 1151 (2006).
14. K.-L. Kratz et al., *Ap. J.*, **662**, 39 (2007).
15. K.-L. Kratz et al., *Ap. J.*, **403**, 216 (1993).
16. K.-L. Kratz et al., *New Astron. Rev.*, **48**, 105 (2004).
17. K.-L. Kratz et al., in *Origin of Elements in the Solar System: Implications of Post-1957 Observations*, Kluwer Academics / Plenum Publishers, Boston/Dordrecht, p. 119 (2001).
18. B. Pfeiffer et al., *Nucl. Phys.*, **A693**, 282 (2001).
19. K.-L. Kratz, *AIP Conf. Proc.*, **819**, 409 (2006).
20. S.E. Woosley and R.D. Hoffman, *Ap. J.*, **395**, 202 (1992).
21. K. Takahashi et al., *Astr. Astroph.*, **286**, 857 (1994).
22. Y.-Z. Qian and S.E. Woosley, *Ap. J.*, 471, 331 (1996).
23. R.D. Hoffman et al., *Ap. J.*, **482**, 951 (1997).
24. C. Freiburghaus et al., *Ap. J.*, **516**, 381 (1999).
25. T.A. Thompson et al., *Ap. J.*, **562**, 887 (2001).
26. K. Farouqi et al., *Nucl. Phys.*, **A758**, 631 (2005).
27. S. Wanajo et al. *Ap. J.*, **636**, 842 (2006).
28. A. Burrows et al., *Nucl. Phys.*, **A777**, 356 (2006).
29. C. Fröhlich et al., *Ap. J.*, **637**, 415 (2006).
30. K. Farouqi et al., *AIP Conf. Proc.*, **819**, 419 (2006).
31. B. Pfeiffer, U. Ott & K.-L. Kratz, *Nucl. Phys.*, **A688**, (2001).
32. K. Farouqi et al., *AIP Conf. Proc. First Stars III*, 2007, in print.
33. Honda et al. *Ap.J.*, **666**, 1189 (2007).
34. C. Travaglio et al., *Ap. J.*, **601**, 864 (2004).
35. Y.-Z. Qian & G.J. Wasserburg, *Phys. Rep.*, **442**, 237 (2007); and Refs. therein.
36. K.-L. Kratz et al., *J. Phys.*, **G14**, 331 (1988).
37. J. Dobaczewski et al., *Phys. Rev. Lett.*, **72**, 981 (1996).
38. B. Pfeiffer, K.-L. Kratz & F.-K. Thielemann, *Z. Physik*, **A357**, 235 (1997).
39. I. Dillmann et al., *Phys. Rev. Lett.*, **91**, 162503 (2003).
40. K.-L. Kratz et al., *AIP Conf. Proc.*, **769**, 1356 (2005).
41. K.-L. Kratz et al., *Hyperfine Interactions*, **129**, 141 (2000).
42. K.-L. Kratz et al., *Phys. Rev.*, **C38**, 105 (1988).
43. K.-L. Kratz, *Nucl. Phys.*, **A417**, 447 (1984).
44. P. Möller et al., *ADNDT*, **66**, 131 (1997).
45. P. Möller et al., *Phys. Rev.*, **C67**, 055802 (2003).
46. J. Engel in *"The Astrophysical Origin of the Heavy Elements"*, World Scientific, **13**, 53 (2004).
47. K.-L. Kratz et al., *Nucl. Phys.*, **A317**, 331 (1979).
48. K.-L. Kratz et al., *Phys. Lett.*, **86B**, 21 (1979).
49. K.-L. Kratz et al., *Proc. 4th Int. Conf. on "Nuclei far from Stability"*, CERN-Report **81-09**, 317 (1981).
50. J.M. Pearson et al., *Phys. Lett.*, **B387**, 455 (1996).
51. K.-L. Kratz et al., *EPJ*, **A25 Supplement 1**, 633 (2005).
52. K.-L. Kratz et al., *Z. Physik*, **A325**, 489 (1986).
53. D. Lunney et al., *Rev. Mod. Phys.*, **75**, 1291 (2003).
54. J. M. Pearson in *"The Astrophysical Origin of the Heavy Elements"*, World Scientific, **13**, 43 (2004).
55. S. Goriely et al., *ADNDT*, **77**, 311 (2001).
56. M. Samyn et al., *Phys. Rev.*, **C70**, 044309 (2004).
57. S. Goriely et al., *Nucl. Phys.*, **A750**, 425 (2005).
58. J. Rikovska Stone, *J. Phys. G: (Nucl. Part. Phys.)*, **31**, R211 (2005).
59. G. Audi et al., *Nucl. Phys.*, **A729**, 3 (2003).
60. P. Möller, private communication 2007; and to be subm. to ADNDT.
61. S. Goriely et al., *Phys. Rev.*, **C75**, 064312 (2007).
62. S. Goriely and M. Arnould, *Astr. Astroph.*, **312**, 327 (1996).
63. F. Montes et al., *arXiv:0709.0417v1 [astro-ph]*, (2007).
64. K. Farouqi, L.I. Mashonkina, K.-L. Kratz et al., to be published.
65. P.S. Barklem et al., *Astr. Astroph.*, **439**, 129 (2005).
66. L.I. Mashonkina et al., *Astron. Rep.*, **51**, 903 (2007).
67. T. Kuroda et al., *arXiv:0707.2255v1 [astro-ph]*, (2007).
68. J. Beun et al., *arXiv:0707.4498v1 [astro-ph]*, (2007).
69. S. Wanajo, *arXiv:0706.4360v2 [astro-ph]*, (2007).
70. H. Ning et al., *arXiv:0708.1748v1 [astro-ph]*, (2007).
71. A. Burrows et al., *Ap. J.*, **655**, 416 (2007).
72. C.L. Fryer at al., *Ap. J.*, **646**, L131 (2007).
73. E. Haseltine, *Discover Magazine*, **23**, No.2 (2002).

Rare Isotopes and Accreting Neutron Stars, X-Ray Bursts, Neutron Star Crust, rp-Process

H. Schatz

National Superconductiong Cyclotron Laboratory, Dept. of Physics and Astronomy, Joint Institute for Nuclear Astrophysics, Michigan State University, East Lansing, MI 48824

Abstract. Rare isotopes from the proton drip line to the neutron drip line are produced on the surface and crust of accreting neutron stars. Understanding their properties is essential to address the open questions raised by a variety of new observations of X-ray binaries, in particular X-ray bursters and transients. This paper provides an introduction to the topic.

Keywords: X-ray bursts, neutron star crust, rp-process
PACS: 26.30.Ca 26.50.+x

INTRODUCTION

Neutron stars in binary systems can accrete matter from a nearby companion star. The composition of the accreted matter depends on the type of companion star and is typically a mix of hydrogen and helium, though systems are known that accrete pure helium. Accretion rates range from 10^{-8}-$10^{-10} M_{\odot}/\mathrm{yr}$ resulting in a local influx of material on the surface at rates of up to 10-100 kg/s/cm^2 (assuming distribution across the entire surface). The gravitational energy release heats the infalling material when it interacts with the neutron star surface leading to bright X-ray emission. Accreting neutron stars are therefore bright astronomical X-ray sources with intrinsic luminosities up to 4 orders of magnitude brighter than the sun. Observationally they represent a subset of X-ray binaries (together with systems containing black holes). For recent reviews of the topic see [1, 2, 3].

Upon arrival on the neutron star the accreted matter is distributed across the neutron star surface, where it accumulates and is compressed and heated by the matter continuously accreted on top of it. As a result, an accreted fluid element experiences steadily increasing pressure and density conditions that lead to a series of nuclear transformations that involve extremely unstable nuclei. After typically hours to days, hydrogen and helium burns unstably in a thermonuclear explosion observed as a type I X-ray burst. Such unstable burning occurs for accretion rates below $3 \times 10^{-8} M_{\odot}$ [4], roughly twice the Eddington accretion rate $\dot{m}_{\mathrm{Edd}} \approx 75 \times 10^4$ g/s/cm^2 or $\approx 1.4 \times 10^{-8}$ $M_{\odot}$/yr for spherical accretion on a typical neutron star. The Eddington accretion rate is defined as the rate where the radiation pressure from the gravitational energy release matches the gravitational binding energy on the neutron star surface. The actual value depends on the opacity and therefore the composition. The fact that the transition from stable to unstable burning occurs near the Eddington accretion rate is probably a coincidence.

The nuclear reaction sequence during an X-ray burst is typically the rapid proton capture process (rp-process) [5, 6] burning the accreted hydrogen and helium into

CP972, *Exotic Nuclei and Nuclear/Particle Astrophysics (II)—Proceedings of the Carpathian Summer School of Physics 2007,* edited by L. Trache and S. Stoica

heavier elements in the mass range $A = 64$, in extreme cases up to $A = 106$ [7, 8, 9]. At most a few percent of the burned matter are ejected, with the bulk of the ashes remaining on the neutron star surface where it is compressed and forms a liquid ocean. After typically a year, residual carbon in the ocean of ashes might ignite explosively and give rise to so called superbursts [10] producing mostly nuclei in the iron and nickel region. The superburst ashes is compressed further and undergoes a series of electron capture reactions and density induced fusion reactions (so called pycnonuclear fusion reactions) [11, 12] for some thousands of years until the accreted fluid element has reached the crust core interface at a depth of typically a few hundred meters. After transitioning through a phase of nuclear pasta, the nuclei dissolve there into a mix of nucleons and other particles.

In recent years, a wide range of new observations of Galactic X-ray binaries became available, and many space based X-ray observatories continue to operate. This stream of observational results has led to the discovery of many new, largely nuclear physics driven phenomena. In addition, the properties of X-ray bursts are now much more precisely determined and their behavior as a function of system parameters such as neutron star spin (typically 300-600 Hz), accretion rate, and accreted composition, has been explored extensively. All these developments have dramatically increased the need to understand the underlying nuclear physics. The nuclear reactions generate energy, that in some cases is directly observable in X-ray bursts and superbursts, but also set the thermal structure of the neutron star crust. Nuclear reactions also determine the composition of the crust, which is essentially composed of the ashes of the various nuclear processes, and determine the composition of the small amounts of material that might be ejected in particularly powerful X-ray bursts. The various processes occurring at different depth are intimately connected. For example, energy generation by electron capture in deeper regions sets the thermal properties of neutron star crust, which then can affect burning processes in shallower layers related to X-ray bursts. At the same time, the ashes of a shallow process represents the initial composition for the deeper process and therefore determines their nature. An example are the so called superbursts that are most likely produced by explosive burning of carbon in the liquid ocean layers of the neutron star. The amount of carbon in the ashes of the normal, much more frequently occurring X-ray bursts determines the nature of the superbursts, and whether they can occur at all. At the same time, electron captures in the deep solid crust (which depend sensitively on the composition of the superburst ashes, on which they start) determine the temperature of the ocean and therefore superburst ignition conditions and recurrence times.

X-RAY BURSTS AND THE RP-PROCESS

X-ray bursts are the most frequent thermonuclear explosions in the universe. The so called type I bursts discussed here are of thermonuclear nature and distinguished from much rarer type II X-ray bursts, which are driven by gravitational energy release associated with a sudden increase in the accretion rate. Type I X-ray bursts typically release 10^{39}-10^{40} erg of energy in 10-100s. There are about 70 X-ray bursting sources known in our Galaxy. Bursts occur once enough matter has been accreted from the companion star to trigger unstable thermonuclear ignition, and repeat with recurrence times from hours to days. Unlike supernovae and most novae, X-ray bursts offer the unique opportunity to frequently observe complete explosive events from the same source. Sources can also be revisited years later once new questions have come up, or if new telescope technology becomes available. The observation of the burst behavior over many years probes the stability of system parameters such as accretion rates. X-ray bursts offer therefore a unique opportunity to study astrophysical thermonuclear explosions.

There are many open questions related to X-ray bursts and advances astrophysics and nuclear physics are needed to address them. Examples include the burst behavior as a function of accretion rate - at high accretion rates above about $\dot{m} = 0.1\dot{m}_{\rm Edd}$ there is a regime with reduced burst frequency contrary to expectations. Uncertainties in the ^{15}O(α,γ) reaction rate have been suspected to contribute to the inability of current models to reproduce this behavior [13, 14]. The occurrence of double peaked bursts in some sources is another puzzle [15]. Among other effects, nuclear physics has also been suggested as a possible explanation [16]. The origin of ms oscillations that have been discovered with the RXTE satellite during some X-ray bursts [2] are also not well understood. While most likely associated with the spin frequency of the neutron star, the challenge is to explain the origin of the surface brightness modulations that would lead to oscillations, and the observed small frequency drift. The most promising models use various types of surface modes that can generate anisotropies in the X-ray emission and propagation of the modes across the neutron star surface can explain the observed frequency drifts [18]. The characteristic frequencies can depend on the composition and therefore on the nuclear physics of X-ray bursts.

The energy released in nuclear reactions during X-ray bursts directly powers the observed burst light curve. The nuclear burning zone is located at a density of 10^5-10^6 g/cm^3 and reaches temperatures up to 1-2 GK. While this is only about 1 m into the neutron star atmosphere, the photon mean free path at such extreme densities is of the order of 10^{-5} cm. Therefore, γ and X-ray radiation released by the nuclear reactions is quickly thermalized and transported to the surface via convection and radiative transfer, where it heats the photosphere to temperatures that lead to thermal X-ray emission with a spectrum resembling roughly a black body. Photons released in the nuclear reactions during an X-ray burst can therefore not be observed directly. Nevertheless, the radiation transport timescale for long bursts is smaller than the burst duration and variations in the nuclear energy release can imprint themselves onto the burst light curve [7, 16, 17]. The overall duration of the frequently observed relatively long bursts (100 s timescale) is most likely set by nuclear reactions in the slow rp-process [7].

Another possible signature of the nuclear reactions in X-ray bursts are the freshly synthesized nuclei. The problem here is that the gravitational binding of a nucleon on the surface of a neutron star is of the order of 200 MeV and therefore about 40 times larger than the energy release of nuclear reactions. Therefore, at most a few percent of matter can be ejected in an X-ray burst. X-ray bursts and the associated processes such as the rp-process are therefore not major nucleosynthesis sites. However, recent theoretical work has demonstrated that a small amount of burned material after being transported to the surface by convection, can be ejected in the frequently observed photospheric radius expansion bursts where radiation pressure can trigger winds [19]. The ejected elements produce potentially observable spectral signatures. This provides strong additional motivation to better understand the nuclear physics during X-ray bursts. In addition, the discovery of superburst (see below) and the observation of crustal cooling in transients [20, 21] has strongly increased the need to understand the composition of the material that is not ejected, but serves as fuel for secondary nuclear processes. These processes can therefore be viewed as indirect signatures of the nucleosynthesis in X-ray bursts.

The nature of the nuclear reaction sequence powering a particular X-ray burst depends sensitively on the accreted composition and other system parameters such as accretion rate or neutron star properties. The most important parameter determining the reaction sequence is the amount of hydrogen available for burning at the time of ignition. Extended stable hydrogen burning via the CNO cycles in between bursts at low accretion rates or the accretion of hydrogen poor matter from the companion star can lead to bursts that burn little or no hydrogen. In this case, helium burns via the 3α reaction into ^{12}C and then into heavier elements. The resulting reaction flow is not just a sequence of α-captures on ^{12}C, as has been frequently assumed in early simulations. At the high temperature reached in X-ray bursts some α-captures quickly reach the Mg region and beyond, where (α,p) reactions start creating hydrogen. The left panel of Fig. 3 in [19] shows a typical reaction flow during the initial stage of a pure He burst. Already a small amount of hydrogen can catalyze a rapid bypass of the slow 12C(α,γ) reaction via ^{12}C(p,γ)^{13}N(α,p)^{16}O [19]. This bypass drastically accelerates the reaction flow towards heavier elements and typically leads to a drastic depletion of ^{12}C, which therefore in many cases is not a major product of helium bursts. In addition, the protons can capture on seed nuclei in the Fe region that were present in the accreted composition.

In many cases however, X-ray bursts ignite in a mixture of hydrogen and helium. In this case, the reaction sequence beyond ^{12}C is governed by the αp process, a sequence of proton captures and (α,p) reactions. The αp process transitions into the rp-process when the Coulomb barrier for α-induced reactions becomes too large. The exact transition point depends on the temperature but is typically somewhere between Mg and Ca. The rp-process is a sequence of rapid proton captures and slower β-decays near the proton drip line that make the process relatively slow. The endpoint of the rp-process depends sensitively on model and parameters, and is typically set by the number of proton capture reactions needed to exhaust the hydrogen [6]. It can therefore vary from system to system, or even from burst to burst. The most extended rp-process is expected for systems that accrete solar composition and have high accretion rates near the Eddington limit so that ignition conditions are reached quickly with little time to burn hydrogen prior to the burst. Earlier one-zone model calculations using ignition conditions from fully one dimensional atmosphere models found that under such conditions the rp-process ends in a Sn-Sb-Te cycle that forms a natural endpoint of the rp-process in X-ray bursts under even the most favorable conditions [7]. Recent fully one dimensional burst models that follow an extended sequence of bursts in principle confirmed this result, but found that subsequent bursts have much less hydrogen to burn as some of it is captured on the ashes of preceeding bursts prior to ignition [8, 9]. As a consequence, the first burst after a long period of quiescence indeed reaches the Sn-Sb-Te cycle, but for later bursts in the sequence the rp-process ends in the $A = 64$ region instead. Recently [9] presented an extensive analysis of the reactions flows as a function of depth in a one dimensional X-ray burst model. One-zone models remain an important tool to study the role of nuclear physics in X-ray bursts due to their computationally less demanding nature, provided they are used with appropriate ignition conditions. Element diffusion effects that have not been included in full X-ray burst models so far could further affect the endpoint of the rp-process, especially at lower accretion rates [23].

The important nuclear physics for hydrogen burning in X-ray bursts are the β-decay rates, masses, and nuclear reaction rates of the neutron deficient rare isotopes along the path of the αp and rp-processes (see Fig. 1 in [3]). A special role play the so called waiting points - nuclei where the otherwise very rapid proton captures are slow or inefficient owing to inverse photodisintegration reactions on a weakly proton bound isotone. These nuclei impede the reaction flow, which has to continue via β-decay or slower α-induced reactions. Waiting points provide accumulation points for abundance and therefore set the final composition, and they delay energy production with possible consequences for burst light curves. The effective delay time at a waiting point depends on masses, decay rates, and capture rates in its vicinity (mostly the waiting point itself and the next two heavier isotones).

Most of the β-decay rates in X-ray bursts have been measured at rare isotope facilities in the past, with the exception of a few isotopes such as ^{96}Cd. However, in many cases error bars are still large, and the data rely on single measurements. Recently, the entire rp-process region up to ^{100}Sn came into reach at the National Superconducting Cyclotron Laboratory (NSCL) at Michigan State University with the commissioning of a new RF Separator . This device effectively provides velocity filtering and provides the clean neutron deficient beams needed for β-decay studies far from stability. In general, experiments can only measure the decay rate under terrestrial conditions. Continuum electron capture is usually not important as electron densities in X-ray bursts are too small. However, in some cases the temperatures during X-ray bursts ($kT \approx 100$ keV) are sufficient to thermally excite nuclei thereby modifying their decay properties [6]. This is particularly important for nuclei with low lying excited states. These corrections have to be applied based on theoretical predictions [24].

The importance of masses in rp-process calculations has been reviewed recently [25]. There has been dramatic progress in our understanding of nuclear masses of neutron deficient nuclei in the rp-process in recent years owing to a number for Penning trap facilities coming on line at various rare isotope facilities world wide. Mass measurements in the critical $N \approx Z$ $A = 64 - 72$ range, where the most important rp-process waiting points are located, performed at Argonne National Laboratory (CPT) [26], CERN (ISOLTRAP) [27], and Michigan State University (LEBIT) [28] have removed major uncertainties in X-ray burst calculations. Improved theoretical calculations of Coulomb shifts [29] have also played an important role and allowed one to take full advantage of the new measurements. With reliable Coulomb shifts, masses of nuclei beyond the $N = Z$ line that are still out of reach for mass measurements can be calculated with reasonable precision (typically 100 keV) from their precisely measured, more stable, mirror nuclei. Other new Penning Trap facilities such as JYFLTRAP [30] and SHIPTRAP [31] have pushed mass measurements of still heavier nuclei closer to the path of the rp-process. As a consequence, X-ray burst model calculations do not rely anymore on uncertain global mass models, but employ a mix of experimental masses, Coulomb shift calculations, and extrapolations. The major impact these ongoing experimental efforts had on X-ray burst model calculations is illustrated in Fig. 1. X-ray burst light curves for a given astrophysical model can now be calculated with much higher reliability. This opens the door for a detailed comparison with observations to test the physics in current X-ray burst

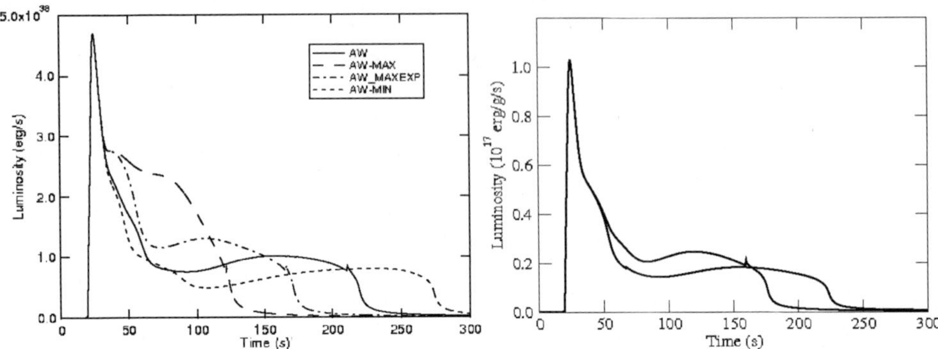

FIGURE 1. X-ray burst light curve model uncertainties due to mass uncertainties in the $A = 64 - 72$ region. Left panel after [29] for uncertainties as of 1995. Right panel with same model using the latest experimental precision masses and Coulomb shifts (see [25]).

models and to constrain system properties such as the accreted composition. The latter is important to determine the Eddington luminosity and therefore to constrain distance and, possibly in the future, the surface gravity and neutron star compactness. First steps in that direction have now been taken [38].

However, the impact of reaction rate uncertainties (and of masses outside the $A = 64 - 72$ region) is not taken into account in Fig. 1. The measurement of proton and α induced reaction rates on unstable nuclei remains a tremendous challenge for the field. Pioneering work has been done in radioactive beam physics, but beam intensities and the types of available beams fundamentally have been too limited to carry out the wide range of measurements needed to understand X-ray bursts and other explosive hydrogen burning scenarios (see [3] for a recent review). Most reaction rates in the rp-process are largely unconstrained experimentally with uncertainties of many orders of magnitude being common. Clearly next generation rare isotope facilities are needed to transition the field into "production mode", where many of the necessary measurements can finally be performed. Until then, there are a number of promising developments at existing facilities that will continue to provide cutting edge results. This includes the TRIUMF/ISAC facility, where a number of new beams of astrophysical importance are being developed. The NSCL is developing a low energy reaccelerator facility based on the gas stopping of fast beams produced by in-flight fragmentation. At this facilities, beams of a wide range of rare isotopes of importance for the rp-process will become available, including beams that cannot be produced at ISOL facilities due to their chemical properties.

In the mean time, indirect techniques are employed at many facilities to constrain rp-process reaction rates. An example are transfer reactions that provide important information on level energies (resonance energies) and ANCs or spectroscopic factors. Rare isotope beam facilities have played an important role in this area. Examples include the measurement of spectroscopic factors in ^{57}Ni, the mirror of the astrophysically important ^{57}Cu relevant for the ^{56}Ni(p,γ) reaction using a ^{56}Ni radioactive beam at ANL [32], or the program at the NSCL to use (p,d) transfer reactions far from stability to populate levels in critical nuclei along the path of the rp-process [33, 34]. Stable beam facilities also play an important role. Multi neutron transfer reactions such as (^{4}He,^{6}He) and (^{4}He,^{8}He) can be used to populate some rp-process nuclei using stable beams. In addition, a number of critical reactions in the CNO region are within reach of stable beam facilities. An example is the ^{15}O(α,γ) reaction, which has been shown to be critical in determining overall burst properties due to its role in regulating hydrogen burning via the CNO cycle prior to burst ignition [13, 14]. A number of experiments at KVI [35] and ANL [36] have been performed that populate the critical resonances and attempted to measure the small α-branching in ^{19}Ne that determines the reaction rate. These experiments have provided important upper limits. More recently the Notre Dame group succeeded with a one-sigma detection of the α-branching [37]. Taking the result at face value, the measurement excludes the very low reaction rate values that have been invoked to explain the puzzling observed burst behavior at relatively large accretion rates, where the burst frequency is drastically reduced compared to expectations [14, 4]. With the new nuclear physics results, alternative explanations for this behavior need to be found. This illustrates nicely the importance of improving the nuclear physics for driving progress in the theoretical understanding of X-ray bursts.

SUPERBURSTS AND CRUST PROCESSES

Superbursts are a new class of X-ray bursts that fundamentally differ from ordinary type I bursts [10]. Superbursts are about three orders of magnitude longer with typical durations of several hours, and therefore release three orders of magnitude more energy. They are extremely rare with an estimated recurrence time of the order of a year. They occur occasionally in systems that otherwise show normal X-ray bursts. So far a total of 13 superbursts has been discovered, in 8 sources. Superbursts match the spectral properties used to classify type I bursts and are therefore likely of thermonuclear origin. The leading model is the ignition of residual carbon in the ashes of the normal bursts that accumulate in the liquid ocean on the neutron star surface during the normal bursting activity [40, 39]. The three orders of magnitude longer recurrence time leads to the accumulation of three orders of magnitude more nuclear fuel and explains the energetics. The ignition layer after such a long accretion period is at a much greater depth compared to ordinary bursts, and the larger heat transport timescale from such a depth to the surface explains the long burst duration. Cumming et al. [41] modeled the heat transport and showed that observed superburst profiles can be used to constrain the ignition column depth y, which is related to the recurrence time t via $y = \dot{m}tg$ with surface gravity g and accretion rate per surface area $\dot{m}$. This is important as it provides a much more reliable handle on the recurrence time without the necessity of many years of continuous observations. Schatz et al. [42] had determined the likely ashes of such a carbon ignition and demonstrated that at peak temperatures around 7 GK nuclear statistical equilibrium is reached resulting in the synthesis of mainly iron group elements. Any nuclei heavier than iron present in the ocean from the rp-process in ordinary bursts are also destroyed and their conversion into iron group nuclei can provide additional energy for the superburst.

However, there are two major open questions related to superburst models based on carbon ignition. Most importantly, it is not understood where the carbon needed to fuel the superburst comes from. X-ray bursts have been shown to only produce very small amounts of carbon [42] and recent studies following a sequence of bursts have found that the little carbon that is made is completely destroyed by subsequent bursts that send heat waves down into the ashes where the carbon fuses with residual helium [8]. The key to the solution of this problem most likely lies in understanding the puzzling behavior of ordinary bursts at high accretion rates - exactly the regime where superbursts occur. One important piece of information is the fraction of energy that is released in bursts as opposed to the persistent X-ray flux during the accretion. Observations show that at high accretion rates this fraction becomes very small, indicating that a mix of stable and unstable burning is occurring. Indeed, stable burning of hydrogen and helium has been shown to produce the necessary amounts of carbon under certain conditions [42]. However, fully one-dimensional X-ray burst models do not yield any stable burning at those accretion rates. Therefore, a better understanding of ordinary X-ray bursts is needed to understand superbursts.

The second major problem is the observed recurrence time of the order of a year, which is about an order of magnitude shorter than what models predict. The problem is the temperature of the liquid ocean where ignition occurs. It is largely determined by the thermal profile in the crust of the neutron star, which in turn is set by electron capture and pycnonuclear fusion reactions that occur as matter is continuously compressed by the ongoing accretion. Electron capture reactions are induced step by step by the rising electron Fermi energy and convert the nuclei into more and more neutron rich species. Eventually so called pycnonuclear fusion reactions are induced as well. In such reactions, the Coulomb repulsion between two nuclei is overcome by their zero point motion in a lattice structure when the density is sufficiently high to bring two neighboring nuclei close enough together. As opposed to thermonuclear reactions, where the Coulomb barrier is overcome by temperature, pycnonuclear fusion reactions are induced by high density and would also occur at zero temperature. Gupta et al. [12] recently demonstrated that taking into account detailed information on the nuclear physics of electron capture reactions, in particular the excitation energy of the final states, leads to an increase in heating compared to previous models. This results in a hotter ocean, which alleviates the superburst problem somewhat but not entirely. Gupta et al. [12] also pointed out that the heat generation depends greatly on the initial composition, which determines whether the associated electron capture reaction sequence passes through regions of neutron rich nuclei characterized by shell- or subshell closures. Would superbursts produce ashes in the $A = 40$ region, the heating could be significantly increased. Recent superburst model calculations seem to indicate that this is a possibility [43].

The sensitivity of superbursts to the thermal profile of the neutron star crust makes them excellent observational probes of neutron star physics [44]. Superburst properties depend on the thermal crust conductivity and the neutrino cooling rate in the neutron star core, which can be an indicator of the presence of exotic phases. However, the nuclear physics of superbursts, X-ray bursts, and crust processes needs to be better understood. This requires the study not only of the neutron deficient rare isotopes in the rp-process, but also the extremely neutron rich nuclei involved in crust processes. These are essentially all nuclei between the proton- and the neutron drip line up to $A \approx 100$. Next generation rare isotope facilities will play a major role in this endeavor and provide a first opportunity to access most of the relevant nuclei for the first time.

ACKNOWLEDGMENTS

This work is supported by the Joint Institute for Nuclear Astrophysics under NSF-PFC grant PHY 02-16723 and by NSF grant PHY 06-06007.

REFERENCES

1. D. Psaltis, arXiv:astro-ph/0410536 (2004).
2. T.E. Strohmayer and L. Bildsten, Compact Stellar X-ray Sources, ed. W. H. G. Lewin and M. van der Klies (Cambridge: Cambridge Univ. Press), arXiv:astro-ph/0301544 (2003).
3. H. Schatz and K.E. Rehm, Nucl. Phys. A777 (2006) 601.
4. J.L. Fisker et al. Ap. J. 665 (2007) 637.
5. R. K. Wallace and S. E. Woosley, Ap. J. Suppl. 45 (1981) 389.
6. H. Schatz *et al.*, Phys. Rep. 294 (1998) 167.
7. H. Schatz *et al.*, Phys. Rev. Lett. 86 (2001) 3471.
8. S. E. Woosley et al., Ap. J. Suppl. 151 (2004) 75.
9. J.L. Fisker, H. Schatz, and F.-K. Thielemann, arXiv:astro-ph/0703311.
10. J. M. M. in't Zandt et al. astro-ph/0407087 (2004).
11. P. Haensel and J.L. Zdunik arXiv:astro-ph/0708.3996v1 (2007).
12. S. Gupta et al. Ap. J. 662 (2007) 1188.
13. J.L. Fisker, J. Görres, M. Wiescher, B. Davids Ap.J. 650 (2006) 332.
14. R.L. Cooper, R. Narayan, Ap. J. 648 (2006) L123.
15. A. Watts and I. Maurer, A&A 467 (2007) L33.
16. J.L. Fisker, F.-K. Thielemann, M. Wiescher, Ap. J. 608 (2004) L61.
17. O. Koike *et al.*, Astron. Astrophys. 342 (1999) 464.
18. for example A. L. Piro and L. Bildsten, Ap. J. 638 (2006) 968.
19. N.N. Weinberg, L. Bildsten, and H. Schatz, Ap. J. 639 (2006) 1018.
20. R. E. Rutledge et al. Ap. J. 577 (2002) 346.
21. E.M. Cackett et al. M.N.R.A.S. 372 (2006) 479.
22. A.M. Amthor et al. arXiv:astro-ph/0608451 (2006).
23. F. Peng, E.F. Brown, and J.W. Truran, Ap. J. 654 (2007) 1022.
24. J. Pruet and G. M. Fuller, Astrop. J. Suppl. 149 (2003) 189
25. H, Schatz arXiv:astro-ph/0607625 (2006).
26. J. A. Clark et al., Phys. Rev. Lett. 92 (2004) 192501
27. D. Rodriguez et al. Phys. Rev. Lett. 93 (2004) 161104
28. P. Schury et al. Phys. Rev. C 75, (2007) 055801.
29. B. A. Brown, et al. Phys. Rev. C65 (2002) 045802
30. A. Kankainen et al. Eur. Phys. J. A29 (2006) 271.
31. M. Block et al. Hyperfine Interactions 173 (2006) 133.
32. K. E. Rehm et al., Phys. Rev. Lett. 80 (1998) 676
33. R. R. C. Clement et al., Phys. Rev. Lett. 92 (2004) 172502
34. D. Galaviz et al. Proceedings of Science, NIC IX, P. 99.100/2006.
35. B. Davids et al. Phys. Rev. C 67 (2003) 065808.
36. K.E. Rehm et al. Phys. Rev. C 67 (2003) 065809.
37. W.P. Tan et al. Phys. Rev. Lett. 98 (2007) 242503.
38. A. Heger et al. arXiv:0711.1195 (2007).
39. T. E. Strohmayer and E. F. Brown, Ap. J. 566 (2002) 1042.
40. A. Cumming and L. Bildsten, Ap. J. 559 (2001) L127.
41. A. Cumming et al. Ap. J. 646 (2006) 429.
42. H. Schatz et al. Nucl. Phys. A 718 (2003) 247.
43. N.N. Weinberg and L. Bildsten, arXiv:astro-ph/0706.3062 (2007).

8. HIGH ENERGY COSMIC RAYS

Cosmic Rays and Cosmology

Tadeus Wibig[a] and Arnold Wolfendale[b]

[a]Physics Department, University of Lodz, Lodz, Poland
Wibig@zpk.u.lodz.pl
[b]Physics Department, Durham University, Durham, UK
a.w.wolfendale@durham.ac.uk

Abstract. There are two major aspects of the relationship of Cosmic Rays to Cosmology. The first is the problem of the foreground to the cosmic microwave background (CMB), caused by cosmic ray (CR) effects in the Galaxy, the CMB being 'seen' against this foreground. The second is the relationship of CR to various phenomena in the Cosmos as a whole, specifically, the question of magnetic fields in the Universe and the likely nature of the sources of CR of the highest energy. Both aspects will be considered here.

Keywords: Cosmic Rays, Cosmology, Energy Densities, Magnetic Fields
PACS: 96.50.S-, 98.80.-k

ENERGY DENSITIES

The energy densities of radiations and of particles have significance in that it is sometimes possible to throw light on 'origins' by their inter-relation.

We start with Energy Densities in the Galaxy. Figure 1 shows the situation 'locally', ie in the region of the solar system.

Energy densities in the Galaxy

	$eVcm^{-3}$
Magnetic field ($B^2/8\pi$)	$\simeq$ 0.5
Gas motion ($< \frac{1}{2} Mv^2 >$)	$\simeq$ 0.5
Cosmic Rays (p..)	$\simeq$ 0.5
Starlight	$\simeq$ 0.5
(Comic microwave background, 2.7K)	$\simeq$ 0.24

But, most $\neq$ f(R) and z

FIGURE 1.

This figure shows both the strength and the weakness of the common assumption of 'equipartition of energy'. If one is uncertain about the origin of Galactic CR then the

CP972, *Exotic Nuclei and Nuclear/Particle Astrophysics (II)—Proceedings of the Carpathian Summer School of Physics 2007,* edited by L. Trache and S. Stoica

equalities in the upper part of the Figure might be thought to offer a clue, viz that CR are Galactic. However, the possibility of chance coincidences must be borne in mind. For example, it is very hard indeed to imagine a reason for the local CR energy density and that of starlight being equal. However, the equality of magnetic fields and CR energy densities is easier to understand – perhaps the magnetic field acts a 'valve', ie CR will accumulate until the equality is met.

A problem arises with the CMB energy density, where there is also equality with the CR value to within a factor 2; a variety of tests show that most CR are Galactic and not Universal (eg 1,2). Nevertheless, we proceed by examining energy densities in Universal (ie Extra-galactic, 'EG') space in an effort to put EG CR in perspective. Figure 2 shows the situation.

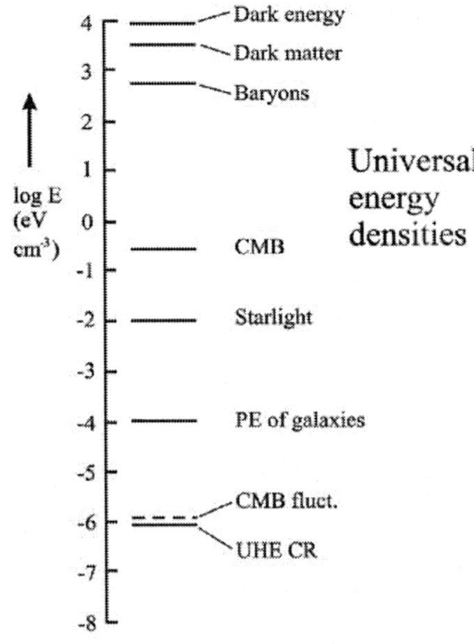

FIGURE 2.

The level marked UHECR is our own estimate based on the assumption that the ankle in the CR spectrum at ~5 x 10^{18} eV marks the transition from mainly Galactic (G-) to mainly Extragalactic (EG-) particles (3). Others (eg 4, and see 5) make the transition occur at a lower energy, perhaps by a factor 10. Thus 'our line' at 10^{-6} eV cm^{-3} could be too low by up to a factor 10. Nevertheless, it is seen that the UHECR energy density is low compared with the other values. For example, it is now a factor of 10^4 lower than that of starlight and, importantly, a factor 3.10^5 lower than that of the CMB. There is thus little guidance from Figure 2 as to the source of the EGCR although the fact that it is only a factor 10^2 below the 'PE of galaxies' may be significant; the argument is that only 1% of the energy released when galaxies form is needed to give the EGCR (ie UHECR-) spectrum.

Importantly, we note that the UHECR energy density is of the same order as that of the fluctuations in the CMB (**not** the CMB itself). At first sight this might be thought to imply that the CR could have an effect on the CMB and affect the interpretation of the observed fluctuations. However, the threshold for CMB gamma- CR interactions is at a high energy, even for very high red-shifts – and the energy content here is too low.

Notwithstanding the above remarks, there **can** be relevance of CR to CMB fluctuations and this is by way of CR-interactions (or something mimicking them) in the Galactic Halo. This is the topic of the next section.

FOREGROUND EFFECTS ON THE CMB RADIATION

A wealth of information of great relevance to Cosmology has come from the detailed analysis of the WMAP (6,7) generally regarded as a view of the Universe at the time when the atoms formed, some 350,000 years after the Big Bang. Although the map has been 'cleaned' for the likely effects of Galactic foreground (8), in our view, the cleaning is incomplete.

In a detailed study (9) we have analysed a number of aspects of the foreground problem. For a start, we have drawn attention to some remarkable symmetries and asymmetries which appear to implicate the Galaxy. For example there appear to be similarities in the Northern and Southern hemispheres (eg Figure 3).

CMB < T > vs Galactic Latitude

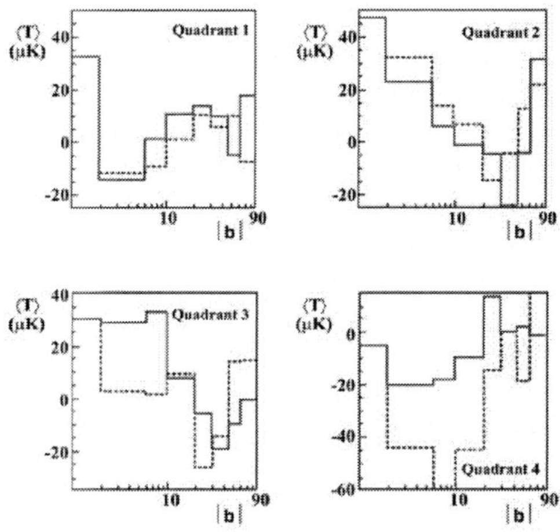

FIGURE 3.

Equally interesting is the frequent coincidence of low CMB temperatures with low CR intensities, and vice versa, the latter being as inferred by us from Gamma Ray Astronomy. Figure 4 gives an example.

CMB ΔT vs CR for

'Hot' (SNR) & 'Cold' (low < I_{CR} >).

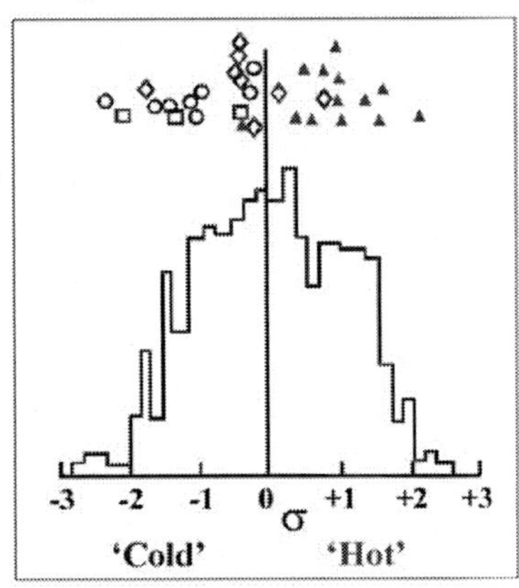

Histogram is N(ΔT_{CMB}) for whole sky

FIGURE 4.

The histogram refers to the distribution of the ΔT_{CMB} values for small areas over the sky and the symbols at the top refer to those areas where we expect high CR intensities (solid triangles) and low I_{CR} (open circles and squares). It is evident that most low I_{CR} regions are 'cold' in ΔT_{CMB} and vice versa.

Clearly, the CMB should not 'know' the nature of the CR intensity in the Galactic Halo. A possible mechanism relates to gentle CR heating of cold dust.

If true, the effect on the interpretation of the CMB properties would be serious. For example, we estimate that $\Omega_m h^2$ could be significantly different from the canonical 0.15; we prefer 0.25±0.05. It is possible that some of the ambiguities present when CMB data are compared with those from other analyses, could be explained in this way. We argue that polarization data are even less reliable from the standpoint of foreground complications.

THE ORIGIN OF EXTRAGALACTIC COSMIC RAYS

In the absence of evidence for an extended magnetic field in a Giant Galactic Halo – which could trap CR of the highest energies (10^{20} eV) – EGCR seem inevitable. The first question is "at what energy do they predominate over CR of Galactic origin?" It would appear natural to identify the transition with the well known ankle in the energy spectrum (eg 3). However, there are dissenters (eg 3) who argue that the particles above 10^{18} eV are mainly EG protons which give an energy spectrum conditioned by losses on the CMB, the ankle being due to the shape of the e^+e^- interaction loss component. Figure 5 shows the rate of energy loss of protons on the contemporary CMB from a variety of authors (see 10 for details).

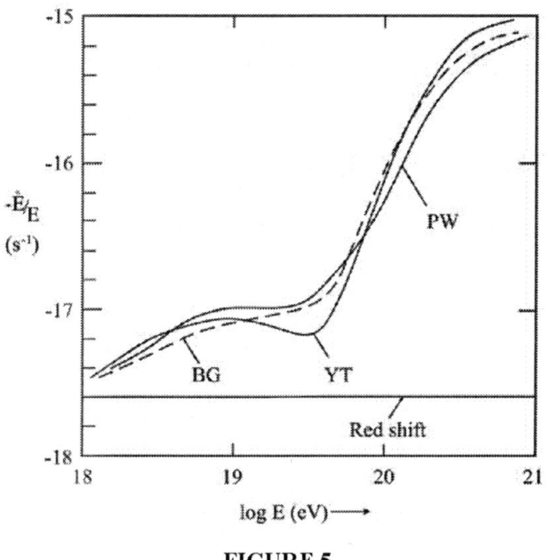

FIGURE 5.

An important matter that now arises concerns the mass composition of the primaries. As Figure 6 shows, only if there is less than 10% 'contamination' by heavier particles will there be a pronounced ankle.

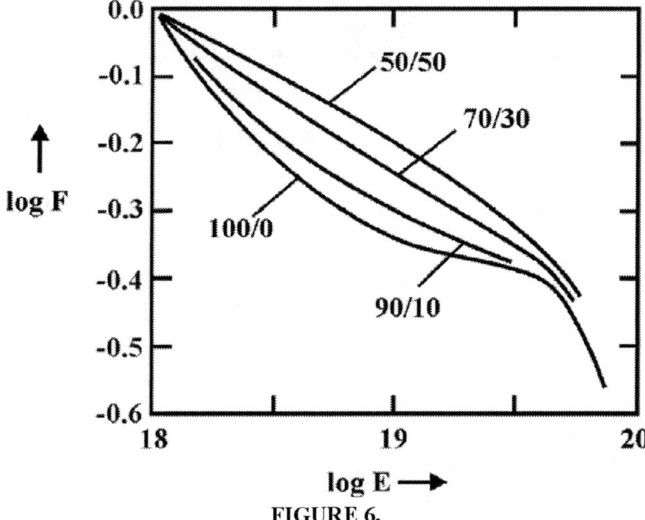

FIGURE 6.

Although it is true that there is some doubt about the mass composition above about 10^{17} eV, largely due to the inevitable uncertainty in the nuclear physics of the CR-air interactions, our survey yielded a mean mass at 10^{19} eV corresponding to ~30% iron. Very recent work with the AUGER observatory appears to confirm this view.

The question now is "what are the actual sources of the EG particles?" There are many possibilities. A promising one some years ago involved colliding galaxies (11) but this has not (yet) been substantiated. We have examined a number of possibilities (5) but without finding compelling evidence for any one. Active galactic nuclei (AGN) or Quasars are 'front-runners' but as yet there is no proof-positive. One difficulty with all the models examined so far is the need for a very high efficiency for producing very high energy particles; essentially, the lack of observation, so far at least, of obvious anisotropies in the arrival directions of the very energetic particles. This means that there is a shortage of 'nearby' particles and, in turn, low intensities of those from further away.

Figure 7 shows one possibility (from 6). The points are from our World Survey.

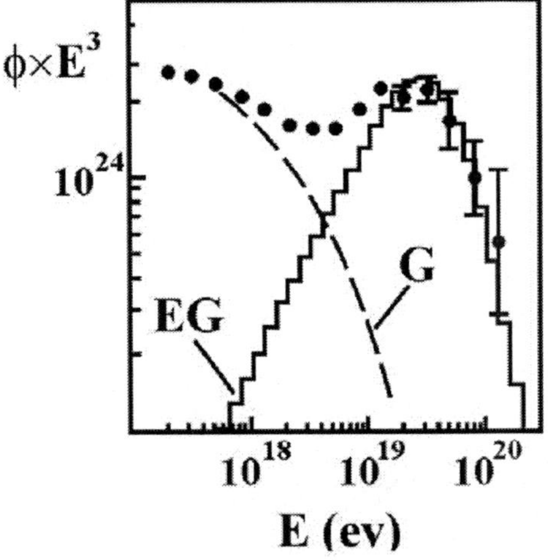

FIGURE 7. Results from [5] for the quasar origin model. G denotes Galactic and EG extragalactic.

THE WAY AHEAD FOR EGCR SEARCHES

There is the hope that the AUGER experiment will provide clues as to origin from observations of anisotropies, and that the mass determination will firm up.

The energy spectrum itself might give further clues as to the nature of the sources from an accurate knowledge of its shape. The hope here is that the effect of other radiation fields such as infra-red might appear. We (13) have drawn attention to the effect of IR in galaxy clusters, where micro-ganss magnetic fields can cause trapping of UHECR accelerated in, for example, AGN in the clusters. Insofar as different clusters have different magnetic fields and IR radiation densities the resulting spectral shape of the spectrum at earth might not be simple, however.

A further possibility, with its attendant complexity, is that the sources are intermittent (e.g. AGN). Thus, because of finite transit times, observed energetic galaxies (ie those observed optically) might not always give detected particles. A further, interesting, possibility is that if, as is possible, our own Galaxy went through an AGN phase, some of the lower energy particles may still be trapped in the Galaxy.

It may be that the cosmos will protect its secret as to UHECR origin somewhat longer.

REFERENCES

1. D.Dodds, A.W.Strong and A.W.Wolfendale, Mon. Not. R. Astr. Soc., 171, 569 (1975).
2. P. Sreckumar et al., Astrophys. J.. 400, L67 (1992).
3. T. Wibig and A.W.Wolfendale, J. Phys. G., 31, 255 (2005).
4. V.S.Berezinsky et al., Astropart. Phys. 21, 617 (2004)
5. T. Wibig and A.W.Wolfendale, J. Phys. G. 34, 1891 (2007)
6. M. Tegmark, A. de Oliveria-Costa and A. Hamilton, Phys. Rev. D., 68, 123523 (2003)
7. C.L.Bennett et al., Ap. J. 583, 1 (2003)
8. C.L. Bennett et al., Ap. J. 583, 148 (2003)
9. T. Wibig and A.W.Wolfendale, Mon. Not.R. Astron. Soc., 360, 236 (2005)
10. T. Wibig and A.W.Wolfendale, 29th Int. Cosmic Ray Conf., Pune, 10, 64 (2006)
11. S.S. Al-Dargazelli et al., J. Phys. G., 22, 1825 (1996)

Cosmic Rays and Global Warming

T. Sloan[a] and A.W.Wolfendale[b]

[a]*Physics Department, University of Lancaster, Lancaster, UK*
[b]*Physics Department, Durham University, Durham, UK*
e-mail: t.sloan@lancaster.ac.uk

Abstract. Some workers have claimed that the observed temporal correlations of (low level) terrestrial cloud cover with the cosmic ray intensity changes, due to solar modulation, are causal. The possibility arises, therefore, of a connection between cosmic rays and Global Warming. If true, the implications would be very great. We have examined this claim in some detail. So far, we have not found any evidence in support and so our conclusions are to doubt it. From the absence of corroborative evidence we estimate that less than 15% at the 95% confidence level, of the 11-year cycle warming variations are due to cosmic rays and less than 2% of the warming over the last 43 years is due to this cause. The origin of the correlation itself is probably the cycle of solar irradiance although there is, as yet, no certainty.

Keywords: Cosmic Rays. Global Warming. Sunspots
PACS: 96.50.S-, 94.20.wq, 98.70.Sa

INTRODUCTION

There are strong politico-social implications in the apparent correlation of cosmic ray (CR) intensity changes with changes in low level cloud cover (CC) – and thereby with mean global temperature [1,2] and it is imperative that the effect be studied. Insofar as there is a possible link of clouds with CR via ionization the correlation cannot be dismissed out of hand. It is not sufficient to say that the energy content of CR is minute in comparison with solar irradiance (SI) and therefore the effect must not be causal; the atmosphere is a highly complex system with subtle properties and unusual sensitivity to small changes is not uncommon.

The Correlation between Cosmic Rays and Cloud Cover

In reference [1] a correlation was demonstrated for 'low clouds' (<3.2 km in altitude) between the changes in the 'CC anomaly' and the changes in the CR count rate (see figure 1 of [1]). Here, 'CC anomaly' means the difference between the mean monthly CC and the time averaged value for the month. A similar effect was also observed in [3] where cloud cover was determined in a completely different way from

CP972, *Exotic Nuclei and Nuclear/Particle Astrophysics (II)—Proceedings of the Carpathian Summer School of Physics 2007,* edited by L. Trache and S. Stoica

that adopted in [1,2]. The CC anomaly in [1] was derived from the ISCCP D2 analysis using the infrared data [4]. It was implied in [1] that a decrease in CR intensity causes a decrease in CC. Since this may not be the case if both effects are correlated to a third variable, it is prudent to look for further evidence of such a causal connection. Such a causal connection would have great significance since, according to [1], it could be the main cause of the presently observed global warming. The mechanism for this depends on the observation of an increase in solar activity over the last century [5]. An increase in solar activity causes a net decrease in CR intensity which, according to the causal connection in [1], causes a decrease in CC. This, in turn, leads to increased solar warming.

The causal connection implied in [1] has been questioned [6] on the grounds that the ISCCP infra red data give different results from the day time low cloud data. In addition the poor quality of the correlation after 1993 is noted (although solar cycles vary). Similar questioning of the conclusions of [1] is reported in [7]. On the other hand, it was shown in [1] that the low IR could top temperature is also strongly correlated with CR, particularly for latitudes below about 45°. Here, we adopt the approach of looking for other manifestations of the causal connection, assuming that it exists, in order to corroborate the effect or otherwise.

The implication of the causal connection proposed in [1] is that CC is strongly influenced by ionizing radiation. Indeed the same group has shown that ionization from radioactive sources can cause aerosol production in a laboratory environment [8]. In our paper, we have examined various incidences of ionizing radiation in the atmosphere to look for consequential changes in CC which would result if the causal connection existed. Firstly, we have examined the correlation between CC and CR intensity as the geomagnetic latitude increases i.e. as the vertical rigidity cut-off (VRCO) decreases allowing the CR intensity to increase. Secondly, we have examined the CC in the vicinity of Chernobyl at the time of the nuclear accident in 1986 to look for an increase after the accident. Thirdly, we have looked at known areas of high atmospheric radon contamination. Fourthly, we have examined the difference in CC over sea where ionization due to radioactivity is small and over land where it is much larger. Fifthly, we have examined the historic records of atmospheric nuclear bomb tests. In all cases we search for evidence favoring ionization leading directly to cloud production.

Latitude Dependence of the 'Effect'

It is well known that the magnitude of the CR time variation, due to the 11 year solar cycle, varies with latitude. More accurately, it is a function of the VRCO, the reason being that the geomagnetic field deflects away more low energy particles as the geomagnetic equator (highest VRCO) is approached. Since this variation falls with increasing primary CR energy, the solar modulation is most severe in the polar regions where lower energy CRs reach the Earth. Hence, if the causal connection proposed in [1] exists, one would expect larger changes in CC, due to solar modulation, in the polar regions than at the Equator. Furthermore it is known that there is a delay of 6-14 months between the decrease in the CR intensity and the increase in the sun spot (SS) number with the even numbered solar cycles showing smaller delays than the odd numbered [9]. Note that the CR count rate is anti-correlated to the SS number.

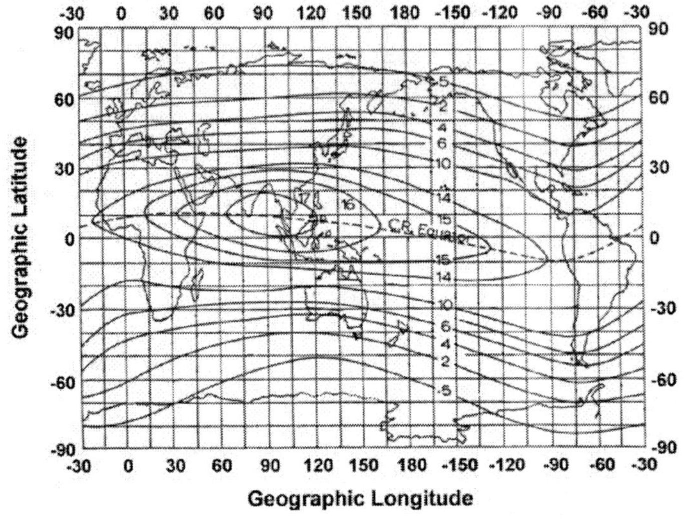

FIGURE 1. Contours of Cosmic Ray Threshold Rigidity, VRCO (the units are GV).

We have studied this effect in some detail using the same data sample as in [1]. Figure 2 shows the CC anomaly for different VRCO bands. The smooth curves in figure 2 show the best fit of the CC anomaly to the mean daily SS number (inverted) with a linearly changing background (the origin of this background trend is not clear). The fit was made using the CERN library fitting programme MINUIT [10] to minimize the value of χ^2 between the measurements and the curve. The errors on the data points were taken from the mean square deviations of independent pairs of neighboring points. The free parameters in the fit were the slope and intercept of the linear background, a multiplicative constant for the monthly averaged daily American sunspot (SS) number [11] and a time shift for the delay between the onset of the *dip* in the CC and the increase in the SS number. The multiplicative constant represents the amplitude of the dip in the CC per unit change in SS number. We take this amplitude as the magnitude of 'the effect'. The fits were rather poor (see figure 2) with values of χ^2 per degree of freedom of from 1.5 to 2.5. However, the fits between 1985 to 1995 were better than this, allowing the amplitude of the dip to be determined in this time range.

The observed dip in figure 2 is similar to that seen in [1] between the years 1985 and 1995. However, the rise in amplitude of this dip with decreasing VRCO expected if the CR hypothesis is valid, is not seen. Furthermore, the dip in CC seen in solar cycle 22 (peaking in 1990) is not evident in solar cycle 23 (peaking in 2000) except in the equatorial region (high VRCO) where the solar modulation is least.

To investigate the effect further and to check that the above result was not due to a latitude dependent 'efficiency' of the cloud production mechanism, the CC was determined in three strips of latitude and the amplitude of the dip in solar cycle 22 was measured from the fit for each, as a function of VRCO. Figure 3 (upper panel) shows that the amplitude of the dip appears to be rather constant with VRCO rather than changing in a manner proportional to the observed CR modulation [12]. The measured value of the delay between the onset of the dip and the change in SS number fluctuates randomly rather than concentrates around a fixed delay (expected to be – 7 months for the CR increase in solar cycle 22). Each latitude band has a median value compatible with zero with an overall mean of -0.9 ± 1.6 months, where the error is the standard error determined from the RMS deviation of the measurements from the mean. This is compatible with the onset of the increase in SS number but incompatible with the arrival time of the CR increase (-7 months). Hence there is an imperfect time correlation between the start of the dip and the change in the CR rate.

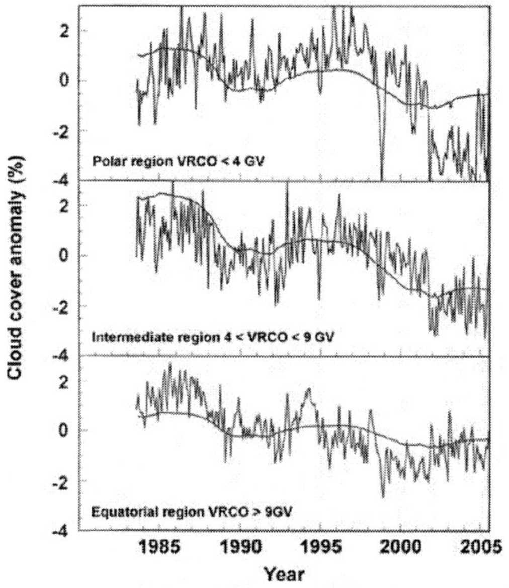

FIGURE 2. Cloud cover anomaly vs year for various bands of vertical rigidity cut off (VRCO)

Neither the amplitude variation with VRCO nor the arrival times shown in figure 3 corroborate the claim of a full causal connection between CR and CC. However, there may be a partial correlation and to identify the part of the distribution in the upper panel of figure 3 which correlates with the CR modulation, a fit was performed of the shape of the neutron modulation curve (the correlated part) and a constant term (the uncorrelated part0 to the measurements. The fit gave the fraction of the distribution correlated with the neutron modulation curve to be 0.012 ± 0.086 i.e. compatible with zero with a value of $\chi^2 = 17.3$ for 16 degrees of freedom. From this it is deduced that less than 15% of the distribution, at the 95% confidence level, belongs to the part correlated with the CR modulation.

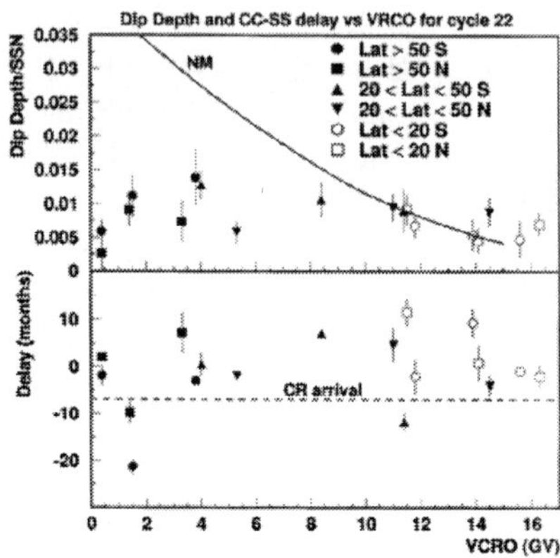

FIGURE 3. Magnitude of the dip (per unit SSN) and of the delay versus VCRO

Sporadic Changes

Rapid changes in the CR intensity occur due to solar flares and related Forbush decreases and the rate of such events increases with solar activity [13]. These events have a rather small energy content and usually last for less than one month. Nevertheless, they give the opportunity of a test of CC – CR correlations. A survey has been given in [14]. If the causal connection proposed in [1] exists then such events would cause fluctuations in CC, the rate of which would increase at times of higher solar activity. We have examined the CC data to search for an increase in fluctuations with mean daily SS number. The fluctuations were measured from the differences of independent pairs of monthly readings and a linear regression line fitted to the variation of these differences with mean daily SS number. The mean fluctuation in global CC in cycle 22 was found to be 0.60%. The slope of the linear regression line was $-6.5 \pm 5.0 \cdot 10^{-4}$% fluctuation per unit SS number i.e. compatible with zero, indicating that the fluctuations change by less than 0.13% over cycle 22 (peak mean daily SS number of 163), at the 95% confidence level – a result in accordance with other recent work [15,16]. This fails to corroborate the causal connection proposed in [1]. However, it is difficult to derive a quantitative upper limit from the observation.

The Chernobyl Nuclear Accident

We now turn to the Chernobyl nuclear reactor accident to see if the released ionization produced an increase in CC.

On the 26 April 1986 there was a nuclear reactor accident at Chernobyl (51.4° N 30.1° E) which released 1.4 10^{19} Bq of radioactivity into the atmosphere during a period of 10 days following the accident [17]. A correlation between CC and ionization from the radioactivity would be expected to produce an increase in CC in the vicinity of Chernobyl following the accident if the causal connection proposed in [1] existed. Figure 4 shows the CC anomaly as a function of time for various regions in the vicinity of Chernobyl. There is no evident increase in the CC following the accident. From the data in figure 4 upper limits at the 95% confidence level on the increase in the CC in the month following the accident are 9.3% (3.6%) in the region of Chernobyl; 50-52.5N, 27.5-32.5E (in the fallout region 45-60 N, 20-35 E). From the published contours of radiation near ground level [18] and an assumed mean height of 0.5 km, the average ionization level for the lowest 3.2 km corresponds to ~ 15 (~ 3) times the CR level in the region of Chernobyl (in the fallout region).

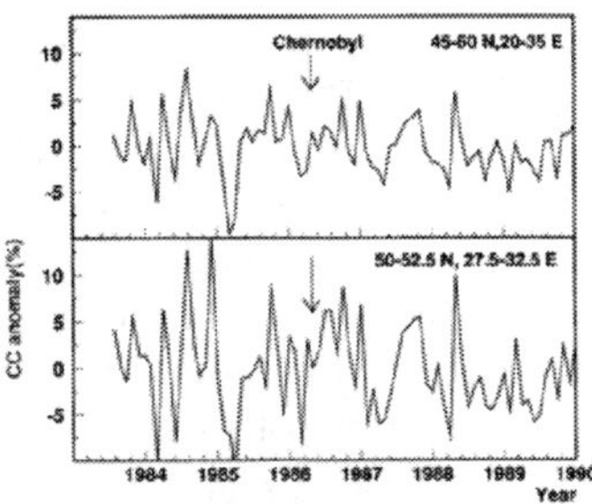

FIGURE 4. CC anomaly versus year of relevance to the Chernobyl Nuclear Accident.

We now proceed to estimate the expected rise in the CC due to the Chernobyl accident if the causal connection from [1] exists. The globally averaged decrease in CC during solar cycle 22 (averaging the dips in figure 4) is 1.3%. The globally averaged modulation in the CR neutron monitor count rate is ~ 10% peak to peak in solar cycle 22 [12]. The muon modulation is about a factor 3 lower than this [19] due to the higher primary energy needed to produce muons. Ionization is also produced from the electromagnetic component of CR whose long term modulation has not been measured. This will depend on π^0 production from CR primary interactions which will have a threshold energy intermediate between those for muons and neutrons. We therefore assume that the total globally averaged solar modulation of the ionization due to cosmic rays is the average of those for muons and neutrons i.e. 7 ± 4%. We estimate that there is a total of 4 ion pairs per second produced by CR in the atmosphere at the altitude of the low cloud cover (<3.2 km). If the causal connection proposed in [1] exists, the rate of change of CC per ion pair per cm^3 per second is then $dCC/dI = 4.6 \pm 2.6\%$ (ion pair cm^{-3} sec^{-1})$^{-1}$.

There is some uncertainty about the time that the ionization from the Chernobyl accident lingered in the atmosphere. However, it must have been a significant fraction of 1 month. Hence it can be seen that, if the causal connection proposed in [1] exists, the CC should have increased to close to 100% in the region of Chernobyl and to several tens of per cent in the fallout region following the accident. Even within the imprecision of these estimated values it can be seen that there is no evidence from the Chernobyl accident to corroborate the causal connection proposed in [1].

Furthermore, from this observation the true value of dCC/dI must be an order of magnitude smaller than the value necessary for the solar modulation to cause the dip in CC observed in [1], in qualitative agreement with the upper limit deduced in section 3.

Radon in India

Radon is an important contributor to the background against which the cosmic radiation is studied and its variation over the earth's surface has been studied in some detail. The average outdoor ionization over the earth is quoted as several times that from cosmic rays [20] and because of the decrease of radon with height (radioactive gases influence atmospheric ionization up to altitudes as high as 2-3 km [21]) and the increase for cosmic rays, they will be of the same order for the 'low cloud region' (mean height ~ 1.5 km).

There are, in India (and no doubt elsewhere), 'hot spots' [22] which are potentially useful for a check on the effect of ionization on CC. Here we use data for two regions: region 1 is centered on 22.4° N 84.5° E and region 2 on 22.2° N 78.9° E. There is a third region at 11.5° N 77.2° E which is unusable due to the effect of the sea which covers about half the 2.5° square containing this position in the ISCCP D2 data. In these 3 regions the indoor radon concentration has been measured to be 50-60 Bq m^{-3}. This is reduced by about a factor 2 outdoors and by a further factor of two at cloud forming altitude [22]. Radon (Rn^{222} from uranium decay) decays by the emission of 5.5 MeV α particles producing roughly 10^5 ion pairs per decay. Hence, assuming that the radon is distributed uniformly throughout the 2.5° square of the ISCCP D2 data, this leads to an ion pair production rate from radon of roughly 1 cm^{-3} sec^{-1}.

If the causal connection proposed in [1] exists with the value of dCC/dI deduced above, the CC would be ~ 2% higher where the radon level is high compared to neighboring regions in which the level is lower by more than a factor of 2. The time averaged excess CC in the squares covering regions 1 and 2 over their neighbors with lower radon levels is measured to be $-0.35\pm0.20\%$ and $0.51\pm0.49\%$ respectively, i.e. each compatible with zero. These imply that the rate of change of CC with ionization change is again lower than the implication of the causal connection proposed in [1]. This result provides supplementary rather than conclusive evidence since many assumptions have been made in deriving it.

Comparison of the CC over Land and Sea

The globally averaged CC was found from the data to be 24.6%. Taking dCC/dI = 4.6 ± 2.6% (ion pair cm^{-3} sec^{-1})$^{-1}$ from the full causal connection proposed in [1] and the rate of production of ions from CR to be 4 ion pair cm^{-3} sec^{-1}, the total CC expected from ionization from CR is 18 ± 10%. This is similar to the globally averaged value, so that if the causal connection exists the majority of the observed CC must be caused by ionization. A substantial part of the ionization which exists over the land is due to radioactive decay of naturally occurring isotopes from ground materials such as uranium (see section 6). Such radioactivity is 1-2 orders of magnitude less over the ocean [23]. Hence, one would expect to see greater CC over land than over the sea if the connection exists.

We proceed to test this hypothesis. In the equatorial region (between latitudes 15° S to 15° N) where we observe a dip in the solar cycle 23 (see section 3), we observe that the mean CC is 12.6% over Africa and 9.9% over South America. In the remainder of this equatorial band, which is predominantly over the ocean, the mean CC is 27.0%. A similar effect was observed in a scan across the Earth between latitudes 45 to 50 °N which includes the region of high atmospheric radioactivity at Innsbruck noted in [23]. The average CC over the ocean was 38% and a factor 2 less over land. There was no abnormal change in the 2.5 degree square containing Innsbruck compared to its neighboring regions.

The change in CC from sea to land in each case is opposite to the expectation from the causal connection proposed in [1] and so this evidence, too, fails to corroborate the connection.

Atmospheric Testing of Nuclear Weapons

There is a wealth of published information on the effect of testing nuclear weapons in the atmosphere [24]. Understandably, one aspect studied was 'the possible influence of nuclear weapons on the weather'. Relevant conclusions from early studies were that 'weapon debris is not effective as a cloud seeding agent' and '...apart from localized effects in the vicinity of the test area, there has been no known influence of nuclear explosions on the weather' [24]. The results for the BRAVO test on 1 March 1954 can be taken as an example. Here a 15 Mt bomb was exploded on the surface and the ensuing fall-out pattern determined. Remote from the immediate local effects of the explosion, specifically at 500 km from "Ground Zero", the radiation level was 20 Rh^{-1} some 16 hours after the explosion. This corresponds to an ionization rate of $\sim 10^7$ ions cm^{-3} sec^{-1} clearly many orders of magnitude greater than the CR level. Even after allowing for a possibly restricted height distribution and radioactive decay the cloud cover should have been complete if the causal connection proposed in [1] existed. There are no reports of such a phenomenon in a carefully recorded test. There is therefore again evidence which fails to corroborate the causal connection proposed in [1].

Conclusions

The dip in amplitude of 1.3% of the low altitude cloud cover noted in reference [1] in solar cycle 22 (peaking in 1990) has also been seen in this analysis. This dip anti - correlates in shape with the observed mean daily sun spot number i.e. correlates with the change in cosmic ray intensity due to solar modulation. A number of attempts are described here to find evidence to corroborate the causal connection between the dip and changes in ionization levels due to changes in the cosmic intensity suggested in [1]. The depth of the dip in solar cycle 22 does not fall as it should with increased vertical cut off rigidity. Nor is the onset of the dip well correlated with the arrival time of the increase in the cosmic ray rate. The dip in the following solar cycle, cycle 23, is only evident in the equatorial regions of the Earth.

We have looked for further evidence linking ionization to cloud formation but have found none. The atmospheric ionization produced by the nuclear accident at Chernobyl gave no observable increase in the mean cloud cover following the accident. The observed regions of high radon concentration in India do not produce observable increases in cloud cover over their neighbors with lower radon levels. No evidence could be found of higher cloud cover over land, where the ionization levels should be higher due to radioactivity, compared to that over the sea where radioactivity should be lower. Nuclear weapons testing in the atmosphere does not produce cloud cover, remotely from the explosion site, which correlates with high ionization levels.

In summary, no corroboration of the claim of a causal connection between the changes in ionization and cloud cover was found in this investigation. From the distribution of the depth of the dip in solar cycle 22 with geomagnetic latitude (the VRCO) we find that, averaged over the whole Earth, less than 15% of the cloud cover change comes from the modulation of the cosmic ray intensity, at the 95% confidence level. In reference [1] it is estimated that if all the cloud cover change in solar cycle 22 were due to cosmic ray modulation then the radiative forcing produced by long term changes in the cosmic ray intensity would be 1.4 Wm^{-2} i.e. the major portion of that necessary to produce the global warming we observe. The upper limit on the effect of cosmic ray forcing of the cloud cover determined here implies that the radiative forcing due to changes in cosmic ray intensity is less than 0.2 Wm^{-2} at the 95% confidence level.

Our results can be used to set a limit on the global temperature rise due to the decrease in cosmic ray intensity of 3.7% since 1964 [5]. The solar modulation of the average global temperature is observed to be 0.1° C peak to peak[25]. If all of this is attributed to the modulation of the ionization from cosmic rays of 7% then the global average temperature change from change in cosmic ray intensity would be 0.06°C. However, our upper limit constrains the temperature rise to be less than 15% of this figure. Hence the mean global temperature rise in the last 43 years due to the change in cosmic ray intensity must be less than 0.008°C i.e. less than 2% of the observed rise.

The lack of a role for cosmic rays in global warming does not imply that the correlation of CR with CC has no interest; something must be responsible for both. The 11-year cycle of solar irradiance [25], by way of a coupling between the upper stratosphere and lower troposphere, [26], is one possibility. Another is more direct – the undoubted role of CR in affecting the electrical conditions in the atmosphere, e.g. [25]. We are impressed by the link between CR and lightning [27] and we ourselves find a stronger dip over land (where lightning is common) in equatorial regions than over the sea (where it is not). Yet another possibility, that relates CR, the stratosphere and coupling to the lower troposphere, is the fact that [14], at a height of 50 km the 'electron production ratio' at middle latitudes varies by a factor 10 over longitude (almost sinusoidally). Such a variation coupled to the troposphere could have significant effects.

REFERENCES

1. N. Marsh and H. Svensmark, 'Low cloud properties influenced by cosmic rays', Phys.Rev. Lett., **85**, 5004 (2000)
2. H. Svensmark, 'Cosmoclimatology:a new theory emerges', New and Reviews in Astronomy and Geophysics, **48**, 18. (2007)
3. R.G. Harrison and D.B. Stephenon, 'Empirical evidence for a nonlinear effect of galactic cosmic rays on clouds', Proc. Roy. Soc. A (doi: 1098/rspa.2005.1628).
4. available at http://isccp.giss.nasa.gov/
5. M.R. Lockwood, R. Stamper and M.N.Wild, 'A doubling of the sun's coronal magnetic field during the past 100 years', Nature, **399**, (1999) 437.
6. J.E. Kristjansson, A. Staple, J. Kristiansen, and E. Kaas, 'A new look at possible connections between solar activity, clouds and climate', Geophysical Research letters, **29**, (2002) 2107.
7. P.M. Udelhofen, and R.D. Cess, 'Cloud cover variations over the United States: An influence of cosmic rays or solar variability?', Geophysical Research Letters, **28**, (2001), 2617.
8. H. Svensmark, J.O.P. Pederson, N.D. Marsh, M.B. Enghoff and U.I. Uggerhoj, 'Experimental evidence for the role of ions in particle nucleation under atmospheric conditions', Proc. Roy. Soc., **A463**, (2006) 385.
9. We are grateful to K. Kudela for the measurement of the CR-SS delay.
10. F. James, CERN Program Library Long Writeup D 506, MINUIT Function Minimization and Error Analysis (available at http://wwwasdoc.web.cern.ch/wwwasdoc.miniut.minmain.html).
11. American Sun Spot numbers, available at http://www.ngdc.noaa.gov/stp/SOLAR/ftpsunspotnumber.html.
12. I. Braun, J. Engler, J.R. Horandel, and J. Milke, 'Solar modulation of cosmic rays in the energy range from 10 to 20 GeV', proceedings of the 29th International Cosmic Ray Conference Pune 2005 **2** 135-138.
13. L.I.Dorman, 'Cosmic Rays:Variations and Space Explorations', (published by North Holland, 1974).
14. P.I.Y. Velinov, L. Mateev and N. Kilifarska, Annales Geophysicae, **23**, 3043 (2005).
15. I.G. Usokin and G. A. Kovaltsov, to be published (2007).
16. J. Calagovic, L. Desorgher, J. Beer, E.O. Fluckiger, and F. Arnold, private communication (2007).
17. Chernobyl Legacy:Health, Enviornment and Socio-Economic Impacts and Recommendations to the Governments of Belarus, the Russian Federation and Ukraine by the Chernobyl Forum:2003-2005.
18. 'The Chernobyl Accident' by UNSCEAR (the United Nations Scientific Community on the Effects of Atomic Radiation) reports published in 1988 and 2000.
19. O.C.Allkofer and H. Jokisch, 'A survey on recent measurements of the absolute vertical cosmic ray muon flux at sea level', Il Nuovo Cimento, **15A**, (1973), 351.
20. J. A. Chalmers, 'Atmospheric Electricity', Clarendon Press (1949).
21. V. Sasikumar, et al., 'atmospheric electrical conductivity variations over different environments', Geophys. J. Int., **122**, 89 (1995).
22. N. Karunakara, et al., 'Indoor and outdoor radon levels and their diurnal variations in the environs of southwest coast of India', Intl. Congress Series, **1276**, 341, (2005).
23. C. E. Junge, 'air chemistry and Radioactivity', (published by Academic Press, new York and London, 1963).
24. S. Glasstone, 'The Effects of Nuclear Weapons', US Energy Commission (1962).
25. J. Lean, G. Rottman, J. Harder and G. Kopp, 'SORCE contributions to new understanding of global change and solar variability', Solar Physics, (2005), **230**, 27-53.
26. J. D. Haigh, 'The role of stratospheric ozone in modulating the solar radiative forcing of climate', Nature, **370**, 544, (1994).
27. V. V. Alexeenko, et al., 'Electric field disturbances in the summer-time atmosphere and associated variations of CR intensity', Proceedings of the 27th international Cosmic Ray Conference (2001), 4161-4164.

The Muon Charge Ratio In Cosmic Ray Air Showers

Heinigerd Rebel[a] and Octavian Sima[b]

[a]*Institut für Kernphysik, Forschungszentrum Karlsruhe, Germany*

[b]*Department of Physics, University of Bucharest, Romania*

Abstract: The muon charge ratio of the lateral muon density distributions in single Extended Air Showers (EAS) is studied on basis of Monte Carlo simulations, in view of proposals to measure this observable in coincidence with EAS observations. Differences of the azimuthal variation of the muon densities of opposite charges and the azimuthal variation of the muon charge ratio appear to be very much pronounced, dependent on the direction of the EAS incidence and the position of the observer in respect to the Earth's magnetic vector. The influence of the geomagnetic field, which induces comparable effects in radio emission from EAS, is obviously of great interest for understanding the shower development.

Keywords: Cosmic rays, extensive air showers, muon charge ratio
PACS: 96.50.s- ; 96.50.sd

INTRODUCTION

The flux, the energy spectrum and the charge ratio of atmospheric muons, resulting from the decay pions and kaons which are produced in collisions of primary cosmic rays with the air molecules, have been studied extensively over wide range of energies at sea level as well as on high mountain altitudes and during balloon ascents. The studies comprise several aspects, in particular, they provide information on the composition of cosmic rays and on characteristic features of the hadronic interaction and of the influence of the geomagnetic field on the primary and secondary components of cosmic radiation. The investigations show that the ratio of the number of positive to negative atmospheric muons $R_\mu^{atm} = N_\mu^+/N_\mu^-$ is experimentally varying between the values of about 1.2 -1.3 [1]. As detailed analyses [2] show, the excess of the positive charge reflects mainly the excess of protons of primary cosmic rays and less eventual small differences of the production of positive and negative

CP972, *Exotic Nuclei and Nuclear/Particle Astrophysics (II)—Proceedings of the Carpathian Summer School of Physics 2007,* edited by L. Trache and S. Stoica

parent particles of the muons. The Earth's magnetic field influences the flux of primary cosmic rays entering the Earth atmosphere leading to the well-known east-west asymmetry and the latitude effect. In addition the geomagnetic field bends the trajectories of the charged particles of the secondary cosmic rays which get curved. The effect is particularly dominant for the muon component which is less affected by Coulomb scatterings processes. This leads to the east-west effect of the muon charge ratio: the observation that the ratio of positive and negative atmospheric muons proves to be different for muons arriving the spectroscopic device from west from those arriving from east direction (see [3]). The finding originates from the fact that in the east-west plane due to different bending, the muons of positive and negative charges have different path lengths from the locus of production to the observer. Hence the decay probability of low energy muons and the attenuation are differently modified.

The features are slightly different for the muon charge ratio in Extensive Air Showers (EAS) whose axis have well defined angles (θ,Φ) of incidence, not only relative to the zenith direction, but also relative to the vector of the magnetic field of the Earth. The total charge ratio integrated over all muons of the EAS is only determined by the hadronic interaction and the decay of the parent particles, and no more by the composition and the dominance of protons in cosmic rays. However, due to various effects the lateral distribution $\rho(r,\varphi)$ of the density of positive and negative muons and its charge ratio $R(r,\varphi)$ vary radially (with the distance r from the EAS centre) as well as azimuthally (with the EAS intrinsic azimuth φ, counted from the direction of the intersection with vertical plane, containing the shower axis, with the observation plane and the normal shower plane, respectively).

In this paper we study, in particular the azimuthal variation of the lateral density distribution of EAS muons of both opposite charges separately and the resulting variation of the muon charge ratio (see also early considerations in ref. [4]). The studies are based on EAS simulations using the Monte Carlo simulation code CORSIKA (vers. 6.0) [5] (E_μ^{thres}=100 MeV). In order to be specific we demonstrate the main features by considering p and Fe induced EAS, with the primary energy of 10^{15} eV, incident with a zenith angle $\theta = 45^\circ$, but comparing different azimuthal directions of incidence, in particular from North and from South. The magnetic inclination of observation locus (Karlsruhe) is adopted to be about 65° with the magnetic field pointing downwards. The studies are in context of various actual [6] and future attempts [7] to measure the muon charge ratio with a spectroscopic device coupled to EAS observation by small detector arrays.

AZIMUTHAL ASYMMETRIES IN THE LATERAL DISTRIBUTIONS OF CHARGED EAS PARTICLES

The origin of azimuthal asymmetries of the lateral distributions of charged EAS particles is mainly attributed to the attenuation and to geometrical effect of showers with inclined incidence. Assuming that the EAS starts from infinity and neglecting any influence of the geomagnetic field for the moment, we have cylindrical symmetry around the shower axis. For inclined showers incident the observation plane, charged particles arriving first ("early" azimuthal region) experience less attenuation than particles arriving later ("late" azimuthal region) due to larger travel distances. Additionally there appear effects from the projection of the normal shower plane to the observational plane. It should be noted that the attenuation of the particle density is largely obscured and counter-balanced to a large extent, when the measured particle densities are reconstructed from the measured energy deposits in scintillation detectors, since this dependence from the angle of particle incidence works in opposite direction [8].

INFLUENCE OF THE GEOMAGNETIC FIELD ON THE AZIMUTHAL LATERAL μ^+ AND μ^- - DISTRIBUTIONS

Figures 1 and 2 display some selective examples of the results for proton induced EAS observed in the observational plane as function of the azimuth angle. Fe induced showers show similar features. In Figs. 1 and 2 as well as in Figs.3-5 azimuth 0° means that the observation point is located towards North from the shower core , 90° means it is located towards East and so.

The asymmetry for EAS incident from various directions can be explained by different relative angles to the Earth magnetic vector. When the geomagnetic field is switched off, the azimuthal μ^+- and μ^--distributions show practically only the variation due to geometric and attenuation effects, for μ^+ and μ^- similar, so that the charge ratio value stays with $R_\mu(r,\varphi) = 1$. Obviously differences in the π^+- and π^-- productions remain rather small at the considered energies. This finding is underlined by the value of the muon charge ratio integrated over all distances and azimuth angles $R_\mu = 1.028 \pm 0.002$ for the cases shown in Fig.2. The influence of the geomagnetic field and the separation of μ^+ and μ^- increase with the path length (slant depth) of the muon trajectories in the atmosphere. Hence the azimuthal R_μ variation gets more pronounced with increasing distances from the shower core, with the threshold of observed muon energies (Fig.3) since muons of higher energies stem from earlier generations, and with the zenith angle of EAS incidence (Fig.4). Recently this feature has been regarded for a separation of positive and negative muons by the Earth's magnetic field [9] in case of very inclined showers of a primary energy high enough so that the muon component observed at ground remains sufficiently intensive for observation.

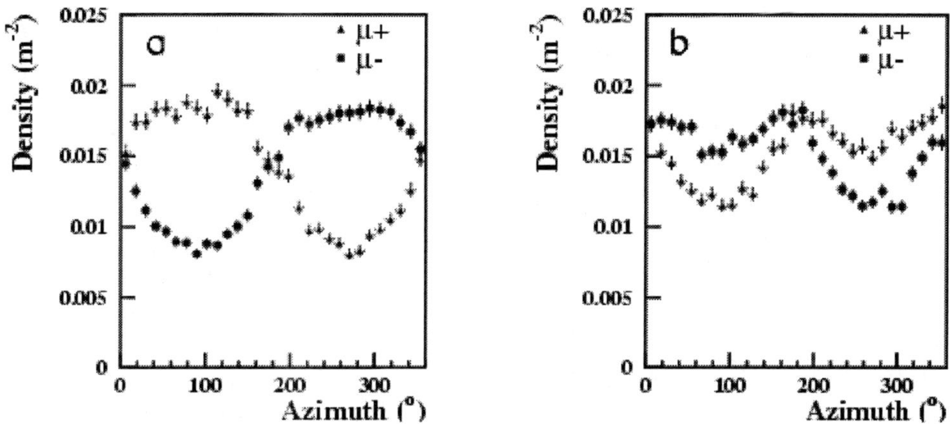

FIGURE 1. The mean azimuthal μ^+ and μ^- - distributions of proton induced inclined EAS ($\theta = 45^\circ$) incident from (a) North and (b) South with the primary energy of 10^{15} eV at a distance of 45m -50 m from the shower axis.

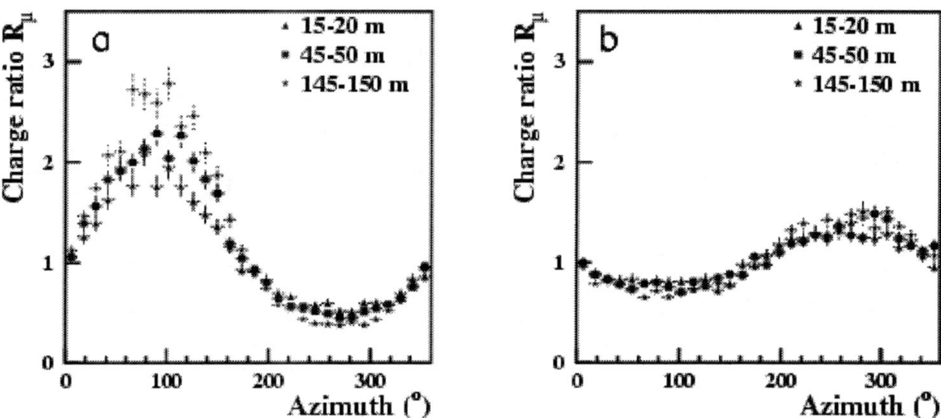

FIGURE 2. Azimuthal variation of the charge ratio $R_\mu(r,\varphi)$ of the mean muon density distribution of proton induced inclined EAS ($\theta = 45^\circ$) incident from (a) North and (b) South with the primary energy of 10^{15} eV at various distances r from the shower axis.

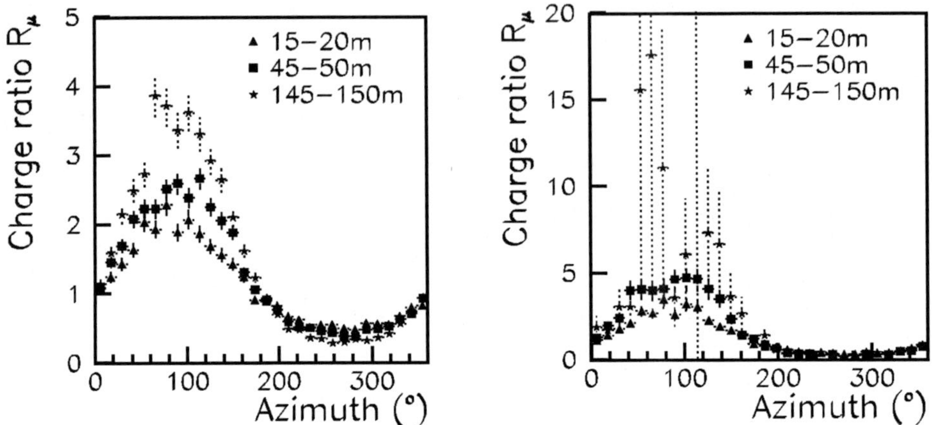

FIGURE 3. Azimuthal variation of the charge ration R_μ of the muon density distribution of proton induced EAS of inclined incidence (θ = 45°) , incident from North with the primary energy of 10^{15} eV, observed at a radial distance of 45-50m. Comparison of different detection thresholds: $E_\mu \geq$ 10 GeV (left), $E_\mu \geq$ 100 GeV (right).

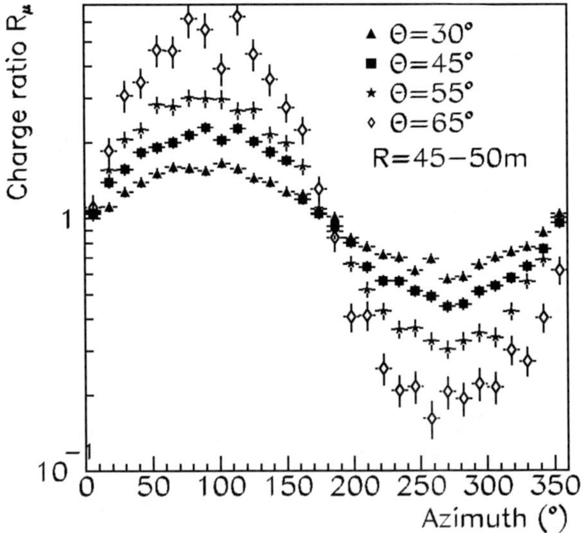

FIGURE 4. Azimuthal variation of the charge ration R_μ of the muon density distribution of proton induced EAS incident with different zenith angles from North with the primary energy of 10^{15} eV, observed at a radial distance of 45m-50m.

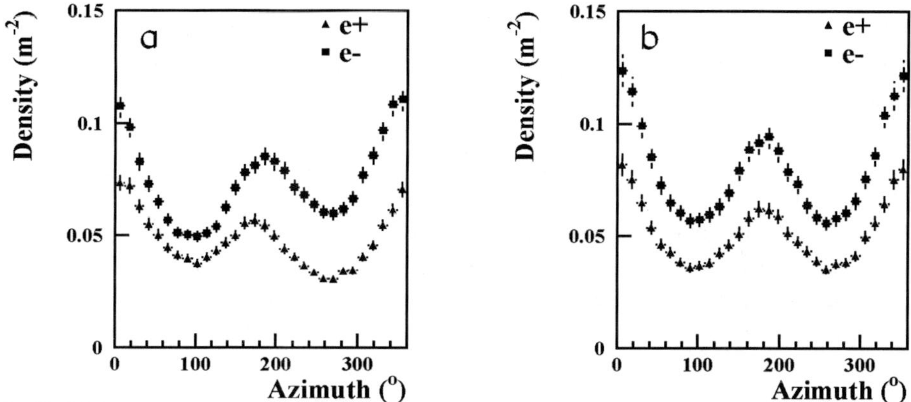

FIGURE 5. The mean azimuthal e^+ and e^- - density distributions of proton induced inclined EAS (θ = 45°) incident from North with the primary energy of 10^{15} eV at a distance of 45m -50 m from the shower axis. The case when the geomagnetic field is ignored, is shown on the right panel (b).

It should be noted that the geomagnetic field also influences the azimuthal EAS electron and positron density distributions, but in a minor extent as compared to the attenuation effects which are dominating there (Fig.5).

ASPECTS FOR STUDIES OF THE MUON CHARGE RATIO IN EAS

Experimental investigations of the azimuthal and radial variations of the charge ratio of the muon density in EAS would need for the determination of R_μ a spectrometer placed within or in an adequate distance from a detector array used for identifying and characterising the incident EAS: the direction of shower incidence, defining also the azimuthal position of the spectrometer and the position of the shower core relative to the location of the spectrometer.

There are different types of spectrometers discriminating the charge of the detected muons in atmospheric muon studies. In many cases magnetic devices are used which, however, suffer from small acceptances and other difficulties at low energies.

The spectrometer WILLI [10] installed in Bucharest determines the muon charge ratio by measuring the life time of the muons stopped in a stack of 16 scintillator detector layers. The life time is different for positive and negative stopped muons as negative muons can be captured in atomic orbits, thus leading to an effectively reduced lifetime, depending on the stopping material. The uncertainties of the efficiencies for the detection of differently charged muons and of the geometrical acceptance, usually affecting magnetic spectrometers, get cancelled out. The detector area of the device is able to be directed in a particular direction to measure the incoming muons under different (zenith and azimuth) angles of incidence. With this apparatus the East –West asymmetry of the charge ratio of the atmospheric muon flux has been measured. The disadvantage of the WILLI method as compared to magnetic spectrometers is that only muons stopped within the electromagnetic calorimeter device, can be differentiated along their charges. In case of atmospheric muons this feature provides no restriction; in case of EAS muons with a spectrum that peaks at 1-2 GeV may need additional absorbers placed in direction of incidence.

It has been proposed [7] to combine the WILLI apparatus with an array of 12 scintillator units, located nearby WILLI, in order to determine the EAS core position relative to the WILLI device and the direction of incidence of showers of the energy region of 10^{14}-10^{15} eV, with providing a trigger for WILLI.

Figure 6 displays schematically a first idea of the geometrical layout of the detector system which is under consideration. Extensive EAS simulations studies are needed to specify the optimal layout parameters:

- *The distance R_{WILLI} of the array centre from WILLI. That determines the radial distance range from the shower axis of the lateral distribution to be efficiently studied*
- *The mutual detector spacing D_{ij} which determines the efficient window of the primary spectrum, the accuracy of core location and of the zenith angle of EAS incidence. It may be wise to keep the installation for these parameters flexible from the very beginning*

As first substantial step for a well founded project proposal studies of the efficiency, of the response and the trigger rate are needed for all possible options. This quest is in addition to a technical concept of an *electronical data acquisition system* with triggering WILLI by a registered EAS. It should be explicitely noted that WILLI is not able to determine the charge ratio of individual showers and is counting only the positive and negative muons, thus forming after observation of many showers an average value.

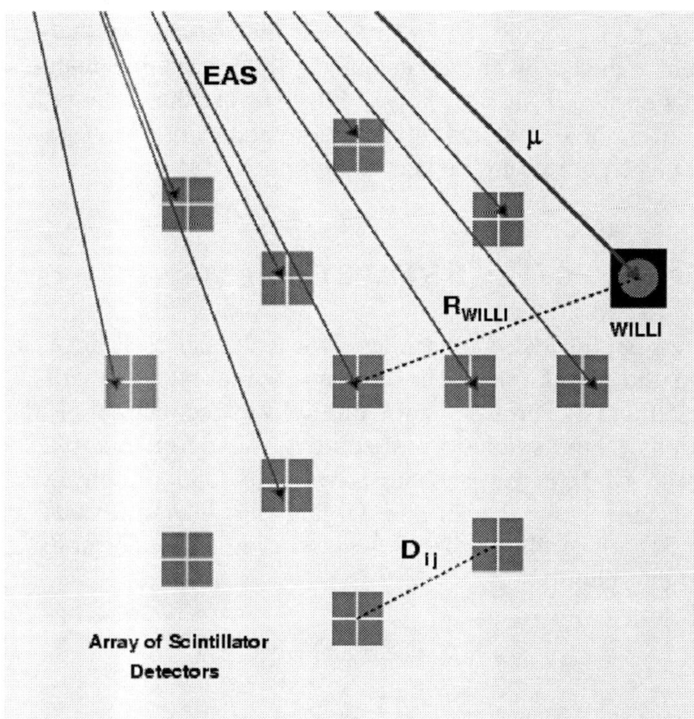

FIGURE 6. First sketch of the geometrical layout of a mini-detector array for triggering the nearby located WILLI [10].

.

CONCLUDING REMARKS

EAS simulations show that the lateral density distributions of the positive and negative muons are varying not only with the (radial) distance from the shower axis, but also with the azimuth relative to the plane of the incident shower. The reasons are different. In addition to the attenuation effects of charged particles of inclined showers in the atmosphere by the variation of the travelling distances, the geomagnetic field affects dominantly the travel of positive and negative muons with deflections in opposite directions. The geomagnetic effects depend on the direction of the EAS axis relative to the Earth's magnetic vector. This leads to an azimuthal variation of the muon charge ratio of the muon density distribution. In the extreme case of very inclined showers (with long slant depths) the Earth field might be used as magnetic separator. Obviously these features which are not yet experimentally explored [6,7] in a systematic way, are of great interest for the understanding of the EAS development. It should be mentioned that some findings for the muon component induced by the geomagnetic field resemble features observed for the radio emission from EAS [11,3]. Furthermore the quantitative results would also provide some detailed information about the hadronic interaction, in particular when observing higher energy muons [3].

ACKNOWLEDGEMENTS

We are grateful to J. Oehlschläger for technical assistance, in particular for preparing the simulations. The authors would like to thank C. Morariu, C. Manailescu and A.Patrascioiu for the help in reconstructing the lateral muon distributions. These studies are in context of experimental plans of the Cosmic Ray Group of NIPNE, Bucharest, we are collaborating with in this field (WILLI-EAS). H.R. thanks the Organising Committee of the Carpathian Summer School of Physics 2007 for providing the opportunity to present these results at the school.

REFERENCES

1. T.Hebbeker and C.Timmermans, Astropart.Phys.18 (2002) 107
2. J.Wentz et al., J. Phys.G:Nucl. Part. Phys. 27 (2001)1699; J.Wentz et al., Phys.Rev. D67 (2003) 073020
3. H.Rebel et al., Report FZKA 7294 Forschungzentrum Karlsruhe 2007; H.Rebel, O.Sima, A.Haungs and J.Oehlschläger, Proc.30th ICRC, Merida, Mexico 2007
4. L.G.Dedenko, G.F. Federova and A.A.Kirillov, Proc. 24th ICRC 1995, Rome, Italy vol. 4, p.309
5. D.Heck et al., Report FZKA 6019 Forschungzentrum Karlsruhe 1998
6. S.Tsuji et al., Proc. 29th ICRC 2005 Pune, India vol.8, p 213 ; ibid vol 6 p.233
7. I.M. Brancus et al., Proc.ISVHECRI 2006, Weihei , China . Nucl.Phys.B (in press) B.Mitrica et al., Proc.30th ICRC, Merida, Mexico 2007, these Proceedings. Letter of Intent , CosmicRay Group IFIN-HH. Bucharest , 19.01.2006.
8. A.Patrascioiu, O Sima and H.Rebel, Internal Report, KASCADE-Grande 2006-01, Forschungszentrum Karlsruhe
9. B. K.Xue and Bo-Qiang Ma, Astropart.Phys.27 (2007) 286 M.Ave, R.A.Vasquez and E.Zas, Astropart.Phys.14 (2000) 91
10. B.Vulpescu et al. Nucl. Instr.Methods A 414 (1998) 1699; B.Vulpescu et al., J.Phys.G:Nucl.Part.Phys. 27(2001)977; I.M.Brancus et al., Nucl. Phys.A271 (2003)1044c
11. A. Saftoiu et al. Proc.30th ICRC, Merida, Mexico 2007

Air Shower Measurements with Radio Antennas: The LOPES Project

Andreas Haungs

Institut für Kernphysik, Forschungszentrum Karlsruhe, 76021 Karlsruhe, Germany
haungs@ik.fzk.de

Abstract. LOPES is set up at the location of the KASCADE-Grande extensive air shower experiment in Karlsruhe, Germany and aims to measure and calibrate radio pulses from Extensive Air Showers. LOPES is designed as a digital radio interferometer using high bandwidths and fast data processing and profits from the reconstructed air shower observables of KASCADE-Grande. First results of the LOPES experiment are very promising for a future large scale application of the technique to detect cosmic rays of highest energies.

Keywords: cosmic rays, extensive air showers, radio emission
PACS: 96.50.sd, 98.70.Sa, 95.55.Jz

INTRODUCTION

A major topic in astroparticle physics is the measurement and the understanding of the source, acceleration and transport of the highest energy cosmic particles. There are novel, promising techniques to boost the energy reach and performance of future detectors for charged cosmic rays and neutrinos. Of particular interest is the detection of particle showers in air and dense media by their radio emissions [1].

The traditional method to study extensive air showers (EAS) is to measure the secondary particles with sufficiently large particle detector arrays. In general these measurements provide only information on the actual status of the air shower cascade on the particular observation level. This hampers the determination of the properties of the EAS inducing primary as compared to methods like the observation of Cherenkov and fluorescence light [2].

In order to reduce the statistical and systematic uncertainties of the detection and reconstruction of EAS, especially with respect to the detection of cosmic particles of highest energies, there is a current methodical discussion on new detection techniques. Due to technical restrictions in past times the radio emission accompanying cosmic ray air showers was a somewhat neglected EAS feature. For a review on the early investigations of the radio emission in EAS in the 60ties see [3]. However, the study of this EAS component has experienced a revival by recent activities, in particular by the LOPES project, and the CODALEMA experiment in France [4].

The main goal of the investigations in Karlsruhe in the frame of LOPES is the 'calibration' of the shower radio emission in the primary energy range of 10^{16} eV to 10^{18} eV. I.e., to investigate in detail the correlation of the measured field strength with the shower parameters, in particular the orientation of the shower axis (geomagnetic angle, azimuth angle, zenith angle), the position of the observer (lateral extension and

CP972, *Exotic Nuclei and Nuclear/Particle Astrophysics (II)—Proceedings of the Carpathian Summer School of Physics 2007,* edited by L. Trache and S. Stoica

polarization of the radio signal), and the energy and mass (electron and muon number) of the primary particle.

EMISSION PROCESS

Recent theoretical studies [5, 6, 7] of the radio emission in the atmosphere are embedded in the scheme of coherent geosynchrotron radiation. Here, electron-positron pairs generated in the shower development gyrate in the Earth's magnetic field and emit radio pulses by synchrotron emission. During the shower development the electrons are concentrated in a thin shower disk ($< 2\,\mathrm{m}$), which is smaller than one wavelength (at 100 MHz) of the emitted radio wave. This situation provides the coherent emission of the radio signal.

In the frame of LOPES, such theoretical and detailed Monte Carlo studies [7, 8] of the radio emission are continued. The simulations lead to expectations of the relevant radio emission at frequencies of 10 MHz to 500 MHz. For showers above a threshold energy of $\approx 5 \cdot 10^{16}\,\mathrm{eV}$ one expects a short, but coherent radio pulse of 10 ns to 100 ns with an electric field strength significantly above the galactic noise and proportionally increasing to the primary energy of the cosmic particle initializing the air shower.

LOPES: GENERAL LAYOUT AND DATA PROCESSING

The basic idea of the LOPES (= LOFAR prototype station) project is to build an array of relatively simple, quasi-omnidirectional dipol antennas, where the received waves are digitized and sent to a central computer. This combines the advantages of low-gain antennas, such as the large field of view, with that one of high-gain antennas, like the high sensitivity and good background suppression. With LOPES it is possible to store the received data stream for a certain period of time, i.e. at a detection of a transient phenomenon like an air shower retrospectively a beam in the desired direction can be formed. To demonstrate the capability to measure air showers with these antennas, LOPES is built-up at the air shower experiment KASCADE-Grande [9]. The air shower experiment provides a trigger of high-energy events and additionally with it's direction reconstruction a starting point for the radio data analysis. KASCADE-Grande is an extension of the multi-detector setup KASCADE [10] (KArlsruhe Shower Core and Array DEtector) built in Germany, measuring air showers in the primary energy range of 100 TeV to 1 EeV with high precision due to the detection of all charged particle types at sea-level, i.e. the electromagnetic, the muonic, and the hadronic shower component. Hence, on the one hand LOPES will profit from the reconstructed air shower observables of KASCADE-Grande, but on the other hand since radio emission arises from different phases of the EAS development, LOPES also will provide complementary information and help to understand the observables measured with the particle detector array of KASCADE-Grande.

In the current status LOPES [11] operates 30 short dipole radio antennas (LOPES-30), data of the first 10 antennas forming LOPES-10 are already analysed.

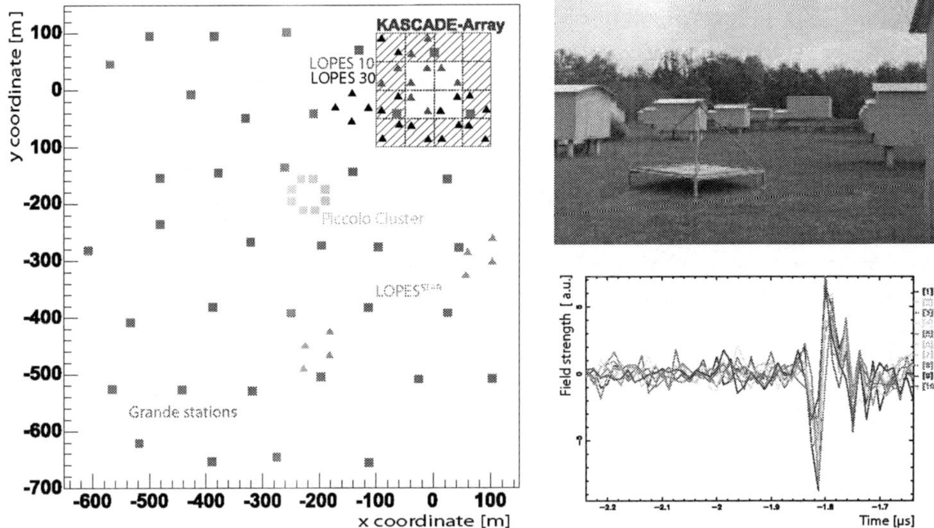

FIGURE 1. Left panel: Sketch of the KASCADE-Grande – LOPES experiments: The 16 clusters (12 with muon counters) of the KASCADE field array, the distribution of the 37 stations of the Grande array, and the small Piccolo cluster for fast trigger purposes are shown. The location of the 30 LOPES radio antennas is also displayed as well as the positions of the newly developed LOPES^{STAR} antennas. Right panel: Photograph of LOPES antennas at the KASCADE field. Raw signals of the individual antennas for one event example. The signals at this stage are prepared for the beam-forming based on shower observables reconstructed by KASCADE-Grande.

The antennas, positioned within or close by the original KASCADE array (fig. 1, left panel), operate in the frequency range of 40 – 80 MHz and are aligned in east-west direction, i.e. they are sensitive to the linear east-west polarized component of the radiation what can be easily changed into the opposite polarization by turning the antennas by 90°. The read out window for each antenna is 0.8 ms wide, centered around the trigger received from the KASCADE array. The sampling rate is 80 MHz. The shape of the antenna and the steel ground screen gives the highest sensitivity to the zenith and half sensitivity to a zenith angle of 45°, almost independent on the azimuth angle. The logical condition for the LOPES-trigger is a high multiplicity of fired stations of the KASCADE array. This corresponds to primary energies above $\approx 10^{16}$ eV; such showers are detected at a rate of ≈ 2 per minute.

The LOPES data processing includes several steps. First, the relative instrumental delays are corrected using a known TV transmitter visible in the data. Next, the digital filtering, gain corrections and corrections of the trigger delays based on the known shower direction (from KASCADE) are applied and noisy antennas are flagged. Then a time shift of the data is done and the combination of the data is performed calculating the resulting beam from all antennas. The geometrical delay (in addition to the instrumental delay corrections) by which the data is shifted, is the time difference of the pulse coming from the given direction to reach the position of the corresponding antenna compared to the reference position. This shift is done by multiplying a phase gradient in the frequency domain before transforming the data back to the time domain. This step includes also a correction for the azimuth and zenith dependence of the antenna gain. Fig. 1, lower right panel shows a particularly bright event as an example. A crucial element of the detection method is the digital beam forming which allows to place a narrow antenna beam in the direction of the cosmic ray event. To form the beam from the time shifted data, the data from each pair of antennas is multiplied time-bin by time-bin, the resulting values are averaged, and then the square root is taken while preserving the sign. We call this the cross-correlation beam or CC-beam. Finally there is a quantification of the radio parameters: Although the shape of the resulting pulse (CC-beam) is not really Gaussian, fitting a Gaussian to the smoothed data gives a robust value for the peak strength, which is defined as the height of this Gaussian. The error of the fit results gives also a first estimate of the uncertainty of this parameter. The finally obtained value ε_ν, which is the measured amplitude divided by the effective bandwidth, is compared with further shower observables from KASCADE-Grande, e.g. the angle of the shower axis in respect to the geomagnetic field, the electron or muon content of the shower, the estimated primary energy or mass, etc.

LOPES 10: FIRST RESULTS

First measurements (6 months data taking) were performed with a set up of 10 antennas. This LOPES-10 [11] data set is subject of various analyses addressing different scientific questions. With a sample asking for high quality events the proof of principle for detection of air showers in the radio frequency range was made [12].

With events falling inside the original $200 \times 200\,\mathrm{m}^2$ large KASCADE array basic correlations of the radio signals with shower parameters are shown. More than 220 events with a clear radio signal and with the shower core inside the KASCADE array could be detected. The analysis of these events concentrates on the correlations of the radio signal with all shower parameters, in particular with the arrival direction and with the shower size, i.e. the primary energy of the shower.

Fig. 2, left panel depicts the dependence of the reconstructed radio pulse height with the primary energy of the cosmic particles. The shown correlation supports the expectation that the field strength increases by a power-law with an index close to one with the primary energy, i.e. that the received energy of the radio signal increases quadratically with the primary energy of the cosmic rays. An index of this power-law close to one serve as a proof of the coherence of the radio emission during the shower development.

Fig. 2, right panel shows the correlation between the normalized reconstructed pulse height of the events with the geomagnetic angle. Normalized here means, that the detected pulse height is corrected for the dependency on the muon number, i.e. up to a large extend, the primary energy. The clear correlation found suggests a geomagnetic origin for the emission mechanisms.

Besides the analyses of events with the core inside the antenna setup, KASCADE-Grande gives the possibility to search for distant events. For each (large) shower triggering KASCADE, the information from the extension of KASCADE, i.e. from the

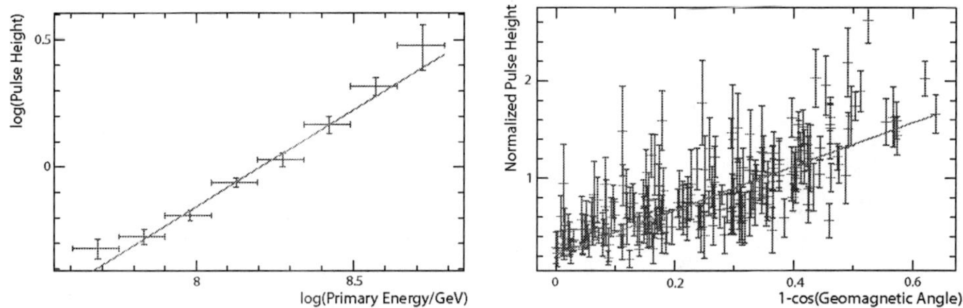

FIGURE 2. Left panel: Average radio pulse height of the detected events (with shower core inside the KASCADE array) plotted versus the primary particle energy as reconstructed by KASCADE [11]. The errors bars show the statistical errors. Right panel: Radio pulse height for individual events normalized with the muon number and distance to the shower axis plotted versus the cosine of the angle to the geomagnetic field [11]. The errors bars show the uncertainty of the Gaussian to the field strength after beamforming.

Grande array, is available (see fig. 1). From that information the shower can be reconstructed even if the core is outside the original KASCADE area, and a radio signal can be searched for events which have distances up to 800 m from the center of the antenna setup. Also for this case several hundred events (372 in 6 months data taking) could be detected, where in particular the lateral behavior of the radio emission is investigated [13]. The functional form of this dependence and also the lateral scaling parameter is of high interest for the further development of the radio detection technique. After linear scaling of the pulse amplitude ε_ν with the primary energy estimated by KASCADE-Grande a clear correlation with the mean distance of the shower axis to the antennas is found (Fig. 3, left panel). Fitting the present data set by assuming an exponential function, the scaling parameter R_0 results to 230 ± 51 m. Such an exponential dependence of signal to distance is expected by detailed simulations of the geosynchrotron effect with a scaling radius of ~ 100 to ~ 800 m, increasing with increasing zenith angle [7].

Further interesting features of the radio emission in EAS were investigated with a sample of very inclined showers [14] and with a sample of events measured during thunderstorms [15]. The first one is of special interest for a large scale application of this detection technique, as due to the low attenuation in the atmosphere also very inclined showers can be detected with high efficiency. This is of great importance if ultrahigh energy neutrinos exist. With LOPES one could show that events above 70° zenith angle still emit a detectable radio signal.

The latter sample is of interest to investigate the role of the atmospheric electric field in the emission process. The LOPES collaboration examines the contribution of an electric field to the emission mechanism theoretically and experimentally. Two mechanisms of amplification of radio emission are considered: the acceleration radiation of the shower particles and the radiation from the current that is produced by ionization electrons moving in the electric field. LOPES data were recorded during thunderstorms, periods of heavy cloudiness and periods of cloudless weather. It was found that during thunderstorms the radio emission can be strongly enhanced (fig. 3, right panel). No

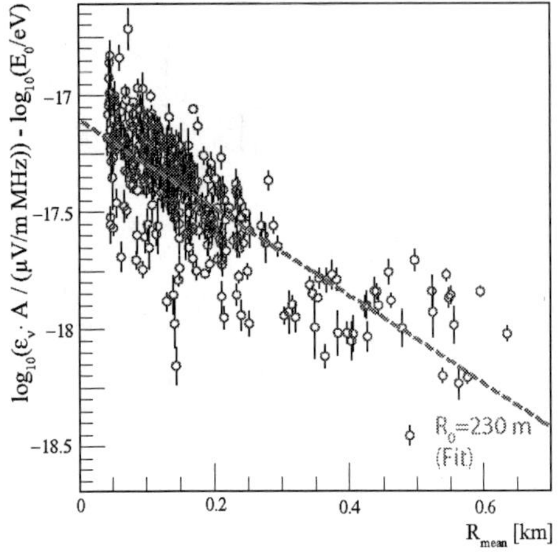

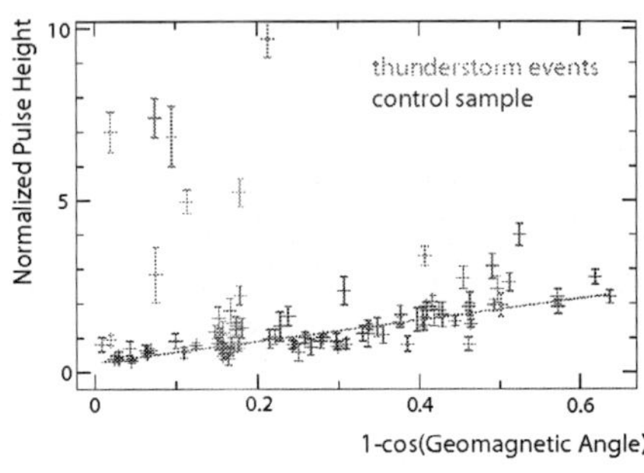

FIGURE 3. Left panel: Correlation of the pulse height corrected for primary energy with the mean distance of the shower axis to the radio antenna system [13]. The line show the result of a fit with an exponential function. Right panel: Normalized pulse height of the events taken at fair weather conditions (control sample) and those detected during thunderstorms plotted against the geomagnetic angle [15].

amplified pulses were found during periods of cloudless sky or heavy cloudiness, suggesting that the electric field effect for radio air shower measurements can be safely ignored during non-thunderstorm conditions [15].

EXTENSION TO LOPES 30

In 2006, 20 additional antennas were installed at KASCADE-Grande. The 30 antenna setup (see fig. 1) provides a larger sensitive area to the radio signal at a single event compared with the original LOPES-10 set-up. This will provide the possibility for a detailed investigation of the lateral extension of the radio signal as LOPES-30 [16] has now a maximum baseline of approximately 260 m.

Each single antenna is absolute amplitude calibrated using a commercial reference antenna [17] of known electric field strength at a certain distance. Each LOPES-30 radio antenna is calibrated at its location inside the KASCADE-Array. The power to be received from the source in calibration mode is compared with the power recorded in the LOPES electronics. The calibration procedure leads to frequency dependent amplification factors representing the complete system behavior (antenna, cables, electronics) in the environment of the KASCADE-Grande experiment (fig. 4, left panel). The calibration results in a spread for the amplification factors of nearly one order of magnitude. The uncertainty of the calibration method can be estimated from repeated measurement campaigns for a single antenna under all kinds of conditions. It results in an systematic uncertainty of the calibration procedure of approximately 25% for the power related amplification factor averaged over the effective frequency range. The correction factors will be applied to the measured signal strength resulting in the true electric field strength

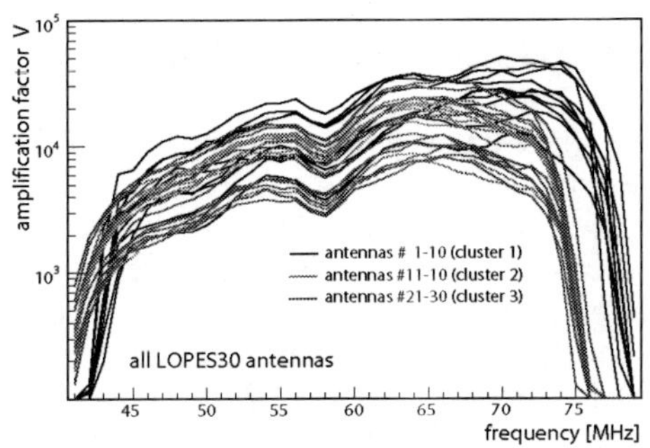

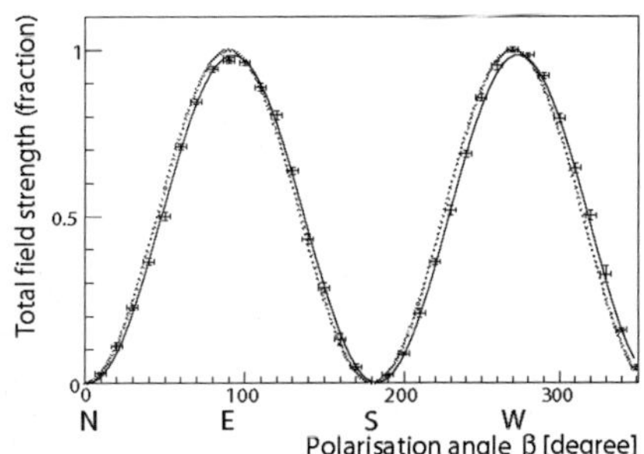

FIGURE 4. Left panel: Frequency dependent amplification factors for all 30 antennas obtained by the amplitude calibration [17]. Right panel: Fraction of the total field strength seen by the linearly polarized LOPES antenna (sensitive to E-W) as a function of the polarization angle β (angle between polarization direction of emitting reference source and receiving LOPES antenna). Points: experimental data by the calibration procedure, data fit (solid line) and pure theoretical sinβ-distribution (dotted line) are also shown [18].

which can be compared to simulated values.

In addition, during LOPES-30 measurements we place emphasis on the monitoring of the environmental conditions by measuring the static electric field and by recording parameters of a nearby weather stations. Atmospheric conditions, in particular E-field variations during thunderstorms influence the radio emission during the shower development, and the measurement of the radio pulses. By monitoring the environmental conditions, and comparing them with the antenna noise level as well as with the detected air shower radio signals, correlations will be investigated and corrected for.

With the data taken by LOPES-30 it is expected to calibrate the radio emission in air showers in the primary energy range from 10^{16} eV to 10^{18} eV, because due to the absolute amplitude calibration a direct comparison of the field strength with the expectations (simulations) is possible. The data of the one year LOPES-30 east-west polarization measurements are currently being investigated.

LOPES POLARIZATION MEASUREMENTS

After one year of measurements of the East-West polarization by all 30 antennas, recently, the LOPES-30 set-up was reconfigured to perform dual-polarization measurements [18]. Half of the antennas have been configured for measurements of the North-South polarization direction. By measuring at the same time both polarization components and by comparing with the expectations, the geosynchrotron effect as the dominant emission mechanism in air showers will be experimentally verified. That the LOPES antenna are sensitive to the polarization of the signal is proved by special calibration campaigns using the reference source (fig. 4, right panel).

LOPES AND THE PIERRE AUGER EXPERIMENT

One of the main goals of the LOPES project is to pave the way for an application of this 're-discovered' air shower detection technique in large UHECR experiments, like the Pierre Auger Observatory. Parallel to the measurements at KASCADE-Grande LOPES follows this aim by optimizing the antenna design for an application at Auger, named as LOPES^{STAR} [19]. Additionally the optimum frequency range, depending on the local noise, and an adapted filtering is investigated. Going in direction of setting up a test array at Auger South, the possibilities of a self-triggering antenna system and an online beam forming are also investigated.

Meanwhile several dual polarized STAR-antennas are in operation at the KASCADE-Grande field (see fig. 1). Also by this system air showers could be detected. These information is used to optimize the self-trigger system of LOPES^{STAR} [20].

In parallel to the activities in Karlsruhe, also the first test set up with LOPES^{STAR} antennas in Argentina at the Auger South experiment has detected first radio signals from extensive air-showers [21].

CONCLUSIONS AND OUTLOOK

LOPES is running and takes continuously data in coincidence with the air shower experiment KASCADE-Grande.

The main goal of the LOPES project is the investigation of the relation between the radio emission from extensive air showers with the properties of the primary particles. Such studies are hoped to pave the way for a large-scale application of a novel technique in future cosmic ray experiments.

With LOPES-10 the proof-of-principle for detection of cosmic particles by radio flashes from extensive air showers could be performed. First results obtained by correlating the observed radio field strength with the shower parameters obtained by the KASCADE measurements appear to be very promising for a more detailed understanding of the emission mechanism from atmospheric showers. Most interesting results are the found quadratic dependence of the radio-power on energy, the correlation of the radio field strength with the direction of the geomagnetic field, the exponential behavior of the lateral decrease of the field strength with an scaling parameter in the order of hundreds of meter, and that except during strong thunderstorms the radio signal is not strongly influenced by weather conditions. In particular, the quadratic dependence on energy will make radio detection to a cost effective method for measuring the longitudinal development of air showers of the highest energy cosmic rays and cosmic neutrinos.

With LOPES-30 we will be able to follow the main goal of the LOPES project: The calibration of the radio emission in extensive air showers for primary energies below 10^{18} eV.

Besides the experimental works done with the present antenna setup the LOPES project aims to improve the theoretical understanding of the radio emission in air showers. Supplementary emission processes like the Cherenkov-Askaryan-effect which plays the dominant role in dense media will be investigated.

The LOPES technology can be applied to existing cosmic ray experiments as well as to large digital radio telescopes like LOFAR and the SKA (square kilometer array), providing a large detection area for high energy cosmic rays. First approaches to use the technique at the Pierre Auger Observatory and at a first LOFAR station are under way.

ACKNOWLEDGMENTS

The author would like to thank the organizers of the school for their kind invitation and all the participants for the lively discussions and the marvelous football match.

REFERENCES

1. D. Z. Besson, *Journal of Physics: Conference Series* **81**, 012026 (2007).
2. A. Haungs, H. Rebel, and M. Roth, *Rep. Prog. Phys.* **66**, 1145 (2003).
3. H.R. Allan, *Prog. in Element. Part. and Cos. Ray Phys.* **10**, 171 (1971).
4. D. Ardouin et al. - CODALEMA coll., *Astropart. Phys.* **26**, 341 (2006).
5. H. Falcke, and P.W. Gorham, *Astropart. Phys.* **19**, 477 (2003).
6. T. Huege, and H. Falcke, *Astronomy & Astrophysics* **412**, 19 (2003).
7. T. Huege, and H. Falcke, *Astropart. Phys.* **24**, 116 (2005).
8. T. Huege, R. Engel, and R. Ulrich, *Astropart. Phys.* **27**, 392 (2007).
9. G. Navarra et al. - KASCADE-Grande coll., *Nucl. Instr. Meth. A* **518**, 207 (2004).
10. T. Antoni et al. - KASCADE coll., *Nucl. Instr. Meth. A* **513**, 429 (2003).
11. A. Horneffer et al. - LOPES coll., *Int. Journ. Mod. Phys. A* **21 Suppl.**, 168 (2006).
12. H. Falcke et al. - LOPES coll., *Nature* **435**, 313 (2005).
13. W.D. Apel et al. - LOPES coll., *Astropart. Phys.* **26**, 332 (2006).
14. J. Petrovic et al. - LOPES coll., *Astronomy & Astrophysics* **462**, 389 (2007).
15. S. Buitink et al. - LOPES coll., *Astronomy & Astrophysics* **467**, 385 (2007).
16. S. Nehls et al. - LOPES coll., *Int. Journ. Mod. Phys. A* **21 Suppl.**, 187 (2006).
17. S. Nehls et al. - LOPES coll., *Proc. of 30th Int. Cosmic Ray Conference ICRC 2007, Merida, Mexico* (2007), in press.
18. P.G. Isar et al. - LOPES coll., *Proc. of 30th Int. Cosmic Ray Conference ICRC 2007, Merida, Mexico* (2007), in press.
19. H. Gemmeke et al. - LOPES coll., *Int. Journ. Mod. Phys. A* **21 Suppl.**, 242 (2006).
20. T. Asch et al. - LOPES coll., *Proc. of 30th Int. Cosmic Ray Conference ICRC 2007, Merida, Mexico* (2007), in press.
21. A. van den Berg et al. - Pierre Auger coll., *Proc. of 30th Int. Cosmic Ray Conference ICRC 2007, Merida, Mexico* (2007), in press.

The Physics of KASCADE-Grande

Gianni Navarra

Dipartimento di Fisica Generale dell'Universita' and INFN, Torino, Italia

Abstract.
The basic experimental information and open questions in cosmic ray studies in the high energy region $10^{15} - 10^{18}$ eV: *knee*, end of galactic radiation, transition to the extragalactic sources, are summarized. It is shown that an extension of the experimental information with the quality achieved in the region of the *knee* ($10^{15} - 10^{16}$ eV) up to 10^{18} eV is required to complete the measurements of the *knee*, identify the transition to the extragalactic component, and put the basis for the interpretation of the data at the highest energies. KASCADE-Grande, now operating at Forschungszentrum Karlsruhe, is shown to fulfill the requirements, both concerning the acceptance and the experimental performances.

Keywords: High Energy Astrophysics, Cosmic Rays, Extensive Air Showers
PACS: 96.50.sd, 98.70.Sa, 95.55.Jz

INTRODUCTION

The study of the region 10^{15} to 10^{18} eV of the cosmic ray spectrum is of main interest for the understanding of the high energy evolution of the galactic radiation, the postulated upper limit of its contribution, and the transition from the galactic to the extragalactic sources (in the usual cosmic ray jargon: the region from the *knee* to the *dip*, see fig. 1). Measurements have to be performed by exploiting the cascades (Extensive Air Showers) that the primaries induce in the atmosphere, and therefore through detectors of the secondaries produced in their interactions.

The statistical limitations of satellite or balloon born detectors can be deduced from fig. 1 [1]: the rate of primaries above 10^{15} eV on a typical square-meter detector being of the order of a few events per month (and moreover, concerning energy determination, the limited depths of the detectors allows the measurement of the energy released into the e.m. (π^0) component just in the first interaction of the incoming particle).

Ground based arrays exploit the detection of the Extensive Air Shower components (e.m., muonic, hadronic, Cherenkov and fluorescent light, radio emission)[1]. They can realize very large collecting areas (depending on the primary energies of interest, and therefore the related fluxes), and have to face experimental problems of different nature. To summarize: a) they are located at fixed atmospheric depth, and therefore are affected by the fluctuations of shower development (enhanced by the steepness of the primary spectrum); b) these are convoluted with the detectors' resolutions, that therefore have to be well under control; c) the energies of interest are above the ones at which we

---

[1] It is interesting to remark that nowadays the indirect and direct measurements match rather reasonably and integrate each other to provide a unified scenario at the transition energies (10^{14} - 10^{15} eV)[1].

CP972, *Exotic Nuclei and Nuclear/Particle Astrophysics (II)—Proceedings of the Carpathian Summer School of Physics 2007,* edited by L. Trache and S. Stoica

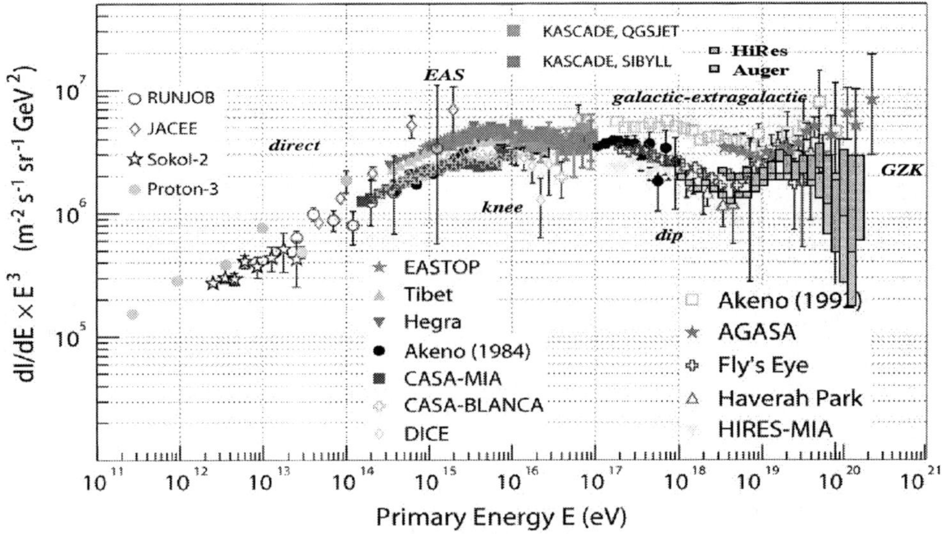

FIGURE 1. The cosmic ray primary spectrum [1] and the peculiarities of the different regions. Auger [19, 24] and HiRes [25] data are included as vertical boxes at 1 sd statistical uncertainties.

have relevant information on hadron cross sections from accelerator experiments, and therefore extrapolations and general checks are required; d) techniques and methods have to be different at different primary energies (due to different required collecting areas), and therefore cross-calibration checks in the overlapping ranges are mandatory; e) usually long exposures are necessary (decades of years of data taking in the field, i.e. far from laboratory conditions) implying severe stability studies.

THE KNEE REGION

The *knee* [2] observed in the region 10^{15}-10^{16} eV (with transition of the exponent of the power law energy spectrum from $\gamma_1 \approx 2.76$ to $\gamma_2 \approx 3.1$) represents a main feature of the cosmic ray primary spectrum, and provides a main opportunity for understanding the high energy galactic radiation. In fact this region is considered to represent both the upper limit to the energy reachable by the commonly accepted galactic sources (shock waves in SNRs) [3] and to the magnetic confinement inside the Galaxy [4, 5][2].

From the experimental point of view, the multi-component air shower arrays of the last generation provide consistent descriptions of the events and the possibility of testing

[2] Both effects imply a rigidity dependence of the knees (cut-off energies) of the spectra of the different elements: $E_k(Z) = Z \cdot E_k(p)$.

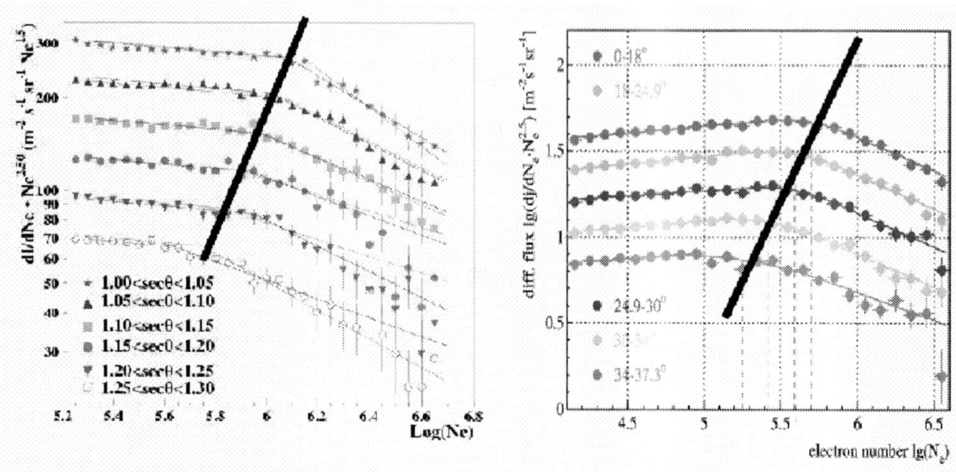

FIGURE 2. Shower size spectra as observed at different atmospheric depths by EAS-TOP (820 gr/cm^2, left) and KASCADE (sea level, right). The dependence of the shower size at the knee on the atmospheric depth is clearly seen, and the two values match (Log(N_e)$\approx$ 5.8) at the same depth (vertical for KASCADE and very inclined for EAS-TOP).

some general features of the interaction and propagation methods used to describe the cascades in the atmosphere. Such possibility can now constrain the old problem of the nature of the *knee*, in connection with possible unknown features of hadronic interactions. Concerning the N_e spectrum, the shift of the *knee* positions vs. zenith angle (i.e. atmospheric depth) is clearly visible and shows that data are compatible with a feature occurring at fixed energy for a given primary nucleus. The EAS-TOP [6] (observation level 820 gr/cm^2) and KASCADE [7] (sea level) data match very well (see fig. 2). The attenuation length of the shower size value at the *knee* $N_{k,e}$ is compatible with the absorption length of EAS obtained from the constant intensity cut [6], and, on another side, the steepening is consistently observed in the muon, hadron and Cherenkov light spectra [7, 8]. Therefore the ensemble of data does not imply any requirement of a sharp change in the character of hadron interactions for its interpretation [9].

The study of the *knee* concerning the evolution of the primary composition is mainly achieved by means of the combined e.m. (N_e) and muon (N_μ) analysis (see fig. 3) [10, 11]. Different experiments based on such method, as KASCADE [10, 12], EAS-TOP (GeV [13], and TeV [11] muons in coincidence with MACRO), CASA-MIA [14], GRAPES [15], Baksan (at muon energy of 230 GeV) [16] agree in the description of the *knee* as due to the steepening of the spectra of the lightest nuclei. The deduced absolute value of the primary mass still depends on the interaction model used for the interpretation[3], but the general explanation, i.e. steepening of the lightest components, is common to all of them (and no one of the models in use can explain the N_e-N_μ relation

[3] The limitations of hadronic interaction models in describing the data in this range are discussed in [12].

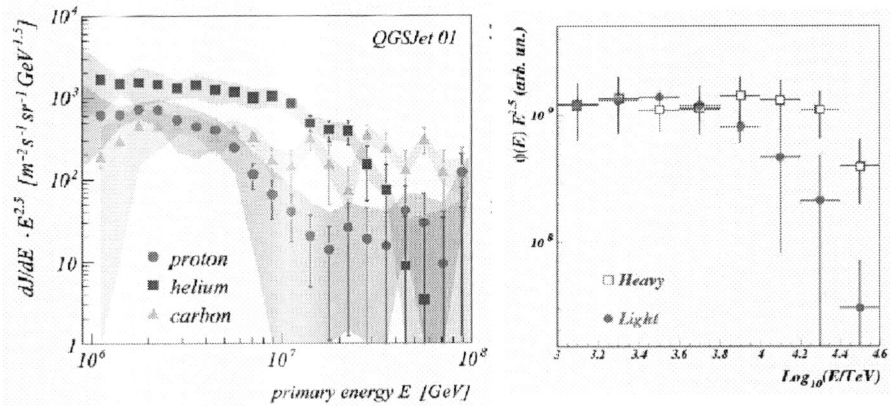

FIGURE 3. Evolution of the primary spectra at the knee from the KASCADE N_e-N_μ analysis (left). Spectra of the Light and Heavy components from EAS-TOP and MACRO data (right).

in the energy range 10^{15}-10^{16} eV without assuming an increasing mass composition [9]). The steepening is observed at increasing primary energies for p, He, and then CNO primaries (with $E_k(p) = (2-3) \cdot 10^{15}$ eV; a definite proof of the A or Z dependence of the breaks of the spectra of the individual elements has still to be obtained). Till now no bending is observed for the iron component.

Similar conclusions are obtained from the Cherenkov light experiments measuring the depth of the shower maximum (see e.g. ref. [8, 17]).

The detection of the iron *knee* (expected at about 10^{17} eV) represents therefore a main goal to complete the experimental information. This is the main aim of the KASCADE-Grande experiment [18], now taking data, and extending the KASCADE N_e-N_μ analysis up to about 10^{18} eV.

FROM GALACTIC TO EXTRAGALACTIC RADIATION

The lack of strong anisotropies related to our Galaxy, and the recent Auger results [19] seem to leave no room for a galactic origin of cosmic rays at the highest energies ($E_0 \approx 10^{20}$ eV). The identification of the transition from the galactic to the extra-galactic contribution is therefore of main interest. Such transition has been historically identified with the *dip* of the spectrum observed above 10^{18} eV and the subsequent *ankle* above 10^{19} eV, and this requires an additional galactic contribution beyond the expected iron *knee* [20]. Rather recently a new approach has been proposed by V.S. Berezinsky and collaborators [21], and with different implementations has attracted much attention [22, 23]. Such interpretation localizes the end of the galactic contribution around 10^{17} eV (as expected for the iron *knee*), and a corresponding rising of the extragalactic component. The analysis of the spectrum is based on the interpretation of the *dip* as due to the energy losses for pair production of primary protons against the cosmic background in the intergalactic space. From the point of view of the composition this

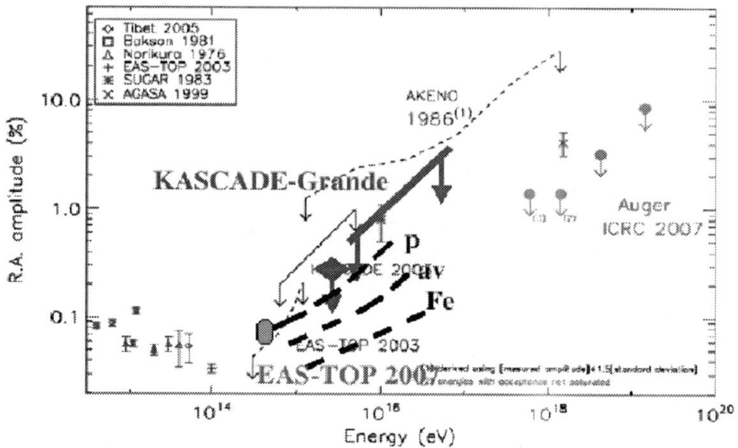

FIGURE 4. Summary of the recent data on the cosmic ray anisotropy over a wide energy range, including general expectations from the galactic diffusion [29], and first KASCADE-Grande limits.

would imply the transition to the extragalactic sources below 10^{18} eV, characterized by a decreasing galactic heavy and an arising extragalactic light component. This at least shows how wide is the possible range for such transition and therefore how different are the implications for galactic accelerators. The solution has of course to come from a detailed study of the primary spectrum and composition covering homogeneously the whole region.

From the experimental point of view, the primary flux at about 10^{18} eV, i.e. the typical threshold of the UHE experiments, is quite debated. Recently [19, 24] the Auger Collaboration exploiting the fluorescence light calorimetric energy calibration and an high (controlled and constant) acceptance from a surface array has measured such flux at the "lowest level" with respect to the existing data (fig. 1). How such level is reached from the *knee*, whether through a steeper spectrum or an harder one followed by a "second knee" (that could represent an additional signature about the galactic component, possibly in correspondence of the iron *knee*) has to be measured.

Energy calibration methods based on hadronic interaction models and cascade simulations in the atmosphere lead to results differing of about 25%. Moreover the analysis of the primary composition at the highest energies becomes more and more uncertain, due to the uncertainties of the hadron interaction models, also due to the required extrapolation of many orders of magnitude in energy. Following the evolution and consistency of the predictions of hadron interaction models with experimental data from 10^{15} to 10^{18} eV represents the main approach to check for their reliability.

A parallel information on the evolutions of the galactic and extragalactic radiations is now reachable from the arrival direction measurements and the study of the evolution of the anisotropies with primary energy. A summary of the status of the anisotropy data is

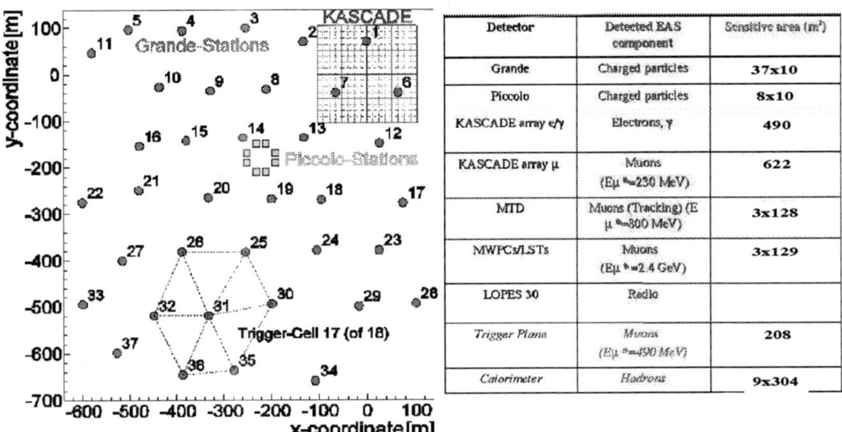

Detector	Detected EAS component	Sensitive area (m^2)
Grande	Charged particles	37x10
Piccolo	Charged particles	8x10
KASCADE array e/γ	Electrons, γ	490
KASCADE array μ	Muons ($E_\mu^{th} \approx 230$ MeV)	622
MTD	Muons (Tracking) ($E_\mu^{th} \approx 800$ MeV)	3x128
MWPCs/LSTs	Muons ($E_\mu^{th} \approx 2.4$ GeV)	3x129
LOPES 30	Radio	
Trigger Plane	*Muons* ($E_\mu^{th} \approx 490$ MeV)	208
Calorimeter	*Hadrons*	9x304

FIGURE 5. Scheme of the KASCADE-Grande array and of its different components.

reported in fig. 4, where the new recent measurements of EAS-TOP [26], KASCADE-Grande [27] and Auger [24] represent significant achievements in sensitivity. The EAS-TOP data provide an indication of an increasing amplitude, together with a change of phase (with respect to 10^{14} eV and to the homogeneous situation in the region 10^{11} - 10^{14} eV) above 4.10^{14} eV, i.e. approaching the *knee*. Anyway an increasing of such anisotropy is expected from the diffusion inside the Galaxy [28, 29]. Establishing such effect would open the possibility of following the evolution of the galactic anisotropy, to match with the Auger measurements at 10^{17} - 10^{18} eV. The observation of a deviation of the properties of the anisotropy from the galactic ones would provide the indication of the appearing of a new component, with different origin and propagation. This would open an additional channel to the study of the transition from the galactic to the extragalactic radiation, and requires a detector of acceptance of the order of the km^2 and high stability (i.e. fluctuations at poissonian level).

KASCADE-GRANDE

In the previous discussions the physical motivations for a detector operating in the energy range from the *knee* to the region of the *giant* arrays have been introduced, resulting in the demand for a detector extending significantly the acceptance of KASCADE, without a significant loss in resolution. These motivations have lead to the KASCADE-Grande program. KASCADE-Grande is made of the complex KASCADE array [30], and of 37 scintillator modules 10 m^2 each ("Grande", from former EAS-TOP experiment) enclosing KASCADE to extend its acceptance of a factor about 10, to 0.5 km^2. A scheme of the whole detector and of its different components is shown in fig. 5. A compact array ("Piccolo") located in a central position has the role of fulfilling the trig-

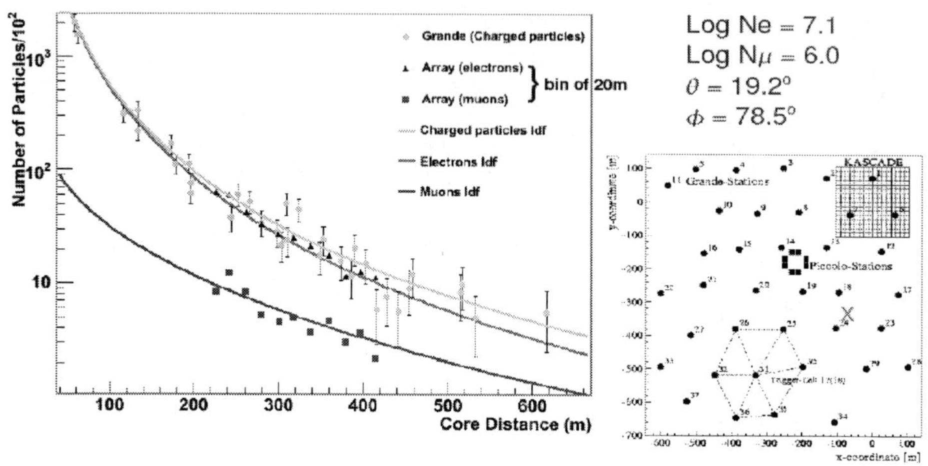

FIGURE 6. An event as observed and reconstructed by the Grande and KASCADE arrays: the information from the different detectors matches very well and covers a wide range of core distances. Core location and main parameters of the event are given.

ger requirements of the two arrays even in the most unfavorable geometrical conditions. "Grande" has a sensitive to acceptance surface ratio of a factor of 10-20 smaller than the KASCADE e.m. array (besides being distributed over a much wider area, which implies more elaborated data transfer techniques). In order to combine the information, its reconstruction accuracies are therefore a main issue to be optimized and checked. This can be done using events with core locations inside the sensitive area of both arrays. The obtained reconstruction accuracies are: 6.4 m on core location, 0.58^o on arrival direction, $\Delta N_e/N_e \approx 0.13$ on shower size, showing the good absolute agreement and response of a distributed array with respect to an "optimum" dense one. Full trigger and reconstruction efficiency for all primaries is reached at Eo = 2.10^{16} eV up to 40°, and therefore continuity checks are possible with the main KASCADE data. As an example, an event as observed by the whole "Grande" and KASCADE elements is shown in fig. 6. The particle density recorded by "Grande" is of course the "total charged particle" component, while KASCADE can detect both the "total charged particle" component, and (through the shielded scintillator modules) select the muon component. It provides thus the muon number information for the composition analysis ("KASCADE-like"). Average lateral distributions of the "total charged particle" and muon components, that represent the basis for the N_e - N_μ analysis are reported in fig. 7, showing a general agreement with the expectations from the simulations (e.g. QGSJET II) over a wide range of distances [27]. A partial sample of the basic N_e - N_μ data for the spectrum and composition analysis as obtained with KASCADE-Grande are shown as a scatter plot in fig. 8 [27]. The primary

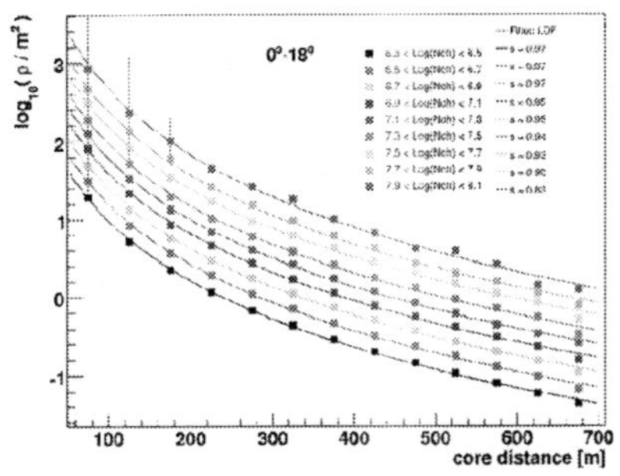

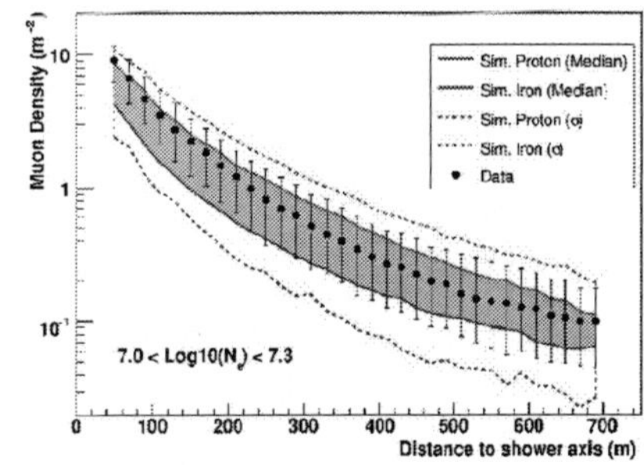

FIGURE 7. Average lateral distributions for the "total charged component" and the muon one, compared to the expectations from QGSJET II.

energy ranges, depending on zenith angles, are indicated. Typically "heavy" primaries are "electron-poor", and are therefore located at the lower edge of the plot; viceversa the "light" ones. Such basic plots summarize at the same time some phenomenological properties of Extensive Air Showers as observed over a wide range of core distances, and the main instruments for the KASCADE-Grande analysis.

We want to stress that the dimensions of the array play a significant role not only on its acceptance, but as well on its performances, concerning the possibilities of: a) checking and comparing interaction and propagation models over a wide range of core distances; b) improving the measurements of production altitudes of the secondaries (and therefore of longitudinal developments and rapidity distributions)[4]; c) using different energy estimators, as S(500) [31, 27].

With my warm thanks to the whole KASCADE-Grande Collaboration and to the organizers for their kind invitation, the excellent site and organization, and their patience!

REFERENCES

1. K.-H. Kampert, *Nuclear Phys. B (Proc. Suppl.)* **165** (2007) 294.
2. G.V. Kulikov and G.B. Khristiansen, *J.E.P.T.* **35(8)** (1959) 441.
3. E.G. Berezhko et al, *J.E.P.T.* **82** (1996) 1.
4. B. Peters, *Proc. 6th ICRC* **3** (1960), 157.
5. G.T. Zatsepin et al., *Izv. Akad. Nauk USSR S.P.* **26** (1962) 685.
6. EAS-TOP Coll. (M. Aglietta et al), *Astropart. Phys.* **10** (1999) 1.
7. R. Glasstetter and J. Hoerandel (KASCADE Coll), *Nuclear Phys. B (Proc. Suppl.)* **75A** (1999) 238.
8. D.V. Chernov et al (Tunka Coll.), *Int. J. Mod. Phys. A* **20** (2005) 6799.
9. G. Navarra et al (EAS-TOP Coll.), *Nuclear Phys. B (Proc. Suppl.)* **151** (2006) 79.

---

[4] Both such measurements are performed by correlating the geometry of the shower with the arrival directions of the muons recorded in the tracking muon detector ("muon tunnel") [27].

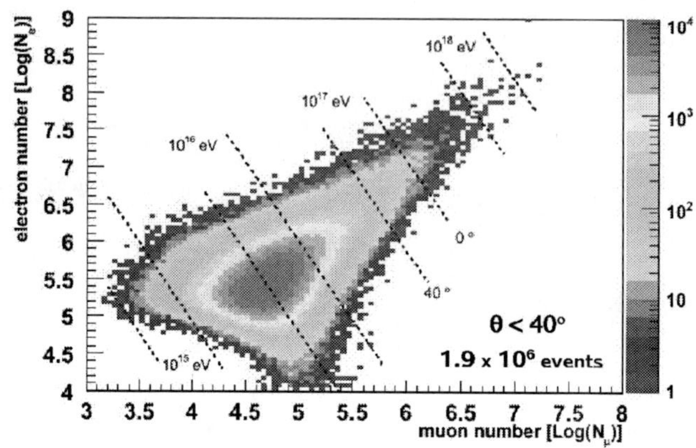

FIGURE 8. Basis for the N_e-N_μ composition analysis in KASCADE-Grande (threshold effects are clearly visible).

10. KASCADE Coll. (T. Antoni et al), *Astropart. Phys.* **16** (2002) 373.
11. EAS-TOP and MACRO Colls. (M. Aglietta et al), *Astropart. Phys.* **22** (2004) 641.
12. KASCADE Coll. (T. Antoni et al), *Astropart. Phys.* **24** (2005) 1.
13. EAS-TOP Coll. (M. Aglietta et al), *Astropart. Phys.* **21** (2004) 583.
14. M.A.K. Glasmacher et al, *Astropart. Phys.* **12** (1999) 1.
15. Y. Hayashi at al, *Proc. 29th ICRC*, **10** (2005) 243.
16. V.B. Petkov et al, *Proc. 29th ICRC*, **1** (2005) 65.
17. N.M. Budnev et al, *Proc. 29th ICRC*, **8** (2005) 255; E.E. Korosteleva et al, *Nuclear Phys. B (Proc. Suppl.)* **165** (2007) 74.
18. KASCADE-Grande Coll: K.-H. Kampert et al, *Nuclear Phys. B (Proc. Suppl.)* **122** (2003) 122; A. F. Badea et al, *Nuclear Phys. B (Proc. Suppl.)* **136** (2004) 384.
19. T. Suomijarvi, these proceedings.
20. A.M. Hillas, *J. Phys. G: Nucl. Part. Phys.* **31/5** (2005) R95.
21. V. S. Berezinsky et al, *Phys. Rev. D* **74**, 043005 (2006), and refs. therein.
22. A.M. Hillas, *Cosmology, Galaxy Formation and Astroparticle Physics on the Pathway to the SKA* Oxford UK (April, 10-12, 2006).
23. D. Allard et al, *Astron. Astrophys.* **443** (2005) L29.
24. The AUGER Collaboration, *Proc. 31st ICRC, Merida* (2007, different papers, in press).
25. R.U. Abbasi et al, Phys. Rev. Lett., **92** (2004) 151101.
26. The EAS-TOP Collaboration, *Proc. 31st ICRC, Merida* (2007, in press).
27. The KASCADE-Grande Collaboration, *Proc. 31st ICRC, Merida* (2007, different papers, in press).
28. V.S. Ptuskin et al, *Astron. Astrophys.* **268** (1993) 726.
29. J. Candia et al, arXiv:astro-ph/0302082v2 25 Apr 2003.
30. T. Antoni et al, KASCADE Coll, *Nucl. Instr. Meth. A* **513** (2003) 490.
31. G. Toma, these proceedings.

The Nature of Light Dark Matter

P. L. Biermann[*,†], L. Caramete[*,**], F. Munyaneza[*] and J. Stasielak[‡]

[*]*MPI for Radioastronomy, Bonn, Germany*
[†]*Dep. of Phys. & Astron., Univ. of Bonn, Germany, Dep. of Phys. & Astr., Univ. of Alabama, Tuscaloosa, AL, USA*
[**]*Institute for Space Studies, Bucharest, Romania*
[‡]*KASI, Daejeon, Korea*

Abstract. For over seventy years we have observed evidence for the presence of dark matter, now clearly shown to be non-baryonic. The microwave background fluctuations show that it is about 0.2 of all density of matter and energy, while baryonic matter is about 0.04; the rest is labelled dark energy, and not even close to an understanding. The sum is just what is required to make the universe geometrically flat. In this overview we discuss one of the attempts to explain the nature of dark matter, light dark matter, also called sterile neutrinos, or right-handed neutrinos, or more generally warm dark matter. Many of the predictions for cold dark matter and warm dark matter are the same, most of modifications are in small scales, in the temporal evolution of structure, and in some of the shapes of structure. We are now convinced that these warm dark matter particles have to be subthermal at injection to get the right free-streaming length in structure formation. Galaxies, their structure as well as their distribution pose a number of tests and riddles, not all of which have yet been passed or solved. We outline and summarize the latest evolution in these discussions, centering now on galaxies, their structure and evolution, as well as large scale distribution. Warm dark matter has survived many critical tests so far, and is clearly worth exploring in much more detail.

Keywords: Dark Matter, First Star Formation, Galaxy, X Rays
PACS: 95.35.+d, 98.35.Ac, 98.70.Qy

INTRODUCTION

Ever since in the early 1930ies Oort first coined the term “dark matter", and Zwicky then proved it to be dominant in its gravitational effects in clusters of galaxies, we have been reminded steadily, how little of the universe we really understand. If we argue that we understand stars, then this is less than one percent of the closure density required to explain the flatness of the universe. With a bold guess we might argue that we understand baryonic mater, but even then it is only 4 percent. The missing baryonic matter is probably in the form of moderately warm to hot gas, so extremely difficult to detect. However, early star formation, galaxies and structure formation are largely influenced by dark matter, of which we have little clue as to its nature.

Most proposals center on the idea that dark matter is the lowest mass stable particle in “supersymmetry", a natural extension of modern particle physics. This suggests that the particles are very massive, and so much more massive than protons. Almost all searches focus on this option, referred to as ‘ńcold dark matter", or CDM. However, CDM has a problem with small scales, as we do not detect a sufficiently large number of small galaxies to match the predictions, while on larger scales CDM works exceedingly well.

Therefore the focus has now shifted a bit to explore ‘ńwarm dark matter", in the

CP972, *Exotic Nuclei and Nuclear/Particle Astrophysics (II)—Proceedings of the Carpathian Summer School of Physics 2007,* edited by L. Trache and S. Stoica

context of this review particles of keV mass, and a simple extension of normal particle physics, with no need to appeal to supersymmetry.

In the following we describe the latest tests and challenges, and center on star formation, galaxies, and large scale structure. For lack of space we just refer to the names of authors, and give only a very limited number of actual references.

DARK MATTER: THE EVIDENCE

Dark matter is detectable only through its gravitational effect, and is required to explain for the following observations:

The disks in galaxies are unstable to perturbations, and require a nearly spherical halo of at least equal mass to keep stable, and so stay flat (Ostriker et al.). In the nearly 100 year old theory of isothermal spheres such a spherical distribution could be simply explained by a huge distribution of gravitating particles, with an outer density law of r^{-2}, where r is the radial distance.

The rotation curves of disk galaxies are flat; while this can be explained in other ways, it is remarkable, that isothermal spheres would precisely give flat rotation curves. In some cases they extend to 50 kpc (Rubin et al.). It is important to note that Stokes' law does not hold in symmetries other than spherically symmetric, and so a radial density law in a disk-like distribution can be constructed, that also gives nearly flat rotation curves (Toomre). The stability argument demonstrates that this is irrelevant.

Early Hubble type galaxies often have huge amounts of hot gas around them, in the case of dominant early Hubble type galaxies in groups, extending throughout the group (such as for NGC5846). The mass in this gas can exceed $10^{10}\,M_{\odot}$. The gravitational potential required to hold this gas confined also requires a dark matter halo (Biermann et al.). Related arguments can be made for clusters of galaxies, where this is much more easily observed (e.g., Forman et al., Böhringer et al., Sarazin et al., Enßlin et al.).

Dark matter is required to explain the motions of galaxies in groups and clusters (e.g. Zwicky). This is the original and first argument, now over 70 years old, and still correct.

Dark matter is required for large scale structure formation in the universe, and to explain the overall flat geometry of the universe; fitting the microwave background fluctuation data gives indeed the most precise numbers for the amount of dark matter (Spergel et al. 2003, 2006 [4]). Many of the key parameters are now known to percent precision, and many talk about high precision cosmology.

It is remarkable how precisely we know some of these key numbers, such as the amount of dark matter, and still have little clue what its nature is.

So, after over 70 years, we still ask: What is dark matter?

Much recent work and many reviews have been published, e.g. by Abazajian, Kusenko, Shaposhnikov, Silk, and many others. Our line of reasoning has been summarized for the Marcel Grossmann meeting (Berlin) and is available (Biermann & Munyaneza 2007 [3]).

DARK MATTER: SOME HISTORY

Sterile neutrinos or right-handed neutrinos, refer to very weakly interacting neutrinos, as they interact with other neutrinos, extremely weakly with photons, and mostly just with gravity.

The concept of dark matter was first proposed by Oort (1932), then by Zwicky (1933, 1937), and again by Swift (1936). Oort was finding undetected stars, and thought so, correctly. Zwicky and Swift detected what we call today non-baryonic dark matter. Pontecorvo (1970) considers sterile neutrinos and introduces the name. Olive & Turner (1982) consider sterile neutrinos in general terms. Dodelson & Widrow (1994) show that sterile neutrinos can be dark matter; the mechanism proposed by them is now ruled out by data. But other mechanisms are now available as production mechanisms (Kusenko, Shaposhnikov). Kusenko & Segre (1997) show that sterile neutrinos can explain pulsar kicks, the observational fact, that an appreciable percentage of all pulsars move at linear velocities faster than 1000 km/s; other mechanisms, such as magnetic field kicks in the explosion, are also possible (Bisnovatyi-Kogan). Interestingly, both these mechanisms utilize the magnetic field and its topology.

PROPOSAL

Most suggestions center on the lightest super-symmetric particle. In large scale structure formation this leads to an overabundance of structure on small scales, such as a multitude of dwarf galaxies around us, and a highly structured dark matter halo around our Galaxy. We will return to this point below.

However, Gilmore et al.[1] find many small dwarf galaxies; also the formation of smaller galaxies might have been hindered by ionization from the original star formation ionization radiation from the Virgo cluster. So smaller galaxies could not form, but their dark matter structure would still be there.

We start with right-handed neutrinos, with a lifetime of

$$\tau = 1.3\,10^{26}\,\mathrm{s}\left(\frac{7\mathrm{keV}}{m_{DM}c^2}\right)^5\left(\frac{0.8\,10^{-9}}{\sin^2\theta}\right) \tag{1}$$

They decay, most importantly: into one active neutrino and a photon. They are never in thermodynamic equilibrium, and so require a chemical potential; probably far sub-thermal. This latter property is a key, as it turns out.

As background information we note that: They can be produced in the right amount (Dodelson & Widrow 1994; Shi & Fuller 1999; Abazijian al. 2001; Dolgov & Hansen 2002; Abazajian 2006); They can explain the baryon asymmetry (Akhmedov al. 1998; Asaka & Shaposhnikov 2005); They can explain the lack of power on small scales in Large Scale Structure Formation (Olive & Turner 1982; Abazajian 2006); They can explain the inner dark matter distribution and spatial shape in galaxies (Belokurov al 2006; Fellhauer al. 2006).

Our recent work started with an argument on pulsar kicks, velocities up to 1000 km/s – see the guitar nebula – (Kusenko 2004). This suggests a mass range of 2 to 20 keV.

In the supernova explosion to form a rotating neutron star, we have a transition from an active left-handed (normal) neutrino to a weakly interacting right-handed neutrino and so prevent scattering: Note that left handed neutrinos have a scattering mean free path during the supernova explosion of a few cm; that explains the 20 second smearing in arrival times of SN 1987A. Our work then continued with a consideration of the early growth of black holes (Munyaneza & Biermann 2005 [2], 2006 [3]) to suggest a range: 12 - 450 keV. Here we used the property that the dark matter particle is a Fermion. At an Aspen meeting September 2005 Kusenko and Biermann noticed a possible overlap in the two independent solutions, an "Eureka-moment". As this suggestion had been considered previously, we had to solve the major conundrum, and remove the key argument against such a proposal, the delayed star formation(Yoshida al. 2003). The solution turned out to be a slightly changed secondary ionization, which in turn increases the molecular hydrogen formation, and so enhances the cooling (Biermann & Kusenko 2007 [5]; Stasielak, Biermann & Kusenko 2007a, 2007b [6]).

FIRST STAR FORMATION

Sterile neutrinos decay into an active neutrino and a photon (of half the mass in energy). After redshift about 40 there is no absorption anymore, at earlier epochs the fraction of ionized atoms is (η fraction of photon energy into ionization)

$$x_e^{(s)} \sim \frac{0.2}{(1+z)^{3/2}} (\frac{\eta}{0.3}) (\frac{m_s}{7\,\mathrm{keV}})^5 (\frac{\sin^2\theta}{0.8\,10^{-9}}) \tag{2}$$

The the formation of H_2 proceeds via the H^- channel. The level of H_2 grows to above the critical threshold of $5\,10^{-4}$ (Tegmark et al. 1997) to provide strong cooling. This gives almost an order of magnitude enhancement, with maximum at redshift 80. In some cases, the corresponding heating could be able to cancel the effect, but only for redshifts < 20 (Mapelli, Ripamonti 2006).

Therefore star formation can be triggered early with many consequences: Early magnetic fields from first massive stars – the fastest process – then cosmic ray driven dynamo process in galactic disks, implies by necessity galactic winds. Reionization from first massive stars at $z > 10$. Energy input from stellar winds and supernovae. Chemical enrichment from first massive stars (Wolf Rayet star winds and supernovae). First dust; First Gamma Ray Bursts; First stellar black holes; First cosmic rays.

Primordial magnetic fields might hinder the secondary ionization: Ionization by secondary electrons, from ionization: The time scale for energy gain is (e.g., Jokipii 2004)

$$\tau_A = \frac{4\kappa}{V_A^2}. \tag{3}$$

Here we have a weakly ionized gas, with the excitation of waves coming from the budding density fluctuations, and the shock waves in the collapse. There is also damping from ion-neutral friction, and other damping (Bykov 2005, Foote & Kulsrud 1979). Here we derive just a limit ignoring both excitation and damping, to derive an extremely safe limit, and use maximal turbulence:

Assuming $\frac{B^2/8\pi}{I(k)k} \simeq 1$ and $I(k) \sim k^{-1}$ to just cancel the energy loss by ionization $B(z=100) < 2\,10^{-8}$ Gauss, corresponding to today

$$B < 4\,10^{-12}\,\text{Gauss}. \tag{4}$$

Work by Dolag et al. (2002, 2005) shows: $B < (0.2 - 1.) \times 10^{-12}$ Gauß from structure formation simulations: Therefore we do have full consistency.

GALAXY MERGING

Galaxies merge, and then a density cusp develops, as energy is transported out, and matter in, rather akin to angular momentum transported out and matter flowing in in accretion disks (Lüst 1952). The inner dark matter law is then a power-law (Moore 1998, Klypin et al. 2002, Navarro, Frenk, & White 1997). At the low end of the phase space distribution one obtains for keV Fermions a degenerate configuration, a dark matter star (Munyaneza & Biermann 2005 [2] 2006 [3]). Such a dark matter star can be eaten relatively quickly by a stellar black hole, which then grows to the full mass of the dark matter star. Thus, depending on the mass of the dark matter star, we can obtain super-massive black holes early in the universe. An alternative would be the direct collapse from a super-massive star, or a very compact star cluster. All concepts allow for an early super-massive black hole of around 10^6 solar masses. It ensues hat our Galactic center black hole could well have grown from dark matter almost exclusively.

X-RAY BACKGROUND

As the sterile neutrino in this model decays, it can produce an X-ray photon; if this photon is not absorbed, then it will produce an emission line; if integrated for cosmological epochs, then due to the increasing redshift an emission wedge. This line is at exactly half the energy of the dark matter neutrino. In the years 2005 - 2007 Boyarsky et al., Riemer-Sørensen et al., Watson et al., and some others have searched for such an emission line, and not found anything. So any search we have to conduct asks: What is the upper limit to an invisible emission line at an energy $m_{DM}c^2/2$? Using then the X-ray background we find that $m_{DM} < 10$ keV. We can also search neighboring clusters of galaxies such as Virgo and Coma, neighboring galaxies such as Andromeda (M31). So far we find only upper limits. We need Large field high spectral resolution spectroscopy, and the satellites CHANDRA and XMM-NEWTON might be able to deepen the upper limits sofar available. It turns out that the Japanese satellite Suzaku is even better suited for such a search for neighboring dwarf elliptical galaxies. This is what M. Loewenstein, A. Kusenko, and P.L. Biermann are trying. The observations have been done, but the data have not yet been fully analyzed.

LYMAN-α FOREST

The Lyman-α forest is due to small density perturbations in the early universe, but at smaller redshifts than the early quasars, and as a consequence so we see narrow absorption lines (McDonald et al. 2000, McDonald et al. 2001, Seljak et al. 2006, Viel et al. 2006). These data suggest that $m_{DM} > 10$ keV; however, using a subthermal particle (Kusenko 2006) modifies this limit to $m_{DM} > 2$ keV.

GALAXIES

The work of the group around G. Gilmore has shown in four different ways, that there is a dearth of structure on small scales (Gilmore et al. 2006, Gilmore et al. 2007 [1], and Wyse & Gilmore 2007 [7]):

The mass of dwarf spheroidal galaxies is limited from below by about $5\ 10^7$ solar masses, a limit in dark matter that is independent of its baryonic content.

The shape of our dark matter halo is very spherical as can be derived from the precessional disruption of incoming dwarf galaxies.

However, as we see one incoming disrupted dwarf galaxy pulled out into a triple stream, and these streams do not laterally mix, the perturbance from any graininess of our dark matter halo must be minimal(Fellhauer et al. 2006).

On the basis of the dynamical friction on globular clusters a similar argument can be constructed (Hernandez & Gilmore 1998; Goerdt et al. 2006).

It follows that the lack of small dwarf ellipticals cannot be simply due to the burst of ionization or heating by the ultra violet radiation emitted by the early star formation burst in the Virgo cluster galaxies, because in that case many dark matter clumps would still be there.

We conclude that all data is consistent with a lower limit to dark matter mass clumps of $5\,10^7\,M_{\odot}$, for a large range of baryonic mass. This may then be an indication of the relevant free streaming length, and so a hint for warm dark matter.

- direct indication for warm dark matter (or formation history? effect of ionization, heating and cooling of satellite galaxy?)

Based on Gilmore et al. (2007) [1], and Wyse & Gilmore (2007) [7] phase space conservation arguments lead to the constraint (Kusenko 2007):

$$m_{DM} < 4\,\mathrm{keV}$$

.

LARGE SCALE STRUCTURE

Starting from the 2MASS survey of the entire sky at 2 microns we constructed a list of nearby black holes, in the following way: First we selected all galaxies within redshift 0.025. Then we limited the sample to all early Hubble type galaxies, up to and including the type Sb. We considered all galaxies for which we could compute the ratio of far-infrared to radio flux densities, and so removed all star burst galaxies. Then we tested with B-V color distribution. Then we calculated the central black hole mass from the correlation between the spheroidal stellar distribution and the black hole mass; the main systematic error in this is the distance for any galaxy closer than approximately the Virgo cluster, because there the peculiar velocities are quite large relative to the Hubble flow, and so redshift derived distances are quite uncertain. Correcting for this effect we find that the black hole masses are on average within a factor of 2 of those directly determined. We also checked whether in this procedure we removed any known active galactic nuclei, and we do; there are some Seyfert galaxies with a later Hubble type, but we rather have too few than too many objects in our list. Finally we checked with the Lauer et al. (2006) survey of the black hole mass function, and our numbers are in full agreement: We have a relatively flat mass function from about $3\,10^6$ solar masses to about 10^8 solar masses, and a much steeper law beyond, to about $3\,10^9$ solar masses. This distribution is quite consistent with what can be expected of black holes grow to a large degree from mergers, in an agglomeration concept a la Press & Schechter (1974), or Silk & Takahashi (1979).

The final sky plot we show in the figure, color-coded for distance. The coordinate system is such, that the Galactic plane is horizontal and the Galactic center is off at the right edge (matching then also on the left), and the Virgo cluster is near the North pole in this graph (slightly to the left). The Cygnus region is about 60 degrees left from the right edge.

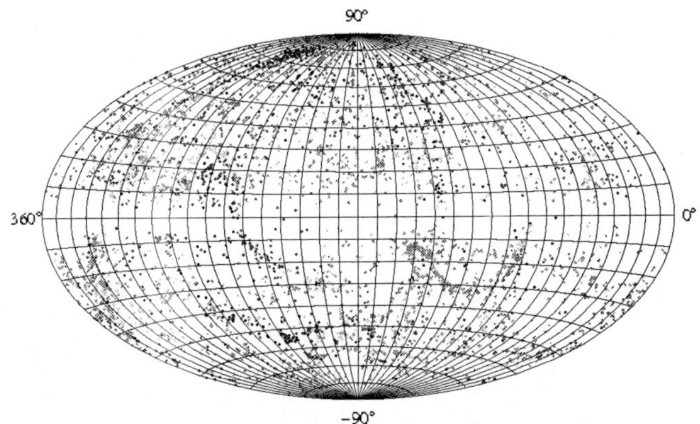

FIGURE 1. The sky in black holes, of mostly $> 10^7$ solar masses. Aitoff projection in galactic coordinates of 5,978 NED candidate sources in the case of a complete sub sample. The choice was made from a complete sample of 10,284 candidate brighter than 0.03 Jy and selected at $z < 0.025$ and 2 micron. The color code is Black, Blue, Green, Orange, Red corresponding to redshifts between 0, 0.005, 0.01, 0.015, 0.02, 0.025

What we notice is that the distribution is not smooth, as expected (de Vaucouleurs 1953, Peebles 1973). However, instead of seeing only planar or straight features, we do find patterns which seem partial circles. Especially near the radio galaxy NGC315 we find a feature which looks like a semi-circle. In order to test whether this feature is actually coherent, we show in the next graph the redshift distribution. The distance is about 70 Mpc, its semi-circular radius is about 30 Mpc, its width is about 5 Mpc, and its velocity dispersion is about 1000 km/s. The feature appears gravitationally bound. The redshift distribution is quite narrow (Fig. 2), and in fact partially broader just because the feature is on either side of the redshift binning near 0.020, so not exactly in the plane of the sky. This filament may be constituted out of two filaments in fact, as suggested by work on the large scale structure near the radio galaxy NGC315 (Ensslin et al. 2001).

Such features cannot be easily explained in cold dark matter or warm dark matter scenarios, and must remain a challenge.

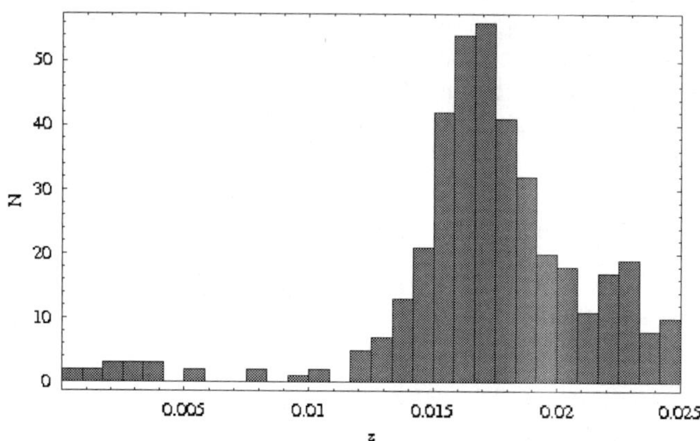

FIGURE 2. Histogram of the redshift distribution for special semicircle feature inside the selection for massive black holes 5,978 NED candidate sources at 2micron; most of these black holes have $> 10^{7}$ solar masses

SUMMARY

We have explored the concept that the dark matter is a right-handed neutrino, or sterile neutrino. For consistency with various observations it would have to be far subthermal (Kusenko 2006).

Such warm dark matter may explain the first star formation; first reionization; first magnetic fields; first cosmic rays; first black holes; first ultra high energy cosmic rays.

It predicts an X-ray line from galaxies, and background.

It may explain pulsar kicks.

The Lyman α-forest is consistent for a subthermal distribution.

It may explain the smoothness of our Galactic halo structure; and the low mass cut-off of dwarf spheroidal galaxies.

However, all of these findings can be explained independently, with a number of different concepts, all plausible. But if we wish to entertain one single concept, then a right-handed neutrino or sterile neutrino can explain it all at once, although some challenges remain ~ 2 keV m_s ~ 4 keV.

ACKNOWLEDGMENTS

PLB acknowledges long and intense discussions with A. Kusenko and Sinziana Păduroiu about this physics. Support for PLB is coming from the AUGER membership and theory grant 05 CU 5PD 1/2 via DESY/BMBF. Support for AK is coming from DOE grant DE-FG03-91ER40662 and NASA ATP grants NAG 5-10842 and NAG 5-13399. Support for FM is coming from the Humboldt Foundation.

REFERENCES

1. "The observed properties of dark matter on small scales", Gilmore et al. (2007); astro-ph/0703308
2. The nature of dark matter, Peter L. Biermann, & Faustin Munyaneza,invited lecture at the Erice meeting June 2006, editors T. Stanev and John Wefel, in press (2007); astro-ph/0702164
3. Dark matter and sterile neutrinos, Peter L. Biermann, & Faustin Munyaneza, invited plenary talk at the 11th Marcel Grossmann meeting, July 2006, editors Ruffini, Jantzen, Kleinert, and Fang, World Scientific, in press (2007); astro-ph/0702173
4. "Wilkinson Microwave Anisotropy Probe (WMAP) Three Year Results:Implications for Cosmology", Spergel, D.N., , (2007); astro-ph/0603449
5. Thermal evolution of the primordial clouds in warm dark matter models with keV sterile neutrinos, Jaroslaw Stasielak, Peter L. Biermann, & Alexander Kusenko, **654**, 290-303 (2007); astro-ph/0606435
6. Sterile neutrinos and structure formation, Stasielak, Jaroslaw; Biermann, Peter L.; Kusenko, Alexander, at the XLVII Cracow School of Theoretical Physics held in Zakopane, Poland, June 2007, to appear in the Acta Phys. Polon. B; arXiv:0710.5431
7. "Observed properties of dark matter on small spatial scales", Wyse, & Gilmore (2007); arXiv/0708.1492

9. NEUTRINO PHYSICS

Nuclear Double-Beta Decay and Derivation of Neutrino Masses

S. Stoica

*Horia Hulubei National Institute for Physics and Nuclear Engineering,
407, Atomistilor Street, P.O. Box MG-6, 077125 Magurele-Bucharest, Romania*

Abstract.
The standard model (SM) of weak interactions incorporates neutrinos as massless particles which are left-handed partners of the charged bosons. However, in the recent past, convincing evidence of massive neutrinos has been obtained in experiments that observe neutrino oscillations. The observation of the neutrino oscillations determines only mass square differences of the neutrino types and thereby sets a lower bound on the largest of the neutrino masses. So, the absolute scale of the neutrino masses remains still unknown. Neutrinoless double beta decay experiments are at present the most sensitive ones for measuring the absolute mass of the electron neutrino. Moreover, these are the only experiments that could answer the fundamental question of whether neutrinos are their own anti-particles, that is whether they are Majorana or Dirac particles. That is why the double-beta decay nuclear process is studied intensively both theoretically and experimentally. In this lecture I give an up-to-date review on this process. After a general introduction in which I stress the importance of the $\beta\beta$ decay study for searching beyond SM physics, I refer to the calculation of nuclear matrix elements involved in $\beta\beta$ decay half-lives and to the to derivation of neutrino mass in connection to other constraints coming from neutrino oscillation experiments and cosmological data.

Keywords: Neutrino,Double-Beta Decay
PACS: 14.60.Pq, 98.80.-k, 98.80.Es

INTRODUCTION

Nuclear $\beta\beta$ decay is a spontaneous decay by which an even-even nucleus transforms into another even-even nucleus by preserving its total mass but changing its charge by two units. It occurs whenever single β decay can not occur due to energetical reasons or if it is highly forbidden by angular momentum selection rules Refs. [1] - [3]. Experimentally, until now, only the $\beta\beta$ decay mode with emission of two neutrinos was measured:

$$(A,Z) \rightarrow (A,Z\pm 2) + 2e^{\mp} + 2(\bar{\nu}/\nu) \qquad (2\nu\beta\beta) \qquad (1.1)$$

This decay mode $(2\nu\beta\beta)$ conserves the lepton number, being allowed within the SM and it has been measured for a dozen of nuclei. It is the rarest decay known at present in nature with half-lives in the range of (10^{18} - 10^{24}) years. It can be viewed as two separate and simultaneous (single) β decays by which two neutrons (protons) are converted into two protons (neutrons) inside the nucleus (See, Fig. 1, left, for the case of two neutron decay). However, the conservation of the lepton number is not a fundamental law in nature and, within GUTs (theories that generalize SM), this symmetry may be violated,

CP972, *Exotic Nuclei and Nuclear/Particle Astrophysics (II)—Proceedings of the Carpathian Summer School of Physics 2007,* edited by L. Trache and S. Stoica

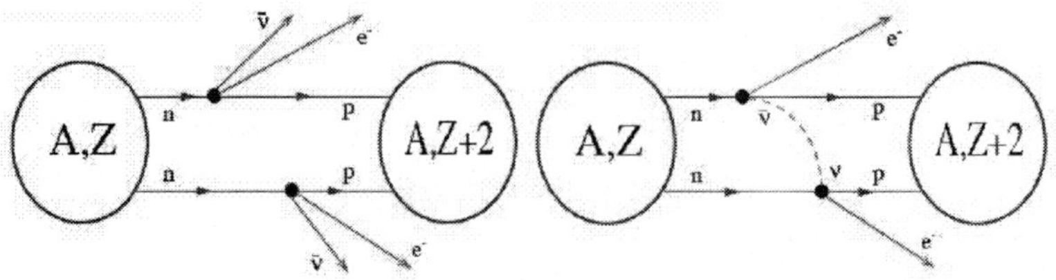

FIGURE 1. Schematic representation of $2\nu\beta\beta$ and $0\nu\beta\beta$ decay modes

resulting in new phenomena beyond the SM. Particularly, one can imagine a $\beta\beta$ decay process without emission of neutrinos:

$$(A,Z) \rightarrow (A,Z\pm 2) + 2e^{\mp} \qquad (0\nu\beta\beta) \qquad (1.2)$$

This process, that violates the lepton number conservation by two units, can occur by exchange of light massive Majorana neutrinos between the two nucleons inside the nucleus (Fig. 1, right). This implies that electron neutrino is a massive Majorana particle. Using the half-life formulas for $0\nu\beta\beta$ and the experimentally non-observability of this decay mode, one can put constraints on the absolute neutrino mass. The prezent stringent constraints on this mass coming from $\beta\beta$ decay is about 0.35 eV. On the other hand, from neutrino oscillation experiments combined with cosmological data constraints one can derive neutrino masses in the range of several meV, for normal hierarchy and several tens of meV for degenerate and inverted hierarchy. But these values are within the aims of the planned next generation of $\beta\beta$ decay experiments (for examples CAMEO [4], EXO [5], MAJORANA [6], MOON [7], NEMO3 [8], etc.) placing these experiments in pole position for deriving an absolute scale for the neutrino mass and deciding on the nature of the neutrino: is it a Dirac or a Majorana particle?

However, there are also other possible mechanisms through which the $0\nu\beta\beta$ decay may occur. For instance, from the non-observability of the $0\nu\beta\beta$ decay one may extract information on the lepton number non-conservation, existence of RH components in the weak interaction, L-R symmetric models ($0\nu\beta\beta$ mediated by heavy RH bosons), Majoron existence (a light or massless boson) or leptoquark existence (bosons carrying both lepton and baryon number)- particles that may also mediate $0\nu\beta\beta$ decay, compositness (possible substructure of quarks and leptons at an energy scale of about TeV), R-parity non-conservation and possiblity that $0\nu\beta\beta$ occurs by exchange of SUSY particles between nucleons, etc. Ref. [3] All these possibilities to constrain the neutrino mass and the new interactions appearing in GUTs demonstrate the broader potential of the $\beta\beta$ decay process to provide us with information for beyond SM physics.

NUCLEAR STRUCTURE METHODS FOR DOUBLE-BETA DECAY

The half-lives for 2ν - and 0ν - $\beta\beta$ decay modes are expressed as follows [2]:

$$\left[T_{1/2}^{2\nu,a}\right]^{-1} = F^{2\nu}|M_{GT}^{2\nu}|^2 \tag{2.1}$$

$$\left[T_{1/2}^{0\nu}\right]^{-1} = C_{mm}\left(\frac{\langle m_{\beta\beta}\rangle}{m_e}\right)^2 + C_{\eta\eta}\langle\eta\rangle^2 + C_{\lambda\lambda}\langle\lambda\rangle^2 + C_{m\eta}\frac{\langle m_{\beta\beta}\rangle}{m_e}\langle\eta\rangle +$$

$$C_{m\lambda}\frac{\langle m_{\beta\beta}\rangle}{m_e}\langle\lambda\rangle + C_{\eta\lambda}\langle\eta\rangle\langle\lambda\rangle \tag{2.2}$$

where $\langle m_{\beta\beta}\rangle$, $\langle\eta\rangle$ and $\langle\lambda\rangle$ are the effective neutrino mass and the right-handed parameters. The coefficients C_{xy} (where $x,y = m,\lambda,\eta$) are products between various NME and the corresponding phase factors F (see, for example Ref. [9].

One of the challenging problems related to the study of $\beta\beta$ decay process is the accurate calculation of the NME that appear in Eqs (2.1) - (2.2).

There are mainly two types of approaches for calculating these NME: i) QRPA-based and ii) Shell Model-based.

Schematically, within QRPA-based methods for describing a particular nucleus, one starts first from a mean field potential and gets the single particle spectrum. Further, one introduces pairing correlations through the BCS procedure and then long range particle-hole (ph) and particle-particle (pp) correlations through the QRPA formalism. These approaches refer to the proton-neutron random phase approximation (pnQRPA), because the long-range correlations are introduced as particle-unlike (i.e. proton-neutron) correlations. and its extensions. Because many $\beta\beta$ decay emitters are nuclei with open shells, QRPA-based methods have been the most used method to perform such calculations. The single particle basis, within these methods, include, usually, 2-3 full shells and one can reproduce the experimental half-lives of the $2\nu\beta\beta$ decay mode for a large number of nuclei, by adjusting the strength (renormalization) parameter of the pp correlations, g_{pp}. However, calculated with this method, the NME manifest a large sensitivity against g_{pp}. To overcome this drawback several improvements of pnQRPA have been advanced. There is a whole history of these developments and we do not intend to describe them in details in this paper but suggest the reader to consult some comprehensive reviews like: Refs. [2]-[3]. However, we remember here shortly several of them. Because pnQRPA is based on the quasi-particle mean field, a certain number of symmetries like particle-number (due to the use of the BCS approximation), the isospin symmetry (due to the use of effective proton and neutron single particle states), angular momentum (when a deformed potential is used) are breaking, several approaches refer to the restoration of these symmetries (see Ref. [2] and references therein). Another type refer to extentions of the pnQRPA method beyond its basic approximation, namely the quasi boson approximation (QBA). On this line two different approaches have been developed. First, the idea that particle-like correlations at the RPA level should also contribute, besides the particle-unlike ones to the $\beta\beta$ decay process, to was exploited in Refs. [10]. In those works the first order corrections beyond QBA have been taken into account in the expressions of the bi-fermion quasiparticle and density-type operators involved in the pnQRPA formalism, by a boson expansion. Then, the corrections were introduced consistenly both in the pnQRPA wave functions (by enlarging the model space with two-boson states) and in the expressions of the $\beta^{\pm}$ operators. As an example we give below the expression of the $M_{GT}^{2\nu}$ obtained within this formalism, that we called SQRPA:

$$M_{GT} = \frac{\langle 0_i || \beta^- || 1_{k_1} \rangle \langle 1_{k_1} | 1_{k'_1} \rangle \langle 1_{k'_1} || \beta^- || 0_f \rangle}{m_e c^2 + Q_{\beta\beta}/2 + \Delta E_1} +$$

$$+ \frac{\langle 0_i || \beta^- || (1_{k_1} 2_{k_2}) 1 \rangle \langle (1_{k_1} 2_{k_2}) 1 | (1_{k'_1} 2_{k'_2}) 1 \rangle \langle (1_{k'_1} 2_{k'_2}) 1 || \beta^- || 0_f \rangle}{m_e c^2 + Q_{\beta\beta}/2 + \Delta E_2} \quad (2.3)$$

where ΔE_1 and ΔE_2 are differences in energy that include the pnQRPA and QRPA energies and the first 1^+ experimental energies in the intermediate nucleus. In the above expression the second row represents the correction added to the initial pnQRPA expression of $M_{GT}^{2\nu}$, due to the higher-order terms.

Several years later, the idea of restoring (partially) the Pauli exclusion principle, which is violated within pnQRPA, leads to the renormalized pnQRPA (RQRPA) approach. Here, higher-order corrections to pnQRPA have been introduced by taking into account the next terms in the commutators of the pair quasiparticle operators involved in the derivation of the pnQRPA equations [11]. The commutator is replaced by its expectation value in the pnQRPA (correlated) ground state and this leads to a renormalization of the relevant operators and of the forwarded- and backwarded-going QRPA amplitudes as well.

A lot of calculations of the NME for $\beta\beta$ decay have been performed with the SQRPA and RQRPA methods, including two- and zero-neutrino modes and transitions to ground and excited final states, transitions for all the $\beta\beta$ decay modes, etc.(see Refs. [12]-[22]). The main improvement these methods brought, was that the large sensitivity of the M_{GT} against g_{pp} was diminished and the g_{pp} parameter was better fixed from $2\nu\beta\beta$ decay calculations to be used further for the $0\nu\beta\beta$ decay modes. This can be seen in Fig. 2 in the case, for instance, of $\beta\beta$ decay of ^{76}Ge, where M_{GT} calculated with SQRPA and RQRPA methods, for two different single particle basis, is represented as function of g_{pp}

In the last time, significant progress has been done in the development of the Shell Model approach [23]. The quantitative and qualitative improvements include i) a significant extension of the valence space: due to the actual computer facilities and the use of the Lanczos tridiagonal construction that made now possible the diagonalization of matrices of dimensionality up to $10^9 - 10^{10}$; ii) construction of new effective NN interactions that include a two-body contribution derived from realistic potentials, which is consistent with two-nucleon data, and three body components incorporating both pure monopole terms and shell-formation properties of the realistic two-body forces. In spite of this, within the ShM approach one has still to severely truncate the model space. Although the calculations may include in principle all the ground state correlations, they are still limited to the $0\hbar\omega$ space.

Calculations performed with ShM include Gamow-Teller and Fermi transitions and thus computation of the NME relevant for $\beta\beta$ decays. Caurier et al. performed for a first time a full pf shell calculation of the NME for the transitions both to ground states and to the 2_1^+ excited states for ^{48}Ca, ^{76}Ge, ^{82}Se, ^{96}Zr, ^{136}Xe, using a Kuo-Brown interaction with monopole corrections [24]. In a recent work [25] the authors used the GXPF1

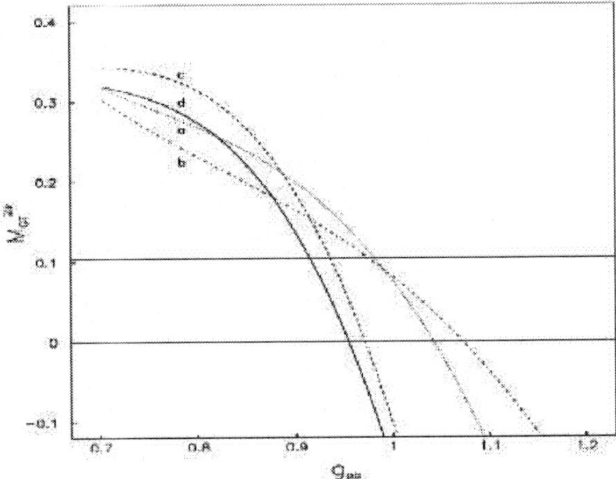

FIGURE 2. The $M_{GT}^{2\nu}$ matrix elements versus g_{pp} calculated with pnQRPA and SQRPA methods for the $\beta\beta$ decay of ^{82}Se. The curves correspond to the calculations performed in the following conditions: (a) SQRPA (s); (b) SQRPA (l); (c) pnQRPA (s); (d) pnQRPA (l). (l) and (s) means at calculations are performed with the large and small s.p. basis, respectively..

interaction, that was tested with good results to correct the previous results for $2\nu\beta\beta$ decay of ^{48}Ca. We obtained:

$$M_{GT}(0^+ \rightarrow 0^+) = 0.0539MeV^{-1} \qquad T_{1/2}(0^+ \rightarrow 0^+) = 3.3 \times 10^{19}y$$

$$M_{GT}(0^+ \rightarrow 2_1^+) = 0.0122MeV^{-3} \qquad T_{1/2}(0^+ \rightarrow 2_1^+) = 8.5 \times 10^{23}y,$$

for transitions to ground states and to the 2_1^+ excited states, respectively.

There are presently still important differences between the values of the NME, for the same $\beta\beta$ decay, calculated with different methods, even of similar type. So, it is very important to understand how large are the uncertainties in their calculation. First, there are uncertainties inherent to each kind of approaches. For instance in the case of QRPA-based methods the uncertainties may come from the residual NN interactions (some approches use schematic zero-range NN interactions, others use interactions derived from different NN potentials, parametrized in different manners), from the renormalization of the residual interactions in the pairing, p-h and mainly p-p channel as well as from the dependence on the size of single particle basis. In the ShM they may come from the reduced dimensionality of the $0\hbar\omega$ space, which is not a good enough approximation, from the missing of the contributions of negative parity intermediate states (which are important, according to the QRPA calculations) or the missing of several spin-orbit partners even in the $0\hbar\omega$ space. Secondly, there are differences between different calculations due to different choice of the model parameters. Thus, one needs to closely compare the various calculations in order to eliminate the errors. Very recently there are debates about the importance of the short-range correlations in treating the neutrino potentials and reasons of uncertainties in calculations of the NME, which were not known until now to be so important (see J. Suhonen, in this volume and [26]). A recent result in such an analysis is related to the sensitivity of the NME against the competition between pairing and proton-neutron pp interactions. After the authors of Ref. [26], this competition implies that only internucleon distances between 2-3 fm contribute and this dependence on separation occurs both in QRPA and ShM. Thus, the proper treatment of the short range correlations turns out to be an important issue related to the accuracy of the NME calculations. Taking advantage of the progress in the ShM approaches, one can proceed by now on to a carreful comparison of the two types of calculations in order to identify and eliminate the eventual systematical errors. The values of the NME for a specific nucleus and decay mode must not depend on a particular nuclear model. We appreciate that a reduction of the uncertainties in the calculation of the NME within 20% is a resonable one.

DERIVATION OF NEUTRINO MASS

Within SM, neutrinos are considered as massless particles which are left-handed partners of the charged bosons. However, in the recent past, neutrino oscillation experiments have convincingly demonstrated that neutrinos are massive and can transform from one flavor into another one. This is the first proved signal that the SM is not an ultimate theory, but only a projection of a more general GUT. The observation of the neutrino oscillations determines only mass square differences of the neutrino types and thereby sets a lower bound on the largest of the neutrino masses. The absolute scale of the neutrino masses remains still unknown and it can be determined from other type of experiments as follows:

i) From precise measurements of β decay of tritium one can derive an average of the neutrino masses:

$$< m_\beta >= \left(\Sigma_i |U_{ei}|^2 m_i^2\right)^{1/2} \tag{3.1}$$

The advantage of these experiments are the following: β decay is superallowed, the energy released and the half-life (12.5 years) are convenient quantities, precisely known, and the NME involved are constant; so, the electron spectrum is determined only by the phase spaces. The inconvenient, however, is the reduced sensitivity. Indeed, the current limits for $< m_\beta >$ come from Troisk [27] and Mainz [28] experiments and are 2.05 eV and 2.3 eV at a 95% CL, respectively.These results are with almost one order of magnitude less that the limits provided by $\beta\beta$ decay experiments. The future KATRIN experiment [29] is planned to reach 0.2 eV at 90% CL. $< m_\beta >$ can also be expressed in terms neutrino oscillation parameters by replacing the $|U_{ei}|$ parameters with their values from the unitary matrix (Eq. 3.3). Further, taking their values compatible with the results from neutrino oscillation experiments [30]-[33], one gets theoretical limits for the average mass parameter: 1) $< m_\beta >\sim 0.012$ eV for NH and $< m_\beta >\sim 0.05$ eV for IH.

ii) From $\beta\beta$ decay experiments another combination of neutrino masses can be extracted:

$$| < m_{\beta\beta} > | = \Sigma_i |U_{ei}|^2 m_i \tag{3.2}$$

In the above formula U_{ei} are the elements of the unitary matrix that connects the neutrino mass eigenvalues and mass flavors:

$$U = \begin{pmatrix} c_{12}c_{13} & s_{12}c_{13} & s_{13}e^{-i\delta} \\ s_{12}c_{23} - c_{12}s_{23}s_{13}e^{i\delta} & c_{12}c_{23} - s_{12}s_{23}s_{13}e^{i\delta} & s_{23}c_{13} \\ s_{12}s_{23} - c_{12}s_{23}s_{13}e^{i\delta} & -c_{12}s_{23} - s_{12}c_{23}s_{13}e^{i\delta} & c_{23}c_{13} \end{pmatrix} \tag{3.3}$$

where $s_{ij} = sin\theta_{ij}$, $c_{ij} = cos\theta_{ij}$ are related to the neutrino mixing angles and δ is the CP violation phase. To calculate the neutrino properties one needs to know all the independent elements of U.

To derive upper limits for the neutrino electron mass, one can assume that $0\nu\beta\beta$ decay occurs only through exchange of light Majorana neutrinos and retain in Eq. (2.2) the first term. Further, one can extracts the neutrino $\beta\beta$ decay mass parameter

$$| < m_{\beta\beta} > | = m_e \frac{1}{\sqrt{T_{1/2}C_{mm}}}, \tag{3.4}$$

where

$$C_{mm} = F^{0\nu} \left(M_{GT} - (\frac{g_v}{g_A})^2 M_F \right)^2 \tag{3.5}$$

Thus, the absolute neutrino mass is related to the non-observability of the $0\nu\beta\beta$ decay and to the uncertainties in the calculation of the NME: M_{GT} and M_F. In Refs. [34]-[35] is made a critical analysis of the values of NME obtained with different methods. In our opinion, if one carefully analyses the results and the ingredients involved in different calculations, one find uncertainties within $50-100\%$. Presently, the most strigent upper limit for $0\nu\beta\beta$ decay is $(0.8-18.3) \times 10^{25}$ y and comes from the Heidelberg-Moscow experiment [36]. In that paper the authors claimed the observation of a positive result for this decay mode. To be valid, this result of a great importance, must be confirmed from other independent sources. Using the average value for the lalf-life reported in that reference and using the NME calculated in [34] one obtains 0.35 eV for the $| < m_{\beta\beta} > |$. This parameter can also be expresssed in terms of neutrino oscillation parameters:

$$| < m_{\beta\beta} > | = |U_{e1}^2 \pm |U_{e2}|^2 (m_1^2 + \Delta m_{12}^2)^{1/2} \tag{3.6},$$

Taking into account the average values of the neutrino oscillation parameters obtained from an analysis of the results of those experiments [30]-[33], one has:

1) $| < m_{\beta\beta} > | \sim$ several meV for NH, 2) $| < m_{\beta\beta} > | \sim$ several tens of meV for IH and 3) $| < m_{\beta\beta} > | \sim 50$ meV for degenerate hierarchy. Some of these values are in the range of the future planned $\beta\beta$ decay experiments.

Besides these experiments, from precision measurements of the cosmic microwave background [37] combined with observations of the large scale structure of the universe and analysis [39] and of the Lyman-alpha spectra (the Lyman alpha forest is the sum of absorption lines arising from the Lyman alpha transition of the neutral hydrogen in the spectra of distant galaxies and quasars [40]) one can derive very stringent limits for the sum of neutrino masses: $\Sigma_i m_i \sim 0.7$ eV or $\Sigma_i m_i \sim 0.4$ eV without or with the inclusion of Lyman-alpha results [41]. Presently, the double beta decay experiments searching for neutrinoless decay mode are the most sensitive ones for measuring the absolute mass of the neutrino. Moreover, these are the only experiments that could answer the fundamental question of whether neutrinos are their own anti-particles, that is whether they are Majorana or Dirac particles. On the other hand, there are stringent constraints coming from cosmological data for the sum of the neutrino masses. A global analysis incorporating results from $\beta\beta$ decay, neutrino oscillation experiments and constraints from cosmology is of great help in deriving neutrino absolute masses.

CONCLUSIONS

In the latest years there were impressive experimental results in favor of the neutrino oscillation interpretation: neutrinos have non-zero mass and mix. This is the first demostrated signal that physics beyond the SM exist. However, there are several fundamental questions that we still do not know the answer: which is the absolute value of the neutrino mass, what is the nature of neutrino: is it a Majorana or a Dirac particle, what is the neutrino mass hierarchy, what is the mass generation mechanism for neutrino?, etc. The double-beta decay experiments are presently the most promissing to answer to some of these questions. Also, the $\beta\beta$ decay process has a broader potential to search for physics beyond the SM. For instance, it can provide us with information about the lepton number non-conservation, existence of RH components in the weak interaction, L-R symmetric models, Majoron or leptoquark existence, compositness, R-parity non-conservation and possibility that $0\nu\beta\beta$ occurs via exchange of SUSY particles between nucleons, etc. The non-observability of the $0\nu\beta\beta$ decay mode can thus be used to also constrain the new interactions appearing in GUTs. Calculation of the NME involved in $\beta\beta$ decay is still a challenge issue. The study of the $2\nu\beta\beta$ decay mode, although does not address fundamental questions, it is nevertheless interesting for testing the nuclear models with which the NME are calculated. Further, these NME are used to derive from $0\nu\beta\beta$ decay half-lives formula neutrino masses. In recent years pnQRPA and its extensions reproduced fairly well the measured half-lives. In addition, the progress of the ShM allows now realistic calculations of the NME with this approach. One of the recent issues related to the accuracy of the NME computation is the manner of calculation of the short-range correlations. By now on, it would be possible a systematical comparison of the results performed with both kind of methods, i.e. pnQRPA and ShM, in order to significantly reduce the uncertainties.

The progress in our knowledge on the neutrino properties is presently focused on the crucial problem of the neutrino mass. Further developments in $\beta\beta$ decay study (both theoretically and experimentally), in neutrino oscillation experiments (particularly the long baseline experiments) and in the precision of cosmological data, will certainly lead in the next future to new exciting results in neutrino physics.

REFERENCES

1. W.C. Haxton and G.J. Stephenson, *Prog. Part. Nucl. Phys.* 12, 409 (1984).
2. J. Suhonen and O. Civitarese, *Phys. Rep.* **3**00 (1998) 123.
3. H.V. Klapdor-Kleingrothaus, *"60 Years of Double-Beta Decay - From Nuclear Physics to Beyond Standard Model Particle Physics"*, World Scientific, Singapore, 2001.
4. Yu. Zdesenko, F. Danevich and V. Tretyakov, *Phys. At. & Nucl.*, **67**, 1974 (2004).
5. M. Danilov et al., *Phys. Lett.* **B 480**, 12 (2000).
6. C. E. Aalseth, et al., *Nucl. Phys. Proc. Suppl.* **138**, 217 (2005).
7. M. Nomachi et al., *Nucl. Phys.* **B 138**, 221 (2005).
8. X. Sarazin et al., *hep-ex/0006031.*
9. K. Muto, E. Bender and H.V. Klapdor, *Z. Phys.* **34**, 177 (1989); Z. Phys. **A 334**, 187 (1989).
10. A. A. Raduta, A. Faessler, S. Stoica and W. A. Kaminski, *Phys. Lett.* **B 254**, 7 (1991); A.A. Raduta, A. Faessler and S. Stoica, *Nucl. Phys.* **A 534**, 149 (1991).
11. J. Toivanen and J. Suhonen, *Phys. Rev. Lett.* **75**, 410 (1995); *Phys. Rev.* **C 55**, 2314 (1997).
12. P. Vogel and M. R. Zirnbauer, *Phys. Rev. Lett.* **57**, 3148 (1986).
13. J. Suhonen, T. Taigel and A. Faessler, *Nucl. Phys.* **A 486**, 91 (1988).
14. A. Staudt, K. Muto and H.V. Klapdor-Kleingrothaus, *Europhys. Lett.* **13**, 31 (1990).
15. S. Stoica and W.A. Kaminski, *Phys. Rev.* **C 47**, 867 (1993); S. Stoica, *Phys. Rev.* **C 49**, 787 (1994).
16. S. Stoica, *Phys. Lett.* **B 350**, 152 (1995); S. Stoica and I. Mihut, *Nucl. Phys.* **A 602**, 223 (1996)
17. J. Schwieger, F. Simcovic and A. Faessler, *Nucl. Phys.* **A 600**, 179 (1996).
18. F. Simkovic, J. Schwieger, M. Veselsky, G. Pantis and A. Faessler, *Phys. Lett.* **B 393**, 267 (1997).
19. A. Bobyk, W.A. Kaminski and P. Zareba, *Eur. Phys. J.* **A 5**, 385 (1999); *Nucl. Phys.* **A 669**, 221 (2000).
20. H. Ejiri and H. Toki, *J. Phys. Soc. Jpn.* **65**, 7 (1996).
21. G. Pantis and F. Simkovic, *Nucl. Phys.* **A 663&664**, 825c (2000).
22. S. Stoica and H. V. Klapdor-Kleingrothaus, *Phys. Rev.* **C 63**, 064304 (2001).
23. E. Caurier, G. Martinez-Pinedo, F. Nowacki, A. Poves and A.P. Zuker, *Rev. Mod. Phys.* **77**, 427 (2005).
24. E. Caurier, A. Poves, and A.P. Zuker, *Phys. Lett.* **B 252**, 13 (1990); E. Caurier, F. Nowacki, A. Poves and J. Retamosa, *Phys. Rev. Lett.* **77**, 1954 (1996).
25. M. Horoi, S. Stoica and B.A. Brown, *Phys. Rev.* **C 75**, 034303 (2007).
26. F. Simkovic, A. Faessler, V.A. Rodin, P. Vogel and J. Engel, *arXiv/0710.2055v2 [nucl-th].*
27. V.M. Lobachev, *Nucl. Phys.* **A 719**, 153 (2003).
28. C. Kraus et al., *Eur. Phys. J.* **C 40]**, 447 (2005).
29. A. Osipovicz et al., *hep-ex/0109033*; J. Bonn, this volume.
30. SNO Collaboration, Q.R. Ahmad, Phys. *Rev. Lett.* **89**, 011301 (2002); 011302 (2002).
31. S. Fukuda et al., *Phys. Rev. Lett.* **86**, 5651 (2001).
32. KamLAND collaboration, K. Eguchi et al., *Phys. Rev. Lett.* **90**, 021802 (2003).
33. CHOOZ collaboration, M. Appollonio et al., *Phys. Lett.* **B 466**, 415 (1999).
34. S. Stoica and H.V. Klapdor-Kleingrothaus, *Nucl. Phys.* **A 694**, 269 (2001).
35. V.A. Rodin, A. Faessler, F. Simkovic, P. Vogel, *Nucl. Phys.* **A 766**, 107 (2006) and erratum, *Nucl. Phys.* **A 793**, 213 (2007).
36. H.V. Klapdor-Kleingrothaus, A. Dietz, H.I. Harney and I.V. Krivosheina, *Mod. Phys. Lett.* **A 16**, 2409 (2001).
37. WMAP collaboration, C.L. Bennet et al. *ApJ, Suppl.* **149** 1 (2003).
38. 2dFGRS collaboration, W.J. Percival et al., *Mon. Nat. Roy. Astron. Soc.* **327**, 1297 (2001).
39. SDSS collaboration, M. Tegmark et al., *Phys. Rev.* **D 69**, 103501 (2004).
40. D.H. Weinberg, R. Dav'e, N. Katz, J.A. Kollmeier, *AIP Conf.Proc.* **666**, 157 (2003).
41. D.N. Spergel et al., *ApJS* **148**, 175 (2003); U. Seljak et al., *Phys. Rev.* **D 71**, 103515 (2005)

The Gran Sasso Laboratory and Neutrinos

Alessandro Bettini[a,b]

[a]*University of Padua - G. Galilei Physics Department- and INFN. Via Marzolo 8 35131 Padova Italy*
Bettini@PD.INFN.IT.
[b]*Laboratorio Subterráneo de Canfranc. Canfranc (Huesca) Spain*

Abstract. After a brief survey of the experimental programme of the INFN Gran Sasso National Laboratory, I summarize the status of neutrino physics. I then focus on two frontier challenges. 1. The possible solution of the mass spectrum hierarchy problem with the observation of neutrinos from a supernova explosion; 2. The establishment of the nature of neutrinos, whether they are Dirac or Majorana particles, with neutrino-less double-beta decay.

Keywords: Underground laboratories; Neutrino mixing; Solar neutrinos; Supernova neutrinos; Dark matter.
PACS: 01.52.+r; 14.60.Pq; 26.65.+t; 97.60.Bw; 95.35.+d.

THE EXPERIMENTS AT LNGS

Underground laboratories are complementary to those with accelerators. They provide the low radioactive background environment necessary to the search for the rare phenomena, an indirect way to explore the highest energy scales. Indeed, physics beyond the standard model has been discovered in underground laboratories, with experiments on neutrinos from natural sources: the Sun and the cosmic rays.

The INFN Gran Sasso Laboratory is located besides the freeway tunnel (10.4 km long) connecting L'Aquila (West) and Teramo (East), at about 6 km from the west entrance, 120 km from Rome. The underground structures consist of three main halls, about 100 m x 20 m x 18 (h) m plus ancillary tunnels providing space for services and small scale experiments, for a total area of 17 300 m^2 and a volume of 180 000 m^3. As important are the facilities on the surface that are fundamental for the research activity

Space allows me only a brief summary of the rich experimental programme of the LNGS. For more details see http://www.lngs.infn.it/. After this summary I shall discuss two issues, Supernova neutrinos and neutrino-less double beta decay.

Dark matter search. No particle in the Standard Model can explain dark matter, the largest fraction of matter in the Universe. However, SUSY, the most popular extension of the model, has it, the neutralino. More generally, dark matter particles of this type are called Weakly Interacting Massive Particles. WIMPs interact only weakly, but, dislike neutrinos, should have masses between several GeV and, say, one TeV. Experiments using all the different frontier approaches for this difficult search are present at LNGS.

DAMA/LIBRA is taking data since 2003 with about 250 kg of hyper-pure NaI crystals. It is the only experiment world wide capable to search for the characteristic

CP972, *Exotic Nuclei and Nuclear/Particle Astrophysics (II)—Proceedings of the Carpathian Summer School of Physics 2007,* edited by L. Trache and S. Stoica

annual modulation of the WIMPs counting rate. Its predecessor, DAMA, is the only experiment that reported a positive signal up to now.

CRESST2 using TeO_2 bolometers measures both the energy deposit, detecting the phonons, and the scintillation light, to discriminate the electromagnetic background.

XENON10 is a 15 kg liquid Xe TPC, measuring both the ionisation charge and the scintillation light. The former is detected by extracting the charges into the gas phase and amplifying them. With its novel technique, it is presently the most sensitive experiment in the world

WARP uses a similar approach with liquid Ar. A 100 litres TPC is under construction. Very interesting results have been already obtained with a small prototype.

Neutrinos from the Sun. BOREXINO has started its data taking by mid May. It detected for the first time the neutrinos from the Be line in the solar spectrum in 100 t of hyper-pure liquid scintillator in the centre of onion-shaped liquid shields[1].

Neutrinos from Supernovae. LVD, a 1000 liquid scintillator detector, is taking data since 1992, in its complete configuration since 2001. It will be discussed later.

Neutrinos from CERN. The **OPERA** experiment, a hybrid emulsion-electronic detector, searches for ν_τ appearance in an initially (almost) pure ν_μ beam from CERN (CNGS). With the approved 5 years run and its almost 2 kt sensitive mass, 15 ν_τ identified events are expected with a background of less than one event.

Neutrino-less double beta decay. The experiments, CUORICINO/CUORE and GERDA will be discussed in detail later on.

Nuclear astrophysics. LUNA and **LUNA2**. The nuclear reactions in the Sun and in the stars take place at very small energies. Consequently, their cross-sections are extremely small and can be measured only underground, to avoid the cosmic rays interference. For the first time LUNA2 and its predecessor LUNA have measured, with small dedicated accelerators, the cross-sections relevant for the solar cycle. This, in particular, excluded the "nuclear solution" of the "solar neutrino puzzle" in 1997.

STATUS OF NEUTRINO PHYSICS

Contrarily to the assumptions of the Standard Model, 1. neutrinos are massive, 2. their stationary states are not the states of definite flavour, 3. flavour lepton numbers are not conserved. Neutrinos have been observed to change flavour in two different ways.

The first phenomenon is neutrino oscillations. It is purely kinematical. It happens both in a vacuum and in matter. It has been discovered as the disappearance of the ν_μs indirectly produced by cosmic rays in the atmosphere. Typical neutrino energies (E) are from sub-GeV to multi-GeV, while the flight lengths (L) are up to a few thousands kilometres.

The probability to observe the flavour state ν_β in a beam initially pure ν_α contains oscillating terms of the type $P_{\alpha\beta} = A(\theta_{12},\theta_{23},\theta_{13})\ \sin^2\left(1.27\left(m_i^2 - m_j^2\right)\left(\mathrm{eV}^2\right)\frac{L(\mathrm{km})}{E(\mathrm{GeV})}\right)$.

The oscillation amplitudes A are functions of the three mixing angles, $\theta_{ij} \in [0,\pi/2]$. The oscillation frequencies are proportional to the differences between the squares of the masses of the eigenstates. We call Δm^2 the square mass difference of the "atmospheric" oscillation. The phenomenon (with a good approximation) is independent on the sign of $m_i^2 - m_j^2$ and, also, on sgn($\pi/2-\theta$□□.

The second phenomenon is the flavour conversion in matter, the Mikehev-Smirnov-Wolfenstein[2] effect. It is a dynamical phenomenon, due to the $\nu_e e$ interaction in the matter. The interaction energy depends on neutrino energy and matter density, and has opposite sign for neutrinos and antineutrinos. While crossing a variable density medium ν_es may reach a critical density layer where a resonant transition to a coherent superposition of the other flavours may happen (level crossing). The phenomenon is the dominant process in the Sun, for above-MeV neutrino energies. We call δm^2 the corresponding square mass difference. δm^2 happens to be much smaller than Δm^2. The matter effect does depend both on the sgn(δm^2) and sgn($\pi/2-\theta_1$□. In a Supernova, due to its much larger density, two level crossings are possible, at δm^2 and at Δm^2.

This evidence tells us that the flavour states are not the mass eigenstates. We call the latter ν_1, ν_2 and ν_3, and m_1, m_2 and m_3 their masses. Flavour states are linear combinations of the eigenstates, $\nu_l = \sum_{i=1}^{3} U_{l,i}\nu_i$, where $l=e,\mu,\tau$ and U is the mixing matrix, which can be written as the product of three rotation matrices including a phase factor, as in the case of quarks, and a fourth diagonal matrix with two more phases. The latter, called Majorana phases, can be absorbed in the wave functions only if neutrinos and antineutrinos are different (Dirac) particles. With $c_{ij}=\cos\theta_{ij}$, and $s_{ij}=\sin\theta_{ij}$ we have

$$U = \begin{pmatrix} 1 & 0 & 0 \\ 0 & c_{23} & s_{23} \\ 0 & -s_{23} & c_{23} \end{pmatrix} \begin{pmatrix} c_{13} & 0 & s_{13}e^{i\delta} \\ 0 & 1 & 0 \\ -s_{13}e^{-i\delta} & 0 & c_{13} \end{pmatrix} \begin{pmatrix} c_{12} & -s_{12} & 0 \\ s_{12} & c_{12} & 0 \\ 0 & 0 & 1 \end{pmatrix} \begin{pmatrix} 1 & 0 & 0 \\ 0 & e^{i\alpha} & 0 \\ 0 & 0 & e^{i\beta} \end{pmatrix}$$

In total there are nine quantities to measure, three masses, three mixing angles and the three phases. The latter, if $\neq 0$ and $\neq \pi$, give *CP* violation effects in the lepton sector. Majorana phases are, even if present, irrelevant in oscillation and matter conversion phenomena; they appear only in double beta decay.

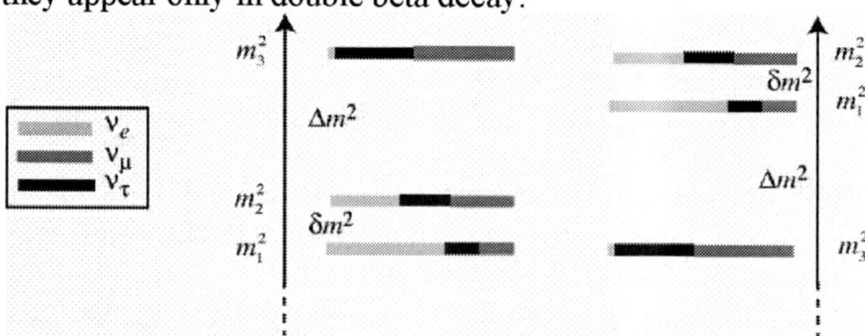

FIGURE 1. Neutrino mass spectrum. Left: "normal" hierarchy; right "inverse" hierarchy

Different groups have performed global fits including all the available data, providing values for the three mixing angles (only an upper limit for θ_{13}) and two squared mass differences. Results can be summarised as follows[3] (1 σ uncertainties):

$$|\Delta m^2| \equiv |m_3^2 - (m_2^2 + m_1^2)/2| = 2600 \pm 180 \text{ meV}^2 \quad \delta m^2 \equiv m_2^2 - m_1^2 = 79.2 \pm 3.6 \text{ meV}^2$$

θ_{12} = 34.1Þ±1.5Þ θ_{23} = 42Þ$^{+4.6Þ}_{-2.5Þ}$ θ_{13} < 8Þ.

Fig. 1 shows schematically the neutrino mass spectrum. It consists of a doublet (m_1 and m_2) and a singlet (m_3). We do not know whether m_3 is larger or smaller than $m_{1,2}$ neither the absolute value of the masses. An indirect upper limit on each of the masses is given by cosmology, approximately m_i<200 meV.

Let us now consider closer the solar neutrinos. Neutrinos are produced by fusion reactions in the core of the star as pure ν_e by charged current (CC) weak interactions. They will then move in matter and interact, again via CC, with the electrons of the medium (one can show that the interactions with protons and neutrons are not relevant for the phenomenon under study). Due to that interaction the neutrino Hamiltonian in the medium differs from that in vacuum and its eigenstates are not the vacuum eigenstates.

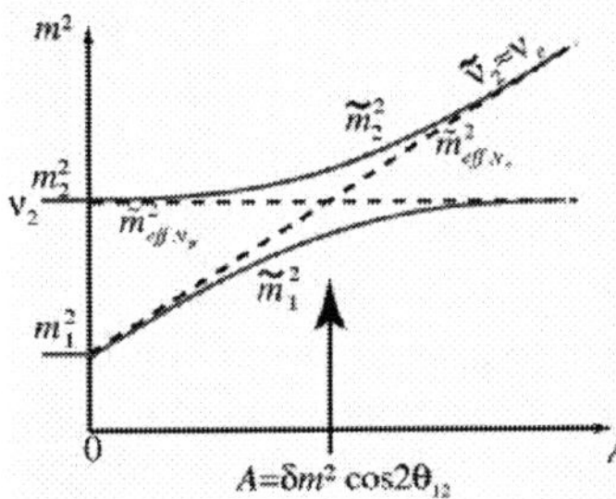

FIGURE 2. Mass eigenvalues vs. $A = 2\sqrt{2}G_F n_e E_\nu$ (n_e is the electron density, E_ν is the neutrino energy)

Fig. 2 shows the two eigenvalues relevant for the Sun as functions of the density, ranging from zero-density, hence in vacuum, to the density in the core. Notice that the upper mass eigenstate $\tilde{\nu}_2$ is at high densities practically identical to the flavour state ν_e. The higher mass eigenvalue gradually decreases with decreasing density. In its route to the exit the neutrino state follows the upper curve and sooner or later encounters a layer in which the density satisfies the level crossing conditions, provided the energy is larger than about one MeV. At the crossing, the particle follows the upper curve, provided that the density variation is not too fast (adiabatic condition). This condition is satisfied if the relevant mixing angle is larger than 1-2°. This is certainly true in the solar case, where the relevant angle is θ_{12}, not necessarily, as we shall see in a Supernova.

Notice that above-MeV neutrinos emitted by the Sun are in a pure mass eigenstate, ν_2. As we shall see a similar situation happens for a Supernova. While the elementary weak interaction processes produce neutrinos in a flavour eigenstate, celestial bodies emit (in certain circumstances) neutrinos as mass eigenstates.

WHAT CAN WE LEARN ON NEUTRINOS FROM A CORE COLLAPSE SUPERNOVA IN OUR GALAXY?

The evolution of isolated iron core massive stars terminates with the collapse of the core, a Supernova explosion. Most of the gravitational energy, an enormous quantity, 3×10^{46} J, is emitted in neutrinos and antineutrinos. The neutrino burst from SN 1987a, exploded in the Large MC, was detected by two water Cherenkov detectors, Kamiokande[4] and IMB[5], and by a liquid scintillator detector, BNO[6]. Even if twenty years ago the sensitive masses were one order of magnitude smaller than the present ones, already the sparse statistics (≈30 events) that have been collected give us hints on neutrino physics. We shall discuss them with the aim to show how much we shall learn from a SN neutrino burst with the now existing detectors.

Neutrinos production in a Supernova evolves in the following phases. In the first phase, called "deleptonisation", the electrons are captured by protons, with production of neutrons, and by nuclei. *n*-reach nuclei become the dominant matter component. Gravitational attraction is no more balanced by the degenerate electron gas. The collapse proceeds rapidly in about 1 ms. An intense purely ν_e burst is produced. Unfortunately, these neutrinos are too few to be detectable by the present detectors (a part for a very near SN). The cooling phases follow, in which $e^+ e^-$ pairs are produced. Initially, in the "accretion" phase, $\bar{\nu}_e$ are produced in the falling *n*-reach matter by $e^+ + n \to \bar{\nu}_e + p$, and ν_e by $e^- + p \to \nu_e + n$. Before thermal equilibrium is reached, in the first second, mainly ν_e and $\bar{\nu}_e$ are produced. The final phase, which lasts ≈30 s, cooling and diffusion of neutrinos outside the inner core takes place. Neutrino-antineutrino pairs of all the flavours are produced via $e^- + e^+ \to \nu_x + \bar{\nu}_x$. ν_μ and ν_τ, having only neutral current interaction, are in equilibrium into a smaller sphere than ν_e. In it the temperature is higher and, as a consequence, the spectrum of ν_μ and ν_τ (and of their antineutrinos) is harder (average energy approximately 25 MeV) than that of ν_e (average energy approximately 12 MeV). There are large uncertainties on these estimates.

Presently (not in 1987), the most sensitive detectors are SeuperKAMIOKANDE, a 22 kt fiducial mass water Cherenkov, and four liquid scintillator detectors: LVD (1080 t), KamLAND (1000 t), BNO (300 t) and BOREXINO (300 t). All experiments are mainly sensitive to $\bar{\nu}_e$ because $\sigma(\bar{\nu}_e + p \to n + e^+)/\sigma(\nu_e + e^- \to \nu_e + e^-) \approx 10$ at the typical 10 MeV energy. In the LS the following neutron capture $n + p \to d + \gamma + 2.2$ MeV is used as a tag. For a SN explosion in the centre of the Galaxy (9 kpc) about 7000 events are expected in SuperK, 300 in LVD and in KamLAND, 100 in BOREXINO and BNO.

SuperK will also provide about 300 elastic scattering events $\nu_x + e \to \nu_x + e$ and $\bar{\nu}_x + e \to \bar{\nu}_x + e$, with information on the direction of the source and on the energy spectrum. LVD and KamLAND have a small sensitivity to electron neutrinos through the charged current process $\nu_e + {}^{12}C \to e^- + {}^{12}N$, followed by ${}^{12}N \to {}^{12}C + e^+ + \nu_e$ (energy threshold at 17.3 MeV). A handful of events are expected in absence of MSW conversion, because ν_e energies would be below the thresholds for a large fraction of their spectra. On the contrary, MSW can make the spectrum harder, with a substantial increase of the yield.

Neutrinos and antineutrinos must cross the high density SN medium before leaving the star. The flavour conversion process is similar to that in the Sun and similar are neutrino energies, but there are important differences. Let us focus on ν_e and $\bar{\nu}_e$. Due to the much larger density, two level crossings, one higher (*HC*) at Δm^2 and one lower (*LC*) at δm^2, are present. The adiabatic condition is guaranteed at the lower crossing because the relevant mixing angle is θ_{12}. On the contrary, at the higher crossing, the mixing angle is θ_{13}, which is unknown. The HC is adiabatic if $|\theta_{13}|$ > about 1°. If not, electron neutrinos do not "see" the higher crossing and continue to the lower, where the transition happens.

Skipping the details, the conclusions are the following. Neutrinos born as ν_e in the SN core leave its surface as ν_3 if $\Delta m^2>0$ and $|\theta_{13}|>1°$ (we call it case a), as ν_2 in the other three cases ($\Delta m^2>0$ with HC non-adiabatic and $\Delta m^2<0$ anyway) (case b). Taking into account that the matter term for antineutrinos is opposite to that of neutrinos, antineutrinos born as $\bar{\nu}_e$ exit as $\bar{\nu}_3$ in case a), as $\bar{\nu}_1$ in case b).

Call Φ_e the $\bar{\nu}_e$ flux detected on Earth and Φ_e^{SN} the original one, and similarly for the other flavours. It can be shown that $\Phi_e=\Phi_{\mu\tau}^{SN}+P_{ee}(\Phi_e^{SN}-\Phi_{\mu\tau}^{SN})$[7]. In case a), being P_{ee} very small, we have $\Phi_e\approx\Phi_{\mu\tau}^{SN}$. No antineutrino born as $\bar{\nu}_e$ reaches us, neutrinos detected as $\bar{\nu}_e$ were originally $\bar{\nu}_{\mu\tau}$ and correspondingly with about twice the energy. In case b), we have $\Phi_e\approx 0.3\,\Phi_{\mu\tau}^{SN}+0.7\,\Phi_e^{SN}$. The spectrum is rather similar to the original one. An analysis by Raffelt and collaborators[8] of the SN1987a data has shown that the observed spectrum is not harder than the expected one before MSW conversion. This points to case b), but statistical and systematic uncertainties do not allow a firm conclusion. However, we see how much we shall learn from a SN now.

Further evidence comes from the observation, by all the experiments, that about 40% of the events are in the first second, while the others are spread over more than 20 s. Quantitative analysis[9] shows that fitting the data with a single time constant is very difficult and leads to values for the SN radius and collapse energy rather different from the expectations. If this conclusion is true, a substantial fraction of the original $\bar{\nu}_e$ burst has reached our detectors without changing to other flavours in the SN. This again points to case b), because in case a) it would be $\Phi_e=P_{ee}\,\Phi_e^{SN}$.

THE NATURE OF NEUTRINOS

In the SM neutrinos are assumed to be Dirac particles. In 1937 E. Majorana proposed[10] a new theory for neutral spinors alternative to the "simple extension of the Dirac equation to neutral particles", "in spite of the fact that is probably not yet possible to ask to the experience to decide". After almost 70 years we do not have the answer to this fundamental question, but we have the means to actively asking a decision to the experience.

Majorana neutrino and antineutrino are two states of the same, massive, particle, distinguished by the helicity. In this case, an extremely rare weak decay can happen, the neutrino-less double beta decay ($0\nu2\beta$). Nuclides stable under β-decay decay via a second order weak process, the two-neutrino double beta decay $N(A,Z)\rightarrow N(A,Z+2)+2e^{\$}+2\bar{\nu}_e$, which is a very rare, but standard process. It has been observed in 10 nuclides with lifetimes of the order of 10^{19} yr.

If Majorana particle, the antineutrino produced at a nucleon can be absorbed by another one as neutrino: the neutrino-antineutrino propagator mediates the $0\nu2\beta$ process $N(A,Z)\rightarrow N(A,Z+2)+2e^{\$}$. The observation of this process would prove the Majorana nature of the electron neutrino. The decay probability is proportional to the square of the so-called "Majorana mass" $M_{ee} = \left|\sum_{i=1}^{3} U_{ei}^{2} m_i\right|$. Notice that the phase factors that can induce cancellations. Fig. 3 shows M_{ee} as a function of the smallest neutrino mass, for the two possible hierarchies, calculated by Feruglio et al.[11] taking into account of the uncertainties in the parameters. $0\nu2\beta$ lifetimes are expected to be larger than 10^{24} yr.

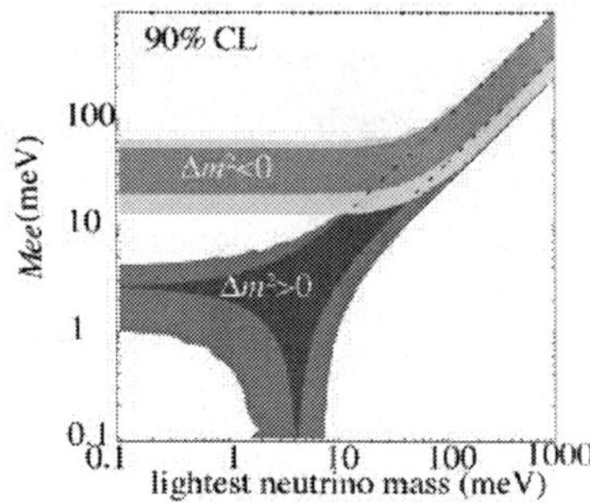

FIGURE 3. "Majorana mass" as a function of the lightest neutrino mass.[11]

The experiments measure (or limit) the decay probabilities, which are proportional to M_{ee}^2 and to the absolute square of the relevant nuclear matrix element. The latter are very difficult to evaluate and have been uncertain by large factors. The situation is improving and recent calculations quote uncertainties around 30%[12], but results from different authors still vary in a wider range.

Experimentally, by far the largest issue is the background. In presence of background the sensitivity to M_{ee} scales with the fourth root of the exposure. Consequently, background free conditions must be achieved in the region of the signal, which is known with high precision, in a range determined by the energy resolution ΔE.

In the "calorimetric" experimental approach, the source coincides with the detector. The energy deposited by the two electrons is accurately (a few ‰ FWHM) measured. Its distribution is a continuous spectrum from $2\nu2\beta$, which is certainly present, and a line after its end-point if $0\nu2\beta$ exists. To this the backgrounds add both continuous and consisting in monochromatic lines. Two techniques have been developed. In the first Germanium, enriched in the active isotope ^{78}Ge, is used to build diodes (GERDA); the second is based on thermal detectors (CUORE). In both cases masses of the order of several 100 kg of extremely radiopure material can be assembled.

The Heidelberg-Moscow experiment at Gran Sasso, the most sensitive so far, used enriched Ge diodes. It took data for 13 years with 11 kg sensitive mass and the lowest background index ever reached, b=0.11/(keV kg yr). A subset of the collaboration reported a positive evidence at 4.2σ level in 2004[13], $T_{1/2}=1.2\times10^{25}$ yr (3σ interval: 0.69-4.18) corresponding to 100 meV<M_{ee}<900 meV, depending on the assumed nuclear matrix element (about 700 meV with that of Rodin et al.[12]). IGEX[14], of a little smaller sensitivity, gives the upper limit M_{ee}<330-1350 meV.

CUORICINO/CUORE. The basic detector elements are TeO_2 cubic crystals of 5 cm side and 790 g mass. The natural abundance of the double beta active isotope, ^{130}Te, is 34%, large enough to make enrichment not necessary. The energy deposit

($Q_{\beta\beta}$=2529 keV) is detected by measuring the temperature increase with a thermal sensor, located on one of the surfaces. To have an accurately measurable ΔT, the specific heat must be as small as possible. Being $C_V \propto (T/T_D)^3$, the crystals are immersed in a thermal bath at a few mK. The FWHM energy resolution ΔE=7 keV has been achieved by CUORICINO at 2 MeV. CUORICINO is one "tower" of crystals, shown in Fig.4a. It contains 11.7 kg of ^{130}Te. It is running since April 2003 and is presently the most sensitive experiment. The background index is b=0.18±0.02. The present limit[15], shown in Fig. 4 and based on a 11.83 kg yr exposure, is $T_{1/2}>3.0\times10^{24}$ yr @ 90% c.l. It corresponds to M_{ee}<160-840 meV, depending on the nuclear matrix element.

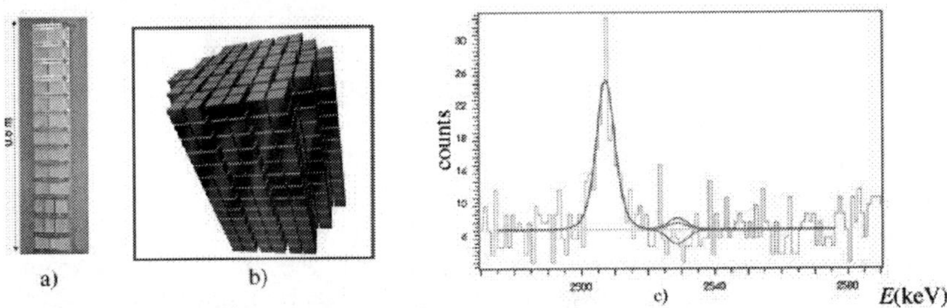

FIGURE 4. a) the CUORICINO tower; b) CUORE; c) CUORICINO spectrum

CUORE[16], the next, already approved, step, will be made of closely packed CUORICINO like towers (988 detectors) for a total ^{130}Te mass of 203 kg. To reach a sensitivity M_{ee}=20-100 meV, the background should be reduced at least to 10^{-2}/(keV kg yr). Special efforts are dedicated to the main problem coming from degraded α and β on and near the surfaces. Data taking is expected to start in 2011.

GERDA. The experiment[17], under construction at LNGS, will use naked Ge crystals assembled into an array suspended in the centre of a liquid Ar volume. This eliminates the problems with the radioactive components near the detectors. The shielding of environmental backgrounds is realised using hyperpure Ar and water. Fig. 5a) is an artist view of GERDA; it is more than 10 m across.

In Phase-I, existing detectors (from HM and IGEX), about 15 kg mass, will be used. They are expected to give the dominant contribution to the background budget (at $Q_{\beta\beta}$=2039 keV), with about 10^{-2} c/(kg keV yr), due mainly to their intrinsic cosmogenic ^{60}Co contamination. The background from the environment, the shields and the set-up components close to the crystals will be reduced to $b=10^{-3}$/(keV kg yr). This will be enough to confirm or refute the Klapdor et al. evidence in an exposure of one year. In phase-II, about 20 kg of new, segmented, detectors (Fig. 5b) will be added aiming at a global background index $b=10^{-3}$/(keV kg yr). As shown in Fig. 5 c) a sensitivity of $T_{1/2}=13\times10^{25}$ yr will be reached with a 100 kg yr exposure, corresponding to M_{ee}=140 meV with the Rodin et al[12] matrix element.

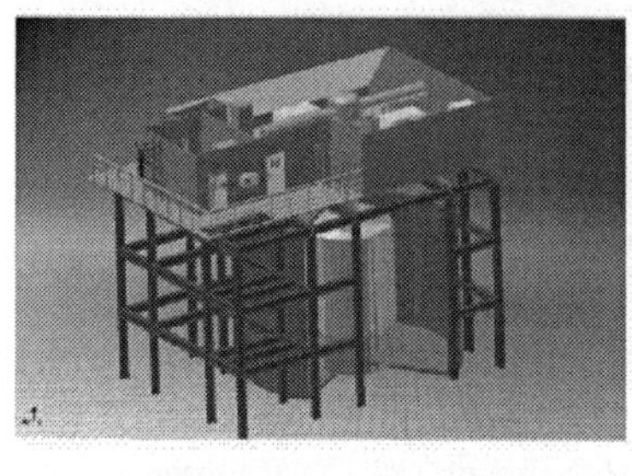

a)

b)

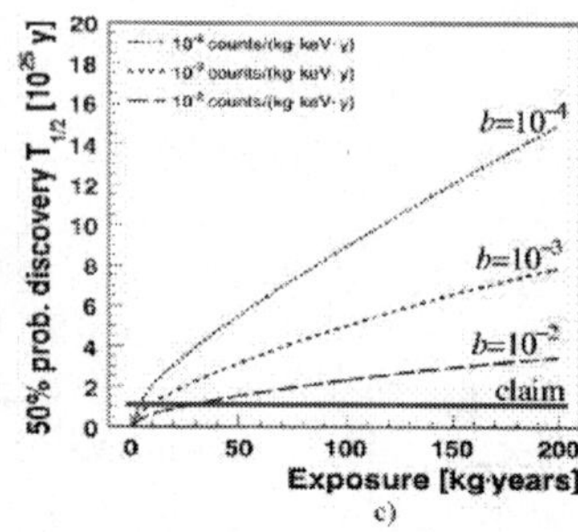

c)

FIGURE 5. a) a view of GERDA; b) segmented crystal with the contacting scheme in its low mass holder; c) sensitivity in half-life, for different background index values ($kg^{-1}keV^{-1}yr^{-1}$)

Background control and reduction (to 1/100 of the present status of the art) is achieved with the following means: 1. accurate calculation of the contributions of all background sources and evaluation of the tolerable specific activity; 2. control and reduction of the Ge exposure on the surface; 3. shielding of γs from the external environment with water and LAr; 4. veto of the μ-induced prompt signals with a plastic scintillator roof and the water used as Cherenkov medium; 5. screening of the materials and the components (of the suspensions, the front-end electronics, the cabling, etc.), by measuring at different sites, by γ-spectroscopy (at LNGS, Baksan, MPIK and Geel) and by ICP-MS (at LNGS, Moscow and Frankfurt University); 6. mechanical design of minimum mass suspension and contact system for the Ge detectors; 7. development of specially designed front-end electronics; 8. geometrical localization and pulse shape analysis to discriminate the backgrounds intrinsic to the Ge diodes.

REFERENCES

1. C. Arpesella et al. astro-ph/0708.2251
2. L. Wolfenstein, *Phys. Rev.* **D17**, 2369 (1978); ibid **D20**, 2634 (1979); S. P. Mikheyev and A. Yu. Smirnov, *Yad. Fiz.* **42**, 1441 (1985); *Nuovo Cimento* **C9**, 17 (1986)
3. G. L. Fogli et al. hep-ph/0506083 (2006) and references therein
4. K. Hirata et al. *Phys. Rev. Lett.* **58** 1490 (1987).
5. R. M. Bionta et al., *Phys. Rev. Lett.* **58** 1494 (1987)
6. E.N. Alekseev, et al., *Phys.Lett.* B **205** 209 (1988)
7. T. K. Kuo and J. Pantaleone; *Phys. Rev.* **D 37** 298 (1988); G. Dutta et al, *Phys. Rev.* **D 61** 013009 (1999); Dinghe, A. Y. Smirnov; *Phys. Rev.* **D 62**, 033007 (2000)
8. M.T. Keil, G.G. Raffelt and H.T. Janka, Astrophys. J. 590 (2003) 971
9. T. J. Loredo and D. Q. Lamb Phys. Rev. **D 65** 063002 (2002); G. Pagliaroli, M. L. Costantini, A. Ianni, F. Vissani astro-ph/0705.4032
10. E. Majorana; *Nuovo Cimento,* **5** (1937) 171
11. F. Feruglio, A. Strumia, F. Vissani; *Nucl. Phys.* **B637** (2002) 345
12. V. A. Rodin et al. *Nucl. Phys.* **A766** (2006) 107; erratum nucl-ex/0706.4304
13. H. V. Klapdor-Kleingrothaus et al.; *Phys. Lett.* **B586** (2004) 198 and references therein
14. C. E. Aalseth et al. Phys. Rev. **D** 73 (2006) 053004
15. C. Arnaboldi et al. *Phys. Rev. Lett.* **95** 142501 (2005); recent results: A. Giuliani, TAUP 2007
16. C. Arnaboldi et al. *Astrop. Phys.* **20** (2003) 91
17. I. Abt et al. hep-ex /04040390 (2004) and http//www.mpi–hd.mpg.de/GERDA/proposal.pdf

Double Beta Decays and Neutrinos - Experiments and MOON -

H. Ejiri

Research Center for Nuclear Physics, Osaka University, Osaka 567-0047, Japan,
Nuclear Science, Czech Technical University, Brehová, Prague, Czech Republic,
National Institute of Radiological Sciences, Chiba, 263-8555, Japan

Abstract. This is a brief review of the present and future experiments of neutrino-less double beta decays ($0\nu\beta\beta$) and the MOON (Mo Observatory Of Neutrinos) project. High sensitivity $0\nu\beta\beta$ experiments are unique and realistic probes for studying the Majorana nature of neutrinos and the absolute mass scale as suggested by neutrino oscillation experiments. MOON aims at spectroscopic $0\nu\beta\beta$ studies with the ν-mass sensitivity of 100-30 meV by means of a super ensemble of multi-layer modules of scintillator plates and tracking detector planes.

Keywords: Double beta decays, neutrino masses, Majorana neutrinos, MOON, lepton number.
PACS: 23.40.-s, 23.40.Bw, 14.60.Pq, 14.60.St

DOUBLE BETA DECAYS AND MAJORANA NEUTRINOS

Neutrinos(ν) are of current interest for particle and astro-nuclear physics. Neutrino properties beyond the standard electroweak theory are studied by high-sensitivity experiments of neutrino-less double beta decays ($0\nu\beta\beta$). They are sensitive and realistic probes for studying the Majorana ν nature, the ν mass spectrum the absolute mass scale as suggested by ν-oscillation experiments, and even the Majorana CP phases. Extensive studies have been carried out on $0\nu\beta\beta$ experiments and theories so far. They are given in recent review articles and references therein [1-7].

Current $0\nu\beta\beta$ experiments use $\beta\beta$ isotopes of the order of 10 kg. Thus the ν-masses to be studied are limited to the 200-300 meV. New-generation $0\nu\beta\beta$ experiments are planned to study the Majorana masses in the quasi-degenerate (QD 100 meV) and the inverted hierarchy (IH 30 meV) ν-mass regions.

Majorana/GERDA, CUORE, EXO and others are calorimetric experiments with $\beta\beta$ isotopes integrated into the detectors. MOON, Super-NEMO and others are spectroscopic experiments with $\beta\beta$ isotopes separated from the detectors. The present and future $0\nu\beta\beta$ experiments are discussed in recent review articles and references there in [1-7].

MOON (Mo Observatory Of Neutrinos) aims at high-sensitivity spectroscopic studies of QD - IH masses of 100-30 meV [1, 3, 8-11].

The $0\nu\beta\beta$ nuclear matrix elements $M^{0\nu}$ are crucial for designing $0\nu\beta\beta$ detectors and for extracting the neutrino mass from the $0\nu\beta\beta$ rate if it is observed. They are discussed in reviews and references therein [1, 2, 4-7].

CP972, *Exotic Nuclei and Nuclear/Particle Astrophysics (II)—Proceedings of the Carpathian Summer School of Physics 2007,* edited by L. Trache and S. Stoica

The present report reviews briefly the present and future 0νββ experiments for Majorana neutrino masses and the MOON project [1,3,10,11]. The present report has been partly given in recent conference report and the review as well [3, 11].

NEUTRINO-LESS DOUBLE BETA DECAYS

Double beta decays within the standard electro-weak model (SM) are followed by two antineutrinos in case of double electron decays, and by two neutrinos in case of double positron decays. Thus they conserve the lepton number L. Double beta decays followed by no neutrinos violate the lepton number conservation law by ΔL=2, and thus are beyond SM. They are called two neutrino ββ decays (2νββ) and neutrino-less ββ decays (0νββ). Schematic diagrams of the ββ decays are shown in Fig. 1.

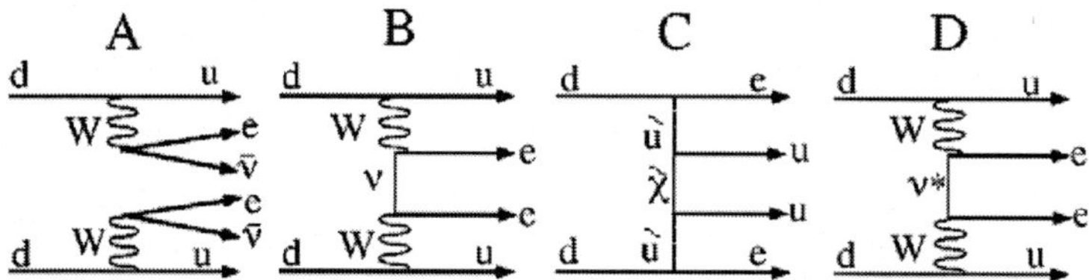

FIGURE 1. Schematic diagrams of *ββ* decays (Ref.1). A: 2νββ. B: 0νββ with a light Majorana ν exchange. C: 0νββ with a SUSY exchange. D: 0νββ with an excited neutrino *ν**.

□

The 2*νββ* process followed by 2 neutrinos are given in Fig. 1A. Decay diagrams for three 0νββ processes with $\Delta L = \pm 2$ are shown in Figs. 1B, 1C and 1D. First we discuss the case of 1B with a light Majorana *ν* exchange between two nucleons (quarks) in a nucleus.

The 0*νββ* transition rate is expressed as

$$T^{0\nu} = G^{0\nu} M^{0\nu} K^{\nu R}, \tag{1}$$

where $G^{0\nu}$ is the kinematical factor for 0νββ and $M^{0\nu}$ is the nuclear matrix element. $K^{\nu R}$ is given in terms of the neutrino mass term and the right-handed weak current terms as

$$K^{\nu R} = [(< m_\nu >) + c < \lambda > + c' < \eta >]^2 \tag{2}$$

where $< m_\nu >$ and $< \lambda >$, $< \eta >$ are the ν-mass term and the right-handed current terms. They are identified by observing the energy and angular correlations of the two β rays. The ν-mass term is characterized by the ββ rays emitted at opposite directions, while the right-handed terms by the ββ rays emitted in the same direction.

The neutrino mass to be studied by neutrino-less double beta decays is an effective Majorana mass $< m_\nu >$. It is expressed by using the mixing coefficient and the Majorana phases of the Majorana neutrinos as

$$< m_\nu > = \Sigma |U_{ei}|^2 m_i \exp(i\alpha_i) \tag{3}$$

The effective mass can be evaluated by using the mass-square differences and the mixing angles, which are obtained from the ν-oscillation experiments. The effective mass in the simple cases of the CP conserving phases is plotted as a function of the lightest mass (m_1 for NH and m_3 for IH) by assuming $\cos(\theta_{13})\approx 1$ in Fig.2.

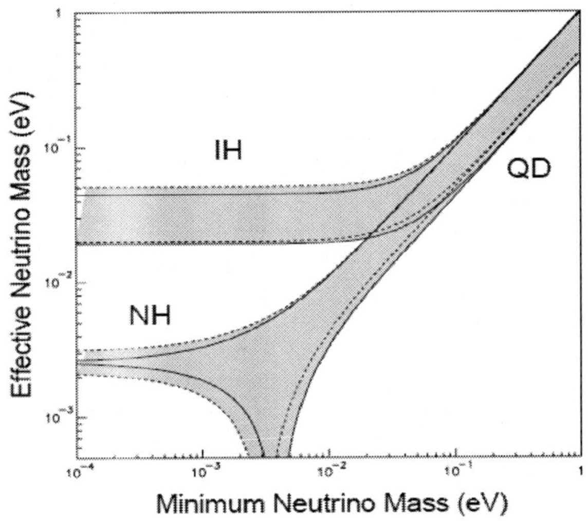

FIGURE 2. The effective masses$< m^{NH} >$ and $< m^{IH} >$ for the NH and IH spectra as a function of the lightest masses of m_1 and m_3, respectively.

Next generation DBD experiments with high mass sensitivity of the order of 100 ~30 meV are of great interest to establish $0\nu\beta\beta$ and the neutrino mass spectrum in case of QD ~ IH mass spectrum. R&D for future DBD experiments with ultra-high mass sensitivity of the order of 2 meV should be encouraged for future experiments in case of NH.

MOON aims at high-sensitivity $0\nu\beta\beta$ experiment with the mass sensitivity of 100 ~ 30 meV in the QD ~ IH region.

THE PRESENT AND FUTURE BB EXPERIMENTS

Experimental studies of $0\nu\beta\beta$ have been made extensively so far and are currently in progress as given in reviews [1−7].

Calorimetric experiments with high E-resolution ^{76}Ge and TeO_2 detectors give stringent upper limits on the $0\nu\beta\beta$ rates. The IGEX - HM data on ^{76}Ge gave limits of around 0.3 ~ 1.4 eV, while claims for the $0\nu\beta\beta$ peak in the HM data were reported. CUORICINO gave the limits of 3 10^{24} y, corresponding to 0.2 - 1.0 eV, for ^{130}Te [12].

Spectroscopic methods have been used for several $\beta\beta$ decays by ELEGANT V and NEMO III groups. Recent NEMO III data gave limits of 4.6 10^{23} y and 0.66 - 2.81 eV for ^{100}Mo [13,14]. CUORICINO and NEMO III are expected to reach the mass region of 0.2 - 0.5 eV in the near future.

Experimental proposals for future ββ experiments with the QD-IH mass sensitivity are as listed in Table 1.

CANDLES is an array of CaF_2 crystals for ββ decays of ^{48}Ca and dark matter. It is based on ELEGANT VI [15].

MAJORANA aims at $0\nu\beta\beta$ studies of ^{76}Ge in the QD-IH mass region on the basis of IGEX [16]. Pulse shape discrimination (PSD) and segmentation of detector (SED) are used to reduce BGs.

GERDA uses ^{76}Ge detectors immersed in liquid argon [17]. It starts with 18 kg enriched Ge crystals at LNGS, and plans to add 20 kg to get the sensitivity of 2 10^{26} y, corresponding to 90 ~ 290 meV, in order to prove or disprove the HM claims.

COBRA uses an array of 1 cm^3 CdZnTe semiconductor detectors at LNGS to study $\beta\beta$ decays of ^{116}Cd and others [18]. The sensitivity with 64 K detector for 5 y will be expected to be around 50-30 meV if BGs are reduced as expected.

CUORE consists of 988 TeO_2 crystals with the net ^{130}Te mass of 203 kg [12]. Reducing the BG rate to 0.01-0.001 per keV kg y, the group aims at the sensitivity of (2 – 6.6)10^{26} y for a 5 y run The mass sensitivity is (35~ 120) meV and (20~ 65) meV in cases fo BG rates of 0.01 and 0.001 per keV kg y, respectively.

EXO studies the $0\nu\beta\beta$ of ^{136}Xe by using the laser technique to identify the decay product [20]. The 1 ~10 ton enriched Xe detectors are used to the get half-life sensitivity of 2 ~41 10^{27} y for a 5 ~10 y run, corresponding to the masses of 60 ~ 15 meV. XMASS for DM can be used for $\beta\beta$ of ^{136}Xe as well [21].

Spectroscopic 0νββ experiments with tracking detectors are complementary to the calorimetric detectors. Spectroscopic experiments with external ββ sources, which are separated from the detectors, can select such ββ nuclide that give the large signal rate and the small BG rate. ^{82}Se, ^{100}Mo and ^{150}Nd are candidates for spectroscopic experiments because of their large $Q_{\beta\beta}$ values and accordingly of the large phase space volumes and the small BG rate. Energy and angular correlations for the two β-rays identify the ν-mass term. The ββ event can be confirmed by measuring at least two ββ isotopes/states with different nuclear matrix elements and different 0νββ peak energies. They are used to identify the ν-mass contribution among the other 0νββ processes.

Using 0νββ isotopes with large $Q_{\beta\beta} \approx$ 3 MeV, the 0νββ signal can be placed well above most RI β–γ rays. ^{82}Se, ^{100}Mo and ^{150}Nd are candidates for future spectroscopic 0νββ experiments.

Two β-ray spectroscopy by tracking detectors make it possible to get the required S/N. Thus natural and cosmogenic RI BGs are negligible in case of ββ sources with realistic ppt-level purity.

Major BGs are contributions from the tail of the 2νββ spectrum into the 0νββ window. They can be reduced to the level required for the mass sensitivity by using detectors with realistic energy resolution of around σ = 2.0-1.5 %.

TABLE 1. Isotopes and detectors to be used for future $\beta\beta$ experiments. SN: nuclear sensitivity [1] in units of 10^{-24} y^{-1} $(eV)^{-2}$

Isotope	A %	$Q_{\beta\beta}$ MeV	S_N	Experiment
^{48}Ca	0.187	4.276	0.11	CANDLES^a
^{76}Ge	7.8	2.039	0.22	MAJORANA^b GERDA^c,
^{82}Se	9.2	2.992	0.86	Super-NEMO^d
^{100}Mo	9.6	3.034	2.02	MOON^e,
^{116}Cd	7.5	2.804	0.90	COBRA^f CAMEO^g
^{130}Te	34.5	2.529	0.73	CUORE^h, COBRA^f
^{136}Xe	8.9	2.467	0.13	EXO^i, XMASS^j
^{150}Nd	5.6	3.368	11.3	DCBA^k

a : [15], b : [16], c : [17], d : [13], e : [3,8-11], f : [18],
g : [19], h : [12] i : [20], j : [21], k : [22]

The SuperNEMO project with 100 kg of ^{150}Nd source is under progress to study the QD masses [13]. The collaboration uses 20 modules, each with 5 kg $\beta\beta$ isotopes. The module is a huge detector ensemble, which consists of two large tracking chambers and many scintillation crystals.

DCBA uses a tracking chamber to study ^{150}Nd $\beta\beta$ decays [22]. The goal is to achieve the half-life sensitivities of 10^{25} y and 10^{26} y with natural and enriched ^{150}Nd sources, respectively.

MOON is an extension of ELEGANT V [23,24]. It is a compact spectroscopic detector with ton-scale $\beta\beta$- isotopes for both the ground and excited 0^+ states with the IH mass sensitivity [3, 8-11]. MOON is used also for real time studies of low-energy solar neutrinos. The $\beta\beta$ and solar-ν transition scheme in ^{100}Mo is shown in Fig. 3.

MOON EXPERIMENT

The MOON detector of the solid scintillator option consists of multi-layer detector modules [3, 11-13]. Each module is composed by a scintillator plate for energy and time, two thin detector planes for position and particle identification, and a thin $\beta\beta$ source film interleaved between the two detector planes. The detector plane made of a thin wire-chamber and that of thin Si-pixel/strip detectors are under consideration.

Individual β rays from $0\nu\beta\beta$ are measured in coincidence by two adjacent scintillator plates to confirm the ν-mass term in the $0\nu\beta\beta$ process. All other modules (layers) are used as active shields to reject BG β-γ rays and neutrons.

Scintillation photons are collected by photomultiplier tubes (PMT) around the scintillator plate. Then the beta ray energy and the time are measured by the scintillator. The $\beta\beta$ vertex point is identified by the position-sensitive detector planes.

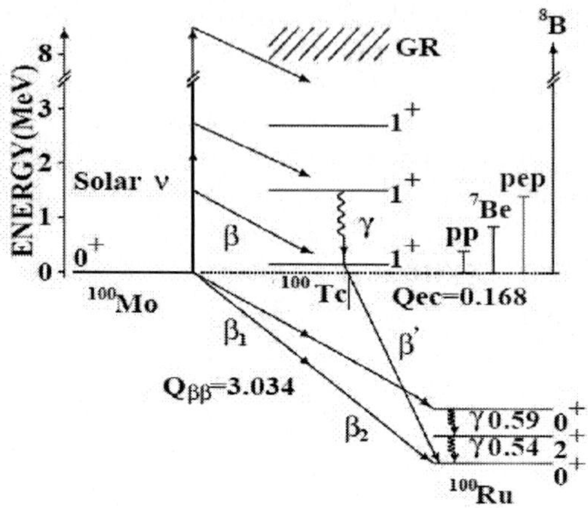

FIGURE 3. Level and transition scheme for ββ and solar- $\tilde{\nu}$.The maximum energy of each solar neutrino source is indicated at the right hand side [1,3].

Then the multilayer detector modules with good energy and position resolutions enable one to select 0νββ signals, reduce BG contributions from the tail of the 2νββ spectrum, and to reject RI-background signals. They can be easily expanded from sub-ton detectors to multi-ton detectors to reach the mass sensitivity even below the IH mass.

The multilayer structure of the scintillator plates makes it realistic to build a compact detector with an order of 1-2m^3 per 10 kg of ββ isotopes to accommodate ton-scale ββ isotopes. The detector is smaller in volume than the SuperNEMO by orders of magnitude.

The mass sensitivity of the MOON experiment is evaluated for ^{82}Se and ^{100}Mo by assuming $M^{0\nu}$=3. They are given in Table 2.

TABLE 2. Neutrino mass sensitivities in unit of meV for MOON with the *N* ton year measurements.

N t y	0.12	0.48	2.0	5.0
^{82}Se	103	63	40	30
^{100}Mo	124	83	56	40

Detection efficiencies for the 0νββ and 2νββ signals are evaluated by Mote Carlo simulation calculations in case of the ββ source thickness of 30 mg/cm^2 and the energy resolution of $\sigma \approx 2.2\%$ [25]. The two β rays are required to be emitted opposite to each other with energy deposits of $E \geq 0.5$ MeV. The QD and IH mass sensitivities of 100 and 40 meV can are achieved by a $N = 0.12$ t y run ($n_{\beta\beta} = 0.06$ t, $T = 2$ y) and N=2 t y ($n_{\beta\beta}$ = 0.48 t, $T = 4$ y) experiments.

Separation of ^{100}Mo is feasible by the centrifugal separation of MoF_6 gas. Using 6000 centrifuges and 40 separation steps, ^{100}Mo isotopes with 85 % are obtained with a rate of 0.1 ton in 1 year.

It is interesting to note that Mo isotopes have a large absorption cross-section for supernova neutrinos as well as that for solar neutrinos. Thus the MOON type detector with natural Mo plates is used for supernova neutrinos [26].

A prototype MOON-1 detector was constructed to prove feasibility of MOON with the IH mass sensitivity [3, 27-29]. It consists of 6-layer plastic-scintillator (PL) plates, each with 53×53×1 cm^3, as shown schematically in Fig. 4. RP-408 (BC-408 equivalent) PL plates were provided by REXON. R&D for other scintillators with better energy resolution is under progress.

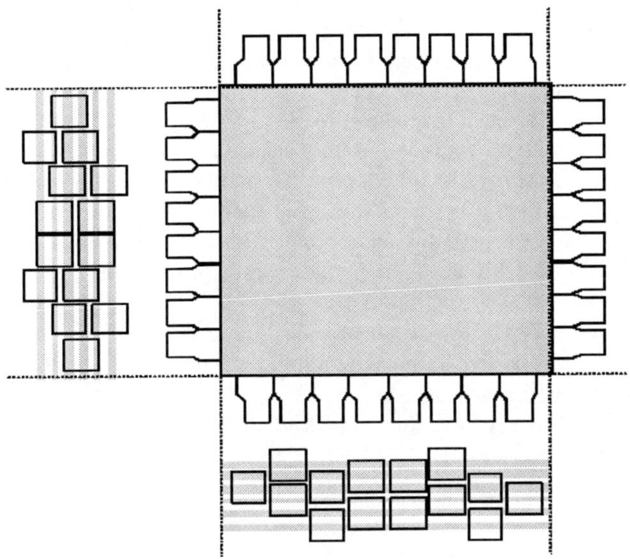

FIGURE 4. A proto-type MOON detector. MOON 1 top and side views. The 4 sides of 6 layers of PL plates are viewed by 56 PMTs.

The 94.5 % enriched ^{100}Mo films with 40 mg/cm^2 are interleaved between the PL plates in the MOON-1 detector. The 6-layer PL plates are viewed by 56 6×6 cm^2 square-type photo-multiplier tubes (PMT). One PMT collects photons from 3 PL plates and the hit PL plate is identified by the PMT hit pattern.

The number of total photo-electrons is 1830 ± 35 for the 976 keV ^{207}Bi conversion electron line, and the energy resolution is σ = 4.8 ± 0.2 %. It corresponds to σ =2.7% at 3 MeV.

The MOON-1 detector was set in the active and passive shields of ELEGANT V [23,24]. The drift chambers were replaced by the MOON 1 multi-layer plates. The measured energy resolution is found to be well reproduced by σ=5.0 $E^{1/2}$% with E in unit of MeV. This leads to the energy resolution of σ =2.9% at 3 MeV, as required for the QD-IH mass of 50 meV in case of ^{82}Se run for $N \approx 2$ t y exposure.

Actually, the energy resolution is quite good in case of a small PL plate. It is σ =3.8 % at 1 MeV (2.2% at 3 MeV). Thus is improved by selecting a small region of the large PL plate. Accordingly we hope to reach the resolution of around σ =2.2 % at $Q_{\beta\beta} \approx$ 3MeV by correcting for the position dependence of the photon collection. Improvement of the resolution is in progress.

CONCLUDING REMARKS

Recently proposed $0\nu\beta\beta$ experiments with calorimetric detectors and those with spectroscopic (tracking) detectors are promising for $0\nu\beta\beta$ studies in the QD-IH mass regions. They are complementary to each other.

MOON is a spectroscopic experiment with multi-layer scintillator plates and tracking detector planes. The good energy and position resolutions, together with the compact multi-layer structure of the detector, make it realistic to study the QD (100 meV) and IH (30 meV) mass regions with $N \approx 0.5$ t y and 2 t y runs, respectively.

It is indispensable to carry out $0\nu\beta\beta$ experiments with different $\beta\beta$ isotopes/states and detection methods (calorimetric and spectroscopic methods) in order to establish the ν-mass $0\nu\beta\beta$ process. MOON with $\beta\beta$ source $\neq$ detector is used to study energy and angular correlations for 2-3 kinds of $\beta\beta$ nuclides and for both the ground and excited states to identify the light ν-mass $0\nu\beta\beta$ process.

Nuclear matrix elements $M^{0\nu}$ are crucial for designing $\beta\beta$ detectors and for extracting the ν-mass from the observed $0\nu\beta\beta$ rate. Theoretical calculations as well as experimental data relevant to $M^{0\nu}$ are important [30,31]

MOON can be used for real-time studies of low energy solar neutrinos [8-11], and the MOON type detector with natural Mo plates is used for studying charged currents of supernova neutrinos [26].

ACKNOWLEDGMENTS

The author thanks the MOON collaborators for very valuable discussions, and Osaka Univ, Univ. Washington, ICU, JASRI, IIAS and NIRS for generous supports.

REFERENCES

1. H. Ejiri, *J. Phys. Soc. Jap.* **74,** 2101-2127 (2005).
2. H. Ejiri, *Phys. Rept.* **338**, 265-351 (2000).
3. H. Ejiri, *Mod. Phys. Lett. A*, **22**, 1277-1291 (2007).
4. J. Vergados, *Phys. Rept.* **361** (2002) 1-56; *Private communication* (2007).
5. J.D. Suhonen and O. Civitarese, *Phys. Rept.* **300**, 123-214 (1998).
6. S. Elliott and P. Vogel, *Ann. Rev. Nucl. Part. Sci.* **52**, 115 (2002) .
7. F. Simkovic and A. Fassler, *Prog. Part. Nucl. Phys.* **48**, 201 (2002).
8. H. Ejiri, et al., *Phys. Rev. Lett.* **85**, 2917-2920 (2000).
9. P. Doe et al., *Nucl. Phys. A,* **721,** 517c -520c (2003).
10. H. Ejiri, *Czechoslovak Journal of Physics B,* **54**, 317-325 (2004).
11. H. Ejiri, *Proc. MEDEX 07, AIP Conference proceedings* 2007.
12. E. Fiorini, Proc. ICNP 2007, Tokyo.
13. F. Piquemal, Proc. ICNP 2007, Tokyo.

14. R. Arnold et al., *Phys. Rev. Lett.* **95**, 182302 1-4 ((2005).
15. S. Umehara et al., *Proc. NDM03 eds. H.Ejiri and I.Ogawa. http://ndm03.phys.sci.osaka-u.ac.jp/proc/index.htm*
I. Ogawa, *Proc. NNR05, ed. H. Ejiri, http://www.spring8.or.jp/ext/en/appeal/nnr05/*
16. C.E. Aalseth et al., *Yad. Fiz.* 67, No. 11 (2004).
17. S. Scheonert, *Proc. Neutrino 2006, Santa Fe*, (2006).
18. K. Zuber, *Phys. Lett. B* 519, 1-7 (2001).
19. G. Bellini et al., *Eur. Phys. J. C* 19 43 (2001).
20. A. Piepke, *Proc. Neutrion 2006, Santa Fe* (2006).
21. S. Moriyama et al., *Proc. Workshop XENON01, Tokyo* (2001).
22. N. Ishihara et al., *Nucl. Instr. Meth. A* 443, 101-107 (2000).
23. H. Ejiri et al., *Nucl. Instrum. Meth. A* **302**, 304-314 (1991).
24. H. Ejiri et al., *Phys. Rev. C,* **63**, 065501 1-7 (2001).
25. T. Shima, *private communication,* (2007).
26. H. Ejiri, J. Engel, and N. Kudomi, *Phys. Lett. B,* **530**, 27-32 (200
27. H. Ejiri, *Prog. Part. Nucl. Phys.* **57**, 153-161 (2006).
28. H. Nakamura et al., *http://arxiv.org/abs/nucl-ex/0609008* (2006).
29. H. Nakamura et al., J. Phys. Soc. Japan, to be published 2007.
30. H. Ejiri: *Nucl. Instr. Meth. Phys. Research,* **503,** 276 -278 (2003).
31. H. Ejiri, *Czechoslovak Journal of Physics B,* **56**, 459-466 (2004).

The Karlsruhe Tritium Neutrino Mass Experiment KATRIN

Jochen Bonn[a]

[a]*Johannes Gutenberg Universitaet Mainz 55099 Mainz Germany*
Jochen.Bonn@uni-mainz,de

Abstract. KATRIN is the next generation tritium neutrino mass experiment presently set up at Forschungszentrum Karlsruhe by the international KATRIN collaboration. The aim is to push the present limit on the neutrino mass[1] from 2eV [1] down to 0.2eV with the hope to detect a finite mass with discovery potential of 0.35eV. The experimental set up and the present status of KATRIN will be given.

Keywords: Neutrino mass, electron spectrometer, high voltage technology
PACS: 14.60.Pq, 07.81.+a, 52.59.Mv

INTRODUCTION

The recent discovery of neutrino oscillation proved that neutrinos mix and can change their flavour. We are convinced that these neutrino oscillations are caused by non-zero masses of neutrinos in contrast to their current description in the Standard Model of particle physics. Unfortunately, oscillation experiments are only sensitive to $\Delta m^2_{ij} = m^2(\nu_i) - m^2(\nu_j)$.

There are different ways to determine the neutrino mass scale:

Cosmology

Information on the absolute scale of the neutrino mass can be obtained from astrophysical observations like the power spectrum of the matter and the energy distribution in the universe at different scales. In most cases they give upper limits on the neutrino mass of several 0.1~eV/c² in some cases non-zero neutrino masses are found illustrating the dependence on the assumptions and the data used to obtain the limits.

---

[1] The experiment is sensitive to the mass of the electron neutrino being a superposition of weak eigenstates. For the mass range in question however the neutrino masses are almost identical.

CP972, *Exotic Nuclei and Nuclear/Particle Astrophysics (II)—Proceedings of the Carpathian Summer School of Physics 2007,* edited by L. Trache and S. Stoica

Neutrinoless Double β Decay

One laboratory way to access the neutrino mass scale is the search for the neutrinoless double β decay. The observable of double β decay is the so-called effective neutrino mass which is a coherent sum over all neutrino mass eigenstates contributing to the electron neutrino with their (complex) mixing matrix elements. The subject was covered in lectures of S. Stoica and J. Suhonen in this summer school.

Direct Neutrino Mass Determination

In these experiments the neutrino mass is determined using the relativistic energy-momentum relationship without further assumptions. Therefore m^2_ν is the observable in most cases.

The non-observation of a dependence of the arrival time on energy of supernova neutrinos from SN1987a gave an upper limit on the neutrino mass of 5.7 eV [1]. Nearby supernova explosions are too rare and well enough understood to allow a further improvement to a sub-eV sensitivity on the neutrino mass. Therefore, the investigation of the electron energy spectrum of a nuclear β-decay is still the most sensitive model-independent and direct method to determine the neutrino mass.

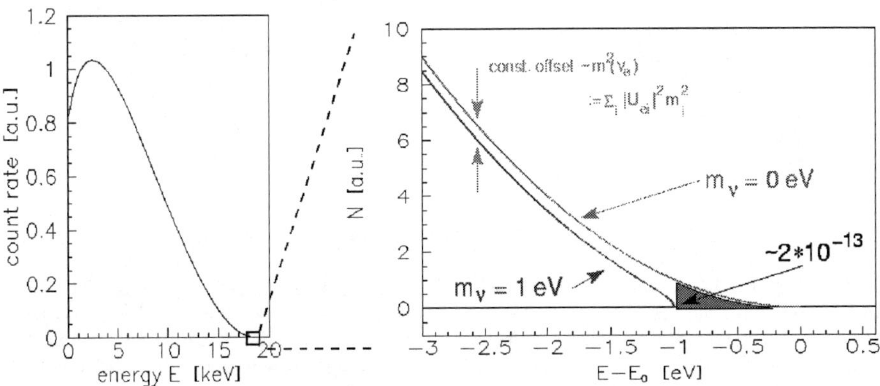

FIGURE 1. Spectrum of tritium β-decay with enlargement of the endpoint region. The figure clearly demonstrates that the main requirement is to cope with the vanishing count rate near the endpoint by providing the strongest possible signal at lowest background. Additionally an energy resolution of a few eV is required to become sensitive to the neutrino mass dependent shape of the β-spectrum.

The fingerprint of the neutrino is the shape of nuclear β-decay spectra. As we all know the neutrino was postulated by W. Pauli to explain the continuous distribution of electron energies without violating the conservation of energy and momentum. The question remains what can we learn about the mass of the neutrino from the shape of the β-spectrum. The influence of the neutrino mass is prominent in the region of the spectrum where the neutrino is non relativistic. At higher neutrino energies the kinetic energy dominates and the relative influence of the mass onto the shape of the spectrum fades away. In other words the influence of the neutrino mass is restricted to the region close to the endpoint. As the count rate is very low in the region of interest the spectrometer used has to combine high resolution, high luminosity and a low background.

NEUTRINO MASS EXPERIMENTS FROM TRITIUM Β-DECAY

Principle of the MAC-E Filter

A major break-through in tritium β-decay experiments was achieved by a new type of spectrometer, the so-called MAC-E-Filter (Magnetic Adiabatic Collimation followed by an Electrostatic Filter). This new type of spectrometer-- based on early work by Kruit [2] -- was developed for the application to tritium β-decay at Mainz and Troitsk independently [3,4].The MAC-E-Filter combines high luminosity at low background and a high energy resolution.

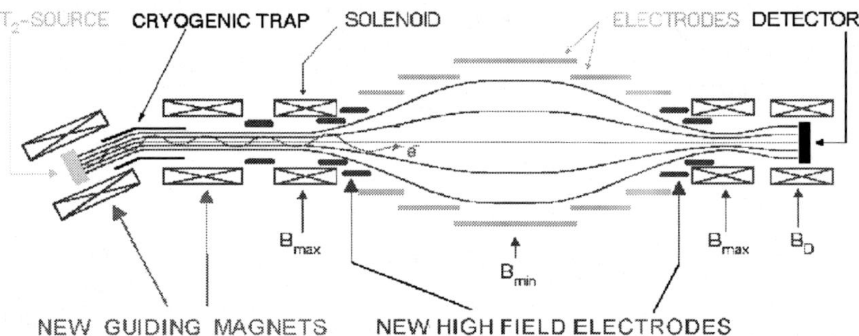

FIGURE 2. The Mainz spectrometer in the version used in phase II of the experiment. The spectrometer itself is the centre part with the two solenoids giving the maximum field B_{max} and the electrodes forming the retarding potential. In the centre the maximum of the retarding potential and the minimum of the B field coincide. The source is decoupled by a cryogenic trap to prevent migration of tritium into the spectrometer.

The principle of the MAC-E-Filter is shown in fig 2. The spectrometer is set up by the two high field solenoids and the retarding electrodes. The source is placed in a strong magnetic field guiding the decay electrons. The gradient force in the fringing field of the source solenoid transforms the energy in the cyclotron motion around the magnetic field line into energy parallel to it. By this adiabatic transformation a large parallel beam of electrons is formed that can be analyzed by the electrostatic filter formed by a set of cylindrical electrodes.

The concept proved to be superior to the magnetic spectrometers used by other groups. The MAC-E filter is well adapted to the problem of investigating the endpoint region of a β-spectrum. The resolution is given by the ratio of the maximum magnetic field B_{max} to the minimum magnetic field B_{min} reached in the analysing plane. The luminosity can reach the full forward solid angle if the source is put into the field maximum. In a real experiment the largest starting angles are cut off by placing the source in a weaker field to reduce energy losses. This reduces the luminosity to values around 60% of the forward solid angle.

A combination of high luminosity and high resolution in general results in a large setup. Comparing the MAC-E filters with the magnetic, momentum analysing spectrometers of similar dimensions the MAC-E filters reach a much better performance at lower demands on mechanical precision and less stringent demands on the parameters of the electric and the magnetic fields. In contrast to magnetic spectrometers there is no way of scattering electrons into spectral regions of higher energy. This is very important for data taking close to the endpoint at relative intensities of 10^{-10} and below.

The present limit on the neutrino mass quoted by:

Particle Data Group m^2_ν=-1.1+/- 2.4 eV^2 m_ν <2eV [1]
based on data from:
Mainz m^2_ν=-0.6+/- $2.2_{stat.}$+/- $2.1_{syst.}$ eV^2 m_ν<2.3V[5]
and Troitsk m^2_ν=-1.9+/- $3.4_{stat.}$+/- $2.2_{syst.}$ eV^2 m_ν<2.5V[6].

THE KATRIN EXPERIMENT

The experiments in Mainz and Troitsk have reached the point at which statistical and systematic uncertainties are about equal. A major improvement can only be achieved by a new, bigger instrument to improve statistics. In addition the systematic uncertainties have to be reduced substantially. To achieve these improvements the dimensions of the MAC-E filter are increased from ~1m in used diameter to~10m, the source strength will be increased by ~ 100 and the resolution will be typically 1 eV.

Experimental Setup

The source will be a windowless gaseous tritium source to avoid the self-charging of a quench condensed source like the one used in Mainz [6]. This effect would limit the source strength in the high resolution experiment. To reduce systematic uncertainties one has to limit data taking to a very narrow region below the endpoint in which energy losses play no role. In addition the experimental parameters have to be very precisely known.

A scheme of the whole experimental setup is shown in Fig. 3: with an enlargement of the source and transport system. One of the most critical points is the reduction of the tritium flow between source and main spectrometer. To keep the background from tritium decaying in the volume of the main below 1 event in 1000 seconds requires an efficiency of differential pumping systems of 10^{14}. The DPS systems are based on turbo molecular pumps (MAG-W2800 Leybold) with an "inner loop" for the DPS1 systems in which the tritium is almost directly pumped back and an "outer loop" in which the tritium will be isotopically enriched. In the CPS system the remaining traces of tritium will be condensed on Ar frost. The system will be warmed up and the tritium will be purged by He gas after data taking for ~2 months.

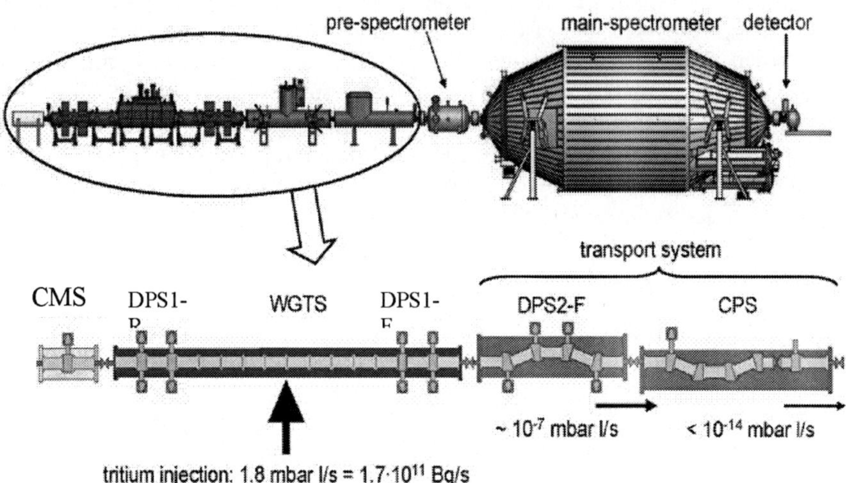

FIGURE 3: Side view of the KATRIN experiment. CMS: control and monitoring system, DPS: differential pumping system, WGTS: windowless gaseous tritium source, CPS: cryogenic pumping system

Control of experimental parameters

Density and the Temperature of the source have to be stable in the 0.1% range. Different systems to control the energy loss are under investigation. One possibility is to use an electron gun in the CMS system and detect the energy loss spectrum in the tritium source. A second possibility is to mix a radioactive source of ^{83m}Kr into the tritium and determine the energy loss spectrum of the mono energetic electron conversion lines. This method will in addition yield information on the electric potential in the source tube. The WGTS will be operated at 30K under standard conditions and at ~ 120K if Kr is added.

Stability and reproducibility of the retarding potential of KATRIN have to be <3ppm including uncertainties in the surface potentials and accuracy and precision of the high voltage. To achieve this, a voltage divider was set up in collaboration with the Physikalisch Technische Bundesanstalt [7]. Long term measurements with the voltage divider showed a drift of 0.6 ppm/month [8]. Tests with the K32 conversion line of ^{83m}Kr [9] are presented in figure 4. As ^{83m}Kr has a rather short half life of 1.8 h it was continuously condensed onto the HOPG (highly oriented pyrolytic graphite). This way the effective half life of the source was given by the mother isotope ^{83}Rb of 86 d. To reach reproducible condition the substrate was cleaned before evaporating the activity by laser ablation. The picture clearly shows that the system was stable over a period of more than one month. For the control of the retarding potential of KATRIN the Mainz MAC-E filter will be used as monitor spectrometer to measure the K32 conversion line of ^{83m}Kr in parallel to tritium data taking. The line position will be shifted to ~18 kV by a bias on the source and the retarding potential of the main spectrometer and the monitor spectrometer will be identical.

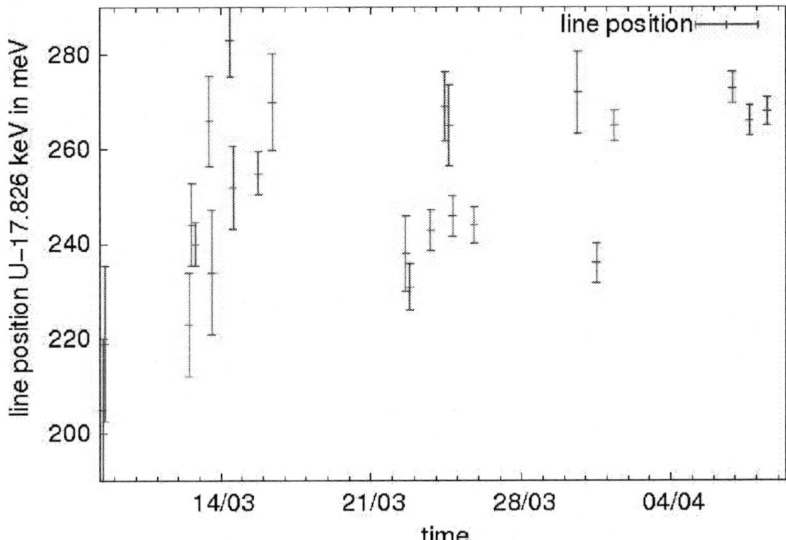

FIGURE 4. Line position of the K32 conversion line of ^{83m}Kr. The full scale of the figure corresponds to the KATRIN request of +/-3ppm. To achieve a "stable" source the ^{83m}Kr was continuously condensed on a HOPG substrate.

Background

The main feature of a MAC-E-Filter, large luminosity and high energy resolution, are connected to one non-negligible disadvantage: The main fraction of the volume of the MAC-E-Filter is put on high voltage of about the retarding potential. Electrons, which are born in this volume e.g. by inelastic scattering, have a significant chance to be accelerated by the electrical potential and counted at the detector. Therefore the pre-spectrometer already limits the electron input rate of the huge main spectrometer but still stringent vacuum conditions have to be fulfilled to suppress background from inelastic scattering within the main spectrometer. Special selection of materials as well as surface cleaning and out-baking at 350°C will allow to reach a residual gas pressure of $<10^{-10}$ mbar. To reduce the size of surfaces inside the vacuum chamber, instead of installing a complex solid electrode system the vacuum vessel itself will be put on high voltage to create the retarding potential.

A second type of background arises from secondary electrons from the walls, which are created by cosmic muons or by environmental radioactivity. Although the magnetic field will prohibit most of these electrons to enter the magnetic flux tube, which is connected to the detector, measurements at Mainz and Troitsk yielded a probability to reach the detector of 10-5 - 10-7, which is critical considering the surface of ~650m² of the KATRIN main spectrometer. To suppress this kind of background a new idea of strong background suppression has been developed and successfully tested at the Mainz spectrometer. It will be also applied to the KATRIN spectrometers: The vessel walls at high potential will be covered by nearly mass less wires, which are put to a slightly more negative potential. Secondary electrons ejected from the vessel wall, will thus be repelled and prohibited from entering the magnetic flux tube which is connected to the detector. This new method resulted at the Mainz spectrometer in a factor 10 reduced background rate. To achieve even higher background suppression factor the KATRIN main spectrometer will be instrumented by a system of 240 double-layer wire electrode modules. Another advantage of the inner wire electrode system is that the retarding voltage can be stabilized more easily.

SENSITIVITY LIMIT OF KATRIN

The detailed simulations of the KATRIN experiment yield the following: A sensitivity of 0.20~eV/c^2 will be achieved with the KATRIN experiment after 3 years of pure data taking. Statistical and systematic uncertainties contribute about equally. This value of 0.20~eV/c^2 corresponds to an upper limit with 90% C.L. in the case that no neutrino mass will be observed. A non-zero neutrino mass of 0.30~eV/c^2 would be detected with 3σ significance, a mass of 0.35~eV/c^2 even with 5σ.

SUMMARY

Among various ways to address the absolute neutrino mass scale the investigation of the shape of β-spectra around the endpoint is the only model-independent method. This direct method is complementary to the search for the neutrino less double β-decay and to the information from astrophysics and cosmology. The investigation of the endpoint spectrum of the tritium β-decay is still the most sensitive direct method. The tritium β-decay experiments at Mainz and Troitsk have been finished yielding an upper limit of about 2eV/c^2. The new KATRIN experiment is being set up Forschungszentrum Karlsruhe by an international collaboration. KATRIN will enhance the sensitivity on the neutrino mass further by one order of magnitude down to 0.2~eV/c^2.

ACKNOWLEDGMENTS

The work by the author for the KATRIN experiment is supported by the German Bundesministerium für Bildung und Forschung, the Deutsche Forschungsgemeinschaft and within the virtual institute VIDMAN by the Helmholtz Gemeinschaft.

REFERENCES

1. W.-M. Yao et al. (Particle data group) J. Phys. G33,1 (2006) and http://pdg.lbl.gov
2. P. Kruit and F.H. Read, *J. Phys. E* 16 (1983) 313
3. A. Picard et al., *Nucl. Instr. Meth. B* 63 (1992) 345
4. V.M. Lobashev et al., *Nucl. Inst. and Meth. A* 240 (1985) 305
5. C. Kraus et al., *Eur. Phys. J. C* 40 (2005) 447
6 V.M. Lobashev et al., *Phys. Lett. B* 460 (1999) 227
7. R. Marx et al. *IEEE Transactions on Instr. and Meas. Vol.50 No.2 (2001)*
8. T. Thümmler, private communication and *PHD thesis Münster (2007)*
9. B. Ostrick, private communication and *PHD thesis Münster (2008)*

Atmospheric ν and Long Baseline ν experiments

Giorgio Giacomelli

Dept of Physics, Univ. of Bologna and INFN Sezione di Bologna
V.le C. Berti Pichat 6/2, Bologna, I-40127, Italy

Abstract. The results obtained by several experiments on atmospheric neutrino oscillations are summarized and discussed. Then the results obtained by different long baseline neutrino experiments are considered. Finally conclusions and perspectives are made.

Keywords: Neutrinos, Neutrino oscillations, Long baseline experiments
PACS: 14.60.Lm, 14.60.St, 11.30.Cp, 13.15.+g

INTRODUCTION

Atmospheric neutrinos are well suited for the study of neutrino oscillations, since they have energies from a fraction of GeV up to more than 100 GeV and they may travel distances L from few tens of km up to 13000 km; thus L/E_ν ranges from $\sim$1 km/GeV to 10^5 km/GeV. Atmospheric neutrinos may study ν oscillations for small Δm^2.

The early water Cherenkov detectors IMB [1] and Kamiokande [2] reported anomalies in the ratio of muon to electron neutrinos, while the tracking calorimeters NUSEX [3] and Frejus [4], and the Baksan [5] scintillator detector did not find any. In 1995 MACRO found a deficit for upthrougoing muons [6]. Then the Soudan 2 experiment [7] confirmed the ratio anomaly. In 1998 Soudan 2 [8], MACRO [9] and SuperKamiokande (SK) [10] reported deficits in the ν_μ fluxes with respect to Monte Carlo (MC) predictions and angular distribution distortions; instead the ν_e distributions were in agreement with non oscillated MCs. These features may be explained in terms of $\nu_\mu \longleftrightarrow \nu_\tau$ oscillations.

The atmospheric neutrino flux was computed by many authors in the mid 1990s [11] and in the early 2000s [12]. The last had improvements, but also a new scale uncertainty.

Several long baseline ν beams were and are operational: KEK to Kamioka (K2K) (250 km), NuMi from Fermilab to the Soudan mine (730 km) and CERN to Gran Sasso(GS) (CNGS) (730 km). The K2K [13] and MINOS [14] experiments obtained results in agreement with the atmospheric ν results. The CNGS beam is under tests: some events were recorded in OPERA [15] and LVD [16] at GS.

Most experiments are disappearance experiments; OPERA is an appearance experiment, searching for ν_τ in a pure ν_μ beam.

Long baseline experiments will give an increasing contribution to neutrino physics.

ATMOSPHERIC NEUTRINO OSCILLATIONS

A high energy primary cosmic ray (CR), proton or nucleus, interacts in the upper atmosphere producing a large number of charged pions and kaons, which decay yielding

CP972, *Exotic Nuclei and Nuclear/Particle Astrophysics (II)—Proceedings of the Carpathian Summer School of Physics 2007,* edited by L. Trache and S. Stoica

muons and ν_μ's; also the muons decay yielding ν_μ and ν_e. The ratio of the numbers of ν_μ to ν_e is $\simeq 2$ and $N_\nu/N_{\overline{\nu}} \simeq 1$. Atmospheric neutrinos are produced at 10-20 km above ground, and they proceed towards the earth.

If ν's have non-zero masses, one considers the *weak flavour eigenstates* $\nu_e,\ \nu_\mu,\ \nu_\tau$ and the *mass eigenstates* $\nu_1,\ \nu_2,\ \nu_3$. Flavour eigenstates are linear combinations of mass eigenstates. For 2 flavour $(\nu_\mu,\ \nu_\tau)$ and 2 mass eigenstates $(\nu_2,\ \nu_3)$ one writes

$$\begin{cases} \nu_\mu = \ \ \nu_2 \cos\ \theta_{23} + \nu_3 \sin\ \theta_{23} \\ \nu_\tau = -\nu_2 \sin\ \theta_{23} + \nu_3 \cos\ \theta_{23} \end{cases} \tag{1}$$

where θ_{23} is the mixing angle. The survival probability of the ν_μ "beam" is

$$P(\nu_\mu \to \nu_\mu) = 1 - P(\nu_\mu \to \nu_\tau) = 1 - \sin^2 2\theta_{23}\ \sin^2\left(\frac{1.27\Delta m^2 \cdot L}{E_\nu}\right) \tag{2}$$

where $\Delta m^2 = m_3^2 - m_2^2$ and L is the distance from ν production to detection.

The simple formula Eq. 2 is modified by additional flavours and by matter effects.

RESULTS FROM ATMOSPHERIC NEUTRINO EXPERIMENTS

After the early indications of an anomaly in atmospheric ν's [1-7], in 1998 Soudan 2, MACRO and SK provided strong indications in favour of $\nu_\mu \longleftrightarrow \nu_\tau$ oscillations [8-10] [17]. Confirming results were presented by long baseline experiments [13] [14].

Soudan 2 used a modular fine grained tracking and showering calorimeter of 963 t located underground in the Soudan Gold mine in Minnesota. The detector was made of 1m×1m×2.5m modules, surrounded by an anticoincidence shield. The bulk of the mass was 1.6 mm thick corrugated steel sheets interleaved with drift tubes [8]. The Soudan 2 double ratio for the zenith angle range $-1 \leq \cos\Theta \leq 1$ is $R' = (N_\mu/N_e)_{DATA}/(N_\mu/N_e)_{MC} = 0.68 \pm 0.11_{stat}$, consistent with ν_μ oscillations.

MACRO detected upgoing ν_μ's via CC interactions $\nu_\mu \to \mu$; upgoing muons were identified with streamer tubes (for tracking) and scintillators (for time-of-flight measurements), points and rectangles in Fig. 1a [18]. Events were classified as:
Upthroughgoing muons from interactions in the rock below the detector of ν_μ with $\langle E_\nu \rangle \sim 50$ GeV. Data were compared with the predictions of the Bartol96 [11], FLUKA, HKKM01 [12] MCs, Fig. 2a. The shape of the angular distribution and the absolute value favoured ν oscillations with $\Delta m^2 = 0.0025$ eV^2 and maximum mixing. The absolute value of the muons is 25% higher than FLUKA and HKKM01 MC predictions . The difference between new and old MC predictions is due to new fits of CR data [21].

Low energy events. Semicontained upgoing muons (IU) come from ν_μ interactions inside the lower apparatus. *Up stopping muons* (UGS) are due to external ν_μ interactions yielding upgoing muons stopping in the detector; the *semicontained downgoing muons* (ID) are due to downgoing ν_μ's interacting in the lower detector; the lack of time information prevents to distinguish between the two samples. An equal number of UGS and ID events is expected. The average parent neutrino energy is 2-3 GeV. The data are compared with MC predictions without oscillations in Figs. 2b,c: they show a uniform deficit over the whole angular distribution with respect to Bartol96 predictions.

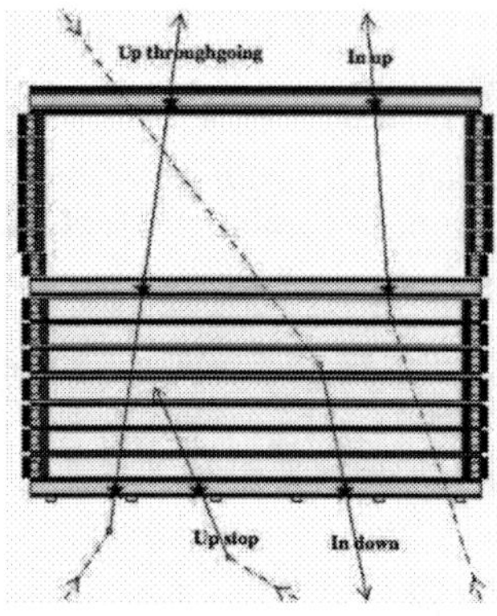

FIGURE 1. a) Cross section of the MACRO detector, sketch of event topologies. b) Schematic layout of the SuperKamiokande detector

$\nu_\mu \longleftrightarrow \nu_\tau$ *against* $\nu_\mu \longleftrightarrow \nu_{sterile}$. Matter effects due to the difference between the weak interaction effective potential for ν_μ with respect to sterile neutrinos, yield different total number and different zenith distributions of upthroughgoing muons. R_{meas} between the events with $-1 < \cos\Theta < -0.7$ and with $-0.4 < \cos\Theta < 0$ was used [9]. The measured ratio, $R_{meas} = 1.38$, is compared with $R_\tau = 1.61$ and $R_{sterile} = 2.03$. One concludes that $\nu_\mu \longleftrightarrow \nu_{sterile}$ oscillations are excluded at the 99.8% c.l.

ν_μ *energy estimate by Multiple Coulomb Scattering (MCS) of upthroughgoing muons.* An estimate of the muon energy was made through their Multiple Coulomb Scattering (MCS) in the absorbers [19]. The ratios Data/MC$_{\text{no osc}}$ as a function of (L/E_ν) are in agreement with $\nu_\mu \longleftrightarrow \nu_\tau$ oscillations [20].

New determination of the oscillation parameters. In order to reduce the effects of systematic uncertainties in the MCs, MACRO used the following three independent ratios (it was checked that FLUKA, HKKM01 and Bartol96 MC yield the same predictions):

(i) High Energy Data: zenith distribution ratio: $R_1 = N_{vert}/N_{hor}$

(ii) High Energy Data: ν_μ energy measurement ratio: $R_2 = N_{low}/N_{high}$

(iii) Low Energy Data: $R_3 = (Data/MC)_{IU}/(Data/MC)_{ID+UGS}$.

The no oscillation hypothesis had a probability P$\sim$3 $\cdot 10^{-7}$ and is ruled out by $\sim 5\sigma$. Fitting the 3 ratios to the $\nu_\mu \longleftrightarrow \nu_\tau$ oscillation formulae, one obtained $\sin^2 2\theta = 1$, $\Delta m^2 = 2.3 \cdot 10^{-3}$ eV^2. Using Bartol96, one adds the information on absolute fluxes:

(iv) High energy data (systematic error $\simeq$17%): $R_4 = N_{meas}/N_{MC}$.

(v) Low energy semicontained muons (scale error 21%): $R_5 = N_{meas}/N_{MC}$.

These leave the best fit values unchanged and improve the significance to 6σ.

SuperKamiokande (SK) is a large cylindrical water Cherenkov detector of 39 m diameter and 41 m height containing 50 kt of water (fiducial mass 22.5 kt); it is seen by 50-cm-diameter inner-facing phototubes (PMTs), Fig. 1b. The 2 m thick outer layer of water acts as an anticoincidence; it is seen by smaller outward-facing PMTs. The detector is located in the Kamioka mine, Japan. Atmospheric neutrinos are detected in SK by measuring the Cherenkov light generated by the charged particles produced in the neutrino CC interactions with water nuclei. Thanks to the high PMT coverage, the

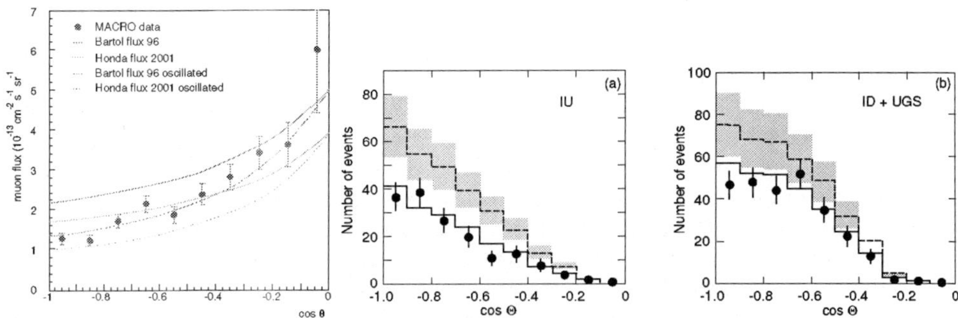

FIGURE 2. left) MACRO upthroughgoing muons compared with the oscillated MC predictions of Bartol96 (solid curve), HKKM01 (dash-dotted line), FLUKA fitted to the new CR fit (dashed curve) and FLUKA with the old CR fit (dotted curve). center) IU and right) ID+UGS events (black points) compared with the non oscillated Bartol96 MC (dashed line) and $\nu_\mu \longleftrightarrow \nu_\tau$ predictions.

experiment detects events of energies as low as ~5 MeV.

The large detector mass allows to collect a high statistics sample of *fully contained* events (*FC*) up to ~5 GeV. The *FC* events yield rings of Cherenkov light on the PMTs. *FC* events can be subdivided into *sub-GeV* and *multi-GeV* events, with energies below and above 1.33 *GeV*. *FC* events include only single-ring events, while *multi-ring* ones (*MRING*) are treated as a separate category. The *partially contained* events (*PC*) are CC interactions with vertex within the fiducial volume and at least a charged particle, typically the μ, exits the detector (the light pattern is a filled circle). *Upward-going muons* (*UPMU*), produced by ν_μ from below interacting in the rock, are subdivided into *stopping* ($\langle E_\nu \rangle \sim 7$ GeV) and *throughgoing muons* ($\langle E_\nu \rangle \sim 70 \div 80$ GeV) [10].

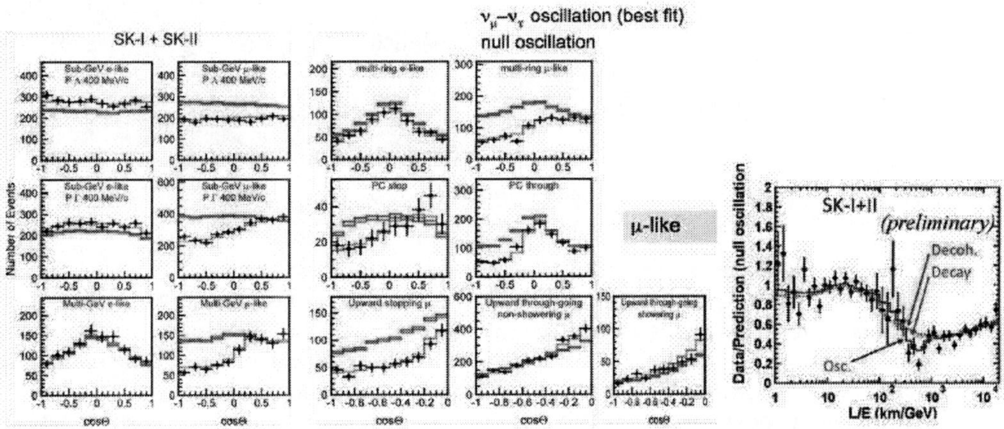

FIGURE 3. Left: SK zenith distributions (black points) for e-like and μ-like sub-GeV, multi-GeV, throughgoing and stopping muons. Solid lines are no oscillation MC predictions. Right: L/E$_\nu$ distribution for μ-like events.

Particle identification is performed using likelihood functions to parametrize the sharpness of the Cherenkov rings, which are more diffused for electrons than for muons. The zenith angle distributions for e-like and μ-like *sub-GeV* and *multi-GeV* events are shown in Fig. 3 left. The MC problem described before exists also in SK [10]: the e-like events were in agreement with the HKKM95 MC predictions for no-oscillations; they are higher than the HKKM01 non oscillated MC. For μ-like events, the new MC predictions are low: to reduce these problems the normalization is a free parameter.

The ratios e-like events/MC do not depend from L/E_ν while μ-like events/MC show a dependence on L/E_ν consistent with an oscillation hypothesis, Fig. 3 right. The overall best fit of SK data corresponds to maximal mixing and $\Delta m^2 = 2.5 \cdot 10^{-3}$ eV^2.

Exotic oscillations. MACRO and SuperK data were used to search for sub-dominant oscillations due to Lorentz invariance violation (or violation of the equivalence principle). In the first case there could be mixing between flavor and velocity eigenstates. 90% c.l. limits were placed in the Lorentz invariance violation parameters $|\Delta v| < 6 \cdot 10^{-24}$ at $\sin^2 2\theta_v$=0 and $|\Delta v| < 4 \cdot 10^{-26}$ at $\sin^2 2\theta_v = \pm 1$ [22]. **Neutrino decay** could be another exotic explanation for neutrino disappearence; no radiative decay was observed [23].

LONG BASELINE NEUTRINO BEAMS AND EXPERIMENTS

Neutrino physics has opened new windows into phenomena beyond the Standard Model of particle physics. Long baseline neutrino experiments may allow further insight into ν physics. The first long baseline ν beam was the KEK to Kamioka (K2K) beam, the 2nd was the Fermilab to the Soudan mine beam (NuMi). CNGS [24] was commissioned in 2006 and started sending neutrinos to the GS Lab.

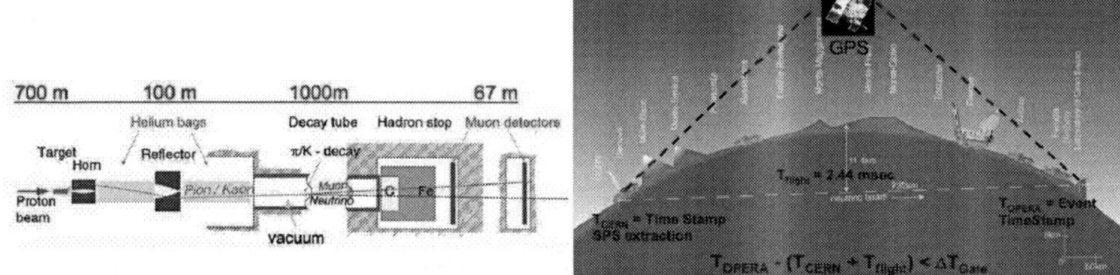

FIGURE 4. Left: The main components of the CNGS neutrino beam at CERN. Right: Sketch of the 730 km neutrino path from CERN to Gran Sasso and GPS time selection of events.

The **CNGS ν_μ beam**. Fig. 4 left shows the main components of the ν_μ beam at CERN [24]. A 400 GeV beam is extracted from the SPS and is transported to the target. Secondary pions and kaons are focused into a parallel beam by 2 magnetic lenses, called horn and reflector. Pions and kaons decay into ν_μ and μ in a decay pipe. The remaining hadrons are absorbed in the hadron stop. The μ's are monitored in 2 detectors.

Fig. 4 right shows the path of the CNGS ν_μ's from CERN to GS. It also shows the synchronization via GPS of the atomic clocks at CERN and GS. The beam is optimised for producing a maximum number of CC ν_τ interactions in OPERA. Fig. 5 left shows the underground layout of the SPS and of CNGS at CERN; Fig. 5 right shows the scheme of the SPS operation during a test run. The mean ν_μ energy is 17 GeV, the $\bar{\nu}_\mu$ contamination

~2%, the ν_e ($\bar{\nu}_e$) < 1% and the number of ν_τ is negligible. The muon beam size at the 2nd muon detector at CERN is $\sigma \sim 1$ m; this gives a ν_μ beam size at GS of $\sigma \sim 1$ km. The first low intensity test beam was sent to GS in August 2006 and 3 detectors (OPERA, LVD and Borexino) obtained their first events. The low intensity CNGS was stable and of high quality. The shared SPS beam sent a pulse of 2 neutrino bursts, each of 10.5 μsec duration, separated by 50 ms, every 12 s. A higher intensity beam expected for October 2006 did not happen because of a water leak at CERN. In September 2007 a 2nd test was successful, but a high intensity run was cancelled because of cooling problems.

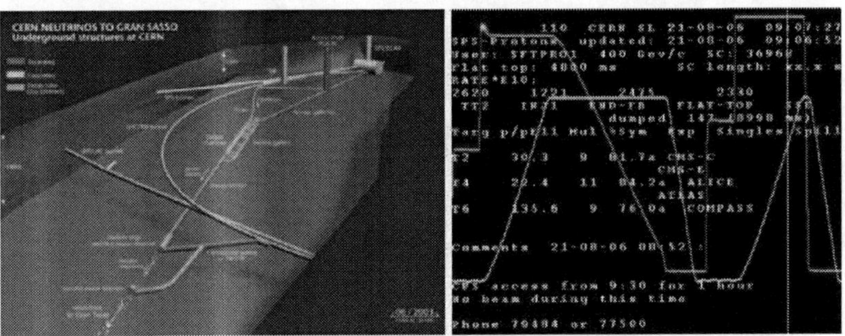

FIGURE 5. Left: Underground layout of the SPS and of the CNGS beam at CERN. Right: Scheme of the SPS operation at CERN during the test runs.

The **K2K** experiment confirmed the results of atmospheric ν_μ oscillations [13].

The **MINOS** experiment on the NuMi low energy neutrino beam is a large magnetised steel scintillator tracking calorimeter, complemented by a similar near detector and a calibration detector. The experiment obtained important results which confirm the atmospheric ν oscillation picture with maximal mixing and $\Delta m^2 = 2.38$ eV^2 [14].

We shall now concentrate on the experiments at GS which may use the CNGS beam.

Borexino is an electronic detector designed to study solar ν_e's, in particular those coming from Be^7 decays in the center of the sun [25]. The important part of the detector is a large sphere, whose central part was filled with liquid scintillator at the beginning of 2007. The outer sphere was started to be filled with water in 2006; during the first run only few meters of water were in the sphere and only few ν_μ events were observed.

LVD is an array of liquid scintillators with a total mass of 1000 t; it is designed to search and study $\bar{\nu}_e$'s from gravitational stellar collapses [16]. LVD plans to be a neutrino flux monitor of the CNGS beam. LVD has 3 identical "towers", each containing 8 active modules; a module has 8 counters of $1 \times 1 \times 1.5\, m^3$, filled with 1.2 t of liquid scintillator. CNGS ν_μ's are observed through the detection of muons produced in neutrino CC interactions in the surrounding rock or in the detector and through the detection of the hadrons produced in neutrino NC/CC interactions inside the detector. In the 2006 test run LVD counted 50-100 muons per day and recorded ~500 events [16].

OPERA [15] is a hybrid-emulsion-electronic detector, designed to search for the $\nu_\mu \longleftrightarrow \nu_\tau$ oscillations in the parameter region indicated by the atmospheric neutrinos, confirmed by the K2K and MINOS experiments. The ν_τ appearance will be made by direct detection of the τ lepton, from ν_τ CC interactions and the τ lepton decay products. To observe the decays, a spatial resolution of ~ 1 μm is necessary; this is obtained in emulsion sheets interspersed with thin lead target plates (Emulsion Cloud Chamber (ECC)). OPERA may also search for the subleading $\nu_\mu \longleftrightarrow \nu_e$ oscillations and make a variety of observations using its electronic detectors. The OPERA detector, Fig. 6, is

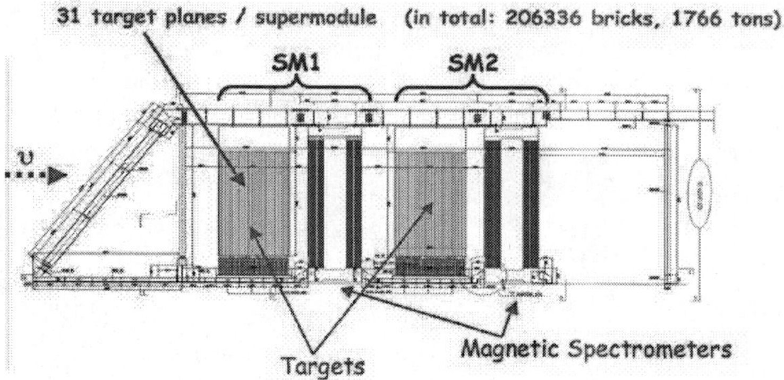

FIGURE 6. Layout of the OPERA detector

made of two identical supermodules, each consisting of a *target section* with 31 target planes of lead/emulsion-film modules ("bricks"), of a scintillator tracker detector and of a muon spectrometer. The final target mass is 1.55 kt.

The first electronic subdetector is an *anticoincidence wall* to separate μ events coming from interactions in OPERA from those in the material and rock before OPERA.

The *target tracker* is made of scintillator strips, each 7 m long and of 25×15 mm^2 cross section. A wavelength shifting fibre of 1 mm diameter transmits the light signals to both ends. The readout is done by 1000 64 channel Hamamatsu PMTs.

The *muon spectrometer* consists of 2 iron magnets instrumented with *Resistive Plate Chambers* (RPC) and *drift tubes*. Each magnet is an 8×8 m^2 dipole with a field of 1.52 T in the upward direction on one side and in the downward direction on the other side. A magnet consists of twelve 5 cm thick iron slabs, alternated with RPC planes.

The *precision tracker* measures the muon track coordinates in the horizontal plane. It is made of drift tube planes, placed in front and behind each magnet and between the 2 magnets. The muon spectrometer has a $\Delta p/p \leq 0.25$ for muon momenta < 25 GeV/c. Two 45° crossed planes of *glass RPC's (XPC's)* are installed in front of the magnets.

The *DAQ system* uses a Gigabit network of 1150 nodes. To match the data of the different subdetectors a "time stamp" is delivered by a clock using the GPS. Also the synchronization with the beam spill is done via GPS. The commissioning of each electronic detector was made with CR muons and with the CNGS at reduced intensity.

Nuclear emulsions and their scanning. The basic target module is a *"brick"*, consisting of 56 lead plates (1 mm thick) and 57 emulsion layers. A brick has a size of 10.2×12.7 cm^2, a depth of 7.5 cm (10 radiation lengths) and a weight of 8.3 kg. Two additional emulsion sheets, the *changeable sheets* (CS), are glued on its downstream face. The bricks are arranged in walls. Within a brick, the achieved spatial resolution is < 1 μm and the angular resolution is $\sim$2 mrad. Walls of target trackers provide the ν interaction trigger and the identification of the brick in which the interaction took place.

The bricks are made by the *Brick Assembling Machine* (BAM), which consists of robots for the mechanical packing of the bricks. The BAM, installed in the GS lab, produces 1 brick every $\sim$2 minutes. The bricks are handled by the *Brick Manipulator System* (BMS), made of two robots, each operating at one side of the detector.

A fast automated scanning system with a scanning speed of $\sim$ 20 cm^2/h per emulsion (44 μm thick) is needed to cope with daily analyses of many emulsions. This is a factor of 10 increase with respect to past systems. For this purpose were developed the *European Scanning System* (ESS) in Europe [26] and the *S-UTS* in Japan [27]. An emulsion is placed on a holder and the flatness is guaranteed by a vacuum system. By adjusting the focal plane of the objective, 16 tomographic images of each field of view are taken at equally spaced depths. The images are digitized, converted into a grey scale, sent to a vision processor and analyzed to recognize sequences of aligned grains. The 3-dimensional structure of a track in an emulsion layer (*microtrack*) is reconstructed by combining clusters belonging to images at different levels. Each microtrack pair is connected across a plastic base to form the *base track*. A set of connected base tracks forms a *volume track*.

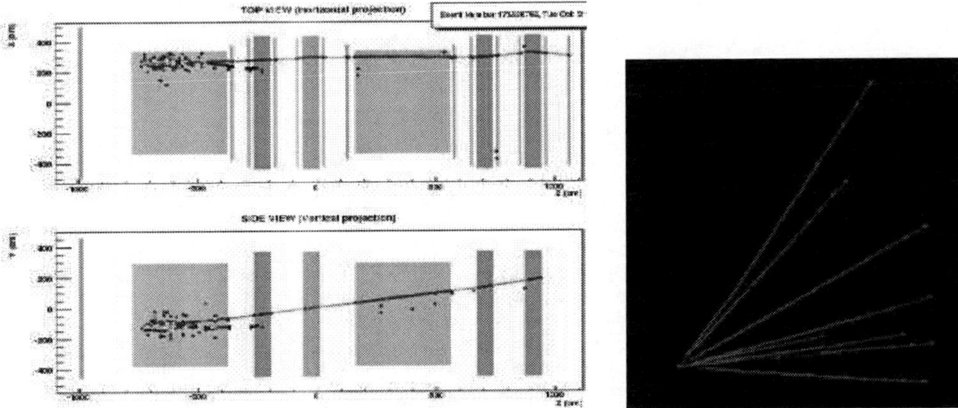

FIGURE 7. Left: display of one event with the ν interacting in a "brick" in the first module, as seen by the electronic detectors; right: the interaction vertex seen in nuclear emulsions.

In the 2006 test run OPERA recorded 319 ν_μ events, consistent with the 300 events expected from the delivered integrated intensity. A fit to the θ angle distribution of these reconstructed events on-time with the beam yielded a mean muon angle of 3.4° in agreement with the value of 3.3° expected for ν_μ originating from CERN and travelling under the earth surface to the GS underground halls. During the same test run, a test of the CS procedure was performed, using an emulsion detector plane inserted in the SM2 target. 9 muons produced by neutrino interactions in the rock surrounding the detector crossed the CS plane surface. The test proved the capability of going from the centimetre scale of the electronic tracker to the micrometric resolution of nuclear emulsions [15]. During the short 2007 test run several interactions in the lead bricks positioned in the target section of SM 1 (>40% of total mass) were detected by the electronic detectors and confirmed by tracks observed in the CS: the event vertexes were observed in the emulsions analysed by microscopes: Fig. 7 left shows an interaction in the electronic detectors, Fig. 7 right the vertex region in the emulsions. This test confirmed the validity of the methods to associate electronic detectors to nuclear emulsions.

Conclusions. The atmospheric neutrino anomaly became, in 1998, the atmospheric neutrino oscillation hypothesis with maximal mixing and $\Delta m^2_{23} \sim 2.4 \cdot 10^{-3}$ eV^2.

It was later confirmed with more data and by the first two long baseline experiments. All experiments agree on maximal mixing, while the Δm^2_{23} are:
Soudan-2 5.2 , MACRO 2.3 , SK 2.5 , K2K 2.7 , MINOS 2.38 $\cdot 10^{-3}$ eV^2.
These results come from disappearance experiments ($\nu_\mu \rightarrow \nu_\mu$). An appearance experiment ($\nu_\mu \longleftrightarrow \nu_\tau$) with ν_τ detection would solve conclusively the situation.

Other more exotic scenarios have been investigate like: Lorentz invariance violation, ν radiative decays, etc. Stringent limits were established for these hypotheses.

Acknowledgments. I acknowledge many colleagues at CERN, LNGS and Bologna. I thank drs. A. Casoni, M. Errico, M. Giorgini and R. Giacomelli for technical help.

REFERENCES

1. R. Becker-Szendy et al., *Phys. Rev.* D **46**, 372 (1992).
2. Y. Fukuda et al., *Phys. Lett.* B **335**, 237 (1994).
3. M. Aglietta et al., 23rd ICRC Proc. **2014**, 446 (1993).
4. K. Daum et al., *Z. Phys.* C **66**, 417 (1995).
5. S. Mikheyev et al., *Phys. Lett.* B **391**, 491 (1997); *Phys. Lett.* B **449**, 137 (1999).
6. S. Ahlen et al., *Phys. Lett.* B **357**, 481 (1995).
7. W. W. M. Allison et al., *Phys. Lett.* B **391**, 491 (1997).
8. W. W. M. Allison et al., *Phys. Lett.* B **449**, 137 (1999); *Phys. Rev.* D **68**, 113004 (2003).
9. M. Ambrosio et al., *Phys. Lett.* B **434**, 451 (1998); *Phys. Lett.* B **478**, 5 (2000); *Phys. Lett.* B **517**, 59 (2001). S. P. Ahlen et al., *Nucl. Instrum. Meth.* A **324**, 337 (1993).
10. Y. Fukuda et al., *Phys. Rev. Lett.* **81**, 1562 (1998); *Phys. Lett.* B **433**, 9 (1998); *Phys. Rev. Lett.* **85**, 3999 (2000); Y. Ashie et al., *Phys. Rev. Lett.* **93**, 101801 (2004); K. Kaneyuki, at HEP2007.
11. V. Agrawal et al., *Phys. Rev.* D **53**, 1314 (1996); M. Honda et al., *Phys. Rev.* D **52**, 4985 (1995).
12. M. Honda et al., *Phys. Rev.* D **64**, 053011 (2001); *Phys. Rev.* D **70**, 043008 (2004); G. Battistoni et al., *Astrop. Phys.* **19**, 269 (2003).
13. D. Kielorezka et al., *Acta Phys. Polon.* B **35**, 1453 (2004).
14. N. Tagg et al., hep-ex/0605058 (2006); A. Weber, Presentation at HEP2007, Manchester.
15. R. Acquafredda et al., *J. Phys.* **8**, 303 (2006); http://operaweb.web.cern.ch/; G. Giacomelli et al., physics/0609045 (2006); E. Barbuto et al., *Nucl. Instr. Meth.* A **525**, 485 (2004).
16. N. Y. Agafonova et al., hep-ex/0710.1536 (2007); http://www.bo.infn.it/lvd; H. Menghetti, at GS.
17. P. Bernardini hep-ex/9809003 (1998); F. Ronga hep-ex/9810008 (1998).
18. M. Ambrosio et al., *Nucl. Instr. Meth. Phys. Res.* A **486**, 663 (2002).
19. M. Ambrosio et al., *Phys. Lett.* B **566**, 35 (2003).
20. M. Ambrosio et al., *Eur. Phys. J.* C **36**, 323 (2004); G. Giacomelli et al., hep-ex/0707.1691 (2007).
21. T. K. Gaisser et al., *Nucl. Phys. B Proc. Suppl.* **118**, 109 (2003).
22. G. Battistoni et al., *Phys. Lett.* B **615**, 14 (2005).
23. S. Cecchini et al., hep-ex/0606037 (2006); *Astrop. Phys.* **21**, 35 (2001).
24. CNGS project: http://proj-cngs.web.cern.ch/proj-cngs/; G. Giacomelli, physics/0703247 (2007).
25. http://borex.lngs.infn.it; M. Pallavicini (A few CNGS events in Borexino) Gran Sasso meeting.
26. N. Armenise et al., *Nucl. Instr. Meth.* A **551**, 261 (2005); M. De Serio et al., *Nucl. Instr. Meth.* A **554**, 247 (2005); L. Arrabito et al., *Nucl. Instr. Meth.* A **568**, 578 (2006); physics/0701192 (2007).
27. T. Nakano, Proc. of Int. Europhys. Conf. on High Energy Physics, Budapest, Hungary, 2001.

Realistic Calculations for Neutrino-Nucleus Reactions Cross sections

V.C. Chasioti, T.S. Kosmas and P.C. Divari

Theoretical Physics Section, University of Ioannina, GR 45110 Ioannina, Greece

Abstract. Inelastic neutrino-nucleus reaction cross sections at low and intermediate neutrino energies are studied. The required many-body nuclear wave-functions are calculated by utilizing a version of the quasi-particle random phase approximation (QRPA) that uses realistic two-body forces. The results presented here refer to the differential, integrated and total cross sections of the neutral-current induced reaction of ν_e with the ^{98}Mo and ^{40}Ar nuclei. These isotopes are considered as promising astrophysical neutrino detection targets in recent neutrino physics research.

Keywords: Neutrino-nucleus reactions; Inelastic cross sections; Quasi-particle random phase approximation; Supernova neutrinos; Solar neutrinos
PACS: 23.40.Bw;25.30.Pt;21.60.Jz;26.30.+k

INTRODUCTION

The neutrino-nucleus reactions offer significant probes to detect or distinguish the neutrinos of several flavors and to study the basic structure of the weak interactions. In particular, several terrestrial experiments, being in operation or planed to operate for the detection of astrophysical neutrinos (solar, supernova, etc.), are good sources of information for the properties of neutrinos and the evolution of stars [1, 2].

In the present work, we examine the role of ^{98}Mo and ^{40}Ar nuclei as neutrino detectors by calculating their neutral current neutrino-induced reaction cross sections at low and intermediate energies ($0 \leq \varepsilon_i \leq 100$ MeV) of the incoming neutrino. The first of the above isotopes, ^{98}Mo (with abundance 24.13% in the natural Mo) was early proposed as an appropriate target for charged-current neutrino-detection experiments of radiochemical type and specifically for measurements of ^{8}B neutrinos [3, 4].

Recently, the MOON (Molybdenum Observatory Of Neutrinos) experiment was proposed, which is a hybrid of $\beta\beta$-decay and solar neutrino experiment. It needs ~ 1 ton of ^{100}Mo but uses natural (non-enriched) Mo which has 9.63% pure ^{100}Mo. This experiment aims at high accuracy studies of the double beta decays, with sensitivity to Majorana neutrino mass of the order of $m_\nu \leq 0.03$ eV, and the charged-current (CC) neutrino spectroscopy of the major components of the pp and ^{7}Be solar neutrinos. Since ^{98}Mo has large abundance in the MOON-target, this experiment may be used as supernova neutrino detection too [1]. The other isotope, ^{40}Ar (its abundance is 99.6% in the natural Ar) is a promising target in direct counting experiments for the detection of atmospheric, solar and supernova neutrinos. So, for example, the final phase of the ICARUS project needs a large volume of liquid Ar [2].

Both of the above isotopes are even-even nuclei and, therefore, their initial and final nuclear states (with good quantum numbers of spin, parity, and isopsin) could be reliably

CP972, *Exotic Nuclei and Nuclear/Particle Astrophysics (II)—Proceedings of the Carpathian Summer School of Physics 2007,* edited by L. Trache and S. Stoica

calculated by adopting the refinement of the QRPA that uses realistic two-body forces [5, 6].

In this work, after a brief description of the main steps of the calculational procedure, we calculate the neutrino-nucleus reaction cross sections coming from both the vector and axial-vector parts of the hadronic currents.

THE QRPA FOR NEUTRAL CURRENT REACTIONS

For neutral current neutrino-nucleus induced reactions studied here, the wave functions for the initial and final nuclear states can be constructed in the context of the pp-nn QRPA [5]. In this method, the mth excited state with total angular momentum J projection M and parity π, denoted by $|J^{\pi}_m M\rangle$, is created by acting on the QRPA ground state with the phononoperator

$$\hat{Q}^{m\dagger}_{J^{\pi}M} = \sum_{\substack{k\leq l \\ \tau}} \left[X^m_{\tau}(kl,J) A^{\dagger}_{\tau}(kl,JM) + Y^m_{\tau}(kl,J) \tilde{A}_{\tau}(kl,JM) \right] \tag{1}$$

where X, Y are the forward and backward going amplitudes. The angular momentum coupled bifermion operators $A^{\dagger}_{\tau}, \tilde{A}_{\tau}$ are written as

$$A^{\dagger}_{\tau}(kl,JM) \equiv (1+\delta_{kl})^{-\frac{1}{2}} \left[a^{\dagger}_{\tau k} a^{\dagger}_{\tau l} \right]^J_M, \quad \left[a^{\dagger}_{\tau k} a^{\dagger}_{\tau l} \right]^J_M = \sum_{m_k, m_l} \langle j_k m_k j_l m_l | JM \rangle a^{\dagger}_{\tau k m_k} a^{\dagger}_{\tau l m_l} \tag{2}$$

and $\tilde{A}_{\tau}(kl,JM) = (-1)^{J-M} A_{\tau}(kl,J-M)$. In neutral-current processes the index τ=1 (-1) for neutrino-proton (-neutron) interactions. In the above notation, $a^{\dagger}$ and a denote quasi-particle creation and annihilation operators.

The next step in using QRPA is the construction of the eigenvalue problem (QRPA equations) which reads

$$\begin{pmatrix} \mathscr{A} & \mathscr{B} \\ -\mathscr{B} & -\mathscr{A} \end{pmatrix} \begin{pmatrix} X^m \\ Y^m \end{pmatrix} = \Omega^m_{J^{\pi}} \begin{pmatrix} X^m \\ Y^m \end{pmatrix}, \tag{3}$$

where $\Omega^m_{J^{\pi}}$ denotes the excitation energy of the nuclear state $|J^{\pi}_m\rangle$. The QRPA-matrices $\mathscr{A}$ and $\mathscr{B}$, are deduced by the matrix elements of the double commutators of $A^{\dagger}$ and A with the nuclear hamiltonian $\hat{H}$ defined as

$$\mathscr{A}_J(\tau k, l; \tau' k', l') = \langle 0^+_{RPA} | [A_{\tau}(k,l,J,M), [\hat{H}, A^{\dagger}_{\tau'}(k',l',J,M)]] | 0^+_{RPA} \rangle, \tag{4}$$

$$\mathscr{B}_J(\tau k, l; \tau' k', l') = \langle 0^+_{RPA} | [A_{\tau}(k,l,J,M), [\tilde{A}_{\tau'}(k',l',J,M), \hat{H}]] | 0^+_{RPA} \rangle \tag{5}$$

Our nuclear hamiltonian $\hat{H}$ includes as field interaction a Coulomb corrected Woods-Saxon potential and as two-body residual interaction the Bonn-C potential. The indices k,l in Eqs. (1)-(5) run over the single particle levels of our valence space which is rather rich with equal number of levels for protons and neutrons [7, 8]. We note that, the particle conservation is separately checked for protons and neutrons.

RESULTS AND DISCUSSION

In the first step of our calculational procedure, we determined the model parameters by following the method described in Ref. [8]: (i) at the BCS level, we fixed the pairing parameters g^p_{pair} (g^n_{pair}), for the proton (neutron) pairing interaction, and (ii) at the QRPA level, we fixed the strength parameters g_{pp} (g_{ph}) for the particle-particle (particle-hole) interaction, separately for each set of multipole states $|J^{\pi}_{m}\rangle$, $m = 1, 2, ...$ The low-lying QRPA energy-eigenvalues obtained this way are well compared with the experimental ones [7, 8].

In the next step, we performed realistic state-by-state calculations for the double differential reaction cross sections, $d^2\sigma/d\Omega d\omega$, where ω=nuclear excitation energy of neutrino scattering off ^{98}Mo and ^{40}Ar nuclei at incoming-neutrino energies $\varepsilon_i \leq 100$ MeV. This neutrino-energy region covers, among other neutrino-sources, the solar and supernova neutrinos we are interested in here. With the aid of these results and by utilizing the method of summing-over-partial-rates, we were able to examine for some dominant sets of multipole states, the dependence of the partial differential cross sections $d\sigma/d\Omega$ on: (i) the energy of the incoming neutrino, and (ii) the scattering angle Φ. Such results are shown in Figs. 1-2, for ^{98}Mo, and in Fig. 3, for ^{40}Ar.

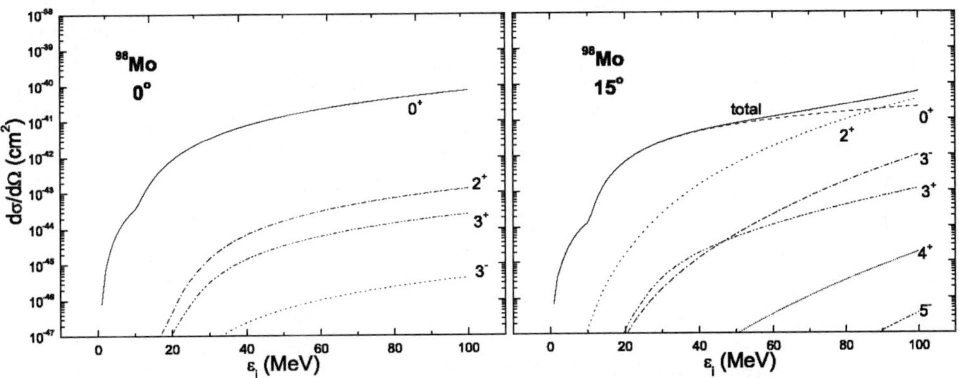

FIGURE 1. Differential cross sections, for the dominant sets of multipole states in the reaction ^{98}Mo$(\nu,\nu')^{98}$Mo*, for two scattering angles: $\Phi = 0^o$ (left) and $\Phi = 15^o$ (right).

In more detail, in Fig. 1 we plot the differential cross sections $d\sigma/d\Omega$ of the reaction ^{98}Mo$(\nu,\nu')^{98}$Mo* versus the incoming neutrino energy ε_i for some sets of multipole states like $J^{\pi} = 0^+, 2^+, 3^+$, etc., which offer dominant contributions. These plots refer to two typical values of the scattering angle, $\Phi = 0^o$ (Fig. 1 left) and $\Phi = 15^o$ (Fig. 1 right). The curve labeled "total" (for $\Phi = 15^o$) represents the total cross section $d\sigma/d\Omega$ obtained by summing over the partial contributions of all sets of multipole states up to $J^{\pi} = 5^{\pm}$. We see that, the total differential cross section essentially originates from the 0^+ multipole states. Another characteristic of these plots is the fact that the dominant multipole contribution changes as the scattering angle and/or the neutrino energy are increased. This feature is more pronounced for $\Phi = 15^o$ (Fig. 1 right side) or higher.

In Figs. 2 and 3, we illustrate the angular variation and the dependence on the initial neutrino energy of the dominant multipole contributions in $d\sigma/d\Omega$. The dominant

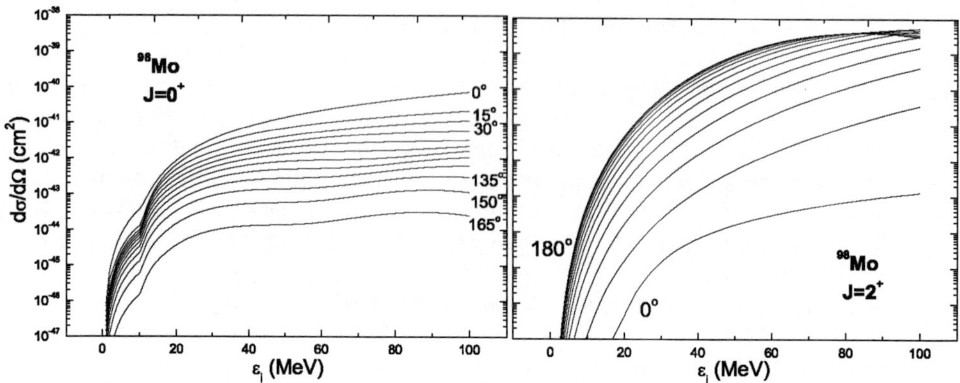

FIGURE 2. Angular dependence of the differential cross section $d\sigma/d\Omega$ of the reaction $^{98}Mo(\nu,\nu')^{98}Mo^*$ (step $\Delta\phi = 15^o$) for the multipoles 0^+ and 2^+.

multipoles for the ^{98}Mo isotope are the $0^+, 2^+$ (Fig. 2), while for ^{40}Ar these are the $0^+, 1^-$ (Fig. 3). In order to study the dependence of $d\sigma/d\Omega$ on the scattering angle, we varied the angle Φ (in the range $0 \leq \Phi \leq \pi$) by choosing a step $\Delta\Phi$=15°, and performed the calculations discussed above. From the plotted results for the 0^+ multipole in Figs. 2 and 3, we conclude that, for both isotopes the differential cross section $d\sigma/d\Omega$, keeps decreasing as the scattering angle increases. For the other multipoles the corresponding behavior reflects the angular dependence of the spherical harmonics which dominate this multipole.

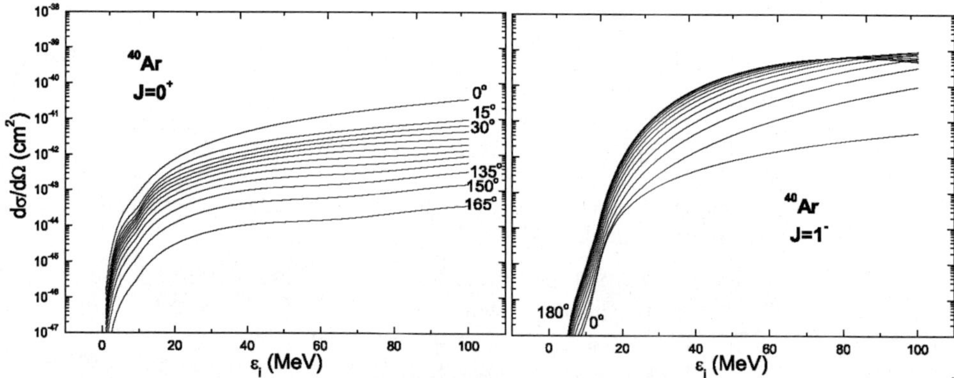

FIGURE 3. Plots showing the angular dependence of the differential cross section $d\sigma/d\Omega$ of the dominant multipoles, 0^+ and 1^-, for the neutral-current reaction $^{40}Ar(\nu,\nu')^{40}Ar^*$

We also mention that, in both nuclei, for the 0^+ multipole the curves corresponding to $\Phi = 180^o$ become negligible (not shown in the figures) and that the differential cross section $d\sigma/d\Omega$ increases slowly but smoothly as ε_i increases. Furthermore, for the 2^+ multipole in ^{98}Mo and 1^- in ^{40}Ar, the differential cross section increases very rapidly up to the incoming neutrino energy $\varepsilon_i \sim 30$ MeV, while for higher energies the increase of

$d\sigma/d\Omega$ becomes moderate. This variation resembles to that found for the cross sections in the charged current ν-nucleus processes [4].

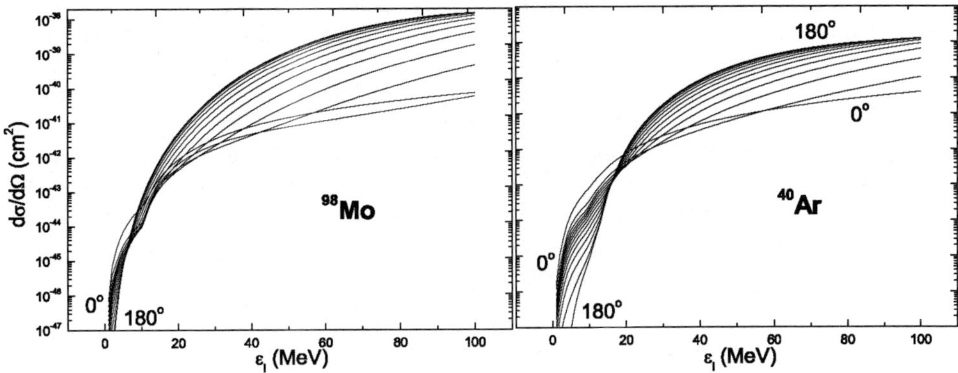

FIGURE 4. Total differential cross sections obtained by summing-over-partial contributions of all multipole states up to J^{π}=5$^{\pm}$ for the reactions: ^{98}Mo$(\nu,\nu')^{98}$Mo* (left) and ^{40}Ar$(\nu,\nu')^{40}$Ar* (right).

In Fig. 4, we show the total differential cross sections for ^{98}Mo (left) and for ^{40}Ar (right) as functions of the incoming neutrino-energy ε_i and for various values of the scattering angle in the range $0 \leq \Phi \leq \pi$, with step $\Delta\Phi = 15^o$. These plots correspond to the sum over the contributions of all excited states given from the solutions of the QRPA equations (up to $\sim$ 45 MeV) for each value of Φ. We remark that, the total differential cross section, for $\varepsilon_i > 20$ MeV generally increases as the scattering angle increases.

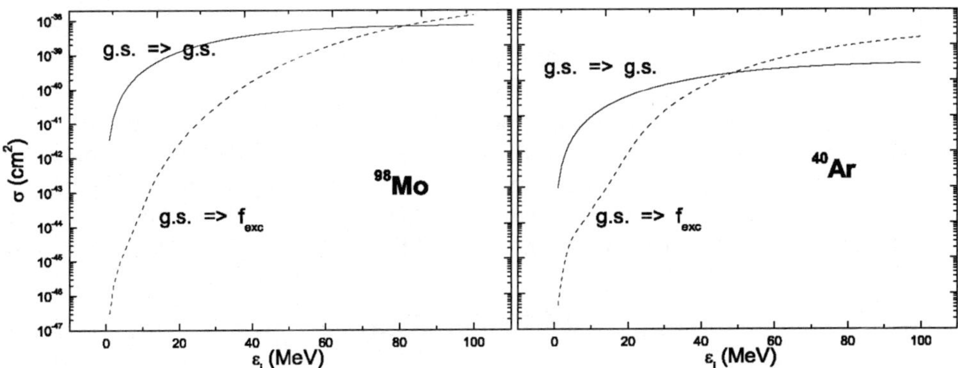

FIGURE 5. Total cross sections for the coherent ($|gs\rangle \to |gs\rangle$) and incoherent ($|gs\rangle \to |f\rangle$, where $|f\rangle$ denotes any excited QRPA state) channels of the reaction ^{98}Mo$(\nu,\nu')^{98}$Mo* (left) and ^{40}Ar$(\nu,\nu')^{40}$Ar* (right).

In Fig. 5 we compare the coherent cross section with the total incoherent one for the reactions ^{98}Mo$(\nu,\nu')^{98}$Mo* (left) and ^{40}Ar$(\nu,\nu')^{40}$Ar* (right). The plots labeled $|gs\rangle \to |f\rangle$ (incoherent cross section) result by integrating $d\sigma/d\Omega$ numerically over angles. In the same way we obtain the total cross sections for the coherent channel. We see that, for low incoming neutrino-energies ($\varepsilon_i < 50$ MeV, for ^{98}Mo and $\varepsilon_i < 35$ MeV for ^{40}Ar), the coherent cross sections dominates but for higher energies the incoherent channel becomes larger than the coherent one.

Before closing, it is worth mentioning that, the cross sections obtained above offer the possibility to explore the nuclear response of ^{98}Mo and ^{40}Ar nuclei to astrophysical (solar, supernova, etc.) neutrino spectra [1, 6]. To this aim, the calculated cross sections must be convoluted with a neutrino distribution which characterizes the neutrino source in question. Thus, for example, in order to estimate the response of ^{98}Mo and ^{40}Ar to supernova neutrinos a two-parameter Fermi-Dirac neutrino distribution could be assumed for the description of the energy spectra of all six neutrino flavors (ν_ℓ, $\tilde{\nu}_\ell$, $\ell = e, \mu, \tau$) emitted during a supernova explosion. Results of this type, for various values of neutrino temperature and degeneracy-parameter (associated with the chemical potential) would be published elsewhere [9].

SUMMARY AND CONCLUSIONS

We carried out realistic calculations of differential, total and integrated cross sections in inelastic scattering of ν_e, and $\bar{\nu}_e$ on ^{98}Mo and ^{40}Ar nuclei at low and intermediate neutrino energies ($0 \leq \varepsilon_i \leq 100$ MeV). This neutrino-energy region covers solar, supernova and the low-energy stellar-collapse neutrinos.

Such reliable results are significant to explore the nuclear response of the studied isotopes to supernova neutrino spectra. Flux averaged cross sections by assuming a two parameter Fermi-Dirac distribution for the supernova neutrino spectra in various values of neutrino temperatures T and degeneracy-parameter α, would be obtained soon.

ACKNOWLEDGMENTS

This research was funded by the program "Heraklitos" of the Operational Program for Education and Initial Vocational Training of the Hellenic Ministry of Education under the 3rd Community Support Framework and the European Social Fund. The present work was partially supported by the ΠΕΝΕΔ No 03ΕΔ807 project of the General Secretariat for Research and Technology of the Hellenic Ministry of Development.

REFERENCES

1. H. Ejiri, *Phys. Reports* **338** 265–351 (2000), and references therein.
2. A. Ereditato, A. Rubbia, *Nucl. Phys. Proc. Suppl.* **154** 163-178 (2006).
3. E. Kolbe and T.S. Kosmas, *Springer Trac. Mod. Phys.* **163** 199–225 (2000).
4. T.S. Kosmas and E. Oset, *Phys. Rev.* **C 53** 1409–1415 (1996); E. Kolbe, *Phys. Rev. C* **54** 1741–1748 (1996).
5. T.S. Kosmas, *Nucl. Phys. A* **683** 443–462 (2001).
6. E. Kolbe, K. Langanke and P. Vogel, *Phys. Rev. D* **66** 013007 (2002);E. Kolbe and K. Langanke, *Phys. Rev. C* **63** 025802 2001.
7. V.Ch. Chasioti and T.S. Kosmas, *Czech. J. Phys.* **52** 467–479 (2002) ; E-print: nucl-th/0202062.
8. V.Ch. Chasioti, T.S. Kosmas and P.C. Divari, *Prog. Part. Nucl. Phys.* **59**, 481–485 (2007).
9. V.Ch. Chasioti, T.S. Kosmas and P.C. Divari, to be submitted in *Nucl. Phys.* **A**.

Neutrinos in Cosmology

Yvonne Y. Y. Wong

Max-Planck-Institut für Physik (Werner-Heisenberg-Institut)
Föhringer Ring 6, 80805 München, Germany

Abstract. I give an overview of the effects of neutrinos on cosmology, focussing in particular on the role played by neutrinos in the evolution of cosmological perturbations. I discuss how recent observations of the cosmic microwave background and the large-scale structure of galaxies can probe neutrino masses with greater precision than current laboratory experiments. I describe several new techniques that will be used to probe cosmology in the future.

Keywords: Neutrino, Cosmology
PACS: 14.60.Pq, 98.80.-k, 98.80.Es

INTRODUCTION

Standard big bang theory predicts some 10^{87} neutrinos per flavour in the visible universe, an abundance second only to the cosmic microwave background photon. Relativistic neutrinos constitute a significant fraction of the total energy density in the epoch of radiation domination. Even in today's matter-dominated universe, a minimum neutrino mass of 0.07 eV established by the solar and atmospheric neutrino oscillations experiments means that relic big bang neutrinos can still contribute at least 0.5% of the total matter density. These neutrinos have a definite impact on cosmic structure formation.

This article provides a brief introduction to the subject of neutrino cosmology, explaining in simple terms the main concepts that allow us to determine neutrino properties with cosmological observations. For more detailed discussions, I refer the reader to the excellent reviews of Refs. [1, 2]. For a review on neutrinos and their connection to big bang nucleosynthesis, see Ref. [3].

THE COSMIC NEUTRINO BACKGROUND

In the first second after big bang, the temperature of the universe is so high that even weak interactions are able to hold neutrinos in thermal equilibrium with the cosmic plasma (photons, electrons, positrons, etc.). When in equilibrium, the neutrinos' phase space density is given by the Fermi–Dirac distribution,

$$f(p,T) = \frac{1}{1+\exp[(E-\mu)/T]}, \tag{1}$$

where the chemical potential μ is expected to be of the same order of magnitude as the universal matter–antimatter asymmetry, $\mu/T \sim 10^{-10}$, i.e., negligible. The number density and energy density per neutrino flavour are given respectively by $n_\nu(T) =$

CP972, *Exotic Nuclei and Nuclear/Particle Astrophysics (II)—Proceedings of the Carpathian Summer School of Physics 2007,* edited by L. Trache and S. Stoica

$g/(2\pi)^3 \int d^3p\, f(p,T)$ and $\rho_\nu(T) = g/(2\pi)^3 \int d^3p\, E\, f(p,T)$, where g counts the internal degrees of freedom. For neutrinos, $g = 2$ for two helicity states.

As the universe expands and cools, the interaction rate drops. At some point the interactions becomes too slow to keep up with the expansion. The neutrinos can no longer be held in thermal equilibrium and decouple from the plasma. A rough estimate for the time at which this decoupling happens is to compare the Hubble expansion rate, $H \sim T^2/m_{\text{planck}}$, to the interaction rate, $\Gamma \sim G_F^2 T^5$. By demanding $\Gamma \sim H$, we can see that neutrinos decouple roughly at a temperature of $T \sim 1$ MeV.

Note that because neutrinos decouple while they are relativistic, they preserve, to an excellent approximation, their relativistic Fermi–Dirac phase space density even after they become nonrelativistic. This fact has important consequences.

Immediately after decoupling the neutrinos have the same temperature as the photons in the plasma. Shortly after, at $T \sim m_e/3$, the process of electron–positron annihilation causes the plasma to reheat. This reheating is, however, not felt by the neutrinos because they have already decoupled. The end result is that the neutrinos are colder than the photons. By demanding entropy conservation in the electron–photon system, one can show that the neutrino temperature is given by

$$T_\nu = \left(\frac{4}{11}\right)^{1/3} T_\gamma, \tag{2}$$

and this expression remains valid until today.

Today, we expect to see a neutrino background of temperature $T \sim 1.95$ K, with a number density of 112 cm^{-3} per flavour. Assuming a neutrino mass $m_\nu \gg T_\nu \sim 10^{-4}$ eV, the energy density per flavour is

$$\Omega_\nu h^2 \equiv \frac{\rho_\nu}{\rho_{\text{crit}}} h^2 \simeq \frac{m_\nu}{93\ \text{eV}}, \tag{3}$$

where ρ_{crit} is the present critical density. By demanding that massive neutrinos do not overclose the universe, i.e., $\Omega_\nu < 1$, one can immediately set an upper bound on the neutrino mass of $m_\nu < 93$ eV [4, 5]. Historically this is first upper bound on the neutrino mass derived from cosmology.

NEUTRINOS AND STRUCTURE FORMATION

In the current theory of structure formation, the universe is initially filled with a nearly smooth density field with small, random fluctuations. The regions of high density act like gravitational potential wells, attracting matter from the surrounding low density regions. Over time, the dense regions form ever denser clumps, while the sparse regions become even sparser—the technical jargon for this is to say that the density perturbations, $\delta \equiv \delta\rho/\rho$, "grow". Eventually, the clumps become so dense that galaxies begin to form within. Of course, the initial small density perturbations must have come from somewhere. The leading theory for this is inflation, in which the density perturbations have their origin in quantum fluctuations of the inflaton field.

Generally, in order to study the evolution of the perturbations, we need to solve the Boltzmann equations. Schematically, the Boltzmann equation for the species i,

$$\frac{D}{Dt} f_i(\vec{x},t) = C[f_i(\vec{x},t)], \tag{4}$$

consists of, on the left hand side, the Liouville operator acting on the phase space density $f_i(\vec{x},t)$ of i, and, on the right hand side, a collision term incorporating all non-gravitational interactions experienced by i. The density perturbations in i are related to f_i via

$$\delta\rho_i = \frac{g_i}{(2\pi)^3} \int d^3p\, E\, \left[f_i(\vec{x},t) - \bar{f}_i(t)\right], \tag{5}$$

where $\bar{f}_i(t)$ denotes the unperturbed phase space density. In Newtonian gravity, the Liouville operator is simply,

$$\frac{D}{Dt} = \frac{\partial}{\partial t} + \vec{v}\cdot\frac{\partial}{\partial \vec{x}} - m\nabla\Phi\cdot\frac{\partial}{\partial \vec{p}}, \tag{6}$$

where Φ is the Newtonian gravitational potential, related to the global density perturbations via the Poisson equation, $\nabla^2\Phi = 4\pi G\, \sum_i \delta\rho_i$. In a general relativistic treatment, the role of the Poisson equation is subsumed by the Einstein equations.

The full $(3+3+1)$-dimensional Boltzmann equations are generally very difficult to solve. One way to solve them is by using particle realisations, i.e., N-body simulations. However, if the perturbations are small, i.e., $\delta \ll 1$, one can solve them perturbatively. Perturbative solutions are easiest when conducted in Fourier space, i.e., $\delta(\vec{x},t) \to \delta_k(t)$, because to first order in δ_k, all modes k evolve independently of each other.

Rather than presenting the full set of equations to be solved, I will briefly summarise some important results from linear (first order) perturbation theory. The growth of density perturbations is essentially governed by:

1. *Background evolution* The optimal growth of structure happens when matter dominates the energy density of the universe (i.e., the epoch of matter domination). Here, the growth rate is proportional to the scale factor, $\delta_k \propto a$. During radiation domination or when dark energy constitutes a significant fraction of the total energy density, the fast expansion rate tends to inhibit growth.
2. *The velocity dispersion* The central premise of structure formation is that matter fall into gravitational potential wells, thus making the wells even deeper. However, if the matter has too much velocity, then this will prevent them from clustering on small scales, thereby hindering the growth of perturbations on these scales. This effect is known as free-streaming suppression, and is the primary reason why neutrinos cannot be the dominant dark matter component: they are born with too much thermal velocity. Current data prefer dark matter that is "cold".
3. *Interactions* If the matter concerned has significant interactions (besides gravitational), then this interaction can also prevent the growth of perturbations in the interacting species. This is because interacting fluids can support sound waves, so that whenever gravity pulls and compresses the fluid, the pressure in the fluid will

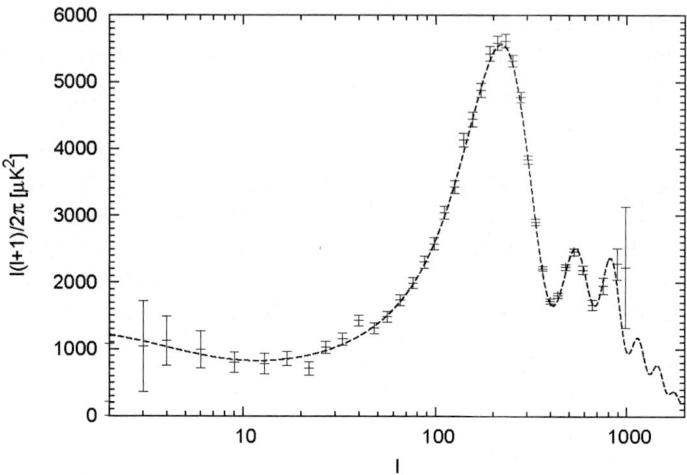

FIGURE 1. The cosmic microwave background temperature anisotropy spectrum for the standard ΛCDM model (see Table 1 for parameter values), with data from the WMAP satellite [8].

push out, setting up acoustic oscillations. This is what happens to baryons before recombination: it is only after decoupling from the photons that the baryons can fall freely into potential wells, and perturbations in their density are allowed to grow.

Probes of Perturbations

From this discussion, it is clear that the density perturbations can tell us a lot about the underlying cosmology. Currently there are two types of probes of density perturbations.

Cosmic microwave background (CMB) anisotropies. Discovered first by the COBE satellite in 1992 [6], these are tiny ($\sim 10^{-5}$) fluctuations around the mean temperature of 2.73 K of the cosmic microwave background. The resolution of COBE satellite was about 7 degrees. Today, the state-of-the-art instrument, the WMAP satellite, can measure the full-sky anisotropies with a 0.5 degree resolution [7, 8]. In addition WMAP also measures the anisotropies in the CMB polarisation [9]. Anisotropies on scales below 0.5 degrees have been measured by an array of ground-based and balloon experiments such as CBI [10], VSA [11], DASI [12], BOOMERanG [13], etc..

These temperature and polarisation fluctuations are usually discussed in terms of a two-point correlation. For example, for the temperature fluctuations,

$$\left\langle \frac{\delta T(\hat{x}_1)}{T} \frac{\delta T(\hat{x}_2)}{T} \right\rangle = \frac{1}{4\pi} \sum_{\ell=2}^{\infty} (2\ell+1) \langle |a_{\ell m}|^2 \rangle P_\ell(\hat{x}_1 \cdot \hat{x}_2), \tag{7}$$

where P_ℓ are the Legendre polynomials, and $C_\ell \equiv \langle |a_{\ell m}|^2 \rangle$ the moment of the multipole ℓ. Figure 1 shows C_ℓ as a function of ℓ, also called the anisotropy spectrum.

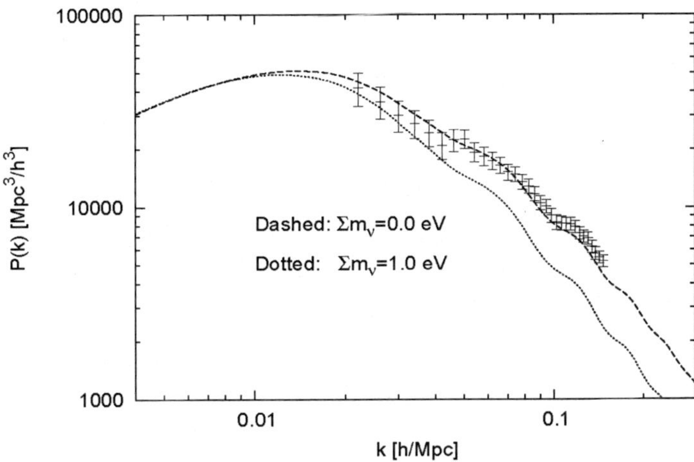

FIGURE 2. The large-scale matter power spectrum. The dashed line denotes the standard ΛCDM model with zero neutrino mass; the dotted line has $\sum m_\nu = 1$ eV. The data points are from the 2dF survey [16].

The physics of these anisotropies may be understood as follows. Before recombination, photons and baryons form a tightly coupled fluid. This means that whenever gravity pulls the baryons into the dark matter potential wells and compresses the fluid, the pressure in the photons will push the fluid out again. This competition between gravity and pressure sets up acoustic oscillations, so that the compressed regions are hotter and the rarefied regions cooler. What we see in the CMB anisotropy spectrum are these acoustic oscillations, just before recombination. After recombination, photons free-stream to infinity, preserving the oscillation pattern in their temperature. This makes the CMB anisotropies are very powerful probe of the baryon density, which determines to the sound speed of the baryon–photon fluid. Furthermore, since the CMB anisotropies can be thought of as a projection of the last scattering surface onto the sky, they also contain information about the geometry of the universe.

Probes of large-scale structure (LSS). The most powerful LSS probes at present are galaxy redshift surveys, with the Two Degree Field Galaxy Redshift Survey (2dFGRS) [14] and the Sloan Digital Sky Survey (SDSS) [15] being the largest to date. Essentially these surveys produce galaxy catalogues. With these catalogues one can calculate the galaxy number density at each point in space, and compute their two-point correlations,

$$\langle \delta(\vec{x}+\vec{r})\delta(\vec{x})\rangle = \int \frac{d^3k}{(2\pi)^3}|\delta_k|^2 \exp(-i\vec{k}\cdot\vec{r}). \tag{8}$$

The quantity $P(k) \equiv |\delta_k|^2$ is the power spectrum. In order to use this power spectrum to constrain cosmology, one makes one further assumption, that the clustering of galaxies traces that of the underlying matter distribution. In technical jargon, we say that the power spectrum reconstructed from galaxy clustering data, $P_{\rm gal}(k)$, is proportional to

TABLE 1. Free parameters of the "vanilla" cosmological model, their estimated values, and 68% credible intervals from a combined WMAP+SDSS analysis [17, 18]. The fluctuation amplitude is expressed in terms of σ_8, defined as the present-day rms fluctuations in a sphere of 8 h Mpc radius.

Parameter	Estimate
Matter density, $\Omega_m h^2$	$0.1272^{+0.0044}_{-0.0043}$
Baryon density, $\Omega_b h^2$	$0.0222^{+0.0007}_{-0.0007}$
Hubble parameter, h	$0.073^{+0.019}_{-0.019}$
Fluctuation amplitude, σ_8	$0.756^{+0.035}_{-0.035}$
Spectral index, n_s	$0.953^{+0.016}_{-0.016}$
Optical depth to reionisation, τ	$0.087^{+0.028}_{-0.030}$

the underlying matter power spectrum $P_{\rm m}(k)$ up to a normalisation or bias factor, i.e., $P_{\rm gal} = b^2 P_{\rm m}(k)$. This assumption should work well on large scales (i.e., small values of k), but is expected to fail on small scales. Where exactly it fails and how to correct for it is a major research question.

Other probes of LSS include weak gravitational lensing, Lyman-α absorption lines, cluster abundance, etc.. However, they all in one way or another rely on nonlinear modelling, which is presently not very well understood. I shall not discuss them here.

Figure 2 shows the LSS power spectrum. It is quite featureless, except for a marked turning point. Inflation generally predicts an initial power spectrum that is power-law like. The present day power spectrum ends up with a turning point because of evolution of the density perturbations: as the universe expands, perturbations with increasing wavelengths, or decreasing k values, "enter" the Hubble radius H^{-1}, i.e., the wavelength becomes shorter than H^{-1} and the mode is in causal contact. The shortest wavelengths enter the Hubble radius during radiation domination; the perturbations in these modes therefore cannot grow. On the other hand, the larger wavelengths enter the Hubble radius late, during matter domination. These modes can grow. The net result is a suppression of the power spectrum at large k values, with a turning point at $k_{\rm tp}$ corresponding to the comoving Hubble radius during matter–radiation equality, $k_{\rm tp} \sim a_{\rm eq} H_{\rm eq}$.

Cosmological Model

The minimum model required to fit all current observational data has six free parameters: the matter density $\Omega_m h^2$ and baryon density $\Omega_b h^2$ describe the energy content of the universe; the Hubble parameter h gives the expansion rate; the initial conditions are specified by the initial power spectrum amplitude A_s and spectral index n_s, such that $P_{\rm ini}(k) = A_s k^{ns}$; lastly, an astrophysical parameter τ, which affects only the CMB anisotropies, gives the optical depth to reionisation. The geometry of the model is flat, i.e., no spatial curvature. Furthermore, the initial conditions are adiabatic, meaning that the density perturbations in various components satisfy the relation $\delta_\gamma/4 = \delta_\nu/4 = \delta_m/3 = \delta_b/3$. This relation is predicted by the simplest kind of inflation

TABLE 2. 95% C.L. upper limit on the sum of neutrino masses $\sum m_\nu$ from various cosmological observations. The acronym SN stands for supernova Ia, HST for Hubble Space Telescope key project, BAO for baryon acoustic oscillations, and Lya for Lyman-α absorption lines. Two classes of models are considered: minimal models "m" with the six basic parameters of Table 1 plus neutrino mass, and extended models with additional free parameters to describe (a) dark energy beyond a cosmological constant, (b) non-adiabatic (isocurvature) initial conditions, and (c) the possible existence of axion hot dark matter.

Reference	95% upper bound [eV]	Model	Data
Spergel *et al.* [17]	0.7	m	WMAP,LSS,SN,HST
Tegmark *et al.* [18]	0.94	m	WMAP,SDSS
Goobar *et al.* [20]	0.6	a	WMAP,LSS,SN,BAO,HST
Seljak *et al.* [21]	0.17	m	WMAP,LSS, SN,BAO,Lya,HST
Ichikawa *et al.* [22]	2.0	m	WMAP
Kristiansen *et al.* [23]	1.43	m	WMAP,Cluster mass function
Zunckel and Ferreira [24]	2.2	b	WMAP,LSS
Hannestad *et al.* [25]	0.65	c	WMAP,LSS,SN,BAO

models, single-field inflation. Table 1 shows the current estimates of these parameters from the CMB experiment WMAP [17] and the SDSS galaxy redshift survey [18].

Of course, these six parameters are not the only ones that can affect cosmology: there are many more. These parameters are not included in most analyses at the moment because they do not change significantly the goodness-of-fit. If we quantify the goodness-of-fit with the best fit χ^2 of the model, then it turns out that adding any one extra parameter does not improve the best fit χ^2 by more than 3 or 4. This improvement is deemed too modest by many practitioners to justify the inclusion of extra parameters, although it is unclear where one should draw the line. Indeed, some of these extra parameters are physically extremely well motivated. Here I discuss the case of neutrino masses.

Constraints on Neutrino Masses

The idea of "weighing neutrinos" with cosmology [19], especially with LSS probes, hinges on an effect known as free-streaming. Neutrinos with sub-eV to eV masses contribute to the dark matter content of the universe: the heavier they are, the more their contribution. However, unlike ordinary cold dark matter, neutrinos are born with a high velocity, and become nonrelativistic only at very late times during structure formation. While relativistic, their high velocity makes them difficult to capture into the dark matter potential wells, especially the ones that are narrow in width. This free-streaming inhibits the growth of the density perturbations on length scales below the free-streaming scale. In the power spectrum, the presence of neutrino mass is manifested in a suppression of power on small scales. Figure 2 shows this suppression for a 1 eV neutrino.

Using this effect many authors have placed bounds on the sum of the neutrino masses. Table 2 shows a selection of recent results. Clearly, the mass bound depends very much on which data sets are used to produce it. In particular, one can obtain an extremely tight bound of $\sum m_\nu < 0.17$ eV if Lyman-α absorption lines are included. However, the Lya data used in [21] suffer from unknown systematic effects, and show a 2σ discrepancy in the fluctuation amplitude when compared with CMB data. This discrepancy pushes the allowed parameter space to a "compromise region" and artificially pushes down the neutrino mass bound. Therefore, one should treat bounds from Lya with extreme caution.

Note also that the neutrino mass bounds are model-dependent: adding to the model parameters that are degenerate with the neutrino mass tends to relax the mass bound.

The bottom line is that precision cosmology can at present constrain the sum of the neutrino masses to order 1 eV. To obtain tighter bounds one would need to make assumptions about unknown factors such as the nature of dark energy and the initial conditions of the primordial density fluctuations.

FUTURE COSMOLOGICAL PROBES

Many cosmological probes have been planned for the next decade. There will be more CMB surveys, including the Planck satellite to be launched in 2008. Galaxy clustering surveys will probe LSS at higher redshifts, enabling us to track the evolution of structure and to probe the neutrino mass with greater sensitivity [26, 27, 28]. New kinds of observations utilising different techniques to probe cosmology will also become available. Here are some examples.

- *Weak gravitational lensing of galaxies* Light travelling from distant galaxies is bent by the foreground matter (both luminous and dark). This causes distortions (magnification and stretching) in the images of these galaxies. By measuring these distortions, one can reconstruct the foreground matter distribution (the lens) [29]. An interesting possibility is to bin the galaxy images by redshift, so that one can even obtain some 3D information about the foreground distribution. Using this technique, and together with future CMB observations, one can expect to probe the sum of the neutrino masses with a 1σ sensitivity of 0.01–0.05 eV [30, 31].
- *Weak lensing of the CMB* Light from the last scattering surface can also be lensed by the foreground matter [32]. Indeed, with the Planck satellite, one can hope to probe $\sum m_\nu$ with sensitivity of 0.1 eV via CMB lensing [33, 34, 35].
- *21 cm* These are emission lines from neutral Hydrogen spin flip at very high redshifts ($z > 50$). The density fluctuations in the primordial Hydrogen gas are imprinted on these emission lines. Thus, if detected they will offer a unique probe of the density perturbations in the "cosmic dark ages" [36]. However, detecting these emission lines require sophisticated radio telescopes, and currently planned facilities cannot look beyond $z = 20$. Thus, 21 cm emission lines are for now better suited for the study of astrophysics, especially reionisation physics.

CONCLUSIONS

Cosmology offers an interesting way to probe many aspects of fundamental physics. Neutrino properties are an excellent example. Using probes of the large-scale structure and of the cosmic microwave background anisotropies, we can already deduce that the sum of the neutrino masses must be less than about 1eV. This limit is already better than those from laboratory experiments (roughly $\sum m_\nu < 6$ eV [37, 38]). Future cosmological observations will do even better: in the most optimistic case, a sensitivity of $\sigma(\sum_\nu) \sim 0.01$ eV may be achievable in the next decade with CMB and weak gravitational lensing observations. With this kind of sensitivity, there is real hope that we may finally be able to measure the neutrino mass with precision cosmology.

ACKNOWLEDGMENTS

I would like to thank the organisers of the Carpathian Summer School of Physics for a most enjoyable meeting.

REFERENCES

1. J. Lesgourgues and S. Pastor, Phys. Rept. **429**, 307 (2006) [arXiv:astro-ph/0603494].
2. S. Hannestad, Ann. Rev. Nucl. Part. Sci. **56**, 137 (2006) [arXiv:hep-ph/0602058].
3. A. D. Dolgov, Phys. Rept. **370**, 333 (2002) [arXiv:hep-ph/0202122].
4. S. S. Gershtein and Y. B. Zeldovich, JETP Lett. **4**, 120 (1966).
5. R. Cowsik and J. McClelland, Phys. Rev. Lett. **29**, 669 (1972).
6. G. F. Smoot *et al.*, Astrophys. J. **396**, L1 (1992).
7. C. L. Bennett *et al.*, Astrophys. J. Suppl. **148**, 1 (2003) [arXiv:astro-ph/0302207].
8. G. Hinshaw *et al.*, Astrophys. J. Suppl. **170**, 288 (2007) [arXiv:astro-ph/0603451].
9. L. Page *et al.*, Astrophys. J. Suppl. **170**, 335 (2007) [arXiv:astro-ph/0603450].
10. B. S. Mason *et al.*, Astrophys. J. **591**, 540 (2003) [arXiv:astro-ph/0205384].
11. K. Grainge *et al.*, Mon. Not. Roy. Astron. Soc. **341**, L23 (2003) [arXiv:astro-ph/0212495].
12. N. W. Halverson *et al.*, Astrophys. J. **568**, 38 (2002) [arXiv:astro-ph/0104489].
13. J. E. Ruhl *et al.*, Astrophys. J. **599**, 786 (2003) [arXiv:astro-ph/0212229].
14. M. Colless *et al.*, arXiv:astro-ph/0306581.
15. J. K. Adelman-McCarthy *et al.*, arXiv:0707.3413 [astro-ph].
16. S. Cole *et al.*, Mon. Not. Roy. Astron. Soc. **362**, 505 (2005) [arXiv:astro-ph/0501174].
17. D. N. Spergel *et al.*, Astrophys. J. Suppl. **170**, 377 (2007) [arXiv:astro-ph/0603449].
18. M. Tegmark *et al.*, Phys. Rev. D **74**, 123507 (2006) [arXiv:astro-ph/0608632].
19. W. Hu, D. J. Eisenstein and M. Tegmark, Phys. Rev. Lett. **80**, 5255 (1998) [arXiv:astro-ph/9712057].
20. A. Goobar, S. Hannestad, E. Mortsell and H. Tu, JCAP **0606**, 019 (2006) [arXiv:astro-ph/0602155].
21. U. Seljak, A. Slosar and P. McDonald, JCAP **0610**, 014 (2006) [arXiv:astro-ph/0604335].
22. K. Ichikawa *et al.*, Phys. Rev. D **71**, 043001 (2005) [arXiv:astro-ph/0409768].
23. J. R. Kristiansen, O. Elgaroy and H. Dahle, Phys. Rev. D **75**, 083510 (2007).
24. C. Zunckel and P. G. Ferreira, JCAP **0708**, 004 (2007) [arXiv:astro-ph/0610597].
25. S. Hannestad *et al.*, JCAP **0708**, 015 (2007) [arXiv:0706.4198 [astro-ph]].
26. M. Takada, E. Komatsu and T. Futamase, Phys. Rev. D **73**, 083520 (2006) [arXiv:astro-ph/0512374].
27. S. Hannestad and Y. Y. Y. Wong, JCAP **0707**, 004 (2007) [arXiv:astro-ph/0703031].
28. F. B. Abdalla and S. Rawlings, arXiv:astro-ph/0702314.
29. M. Bartelmann and P. Schneider, Phys. Rept. **340**, 291 (2001) [arXiv:astro-ph/9912508].
30. Y. S. Song and L. Knox, arXiv:astro-ph/0312175.
31. S. Hannestad, H. Tu and Y. Y. Y. Wong, JCAP **0606**, 025 (2006) [arXiv:astro-ph/0603019].
32. A. Lewis and A. Challinor, Phys. Rept. **429**, 1 (2006) [arXiv:astro-ph/0601594].
33. M. Kaplinghat, L. Knox and Y. S. Song, Phys. Rev. Lett. **91**, 241301 (2003) [arXiv:astro-ph/0303344].
34. J. Lesgourgues *et al.*, Phys. Rev. D **73**, 045021 (2006) [arXiv:astro-ph/0511735].
35. L. Perotto *et al.*, JCAP **0610**, 013 (2006) [arXiv:astro-ph/0606227].
36. S. Furlanetto, S. P. Oh and F. Briggs, Phys. Rept. **433**, 181 (2006) [arXiv:astro-ph/0608032].
37. V. M. Lobashev, Nucl. Phys. A **719**, 153 (2003).
38. C. Kraus *et al.*, Eur. Phys. J. C **40**, 447 (2005) [arXiv: hep-ex/0412056].

PART II: COMMUNICATIONS

Extracting the Asymptotic Normalization Coefficients in Neutron Transfer Reactions to Determine the Astrophysical Reaction Rates for ^{22}Mg(p,γ)^{23}Al and ^{17}F(p,γ)^{18}Ne

T. Al.-Abdullah[a], F. Cârstoiu[b], X. Chen[a], H. Clark[a], C. Fu[a], C. A. Gagliardi[a], Y.-W. Lui[a], A. M. Mukhamedzhanov[a], G. Tăbăcaru[a], Y. Tokimoto[a], L. Trache[a], R. E. Tribble[a], and Y. Zhai[a]

[a]*Cyclotron Institute, Texas A&M University, College Station, TX 77843, U. S. A.*
[b]*National Institute for Nuclear Physics and Engineering, P.O. Box MG-6, Bucharest, Romania*

Abstract. It was expected that γ rays from ^{18}F and ^{22}Na would be observed in massive ONe novae, but neither has been to date. The ^{17}F(p,γ)^{18}Ne and ^{22}Mg(p,γ)^{23}Al reactions may play important roles in the production and destruction of ^{18}F and ^{22}Na. We have determined the asymptotic normalization coefficients for ^{18}Ne and ^{23}Al through measurements of peripheral neutron transfer reactions in the mirror nuclear systems. We use the ANCs to calculate the astrophysical *S*-factors and reaction rates for these two reactions.

Keywords: ONe novae; HCNO cycle; NeNa cycle; asymptotic normalization coefficients.
PACS: 26.50.+x, 25.40.Lw, 25.70.Hi.

INTRODUCTION

Massive ONe novae are common events in the Galaxy. The theoretical and astrophysical analysis of the spectroscopic lines reveals that their probability to occur accounts for approximately one-third of all the recently detected novae [1]. In these events, the H-burning sequence of reactions by means of HCNO and NeNa cycles can yield substantial concentrations of ^{18}F and ^{22}Na. However, observations using space-based γ-ray telescopes have not found the signatures of these γ-rays. The evolution of the nucleosynthesis in the *rp*-process may give an explanation for the missing flux of ^{18}F and ^{22}Na γ-rays.

^{17}F(p,γ)^{18}Ne is an important nuclear reaction to understand the absence of 511 keV lines following ^{18}F decay. The rate of this reaction may influence the abundances of ^{18}F, ^{18}Ne, ^{17}F, and ^{15}O, and explain the transition sequence from the HCNO cycle to the *rp*-process. Estimates indicate that direct capture to the bound states in ^{18}Ne, especially the first two 2^+ states, dominates the rate at typical nova temperatures, $T_9 <$ 0.4 [2,3]. ^{22}Na is synthesized in novae through both the cold and hot NeNa cycles. In the hot NeNa cycle, the conventional path to produce ^{22}Na is ^{20}Ne(p,γ)^{21}Na(p,γ)^{22}Mg ($\beta^+\nu$)^{22}Na. In explosive conditions, a fast ^{22}Mg(p,γ)^{23}Al reaction may bypass the production of ^{22}Na [4].

CP972, *Exotic Nuclei and Nuclear/Particle Astrophysics (II)—Proceedings of the Carpathian Summer School of Physics 2007*, edited by L. Trache and S. Stoica

No direct measurements of the $^{17}F(p,\gamma)^{18}Ne$ and $^{22}Mg(p,\gamma)^{23}Al$ reactions at stellar energies have been performed to date. Here we report measurements of the asymptotic normalization coefficients (ANCs) for ^{18}Ne and ^{23}Al from peripheral neutron transfer reactions in the mirror nuclear systems. The ANCs are used to calculate the corresponding astrophysical reaction rates.

NEUTRON TRANSFER REACTIONS

Experiments were carried out at the Texas A&M University K500 Superconducting Cyclotron to measure the neutron transfer reactions $^{13}C(^{17}O,^{18}O)^{12}C$ and $^{13}C(^{22}Ne,^{23}Ne)^{12}C$. In addition, the reaction $^{13}C(^{12}C,^{13}C)^{12}C$ was studied in order to determine the ANC for $^{13}C \leftrightarrow {}^{12}C + n$, which is needed in the analysis of the other two reactions. The elastic scattering cross section for the incoming and outgoing channels of each reaction was also measured in order to obtain optical model parameters. The energy of the beams was about 12 MeV/u to ensure that the reactions are peripheral. Reaction products were measured with the MDM magnetic spectrometer combined with the modified Oxford detector [5].

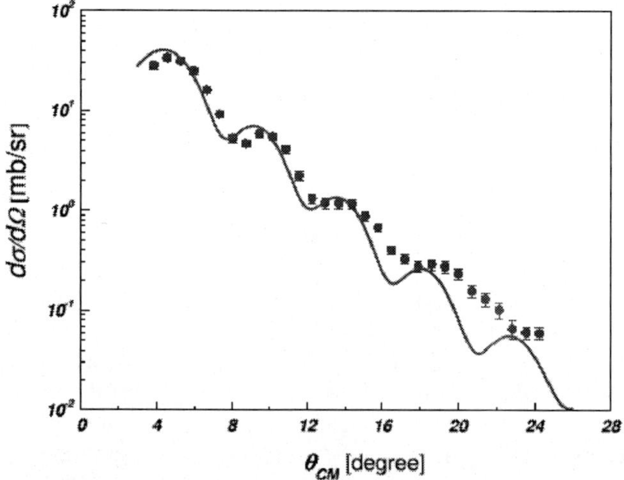

FIGURE 1. Angular distribution for the ground state of ^{23}Ne from the $^{13}C(^{22}Ne,^{23}Ne)^{12}C$ reaction. The curve shows a DWBA calculation that has been normalized to the data to infer the ANC.

The $^{13}C(^{22}Ne,^{23}Ne)^{12}C$ reaction was measured over the angular range 3-32^0 in the center-of-mass system. Figure 1 shows the angular distribution that was obtained for the ^{23}Ne ground state. The first excited state was also measured. DWBA descriptions of the transfer reaction cross sections were calculated using PTOLEMY. The ^{23}Ne states were treated as a $d_{5/2}$ or $s_{1/2}$ neutron coupled to the ground state of ^{22}Ne. The calculations were performed using the optical potential obtained from $^{22}Ne+^{13}C$ scattering for the entrance channel and the optical potential extracted from $^{22}Ne+^{12}C$ scattering for the outgoing channel. The comparison between the data and the calculations is shown in Fig. 1. The ANCs $C^2_{5/2}$ and $C^2_{1/2}$ were obtained by fitting the calculations to the data. Figure 1 shows that a phase shift was observed between the data and the DWBA calculation. A contribution was included in the uncertainty to account for this. We find $C^2_{5/2} = 0.86 \pm 0.08 \pm 0.12$ fm^{-1} and $C^2_{1/2} = 18.2 \pm 1.8 \pm 3.8$ fm^{-1}. These experimental results are consistent with previous predictions of 0.71-0.81 fm^{-1} [6] and 16.3-18.5 fm^{-1} [7], respectively.

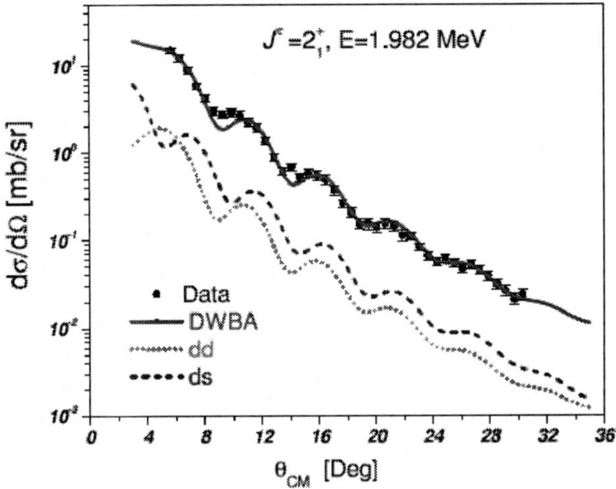

FIGURE 2. The angular distribution for the first excited state in ^{18}O, populated in ^{13}C(^{17}O,^{18}O)^{12}C. The curve is the DWBA best fit, which includes contributions from *dd* and *ds* components.

The ^{13}C(^{17}O,^{18}O)^{12}C reaction was measured over the angular range 5-31^0 in the center-of-mass system. Angular distributions were obtained for the 0^+_1, 2^+_1, 4^+_1, and 2^+_2 states in ^{18}O. In our DWBA calculations, the 2^+ states result from the coupling of $1d_{5/2}$ or $2s_{1/2}$ neutrons to the $5/2^+$ ground state of ^{17}O, while the 0^+ and 4^+ states can only be obtained from adding a $1d_{5/2}$ neutron to the $5/2^+$ core configuration. For the first 2^+ state, the ANCs were obtained by fitting the data to a sum of the calculated DWBA angular distributions for the (*dd*) and (*ds*) configurations, as shown in Fig. 2. The resulting ratio of the spectroscopic factors agrees with previous measurements [8,9]. We find $C^2(dd) = 2.06 \pm 0.21$ fm^{-1} and $C^2(ds) = 6.55 \pm 0.69$ fm^{-1} for the first 2^+ state, which are in reasonable agreement with theoretical estimates [10]. A similar procedure was followed for the second 2^+ state. The 0^+ and 4^+ states were treated like the states in ^{23}Ne.

ASTROPHYSICAL CONSEQUENCES

The measured ANCs for ^{23}Ne were used to calculate the corresponding ANCs in ^{23}Al. An additional 3% uncertainty was included to account for possible charge-symmetry breaking effects. The ^{23}Al ANCs were then used to calculate the reaction rate for ^{22}Mg(p,γ)^{23}Al. The astrophysical *S*-factor for direct capture to the ground state, which is shown in Fig. 3, was calculated using RADCAP [11]. We find *S*(0) =

960 ± 110 eV b, which is nearly 50% larger than a previous estimate [12]. The total reaction rate, which also includes a contribution from resonant capture through the first excited state, is shown in Fig. 3. Our result is similar to Wiescher's estimate [4] and approximately two times larger than Caggiano's determination [12]. The difference is mainly due to the binding energy of the proton in ^{23}Al and the ANCs that are used in the reaction rate calculations.

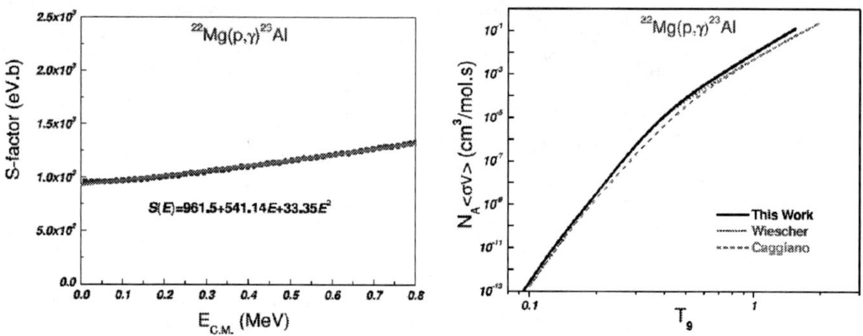

FIGURE 3. The astrophysical S-factor for direct capture to the ^{23}Al ground state determined from the ANC (left) and total reaction rate (right) for ^{22}Mg(p,γ)^{23}Al. The reaction rate is compared to previous estimates by Wiescher [4] and Caggiano [12].

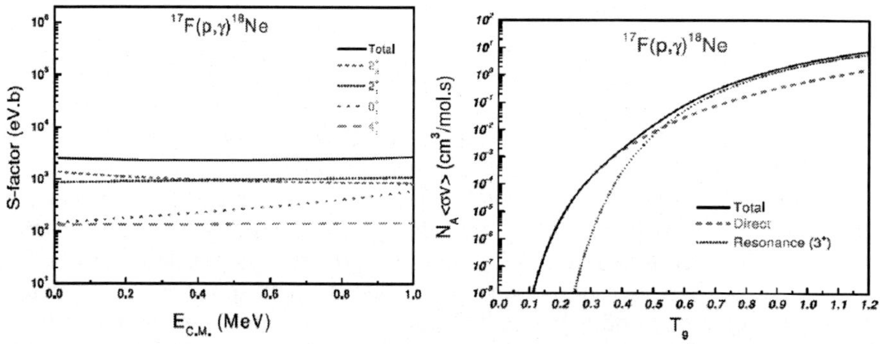

FIGURE 4. The S-factor components (left) and total reaction rate (right) for ^{17}F(p,γ)^{18}Ne. The direct capture contribution is calculated from the ANCs measured here.

The ANCs of the low-lying states in ^{18}Ne were determined from those of their corresponding mirror states in ^{18}O using a procedure identical to that for ^{23}Al. The ANCs were then used to calculate the direct capture S factors shown in Fig. 4. The dominant contributions to the total astrophysical S factor arise from the (ds) components of the two 2^+ states. We find $S(0)$ = 2.5 ± 0.4 keV b, which is consistent with the previous estimate 2.9 ± 0.4 keV b reported in [3] and smaller than 3.5 keV b calculated in [2]. This is the first time to evaluate the direct capture rate of ^{17}F(p,γ)^{18}Ne through the measurement of the corresponding ANCs. The present results show that the thermonuclear reaction rate is dominated by the direct capture component for T_9 = 0.2-0.4, as shown in Fig. 4. The resonant contribution is more significant for temperatures $T_9 > 0.5$ that characterize X-ray bursts or neutron stars.

The astrophysical reaction rates found here can be inserted into the nuclear reaction rate library to study their influence on the abundance of CNO, NeNa, and MgAl materials in novae.

ACKNOWLEDGMENTS

This work was supported in part by the U.S. DOE under grant DE-FG02-93ER40773 and the Robert A. Welch Foundation under grant A-1082.

REFERENCES

1. R.D. Gehrz, J.W. Truran, R.E. Williams, and S. Starrfield, *Publ. Astron. Soc. Pac.* **110**, 3 (1998).
2. D.W. Bardayan *et al., Phys. Rev.* C **62**, 055804 (2000).
3. A. Garcia *et al., Phys. Rev.* C **43**, 2012 (1991).
4. M. Wiescher *et al., Nucl. Phys.* **A484**, 90 (1988).
5. D.H. Youngblood *et al., Nucl. Instrum. Meth.* **A361**, 539 (1995).
6. N.K. Timofeyuk and P. Descouvemont, *Phys. Rev.* C **71**, 064305 (2005).
7. N.K. Timofeyuk and P. Descouvemont, *Phys. Rev.* C **72**, 064324 (2005).
8. T.K. Li, D. Dehnhard, R.E. Brown, and P.J. Ellis, *Phys. Rev.* C **13**, 55 (1976).
9. R.L. Lawson, F.J.D. Serduke, and H.F. Fortune, *Phys. Rev.* C **14**, 1245 (1976).
10. N.K. Timofeyuk (private communication).
11. C.A. Bertulani, *Comp. Phys. Comm.* **156**, 123 (2003).
12. J.A. Caggiano *et al., Phys. Rev.* C **64**, 025802 (2001).

Astrophysical S-factor of the $^{12}N(p,\gamma)^{13}O$ Reaction Determined from the (^{12}N,^{13}O) Proton Transfer Reaction

A. Banu[a], T. Al.-Abdullah[a], V. Burjan[b], F. Cârstoiu[c], C. Fu[a], C. A. Gagliardi[a], M. McCleskey[a], G. Tăbăcaru[a], L. Trache[a], R. E. Tribble[a], and Y. Zhai[a]

[a]*Cyclotrone Institute, Texas A&M University, College Station, TX 77843, U. S. A.*
[b]*Institute of Nuclear Physics, Czech Academy of Sciences, Prague-Řež, Czech Republic*
[c]*National Institute for Nuclear Physics and Engineering, P.O. Box MG-6, Bucharest, Romania*

Abstract. The reaction cross section of the radiative proton capture on the drip line nucleus ^{12}N was studied at the Cyclotron Institute at Texas A&M University. We have employed the indirect Asymptotic Normalization Coefficient method to determine the direct radiative component of the astrophysical S-factor of the $^{12}N(p,\gamma)^{13}O$ reaction. A value of 0.31(5) keV·b was found at zero energy. Interference between direct and resonant captures leads to a further enhancement of a factor of two. The reaction was investigated in relation to the evolution of hydrogen-rich massive objects, Population III stars, because of the role that it may play in the hot pp chain nuclear burning processes occurring in such stars.

Keywords: Population III stars, radiative proton capture, ANC, DWBA
PACS: 25.40.Lw, 25.60.Bx, 25.60.Je, 26.50.+x

INTRODUCTION

From modern cosmology we have learned that the first stars in the Universe have formed at the end of the so-called "dark ages", a few 10^8 years after the Big Bang. Supernova explosions that ended the lives of the first stars were responsible for the initial enrichment of the intergalactic medium with heavy elements and consequently had important effects on subsequent galaxy formation. Therefore understanding the properties of the first stars and what was their impact in driving early cosmic evolution is one of the key problems in modern cosmology.

The most fundamental question about the first stars is how massive they typically were. Numerical simulations of the collapse of the primordial gas indicate that the first stars were predominantly very massive with masses bigger than hundreds of solar masses. Currently we do not have direct observational constrains on any of the properties of the first stars because not a single metal-free star has ever been detected. Since we have never observed directly Population III stars, their study might appear to be rather academic. The motivation for our measurement of the radiative proton capture reaction $^{12}N(p,\gamma)^{13}O$ in relation to the fate of the first stars was based on two such studies.

CP972, *Exotic Nuclei and Nuclear/Particle Astrophysics (II)—Proceedings of the Carpathian Summer School of Physics 2007,* edited by L. Trache and S. Stoica

In 1986, Fuller et al. [1] addressed the classic problem of the evolution of supermassive non-rotating Population III stars with $M > 10^5 \cdot M_{\odot}$: given that such an object has formed and quasi-statically contracted to the point of dynamical instability, is the nuclear energy generated in the subsequent collapse enough to blow up the star? The modeled two possible scenarios and concluded that non-rotating supermassive stars with zero metallicity will never explode but collapse into black holes.The smallest metallicity needed for an explosion was found to be Z = 0.005. For the failed explosions it is reasoned that in the short time scales of the collapse insufficient amounts of ^{12}C and other heavier isotopes are produced by the $3\alpha \rightarrow {}^{12}$C process. Three years later, Wiescher et al. [2] proposed alternative ways to bypass the 3α reaction and produce CNO material — the hot pp chains and rap-processes — shown below, with ^{12}N(p,γ)^{13}O being an important branching point in these chains:
pp-IV: ^{7}Be(p,γ)^{8}B(p,γ)^{9}C(β+ν)^{9}B(p)^{8}Be(α)^{4}He
pp-V: ^{7}Be(α,γ)^{11}C(β^{+}ν)^{11}B(p,2α)^{4}He
rap-I: ^{7}Be(p,γ)^{8}B(p,γ)^{9}C(α,p)**^{12}N(p,γ)^{13}O**(β^{+}ν)^{13}N(p,γ)^{14}O
rap-II: ^{7}Be(α,γ)^{11}C(p,γ)**^{12}N(p,γ)^{13}O**(β^{+}ν)^{13}N(p,γ)^{14}O
rap-III: ^{7}Be(α,γ)^{11}C(p,γ)^{12}N(β^{+}ν)^{12}C(p,γ)^{13}N(p,γ)^{14}O
rap-IV: ^{7}Be(α,γ)^{11}C(α,p)^{14}N(p,γ)^{15}O.

The outcome of Wiescher's et al. study is that for densities in excess of 1 g·cm^{-3}, temperatures of (2-3) × 10^8 K are sufficient to process material from the pp chains to CNO nuclei. The question to be answered is whether the rap-processes can lead to the production of sufficient amounts of CNO material that may also result in an explosion of supermassive objects with lower metallicity than proposed by Fuller's et al. study. Taking into account that only primordial abundances of ^{2}H and ^{3}He are available to produce CNO material through the hot pp chain and rap-process scenario, Wiescher et al. came up with a mass fraction of CNO material of $\sim 3 \times 10^{-4}$, much smaller than the threshold in CNO nuclei of 5×10^{-3} quoted by Fuller et al. as needed to trigger an explosion of non-rotating supermassive stars. However, the energy release by the rap-processes was not taken into consideration in the original hydrodynamical calculations of the collapse. Wiescher et al. suggested that this energy release may moderate or even halt the collapse. Hence the inversion of the collapse may be possible at much smaller CNO abundances.

More recent simulation studies [3] in massive metal-free stars between 120 and 1000 solar masses indicated that a metallicity of $\sim 1 \times 10^{-9}$ is sufficient to stop the contraction and supply energy by the hot CNO cycle for hydrogen burning. The new experimental information on the ^{12}N(p,γ)^{13}O reaction reported here might also play an important role in the nucleosynthesis of these lower mass Population III stars.

ANC AS INDIRECT METHOD TO STUDY ^{12}N(P,γ)^{13}O

Radiative proton capture reactions at stellar energies are very peripheral processes which can be easily understood by the presence of the Coulomb barrier. Hence to study the direct radiative proton capture on ^{12}N nucleus, we have used the proton transfer reaction ^{12}N(^{14}N,^{13}O)^{13}C at around 10 MeV/nucleon, which is also a peripheral process but it happens at smaller core-proton distances and thereby it has a much larger cross section.

We have applied what has become known as the Asymptotic Normalization Coefficient (ANC) method [4] to determine the amplitude of the tail of the overlap integral of the ground state wave function of ^{13}O into the two-body channel ^{12}N + p. The normalization of the wave function of the last proton determines the astrophysical *S*-factor for the charged particle radiative capture.

In Figure 1 it is illustrated that the reaction rate of ^{12}N(p,γ)^{13}O is actually determined by both the direct and resonant captures to ground state and to the first excited state in ^{13}O, respectively. We report here only on the direct capture component which we have measured.

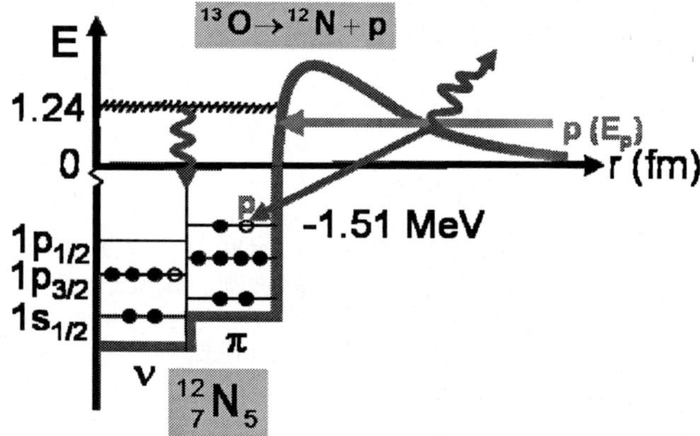

FIGURE 1. This cartoon illustrates the two components of the radiative proton capture process: the direct capture to the ground state in ^{13}O bound by 1.51 MeV and a resonant capture to the first excited state in ^{13}O at E_x = 2.75 MeV.

Figure 2 shows schematically the direct capture reaction vertex (a) and the reaction vertices in the proton transfer (b).

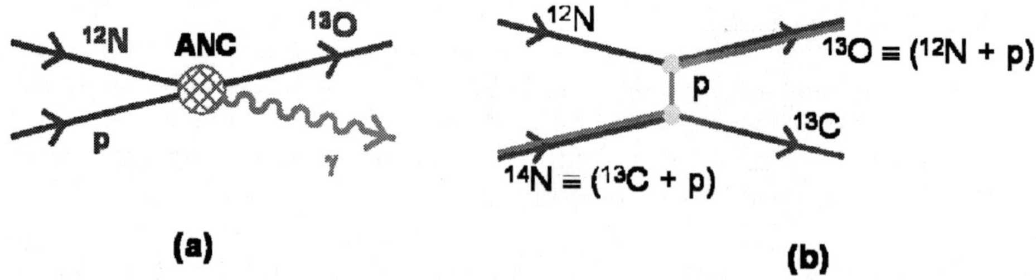

FIGURE 2. The reaction vertex in the direct radiative proton capture ^{12}N(p,γ)^{13}O determined by the ANC (a); the reaction vertices in the proton transfer ^{12}N(^{14}N,^{13}O)^{13}C in which one of the reaction vertex is the same one that appears in the direct capture process.

We can extract thc ANC of ^{13}O by fitting the measured angular distributions with calculated Distorted Wave Born Approximation (DWBA) cross sections as shown in the following mathematical expression:

$$\sigma_{\exp} = \left(C^{13O}_{p1/2}\right)^2 \left\{ \left(\frac{C^{14N}_{p1/2}}{b^{13O}_{p1/2} b^{14N}_{p1/2}} \right)^2 \sigma^{DW}_{\frac{1}{2}\frac{1}{2}} + \left(\frac{C^{14N}_{p3/2}}{b^{13O}_{p1/2} b^{14N}_{p3/2}} \right)^2 \sigma^{DW}_{\frac{1}{2}\frac{3}{2}} \right\}$$

Here we have taken into account the mixing orbital configurations of the structure of ^{14}N target and ^{13}O ejectile — ${}^{13}_{6}C_7(1/2^-) \otimes \pi[1p_{1/2} + 1p_{3/2}]$ and ${}^{12}_{7}N_5(1^+) \otimes \pi[1p_{1/2} + 1p_{3/2}]$, respectively. The ANC's of the ^{14}N were measured before [5] and the *b*'s coefficients represent the ANC's of the normalized single particle wave functions used in the DWBA analysis.

PROTON TRANSFER EXPERIMENT (^{12}N,^{13}O)

A 12 MeV/nucleon ^{12}N radioactive beam was produced via the reaction ^{1}H(^{12}C,^{12}N)n using a 23 MeV/nucleon ^{12}C primary beam delivered by the K500 Superconductive Cyclotron at Texas A&M University, and separated by the Momentum Achromat Recoil Spectrometer (MARS) [6]. The ^{12}N beam was produced at an average rate of 200 kHz and had a purity of 99.8%. It bombarded a secondary melamine ($C_3N_6H_6$) target.

The detection system consisted of four Si detector telescopes for particle identification, the front ΔE detectors being position sensitive strip detectors. The telescope arrays were positioned at angles covering a range from 8° to 35° to measure simultaneously transfer reaction products and the elastic scattering processes in order to get/check the Optical Model Potential (OMP) parameters used in the incoming channel in the DWBA analyis.

RESULTS

Figure 3 (a) shows the measured elastic angular distribution (in the laboratory reference system) from which the OMP parameters were extracted using a double folding potential with JLM effective interaction [7] whereas in Figure 3 (b) is shown the measured angular distribution plotted in the centre of mass system. Details on data analysis procedure are to be found in our forthcoming paper [8].

We have found the ANC of ^{13}O to be 2.8(4) fm^{-1} (its squared value) from which we have calculated the direct capture component of the astrophysical *S*-factor for the ^{12}N(p,γ)^{13}O reaction to be 0.31(5) keV·b at zero energy. However, the total astrophysical *S*-factor is determined by the interference of the capture amplitudes of both direct and resonant reactions which was found to be constructive [9] resulting in an enhancement of a factor of 2 in the astrophysical *S*-factor.

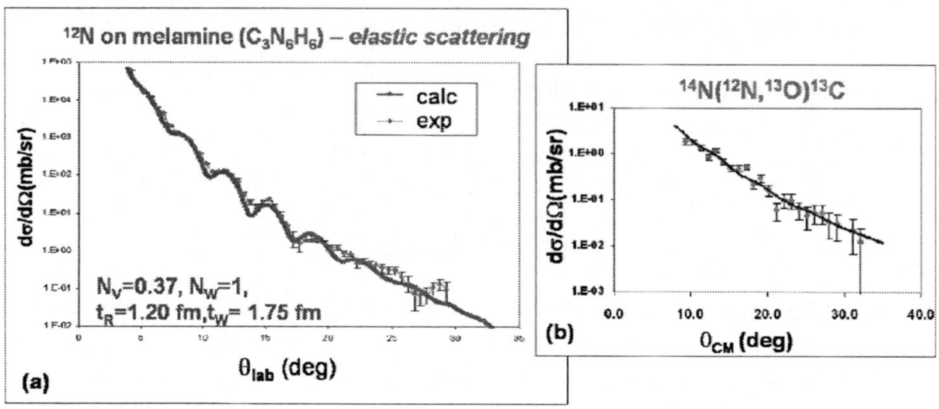

FIGURE 3. Measured angular distributions for the elastic scattering on the melamine target (a) and for the proton transfer reaction (b).

ACKNOWLEDGMENTS

This work was supported in part by the U.S. DOE under grant DE-FG02-93ER40773 and the Robert A. Welch Foundation under grant A-1082.

REFERENCES

1. G. M. Fuller, S.E. Woosley, T. A. Weaver, *APJ* **307**, 675-686 (1986).
2. M. Wiescher, J. Görres, S. Graff, L. Buchmann, F.-K. Thielemann, *APJ* **343**, 352-364 (1989).
3. A. Heger, S. E. Woosley, I. Baraffe, and T. Abel, "Evolution and Explosion of Very Massive Primordial Stars" in *Lighthouses of the Universe: The Most Luminous Celestial Objects and Their Use for Cosmology-2001,* edited by M. Gilfanov et al., Proceedings of the MPA/ESO/MPE/USM Joint Astronomy Conference, Springer-Verlag, NY, 2002, pp. 369-375.
4. A. M. Mukhadmedzhanov, C. A. Gagliardi, R. E. Tribble, N. K. Timofeyuk, *Phys. Rev. C.* **51**, 3472-3478 (1995).
5. L. Trache et al., *Phys. Rev. C* **58**, 2715-2719 (1998).
6. R. E. Tribble, R. H. Burch, C. A. Gagliardi, *Nucl. Instr. Meth.* **A285**, 441-446 (1989).
7. L. Trache et al., *Phys. Rev. C.* **61**, 024612 (2000); F. Carstoiu et al., *Phys. Rev. C* **70**, 054610 (2004).
8. A. Banu et al. (to be published)
9. A. M. Mukhadmedzhanov (private communication).

Secondary Cosmic Ray Particles due to GCR Interactions in the Earth's Atmosphere

G. Battistoni*, F. Cerutti†, A. Fassò**, A. Ferrari†, M.V. Garzelli *, M. Lantz‡, S. Muraro*, L.S. Pinsky§, J. Ranft¶, S. Roesler† and P.R. Sala*

*University of Milano, Department of Physics, and INFN, Milano, Italy
†CERN, Geneva, Switzerland
**SLAC, Stanford, CA, US
‡Chalmers University, Department of Fundamental Physics, Goteborg, Sweden
§University of Houston, Department of Physics, Houston, TX, US
¶Siegen University, Fachbereich 7 - Physik, Siegen, Germany

Abstract. Primary GCR interact with the Earth's atmosphere originating atmospheric showers, thus giving rise to fluxes of secondary particles in the atmosphere. Electromagnetic and hadronic interactions interplay in the production of these particles, whose detection is performed by means of complementary techniques in different energy ranges and at different depths in the atmosphere, down to the Earth's surface.

Monte Carlo codes are essential calculation tools which can describe the complexity of the physics of these phenomena, thus allowing the analysis of experimental data. However, these codes are affected by important uncertainties, concerning, in particular, hadronic physics at high energy. In this paper we shall report some results concerning inclusive particle fluxes and atmospheric shower properties as obtained using the FLUKA transport and interaction code. Some emphasis will also be given to the validation of the physics models of FLUKA involved in these calculations.

Keywords: GCR EAS, inclusive fluxes of secondary particles, Monte Carlo models and codes
PACS: 98.70.Sa, 13.85-t, 13.75-t

INTRODUCTION: EAS DEVELOPMENT AND DETECTION

GCR interactions in the Earth's atmosphere, induced by particles with energies high enough (E $\gtrsim 10^{14} - 10^{15}$ eV), may originate Extended Air Showers (EAS). Both the EM and the μ component of EAS can be detected and used to infer properties concerning GCR primary spectrum energy and composition. At present, a special emphasis is put in the investigation of Very High and Ultra High Energy Cosmic Rays (VHECR and UHECR), through dedicated experiments devoted to the detection of EAS originated by these particles, due to the still open questions concerning the shape of the spectrum (2nd knee, ankle), the composition (γ, p, heavy ions), and the origin of cosmic rays (galactic / extragalactic), their transport, acceleration and the GZK cut-off.

Information on GCR primary flux and composition are inferred from the experimental observables by using Monte Carlo (MC) simulations. In particular, while the fluorescence technique, used by many experiments, allows with some uncertainties to estimate the EM energy deposited in the atmosphere, the total primary energies can be obtained only

[1] Corresponding author, e-mail: *Maria.Garzelli@mi.infn.it*

CP972, *Exotic Nuclei and Nuclear/Particle Astrophysics (II)—Proceedings of the Carpathian Summer School of Physics 2007,* edited by L. Trache and S. Stoica

after estimating the missing energy carried by other shower components (μ, nucleons...), and this requires indeed MC simulations. Even the search for particular estimators, such as the S variable, allows to obtain primary energy information in a way nearly independent of composition [1], but needs MC for calibrations.

In particular, in the AUGER experiment [2] information on primary energies are inferred by hybrid measurements and the fluorescence technique as far as the EM sector is concerned (with an uncertainty amounting to $\sim$ 20%), whereas the number of μs at the Earth's surface, N_μ, is still related to total primary energies by means of MC simulations. On the other hand, in the KASCADE and KASCADE-Grande experiments [3] information on primary energies are inferred from measures of μ and e with the aid of MC shower simulations. Both in AUGER and in KASCADE / KASCADE-Grande, as well as in many other EAS experiments, MCs are heavily needed to analyse data in terms of GCR mass composition. At present there are however still important uncertainties, since different MC models produce important differences on the interpretation of the same data. The critical ingredients in these MC codes are the physical models for hadronic and nuclear interaction, and their implementation.

In this work we summarize some of the results which can be obtained by MC simulation of EAS, giving examples taken from the use of the FLUKA code.

MC SIMULATION OF EAS

EAS simulations can be performed by means of MC codes, which include hadronic and EM modules. This method leads, in a straightforward way, to account for fluctuations in the evolution of different showers initiated by primaries of the same type and energy.

The GCR primary spectrum covers a wide energy range, extending on many orders of magnitude. This fact implies that it is very difficult to have in a MC code a single hadronic model able to describe primary and secondary hadron interactions in the Earth's atmosphere. Therefore in many cases different models, differing according to the energy range under study, have to be merged together.

At present the most diffused MC package for GCR induced shower simulation is CORSIKA [4]. As far as its hadronic sector is concerned, it distinguishes between high energy (E $\gtrsim$ 100 - 200 GeV) and low energy (E $\lesssim$ 100 - 200 GeV) hadronic models. In the present version, CORSIKA offers the choice, at high energy, among models like QGSJET, QGSJET-II, SIBYLL, DPMJET-2.55, NEXUS, EPOS, whereas, at low energy, GHEISHA, UrQMD, FLUKA are available. Other packages for GCR EAS simulation also exist. We just mention, among the others, the AIRES [5] and the Cosmos [6] codes.

In this paper, we focus instead on the performances of the FLUKA multipurpose code [7]. FLUKA is a fully integrated, high precision, transport and interaction code, without the need of invoking other external packages. Besides its use in the CORSIKA package at energies $\lesssim$ 100 - 200 GeV, FLUKA has recently been used also as a stand-alone code, complemented by the DPMJET-III code [8], for the simulation of EAS at energies < 10^{16} eV. FLUKA simulations can, in principle, lead to results different from the ones made by FLUKA+CORSIKA, due to the implementation of different EM and hadronic models. FLUKA contains a highly accurate EM interaction model [9, 10], not included in CORSIKA. As far as the hadronic sector is concerned, the following models

are available in FLUKA:
- low energy (E < 20 TeV) h + A reactions can be simulated by means of the PEANUT (PreEquilibrium Approach to NUclear Thermalization) module, originally worked out for energies < tens of GeV [11], and recently improved and extended to higher energies [12]. This extension has superseded a previously existing DPM + INC model, working at E > 5 GeV.
- high energy (E > 20 TeV) h + A reactions can be described by means of the interface to the DPMJET-2.5 or III codes.
- A + A reactions can be simulated by means of an interface to the RQMD2.4 code, written in Frankfurt and modified for insertion in FLUKA [13], for E < 5 GeV/A, and by means of the interface to the DPMJET-2.5 or III codes for E > 5 GeV/A. Alternatively, the superposition model is also available to roughly describe heavy-ion interactions.

Constraints on theoretical models for particle interactions can (or will) be obtained by data collected at collider experiments, but, unfortunately, diffractive cross-sections are difficult to be measured at colliders. Furthermore, present and near future accelerators do not and will not allow to test the physics of particles at the highest GCR energies (E > 1.5 10^{18} eV, GCR spectrum tail). Thus, MC models used in the analysis of EAS data are still affected by important uncertainties, concerning in particular the hadronic sector. Other constraints/checks on low-energy models, besides the ones from accelerator data, can come from astrophysical measurements of inclusive particle fluxes at different atmospheric depths, as performed by balloon-borne experiments and satellites. We emphasize that for an accurate description of VHECR and UHECR EAS also low-energy physics is important, e.g. the larger is the distance from the shower core and the shower inclination, the stronger is the required accuracy of the adopted low energy models.

Validation of the FLUKA Models at Low Energies

Validation of the FLUKA models at low energies has been performed by means of comparisons of their theoretical predictions with data collected at the accelerators. Data concerning inclusive particle fluxes in the atmosphere down to the Earth's surface, have been then used to assess the FLUKA performance on cosmic ray physics.

While the first issue can be addressed by using the version of the code officially distributed and available on the web, the second one requires the inclusion of further (geometrical, geomagnetical, astrophysical) assumptions and information. These elements have been inserted by means of the FLUKA user routines.

Validation of FLUKA using accelerator data. Data recently collected by the HARP and the NA49 experiments have been nicely reproduced by means of FLUKA [10], Many others validations of the code, referred to previously published data, can be found in literature. Furthermore, data from the SPS collider, Tevatron and RHIC have been used to successfully validate the DPMJET2.5 and III codes [14].

Validation of FLUKA using astrophysical data from satellites and balloon-borne experiments. Inclusive particle fluxes in the Earth's upper and lower atmosphere, induced by GCR propagation and interactions, have been measured by means of satellites

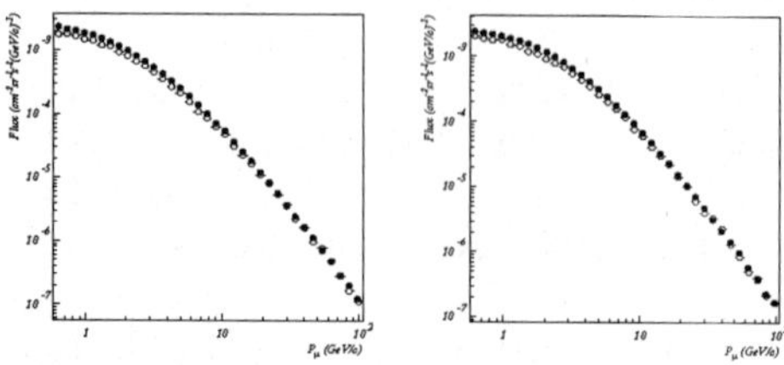

FIGURE 1. Calculated inclusive fluxes of μ^- (*left panel*) and μ^+ (*right panel*) at the top of Mt. Norikura ($\sim$ 2700 m a.s.l.) as obtained by FLUKA simulations (open symbols) vs. exp. data [15] from BESS'99 (full symbols). Other examples of FLUKA benchmarks in GCR physics can be found in [16].

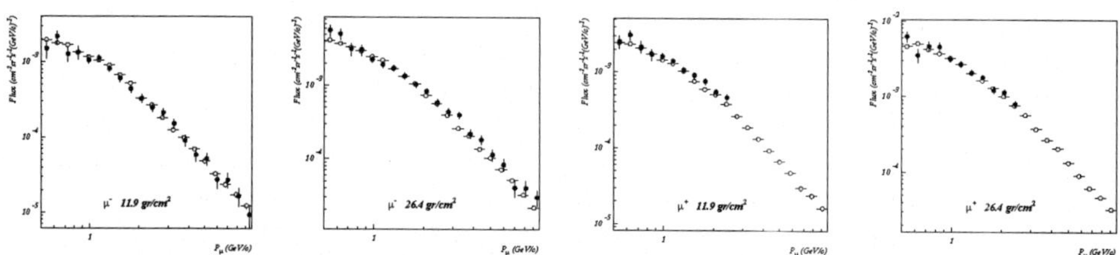

FIGURE 2. Calculated inclusive fluxes of μ^- (*left panels*) and μ^+ (*right panels*) at atmospheric depths of 11.9 g/cm^2 (*odd panels*) and 26.4 g/cm^2 (*even panels*), respectively, as obtained by FLUKA simulations (open symbols) vs. exp. data [17] from BESS 2001 (full symbols).

and balloon-borne experiments such as BESS. Different sets of data, taken in different atmospheric and geomagnetic conditions, have been analyzed and reproduced by FLUKA, under proper assumptions concerning GCR primary spectrum, solar modulation, atmospheric and geomagnetic models. As examples, data concerning μ collected at the top ($\sim$ 2700 m a.s.l.) of Mt. Norikura (geomagnetic cut-off $\sim$ 11.2 GV) in 1999, and at two different atmospheric depths (11.9 and 26.4 g/cm^2) in a balloon flight campaign over Ft. Sumner (geomagnetic cut-off $\sim$ 4.2 GV) in 2001, are presented in fig. 1 and 2, respectively, and compared to the theoretical fluxes expected on the basis of FLUKA simulations. The agreement between the experimental and theoretical results is quite nice and increases at increasing energies. Furthermore, atmospheric μ charge ratios in the energy range from $\sim$ 20 GeV up to $\sim$ 3 TeV have been detected by the L3+C experiment at CERN ($\sim$ 450 m a.s.l.), that has provided also μ arrival directions. FLUKA has successfully reproduced also these data [18].

Proton and heavier ion inclusive fluxes have also been obtained by FLUKA theoretical simulations and compared to experimental data available from BESS. In these cases, the results of the simulations turn out to be very sensitive to a correct treatment of the geomagnetic cut-off, of the transport of particles in the upper layers of the atmosphere (trapping of particles and their recirculation), and to the accuracy of the theoretical modelling of low-energy A-A interactions.

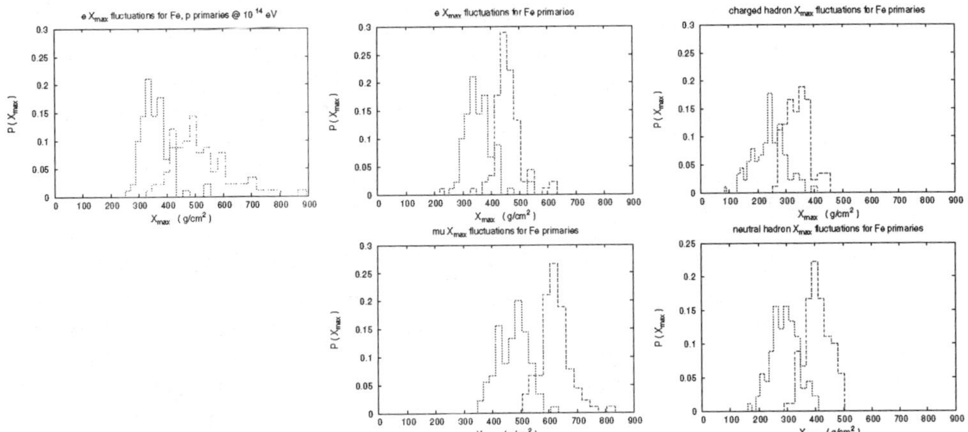

FIGURE 3. *Left panel*: fluctuations in the distribution of e $X_{\max}$ for a few hudreds of vertical showers induced by p (*dot-dashed green line*) and Fe ions (*dotted red line*) at 10^{14} eV energy. *Central and Right panels*: fluctuations in the distribution of e $X_{\max}$ (*central upper panel*), μ $X_{\max}$ (*central lower panel*), charged hadron $X_{\max}$ (*right upper panel*) and neutral hadron $X_{\max}$ (*right lower panel*) for a few hudreds of vertical showers induced by Fe ions at 10^{14} eV (*dotted red line*) and 10^{15} eV (*dashed blue line*) energies.

Predictions of the FLUKA Models at High Energies: EAS

In the following we shall report about some tests of EAS simulation at energies 10^{14} - 10^{15} eV, performed by means of FLUKA, complemented by DPMJET-III for the description of A-A reactions at E > 5 GeV/A and h-A reactions at E > 20 TeV. Vertical showers, induced by primaries of different mass (p and Fe ions), have been considered.

The $X_{\max}$ position, i.e. the atmospheric depth where the shower reaches its maximum development (in terms of EM energy deposition), is a good probe of GCR composition. In fact, since the inelastic scattering cross-section for p-Air interactions is lower than the A-Air ones, EAS induced by Fe primaries have, on the average, vertical profile maxima shifted towards lower atmospheric depths than p induced showers. e and μ fluences as a function of the atmospheric depth ρ for p and Fe induced showers at 10^{14} and 10^{15} eV, as calculated with FLUKA + DPMJET-III, are shown in ref. [10]. In general, different codes, based on different models, due to different modelings of hadronic and EM processes, give different results for $X_{\max}$. Anyway, to disentangle the effects of different models and different primary composition, also more specific variables can be considered. Among the others, the study of $X_{\max}$ fluctuations [19] for primaries at a fixed energy has been proposed for this purpose. As an example, $X_{\max}$ fluctuations for Fe induced showers have been computed from our simulations, considering the e, μ, charged hadron and neutral hadron vertical shower profiles, and are shown in the central and right panels of fig. 3. In general, the average muon $X_{\max}$ is located at larger ρs with respect to the electron one, since μs penetrate more deeply. The charged hadrons undergo strong absorption effects, thus their $X_{\max}$ are located at lower ρs with respect to the ones of neutral hadrons. Increasing energies lead, in general, to a shift of $X_{\max}$ to higher ρs for all considered profiles, even if, in all cases, the fluctuation profiles at

10^{14} and 10^{15} eV overlap. That means that reconstructing the energy of GCR primaries out of this method is not possible, since showers of energies that differ even by a factor 10 can give rise, in some cases, to the same position of X_{max}. An example from our simulations of the sensitivity of X_{max} fluctuations to the GCR primary composition is shown in the left panel of fig. 3. Not only the position of maxima but also the shape and broadness of the fluctuation profiles change according to the GCR masses, as can be seen by comparing the cases of p and Fe induced EAS at a fixed energy (10^{14} eV for the cases considered in the figure).

Besides the vertical spreads of particle distribution, also lateral spread can be calculated. In particular, in case of EM particles, the lateral spread is due to the effect of multiple Coulomb scattering, while, in case of μs, the lateral spread is mainly correlated with the p_T distribution of charged π and K, from the decay of whom μs originate (we are not mentioning here μs from the decay of short-lived charmed particles, which are also considered in FLUKA and other codes). Detailed maps of e and μ fluences in the atmosphere, down to the Earth's surface, which allow to appreciate both the average vertical profile and the lateral spread of each component, as calculated by FLUKA for Fe induced showers, can be found in [10].

REFERENCES

1. H. Rebel *et al.*, Proc. ICRC2005 **6** (2005), 297–300; I.M. Brancus *et al.*, Proc. ICRC2005 **6** (2005), 361–364; G. Toma for the KASCADE-Grande collaboration, these proceedings
2. T. Suomijarvi, these proceedings
3. G. Navarra, these proceedings
4. D. Heck *et al.*, Report FZKA 6019 (1998), 1–90; talks presented at the Vihkos CORSIKA School 2005, Lauterbad, Germany, May 31 - June 5 2005, available on the web
5. S.J. Sciutto, `http://www.fisica.unlp.edu.ar/auger/aires/`
6. K. Kasahara, `http://cosmos.n.kanagawa-u.ac.jp/cosmosHome/index.html`
7. A. Ferrari *et al.*, CERN Yellow Report 2005–10 (2005), INFN/TC_05/11, 1–387; A. Fassò *et al.*, Proc. CHEP2003 Conference, La Jolla, CA, US, March 24 - 28, 2003, (paper MOMT005) eConf C0303241 (2003), hep-ph/0306267; F. Ballarini *et al.*, Adv. Space Res. **40** (2007), 1339–1349
8. S. Roesler *et al.* Proc. MonteCarlo 2000 Conference, Lisboa, October 23 - 26 2000, Springer Verlag eds. (2001), 1033–1038
9. A. Fassò *et al.*, Proc. MonteCarlo 2000 Conference, Springer Verlag eds. (2001), 159–164
10. G. Battistoni *et al.*, Proc. ISVHECRI 2006 Symposium, Weihai, China, August 15 - 22 2006, to be published in Nucl. Phys. B (Proc. Suppl.), hep-ph/0612075
11. A. Ferrari and P.R. Sala, Proc. Workshop on Nuclear Reaction Data and Nuclear Reactors Physics, Design and Safety, Trieste, Italy, April 1996, A. Gandini and G. Reffo eds., **2** (1998), 424–532
12. G. Battistoni *et al.*, Proc. 11th Int. Conf. on Nuclear Reaction Mechanisms, Varenna, Italy, June 12 - 16 2006, Ric. Scient. ed Ed. Perm. Suppl. **126**, E. Gadioli ed. (2006), 483–495
13. H. Sorge *et al.*, Nucl. Phys. A **498** (1989), 567–576; G. Battistoni *et al.*, Braz. J. Phys. **34** (2004), 897–900; V. Andersen *et al.* Adv. Space Res. **34** (2004), 1302–1310
14. J. Ranft, Proc. HSS06 Workshop, September 6 - 8 2006, FNAL, Illinois, US, AIP Conf. Proc. **896** (2007), 102–111
15. K. Abe *et al.* Phys. Lett. B **564** (2003), 8–20; erratum in Phys. Lett. B **645** (2007), 472–475
16. S. Muraro, Ph.D. Thesis (2006), Milano; G. Battistoni *et al.*, Astropart. Phys. **12** (2000), 315–333.
17. M. Motoki *et al.* Astropart. Phys. **19** (2003), 113–126, astro-ph/0205344
18. G. Battistoni *et al.*, Proc. NOW 2006 Workshop, September 9 - 16 2006, Conca Specchiulla, Italy, to be published in Nucl. Phys. B(Proc. Suppl.).
19. R. Engel, Summary Talk at the ICRC2007 Conference, Merida, Mexico, July 3 - 11 2007, to be published in the Conf. Proc.

Neck Influence on Fission Paths

Radu A. Gherghescu and D. N. Poenaru

Horia Hulubei National Institute of Physics and Nuclear Engineering (IFIN-HH), P O Box MG-6, RO 077125 Bucharest-Magurele, Romania

Abstract.
The neck region generates a microscopic potential, derived in correlation with the necking region within the fission-like shape on the potential theory basis. The whole microscopic potential is of the two-center type, yielding the evolution of proton and neutron level schemes from one parent to two completely separated fragment nuclei. The shell corrections are calculated using the neck in single-particle levels. The total deformation energy is obtained from the macroscopic-microscopic method. As an application, dynamic calculation is performed for the fission of ^{236}Pu, using the multidimensional minimization within the total space of deformation of two spheroids joined by a smoothed necking region.

Keywords: shell model, fission, shell and pairing corrections
PACS: 25.70.Jj, 21.60.Ev, 21.60.Cs

INTRODUCTION

Neck influence on fission process is studied since early articles on the basis of macroscopic-microscopic method [1]. Necking appears in its most common form in the neck rupture scenario [2]. A general approach derives from the widely used Brosa model [3], where a dumbel-like shape develops a middle neck whose characteristics and rupture are determined by the balance between Coulomb and nuclear part of the macroscopic energy. A first attempt to link the neck to the binary type potential within the two-center shell model has been made in [4]. In this case the necking potential is defined by means of a geometry dependent frequency. The study of symmetric and almost-symmetric systems showed that a decreasing ε (growing neck) leads the system to fall into the fission type valley [5]. One shall introduce a microscopic necking potential on the basis of the potential theory, constructed in such a way as to fit the two-center deformed oscillator potential formed by two Nilsson-type wells. The influence of the neck on the total deformation energy is presented for ^{236}Pu and the macroscopic-microscopic method is employed.

THE NECK POTENTIAL

The binary fission like configuration is described by two intersected spheroids smoothly joined by a spherical necking surface, expressed in cylindrical coordinates by:

CP972, *Exotic Nuclei and Nuclear/Particle Astrophysics (II)—Proceedings of the Carpathian Summer School of Physics 2007,* edited by L. Trache and S. Stoica

$$\rho(z) = \begin{cases} \rho_H(z) = & [b_H^2 - \chi_H^2 z^2]^{1/2} & -a_H \le z \le z_{cH} \\ \rho_g(z) = & \rho_3 - [R_3^2 - (z-z_3)^2]^{1/2}, & z_{cH} \le z \le z_{cL} \\ \rho_L(z) = & [b_L^2 - \chi_L^2 (z-R)^2]^{1/2}, & z_{cL} \le z \le R + a_L \end{cases} \quad (1)$$

where a_H, b_H and a_L, b_L are the semiaxes of the heavy (H) and light (L) fragment, $\chi_i = b_i/a_i$ and the neck surface $\rho_g(z)$ is generated by the rolling sphere around the symmetry axis, with the center in (ρ_3, z_3) and a radius R_3. The parameter R is the distance between centers, z_H and z_L are the tangent points of the necking sphere with the heavy and the light fragment respectively. Each of the potentials follows the shape described by $\rho_H(z)$ and $\rho_L(z)$ respectively. One shall try to fi nd a potential which corresponds to $\rho_g(z)$ describing the neck. In the spheroid region, an attractive oscillator-type force confi nes the nucleons within the two convex surface bordered volumes. A rejective force directed from outside towards the neck surface is the result of the necking nucleon average nuclear fi eld:

$$\vec{F}(\vec{r}) = -C\vec{r} \quad (2)$$

According to the potential theory, this radially outward force corresponds to a potential function $\phi(r)$ through the relation:

$$\int_0^r \vec{F} d\vec{r} = \varphi(r) - \varphi(0) \quad (3)$$

where $\vec{r}$ describes the necking surface. When using the eq. (2) one obtains the general form of the potential function:

$$\int_0^r \vec{F} d\vec{r} = \int_0^r -C\vec{r} d\vec{r} = -\frac{Cr^2}{2} \quad (4)$$

Up to a constant V_c, the neck potential must have an expression like:

$$V_{g1}(r) = V_c - \frac{Cr^2}{2} \quad (5)$$

In order to determine the constant V_C, we appeal to the equipotentiality condition on the nuclear surface:

$$V_{g1} |_{S_g} = V_c + \varphi(r) |_{S_g} = V_0 \quad (6)$$

where V_0 is the potential value on the nuclear surface. The potential constant comes out from:

$$V_c = V_0 - \varphi(r) |_{S_g} \quad (7)$$

once one has the form of $\varphi(r)$. From an oscillator type force one has to comply to a constant form of $C = m_0 \omega_g^2$, so that:

$$\vec{F}(\vec{r}) = -m_0 \omega_g^2 \vec{r} \quad (8)$$

which gives the potential function fi nal form:

$$\varphi(r) = -\frac{m_0 \omega_g^2 r^2}{2} \tag{9}$$

When one replaces the shape function $r(\rho, z)$ by the neck generating function of the geometry in cylindrical coordinates, we obtain:

$$\varphi(\rho, z) = -\frac{m_0 \omega_g^2}{2} [(\rho - \rho_3)^2 + (z - z_3)^2] \tag{10}$$

This is a potential function centered in the middle of a sphere at (ρ_3, z_3), which is tangent all the time at the spheroidal fragments. The frequency ω_g defi ning the strength of the neck potential is obtained again from the equipotentiality condition, and the value for the constant results as $V_c = 2V_0$.

The outer neck potential (outer region down to the nuclear neck surface) is then:

$$V_{g1}(r) = 2V_0 - \frac{m_0 \omega_g^2}{2} [(\rho - \rho_3)^2 + (z - z_3)^2] \tag{11}$$

The inner neck potential is constant. The fi nal form of the potential describing the necking binary shapes appropiate for fi ssion:

$$V_{osc}(\rho, z) = \begin{cases} V_H(\rho, z) = & \frac{1}{2} m_0 \omega_{\rho H}^2 \rho^2 + \frac{1}{2} m_0 \omega_{zH}^2 (z + z_1)^2, v_H \\ V_g(\rho, z) = & \begin{cases} 2V_0 - [\frac{1}{2} m_0 \omega_g^2 (\rho - \rho_3)^2 + \frac{1}{2} m_0 \omega_g^2 (z - z_3)^2] & , v_{g1} \\ V_0 & , v_{g2} \end{cases} \\ V_L(\rho, z) = & \frac{1}{2} m_0 \omega_{\rho L}^2 \rho^2 + \frac{1}{2} m_0 \omega_{zL}^2 (z - z_2)^2, v_L \end{cases} \tag{12}$$

where v_H and v_L are the regions on the heavy and light fragment side except the neck region and v_{g1} and v_{g2} are the outer and inner regions of the neck.

TOTAL HAMILTONIAN AND DEFORMATION ENERGY

The total Hamiltonian contains also the spin-orbit and Ω^2 term, so that the three-dimensional Schroedinger equation is:

$$[-\frac{\hbar^2}{2m_0} \Delta + V_{osc}(\rho, z) + V_{\Omega s} + V_{\Omega^2}] \Phi = E \Phi \tag{13}$$

A working basis for such binary systems has been obtained in [6] for two intersected spheroids without neck. The matrix elements for the necking spin-orbit part are obtained from the creation and anihilation operators:

$$\Omega s(neck) \quad = \quad \frac{1}{2} [\Omega^+(v_g) s^- + \Omega^-(v_g) s^+] + \Omega_z(v_g) s_z \tag{14}$$

where:

$$\Omega^+(v_{g1}) = -e^{i\varphi}\left[\frac{\partial V_g(\rho,z)}{\partial\rho}\frac{\partial}{\partial z} - \frac{\partial V_g(\rho,z)}{\partial z}\frac{\partial}{\partial\rho} - \frac{i}{\rho}\frac{\partial V_g(\rho,z)}{\partial z}\frac{\partial}{\partial\varphi}\right]$$
$$= -e^{i\varphi}\left[m_0\omega_g^2(\rho-\rho_3)\frac{\partial}{\partial z} - m_0\omega_g^2(z-z_3)\frac{\partial}{\partial\rho} - \frac{i}{\rho}m_0\omega_g^2(z-z_3)\frac{\partial}{\partial\varphi}\right]$$
$$\Omega^-(v_{g1}) = -e^{i\varphi}\left[\frac{\partial V_g(\rho,z)}{\partial\rho}\frac{\partial}{\partial z} - \frac{\partial V_g(\rho,z)}{\partial z}\frac{\partial}{\partial\rho} + \frac{i}{\rho}\frac{\partial V_g(\rho,z)}{\partial z}\frac{\partial}{\partial\varphi}\right]$$
$$= e^{-i\varphi}\left[m_0\omega_g^2(\rho-\rho_3)\frac{\partial}{\partial z} - m_0\omega_g^2(z-z_3)\frac{\partial}{\partial\rho} + \frac{i}{\rho}m_0\omega_g^2(z-z_3)\frac{\partial}{\partial\varphi}\right]$$
$$\Omega_z(v_{g1}) = -\frac{i}{\rho}\frac{\partial V_{g1}}{\partial\rho}\frac{\partial}{\partial\phi} = im_0\omega_g^2\frac{\rho-\rho_3}{\rho}\frac{\partial}{\partial\varphi} \tag{15}$$

The total matrix of the Hamiltonian is then diagonalized separately for protons and neutrons and the level scheme evolution from one parent potential well to two separated spheroidal potentials is obtained. The single necked binary particle levels are input data for the Strutinsky method in order to calculate the proton and neutron shell correction for the fi ssion confi guration.

The macroscopic-microscopic method is applied. The liquid drop part E_{LDM} is added to the shell correction E_{shell} and the total deformation energy E_{def} is obtained:

$$E_{def} = E_{LDM} + Eshell \tag{16}$$

The Coulomb E_C and Yukawa-plus-exponential energy E_Y are the only macroscopic terms which are deformation dependent. The total deformation energy is calculated in the four-dimensional space of deformation which comprises: the two semiaxis ratios of the spheroids $\chi_H = b_H/a_H$, $\chi_L = b_L/a_L$, the neck radius R_3 and the distance between centers R. The inertia tensor is obtained with the Werner-Wheeler approximation and the action integral is minimized over the deformation space.

^{236}PU FISSION - RESULTS AND DISCUSSION

The neck influence on fi ssion barriers has been studied on the particular case of ^{236}Pu fi ssion. The mass distribution of the fi ssion fragments for this nucleus has been measured among others by Dematte, Wagemans et al. [7]. A peak is reported around heavy fragment mass of about 141 amu. The results of the present calculations show that the neck can be in part responsible for such mass asymmetry in spontaneous fi ssion.

The results of the minimization of the action integral using Werner-Wheeler mass inertia are summarized on a contour map. The fi nal fi ssion path is plotted on the corresponding contour map in fi g. 1. It starts with low R_3 values close to parent confi guration, goes through larger neck radii and reaches the scission point ($R - R_i \approx 13$ fm) at $R_3 \approx 6.5$ fm. The valley around a pair (A_H, A_L) of (144, 92) has been obtained and is very close to the result reported in [7]. Other pairs like (142, 94) and (140, 96) are also belonging to this fi ssion valley. Their probability of occurence is at least partially due to the shell correction minima generated by the introduction of a smooth microscopic neck potential between fragments.

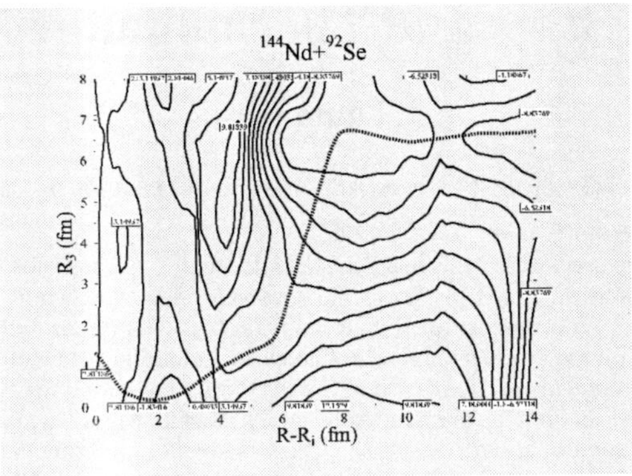

FIGURE 1. Dynamical fission path in the (R, R_3) space of deformation for the fission of ^{236}Pu.

CONCLUSIONS

A microscopic neck potential is introduced in the deformed two–center shell model. Calculations are based on potential theory under the assumption of generating the fi ssion–like shape of two spheroids smoothly linked by a spherical neck surface. The neck potential introduces an oscillator and a spin–orbit component. The level scheme is directly influenced by the geometry of the neck and the shell corrections depend on the radius of the neck generated sphere. Calculations have been applied to the fi ssion of ^{236}Pu. An asymmetric valley appears at $^{144}Nd+^{92}Se$ fi ssion channel, exclusively due to the microscopic neck potential through the shell corrections. This result shows that the neck is at least partially responsible for asymmetric fragment fi ssion valleys.

ACKNOWLEDGMENTS

This work was supported by Ministry of Education and Research, Bucharest within the "IDEI" programme.

REFERENCES

1. M. Bosterli, E. O. Fiset, J. R. Nix and J. L. Norton, Phys. Rev. C5 (1972) 1050.
2. K. T. R. Davies, R. A. Managan, J. R. Nix and A. J. Sierk, Phys. Rev. C16 (1977) 1890.
3. U. Brosa and S. Grossmann, Z. Phys. A310 (1983) 177.
4. J. Maruhn and W. Greiner, Z. Phys. 251 (1972) 431.
5. G. G. Adamian, N. V. Antonenko, S. P. Ivanova and W. Scheid, Nucl. Phys. A646 (1999) 29.
6. R. A. Gherghescu, Phys. Rev. C67: 014309 (2003).
7. L. Dematte, C. Wagemans, R. Barthelemy, P. D. D'hondt and A. Deruytter, Nucl. Phys. A617 (1997) 331.

Barium Stars and Thermohaline Mixing

Laura Husti

Dipartimento di Fisica Generale, Università di Torino, Via Pietro Giuria 1, 10125 Torino, Italy

Abstract. Barium stars are formed in binary systems through mass transfer from the carbon and s-element rich primary in the AGB phase, to the secondary star which is in a less evolved evolutionary stage. The mixing of the accreted material from the AGB donor with the envelope of the secondary results in a dilution of the s-element abundances. Dilution in red giants is explained by the occurence of the first dredge up, while in case of dwarfs thermohaline mixing would determine it. A comparison between the theoretical predictions of the AGB stellar models and the spectroscopical observations of a large sample of barium stars has been made. Dilution due to thermohaline mixing was taken into account when searching for best fits of the observational data. The importance of thermohaline mixing in barium dwarfs is discussed.

Keywords: Stars: abundances - Stars: Barium - Stars: AGB - Nucleosynthesis: s elements
PACS: 97.10.Cv, 97.10.Tk, 97.20.Jg, 97.30.Hk, 98.62.Bj, 98.80.Ft

INTRODUCTION

Barium stars are particular giants, subgiants or dwarf stars of Population I [1, 2, 3, 4]. Their spectra present enhanced lines of Ba II (4554 A), Sr II (4077 A, 4215 A) and the CH G band.

Barium (as strontium) lies on the s-process path and is formed at neutron densities of 10^5-10^{11} free neutrons/cm^3. The conditions for such neutron densities are encountered in the thermal pulsing AGB phase of low mass stars [5, 6].

Since barium is expected to form in TP-AGB stars, and the barium stars are main sequence dwarfs, subgiants or red giants, the most plausible scenario on their formation is mass transfer in a binary system [7]. The more massive star evolved first to the AGB phase and polluted the less evolved one through stellar winds, making the nowadays observed barium star and then evolving to a white dwarf. This hypothesis is sustained by the observed variations in radial velocity of most barium stars [8, 9], consistent with the presence of companions that have masses expected for white dwarfs and by the detection of an ultraviolet excess in some cases [10, 11, 12].

We present some theoretical element abundance distributions of barium stars emphasizing the consequences of binarity.

EVOLUTION OF LOW MASS STARS

In stars less massive than $\sim 5\ M_\odot$, core hydrogen burning starts on the zero age main sequence and goes on until H is exhausted in the core over a mass fraction close to 10%. Then the contraction of the star begins due to the unbalanced gravitational force until a H-shell surrounding the He core is ignited. Afterwards, the core continues to shrink,

CP972, *Exotic Nuclei and Nuclear/Particle Astrophysics (II)—Proceedings of the Carpathian Summer School of Physics 2007,* edited by L. Trache and S. Stoica

while the stellar radius increases to carry out the energy produced by the H-burning shell. As a consequence of the envelope expansion, the stellar representative point in the H-R diagram moves to the red and to increasing luminosities, climbing a track called the red giant branch (RGB). This phase lasts for 100 to 500 million years.

While the star ascends along the RGB, convection penetrates into regions that during the main sequence had already experienced partial C-N processing and whose abundance of light elements has been modified by proton captures. This process, called first dredge-up, mixes the pre-existing layered structure homogenizing about 80% of the mass.

In stars below a critical mass $M_{\rm Hef}$ (the mass at helium flash), the conditions to ignite He are reached when the stellar interior is strongly degenerate. $M_{\rm Hef}$ spans a range of 1.7 to 2.5 $M_{\odot}$, for different choices of the mixing algorithm. When He is ignited in degenerate conditions, a thermal runaway, called core He-flash, develops. After a number of flashing episodes, which burn some He into C, the electron degeneracy is removed without modification in the envelope composition. The start of core He burning forces the representative point of the star to move toward the blue and to lower luminosities in the H-R diagram. Then He burning proceeds quietly up to central He exhaustion, into C (20% in mass) and O (80% in mass). In stars with masses above $M_{\rm Hef}$ electron degeneracy at core He ignition is avoided and core He burning occurs quietly up to exhaustion. This phase is similar to that of main sequence and is called horizontal branch (HB).

At core He exhaustion, the star gets the energy from a He-burning shell and the contraction of the C-O core. The representative point follows a track that asymptotically approaches the RGB and is therefore called the asymptotic giant branch (AGB). During the early phases (E-AGB), for all stars more massive than ~ 3.5 $M_{\odot}$ the energy output from the He-shell forces the star to expand and cool so that the H shell becomes inactive. The base of the convective envelope penetrates beyond the H-He discontinuity mixing H-burning ashes with the envelope. This new mixing episode is called second dredge-up.

In the AGB stage the star is characterized by a degenerate C-O core, by two alternatively burning shells (of H and He), separated by a He intershell and by an extended convective envelope (of a few AU). As pressure and temperature decrease, He burning in the shell is extinguished and the star begins to contract again. Due to compression, the temperature increases until the H-shell is ignited while the He-shell remains inactive.

The He produced by H burning sinks onto the He intershell, so that the mass of the intershell increases and attains higher densities and temperatures. When the conditions for He fusion are reached, it starts burning almost explosively for a short period of time generating a so called thermal pulse (TP). In all the He-rich region convection takes place of the radiative state. The envelope is expanded to output the energy resulted from partial He burning so that the H-shell is cooled and extinguished. For low mass AGB stars (M ≤ 3 $M_{\odot}$), the pulse lasts for about 100 years, while just 25% of the He is burned into C via the $3-\alpha$ reaction as there is not enough time to capture another α particle and form oxygen.

After the thermal pulse, the star begins to shrink and the convective envelope penetrates below the H-He discontinuity and allows the mixing of freshly synthesized nuclei with the envelope in the phase of the third dredge-up (TDU). Then H is reignited and the cycle of the alternatively burning shells continues. The whole thermal pulsing phase lasts a few million years, thermal pulses become more and more frequent, the star becomes unstable and loses mass due to the pulsation and to radiation pressure. The stellar wind will finally remove the entire envelope and the nucleus will remain as a cooling white dwarf surrounded by a planetary nebula.

For a more detailed description of low mass stellar evolution we refer to Busso, Gallino & Wasserburg [13].

AGB PREDICTIONS FOR BARIUM STARS

The main neutron source in low mass AGB stars is the $^{13}C(\alpha,n)^{16}O$ reaction. During the third dredge up, at the contact of the H-rich convective envelope and the He-rich radiative zone, a small amount of protons is assumed to penetrate from the envelope into the top layers of the He intershell. At H reignition, these protons are captured by the large abundance of ^{12}C, giving rise to a ^{13}C pocket, via the reaction chain $^{12}C(p,\gamma)^{13}N(\beta^+\nu)^{13}C$, eventually followed by $^{13}C(p,\gamma)^{14}N$ in case some extra protons survive the capture on the abundant ^{12}C. Later on, ^{13}C burns radiatively in the interpulse phase at $T \sim 0.9 \times 10^8$ K, thus releasing neutrons which are used for the build up of s-process elements. The s-rich pocket is then engulfed by the following convective thermal instability. We consider a large range of ^{13}C pocket efficiencies [13]. The standard ST case corresponds to a mass pocket of 9.37×10^{-4} $M_\odot$ containing a total mass of ^{13}C atoms of 4.69×10^{-6} $M_\odot$ distributed in three subzones. Starting from the top of the pocket, the masses of the three subzones are 7.5×10^{-6} $M_\odot$, 5.3×10^{-4} $M_\odot$, and 4.0×10^{-4} $M_\odot$; each subzone contains a mass fraction $X_{\rm eff}(^{13}C) = 1.5 \times 10^{-2}$, 6.37×10^{-3}, and 3.0×10^{-3}, respectively. Here we consider the effective ^{13}C mass fractions, defined as $X(^{13}C) - 13/14\ X(^{14}N)$. In fact, during the subsequent release of neutrons by the reaction $^{13}C(\alpha,n)^{16}O$, any ^{14}N present in the zone would act as a major neutron poison via the resonant reaction $^{14}N(n,p)^{14}C$. For a general discussion of the choice of ^{13}C pocket and its profile we refer to Gallino et al. [14]. This ST case was used by Arlandini et al.[15] and was shown to best reproduce the solar system main component with low mass AGB stellar models of half solar metallicity.

The H burning ashes contain ^{14}N from the original CNO nuclei. During each convective thermal instability in the He shell, all the ^{14}N nuclides are converted to ^{22}Ne by $^{14}N(\alpha,\gamma)^{18}F(\beta^+\nu)^{18}O$, followed by $^{18}O(\alpha,\gamma)^{22}Ne$. At the peak temperature reached at the base of the thermal pulse, the $^{22}Ne(\alpha,n)^{25}Mg$ reaction is partly activated, giving rise to a small neutron exposure. The contribution of this second neutron source to the total neutron irradiation is small, but important in some branchings.

The free neutrons provided by the two neutron sources are captured on iron seeds and give rise to the s-elements. The third drege up takes to the surface the fresh C made by partial He burning in the thermal pulse and the s-elements. The C and s-process enhanced material is diluted after the mass transfer, into the envelope of the star observed today. We consider the dilution factor dil $= \log(M^{\rm env}_{\rm obs}/M^{\rm AGB}_{\rm transf})$, where $M^{\rm env}_{\rm obs}$ is the mass of the convective envelope after mass transfer and $M^{\rm AGB}_{\rm transf}$ is the mass transfered from the AGB. This dilution will affect the observed abundance of each element in the envelope of the barium stars according to the formula:

$$\left[\frac{X}{Fe}\right]^{obs} = \log\left(10^{\left[\frac{X}{Fe}\right]^{ini}} \cdot (1 - 10^{-dil}) + 10^{\left[\frac{X}{Fe}\right]^{AGB}} \cdot 10^{-dil}\right)$$

Dilution will lower the abundances of the elements, but will not change the shape of the curve above iron.

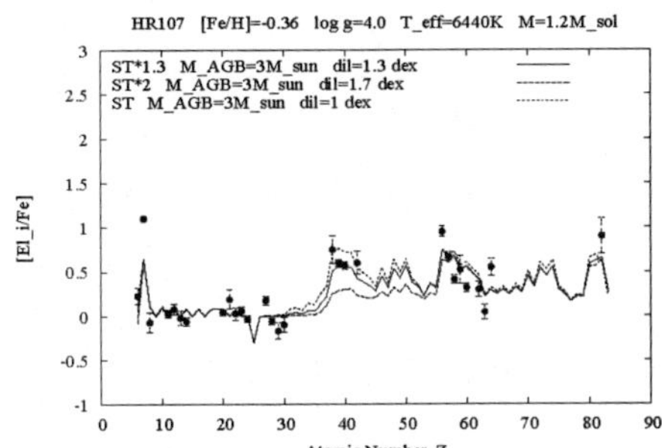

FIGURE 1. Star HR 107. Choice of the ^{13}C-pocket. On top the observational parameters of the star are reported: metallicity, logarithm of the surface gravity, effective temperature and the estimated mass from isochrones. In the inset we report the parameters that determine the theoretical abundance: ^{13}C-pocket efficiency, initial mass of the AGB donor, and the dilution factor we obtain.

We considered the sample of 26 barium stars from Allen & Barbuy [16]. In order to obtain a fit, we search for the ^{13}C pocket efficiency that would give the right trend of the elements in s-peaks. In Fig.1 we fit the observed values of the elements using a ST $\times$ 1.3 ^{13}C pocket. We see that we cannot fit the observations neither using a ST nor a ST $\times$ 2 ^{13}C pocket. These would fit well Pb and the elements of the Ba peak, but not those of the Sr peak. We usualy found good fits using a unique choice of the ^{13}C pocket, for each star.

THERMOHALINE MIXING IN BARIUM STARS

Subgiants and red giants have large convective envelopes and many of them have already suffered first dredge up, so we expect large dilution factors in their cases (dil $\sim$ 1 dex). Main sequence dwarfs have small convective envelopes and we expect smaller dilutions than in case of giants. Although, for some dwarf barium stars we obtain larger dilution factors than expected, suggesting that thermohaline mixing may play a part (see Fig 1).

Thermohaline mixing is a process that occurs when the mean molecular weight of the stellar material increases towards the surface. The heavier material transfered from the AGB will sink until the difference in molecular weight will disappear.

For most barium dwarfs we obtained good fits using the same ^{13}C pocket with initial masses of the AGB of 1.5, 2 and 3 $M_{\odot}$. In Fig.2 we have ploted the fits we obtained for star HD 89948. When using a 1.5 $M_{\odot}$ AGB model, the expected dilution factor is 0.1 dex, while in case of the 3 $M_{\odot}$ model, dil = 0.5 dex. The 1.5 $M_{\odot}$ case would need no thermohaline mixing to explain the dilution, the 3 $M_{\odot}$ model would need some thermohaline mixing in order to get the expected dilution factor.

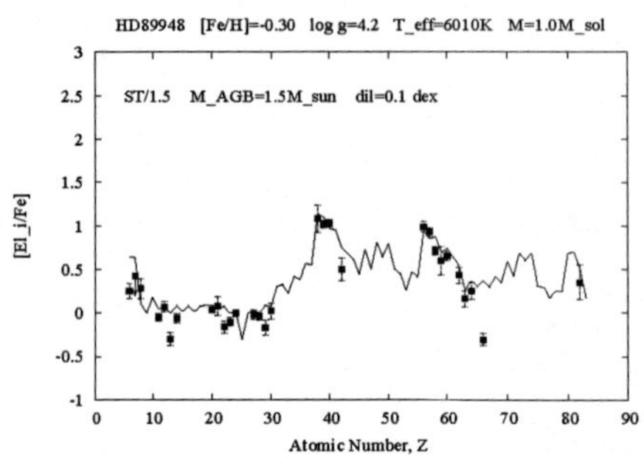

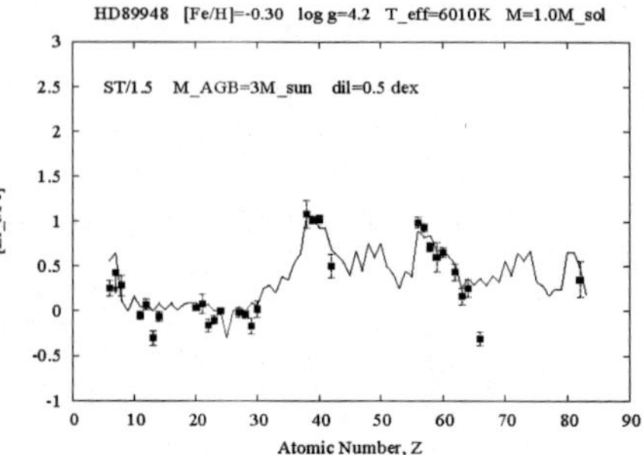

FIGURE 2. Star HD 89948. Initial mass of the AGB donor = 1.5 $M_{\odot}$ (left panel). Initial mass of the AGB donor = 3 $M_{\odot}$ (right panel).

CONCLUSIONS

Barium stars are extrinsic AGBs, so the dilution of the s-process enriched material into the envelope of the observed star must be considered. In order to get a good fit we have to choose a proper ^{13}C-pocket for each star. For some stars, we obtain good fits for all the initial masses of the AGB star, using the same ^{13}C-pocket, but with different dilution factors. In these cases, a good knowledge of thermohaline mixing would help us choose the suitable initial mass of the AGB donor.

ACKNOWLEDGMENTS

I thank prof. Roberto Gallino for guiding my PhD research.

REFERENCES

1. Bidelman, W.P., Keenan, P.C., 1951, ApJ, 114, 473
2. Bidelman, W.P., 1981, AJ, 86, 553
3. Tomkin, J. et al., 1989, A&A, 219, L15
4. North, P. et al., 1994, A&A, 468, 679
5. Iben, I., Jr., Truran, J.W., 1978, ApJ, 220, 980
6. Iben, I., Jr., Renzini, A., 1983, ARAA, 21, 271
7. McClure, R.D., Fletcher, J.M., Nemec, J.M., 1980, ApJ, 238, L35
8. McClure, R.D., 1983, ApJ, 268, 264
9. McClure, R.D., Woodsworth, A.W., 1990, ApJ, 352,709
10. Böhm-Vitense, E., 1980, ApJ, 239, L79
11. Böhm-Vitense, Johnson, H.R., 1985, ApJ, 293, 288
12. Böhm-Vitense, E., 2000, ApJ, 533, 969
13. Busso, M., Gallino, R., Wasserburg, G.J., 1999, Annu. Rev. Astron. Astrophys. 37, 239
14. Gallino, R. et al., 1998, ApJ, 497, 388
15. Arlandini, C. et al., 1999, ApJ, 525, 886
16. Allen, D.M., Barbuy, B., 2006, A&A, 454, 895

Quantum Entanglement in Open Systems

Aurelian Isar

Institute of Physics and Nuclear Engineering, Bucharest-Magurele, Romania

Abstract. In the framework of the theory of open systems based on completely positive quantum dynamical semigroups, the master equation for two independent harmonic oscillators interacting with an environment is solved in the asymptotic long-time regime. Using the Peres–Simon necessary and sufficient condition for separability of two-mode Gaussian states, we show that the two non-interacting systems become asymptotically entangled for certain environments, so that in the long-time regime they manifest non-local quantum correlations. We calculate also the logarithmic negativity characterizing the degree of entanglement of the asymptotic state.

Keywords: Open systems, quantum entanglement, separable states, logarithmic negativity
PACS: 03.65 Yz, 03.67 Mn

INTRODUCTION

Quantum entanglement represents the physical resource in quantum information science which is indispensable for the description and performance of such tasks like quantum teleportation, quantum cryptography and quantum computation [1]. When two systems are immersed in an environment, then, besides and at the same time with the quantum decoherence phenomenon, the external environment can also generate a quantum entanglement of the two systems [2]. The structure and properties of the environment may be such that not only the two systems become entangled, but also such that a certain amount of entanglement survives in the asymptotic long-time regime [2].

In the framework of the theory of open systems based on completely positive quantum dynamical semigroups, we investigate the existence of the asymptotic entanglement for a subsystem composed of two identical harmonic oscillators interacting with a common environment. We are interested in the correlation effect of the environment, therefore we assume that the two systems are initially independent. Using the Peres-Simon necessary and sufficient condition for separability of two-mode Gaussian states [3, 4], we show that for certain classes of environments the initial state evolves asymptotically to an equilibrium state which is entangled, i.e. there exist non-local quantum correlations for the bipartite states of the subsystem, while for other environments, the entanglement is suppressed and the asymptotic state is separable. We calculate also the logarithmic negativity characterizing the degree of entanglement of the asymptotic state.

EQUATIONS OF MOTION FOR TWO INDEPENDENT HARMONIC OSCILLATORS

The time evolution of an open system can be described by a Markovian, completely positive quantum dynamical semigroup and in the Heisenberg picture the corresponding

CP972, *Exotic Nuclei and Nuclear/Particle Astrophysics (II)—Proceedings of the Carpathian Summer School of Physics 2007,* edited by L. Trache and S. Stoica

master equation for an operator A has the form [5, 6]:

$$\frac{dA(t)}{dt}=\frac{i}{\hbar}[H,A(t)]+\frac{1}{2\hbar}\sum_j(V_j^\dagger[A(t),V_j]+[V_j^\dagger,A(t)]V_j). \tag{1}$$

Here, H denotes the Hamiltonian of the open quantum system and $V_j,V_j^\dagger$, which are operators defined on the Hilbert space of H, represent the interaction of the open system with the environment. We are interested in the set of Gaussian states, therefore we introduce quantum dynamical semigroups that preserve that set. Then H is taken to be a polynomial of second degree in the coordinates x,y and momenta p_x,p_y of the two quantum oscillators and $V_j,V_j^\dagger$ are taken polynomials of only first degree in these canonical observables [7]:

$$V_j=a_{xj}p_x+a_{yj}p_y+b_{xj}x+b_{yj}y, \tag{2}$$

where $a_{xj},a_{yj},b_{xj},b_{yj}\in\mathbf{C},\ j=1,2,3,4$. The Hamiltonian H of the two uncoupled identical harmonic oscillators of mass m and frequency ω is given by

$$H=\frac{1}{2m}(p_x^2+p_y^2)+\frac{m\omega^2}{2}(x^2+y^2). \tag{3}$$

The dynamical semigroup implies the positivity of the following matrix formed by the scalar products of the four vectors $\mathbf{a}_x,\mathbf{b}_x,\mathbf{a}_y,\mathbf{b}_y$, whose entries are the components $a_{xj},b_{xj},a_{yj},b_{yj}$, respectively:

$$\frac{1}{2}\hbar\begin{pmatrix}(\mathbf{a}_x\mathbf{a}_x) & (\mathbf{a}_x\mathbf{b}_x) & (\mathbf{a}_x\mathbf{a}_y) & (\mathbf{a}_x\mathbf{b}_y)\\ (\mathbf{b}_x\mathbf{a}_x) & (\mathbf{b}_x\mathbf{b}_x) & (\mathbf{b}_x\mathbf{a}_y) & (\mathbf{b}_x\mathbf{b}_y)\\ (\mathbf{a}_y\mathbf{a}_x) & (\mathbf{a}_y\mathbf{b}_x) & (\mathbf{a}_y\mathbf{a}_y) & (\mathbf{a}_y\mathbf{b}_y)\\ (\mathbf{b}_y\mathbf{a}_x) & (\mathbf{b}_y\mathbf{b}_x) & (\mathbf{b}_y\mathbf{a}_y) & (\mathbf{b}_y\mathbf{b}_y)\end{pmatrix}. \tag{4}$$

For simplicity we take this matrix of the following form, where all coefficients $D_{xx},D_{xp_x},...$ and λ are real quantities:

$$\begin{pmatrix}D_{xx} & -D_{xp_x}-i\hbar\lambda/2 & D_{xy} & -D_{xp_y}\\ -D_{xp_x}+i\hbar\lambda/2 & D_{p_xp_x} & -D_{yp_x} & D_{p_xp_y}\\ D_{xy} & -D_{yp_x} & D_{yy} & -D_{yp_y}-i\hbar\lambda/2\\ -D_{xp_y} & D_{p_xp_y} & -D_{yp_y}+i\hbar\lambda/2 & D_{p_yp_y}\end{pmatrix}. \tag{5}$$

It follows that the principal minors of this matrix are positive or zero. From the Cauchy-Schwarz inequality the following relations for the coefficients defined in Eq. (5) hold (from now on we put $\hbar=1$):

$$D_{xx}D_{yy}-D_{xy}^2\geq 0,\ D_{xx}D_{p_xp_x}-D_{xp_x}^2\geq\frac{\lambda^2}{4},\ D_{xx}D_{p_yp_y}-D_{xp_y}^2\geq 0,$$
$$D_{yy}D_{p_xp_x}-D_{yp_x}^2\geq 0,\ D_{yy}D_{p_yp_y}-D_{yp_y}^2\geq\frac{\lambda^2}{4},\ D_{p_xp_x}D_{p_yp_y}-D_{p_xp_y}^2\geq 0. \tag{6}$$

We introduce the following 4×4 bimodal covariance matrix:

$$\sigma(t) = \begin{pmatrix} \sigma_{xx} & \sigma_{xp_x} & \sigma_{xy} & \sigma_{xp_y} \\ \sigma_{xp_x} & \sigma_{p_xp_x} & \sigma_{yp_x} & \sigma_{p_xp_y} \\ \sigma_{xy} & \sigma_{yp_x} & \sigma_{yy} & \sigma_{yp_y} \\ \sigma_{xp_y} & \sigma_{p_xp_y} & \sigma_{yp_y} & \sigma_{p_yp_y} \end{pmatrix}. \tag{7}$$

By direct calculation we obtain (T denotes transposition) [7]:

$$\frac{d\sigma}{dt} = Y\sigma + \sigma Y^{\mathrm{T}} + 2D, \tag{8}$$

where

$$Y = \begin{pmatrix} -\lambda & 1/m & 0 & 0 \\ -m\omega^2 & -\lambda & 0 & 0 \\ 0 & 0 & -\lambda & 1/m \\ 0 & 0 & -m\omega^2 & -\lambda \end{pmatrix} \tag{9}$$

and D is the matrix of the diffusion coefficients

$$D = \begin{pmatrix} D_{xx} & D_{xp_x} & D_{xy} & D_{xp_y} \\ D_{xp_x} & D_{p_xp_x} & D_{yp_x} & D_{p_xp_y} \\ D_{xy} & D_{yp_x} & D_{yy} & D_{yp_y} \\ D_{xp_y} & D_{p_xp_y} & D_{yp_y} & D_{p_yp_y} \end{pmatrix}. \tag{10}$$

The time-dependent solution of Eq. (8) is given by [7]

$$\sigma(t) = M(t)(\sigma(0) - \sigma(\infty))M^{\mathrm{T}}(t) + \sigma(\infty), \tag{11}$$

where $M(t) = \exp(tY)$. The matrix $M(t)$ has to fulfil the condition $\lim_{t\to\infty} M(t) = 0$. In order that this limit exists, Y must only have eigenvalues with negative real parts. The values at infinity are obtained from the equation [7]

$$Y\sigma(\infty) + \sigma(\infty)Y^{\mathrm{T}} = -2D. \tag{12}$$

ENVIRONMENT INDUCED ENTANGLEMENT

The two-mode Gaussian state is entirely specified by its covariance matrix σ (7), which is a real, symmetric and positive matrix with the following block structure:

$$\sigma = \begin{pmatrix} A & C \\ C^{\mathrm{T}} & B \end{pmatrix}, \tag{13}$$

where A, B and C are 2×2 matrices. A and B denote the symmetric covariance matrices for the individual reduced one-mode states, while the matrix C contains the cross-correlations between modes. The entries of the covariance matrix depend on Y and D and can be calculated from Eq. (12). To simplify the calculations, we consider environments for which $D_{xx} = D_{yy}$, $D_{xp_x} = D_{yp_y}$, $D_{p_xp_x} = D_{p_yp_y}$ and, in addition, $D_{xp_y} = D_{yp_x}$. Then both unimodal covariance matrices are equal, $A = B$ and the entanglement matrix C

is symmetric. With the chosen coefficients, we obtain the following elements of the asymptotic entanglement matrix C:

$$\sigma_{xy}(\infty) = \frac{m^2(2\lambda^2+\omega^2)D_{xy}+2m\lambda D_{xp_y}+D_{p_xp_y}}{2m^2\lambda(\lambda^2+\omega^2)}, \tag{14}$$

$$\sigma_{xp_y}(\infty) = \sigma_{yp_x}(\infty) = \frac{-m^2\omega^2D_{xy}+2m\lambda D_{xp_y}+D_{p_xp_y}}{2m(\lambda^2+\omega^2)}, \tag{15}$$

$$\sigma_{p_xp_y}(\infty) = \frac{m^2\omega^4D_{xy}-2m\omega^2\lambda D_{xp_y}+(2\lambda^2+\omega^2)D_{p_xp_y}}{2\lambda(\lambda^2+\omega^2)}. \tag{16}$$

The elements of matrices A and B are obtained by taking $x=y$ in Eqs. (14) – (16).

With these quantities we calculate the determinant of the entanglement matrix:

$$\det C = \frac{1}{4\lambda^2(\lambda^2+\omega^2)}\left[(m\omega^2D_{xy}+\frac{1}{m}D_{p_xp_y})^2+4\lambda^2(D_{xy}D_{p_xp_y}-D^2_{xp_y})\right]. \tag{17}$$

The Gaussian states with $\det C \geq 0$ are separable states, but for $\det C < 0$, it may be possible that the asymptotic equilibrium state is entangled. In order to investigate whether an external environment can actually entangle the two independent systems, we use the partial transposition criterion [3, 4]: a state is entangled if and only if the operation of partial transposition does not preserve its positivity. Simon [4] obtained the following necessary and sufficient criterion for separability: $S \geq 0$, where

$$S \equiv \det A\,\det B+(\frac{1}{4}-|\det C|)^2-\mathrm{Tr}[AJCJBJC^{\mathrm{T}}J]-\frac{1}{4}(\det A+\det B) \tag{18}$$

and J is the 2×2 symplectic matrix

$$J=\begin{pmatrix} 0 & 1 \\ -1 & 0 \end{pmatrix}. \tag{19}$$

In order to analyze the possible persistence of the environment induced entanglement in the asymptotic long-time regime, we consider the environment characterized by the following values of its parameters: $m^2\omega^2D_{xx}=D_{p_xp_x}$, $D_{xp_x}=0$, $m^2\omega^2D_{xy}=D_{p_xp_y}$. In this case the Simon expression (18) takes the form:

$$S=\left[\frac{m^2\omega^2(D^2_{xx}-D^2_{xy})}{\lambda^2}+\frac{D^2_{xp_y}}{\lambda^2+\omega^2}-\frac{1}{4}\right]^2-4\frac{m^2\omega^2D^2_{xx}D^2_{xp_y}}{\lambda^2(\lambda^2+\omega^2)}. \tag{20}$$

For environments characterized by such coefficients that the expression (20) is negative, the asymptotic final state is entangled. In particular, if $D_{xy}=0$, we obtain that $S<0$, i.e. the asymptotic final state is entangled, for the following range of values of the coefficient D_{xp_y} characterizing the environment [8]:

$$\frac{m\omega D_{xx}}{\lambda}-\frac{1}{2}<\frac{D_{xp_y}}{\sqrt{\lambda^2+\omega^2}}<\frac{m\omega D_{xx}}{\lambda}+\frac{1}{2}, \tag{21}$$

where the coefficient D_{xx} satisfies the condition $m\omega D_{xx}/\lambda \geq 1/2$, equivalent with the unimodal uncertainty relation. If the inequalities (21) are not fulfilled, then $S \geq 0$ and therefore the asymptotic final state is separable.

We apply the measure of entanglement based on negative eigenvalues of the partial transpose of the subsystem density matrix. In the case of a Gaussian density operator, the negativity is completely defined by the symplectic spectrum of the partial transpose of the covariance matrix. Logarithmic negativity, defined by $E = -\frac{1}{2}\log_2[4f(\sigma)]$, where

$$f(\sigma) = \frac{1}{2}(\det A + \det B) - \det C - \sqrt{\left[\frac{1}{2}(\det A + \det B) - \det C\right]^2 - \det \sigma}, \tag{22}$$

determines the strength of entanglement for $E > 0$; if $E \leq 0$, then the state is separable. We obtain

$$E = -\frac{1}{2}\log_2\left[4\left(\frac{m\omega D_{xx}}{\lambda} - \frac{D_{xp_y}}{\sqrt{\lambda^2+\omega^2}}\right)^2\right]. \tag{23}$$

We see that logarithmic negativity does not depend on the initial Gaussian state, it only depends on the diffusion and dissipation coefficients characterizing the environment.

SUMMARY

In the framework of theory of open systems based on completely positive quantum dynamical semigroups, we investigated the existence of the asymptotic quantum entanglement for a system composed of two uncoupled identical harmonic oscillators interacting with an environment. Using the Peres-Simon necessary and sufficient condition for separability of two-mode Gaussian states, we have shown that for certain classes of environments the initial state evolves asymptotically to an equilibrium state which is entangled, while for other values of the coefficients describing the environment, the asymptotic state is separable. We calculated also the logarithmic negativity characterizing the degree of entanglement of the asymptotic state.

ACKNOWLEDGMENTS

The author acknowledges the financial support received within Project CEEX 68/2005.

REFERENCES

1. M. A. Nielsen, I. L. Chuang, *Quantum Computation and Quantum Information,* Cambridge Univ. Press, Cambridge, 2000.
2. F. Benatti, R. Floreanini, Int. J. Mod. Phys. B **19**, 3063 (2005); J. Phys. A: Math. Gen. **39**, 2689 (2006).
3. A. Peres, Phys. Rev. Lett. 77, 1413 (1996).
4. R. Simon, Phys. Rev. Lett. **84**, 2726 (2000).
5. G. Lindblad, Commun. Math. Phys. **48**, 119 (1976).
6. A. Isar, A. Sandulescu, H. Scutaru, E. Stefanescu, W. Scheid, Int. J. Mod. Phys. E **3**, 635 (1994).
7. A. Sandulescu, H. Scutaru, W. Scheid, J. Phys. A: Math. Gen. **20**, 2121 (1987).
8. A. Isar, J. Russ. Laser Res. **4**, 133 (2007).

Radio Emission in Atmospheric Air Showers Measured by LOPES-30

P.G. Isar[a,b]

[a] *Institut für Kernphysik, Forschungszentrum Karlsruhe, Germany*
[b] *On leave of absence from Institute of Space Science, Bucharest, Romania*
gina.isar@ik.fzk.de

Abstract. When Ultra High Energy Cosmic Rays (UHECR) interact with particles in the Earth's atmosphere, they produce a shower of secondary particles propagating towards the ground. These relativistic particles emit synchrotron radiation in the radio frequency range when passing the Earth's magnetic field. The LOPES (LOFAR Prototype Station) experiment investigates the radio emission from these showers in detail and will pave the way to use this detection technique for large scale applications like in LOFAR (Low Frequency Array) and the Pierre Auger Observatory. The LOPES experiment is co-located and measures in coincidence with the air shower experiment KASCADE-Grande at Forschungszentrum Karlsruhe, Germany. LOPES has an absolute amplitude calibration array of 30 dipole antennas (LOPES-30). After one year of measurements of the single East-West polarization by all 30 antennas, recently, the LOPES-30 set-up was configured to perform dual-polarization measurements. Half of the antennas have been configured for measurements of the North-South polarization. Only by measuring at the same time both, the E-W and N-S polarization components of the radio emission, the geo-synchrotron effect as the dominant emission mechanism in air showers can be verified. The status of the measurements, including the absolute calibration procedure of the dual-polarized antennas as well as analysis of dual-polarized event examples are reported.

Keywords: Cosmic Ray Air Showers, New Detection Technique
PACS: 90

INTRODUCTION

The LOFAR (LOw Frequency ARray) PrototypE Station - LOPES experiment [1] is operating in the 40-80 MHz frequency range. The KASCADE-Grande [2] experiment (an extension of the KArlsruhe Shower Core and Array DEtector KASCADE) provides the trigger information and well-reconstructed parameters of the air shower properties in the energy range from a few PeV to 1 EeV. The LOPES-30 configuration is an extension of the initially installed 10 LOPES antennas (LOPES-10) by the addition of 20 dipole antennas, which have an absolute amplitude calibration providing a larger sensitive area and a large baseline to the radio signal on a single event basis. This provides the possibility for a detailed investigation of the lateral extension of the pulsed radio signal. Moreover, the antenna number is high enough for a new configuration sensitive to both, the East-West and North-South polarization directions.

CP972, *Exotic Nuclei and Nuclear/Particle Astrophysics (II)—Proceedings of the Carpathian Summer School of Physics 2007,* edited by L. Trache and S. Stoica

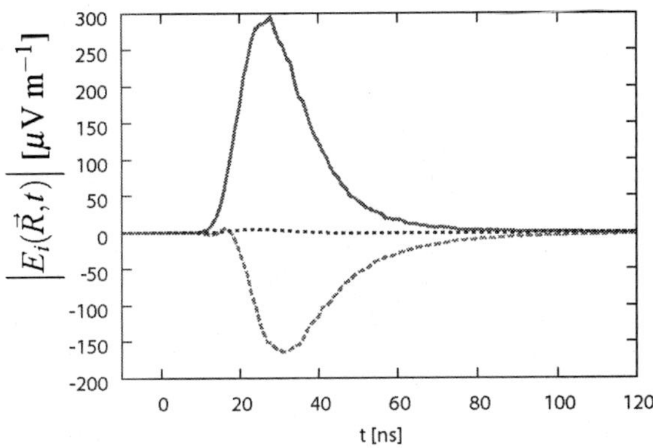

FIGURE 1. Raw simulated pulses in the individual linear polarization components of a 10^{17} eV vertical air shower. The solid line represent the E-W polarization component, dashed line N-S and dotted line the vertical component respectively.

MOTIVATION FOR PERFORMING POLARIZATION MEASUREMENTS

For the initial measurements, all 30 antennas of the LOPES experiment were equipped with dipoles in E-W direction, measuring the single polarization of the radio emission only. Recently, for recording the full radio signal, LOPES-30 has been reconfigured to perform dual-polarization measurements. The radio emission generated by the geosynchrotron mechanism is expected to be highly linearly polarized [3]. As predicted by the sophisticated Monte Carlo simulation for the radio emission, the signal is usually present in both polarization components which strengths strongly depends on the position of the observer relative to the incoming air shower. In addition, the polarization is directly related with the shower azimuth for a given zenith angle. Knowledge of these polarization characteristics of the radio emission are mandatory for the interpretation of experimental measurements, which can directly verify the geomagnetic origin of the radio emission from atmospheric air showers. Fig.1 shows a simulated raw (unlimited bandwidth) pulse arriving at a distance of 200 m to the N-W from the center of a 10^{17} eV vertical shower. The N-S and E-W polarization components are of similar strength and arrive mostly synchronously, as expected for a linearly polarized pulse. For inclined air showers, a vertical polarized component can occur as well.

THE LOPES-POL CONFIGURATION

Since end of 2006, LOPES is performing polarization measurements [8]. Half of the 30 antennas are configured for the East-West and North-South polarization measurements respectively, while five of them are configured as dual-polarization antennas at the same place, measuring at the same time both, the N-S and E-W polarization directions. The actual configuration is depicted in Fig.2.

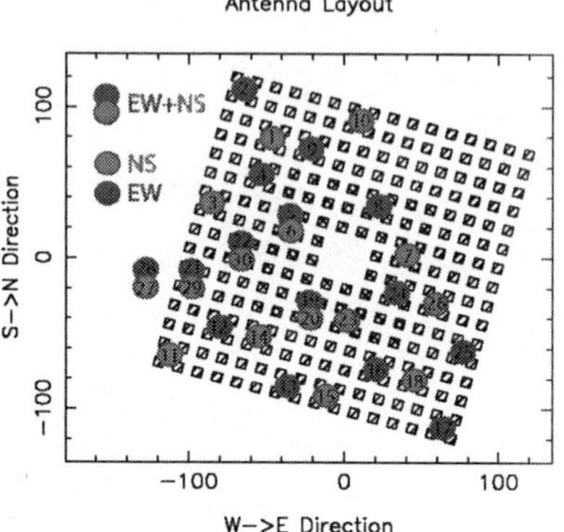

FIGURE 2. The antenna layout at the KASCADE array. The all antenna setup is divided in 15 E-W and 15 N-S polarized channels. Five antennas are sensitive to both polarization directions at the same station.

Since December 2006, the LOPES antennas are triggered by the original KASCADE and by the KASCADE-Grande array. Within this new additional trigger source, the LOPES experiment can therefore benefit from the extended detection area making possible the analysis of large events at large distances with a much better accuracy [9].

POLARIZATION SENSITIVITY

The array of 30 digital radio antennas have an absolute amplitude calibration in order to estimate the electric field strength of the short radio pulse generated in air showers [5, 6]. To perform the calibration, a commercial calibrated radio source is used as emitter and LOPES antennas are measuring within an artificially triggered event. For a calibration campaign, the reference source is placed at 11m above each LOPES antenna. It is a movable biconical reference antenna which is linearly polarized and has a nearly constant directivity close to its principal axis.

Fig.3 shows the fraction of the total field strength seen by a linearly dual-polarized LOPES antenna as a function of the angle β between the polarization axes of the reference source and the LOPES antenna. Each polarization direction is measured individually during the same campaign varying the polarization angle of the reference antenna in steps of 10 degrees by rotating the reference source in vertical position. The results obtained prove the expected polarization sensitivity (and insensitivity) of both, the E-W and N-S oriented LOPES antennas.

FIRST EVENTS

As an example, (left panel at fig.4) we display an event detected by LOPES in the new polarization configuration in December 2006. The KASCADE shower reconstruction results in a primary energy of $E_0 \approx 10^{18}$ eV, a geomagnetic angle (the angle between the

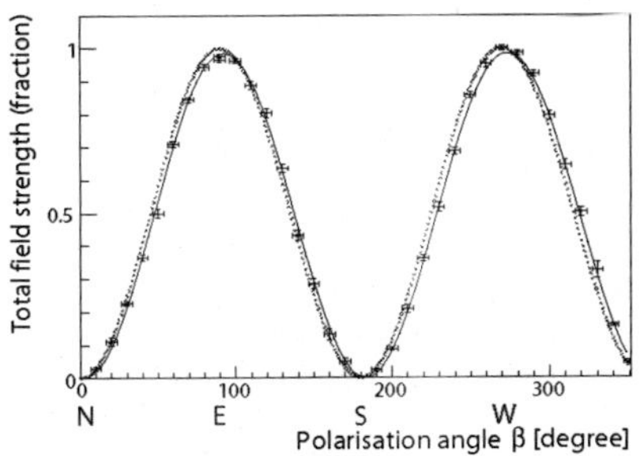

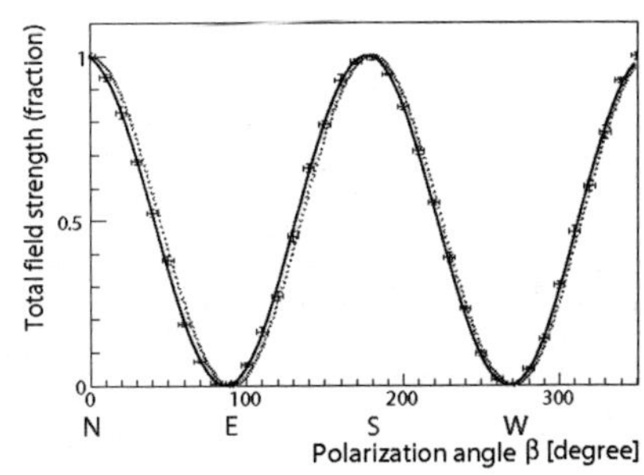

FIGURE 3. Left panel: Fraction of the total field strength seen by the linearly polarized LOPES antenna (sensitive to E-W) as a function of the polarization angle β. Points: experimental data, solid line: data fit by a sinβ-function, dotted line: theoretical sinβ-distribution (normalized to the value at β=270°). Right panel: An antenna sensitive to the N-S polarization direction of the radio signal. The dotted curve is normalized to the value at β=180°.

shower axis and the Earth magnetic field) of 83°, an azimuth angle of 51° (i.e. coming from North-East), and a zenith angle of 66°. The shower shows clearly signals in both polarization directions. The figure shows the value of the reconstructed CC-beam [10] which describes the average electric field strength for the mean distance of the antennas from the shower axis.

As another example, (right panel at fig.4) we display an event detected in December 2006 with a primary energy of $E_0 \approx 3 \cdot 10^{17}$ eV, a geomagnetic angle of 77°, an azimuth angle of 333° (i.e. coming from North, North-West), and a zenith angle of 54°. Here, a clear signal is recorded only in the E-W polarization direction without a significant signal contribution in the N-S polarization direction.

These two selected events clearly show the capabilities of the LOPES experiment in recording the radio emission in both, the E-W and N-S polarization components [8]. The new antenna configuration allows a full detection of the pulsed radio signal generated by the cosmic ray air showers in the Earth's atmosphere.

SUMMARY

In its current configuration, the LOPES experiment is performing polarization measurements; measuring at the same time both, the E-W and N-S polarization components allows a much more detailed analysis of the radio events than with the East-West polarisation measurements alone.

Providing calibrated data, the LOPES experiment allows a direct comparison with theoretical predictions for the first time. A direct verification of the geosynchrotron effect as the dominant emission mechanism in air showers is possible now.

The current data set of the dual-polarized events are being under detailed investigation; correlations between the radio pulse heights (as seen in Fig.4) and different shower parameters (like shower size, arrival direction, geomagnetic angle, primary energy) are

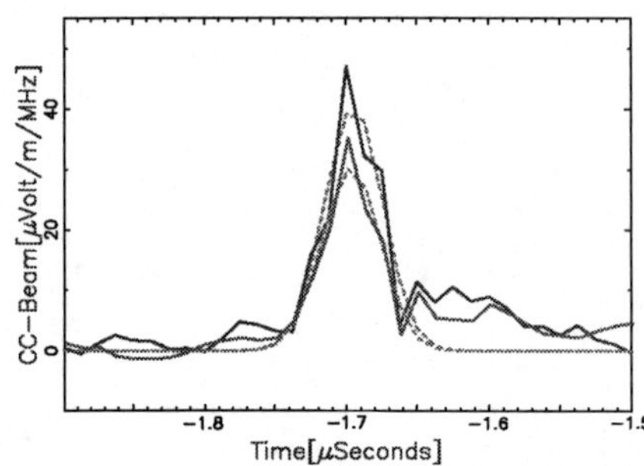

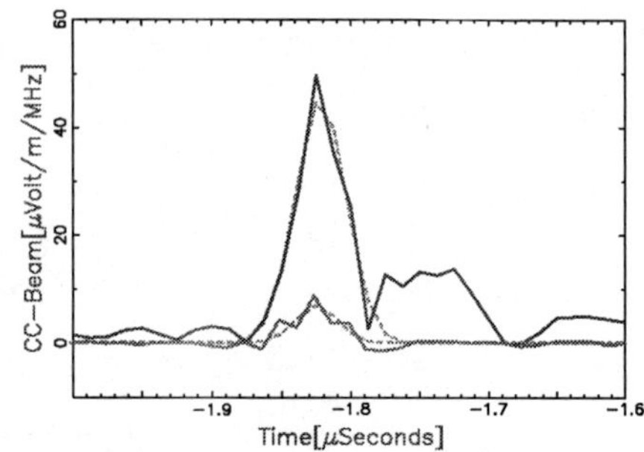

FIGURE 4. Left panel: The Cross Correlation(CC)-beams for a dual-polarized event (E-W component: upper signal, N-S component: lower signal) of a primary energy of 10^{18} eV coming from North-East. The full lines indicate the CC-beams and the dotted lines the Gaussian fits respectively. One can clearly see very high radio signal recorded in both polarization directions. Right panel: The same Cross Correlation(CC)-beam for a dual-polarized event of a primary energy of $3 \cdot 10^{17}$ eV coming from North, North-West. One can cleraly see very high signal recorded in the E-W polarization direction only.

carefully analyzed for each polarization direction individually.

ACKNOWLEDGEMENTS

The author would like to thank the German DAAD-Helmholtz grant system for the excellent research opportunity in the frame of the LOPES project. Thanks to the LOPES collaboration for all helpful contributions as well as to the organizing committee of the Carpathian summer school for the opportunity to attend the school with all nice discussions during the lectures and in the mountains.

REFERENCES

1. H. Falcke et al., Nature 435, 313-316 (2005)
2. A. Haungs et al., 30th ICRC Proc. Merida, Mexico, in press (2007)
3. T. Huege & H. Falcke, Astropart.Phys. 24 116-136 (2005)
4. T. Huege, R. Ulrich and R. Engel, Astropart.Phys. 27 392-405 (2007)
5. S. Nehls et al., ARENA Proc. 2005, Int.J.Mod.Phys. A21S1, 187 (2006)
6. S. Nehls et al., 30th ICRC Proc. Merida, Mexico, in press (2007)
7. P.G. Isar et al., ARENA Proc., in press (2006)
8. P.G. Isar et al., 30th ICRC Proc. Merida, Mexico, in press (2007)
9. W. Apel et al., Astropart. Phys. 26, 332 (2006)
10. A. Horneffer et al., ARENA Proc. 2005, Int.J.Mod.Phys. A21S1, 168 (2006)
11. A. Horneffer et al., 30th ICRC Proc., Merida, Mexico, in press (2007)

Pionic Fusion Experiments at Subthreshold Energies

L. Joulaeizadeh*, J. Bacelar*, I. Gašparić† and H. Löhner*

*KVI, University of Groningen, Groningen, The Netherlands
†Ruder Bošković Institute, Zagreb, Croatia

Abstract. In order to study the role of pions and clustering phenomena in nuclei, two experiments have been performed using the AGOR accelerator facility. In collisions of two nuclei a pion and a fused nucleus were produced. The examined reactions were $^4He(^3He,\pi^0)^7Be$ and $^6Li(^4He,\pi^0)^{10}B$ at beam energies about 10 MeV above the coherent pion production threshold (256 MeV and 236.4 MeV, respectively). Since the available energy is well below the pion production threshold in an elementary nucleon-nucleon process, a highly coherent mechanism is needed. We identified the reaction by measuring the fused system in the magnetic spectrometer and the produced neutral pions in the Plastic Ball detection system with large acceptance. Our experimental setup provided the exclusive cross sections by identifying all products in overdetermined kinematics. Here we present the preliminary results of the ongoing analysis for the second reaction. About 700 events fulfilling the kinematical conditions for an outgoing ^{10}B and a π^0 decaying with large opening angle have been selected. Angular distribution of neutral pions will be discussed.

Keywords: Pionic fusion, Neutral pion production, Subthreshold energy.
PACS: 25. Nuclear reactions, Specific reactions - 25.60.Pj Fusion reactions - 23.20.En Angular distribution and correlation measurements.

INTRODUCTION

In the extreme limit of a nuclear reaction, the total available center-of-mass energy may be concentrated into the pion field. The pion is emitted and the colliding nuclei fuse into a united nucleus in a specific bound state which is, among others, defined by the conservation of isospin. In this way, pions can also be produced at subthreshold energies, i.e. at energy per nucleon less than 280(290) MeV which is the threshold energy for neutral (charged) pion production in a free nucleon-nucleon collision. So a highly coherent mechanism is required if the production of a single pion in a nuclear reaction demands a significant fraction of the available energy. We are in fact aiming to understand this coherent reaction mechanism.

There are some theoretical models aiming to describe the scarce experimental data on pionic fusion at subthreshold energies [1, 2] and all models are in favor of clustering phenomena in nuclei [3] since results of cluster models are closer to the exsisting experimental data but still understimate the data [4]. In the cluster model, in addition to the nucleon-nucleon interaction, the strong clustering correlations of two colliding nuclei are taken into account. It was concluded that high-energy and high-momentum parts of clustering components in the final nucleus cooperate with the relative motion in the entrance channel to permit coherent pion production.

We have done two pionic fusion experiments at KVI using the AGOR facility. The

CP972, *Exotic Nuclei and Nuclear/Particle Astrophysics (II)—Proceedings of the Carpathian Summer School of Physics 2007,* edited by L. Trache and S. Stoica

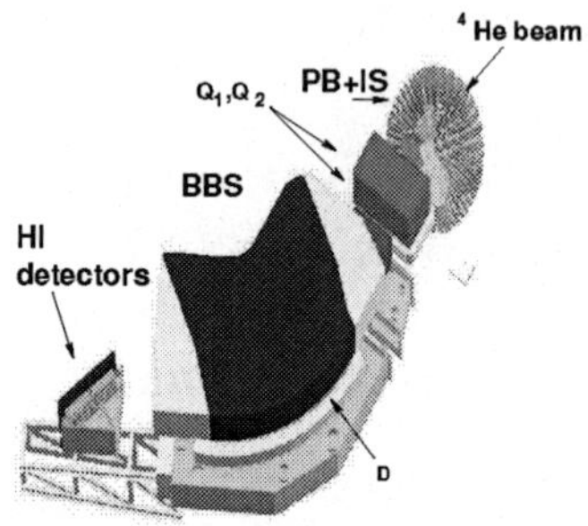

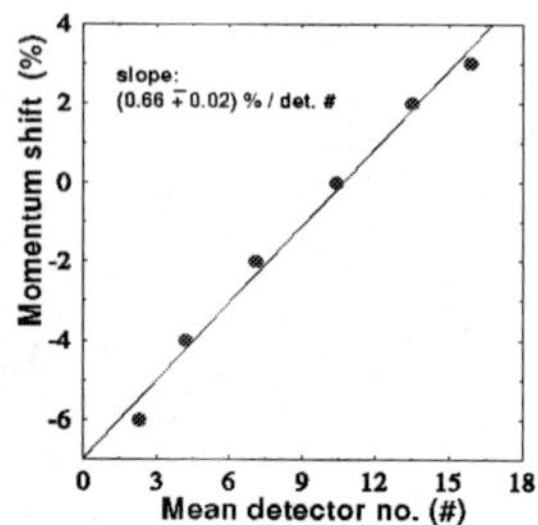

FIGURE 1. Left: Scheme of the experimental setup consisting of BBS, PB, IS and HI detectors. $Q_{1,2}$ and D represent quadrapole and dipole magnets of the BBS, respectively. Right: Momentum shift measured by detector position in the HI array.

first experiment used an 85.3 AMeV ^3He beam on a 130 mg/cm^2 liquid ^4He target. The final state is $^7Be + \pi^0$ and the beam current was on average about 0.3 nA. In the second experiment we used an 59.1 AMeV ^4He beam on a 2 mg/cm^2 solid ^6Li target. The final state in this case is $^{10}B + \pi^0$ and the beam current was on average 3 nA. Both experiments were carried out at about 10 MeV above the coherent threshold energy of pion production.

EXPERIMENTAL SETUP

In a novel experimental approach, we study the pionic fusion process by exclusive pion production in overdetermined kinematics. This requires a suitable combination of detector systems that allow to measure the emitted pion and the residual nucleus. Only in this way one can guarantee clean and truly exclusive data providing also the angular distribution of pions.

For the detection of low momentum neutral pions, we used the spherically symmetric Plastic Ball detector (PB) [5]. The Plastic Ball is a highly segmented 4π detector system consisting of 552 plastic/CaF_2 phoswich scintillators. In order to improve the detector response for photons, the Plastic Ball is equipped inside the hollow sphere of 50 cm diameter with one hemisphere of 64 CsI(Tl) scintillators, the Inner Shell (IS), acting as active converter with a thickness of 5 cm equivalent to about 2.5 radiation lengths. Also for the detection and identification of ions with Z in the range of 4-10 and typical energies of 10-30 AMeV, a specially designed array of Heavy Ion detectors (HI) consisting of 60 CsI/plastic phoswich scintillators has been used. This array was placed in the vacuum chamber just in front of the nominal focal plane of the Big Bite Spectrometer (BBS), figure 1-left. In commissioning runs with the HI detectors we have performed a test experiment with various target-projectile combinations and measured the elastic peak. We correlated the momentum shift in the BBS with the individual HI detector position by applying a systematic change of the dipole magnetic field in the BBS. As can be seen in figure 1-right, a good linearity with slope of 0.66%/detector unit has been obtained. For the calibration of the PB and IS detectors, we used high energy

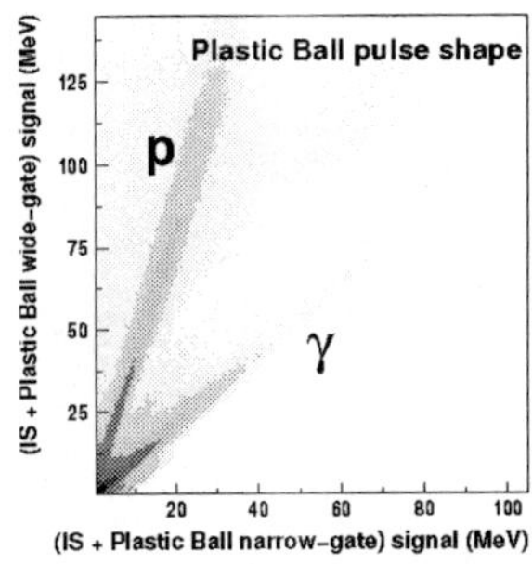

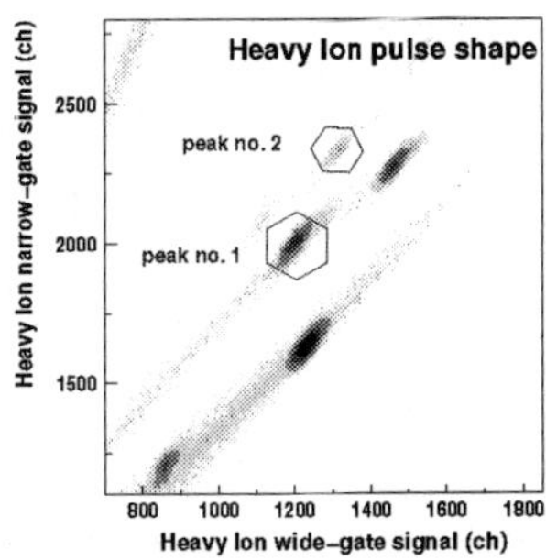

FIGURE 2. Left: Calibrated PB pulse shape. Lines labeled p and γ represent protons and photons, respectively. Right: HI pulse shape and identification of isotopes.

protons in our pionic fusion data for the forward part of the PB and high energy cosmics for the IS and backward part of the PB. In the next section, results of the $^{6}Li(^{4}He,\pi^{0})^{10}B$ reaction will be discussed.

RESULTS

The result of the calibration procedure for the Plastic Ball is shown in the pulse shape spectrum in figure 2-left. The y-axis shows the sum of the IS, CaF_2 and plastic light output using a wide-gate integration of the PB signal within 944 ns. The x-axis shows the sum of the IS light output and the plastic light output using a narrow-gate integration of the PB signal within 133 ns. In order to identify ions in the HI detector (figure 2-right), we used the HI Time Of Flight (TOF) information which is the time difference between ion and photon detection. The HI TOF is proportional to the m/q ratio for every particle. As well as ^{10}B, produced in the pionic fusion reaction, as ^{6}Li, produced in the elastic interaction of ^{4}He with ^{6}Li, have the same m/q ratio. Therefore the TOF information is not sufficient to identify ^{10}B among the peaks in figure 2-right.

In figure 2-right, the y-axis (x-axis) shows the light output of plastic scintillator (total light output) from the HI detector using a narrow (wide) gate integration within 117 (1115) ns. Although in case of ^{6}Li the momentum acceptance suppresses ^{6}Li in the focal plane of the BBS, we aimed at a unique identification of the fusion product. We used other available informations like the light output of the scintillators for different produced particles. Horn et al. [6, 7] have calibrated plastic and also CsI scintillators using different known particles in various regions of energy taking into account the strong quenching effects. We compared the relative light output of the HI detector for different particles with the relative light output of the Horn detector setup. As shown in figure 3, the relative light output of different particles from the plastic scintillator of the HI detector in our data is in good agreement with Horn for the known particles like p or ^{7}Be. Still we can not distinguish ^{10}B from ^{6}Li due to the fact that they have almost the same light output from the plastic scintillator in the region of energy in which they have been produced. If we compare the CsI light output of our data and that of Horn (figure 3), we can easily conclude that the full circle in figure 3, which belongs to peak number

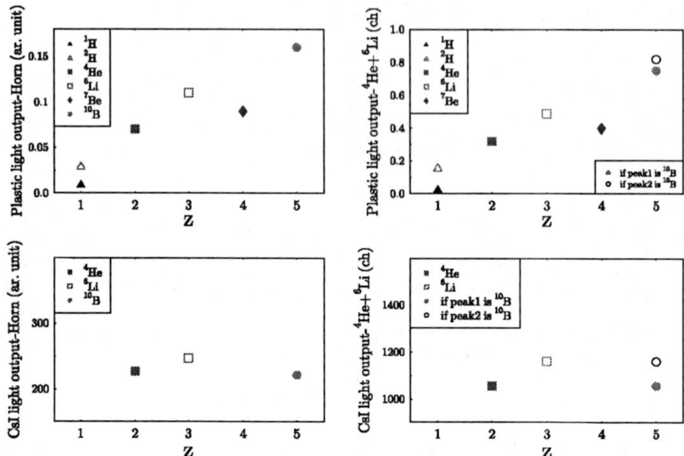

FIGURE 3. Left-top (bottom): Plastic(CsI) light output of the produced ions in $^6Li(^4He,\pi^0)^{10}B$ reaction using Horn setup parameters. Right-top (bottom): Plastic(CsI) light output of the produced particles using the AGOR facility.

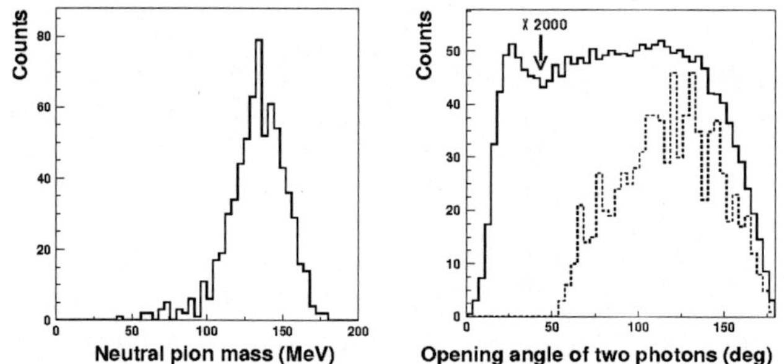

FIGURE 4. Pion mass (Left) and opening angle of two photons (Right) from neutral pion decay without (solid line) and with (dashed line) condition on the ^{10}B production.

1 in the pulse shape of the HI detector, is ^{10}B. In addition, the momentum distribution of peak number 1 in the HI detector is in agreement with the momentum distribution of ^{10}B from the kinematical calculations. The identification of ^{10}B is consistent with the TOF, energy loss, quenching of the energy in the scintillators and momentum distribution.

Examining events in the PB which were registered every time a ^{10}B nucleus was identified in the BBS, we have seen a clear enhancement of two-photon events. Furthermore these two-photons are primarily emitted with large opening angles as expected for the two-photon decay of π^0's, moving with small velocity in the laboratory (figure 4). In figure 4-right, the solid line represents the (down-scaled) two-photon production without any condition on fusion product, and the dashed line is related to the two-photon production in coincidence with ^{10}B. If we select those two-photon events with high opening angle requiring a coincidence with ^{10}B and calculate the two-photon invariant mass, we

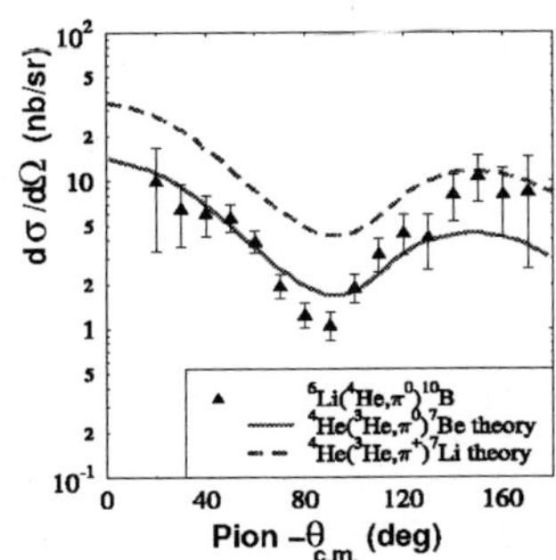

FIGURE 5. Angular distribution of the $^6Li(^4He,\pi^0)^{10}B$ reaction cross section at the incident energy $E_{lab}/nucleon = 59.1$ MeV.

obtain a peak around 135 MeV, which is the neutral pion mass (figure 4-left). After selection of pions and normalizing the number of produced pions to the beam current, target thickness, efficiencies of detector setup and life time of the data taking, the differential cross section is obtained (figure 5). The general behavior of the differential cross section is as predicted for $^4He(^3He,\pi^0)^7Be$ and $^4He(^3He,\pi^+)^7Li$ reactions taking clustering correlations into account. We can estimate that the total cross section of neutral pion production for the $^6Li(^4He,\pi^0)^{10}B$ reaction is as expected from the extrapolation of the total cross section for different reactions using clustering correlations [3].

SUMMARY

Two pionic fusion experiments have been performed using the AGOR facillity. The used experimental setup provides exclusive measurements by identifying all the products overdetermining the kinematics. Angular distribution of produced neutral pions has been obtained. The general behaviour of the cross section is in good agreement with calculations based on clustering correlations for the reactions with lighter target and projectiles. Further analysis is being performed.

We thank the AGOR accelerator staff for providing high-quality beam. The support of P. Dendooven and H. Timersma in operating the LHe target is gratefully acknowledged. This work was performed as part of the research program of the Stichting voor Fundamenteel Onderzoek der Materie (FOM) with financial support from the Nederlandse Organisatie voor Wetenschappelijk Onderzoek (NWO).

REFERENCES

1. A. Volya, S. Pratt, V. Zelevinsky, *Phys. Rev. C* **59**, 305 (1999).
2. P. Navratil, B.R. Barrett, *Phys. Rev. C* **54**, 2986 (1996).
3. T. Kajino, H. Toki, K. Kubo, *Phys. Rev. C* **35**, 1370 (1987).
4. L. Bimbot et al., *Phys. Lett. B* **114**, 311 (1982).
5. A. Baden et al., *Nucl. Instrum. Methods* **203**, 189 (1982).
6. D. Horn et al., *Nucl. Instr. Meth. A* **320**, 273 (1992).
7. D. Fox et al., *Nucl. Instr. Meth. A* **374**, 63 (1996).

Spin-Isospin Excitations and Muon Capture in Nuclei

V. A. Kuz'min* and T. V. Tetereva†

*Joint Institute for Nuclear Research, Dubna, Russia
†Skobeltsyn Institute of Nuclear Physics, Lomonosov Moscow State University, Moscow, Russia

Abstract. The muon capture in nuclei is considered as the tool for studing the weak axial vector coupling and the nuclear particle-hole residual interactions.

Keywords: beta-decay, Gamow-Teller strength function, residual interaction, muon capture
PACS: 21.60.Jz, 23.40.-s, 25.30.Mr

INTRODUCTION

The investigations of muon capture in nuclei could be very useful for the studies of two entangled problems: i) the determination of renormalization of weak axial-vector interaction in nuclear medium [1] and ii) the distribution of the strength of Gamow-Teller (GT) transitions in nuclei [2]. The product of the constant of axial-vector coupling, g_A, of weak nucleon current with the reduced probability of GT transition, $B(\mathrm{GT})$, determines the response of the nucleus in any weak interaction process. The examples important for nuclear astrophysics are the β-decay and the neutrino absorption in nuclei.

The ft values of allowed β-decay are connected with $B(\mathrm{GT})$ by simple ratio [3]

$$ft = 6166(3)/\left(g_V^2 B(\mathrm{F}) + g_A^2 B(\mathrm{GT})\right). \tag{1}$$

The numerical value in nominator is quoted from [4]. From ft values one can extract only the product $g_A^2 B(\mathrm{GT})$, and the independent determination either g_A or $B(\mathrm{GT})$ is necessary. It is common procedure to extract $B(\mathrm{GT})$ from the cross sections of charge-exchange (CEX) reactions such as (p,n). But the reliable determination of $B(\mathrm{GT})$ from the cross sections of CEX reactions is possible for $B(\mathrm{GT}) \geq 0.4$ only [5]. The corresponding β-decay values are $\log(ft) < 4.2$ for $|g_A/g_V| = 1.0$ or $\log(ft) < 4$ if $|g_A/g_V| = 1.26$. Another way to determine g_A from the experimental ft values is to use in Eq. (1) the $B(\mathrm{GT})$ calculated within certain nuclear model. The use of $B(\mathrm{GT})$'s, obtained within multiparticle shell model, to fit of allowed β-decay rates of sd-shell nuclei gives $g_A/g_V = -0.95$ [6]. In complex nuclei the attractive particle-particle residual interaction reduces greatly the strength of σt^+ transitions and allows one to describe ft values of β^+-decay of proton-rich nuclei with $g_A/g_V = -1.0$ or even with unquenched value $g_A/g_V = -1.26$ [7]. Later it was shown [8] that particle-particle interaction accelerated the Gamow-Teller β^--decay of neutron rich nuclei too much. The main difficulty in nuclear structure calculations related to the β decay is the determination of the strength of GT transitions participating in β-decay. The distribution of the strength of single-particle transition operator on the excitation energy (strength function) is mainly

CP972, *Exotic Nuclei and Nuclear/Particle Astrophysics (II)—Proceedings of the Carpathian Summer School of Physics 2007,* edited by L. Trache and S. Stoica

TABLE 1. The $\Lambda^{tot}(^{90}Zr)$ (in $10^5\,s^{-1}$) calculated with different sets of the constants of effective particle-hole residual interactions and $g_A = -1.26 g_V$.

g_P	$\kappa_1^{LJ} \cdot A =$	0.0	−0.23	−0.33	−0.43	−0.33	−0.33
g_A	$\kappa_1^{J} \cdot A =$	0.0	−0.42	−0.42	−0.42	−0.21	−0.63
2.0		137.9	95.99	90.37	86.89	91.31	90.33
4.0		132.8	92.42	87.06	83.73	88.00	87.02
6.0		128.3	89.29	84.15	80.95	85.08	84.11
8.0		124.5	86.59	81.63	78.54	82.57	81.59
10.0		121.3	84.33	79.51	76.50	80.45	79.47
12.0		118.7	82.50	77.78	74.84	78.72	77.75
14.0		116.7	81.10	76.46	73.54	77.39	76.42

determined by the interaction between particle-hole $(1p-1h)$ configurations. Also, it was shown that the effect of *quenched* (or *missed*) GT strength revealed in (p,n) reactions [2] should be result of the interaction between $1p-1h$ excitations, only [9, 10]. It means that nuclear residual forces must give rise to strong interaction between particle-hole pairs in which particle and hole are from different major shells. For GT transitions the interaction between $\Delta\mathcal{N} = 0$ and $\Delta\mathcal{N} \geq 2$ particle-hole states is the most important one. Here $\mathcal{N}$ is the principal quantum number of the shell. Presently the effect of missed GT strength is the main evidence of such particle-hole interaction, therefore it is interesting to see how it will work in others spin-isospin processes.

In this contribution we show that ordinary muon capture (OMC) $\mu^- + A(Z,N) \rightarrow B(Z-1,N+1) + \nu_\mu$ is useful for studying both the renormalization of weak axial coupling and the properties of spin-isospin particle-hole residual interaction in nuclei.

MUON CAPTURE AND AXIAL COUPLING

As a tool for determination of weak nucleon couplings muon capture has some advantages over β-decay. They are due to a large energy release of muon capture. First one is that we should not worry about the precision calculation of the transition strength within the beta-decay window because whole excitation spectrum of product nucleus participate in muon capture. The second advantage is that the energy of nuclear transition enters into muon capture rate, Λ_f, through neutrino energy, $E_\nu \approx (m_\mu + M_i - M_f)c^2 - E_f - \varepsilon_{1S}$, only. Here, m_μ, $M_{i,f}$, ε_{1S}, and E_f are the masses of muon, of initial and final nuclei, muon binding energy and excitation energy of product nucleus respectively. The influence of the error of the energy of final state, ΔE_f, on calculated OMC rate is reduced by a large neutrino energy $\frac{\Delta\Lambda_f}{\Lambda_f} \sim \frac{\Delta E_f}{E_\nu} \ll \frac{\Delta E_f}{E_f}$, and the requirements to the accuracy of nuclear structure calculations may be relaxed considerably. A general distribution of the transition strength over the excitation energy is requested for the calculation of the rate of muon capture.

The Table 1 presents the results of calculations of total muon capture rates in ^{90}Zr for several values of weak induced pseudoscalar coupling, g_P. The excitation energies and transition amplitudes were calculated within quasiparticle Random Phase Approx-

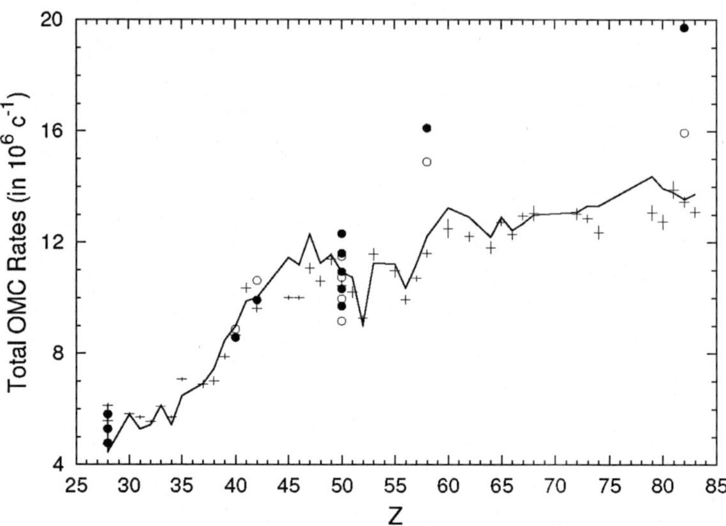

FIGURE 1. Total rates of OMC in complex nuclei. The experimental data are shown by crosses. Black and light circles present the results of calculations with $g_A(q^2=0) = -1.26\,g_V(q^2=0)$.

imation. The Hamiltonian of nuclear model, employed residual interactions and details of calculation are described in [11]. The energies of bound and quasi-bound single-particle states in Saxon-Woods shell potential were used. The κ_1^{LJ} and κ_1^{J} are the effective constants of particle-hole spin-multipole and multipole nuclear residual interactions respectively. The interaction which mixes $(\Delta\mathcal{N}=0)$ and $(\Delta\mathcal{N}\geq 2)$ particle-hole states was used in calculations. The comparison of the rates in fourth, sixth and seventh columns shows that Λ^{tot} does not practically depend on multipole residual interactions. Is is shown in [10] that the calculations with $\kappa_1^{01}/A = -0.33$ reproduces correctly the position of collective GT state and distribution of σt^- transition strength obtained from cross section of ^{90}Zr$(p,n,)^{90}$Nb in [12]. It is expected that $6.0 < g_P/g_A 8.0$ [13]. The experimental rate of OMC in the natural mix of zirconium isotopes is $(86.6\pm 0.8)\cdot 10^5\,\mathrm{s}^{-1}$ [14]. The rates in table show that the calculation even with "free" g_A value will slightly underestimate the experimental Λ^{tot}, because ^{90}Zr is the lightest stable isotope.

Total rates of OMC in several spherical complex nuclei are presented on the Fig. 1. Solid line connects the phenomenological estimations by Goulard-Primakoff formula with parameters fitted in [14]. The circles present Λ^{tot} calculated with different residual interactions. The results, obtained with particle-hole residual interaction which mixes $(\Delta\mathcal{N}=0)$ and $(\Delta\mathcal{N}\geq 2)$ particle-hole states are shown by the black circles. The light circles show the results of calculations with the interaction without this property [11]. The Figure shows that the theory describes reasonably the experimental Λ^{tot} in nuclei with $Z\leq 50$ $(A\leq 130)$ and goes above experiment for larger Z and A (the circles corresponding to ^{140}Ce and ^{208}Pb). In order to bring the theory to the experiment one can reduce the axial coupling, but this renormalization will be clearly model dependent.

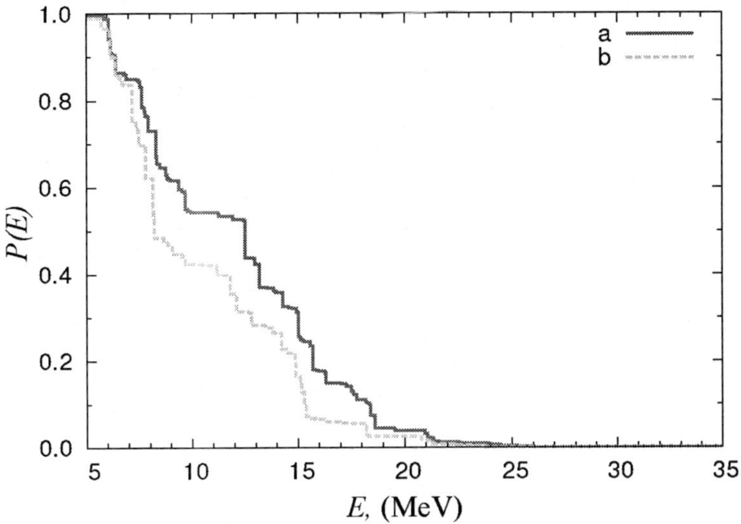

FIGURE 2. Normalized rates of muon capture in ^{208}Pb calculated with different residual interactions.

MUON CAPTURE AND RESIDUAL INTERACTION

It was said above, that the total rates of nuclear muon capture are mainly determined by spin-multipole transitions caused by the axial-vector part of weak current. So, one can expect to reduce the effect of possible in-medium modifications of g_A and reveal the dependence on the particle-hole interaction by considering the theoretical energy distribution of muon capture rates normalized by theoretical total rate

$$P(E) = 1 - \frac{1}{\Lambda^{\text{tot}}} \sum_{f:E_f \leq E} \Lambda_f. \tag{2}$$

The Fig. 2 shows $P(E)$ calculated with two residual interactions. The line "a" shows the rates obtained with interaction with strong interaction between ($\Delta\mathcal{N} = 0$) and ($\Delta\mathcal{N} \geq 2$) excitations and line "b" corresponds to the interaction without this mixing. According to Fig. 2 the interaction "a" produces more high excited states than the interaction "b". And the question is how to test this difference by the experimental data.

If we assume that neutron leaves the nucleus each time when it has sufficient energy (all decay channels except the neutron one are neglected), we can estimate the probability distribution of multiplicity of neutrons released after muon capture. The Table 2 contains the estimations of probabilities of emission k neutrons after muon capture in ^{208}Pb. As an example, the Table shows the probabilities of reactions $^{209}\text{Bi}(\mu, \nu\, kn)^{209-k}\text{Pb}$ determined in [15]. In this paper no proton decay and no alpha decay after muon capture in ^{209}Bi were observed. The listed energies of separation of k neutrons, S_{kn}, are reckoned from the ground state of target nucleus. If we assume that the last odd proton in ^{209}Bi

TABLE 2. Neutron multiplicity pattern for muon capture in ^{208}Pb.

Interaction or S_{kn}	Probability of k neutrons emission (in %)			
	$k=0$	$k=1$	$k=2$	$k\geq 3$
"a"	38	44	17	1
"b"	55	38	6	1
S_{kn}, (MeV)		9.305	16.147	22.650
Experiment on ^{209}Bi [15]	5	47	29	19
S_{kn}, (MeV)		5.092	12.459	19.197

does not participate in muon capture some observations can be made. The one neutron emission has the largest probability at the experiment as well as in calculation "a". The low probability of emission $k > 3$ neutrons show that there is definite lack of high excited states in the calculations with both interactions.

CONCLUSION

The calculated distribution of neutron multiplicities in reaction $^{208}\mathrm{Pb}(\mu, \nu\, kn)^{208-k}\mathrm{Tl}$ depends on the nuclear spin-isospin residual interaction in particle-hole channel.

The experimental pattern will be a strong test for nuclear spin-isospin residual interaction, and therefore for the effect of missed Gamow-Teller strength.

REFERENCES

1. I. S. Towner, and J. C. Hardy, in *The Nucleus as a Laboratory for Studying Symmetries and Fundamental Interactions,* Eds. E. M. Henley and W. C. Haxton, World Scientific Pobl. Co, New Jersey, London, Hong Kong, 1995, pp. 183–249.
2. F. Osterfeld, Rev. Mod. Phys. **64,** 491 (1992).
3. A. Bohr, and B. R. Mottelson, *Nuclear Structure, Vol. 1,* W. A. Benjamin, Inc., New York, Amsterdam, 1969.
4. E. K. Warburton, Phys. Rev. C **33,** 303 (1986).
5. S. A. Austin, N. Anantaraman, W. G. Love, Phys. Rev. Lett. **73,** 30 (1994).
6. B. A. Brown, B. H. Wildenthal, At. Data and Nucl. Data Tabl. **33,** 347 (1985).
7. V. A. Kuz'min, V. G. Soloviev, Nucl. Phys. A **486,** 118 (1988).
8. V. A. Kuz'min, T. V. Tetereva, Yad. Fiz. **56,** no. 8, 76 (1993) [Phys. Atom. Nucl. **56,** 1035 (1993)].
9. K. Junker, V. A. Kuz'min, and T. V. Tetereva, Eur. Phys. J. A **5,** 37 (1999).
10. V. A. Kuz'min and T. V. Tetereva, Fiz. Elem. Chast. At. Yadra **37,** 493 (2006) [Phys. Part. Nucl. **37,** 268 (2006)]
11. V. A. Kuz'min, T. V. Tetereva, K. Junker, and A. A. Ovchinnikova, J. Phys. G: Nucl. Part. Phys. **28,** 665 (2002).
12. T. Wakasa, *et al.,* Phys. Rev. C **55,** 2909 (1997).
13. T. P. Gorringe, H. W. Fearing, Rev. Mod. Phys. **76,** 31 (2004).
14. T. Suzuki, D. F. Measday, and J. P. Roalsvig, Phys. Rev. C **35,** 2212 (1987).
15. D. F. Measday, T. J. Stocki, and H. Tan, Phys. Rev. C **75,** 045501 (2007).

Recent Applications of the THM to the AGB Star Nucleosynthesis

M. La Cognata*, C. Spitaleri*, R. Tribble†, T. Al-Abdullah†, A. Banu†, S. Cherubini*, V. Crucillà*, C. Fu†, V. Goldberg†, M. Gulino*, L. Lamia*, A.M. Mukhamedzhanov†, R.G. Pizzone*, S.M.R. Puglia*, G.G. Rapisarda*, S. Romano*, M.L. Sergi*, G. Tabacaru†, L. Trache†, A. Tumino* and Y. Zhai†

*Istituto Nazionale di Fisica Nucleare, Laboratori Nazionali del Sud and Università degli Studi di Catania, Catania, Italy
†Cyclotron Institute, Texas A&M University, College Station, TX 77843, USA

Abstract. The THM has been applied to the study of reactions involved in the fluorine production in AGB stars with the aim of extending these measurements down to the Gamow peak. Fluorine abundance is important since it allows to constrain mixing models from the comparison of the observed fluorine abundances with the ones predicted by models. In particular the study focuses on the $^{15}N(p,\alpha)^{12}C$ and the $^{18}O(p,\alpha)^{15}N$ reactions which can influence fluorine yield as they are part of ^{19}F production (and destruction) network.

Keywords: Nuclear reactions, indirect techniques, AGB star nucleosynthesis
PACS: 24.10.-i Nuclear reaction models and methods – 26.20.+f Hydrostatic stellar nucleosynthesis

INTRODUCTION

An open question in nuclear astrophysics is ^{19}F abundance in the Milky Way [1]. Even its production sites are still a debated topic: it has been shown that this isotope is synthesized in type II supernovae explosions as well as in Wolf-Rayet stars and during the AGB phase of stellar evolution [1]. In particular it was shown that it is not possible to explain the observed abundances of ^{19}F in Milky Way giants without taking into account its production in the AGB phases [2]. Large efforts were then devoted to understand how ^{19}F nucleosynthesis proceeds in AGB's since fluorine abundances observed in giants can constrain AGB star models [2]. Anyway large uncertainties characterize ^{19}F nucleosynthesis since experimental ^{19}F abundances are not reproduced by current AGB models [2].

In this context it has been shown that a key isotope for ^{19}F production in AGB stars is ^{15}N and a crucial role is played by the $^{15}N(p,\alpha)^{12}C$ reaction, removing both proton and ^{15}N nuclei from ^{19}F production chain in the AGB intershell environment. The available reaction rates give at least an 8% uncertainty in the fluorine surface abundance [2], due to a factor 2 difference between NACRE [3] and CF88 [4] reaction rates usually employed in calculations.

On the other hand the $^{18}O(p,\alpha)^{15}N$ reaction can affect fluorine yield from AGB stars since it produces ^{15}N nuclei which are later burnt to ^{19}F through the $^{15}N(\alpha,\gamma)^{19}F$ cap-

CP972, *Exotic Nuclei and Nuclear/Particle Astrophysics (II)—Proceedings of the Carpathian Summer School of Physics 2007,* edited by L. Trache and S. Stoica

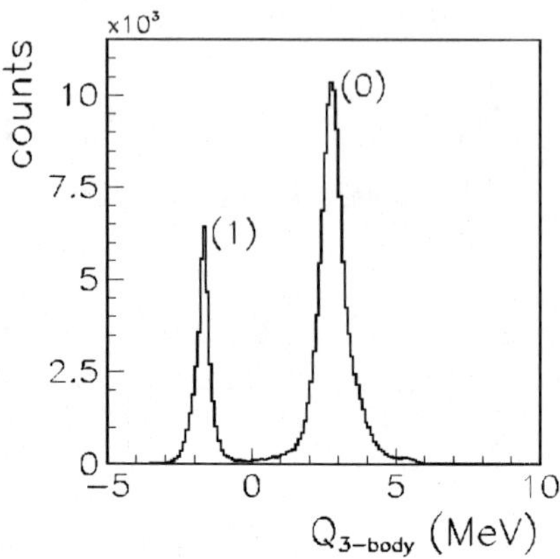

FIGURE 1. Q–value spectrum for the three-body process.

ture reaction during a thermal pulse [5]. Though this reaction has been extensively studied both in direct [6] and indirect [7] ways, the reaction parameters of some important resonances are not well established yet. In particular, the width and the strength of the 656 keV resonance are poorly known, though this resonance provide a non negligible contribution to the reaction rate both at high and low temperatures (around about 10^8 K), being about 300 keV wide [3].

An additional relevant source of uncertainty is the enhancement of the S-factor due to the electron screening effect. The target nuclei and the projectiles are usually in the form of neutral atoms or molecules and ions, respectively. Because of the electron clouds surrounding the interacting nuclei, the projectile effectively sees a reduced Coulomb barrier. This, in turn, leads to a higher cross section for screened nuclei, $\sigma_s(E)$, compared to the cross section one would get in the case of bare nuclei.

The Trojan-horse Method (THM) is a powerful technique allowing to measure the low-energy trend of the astrophysical S(E)-factor by selecting the quasi-free (QF) contribution to a suitable three-body process. This is because the QF reaction is studied at energies well above the Coulomb barrier thus it does not prevent the exploration of the astrophysically relevant energy region. For the same reason the bare nucleus cross section is deduced, which is the relevant input to astrophysical codes. More details about the THM are given in [8].

THE ^{15}N$(p,\alpha)^{12}$C REACTION

The experiment was performed at Texas A&M University Cyclotron Institute. Further details on the experiment and the data analysis are reported in [9]. The ^{15}N$(p,\alpha)^{12}$C S(E)-factor has been deduced from the ^{2}H(^{15}N,α^{12}C)n three-body process. Before proceeding to the extraction of the astrophysical factor two steps were necessary: (i) by investigation of the kinematics of the three-body reaction the α_0 channel (which is the relevant one for astrophysics [10]) was disentangled from others; (ii) from the study of

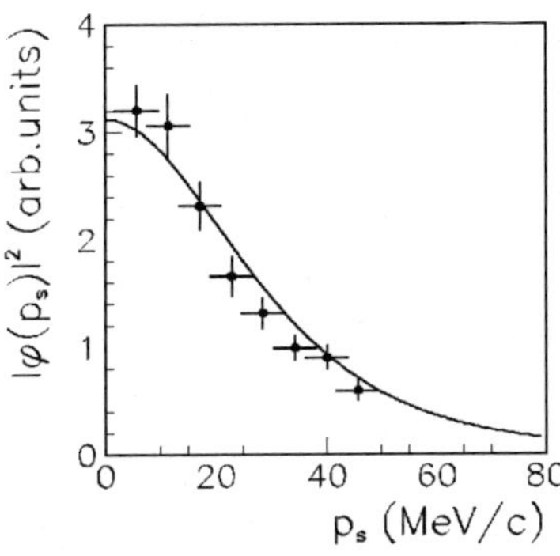

FIGURE 2. The momentum distribution deduced from the present data set compared with the behaviour of the Hulthén function.

the three-body reaction dynamics and in particular from the analysis of the experimental neutron momentum distribution the QF contribution to the total cross section was selected.

In order to single out the contribution of the ^{2}H(^{15}N,α_0^{12}C)n process a number of conditions was imposed to the data set. In detail, coincidences are selected by introducing a window on the Time-to-Amplitude Converter (TAC) parameter and the carbon locus was identified in the $\Delta E - E$ spectra. An additional cut was introduced in the $E_{^{12}C}$ vs. E_α 2D spectrum to single out the kinematic locus of the three-body reaction. Finally the $Q-$value spectrum of the three-body process was deduced through energy conservation (see figure 1). Two peaks showed up, corresponding to the ^{2}H(^{15}N,α_0^{12}C)n channel (the one at higher energy, $Q_{3-body} = 2.74$ MeV) and to the ^{2}H(^{15}N,α_1^{12}C)n channel ($Q_{3-body} = -1.7$ MeV). The absence of additional peaks and the good agreement between the theoretical and experimental $Q-$values made us confident that additional reaction channels had been disentangled from the ^{2}H(^{15}N,α_0^{12}C)n coincidence yield and that an accurate calibration procedure had been performed. Only the events belonging to the high energy peak were retained for further analysis.

The study of the reaction mechanisms feeding the exit channel was then necessary to disentangle the QF component from others. In fact the analysis of the experimental results is often complicated by the presence of other reaction mechanisms yielding the same particles in the final state, e.g. sequential decay (SD) [11]. In the present analysis it was found that in the chosen phase space region (within a neutron momentum window $|\vec{p}_s| < 40$ MeV/c) SD processes gave negligible contributions to the coincidence yield in the astrophysically relevant energy region. Indeed, if direct break-up occurs, the experimental momentum distribution should resemble the one neutron has inside deuteron, given by the Fourier transform of the intercluster wave function in the PWIA approach. The momentum distribution deduced from the present data set is displayed in figure 2. The theoretical one, the square of the Hulthén wave function in momentum space, is superimposed to the data (black line in figure 2). The normalization constant, which is the only adjustable parameter, was fitted to the experimental distribution. The nice agreement between the two demonstrated that in the $|\vec{p}_s| < 40$ MeV/c momentum

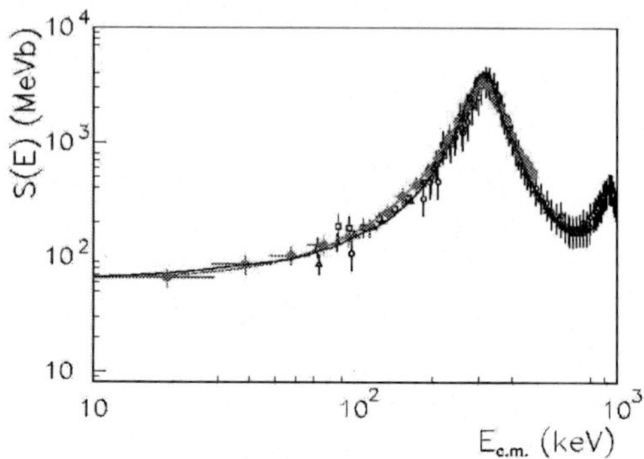

FIGURE 3. Astrophysical S(E)-factor extracted via THM (red dots) compared with direct data (black symbols) from [12].

window the QF contribution was dominant thus neutron could be regarded as a spectator to the ^{15}N$(p,\alpha)^{12}$C subreaction.

The astrophysical S-factor was finally deduced and compared to the ones in literature [12, 13]. Normalization to direct data was deduced by equating the areas subtended by the 314 keV peak, due to the 12.44 MeV 1^{-} ^{16}O excited state, in both data sets. The results for the astrophysical S(E)-factor are shown in figure 3. THM and direct data are displayed as full red and black symbols respectively. The red and black lines represent instead Breit-Wigner best-fits to THM and direct data from [12]. A good agreement is apparent in the energy region where direct data are available.

THE ^{18}O$(p,\alpha)^{15}$N REACTION

The ^{18}N$(p,\alpha)^{15}$N S(E)-factor has been deduced from the analysis of the ^{2}H(^{18}O,α^{15}N)n three-body process (see [14] for a more detailed discussion). The experimental procedure is similar to the one reviewed in the previous section. The experiment was performed at the Texas A&M University Cyclotron Institute. After detector calibration, the first step in the analysis was the selection of the ^{2}H(^{18}O,α^{15}N)n channel. To this purpose the three-body kinematics was studied in order to disentangle the contribution of the channel of interest from others. The Q-value spectrum was deduced, showing a single peak, which demonstrated that the ^{2}H(^{18}O,α^{15}N)n channel only contributed to the three-body coincidence yield.

A three-body process can proceed through different reaction mechanisms besides the quasi-free one as already discussed. In the THM approach, the QF mechanism has to be disentangled in order to extract the cross-section for the two-body process. This is accomplished by analyzing the reaction dynamics. A quantity which turns out to be strongly dependent on the reaction mechanism is the shape of the momentum distribution of participant - spectator relative motion in the THM nucleus (here, $p-n$ in the deuteron). Indeed, only if the deuteron break-up process is direct and neutron is emitted without interacting with the ^{18}O$+p$ subsystem, does the neutron momentum

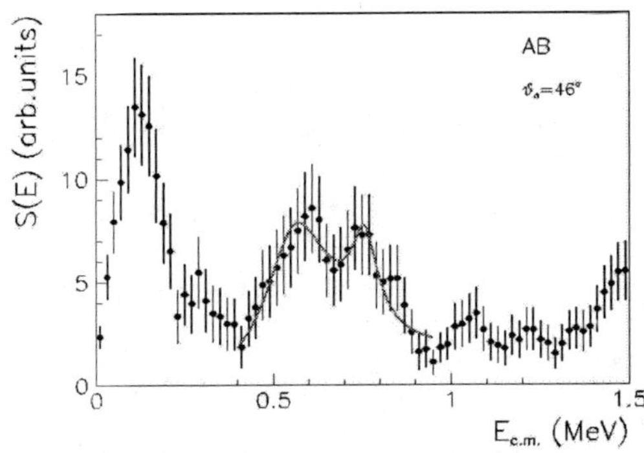

FIGURE 4. Indirect astrophysical S-factor extracted from the A-B coincidences (black dots). The red line represents a fit of the two levels at 656 *keV* and 799 *keV* including interference effects.

distribution keep the same shape as inside d. As in the previous case, a good agreement was found between the theoretical and experimental momentum distributions, making us confident about the presence of the QF mechanism, which was the dominant process at the energies of interest.

Finally, the astrophysical S(E)-factor was extracted for each available $\theta_{c.m.}$. In figure 4 a typical spectrum obtained for $\theta_{c.m.} = 46^\circ$ is displayed. Similar spectra were derived for other $\theta_{c.m.}$ angles in the $30^\circ - 150^\circ$ range. The resonance parameters were deduced for each $\theta_{c.m.}$ angle in this preliminary analysis by fitting the THM S(E)-factor (in arbitrary units) with the sum of two Breit-Wigner functions, taking into account the 656 keV and 799 keV resonances, plus an interference term. Regarding the strength of the 656 keV resonance, our approach lead to: $(\omega\gamma)_{656}/(\omega\gamma)_{799} = 14.8 \pm 2.5$ (where the cited error is only statistical), to be compared with ~ 0.5 [3], ~ 3 [6] and ~ 0.4 [15]. Since the ratio of the two strength is evaluated, no systematic uncertainties from the normalization procedure can affect the deduced $(\omega\gamma)_{656}$.

REFERENCES

1. A. Renda, et al., *Mon. Not. R. Astron. Soc.* **354**, 575 (2004)
2. M. Lugaro, et al., *Ap. J.* **615**, 934 (2004)
3. C. Angulo, et al., *Nucl. Phys.* **A656**, 3 (1999)
4. G.R. Coughlan, and W.A Fowler, *At. Data Nucl. Data Tables* **40**, 283 (1988)
5. A. Jorissen, et al., A&A **261**, 164 (1992)
6. H. Lorentz-Wirzba, et al., *Nucl. Phys.* **A313**, 346 (1979)
7. A.M. Champagne, and M.L. Pitt, *Nucl. Phys.* **A457**, 367 (1986)
8. C. Spitaleri, et al., this volume
9. M. La Cognata, et al, *Phys. Rev. C*, in press (2007)
10. C. Rolfs, and W.S. Rodney, *Nucl. Phys.* **A235**, 450 (1974)
11. M. La Cognata, et al., *Phys. Rev. C* **72**, 065802 (2005)
12. A. Redder, et al., *Z. Phys.* **A305**, 325 (1982)
13. A. Schardt, et al., *Phys. Rev.* **86**, 527 (1952)
14. M. La Cognata, et al., *J. Phys. G Nucl. Part. Phys.*, in press (2007)
15. K. Yagi, et al., *J. Phys. Soc. Japan*, **17** 595 (1962)

Cosmic Ray Acceleration in Astrophysical Environments

A. Meli and A. Mastichiadis

Department of Physics, National University of Athens, Panepistimiopolis, Zografos 15783, Greece

Abstract. Cosmic ray acceleration in astrophysical shocks is important. We discuss these shocks' properties and we perform Monte Carlo simulations studying their efficiency in accelerating particles (i.e. protons or electrons) up to very high energies with an application to astrophysical regions such as Super Novae, Active Galactic Nuclei hot spots and Gamma Ray Bursts. The efficiency of the acceleration mechanism at shocks varies in regard to the inclination of the magnetic field to the shock normal (e.g. subluminal shocks, superluminal shocks). This fact could have consequences to the contribution of the very high energy cosmic rays to the observed cosmic ray spectrum and the relevant neutrino and gamma-ray emission.

Keywords: cosmic rays, shocks, acceleration mechanism
PACS: 95.85.Ry, 98.70.-f, 98.70.Sa, 98.70.Rz, 98.54.-h

INTRODUCTION

It is the work of a number of authors [10, 4, 5, 1, 6] which established in principle the basic mechanism of particle diffusive namely Fermi acceleration in non-relativistic shocks. Astrophysical collisionless plasma shocks can accelerate charged particles into high energies resulting into the observed cosmic ray spectrum. It is widely accepted today that this process can account for the majority of the energetic, non-thermal particle distributions that we infer in various astrophysical environments. Considerable work has been done analytically and numerically on this subject, however, questions still need to be answered on issues concerning the very nature of this mechanism especially at relativistic shock speeds. We will discuss the properties of astrophysical shocks and we will present Monte Carlo simulations studying the shocks' efficiency in accelerating particles (i.e. hadrons or leptons) up to very high energies. These simulations include subluminal and superluminal shock acceleration with an aim to draw conclusions on the primary spectra of the relevant astrophysical sources such as Supernovae Remnants, Active Galactic Nuclei (AGN) hot spots and Gamma Ray Bursts (GRBs). In the superluminal shocks the particles are accelerated in a shock drift, since there is no possibility to transform into the de Hoffman-Teller (HT) frame [8], where $\vec{E} = \vec{0}$. Further to that, it is known that for the transformations from the normal shock frame (NSH) frame to the HT frame, we need to boost by a V_{HT} speed along the shock frame, where $V_{HT} \leq V_{NSH} \tan\psi$. Due to this physical causality, this transformation is only possible if V_{HT} is less or equal to the speed of light, $c = 1$. This means that when $\tan\psi \leq 1$, then one treats the subluminal shock transformation case. Therefore, for all other cases where the inequality of the HT fails to hold, due to a very high inclination in combination with a high shock velocity, the superluminal shock condition is applied.

CP972, *Exotic Nuclei and Nuclear/Particle Astrophysics (II)—Proceedings of the Carpathian Summer School of Physics 2007,* edited by L. Trache and S. Stoica

NUMERICAL SIMULATIONS

Our codes simulate the particle transport equation, which in our case is for mild and relativistic shock velocities. The appropriate time independent Boltzmann equation is given by $\Gamma(V+\upsilon\mu)\frac{\partial f}{\partial x}=\frac{\partial f}{\partial t}|_c$, where V is the fluid velocity, υ the velocity of the particle, Γ the Lorentz factor of the fluid frame, μ the cosine of the particle's pitch angle and $\frac{\partial f}{\partial t}|_c$ the collision operator. The frames of reference used in the code are the local fluid frame, the normal shock frame and the de Hoffman-Teller frame as we mentioned in the previous section. Pitch angle is measured in the local fluid frame, while the value x is the distance of the particles to the shock front, where the shock is assumed to be placed at $x=0$. In order for the above equation to be solved by the Monte Carlo technique, we assume that first, the collisions represent scattering in pitch angle and secondly, the scattering is elastic in the fluid frame. The simulation begins by injecting 10^4 particles far upstream. First order Fermi (diffusive) shock acceleration can be simulated provided that the particles' guidance center undergoes multiple scatterings with the assumed magnetized media and in each shock-crossing the particles gain an amount of energy. Provided multiple scattering between the upstream and downstream regions of the shock occurs, the particles gain energy in each shock-crossing so in that way diffusive shock acceleration is simulated. The mean free path λ is calculated in the respective fluid frames (i.e. upstream or downstream) assuming a momentum dependence to the mean free path of the particle such as, $\lambda=\lambda_o\cdot p$, where $\lambda_o=10\cdot r_g$, which is related to the spatial diffusion coefficient, κ. Specifically for relativistic oblique shocks, in the shock normal which lies in the x direction, the diffusion coefficient κ, equals to $\kappa_{\parallel}cos^2\psi$, where $\kappa_{\parallel}=\lambda v/3$ and ψ is the inclination of the shock normal to the magnetic field lines. A pitch angle diffusion algorithm is used (e.g. see [12]).

The results shown here are given at the downstream side of the shock frame. It is important to note that the flow into and out of the shock discontinuity is not along the shock normal, but a transformation is possible into the shock frame to render the flows along the normal (e.g. [3]) therefore we start off the simulations by this frame. During our simulations continuous Lorentz transformations are performed from the local fluid rest frames to the shock frame to check for shock-crossings. Furthermore, for the superluminal shock conditions (see [12]) we consider a helix-trajectory motion of each test-particle of momentum p, in the fluid frames upstream or downstream respectively, where the velocity coordinates (v_x, v_y, v_z) of the particle are calculated in a three dimensional space. In principle, we follow the helix-trajectory of each particle in time t, where t is the time from detecting the shock intersection at x, y, z. Further details of this method can be found in the work mentioned above. All particles leave the system if they escape far downstream at the spatial boundary r_b, or alternatively, if they reach a specified maximum energy E_{max}, for computational convenience, even though other physical parameters describing particle energy loss would need to be taken into account [16]. The downstream spatial boundary required, can be initially estimated from the solution of the convection-diffusion equation in a non-relativistic, large-angle scattering approximation in the downstream plasma which gives the chance of return to the shock as, $exp(-V_2 r_b/x_i)$. In fact, we have performed many test runs with different spatial boundaries to investigate the effect of the size of the acceleration region on the

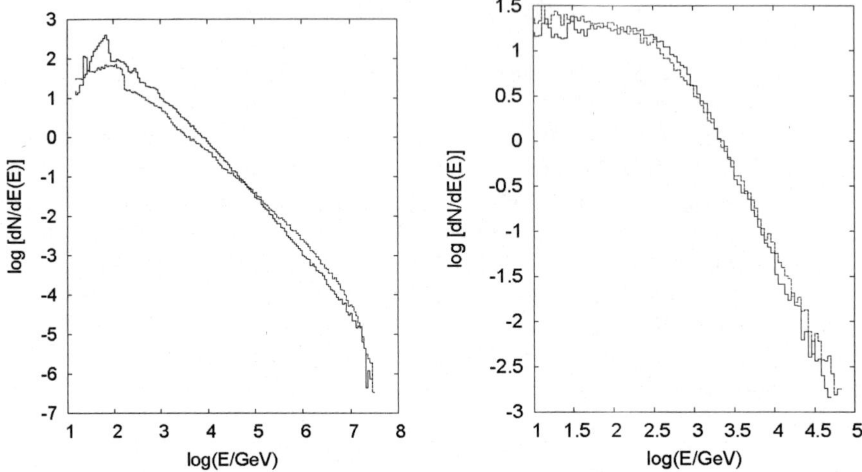

FIGURE 1. Left: Two subluminal shock spectra, one for a non-relativistic shock of a gamma factor equal to 1.0001 (i.e. V=0.01c) and the other for a gamma equal to 5.0 (i.e. V=0.98c) with an inclination angle of 89° and 39° respectively. One can see the almost perfect superposition of the spectra, showing the equivalence of acceleration efficiency between the non-relativistic *perpendicular* shock and the relativistic subluminal oblique one. Right: Two superluminal shock spectra for a gamma factor of 500, one for an inclination angle of 50° and the other for an inclination of 89°. The spectra reach a maximum of $\sim 10^5$GeV for protons. The cut-off in the energy is due to the stream flow sweeping the particles downstream, limiting their chance of returning to the shock.

spectrum, as well as to find a region where the spectrum is size independent. We note that in the pitch angle diffusion case, the inherent anisotropy due to the high downstream sweeping effect may greatly modify this analytical estimate.

RESULTS

We performed a series of Monte Carlo simulations and we show two representantive graphs in figure 1. Specifically, two subluminal shock spectra are shown one for a non-relativistic shock of a Lorentz gamma equal to 1.0001 (i.e. V=0.01c) and the other for a Lorentz gamma equal to 5.0 (i.e. V=0.98c) with an inclination angle of 89° and 39° respectively. One can see the almost perfect superposition of the spectra, showing the equivalence of the acceleration efficiency between the non-relativistic *perpendicular* shock and the relativistic subluminal oblique one. The above correlation of efficiency between non-relativistic perpendicular shocks and mild relativistic oblique ones, applies to many other shock spectra we simulated, but as long as the non-relativistic shocks are perpendicular and the "Jokipii limit" applied for the particle diffusion. Further to that, the reader refers to [14] where it is shown that non-relativistic perpendicular shocks are efficient accelerators for cosmic rays reaching energies as high as $\sim 10^8$GeV, because the acceleration in those shocks seems to be faster under certain diffusion conditions [9].

For the case shown here, the maximum energy corresponds to $\sim 10^{7.5}$GeV for protons. One further result of the simulations is that for shock speeds below a Lorentz factor of Γ=5 (i.e. V=0.98c), as higher the inclination of the shock as greater its efficiency to accelerate particles up to $\sim 10^{7.5}$GeV (i.e. protons, iron nuclei). Moreover, as [12, 15] have shown, subluminal highly relativistic shocks are more efficient accelerators than the non-relativistic shocks, resulting in an average cosmic ray energy gain of $\sim \Gamma^{3.4}$ after two complete shock-cycles. For this case, Γ is the boost factor of the shock, while in the simulations the test-particles are considered already relativistic with an initial Lorentz gamma, $\gamma = \Gamma+100$. Furthermore, in the same figure on the right side one sees the two superluminal shock spectra for a gamma factor of 500, one for an inclination angle of 50° and the other for an inclination of 89°. Simulations show that the inclination of the shock in the superluminal cases does not change the results regarding the spectral shape and the efficiency of the acceleration. The 'early' cut-off in the spectra is due to the flow, sweeping the particles downstream limiting their chance to return to the shock, diminishing the chances for these shocks to be efficient cosmic ray accelerators. We note here that one may see the smooth power-law formation of the above spectra in contrast to the subluminal spectra formations in e.g. [12, 15]. Simulations show that even for very high Lorentz boost factors the particles accelerated in superluminal shocks, do not reach very high energies (e.g. $\sim 10^5$GeV for protons) since almost all of them are advected downstream, after their helix trajectory performs a shock-cycle crossing. All the simulated superluminal shock spectra can be well fitted by a power-law with a spectral index ranging between $\sim 2.0 - 2.3$. Nevertheless our simulations show that subluminal relativistic shocks appear flatter as higher the Lorentz boost factor of the shock.

DISCUSSION AND CONCLUSIONS

We discussed about the mechanism of diffusive shock acceleration and we performed simulations in order to test the behaviour and the efficiency of subluminal and superluminal relativistic shocks in a quantitative comparison to non-relativistic ones. It is shown that superluninal shocks are not as efficient first order Fermi cosmic ray accelerators as the subluminal ones. Superluminal shocks can be well fitted by a power-law with a spectral index ranging between $\sim 2.0 - 2.3$ in contrast to the relativistic subluminal ones which give flatter spectra. Specifically, the superluminal shocks are efficient in accelerating particles up to a maximum energy of around 10^5GeV (i.e. protons or iron nuclei). On the other hand, the subluminal relativistic oblique shocks are better accelerators reaching maximum energies between 10^7GeV and 10^{10}GeV for higher Lorentz boost shock factors. Further to that, we refer the reader to [2, 17, 15] where one can see spectral features developing for very high subluminal relativistic shocks, indicating a sensitivity connecting to the kinematical details of the particle diffusion while crossing the relativistic shock front.
Moreover, we showed that perpendicular non-relativistic shocks seem to be as efficient accelerators as subluminal mild relativistic ones, as long as the "Jokipii limit" [9] for the particle diffusion was used. The consequences of the results presented here are discussed in [15] where the possible contribution of AGN and GRBs sources to the ultra high en-

ergy cosmic rays (UHECR) flux is calculated, as well how the simultaneous radiative losses could alter the final cosmic ray energy spectrum output [16]. We note that simulation calculations show that for sources where relativistic superluminal shocks occur, a contribution to the UHECR flux is not significant, using current neutrino flux limits from measurements to normalise the spectrum to the highest possible contribution. In the case of sources where relativisitc subluminal shocks take place, the spectrum could be normalised to the cosmic ray flux above the ankle, $E_{\rm min} = 5 \cdot 10^9$ GeV. Comparison with observation seems to suggest that subluninal shocks in AGN are the most likely source of UHECR. On the other hand, if one assumes that UHECR are accelerated in the external shocks of the early expansion phase in GRB, their spectra would resemble the AGN cosmic ray spectra and a contribution from these sources to the UHECR flux could be possible.

ACKNOWLEDGMENTS

The project is funded by the European Social Fund and National Resources (EPEAEK II) PYTHAGORAS.

REFERENCES

1. Axford, W.I., Leer, E. and Skadron, G. 1978, 15th ICRC Plovdiv, Bulgaria, Proc. 11, 132
2. Baring, M. G., 2004, Nucl.Phys.B.Suppl. 136, 198
3. Begelman, M. C. and Kirk, J. G., 1990, ApJ 353, 66
4. Bell, A. R., 1978a, MNRAS 182, 147
5. Bell, A. R., 1978b, MNRAS 182, 443
6. Blandford, R.D. and Ostriker, J.P., 1978, ApJLet 221, L29
7. Dingus, B. L., 1995, APSS 231, 187
8. de Hoffmann, F. and Teller, E., 1950, Pr 80, 692
9. Jokipii, J. R., 1987, ApJ 313, 842
10. Krymsky, G. F., 1977, Dokl.Acad.Nauk. SSSR 234, 1306
11. Machalski, J., Chyży, K. T., Stawarz, Ł. and Kozieł, D., 2007, A&A 462, 43
12. Meli, A. and Quenby, J. J., 2003b, APh 19, 649
13. Meli, A., Quenby J. J., Becker, J., 2005, ICRC 3, 229
14. Meli, A. and Biermann, P. L., 2006, A&A 454, 687
15. Meli, A., Becker, J., Quenby, J. J., 2007, submitted in A&A
16. Meli, A. and A. Mastichiadis, to be submitted in A&A
17. Stecker, F. W., Baring, M. G. and Summerlin, E. J., 2007, astro-ph/0707.4676

Landau-Zener Effect In The Time-Dependent Hartree-Fock-Bogoliubov Equations

M. Mirea

Institute of Physics and Nuclear Engineering, P.O. Box MG-6, Bucharest-Magurele, Romania

Abstract. If the Hartree-Fock-Bogoliubov equations of motion are solved in the case of seniority one nuclear systems, the unpaired nucleon remains on the same orbital. The blocking effect hinders the possibility that the unpaired nucleon skips from one orbital to another. A new set of coupled channel equations based on the time dependent Hartree-Fock-Bogoliubov formalism is deduced. This system includes a mechanism similar to the Landau-Zener effect, allowing that the unpaired nucleon jumps from one level to another.

Keywords: Landau-Zener effect; Time-dependent Hartree-Fock-Bogoliubov equations.
PACS: 24.10.Eq. Coupled channel and distorded wave models.

INTRODUCTION

The Time Dependent Hartree-Fock-Bogoliubov (TDHFB) method allows a description of the dynamics of the nucleus. In the case of even-even system, this method represents a tool to estimate the dissipation during disintegration processes [1]. Unfortunately, for seniority one nuclear systems, during the deformation of the nucleus from its initial state up to scission, the unpaired nucleon remains located on the same orbital. The blocking effect hinders the possibility to skip from one orbital to another. This behavior leads to the unpleasant feature that a system, even moving infinitely slowly, could not end up in its ground state. On another hand, in a single-particle description, the Landau-Zener effect allows that the unpaired nucleon skips from one level to another in some avoided level crossing regions. The probabilities that the unpaired nucleon ends up in different final states can be computed by using a system of coupled channel equations that characterizes the microscopic motion. A way to introduce a similar behavior for the unpaired nucleon in superfluid systems is investigated. The TDHFB equations will be generalized in order to include the Landau-Zener effect. The classical TDHFB equations and the equations that govern the Landau-Zener effect will be obtained as particular cases of new equation.

LANDAU-ZENER EFFECT

The single-particle levels are function of the deformation parameters that characterize the shape of a nucleus. Levels characterized by the same quantum numbers associated to some symmetry of the system cannot cross and exhibit avoided

CP972, *Exotic Nuclei and Nuclear/Particle Astrophysics (II)—Proceedings of the Carpathian Summer School of Physics 2007,* edited by L. Trache and S. Stoica

level crossings. The transition probability of one nucleon from one adiabatic level to another adiabatic level is strongly enhanced in an avoided crossing region. This promotion mechanism is known as the Landau-Zener effect. In Fig. 1 the avoided crossing between two adiabatic levels e_k and e_i is displayed. The diabatic levels are ε_m and ε_j. If the variation of the generalized coordinate is produced slowly and the nucleon is initially located on the level e_i, after the passage of the avoided crossing region, the nucleon will practically remain on the same adiabatic level. In this case, the motion is adiabatic and the nucleon follows the adiabatic state e_i. If the variation of the generalized coordinate is produced suddenly, then the nucleon will skip with a great probability on the level e_k. The motion is diabatic, and the nucleon follows the diabatic state ε_j. A formalism can be used to obtain the promotion probability. Assuming an n-state approximation, the wave function of the unpaired nucleon can be formally expanded in a basis of n diabatic wave functions $\varphi_i(r)$ as:

$$\Psi(r,t)=\sum_i^n c_i(t)\varphi_i(r)\exp\left(-\frac{i}{\hbar}\int_0^t \varepsilon_{ii}dt\right)$$

where the matrix elements with the diabatic states are abbreviated as follows,

$$\varepsilon_{ij}=\langle\varphi_i|H|\varphi_j\rangle$$

where H is the Hamiltonian and c_i amplitudes. Inserting ψ in the time-dependent Schrödinger equation,

$$\langle\varphi_i|H-i\hbar\frac{\partial}{\partial t}|\Psi\rangle=0$$

the following system of coupled equations is obtained:

$$\dot{c}_i=\frac{1}{i\hbar}\sum_{j\neq i}^n c_j\varepsilon_{ij}\exp\left(-\frac{i}{\hbar}\int_0^t(\varepsilon_{jj}-\varepsilon_{ii})dt\right) \quad (1)$$

with $p_i=|c_i|^2$ being the probability to find the unpaired nucleon on the level i.

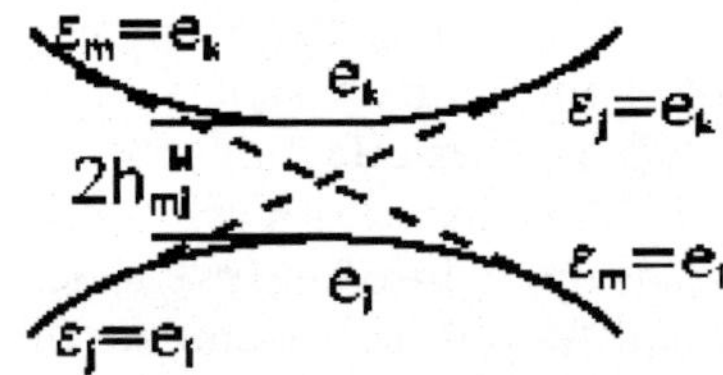

FIGURE 1: Ideal avoided crossing region.

SUPERFLUID SYSTEMS

Using quasiparticle creation and annihilation operators α_k^+ and α_k

$$\alpha_{\bar{k}}^+=u_k a_k-v_k a_{\bar{k}}^+;\quad \alpha_k^+=u_k a_{\bar{k}}+v_k a_k^+;$$

$$\alpha_{\bar{k}}=u_k^* a_k^+-v_k^* a_{\bar{k}};\quad \alpha_k=u_k^* a_{\bar{k}}^++v_k^* a_k,$$

it is possible to construct the operators that help us to promote the nucleon from one diabatic level to another as displayed in Fig. 2. The two situations plotted in Fig. 2 can be modeled. Here a_k^+ and a_k denote operators for creating and destroying a particle in the state k, respectively. The state characterized by a bar signifies the time-reversed partner of a pair. The parameters v_k and u_k are the occupation and vacancy amplitudes, respectively. Because only the relative phase between the parameters u_k and v_k matters, in the following u_k is considered as a real quantity and v_k a complex one.

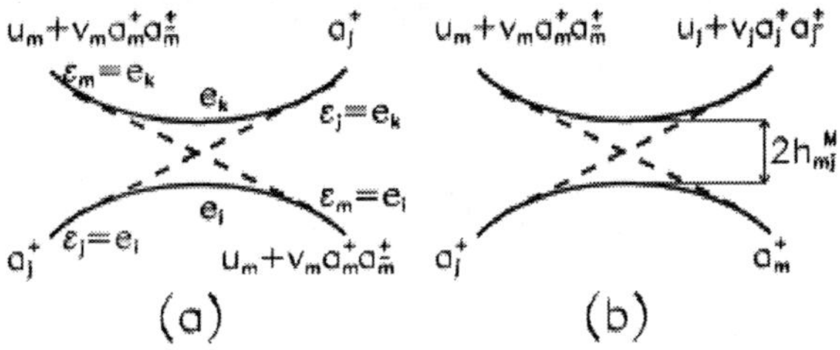

FIGURE 2: Possible transition states in an avoided crossing region for an unpaired nucleon in the superfluid model.

In order to obtain the equations of motion, we shall start from the variational principle taking the Lagrangian as:

$$L=\left\langle \phi \middle| H - i\hbar\frac{\partial}{\partial t} + H' - \lambda N \middle| \phi \right\rangle$$

(2)

and assuming the many-body state formally expanded as a superposition of time dependent BCS seniority one diabatic wave functions

$$|\phi\rangle = \sum_i c_i(t) a_i^+ \prod_{j\neq i} (u_j(t) + v_j(t) a_j^+ a_{\bar{j}}^+)|0\rangle$$

The Lagrangian contains several terms. The first one is the many body Hamiltonian with pairing residual interactions:

$$H = \sum_{k>0} \varepsilon_k (a_k^+ a_k + a_{\bar{k}}^+ a_{\bar{k}}) - G \sum_{k,l>0} a_k^+ a_{\bar{k}}^+ a_{\bar{l}} a_l$$

The residual interactions between diabatic levels characterized by the same quantum numbers is

$$H' = \sum_{mj} h_{mj} \left[\left(u_m a_m - v_m a_{\bar{m}}^+ \right)\left(u_j^* a_j^+ - v_j^* a_{\bar{j}} \right) + \left(u_j a_j - v_j a_{\bar{j}}^+ \right)\left(u_m^* a_m^+ - v_m^* a_{\bar{m}} \right) \right].$$

The sum runs over n, the total number of avoided crossings between adiabatic levels m and j, for any pair (mj).. The number of particle operators is:

$$N = \sum_{k>0} (a_k^+ a_k + a_{\bar{k}}^+ a_{\bar{k}})$$

After some calculations we obtain:

$$\left\langle \phi \middle| H - i\hbar\frac{\partial}{\partial t} + H' - \lambda N \middle| \phi \right\rangle$$

$$= \sum_m |c_m|^2 \left\{ 2 \sum_{k\neq m, k>0} |v_k|^2 (\varepsilon_k - \lambda) + (\varepsilon_m - \lambda) - G \left| \sum_{k\neq m, k>0} u_k v_k \right|^2 \right\} - \qquad (3)$$

$$i\hbar \sum_m |c_m|^2 \left[\sum_{k\neq m, k>0} \frac{1}{2}\left(v_k^* \dot{v}_k - \dot{v}_k^* v_k \right) \right] - i\hbar \sum_{m>0} c_m^* \dot{c}_m + \frac{1}{2} \sum_{mj} h_{mj} \left(c_m^* c_j + c_j^* c_m \right)$$

where terms like

$$\sum_m |c_m|^2 \sum_k v_k^4$$

are neglected. The independent parameters are v_k, v_k^*, c_m and c_m^* are subject of conditions:

$$u_k^2 + |v_k|^2 = 1; \quad \sum_{m>0} |c_m|^2 = 1.$$

In the following, the calculations are made in a way similar to that given in Ref. [2]. The equation (3) is derived with respect v_l, v_l^*, c_m and c_m^* and set to zero. After some calculations, a set of four equations is obtained [3]:

$$i\hbar\dot{\rho}_l = \sum_m p_m \left\{ \kappa_l \Delta_m^* - \kappa_l^* \Delta_m \right\}$$

$$i\hbar\dot{\kappa}_l = \sum_m p_m \left\{ (2\rho_l - 1)\Delta_m - 2\kappa_l \left[\varepsilon_l(t) - \lambda(t) \right] \right\}$$

$$i\hbar\dot{p}_m = \sum_{mj} h_{mj} \left(S_{mj} - S_{jm} \right)$$

$$i\hbar\dot{S}_{jm} = S_{jm} \left\{ -\frac{1}{G}\left(|\Delta_m|^2 - |\Delta_j|^2 \right) + \left[\varepsilon_m(t) - \varepsilon_j(t) \right] \right.$$

$$\left. -\frac{1}{2}\left[-\frac{\rho_m}{\kappa_m} + 2\kappa_m^* + \frac{\rho_j}{\kappa_j} - 2\kappa_j^* \right] \sum_l p_l \Delta_l - \frac{1}{2}\left[-\frac{\rho_m}{\kappa_m^*} + 2\kappa_m + \frac{\rho_j}{\kappa_j^*} - 2\kappa_j \right] \sum_l p_l \Delta_l^* - \right\}$$

$$+ \sum_{k\neq m, j} \left[h_{mk}(t) S_{jk} - h_{jk}(t) S_{km} \right] + h_{jm}(t)(p_j - p_m)$$

(4)

where the next notations are used

$$\Delta_m = G \sum_{\substack{k\neq m \\ k>0}} \kappa_k; \quad \Delta_m^* = G \sum_{\substack{k\neq m \\ k>0}} \kappa_k^*; \quad \kappa_k = u_k v_k; \quad \rho_k = |v_k|^2;$$

$$p_m = |c_m|^2; \quad s_{jm} = c_j^* c_m.$$

A system of differential equations for coupled channels is obtained, p_m being the probability to find the unpaired nucleon on the bocking level m. The Eqs. (4) include the Landau-Zener effect in the TDHFB equations.

GENERALIZATION

If the blocking level is neglected, the system (4) reduces to:

$$i\hbar\dot{\rho}_l = \kappa_l\Delta^* - \kappa_l^*\Delta$$
$$i\hbar\dot{\kappa}_l = (2\rho_l - 1)\Delta - 2\kappa_l[\varepsilon_l(t) - \lambda(t)]$$

the well known TDHFB equations [2,4]. On another hand if the pairing is neglected, the third equation of the system (4) can be written:

$$i\hbar\left(\dot{c}_m c_m^* + \dot{c}_m^* c_m\right) = \sum_j h_{jm}\left(c_j c_m^* - c_j^* c_m\right)$$

Introducing explicitly the time dependence of the amplitudes c_m

$$c_i = c_{0i}\exp\left(-\frac{i}{\hbar}\int_0^t \varepsilon_i(t)dt\right)$$

the next relation is obtained:

$$i\hbar\left(\dot{c}_{0m}c_{0m}^* + \dot{c}_{0m}^* c_{0m}\right) =$$
$$\sum_j h_{jm}\left(c_{0j}c_{0m}^*\exp\left(-\frac{i}{\hbar}\int_0^t(\varepsilon_j - \varepsilon_m)dt\right) - c_{0j}^* c_{0m}\exp\left(\frac{i}{\hbar}\int_0^t(\varepsilon_j - \varepsilon_m)dt\right)\right)$$

The last relation is an equivalent form of the Landau-Zener Eq. (1) obtained in the frame of the single-particle model. Furthermore, if the pairing interaction is neglected, ρ and κ can be either zero or one, and the fourth equation of the system (4) reduces to

$$i\hbar\left(\dot{c}_m c_j^* + \dot{c}_j^* c_m\right) = c_j^* c_m\left(\varepsilon_m - \varepsilon_j\right) + \sum_{k \neq m,j}\left(h_{mk}c_j^* c_k - h_{jk}c_k^* c_m\right) + h_{jm}\left(c_j^* c_j - c_m^* c_m\right)$$

After introducing the exponential dependence, the next relation emerge:

$$i\hbar\left(\dot{c}_{0m}c_{0j}^* + \dot{c}_{0j}^* c_{0m}\right)\exp\left(-\frac{i}{\hbar}\int(\varepsilon_m - \varepsilon_j)dt\right)$$
$$= \sum_{k \neq m,j}\left[h_{mk}c_{0j}^* c_{0k}\exp\left(-\frac{i}{\hbar}\int(\varepsilon_k - \varepsilon_j)dt\right) - h_{jk}c_{0k}^* c_{0m}\exp\left(-\frac{i}{\hbar}\int(\varepsilon_m - \varepsilon_k)dt\right)\right]$$
$$+ h_{jm}\left(c_{0j}^* c_{0j} - c_{0m}^* c_{0m}\right),$$

that is another form of the Landau-Zener relation (1). So, the Landau-Zener equation for single particle systems and the TDHFB equations are two particular cases of the system (4). So, Eqs. (4) represents a generalization of the TDHFB equations in the case of seniority one nuclear systems.

REFERENCES

1. M. Mirea, L. Tassan-Got, C. Stephan and C.O. Bacri; *Nucl. Phys.* **A735**, 21 (2004).
2. J. Blocki and H. Flocard, *Nucl. Phys.* **A273**, 45 (1976).
3. M. Mirea; *Mod. Phys. Lett.* **A18**, 1809 (2003).
4. S.E. Koonin and J.R. Nix, *Phys. Rev.* **C13**, 209 (1976).

Asymmetry of Charge Ratio for Low Energetic Muons

Bogdan Mitrica

National Institute for Physics and Nuclear Engineering, Horia Hulubei, Bucharest, Romania
mitrica@nipne.ro

Abstract. The WILLI calorimeter, installed in NIPNE Bucharest, is operated since several years for measuring charge ratio of atmospheric muons at low energies (E < 1 GeV), particularly exploring its directional dependence. Measurements of the muon flux are compared with Monte-Carlo simulation with CORSIKA code. Recently WILLI detector is proposed to get extended with a by mini-array of 12 scintillators (WILLI-EAS) in order to measure muon charge ratio of the muon density of EAS lateral distributions. Experimental studies of this kind could provide detailed information on the shower development under the influence of the geomagnetic field and also on features of the hadronic interaction.

Keywords: muon charge ratio ;

PACS: 98,70 Sa

INTRODUCTION

Muons originate from the decay of charged pions and kaons produced by cosmic rays in the atmosphere:

$$\pi^{\pm} \rightarrow \mu^{\pm} + \nu_{\mu} (\bar{\nu}_{\mu}) \quad 100.0\ \% \text{ lifetime } 2.6 \cdot 10^{-2}\ \mu s \tag{1}$$

$$K^{\pm} \rightarrow \mu^{\pm} + \nu_{\mu} (\nu_{\mu}) \quad 63.5\% \quad \text{lifetime } 1.2 \cdot 10^{-2}\ \mu s \tag{2}$$

The muons decay subsequently with a larger life time in:

$$\mu^{+} \rightarrow e^{+} + \nu_{e} + \bar{\nu}_{\mu} \tag{3}$$

$$\mu^{-} \rightarrow e^{-} + \bar{\nu}_{e} + \nu_{\mu} \tag{4}$$

Measurements of muon flux and muon charge ratio at low energies (<1 GeV / c) are important for the understanding of neutrino flux and the hadronic interaction models. The muon charge ratio R_{μ}, whose value is governed by the proton excess in the flux of the primary cosmic rays, is dependent of the direction of observation, since the muon tracks are influenced by the geomagnetic field and differently bent in East and West direction. This influence leads to the well known East-West effect of the muon charge ratio of atmospheric muons (i.e with unspecified EAS origin) [1].

In contrast the charge ratio of the muon density of single EAS registered with well specified observation conditions (energy, direction of incidence, distance from the shower axis etc.) has been seldom considered and is experimentally unexplored. Recently the features of the charge ratio of the density of the EAS muon component have been extensively studied on basis of Monte Carlo simulations [2] revealing that

CP972, *Exotic Nuclei and Nuclear/Particle Astrophysics (II)—Proceedings of the Carpathian Summer School of Physics 2007,* edited by L. Trache and S. Stoica

the radial and azimuthal muon density distributions of EAS observed by surface detector are strongly influenced by the magnetic field of the Earth.

MUON CHARGE RATIO

The WILLI detector [3, 4] installed for studying the muon charge ratio of atmospheric muons discriminates positive and negative muons by measuring the life time of muons stopped in the detector layers: Stopped positive muons decay with a lifetime of 2.2 μs, while negative muons are captured in atomic orbits, thus leading to an effectively smaller lifetime depending on the stopping material. From the adjustment of the measured decay curve to the actual mean value of the life time the value of Rμ can be deduced. Details of the method are described elsewhere [3]. It largely avoids the uncertainties due to different detection efficiencies and geometrical acceptances of muons of different charges which affect the results of magnetic spectrometers. In the present configuration [5] the WILLI setup is able to get directed to a pre-chosen direction of muon incidence (zenith and azimuth angle) for studies of the East-West effect of atmospheric muons e.g.(see Figure 1).

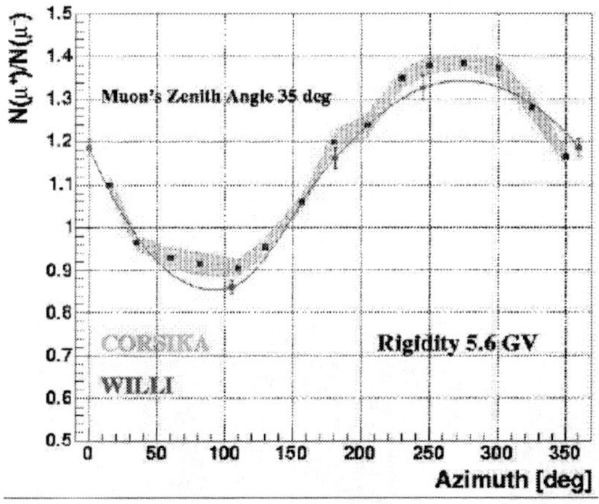

FIGURE 1. The azimuth dependence of the charge ratio of atmospheric muons compared with simulation result [5]

The results shows that CORSIKA simulations reproduce well the azimuthal variation i.e. the East-West effect as observed by WILLI. The Okayoma group reported a less pronounced azimuth dependence [6] but for muon with energies above 1 GeV. The anisotropy can be attributed to the anisotropy of primary proton flux caused by the geomagnetic cut-off, as well as to the geomagnetic influence on the atmospheric muon propagation [7].

Fig.2 gives a comparison of the muon charge ratio measured for vertical muons and for muons measured with rotatable WILLI, inclined at 45°, separately in the East and West direction displaying a good agreement.

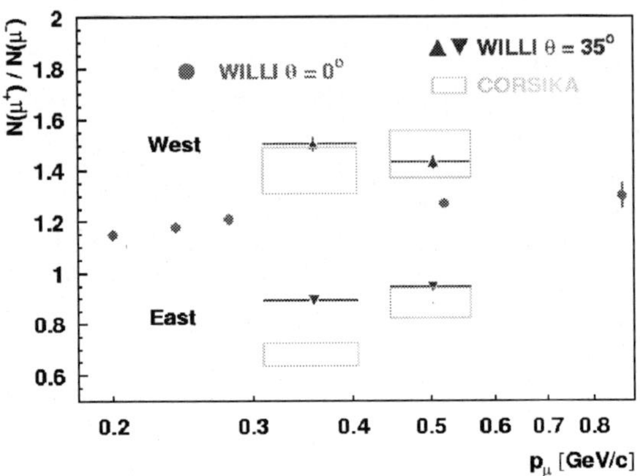

FIGURE 2. The energy dependence of the muon charge ratio, measured with WILLI

WILLI-EAS

The application of the WILLI device for experimental studies of the features of the radial and azimuthal variation of the charge ratio of the muon density distribution in EAS needs a link to the EAS registration by a nearby located detector array (see Fig.3), specifying the characteristics (core location, direction of incidence etc.) of the triggering EAS.

FIGURE 3. Sketch of the geometrical layout of a mini-array for triggering the nearby locatedWILLI device for muon detection by registering EAS

The necessary simulation studies of the performance of such a mini-array and of the expected trigger rate (roughly estimated to be in the order of 1 event/min) are in progress. In particular, the layout studies must also include an adequate electronic system for triggering and data acquisition since only one stopped muon per shower can be handled in the present system. On basis of the studies of ref. [2] we investigate, how the finite angular acceptance of the WILLI spectrometer, positioned at a particular accurately defined distance from the shower core and observing muons from a particular direction will affect the pronounced predicted variation of the charge ratio of the observed muon density.

For devising the mini-array (intended to be set up with 12 scintillator units (formerly used in the Central Detector of KASCADE in Forschungszentrum Karlsruhe [8]) special attention must be paid to the accuracy of the core location since inaccuracies of the core location are propagated in the azimuthal (and radial) position of the WILLI device with respect to the incoming shower and would smear out azimuthal variations of R_μ.

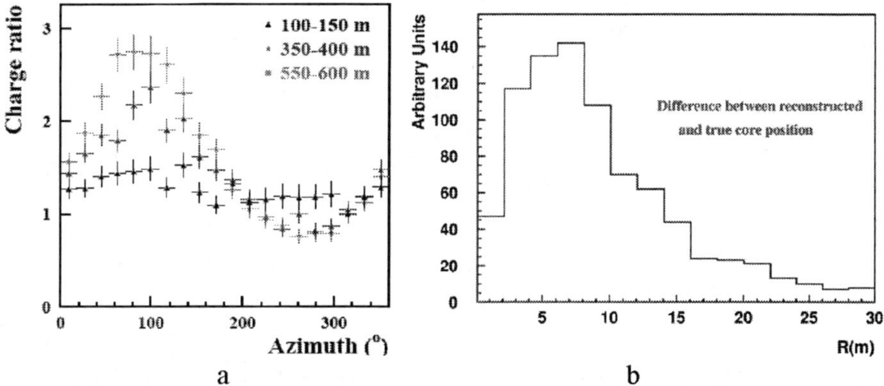

FIGURE 4. (a) Dependence of the charge ratio on the azimuthal position of WILLI around shower core for various radial ranges. Proton showers ($E=10^{15}$ eV, $\theta=45°$) coming from North. WILLI is oriented parallel to the shower axis [9]. (b) Difference between the reconstructed and the true core position as simulated for a rectangular mini-array consisting of 12 scintillator plates

Simulation studies of these questions have been started [11] and preliminary results are displayed in Fig.4. In Fig.4(a) is represented the dependence of the charge ratio on the azimuthal position of WILLI around shower core for various radial ranges for proton showers ($E=10^{15}$ eV, $\theta=45°$) coming from North. In Fig.4(b) is represented the difference between the reconstructed and the true core position as simulated for a rectangular (20 X 10 m) mini-array consisting of 12 scintillator plates. Further careful performance studies, optimising the layout of the planned mini-array, are needed.

CONCLUSIONS

The rotatable system WILLI allows studies of the charge ratio of muons with different angles-of-incidence, enabling to measure the East-West effect by the influence of the Earth's magnetic field. Measurements of the muon flux are in progress.

The WILLI detector is going to be extended with an array of scintillator modules in order to determine the mean value of the charge ratio in single showers, indicating the influence of the geomagnetic field on the propagation of the muon component.

ACKNOWLEDGMENTS

The author would like to thank the colleagues participating in the described studies, in particular Prof. Dr. O. Sima of the University of Bucharest, Prof. Dr. H. Rebel of Forschungsztentrum Karlsruhe,Germany, Dr. Iliana Brancus, D.E. Mirel Petcu, D.P, Gabriel Toma, D.P. Alexandra Saftoiu, D.M. Marin Duma and D.E. Marian Petre.

These studies have been supported by the Romanian Authority for Scientific Research in the project CEEX 05-D11-79 and in the Program Nucleu.

REFERENCES

1. O.C. Allkofer et al.: *Journ. Geophys. Research* 90 (1985) 3537..
2. H.Rebel and O.Sima : Rom. Rep.Phys. 59 (2007)609; H.Rebel, O.Sima, A.Haungs, J.Oehlschlaeger, Proc. of 30th ICRC, Merida, Mexico 2007
3. B.Vulpescu et al.: Nucl Instr. and Meth.Phys.Res.A414 (1998) 205
4. B. Vulpescu et al.: *J. Phys. G: Nucl. Part. Phys.* 27 (2001) 977.
5. I.M. Brancus et al.: *Nucl. Phys. A* 271 (2003) 1044c.
6. S.Tsuji et al.: Proc. 29th ICRC2005, p.211, vol.6, Pune, India
7. J.Wentz et al.: Phys.Rev.D67, 073020 (2003)
8. T.Antoni et al.-KASCADE Collaboration :Nucl.Instr. and Meth.Phys.Res. A513 (2003)490
9. B. Mitrica et al. : Proc. of 30th ICRC, Merida, Mexico 2007

$\alpha-$halflife Estimates for the Heaviest Elements

I.Silişteanu , A.Neacşu , A.O.Silişteanu and M. Rizea

Horia Hulubei Institute of Physics and Nuclear Engineering, RO-077125, Bucharest-Magurele,Romania

Abstract. The microscopic shell model theory of α-decay is reviewed and resonance formulas for α-rates of heaviest deformed nuclei are derived within the framework of multichannel resonance scattering approach. The very sharp α-resonances observed in current reactions with accelerated heavy ions may be interpreted as levels of reaction product and such a connection contributes directly to decay and in-beam studies of the nuclear structure. The reliability of results is proved by a comparison with the decay data and relevant approaches.

Keywords: SHE; α-decay; coupled channels analysis
PACS: 25.70.Jj; 27.90.+b

INTRODUCTION

In reactions with heavy heavy ions were detected the α-chains of decays which matched very well the signature expected for the decay of superheavy elements (SHE). A series of measurements established that the SHE with Z up to 118 are sufficiently bound against fission to preferentially decay by α-emission [1, 2, 3, 4, 5]. Significant progress in the experimental SHE research, raised a natural question of how to gain access to the basic ground-state properties of these nuclei- their masses, lifetimes, moments, spins and radii. This involves developing and applying theories and methods of prediction, analysis and interpretation of specific data, with the aim of achieving a deeper understanding of the physical nature of quantum many nucleon systems. The progress made thus far in such theories and the availability of faster and larger computers, do enable us to investigate the nuclear properties of new SHEs.

Progress in obtaining the most complete information on the development of the clustering and fine structure in heaviest nuclei is being made on several ways: 1) Improving the structure models in order to describe essential features; 2) Extending the range of applicability of reaction models; 3) Including microscopic information in coupled channel reaction models. In this work we try to combine these methods to obtain the α-scattering wave functions in the vicinity of a multichannel resonance by solving the radial Schrodinger equation for coupled channels with outgoing asymptotes and resonance conditions.

DECAY OF A RESONANCE LEVEL

The first many-body description of α-decay[6] was an application of the general theory of nuclear reactions[7]. This, combined with shell model, which was introduced at the same time, opened the possibility of performing microscopic shell model calcula-

CP972, *Exotic Nuclei and Nuclear/Particle Astrophysics (II)—Proceedings of the Carpathian Summer School of Physics 2007,* edited by L. Trache and S. Stoica

tions for the nucleon clustering into the α-particle and its subsequent emission [8, 9]. To overcome some disadvantages of the R-matrix theory, there have been formulated alternative approaches [12, 13, 14, 15] in the dynamical reaction theories [10, 11]. All these theories are reviewed in [16, 17].

The fine structure has been observed in the α-decay of SHE[1, 2, 18]. The usual way to address the fine structure and coupling between the intrinsic degrees of freedom and relative motion is to numerically solve the coupled channel equations, including[19] all the relevant channels[20, 21]. Thus, the total decay width for the multichannel decay of the state k into a set of $\{n\}$ different channels is

$$\Gamma^k = \sum_n \Gamma_n^k, \quad where: \tag{1}$$

$$\Gamma_n^k = 2\pi \left| \frac{\int_{r_{\min}}^{r_{\max}} I_n^k(r) u_n^0(r) dr}{\int_{r_{\min}}^{r_{\max}} I_n^k(r) u_n^k(r) dr} \right|^2 . \tag{2}$$

In Eq. (9) $I_n^k(r)$ is the particle (cluster) formation amplitude (FA) and $u_n^k(r)$ and $u_n^0(r)$ are the solutions of the systems of differential equations

$$\left[\frac{\hbar^2}{2m} \left(\frac{d^2}{dr^2} - \frac{l(l+1)}{r^2} \right) - V_{nn}(r) + Q_n \right] u_n^0(r) + \sum_{m \neq n} V_{nm}(r) u_m^0(r) = 0 \tag{3}$$

$$\left[\frac{\hbar^2}{2m} \left(\frac{d^2}{dr^2} - \frac{l(l+1)}{r^2} \right) - V_{nn}(r) + Q_n \right] u_n^k(r) + \sum_{m \neq n} V_{nm}(r) u_m^k(r) = I_n^k(r) \tag{4}$$

The radial motion of the fragments at large and small separations, respectively is described in terms of the reduced mass m, the kinetic energy of emitted particle $Q_n = E - E_T - E_P$, the FA $I_n^k(r)$, l_n are the angular momenta in channel n, and the interaction potential $V_{nm}(r)$. The FA is the antisymmetrized projection of the parent wave function (WF) $| \Psi_k >$ on the channel WF $| n\rangle = | \left[\Phi_D(\eta_1) \Phi_P(\eta_2) Y_{lm}(\hat{r}) \right]_n \rangle$:

$$I_n^k(r) = r \langle \Psi_k | \mathscr{A} \{ \left[\Phi_D(\eta_1) \Phi_P(\eta_2) Y_{lm}(\hat{r}) \right]_n \} \rangle \tag{5}$$

where $\Phi_1(\eta_1)$ and $\Phi_2(\eta_2)$ are the internal (space-spin) wave functions of the fragments, $Y_{lm}(\hat{r})$ is the wave function of the angular motion, r connects the centers of mass of the fragments, and the symbol $<|>$ means integration over the internal coordinates and angular coordinates of relative motion. The diagonal elements V_{nn} of the potential are:

$$V_n(r) = V_{nn}(r) = V_0^{nucl.}(r) + V_0^{Coul.}(r) \tag{6}$$

For the nuclear potential we use the Woods-Saxon parametrization (V_0, r_0, a):

$$V_0^{nucl}(r) = V_0^{nucl}(r) = -V_0/[1 + \exp((r - R_0)/a)] \,, R_0 = r_0(A_P^{1/3} + A_T^{1/3}), \tag{7}$$

while the Coulomb potential is taken of the usual form:

$$V_0^{Coul}(r) = \{ \begin{array}{ll} (Z_P Z_T e^2/2R_c)[3 - (r/R_c)^2], & r \leq R_c \\ Z_P Z_T e^2/r, & r > R_c. \end{array} \}, R_c = r_c(A_P^{1/3} + A_T^{1/3}) \tag{8}$$

The matrix elements V_{nm} of the coupling Hamiltonian consist also of nuclear and Coulomb components. The nuclear component can be generated by changing the target radius in the nuclear potential to a dynamical operator[22]

$$R_0 \rightarrow R_0 + \hat{O} = R_0 + \beta_2 R_T Y_{20} + \beta_4 R_T Y_{40} \tag{9}$$

where $R_T = r_0 A_T^{1/3}$ and β_2, β_4 are the quadrupole and hexadecapole deformation parameters of the deformed target nucleus, respectively. The nuclear coupling term is

$$V^{nucl}(r, \hat{O}) = \; -V_0/(1+\exp((r-R_0-\hat{O})/a)). \tag{10}$$

and the coupling matrix elements between rotational states states$|\, n >=|\, I0 >$ and $|\, m >=|\, I^{'}0 >$ are

$$V_{nm}^{Coul.}(r) = 3Z_D Z_p R_D^2/5r^3 \left[5(2I+1)(2I^{'}+1)/4\pi\right]^{1/2} \left(\beta_2 + (2/7)(5/\pi)^{1/2}\beta_2^2\right) \left(C_{000}^{I2I^{'}}\right)^2$$

$$+3Z_D Z_p R_D^2/9r^5 \left[9(2I+1)(2I^{'}+1)/4\pi\right]^{1/2} (\beta_4 + (9/7)\beta_2^2) \left(C_{000}^{I4I^{'}}\right)^2. \tag{11}$$

If the all exit channels are open the boundary conditions should be:

$$u_n^0\,(r \rightarrow 0) = 0; \quad u_n^0(r \rightarrow \infty) \rightarrow i/2(k_n/\pi Q_n)^{1/2}\{\delta_{nm}\exp\left[-i(k_n - l\pi/2)\right]$$

$$-S_{nm}\exp\left[i(k_n r - l\pi/2)\right]\} \tag{12}$$

$$u_n^k(r \rightarrow 0) = 0; \qquad u_n^k(r \rightarrow \infty) = 0 \tag{13}$$

where $\hbar k_n = (2mQ_n)^{1/2}$, S_{nm} is the scattering matrix., $u_n^{(\pm)}(r) = G_n(r) \pm iF_n(r)$ and F_n and G_n are the regular and irregular Coulomb functions. The lower limit in integrals (1) is an arbitrary small radius $r_{\min} > 0$, while the upper limit $r_{\max}$ is close to the first exterior node of $u_n^0(r)$. To avoid the usual ambiguities encountered in formulating the potential for the resonance tunneling of the spherical barrier we iterate directly the nuclear potential in equations of motion. The solutions $u_n^0(r\ )$may be matched to the boundary conditions at two values of r large enough so the terms $V_{nm}$are negligible. A special type of eigenvalue solution will be considered here for which the behavior of solution in each separate channel is similar to that of $G_n$in the one channel problem[23, 24]. The ratio[20, 21]

$$S_n^k = \Gamma_n^k/\Gamma_n^{o.b.} = T_n^{o.b.}/T_n^k \tag{14}$$

is a measure of global contribution of shell effects, finite sizes of nucleons and α-particle to the one-body width approximation (where α is a pointlike particle and all these effects are merely neglected).

DETAILS OF THE CALCULATIONS

In present estimates of decay rates we take into the account the shell model (SM) and "one-body" resonance (R) of CFAs, couplings between the relative motion of the fragments and several nuclear collective motions, the effects of non-linear couplings to all orders, the self-consistence of scattering potential and convergence of resonance scattering solution, the measured Q_α values. In the SM case we simply put in Eq.(4) $\mid \Psi_k >= \Psi_{\{\alpha\}} = (A!)^{-1/2} \det \parallel \Psi_{\{\alpha_i\}}(r_i) \parallel$, i=1,A. For the all members of α-decay chains of $^{294}118$ and $^{290-291}116$ we estimate the α-decay rates using: the SMCFA constructed from the s.p. orbits[26] (protons: $1i_{13/2}, 2f_{7/2}, 3p_{3/2}$, neutrons: $2g_{7/2}, 3d_{5/2}$, $3d_{5/2}$)and RCFA[21, 25, 23, 24]; the deformation parameters [27]in Eq.(????x), and the measured Q_α-values[5]. The numerical code for emission rates is written in Fortran complex, double precision and uses the published codes:[22, 23, 28, 29]

RESULTS

Our results for different model approximations are shown in Table 1. For all the α-channels of $^{294}118$, $^{290}116$ nuclides, we can observe a reasonable agreement of our model results (with the screening energy) with empirical estimates[5] (without the screening energy). In Fig.1 we compare the shell model and empirical[5] estimates. With the screening corrections added the empirical halftime estimation becomes shorter (~by a factor 2). In fact, the halftimes [5] should be overestimated.

SUMMARY AND OUTLOOK

Systematic calculations of spectroscopic properties of SHEs have become feasible, as has been shown on the examples of ground state decay of $^{294}118$ and $^{290}116$ nuclides and their α-descendants. The emission rates of SHE can be enhanced due to the collective

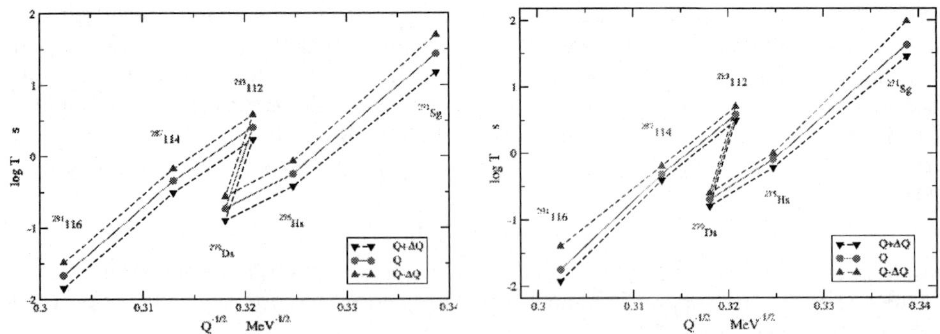

FIGURE 1. The Gamow plot of calculated (our shell model-left and empirically [5]-right) α-half-lives for the $^{291}116$ nuclide and its descendants. The Q_α-values are measured ones[5].

TABLE 1. Decay properties of nuclei produced by Yu. Ts. Oganessian et al.[5]

Nuclide	Q_α MeV	$T^{\alpha}_{1/2}$ Empiric estimates* Ref.[5] Q_α	$T^{\alpha}_{1/2}$ Empiric estimates* This work $Q_\alpha + E_{scr}$	$T^{\alpha}_{1/2}$ Microscopic calculations Shell model† $Q_\alpha + E_{scr}$	$T^{\alpha}_{1/2}$ Microscopic calculations Superfl. model** $Q_\alpha + E_{scr}$	S^k_n $^{294}118$
$^{294}118$	$11.81 \pm .07$	$.89^{+1.07}_{-0.31} ms$	$0.48^{+0.21}_{-0.15} ms$	$0.72^{+1.98}_{-0.27} ms$	$.13^{+.49}_{-.35} ms$	$.4\ 10^{-3}$
$^{290}116$	$10.95 \pm .09$	$7.1^{+3.2}_{-1.7} ms$	$15.15^{+10.50}_{-6.14} ms$	$21^{+16}_{-8} ms$	$2.68^{+1.83}_{-1.08} ms$	$.2\ 10^{-3}$
$^{286}114$	$10.18 \pm .09$	$.13^{+0.04}_{-0.02} s$	$0.40^{+0.31}_{-0.13} s$	$0.28^{+0.17}_{-0.13} s$	$0.02^{+0.01}_{-0.01} s$	$.1\ 10^{-2}$
$^{282}112$	$10.25 \pm .25$ ‡		$60.49^{+231.90}_{-47.25} ms$	$40^{+154.2}_{-31.92} ms$	$3.23^{+11.73}_{-2.49} ms$	$.5\ 10^{-3}$
$^{290}116$	$11.00 \pm .08$	$7.1^{+3.2}_{-1.7} ms$	$11.34^{+6.7}_{-4.18} ms$	$12.70^{+6.9}_{-10.4} ms$	$12.34^{+7.0}_{-4.4} ms$	0.20
$^{286}114$	$10.33 \pm .06$	$.13^{+.04}_{-.02} s$	$.15^{+.07}_{-.04} s$	$.14^{+.17}_{-.01}$ s	$.10^{+.05}_{-.03} s$	0.17
$^{291}116$	$10.89 \pm .07$	$18^{+22}_{-6} ms$	$21^{+11}_{-7} ms$	$27^{+14}_{-5} ms$	$1.12^{+5.5}_{-3.6} ms$	$.9\ 10^{-2}$
$^{287}114$	$10.16 \pm .06$	$0.48^{+0.16}_{-0.09} s$	$0.45^{+0.21}_{-0.14} s$	$0.62^{+0.2}_{-0.19} s$	$16.18^{+4.23}_{-5.01} ms$	$.5\ 10^{-2}$
$^{283}112$	$9.67 \pm .06$	$3.8^{+1.2}_{-0.2} s$	$2.56^{+1.29}_{-0.85} s$	$3.95^{+1.85}_{-1.11} s$	$59.12^{+18.25}_{-24.02} ms$	$.3\ 10^{-2}$
$^{279}110$	$9.84 \pm .06$	$0.2^{+0.05}_{-0.04} s$	$0.18^{+0.09}_{-0.16} s$	$0.10^{+0.06}_{-0.03} s$	$8.17^{+3.76}_{-2.55} ms$	$.2\ 10^{-2}$
$^{275}108$	$9.44 \pm .06$	$0.19^{+0.22}_{-0.07} s$	$0.56^{+0.29}_{-0.19} s$	$0.26^{+0.15}_{-0.08} s$	$25.41^{+12.66}_{-8.17} ms$	$.2\ 10^{-1}$
$^{271}106$	$8.67 \pm .08$	$114^{+144}_{-36} s$	$27.33^{+22.68}_{-12.28} s$	$68^{+51}_{-22} s$	$1.373^{+1.119}_{-.611} s$	$.4\ 10^{-2}$

* -Viola-Seaborg formula[30] with parameters [31, 32]
† -Shell-model formation amplitude [19, 20]
** -Superfluid model formation amplitude [33]
‡ - The limit values include all the predictions:[34, 35, 36, 37, 38]

behavior or forbidden by internal selection rules. The agreement with experiment supports the basic correctness of our unified picture for heavy-ion and α-resonances and gives us some guaranty for reasonable predictions to the SHE. Studies of the α-decay of SHE have been offering unique insights into the structures of nuclei as seen in earlier and recent works, such as[39, 40, 41, 42, 43, 44, 45].

ACKNOWLEDGMENTS

We thank to Profs. Yu.Ts. Oganessian, V.K. Utynkov, S. Hofmann, G. Münzenberg, W. Scheid, S. Mişicu, A. Şandru A. Sandulescu for many stimulating discussions. This work was supported in part from Contracts CEX05-D08-10, PN2-71-112.

REFERENCES

1. S. Hofmann et al., Zeit. Phys. **A354**, 229 (1996).
2. S. Hofmann and G. Münzenberg, *Rev. Mod. Phys.* **72**, 733 (2000).
3. Yu. Ts. Oganessian et al, Phys. Rev. **C 69**, 021601 (2004).
4. Yu. Ts. Oganessian et al, Phys. Rev. **C 69**, 054607 (2004).
5. Yu. Ts. Oganessian et al, Phys. Rev. **C 74**, 044602(2006).
6. R.G.Thomas, Prog. Theor.Phys., **12**, 253(1954).
7. T.Teichmann, E.P.Wigner,Phys.Rev., **79**, 30 (1952).
8. H.J.Mang,Phys.Rev., **119**, 1069(1960).
9. H.J.Mang, J.O.Rasmussen, Kgl.Danske Videnscab Selskab,Mat.Fys.Skifter 2 **3**,(1962).
10. G.Breit, Encyclopedia of Physics, Vol.XLI, Nuclear Reactions II: Theory,Springer-Verlag,1959.
11. H.Feshbach,Ann.Phys.5 (NY), **357**,(1958).
12. K.Harada, E.A.Rauscher,Phys.Rev., **169**, 818(1968).
13. S.G.Kadmensky, V.E.Kaletchis,Yad.Fiz., **12**,70(1970).
14. J.Schlitter, Nucl.Phys.,**A 211**, 96(1973).
15. A.Sandulescu, I.Silisteanu, R.Wuensch, Nucl.Phys.**A305** 205(1978).
16. R.G. Lovas et al, Phys. Rep. **294**, 265(1998).
17. P.E.Hodgson and E. Betak, Phys. Rep. **374**,89 (2003).
18. P. Reiter et al, Phys.Rev.Lett. **82**, 509(1999).
19. I. Silişteanu, W.Scheid and A.O.Silişteanu, Rom.Rep.Phys. **57**, 4(2005).
20. I. Silişteanu , W. Scheid and A. Sandulescu, Nucl. Phys **A679**,317(2001).
21. I. Silişteanu, W.Scheid, Phys.Rev. **51**, 2023 (1995).
22. K.Hagino, N.Rowley, A.T.Kruppa, Comp.Phys.Commun.**123** ,143(1999).
23. M. Rizea, Comp. Phys. Commun. **143**,83(2002).
24. M. Rizea, Contribution to this volume
25. I. Silisteanu et al, Proc. CSSP-2005,World Sci.2006, Eds. Stoica, Trache, Tribble), pag.423.
26. A.T.Kruppa et al. Phys.Rev.**C61** 034313 (2000).
27. P. Möller et al, At. Data and Nucl. Data Tab. **66**, 131 (1997).
28. J.A. Christley and I.J. Thomson, Comp.Phys.Commun. 79,143(1994).
29. I. J. Thompson and A. R. Barnett, Comput. Phys. Commun. 36 (1985) 363-372
30. V. E. Viola and G. T. Seaborg, J. Inorg. Nucl. Chem. **28**, 741 (1966).
31. A. Sobiczewsky, Z. Patyk, and S Cwiok, Phys. Lett. **B 224**, 1(1989).
32. Z. Patyk and A. Sobiczewski, Nucl. Phys. **A354**, 229(1996).
33. F. Barranco, R.A. Broglia and G.F. Bertrech, Phys. Rev. Lett 60,507(1988).
34. S. Gorieely et al, Phys.Rev. **C 66**, 024326(2002).
35. S.Typel and B.A. Brown Phys.Rev **C 67**, 034313 (2003).
36. J.F.Berger, D.Hirata, M.Girod, Acta Phys. Pol. **B 34**, 1909 (2003).
37. S. Cwiok, P.H.Heenen,W. Nazarewicz, Nature **433**, 705 (2005).
38. M.Warda and al, Intern.J.Mod.Phys.E**15**,101(2007).
39. V.Yu.Denisov and H.Ikezoe, Phys. Rev. **C 72**, 064613 (2005).
40. P.Mohr, Phys. Rev **C 73**, 031301 (2006).
41. P.Roy.Chowdhury,et al, Phys.Rev. **C73**, 014612(2006).
42. Chang Xu and Zhongzhou Ren, Phys. Rev. **C 74**, 014304(2006).
43. Takatoshi Ichikawa et al., Phys. Rev. **C 71**, 044608 (2005).
44. M. Bender, et al , Phys.Rev. **C60**,034304(1999).
45. Z.A. Dupre, T.J. Bürvenich, Nucl.Phys. **A 767**,81(2006).

Nuclearite Detection with the ANTARES Neutrino Telescope

G. E. Păvălaş* and V. Popa*,†

*Institute for Space Sciences, RO 077125 Bucharest - Măgurele, Romania
†for the ANTARES Collaboration

Abstract.
We discuss the possibility to search for cosmic ray nuclearites using large volume neutrino telescopes. We present a short review of the nuclearite expected properties, focussing on their interaction with the Earth's atmosphere and sea water, and on the specific energy loss mechanisms that could make them detectable in undersea neutrino detectors. After a brief description of the ANTARES telescope, currently under deployment in the Mediterranean Sea, we will discuss its sensitivity to down-going nuclearites, and present some results from the Monte Carlo simulations of such events.

Keywords: Quark mater, nuclearites, neutrino telescopes, Cosmic rays interactions
PACS: 21.65.Qr,14.80,-j,95.55.Vj,96.50.sf

INTRODUCTION

Strange Quark Matter (SQM) could be the ground state of quantum chromodinamics [1]. It was hypothesized that "nuggets" of SQM, having masses ranging from heavy nuclei to macroscopic values could be present in the cosmic radiation (the so-called *nuclearites*) [2]. The SQM should have a relatively small positive electric charge, eventually neutralized by an electron cloud. If the size of the SQM is large enough (masses $M \gtrsim 10^7$GeV), some of the electrons could be in chemical equilibrium inside the quark core [3]. Nuclearites larger than 1 Å($M \geq 8.4 \times 10^{14}$GeV) would contain inside all electrons and thus would be completely neutral [2]. A qualitative picture of nuclearites may be found in [4].

Nuclearites are supposed to be relics from the first moments of the Universe, but could be produced also in violent astrophysical processes, as during the pre-colapse of binary strange star systems [5].

The main aim of the future very large volume neutrino telescopes is to detect up-going high energy neutrinos from astrophysical sources; but the search for exotic particles as magnetic monopoles, nuclearites or Q-balls could be also performed, as secondary goals.

STRANGE QUARK MATTER BASIC PROPERTIES

Bulk SQM represents a bound and stable system of u, d and s quarks, in (nearly) equal proportions [1]. In lumps of SQM, a small deficit of s quarks is expected, due to their larger chemical potential, yielding to a net positive electric charge, (partially) neutralized by electrons, either in weak equilibrium inside the SQM, or forming an

CP972, *Exotic Nuclei and Nuclear/Particle Astrophysics (II)—Proceedings of the Carpathian Summer School of Physics 2007,* edited by L. Trache and S. Stoica

external electronic cloud [2]. This makes the properties of SQM nuggets changing significantly according their mass.

Low mass lumps of SQM: *strangelets*

It is presumed that SQM could be stable from masses larger than about 300 GeV; small nuggets (with masses roughly in the TeV region) would behave like super-heavy nuclei, with anomalously low Z/A ratios; those states are known as "strangelets". As they are carrying a net electric charge, they are expected to be accelerated by astrophysical fields, and thus be present in cosmic rays as relativistic objects. There are different structure models, eg. [3, 6, 7], but from the point of view of the experimental searches, their predictions do not differ significantly. This is not the case in what concerns the models describing the interactions of strangelets with the normal matter, in particular with the Earth's atmosphere; as an example, we cite two opposite scenarios: one, based on a spectator - participant picture, assumes a decrease of the mass in average with the nitrogen mass after each interaction [8]; an other, [9] is based on the mass accretion idea, assuming the strangelet increases with the same amount after each interaction in the air. Considering the flux predictions in Ref. [5], those two models would lead to very different fluxes at the level of high altitude experiments; the recent results from SLIM (a passive detector that was exposed at the Chacaltaya Cosmic Ray Lab) exclude on this basis the accretion model [10]. Strangelets, as defined above, cannot reach the Earth's surface, and thus are not relevant for underwater experiments.

Large mass lumps of SQM: *nuclearites*

Much heavier lumps of SQM (masses $M \gtrsim 10^{10}$GeV) would be electrically neutral; even if part of the electrons would be still in the outer electronic cloud, as they would be in gravitational equilibrium in the galactic halo traveling with typical speeds $\beta = v/c \simeq 10^{-3}$, they would not be ionized in interaction with matter. Such very heavy SQM nuggets are known as "nuclearites". In [2] it was postulated that elastic collisions with the atoms and molecules of the traversed medium are the only relevant energy loss mechanism,

$$\frac{dE}{dx} = -\sigma \rho v^2, \tag{1}$$

where ρ is the density of the traversed medium, v is the nuclearite velocity and σ is its geometrical cross section:

$$\sigma = \begin{cases} \pi(3M/4\pi\rho_N)^{2/3} & \text{for } M \geq 8.4 \times 10^{14} \text{ GeV} \\ \pi \times 10^{-16}\text{cm}^2 & \text{for lower nuclearite masses} \end{cases}, \tag{2}$$

with $\rho_N = 3.6 \times 10^{16}$ g cm^{-3}. The mass limit in Eq. 2 corresponds to a radius of the strange quark matter of 1 Å. As finite SQM is always positively charged [2], the cross section for nuclearites with $M < 8.4 \times 10^{14}$ GeV is determined by their electronic cloud.

A phenomenological upper limit of the flux of nuclearites trapped inside the Galaxy hallo could be obtained assuming that they contribute to most of the dark matter in it [2]:

$$\Phi_{max} = \frac{\rho_{DM} v}{2\pi M}, \tag{3}$$

where $\rho_{DM} \simeq \rho_{hallo} \simeq 2 \div 13 \times 10^{-25}$ g cm^{-3} [11], v is the nuclearite speed and M its mass. For nuclearites trapped inside the Galaxy disk, the limit given by Eq. 3 would be about one order of magnitude higher, as in this case $\rho_{DM} \simeq \rho_{disk} \simeq 3 \div 12 \times 10^{-24}$ g cm^{-3} [12].

Assuming a nuclearite of mass M enters the atmosphere with an initial (nonrelativistic) velocity v_0, after crossing a depth L it will be slowed down to

$$v(L) = v_0 e^{-\frac{\sigma}{M}\int_0^L \rho dx} \tag{4}$$

where ρ is the air density at different depths. Considering the parametrization of the standard atmosphere from [13]:

$$\rho(h) = ae^{-\frac{h}{b}} = ae^{-\frac{H-L}{b}}, \tag{5}$$

where the fit constants are $a = 1.2 \times 10^{-3}$g cm^{-3} and $b \simeq 8.57 \times 10^5$ cm; $H \simeq 50$km is the total height of the atmosphere, the integral in Eq. 3 may be solved analytically.

The propagation of nuclearites in sea water is described also by Eq. 4, assuming $\rho \simeq 1$ g cm^{-3} and substituting v_0 with the speed value at the Earth surface. Nuclearites moving into the water could be detected because of the black-body radiation emitted by the expanding cylindrical thermal shock wave [2]. The luminous efficiency (defined as the fraction of dissipated energy appearing as light) was estimated, in the case of water, to be $\eta \simeq 3 \times 10^{-5}$ [2]. One may then compute the number of visible photons emitted per unit path length as:

$$\frac{dN_\gamma}{dx} = \eta \frac{dE/dx}{\pi\ (eV)}, \tag{6}$$

assuming the average energy of visible photons π eV. Very heavy nuclearites ($M \gtrsim 10^{22}$GeV) could traverse the Earth and reach detectors from below too; due to the very small flux allowed by the condition Eq. 3, we are not considering this possibility here. The best present limit for a flux of downgoing nuclearites was reported by the MACRO experiment, at Gran Sasso [14].

THE ANTARES NEUTRINO TELESCOPE

ANTARES is a deep sea neutrino telescope, currently being deployed in the Mediterranean sea, south of Toulon [15]. Fig. 1 shows a drawing of the detector. It is due to become fully operational early in 2008.

We have performed detailed Monte Carlo simulations of nuclearite detection in ANTARES, assuming only the down-going part of an isotropic flux of nuclearites (outside the Earth's atmosphere) and an initial velocity (before entering the atmosphere)

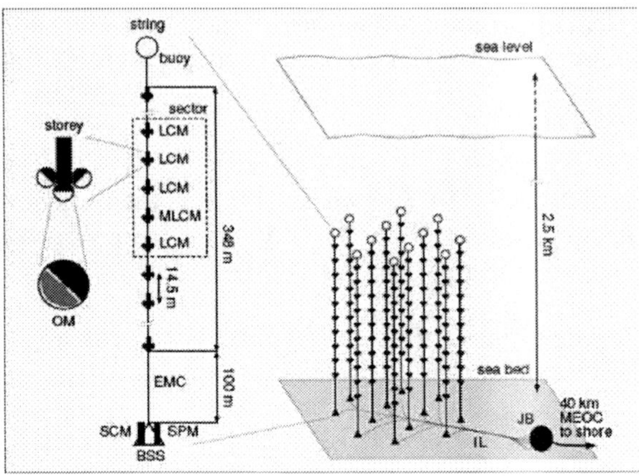

FIGURE 1. A sketch of the ANTARES neutrino telescope

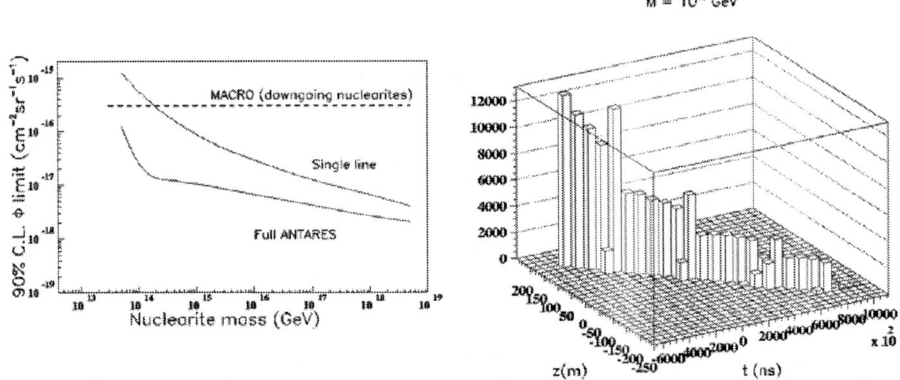

FIGURE 2. Some results of the MC simulation of nuclearite detection in ANTARES. The sensitivities accessible in one year of data tacking with a single line and with the full detector, compared with the MACRO final limit for a down-going flux of nuclearites (left panel) and some characteristics of an event induced by a nuclearite (right panel). See text for details.

$\beta = 10^{-3}$. We considered as "good events" those in which the light produced by the passage of the nuclearite inside the sensitive volume of the detector is "seen" by at least 3 photomultipliers, situated in different optical modules (different ANTARES strings or different storeys of the same string). The left panel of Fig. 2 shows the sensitivity reachable in one year of data tacking by a single ANTARES string and the full detector, in therms of 90% C.L. down-going nuclearite flux limits. The final MACRO limit is also shown. The dependence of the ANTARES sensitivity on the nuclearite mass is due to the solid angles in which nuclearites with different masses could reach the detector with enough residual velocity to produce a detectable signal.

There are different background sources: the decays of ^{40}K present in the sea water (yielding to a constant rate of about 27 kHz), the bioluminiscence (characterized by large fluctuations in time) and the down-going atmospheric muons. The first two contributions produce uncorrelated signals, while the down-going muons are relativistic, thus the muon events are about 3 orders of magnitude shorter than those induced by nuclearites. In the right panel of Fig. 2 we show some characteristics of an event corresponding to a down-going 10^{15}GeV. The "z" axis denotes the depth (the 0 value represents the center of ANTARES), and the t axis is the time of the hits (here 0 is the moment of the hits in the middle plane of the detector); the vertical axis is the number of photons reaching the photomultipliers (integrated over time slices of 50 μs).

CONCLUSIONS

Large volume neutrino telescopes could be very efficient in the search for nuclearites (with masses larger than few 10^{13} GeV) in cosmic radiation; they could, at least, improve with orders of magnitude the present flux limits. Special software triggers have to be developed. They should be efficient in removing the backgrounds, but permissive enough not to reject a possible rare event, tacking into consideration the large amount of data that has to be analyzed, due to the time scale of a nuclearite event (typically of the order of few ms).

ACKNOWLEDGMENTS

The authors are grateful to their colleagues from the ANTARES collaborations, for many fruitful discussions.

REFERENCES

1. E. Witten, *Phys. Rev.* **D30** (1984) 272.
2. A. De Rújula and S. L. Glashow, *Nature* **312** (1984) 734.
3. M. Kasuya, T. Saito and M. Yasuè, *Phis. Rev.* **D47** (1993) 2153.
4. D. Bakari et al., hep-ex/0004019.
5. J. Madsen, *Phys. Rev.* **D71** (2005) 014026; astro-ph/0411538.
6. J. Madsen, *Phys. Rev. Lett.* **87** (2001) 172003.
7. H. Heselberg, *Phys. Rev.* **D48** (1993) 1418.
8. G. Wilk and Z. Wlodarczyk, *Heavy Ion Phys.* **4** (1996) 395.
9. S. Banerjee et al., *Phys. Rev. Lett.* **85** (2000) 1384.
10. G. Giacomelli et al., hep-ex/0702050.
11. E.I. Gates et al., *Ap. J.* **449** (1995) L133.
12. G. Gilmore et al., *Ann. Rev. Astron. Astrophys.* **27** (1989) 555.
13. T. Shibata, *Prog. Theor. Phys.* **57** (1977) 882.
14. M. Ambrosio et al., *Eur. Phys. J.* **C26** (2002) 163, updated in G. Giacomelli and L. Patrizii, hep-ex/0506014.
15. J.A. Aguilar et al., *Nucl. Instr. Meths. in Phys. Res.* **A570** (2007) 107.

Folding Description of the α-decay Fine Structure to 2^+ States in Vibrational and Transitional Nuclei

S. Peltonen*, D.S. Delion† and J. Suhonen*

*Department of Physics, University of Jyväskylä, POB 35, FIN-40351, Jyväskylä, Finland
† "Horia Hulubei" National Institute of Physics and Nuclear Engineering, 407 Atomistilor, Măgurele-Bucharest, 077125, Romania

Abstract. We analyze α-decays to ground and 2^+ vibrational states in even-even nuclei by using a coupled channels formalism. A satisfactory agreement with existing experimental data was obtained by using the vibrational strength as a free parameter. It turns out that the inverse of this parameter is proportional to the logarithm of the hindrance factor squared. More details of our work has been published in [1]

Keywords: alpha decay, fine structure, vibrational levels, decay widths
PACS: 23.60.+e, 24.10.Eq, 25.70.Ef

Theoretical Background

For full theoretical background see [1]. We assume that the interaction between the two nuclei can be written as

$$V(\mathbf{R},\alpha_2) = \sum_{\lambda=0,2} \overline{V}_\lambda(R)\mathscr{Y}_\lambda(\Omega,\alpha_2)\ , \tag{1}$$

The monopole part of the interaction is given by

$$\begin{aligned} \overline{V}_0(R) &= v_a V_0(R)\ , \quad R > R_m \\ &= c(R-R_0)^2 - v_0\ , \quad R \leq R_m\ , \end{aligned} \tag{2}$$

where V_0 is the monopole part of the nuclear plus Coulomb interaction. The second line is the repulsive core mimicking the Pauli effect and by which we fix the energy of the first resonant state to the experimental Q-value, as done in ref. [2].

The $\lambda = 2$ form factor is the linear term in the expansion of the nuclear surface, i.e.

$$\overline{V}_2(R) = -v_2(R-R_0)\frac{d\overline{V}_0(R)}{dR}\ . \tag{3}$$

The matrix elements of the coupled system of differential equations for radial components of the solution have their only off-diagonal non-vanishing values given by

$$\overline{V}_2(R)\langle\mathscr{X}_2|\mathscr{Y}_2|\mathscr{X}_0\rangle \equiv -C_v(R-R_0)\frac{d\overline{V}_0(R)}{dR}\ , \tag{4}$$

depending upon a constant $C_v = v_2\langle\Phi_2||\alpha_2||\Phi_0\rangle/\sqrt{4\pi}$, that we use as a free parameter.

CP972, *Exotic Nuclei and Nuclear/Particle Astrophysics (II)—Proceedings of the Carpathian Summer School of Physics 2007*, edited by L. Trache and S. Stoica

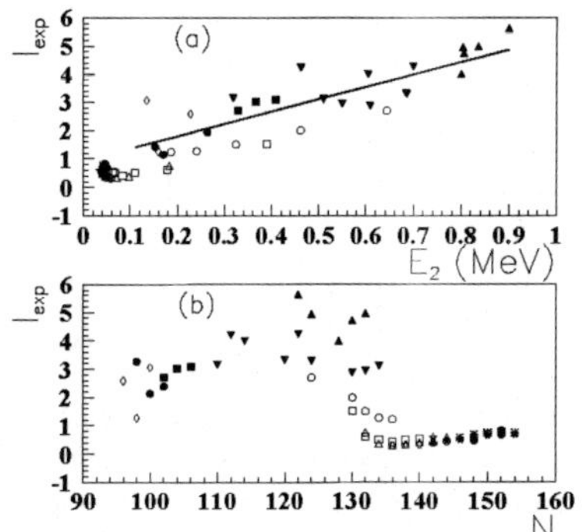

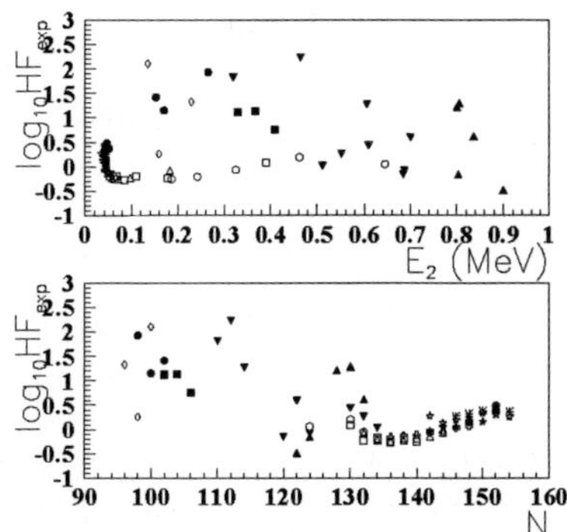

FIGURE 1.(a) Experimental values of the logarithm of the intensity ratio (5) versus the 2^+ energy E_2. Different symbols denote different isotopic chains. The straight line shows the fit of equation 8. (b) Experimental values of the logarithm of the intensity ratio (5) versus the neutron number.

FIGURE 2. (a) Experimental values of the logarithm of the hindrance factor (6) versus the 2^+ energy E_2. Different symbols denote different isotopic chains. (b) Experimental values of the logarithm of the hindrance factor (6) versus the neutron number.

Analysis of Experimental Data

Complete analysis can be found in [1]. We will investigate the so-called fine structure in α-decays to 2^+ states in even-even nuclei. It is defined as the logarithm of the ratio of Γ_0 to Γ_2, where Γ_0 (Γ_2) is the decay width (or intensity) to ground state (2^+ state), i.e.

$$I \equiv \log_{10}\frac{\Gamma_0}{\Gamma_2} \,. \tag{5}$$

In order to remove the effect of the Coulomb barrier and to compare the corresponding formation probabilities one defines the hindrance factor (HF) as

$$\mathrm{HF} \equiv \frac{\Gamma_0}{\Gamma_2}\frac{P_2}{P_0} \,, \tag{6}$$

where P_J is the standard penetrability

$$P_J \equiv \frac{2\kappa_J R}{G_J^2(\kappa_J R)} \,, \tag{7}$$

in terms of the irregular Coulomb function $G_J(\kappa_J R)$. For R we used in our calculations the touching radius, i.e. $R = 1.2(A_D^{1/3} + A_\alpha^{1/3})$. As it was shown in Ref. [2] the relation (6) gives results, that are close to the standard Rasmussen definition [3].

In figure 1 (a) we plot the logarithm of the intensity ratio I (5) versus the energy E_2. All values of the rotational region with $E_2 \leq 0.1$ MeV are concentrated around $I \approx 0.5$, while for the other nuclei one notices an increasing trend, which is best fitted by the following linear ansatz

$$I_{\rm fit} = 4.361\, E_2 + 0.924\ , \quad \sigma = 0.774\ , \tag{8}$$

shown by the solid line. Here σ denotes the mean error. In figure 1 (b) we plot the same quantity against the neutron number. The dependence of the experimental values $\log_{10}{\rm HF}_{\rm exp}$ (6) on E_2 is given in figure 2 (a) and on neutron number in figure 2 (b).

Next we analyzed the dependence of the fine structure (5) and hindrance factors (6) on various parameters of our model. Our model has four essential parameters: v_a and C_v give the attractive part of the potential, and c and v_0 define the repulsive core. In figure 3 we plot the dependence of the logarithm of the half life $T = \hbar \ln 2/\Gamma$ on the parameters v_a and C_v. One can see that it strongly depends on v_a while the dependence upon C_v is very weak.

Our goal is to investigate the fine structure given by (5). It mainly depends upon the vibrational parameter C_v, defining the coupling strength between the two considered channels with $J = 0,\ 2$. This can be seen in figure 4. We simplify the calculation by taking $v_a = 1$, corresponding to a "pure" α-cluster model. Therefore our calculations will provide the components $G_J(\kappa_J R)$ of the wave function giving the penetration factor P_J in (7). Actually the α-cluster exists on the nuclear surface with some probability $v_a < 1$ and therefore the ratio between theoretical and experimental half lives

$$S = \frac{T_{\rm th}}{T_{\rm exp}}\ , \tag{9}$$

will provide us this α-particle preformation probability, or the spectroscopic factor [4].

In ref. [2] we have shown that the parameters c and v_0 of the repulsive core, used to determine the Q-value, are related to each other. This, and the above analysis means that our calculation will depend on only two parameters, namely the coupling parameter C_v and the repulsive strength c.

We have analyzed the α-decay fine structure I (5) for emitters with evaluated experimental data on both Γ_0 and Γ_2. Our calculations show that I can be reproduced for vibrational nuclei only with a strong repulsive core. Here the hindrance factor is estimated as a ratio between the integral of the corresponding preformation amplitudes squared, computed in the region beyond the last maximum.

A plot of the dependence of the computed I of (5) on the vibrational coupling strength C_v for the α-decay to ^{196}Po for the adopted value of the repulsive strength can be found in [1] as figure 7 (a). There one observes a decreasing behaviour for $C_v \leq 0.15$, as already seen in figure 4, followed by a plateau and slow increase.

By using only one parameter C_v with $c = 2000$ MeV fm^{-2} it was possible to reproduce the experimental values $I_{\rm exp}$ of most measured vibrational nuclei. The results of our calculations are given in Table I of [1].

Concerning the last six transitional nuclei in Table I it turned out that their decay properties are closer to rotational nuclei investigated in our previous work [2], where we used a soft repulsive core. It turns out that only by using a rather small repulsive strength $c = 50$ MeV fm^{-2} it is possible to reproduce their fine structure.

It is interesting to witness the fact that the inverse of the vibrational coupling strength, $1/C_v$, is proportional to the square of the logarithm of the HF for all analyzed transitions, as shown in figure 5. Here the solid line represents the fit

$$\frac{1}{C_v} = 49.122\,(\log_{10}\mathrm{HF})^2\ , \quad \sigma = 15.636. \tag{10}$$

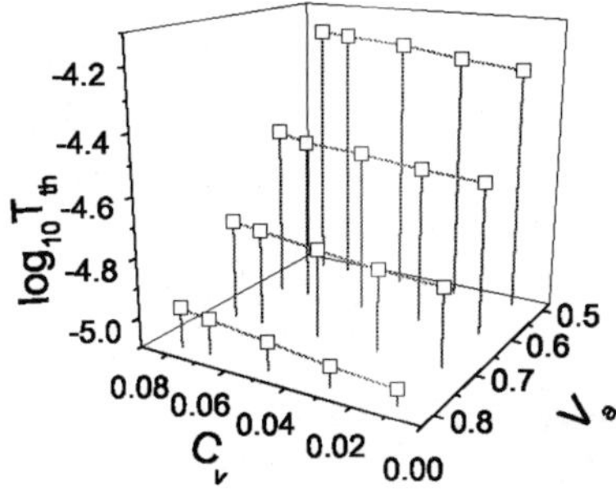

FIGURE 3. (Color online) Logarithm of the half-life (in seconds) as a function of the attraction parameters: double folding strength parameter v_a and vibrational coupling strength C_v.

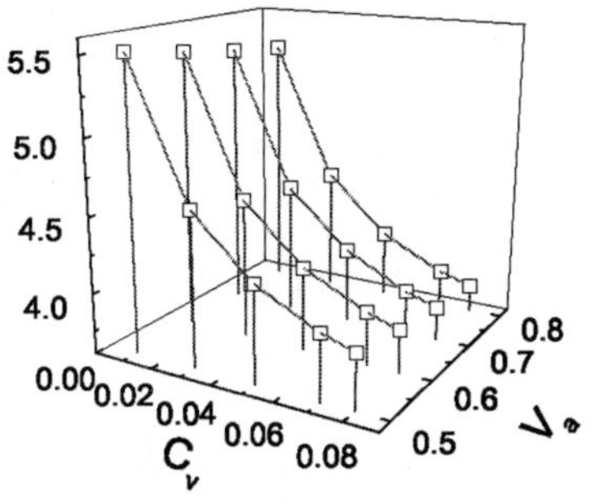

FIGURE 4. (Color online) Logarithm (5) of the intensity ratio as a function of the attraction parameters: double folding strength parameter v_a and vibrational coupling strength C_v.

Systematic Predictions

The relation (8) is able to describe the logarithm of the intensity ratio (5) within a factor of five. At the same time from figure 5 one sees that the relation (10), connecting the vibrational strength and HF, is fulfilled with a much better accuracy for all emitters. By using the definitions (5) and (6) one derives from the quadratic fit of (10) the following relation

$$I_{\mathrm{q}} = \frac{1}{\sqrt{49.122\,C_v}} - \log_{10}\frac{\mathrm{P}_2}{\mathrm{P}_0}\ . \tag{11}$$

On the other hand, the I_{q} provided by our calculation depends on the vibrational strength C_v as depicted in figure 7 (a) of [1]. Thus it is possible to simultaneously determine I_{q} and C_v, i.e. to obtain selfconsistency between the above relation and the value obtained as a result of the coupled channels calculation. This way we predict the fine structure I_{q} for several α-decays from vibrational nuclei. The results are given in the Table II of [1].

Finally, in figure 6, we give the values of the spectroscopic factor (9) versus the neutron number. We emphasize that one can reproduce the experimental half lives by using a strength v_a in (2), that is inversely proportional to $\log_{10}S$.

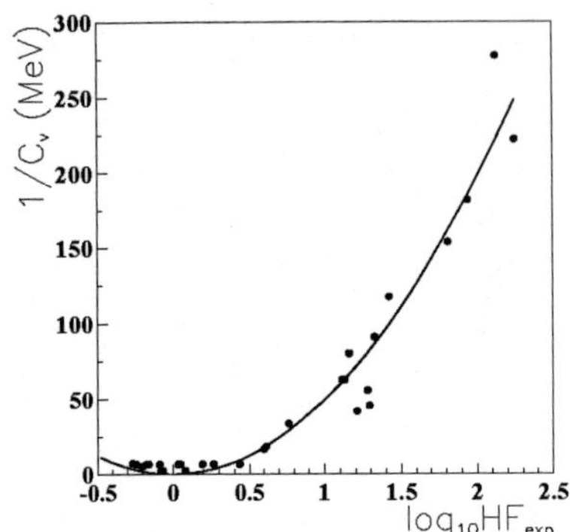

FIGURE 5. Inverse of the vibrational coupling parameter as a function of the logarithm of the experimental HF. The parabola is the fit given by (10).

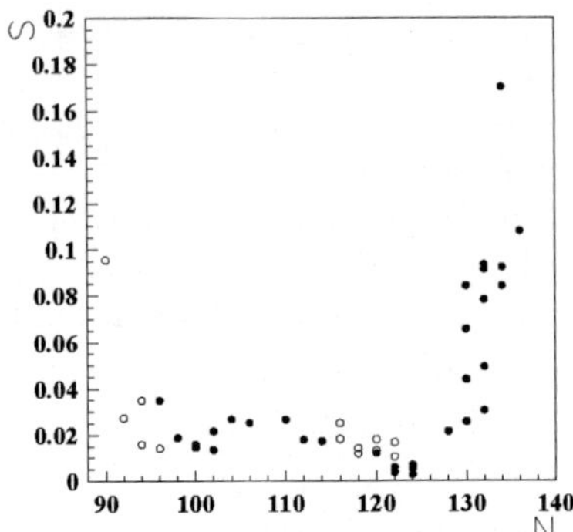

FIGURE 6. The spectroscopic factor (9) versus the neutron number for transitions in Table I (dark circles) and Table II (open circles) of [1].

Conclusions

We have analyzed the α-decay fine structure (5) for decays to 2^+ states and the corresponding hindrance factors (6) in vibrational-transitional nuclei. We have used a stationary coupled channels formalism to describe 2^+ states in even-even nuclei in terms of a vibrational coupling between the daughter nucleus and the α-particle. This interaction was computed as a derivative of the initial potential, simulated by a double folding procedure that uses M3Y plus Coulomb forces for the attractive part. The internal part was given by a parabolic repulsion, simulating the fact that an α-particle exists only on the nuclear surface. We reproduce the experimental Q-value by adjusting the depth of the repulsive interaction. As a result a pocketlike potential was obtained.

The α-decaying state was identified as the zero-nodes eigenstate inside the pocket-like potential. This interpretation is based on microscopic calculations. The analysis of experimental data showed that the logarithm of the intensity ratio depends linearly on the 2^+ energy. Our calculations show that a strong repulsive core is able to reproduce experimental data. By using one parameter, namely the vibrational coupling strength C_v, it was possible to reproduce experimental fine structure. We discovered a remarkable dependence (10) between the inverse of this parameter and the hindrance factor squared. Based on this we made systematic predictions for several vibrational α emitters.

REFERENCES

1. S. Peltonen, D. S. Delion and J. Suhonen, *Phys. Rev.* **C75**, 054301 (2007).
2. D.S. Delion, S. Peltonen, and J. Suhonen, *Phys. Rev.* **C73**, 014315 (2006).
3. J.O. Rasmussen, *Phys. Rev.* **113**, 1593 (1959).
4. P. Mohr, *Phys. Rev.* **C73**, 031301(R) (2006).

Observational Constraints on the Lepton Asymmetry and Radiation Energy Density

L.A. Popa and A.Vasile

ISS Institute for Space Sciences Bucharest-Magurele, Ro-077125 Romania

Abstract. By using most of the present Cosmic Microwave Background (CMB) and Large Scale Structure (LSS) measurements we set bounds on the radiation content of the Universe and neutrino properties. We consider lepton asymmetric cosmological models parametrized by the neutrino degeneracy parameter ξ_ν and the variation of the relativistic degrees of freedom, $\Delta N_{\rm eff}^{\rm oth}$, due to possible unknown physical processes during BBN.
We found $0 < \xi_\nu < 0.788$ (2σ) that implies a lepton asymmetry of the neutrino background $0.001 < L_\nu < 0.661$. We also found $0.079 < \Delta N_{\rm eff}^{\rm oth} < 2.555$ (2σ) and the contribution to the total number of relativistic degrees of freedom $-1.107 < \Delta N_{\rm eff} < 3.482$.
These results bring an important improvement over the same parameters obtained by using the WMAP data alone, showing that the LSS measurements are useful tools in constraining the radiation content of the Universe during BBN.

Keywords: CMBR theory, dark matter, cosmological neutrinos, gravitational lensing
PACS: 98.80.Bp 98.80.Es 98.80.Ft

INTRODUCTION

The primordial light element abundances predictions in the standard theory of the Big Bang Nucleosynthesis (BBN) [1, 2, 3, 4] depend on both baryon-to-photon ratio, η_B, and the radiation energy density at the BBN epoch usually parametrized by the number of relativistic degrees of freedom, $N_{\rm eff}$.
Meanwhile, the number of neutrino flavors have been fixed by Z^0 decay width [4] and the precision WMAP cosmic microwave background (CMB) temperature and polarization measurements [5] strengthen the limit on the neutrino mass to $\sum m_\nu \leq 0.23$ eV and lead to the determination of the baryon abundance $\eta_B = 6.14 \times 10^{-10}(1.00 \pm 0.04)$.

At the same time, more phenomenological extensions to the standard the neutrino sector have been studied, the most natural being the consideration of the lepton asymmetry, L_ν, parametrized by the degeneracy parameter ξ_ν. The present lepton asymmetry of the neutrino background is given by [6]:

$$L_\nu = \frac{1}{12\xi(3)} \frac{T_{\nu_0}}{T_{\rm cmb}} [\xi_\nu^3 + \pi^2 \xi_\nu], \quad \xi_\nu = \frac{\mu_\nu}{T_{\nu_0}}. \tag{1}$$

In the above equation μ_ν is the neutrino chemical potential, T_{ν_0} is the present temperature of the neutrino background, $T_{\nu_0}/T_{\rm cmb} = (4/11)^{1/3}$, and $\zeta(3)$ is the Riemann function of 3. In the early Universe, when the temperatures are very high, the non-perturbative (sphaleron) processes are likely to have occurred, converting lepton asymmetry to baryonic asymmetry. This leads to the assumption that the magnitudes of baryon and lepton

CP972, *Exotic Nuclei and Nuclear/Particle Astrophysics (II)—Proceedings of the Carpathian Summer School of Physics 2007,* edited by L. Trache and S. Stoica

asymmetries are of the same order.

There are many models in which higher orders of magnitude in the lepton asymmetry have been assumed. Unlike the baryonic asymmetry, the lepton asymmetry is not directly observable and so far only upper bounds have been obtained. The most stringent limits on the degeneracy parameter are obtained from BBN that gives $|\xi| < 0.07$ for each neutrino flavor in the case of no neutrino mixing [7, 8] and $|\xi_{\nu_e}| < 0.2$ and $|\xi_{\nu_\mu,\nu_\tau}| < 2.6$ for the case of mixing [9].

The effect of the lepton asymmetry on the CMB anisotropy and LSS matter power spectra is mainly related to the fact that the lepton asymmetry increases the energy density in the relativistic particles relatively to the standard case. This extra energy density can be parametrized by the variation of the number of relativistic degrees of freedom, $\Delta N_{\rm eff}$, that can be expressed as a function of the degeneracy parameter only:

$$\Delta N_{\rm eff}(\xi_\nu) = 3\left[\frac{30}{7}\left(\frac{\xi_\nu}{\pi}\right)^2 + \frac{15}{7}\left(\frac{\xi_\nu}{\pi}\right)^4\right]. \tag{2}$$

On the other hand, new low-energy physical processes such as active-sterile neutrino mixing, the existence of new particles as axions and gravitons, the time variation of the physical constants and many other non-standard scenarios [10] can contribute to the standard radiation energy density at BBN epoch. The extra energy density can be splitted in two distinct contributions, first due to lepton asymmetry of the neutrino background and second due to the extra contributions from other (unknown) processes, that can be parametrized as:

$$\Delta N_{\rm eff} = \Delta N_{\rm eff}(\xi) + \Delta N_{\rm eff}^{\rm oth}. \tag{3}$$

The aim of this paper is to jointly constrain the neutrino lepton asymmetry, L_ν, and the extra contributions to the radiation energy density, parametrized by $\Delta N_{\rm eff}$, by using most of the existing CMB and LSS measurements.

LEPTON ASYMMETRIC COSMOLOGICAL MODELS

We perform our analysis in the framework of the extended standard ΛCDM model cosmological model described by $6+3$ free parameters:

$$\Theta = (\underbrace{\Omega_b h^2, \Omega_{dm} h^2, \theta_s, \tau, n_s, A_s,}_{standard} f_\nu, \xi_\nu, \Delta N_{\rm eff}^{\rm oth}). \tag{4}$$

Here $\Omega_b h^2$ and $\Omega_{dm} h^2$ are the baryonic and dark matter energy density parameters, θ_s is the ratio of the sound horizon distance to the angular diameter distance, τ is the reionization optical depth, n_s is the scalar spectral index of the primordial density perturbation power spectrum and A_s is its amplitude at the pivot scale $k_* = 0.05$ hMpc^{-1}. The additional three parameters denote the neutrino energy density fraction f_ν, the neutrino degeneracy parameter ξ_ν and the extra contributions from other (unknown) processes $\Delta N_{\rm eff}^{\rm oth}$.

Observational implications of lepton asymmetry for density perturbations have been discussed in literature [11, 12, 13, 14]. We applied them to modify the Boltzmann Code for Anisotropies in the Microwave Background (CAMB) [15, 16, 17] to compute the CMB temperature and polarization anisotropies power spectra and LSS matter density fluctuations power spectra for the case of three degenerate neutrinos/antineutrinos with mass m_ν and degeneracy parameter ξ_ν.
The measured neutrino mixing parameters imply that neutrinos reach approximate chemical potential equilibrium before BBN epoch so that we consider in our computation that all three flavors of neutrinos/antineutrinos have the same degeneracy parameter ξ_ν. We also consider for simplicity that all there neutrino/antineutrino flavors have the same mass m_ν.

When the Universe was hot enough, neutrinos and antineutrinos of each flavor behave like relativistic particles with Fermi-Dirac phase space distributions:

$$f_\nu(q) = \frac{1}{e^{E_\nu/T_\nu - \xi_\nu} + 1}, \quad f_{\bar{\nu}}(q) = \frac{1}{e^{E_{\bar{\nu}}/T_\nu - \xi_{\bar{\nu}}} + 1}, \tag{5}$$

where $E_\nu = \sqrt{q^2 + a^2 m_\nu}$ is the one flavor neutrino/antineutrino energy with mass m_ν, degeneracy parameter ξ_ν and the comoving momentum $q = ap$. Hereafter, a is the cosmological scale factor ($a_0 = 1$ today).
We compute the mean energy density and pressure of one species of massive degenerated neutrinos and antineutrinos as:

$$\rho_\nu + \rho_{\bar{\nu}} = (k_B T_\nu)^4 \int_0^\infty \frac{d^3q}{(2\pi)^3} q^2 E_\nu (f_\nu(q) + f_{\bar{\nu}}(q)), \tag{6}$$

$$3(P_\nu + P_{\bar{\nu}}) = (k_B T_\nu)^4 \int_0^\infty \frac{d^3q}{(2\pi)^3} \frac{q^2}{E_\nu} (f_\nu(q) + f_{\bar{\nu}}(q)), \tag{7}$$

and the neutrino and antineutrino number densities as:

$$n_\nu + n_{\bar{\nu}} = (k_B T_\nu)^4 \int_0^\infty \frac{d^3q}{(2\pi)^3} q^2 (f_\nu(q) + f_{\bar{\nu}}(q)). \tag{8}$$

We re-write in CAMB the expressions for the energy density and the pressure in the relativistic and non-relativistic limits, for the degenerate case, as described in [14]. In order to compute the perturbed quantities we follow the standard procedure [17, 18, 19] and expand the phase space distribution function of neutrinos and antineutrinos into homogeneous and perturbed inhomogeneous components.

ANALYSIS

We use the CosmoMC Monte Carlo Markov Chain (MCMC) public package [20] modified for our extended $6+3$ parameter space to sample from the posterior distribution giving the following experimental datasets.
The Cosmic Microwave Background (CMB): For our analysis we use the WMAP 3-year data [21, 22] and the CMB data from Boomerang [23, 24], ACBAR [25] and CBI [26]

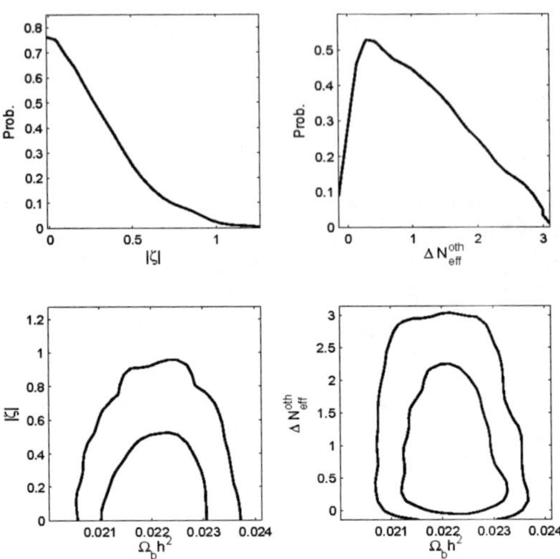

FIGURE 1. Top: the marginalized likelihood probability distribution functions for ξ_ν and $\Delta N_{\rm eff}^{\rm oth}$. Bot tom: 2D marginal 68% and 95% allowed regions for the indicating pairs of cosmological parameters.

experiments. The WMAP temperature and polarization datasets can constrain alone the basic set of the standard ΛCDM cosmological model parameters. As ACBAR and CBI measurements probe the amplitude of CMB temperature anisotropy power spectrum at small scales (l<900), they have the potentiality to complement the WMAP data, improving the analysis by better constraining the baryon energy density and the shape of the primordial power spectrum.

Large Scale Structure (LSS): We also use the large scale structure matter power spectrum from galaxy clustering data of the Sloan Digital Sky Survey (SDSS) [27] and Two-degree Field Galaxy Redshift Survey (2dFGRS) [28].

Type Ia Supernovae (SNIa): We also use Type 1A supernovae data [29, 30] which improves the degeneracy between Ω_m and Ω_Λ. In the computation we used gaussian priors on the Hubble constant $H_0 = 72 \pm 8$ km s^{-1}Mpc^{-1} as given by the HST key project [31] and on $\Omega_b h^2$ as given by the BBN constraints.

The top panels from Figure 1 present the marginalized likelihood probability distribution functions obtained for ξ_ν and $\Delta N_{\rm eff}^{\rm oth}$. In the bottom panels from the same figure we present the 2D marginal 68% and 95% allowed regions for the indicating pairs of cosmological parameters.

From the likelihood probability distributions we compute the mean expectation values of ξ_ν and $\Delta N_{\rm eff}^{\rm oth}$ and their errors. We found from our analysis a value of the degeneracy parameter $\xi_\nu = 0.299^{+0.489}_{-0.299}$ that implies a lepton asymmetry of the neutrino background $L_\nu = 0.242^{+0.419}_{-0.241}$[1]. We also found the contribution to the radiation energy density from

[1] All errors are obtained at 95% CL

other (unknown) physical processes $\Delta N_{\rm eff}^{\rm oth} = 1.11^{+1.445}_{-1.031}$ and the contribution to the total number of relativistic degrees of freedom $\Delta N_{\rm eff} = 1.187^{+2.295}_{-2.294}$. These results bring an important improvement over the same parameters obtained by using the WMAP data alone [13], showing that the LSS measurements are useful tools in constraining the radiation content of the Universe during BBN.

REFERENCES

1. R. V. Wagoner et al, *Ap.J.*, **148**, 3, 1967
2. K. A.Olive et al, *Phys. Rept*, **333**, 389, 2000 [astro-ph/9905320]
3. S. Burles et al, *Ap.J.*, **552**, L1, 2001 [astro-ph/0010171]
4. S. Eidelman et al, *Phys. Lett. B*, **592**, 1, 2004
5. D. N. Spergel et al, *Ap.J.Suppl*, **170**, 377, 2007
6. H. Kang, G. Steigman, (Nucl.Phys), **B 372**, 494, 2002
7. A. D. Dolgov, S. H. Hansen,S. Pastor, S. T. Petcov,G. G.Raffelt, D. V. Semikoz, *Nucl.Phys.*, **B632**, 363, 2002
8. P. D. Serpico and G. G. Raffelt, *Phys.Rev.D*, **71**, 127301, 2005
9. S. H. Hansen, G. Mangano, A. Melchiorri, G. Miele and O. Pisanti, *Phys.Rev.D*, **65**, 023511, 2002
10. S. Sarkar, *Rept.Prog.Phys*, **59**, 1493, 1996 [hep-ph/9602260]
11. J. Lesgourgues, S. Pastor, *Phys. Rev. D* **60** 103521, 1999 [hep-ph/9904411]
12. M. Orito, T. Kajino, G. J. Mathews, Y. Wang, *Phys. Rev. D*, **65** 123504, 2002 [astro-ph/0203352]
13. M. Lattanzi, R. Ruffini,G. V. Vereshchagin, *Phys. Rev. D* **72** 063003, 2005 [astro-ph/0509079]
14. K. Ichiki, M. Yamaguchi and J. Yokoyama, *Phys.Rev D*, **vol.75**, Issue 8, id. 084017, 2007
15. W. Hu,*Phys.Rev.D*,**62** 043007, 2000 [astro-ph/0002238]
16. A. Challinor and A. Lewis, *Phys. Rev. D*, **71** 103010, 2005 [astro-ph/0502425]
17. A. Lewis,A. Challinor and A. Lasenby, *Ap.J.* **538** 473, 2000 [astro-ph/9911177] [2]
18. C. P. Ma, E. Bertschinger, *Ap.J.*, **455** 7M, 1995
19. U. Seljak, M. Zaldarriaga, *Ap.J.* **469** 437S, 1996
20. A. Lewis and S. Bridle, *Phys. Rev. D* **66** 103511 [3]
21. G. Hinshaw et al (WMAP Collaboration), *Ap.J.Suppl.*, **170**, 288 , 2007 [astro-ph/0603451]
22. L. Page et al (WMAP Collaboration), *Ap.J.Suppl.*,**170**,335, 2007 [astro-ph/0603450]
23. C. B. Netterfield et al., *Ap.J.*, **571**, 604,2002, astro-ph/0104460
24. C. J. MacTavish et al., *Ap.J.*, **647**, Issue 2, pp. 799-812, 2006
25. C. I. Kuo et. al., *Ap.J.*, **600**, 32, 2004 [astro-ph/0212289]
26. A. C. S Readhed et al., *Ap.J.*, **609**, Issue 2, 2004, pp. 498–512
27. M. Tegmark et al [SDSS Collaboration], *Phys. Rev. D*, **69** 103501, 2004 [astro-ph/0310725]
28. S. Cole et al [2dFGRS Collaboration], *Mon.Not.Roy.Astron.Soc.*, **362**, 505, 2002 [astro-ph/0501174]
29. A. G. Riess et al., *Ap.J.*,**607**, Issue 2, pp.665-687, 2004, astro-ph/0402512
30. P. Astier et al., *A &A*, **447**, pp. 31-48, 2006
31. W. L. Freedman et al., *Ap.J.*, **553**, 47, 2001 [astro-ph/0012376]

[2] http://camb.info

[3] http://cosmologist.info/cosmomc

Numerical Modeling of Scission Neutrons Emitted During Nuclear Fission

M. Rizea* and N. Carjan†

*"Horia Hulubei" National Institute of Physics and Nuclear Engineering P.O. Box MG-6, Bucharest, Romania
†Centre d'Etudes Nucleaires de Bordeaux - Gradignan, UMR 5797, CNRS/IN2P3 - Universite Bordeaux 1, BP 120 Le Haut Vigneau, 33175 Gradignan Cedex, France

Abstract. We have considered a problem related to nuclear fission, namely the emission of scission neutrons. To study this process we have used the sudden approximation and also a dynamical approach. Numerically, this implies the solution of the Schrödinger equation, both stationary and time-dependent. To describe deformed nuclear shapes, we have used a model based on cylindrical coordinates. The Hamiltonian is discretized by using finite difference approximations of the derivatives. The eigenstates of the two-dimensional stationary Schrödinger equation are obtained as solutions of an algebraic eigenvalue problem, using an appropriate procedure (Arnoldi). In the sudden approach, the physical characteristics of the phenomenon are obtained from the wavefunctions before and after scission. In the more realistic evolutional (diabatic) model, the initial wavefunctions are propagated in time, solving the temporal Schrödinger equation by a Crank-Nicolson scheme.

Keywords: nuclear fission, scission neutrons, sudden approximation, diabatic approach, two-dimensional Schrödinger equation, deformed nuclear shapes, single particle eigenstates
PACS: 02.60.Lj; 02.70.Bf; 03.65.Ge; 25.85.Ec

INTRODUCTION

The emitted neutrons during fission can be differentiated as follows: the scission neutrons (10^{-21} - 10^{-20} s), the prompt neutrons (10^{-18} - 10^{-16} s) and the delayed neutrons ($> 10^{-16}$ s). Most fission neutrons belong to the last two categories, which have been extensively studied. However, it was recongnized that the scission neutrons should also be investigated, since they influence the fission dynamics and are key ingredients in the modelling of this process (see Refs.[1] - [4]).

In order to study the scission neutrons we have started with **the sudden approximation**. In this approximation the scission process is defined by a sudden transition between two very different nuclear configurations. The scission neutrons are characterized by their wave functions just-before-scission and immediately-after-scission. Numerically, these wave functions are the bound states of the bi-dimensional stationary Schrödinger equation with different deformation parameters for the nuclear shape description and, consequently, with different potentials. The potential is assumed to be axially symmetric (without other constraints on the shape). To obtain the bound states, one has to solve the eigenvalue problem:

$$\mathscr{H}\Psi = E\Psi \tag{1}$$

where $\mathscr{H}$ is the Hamiltonian (in cylindrical coordinates).

CP972, *Exotic Nuclei and Nuclear/Particle Astrophysics (II)—Proceedings of the Carpathian Summer School of Physics 2007*, edited by L. Trache and S. Stoica

THE BI-DIMENSIONAL SCHRÖDINGER EQUATION

The total wavefunction has two components, corresponding to **spin-up** and **spin-down** as follows

$$\Psi(\rho,z,\phi)=f_1(\rho,z)e^{i\Lambda_1\phi}|\uparrow>+f_2(\rho,z)e^{i\Lambda_2\phi}|\downarrow>, \quad (2)$$

$$\Lambda_1=\Omega-\frac{1}{2},\ \ \Lambda_2=\Omega+\frac{1}{2}.$$

Ω is the projection of the total angular momentum along the symmetry axis. The Hamiltonian has also two components: H_1 and H_2, of the form

$$H_1\Psi=O_1f_1-2K(S_af_2+S_cf_1), \quad (3)$$

$$H_2\Psi=O_2f_2-2K(S_bf_1+S_df_2). \quad (4)$$

The operators O_1,O_2 contain the Laplacean:

$$O_{1,2}=-\frac{\hbar^2}{2\mu}(\Delta-\frac{\Lambda_{1,2}^2}{\rho^2})+V(\rho,z),\ \ \Delta=\frac{1}{\rho}\frac{\partial}{\partial\rho}+\frac{\partial^2}{\partial\rho^2}+\frac{\partial^2}{\partial z^2}.$$

$\hbar$ is the Planck constant, μ is the reduced mass, V is the potential, K is a constant and

$$S_a=\frac{\partial V}{\partial\rho}\frac{\partial}{\partial z}-\frac{\partial V}{\partial z}\left(\frac{\partial}{\partial\rho}+\frac{\Lambda_2}{\rho}\right),\ S_b=-\frac{\partial V}{\partial\rho}\frac{\partial}{\partial z}+\frac{\partial V}{\partial z}\left(\frac{\partial}{\partial\rho}-\frac{\Lambda_1}{\rho}\right),$$

$$S_c=\frac{\partial V}{\partial\rho}\frac{\Lambda_1}{\rho},\ \ S_d=-\frac{\partial V}{\partial\rho}\frac{\Lambda_2}{\rho}.$$

To describe the nuclear shape we have used the Cassini parametrization, appropriate for strong deformations. It is defined by the relations (see Ref.[5]):

$$\rho=\frac{1}{\sqrt{2}}\left[Q(x)-R^2(x)(2x^2-1)-\varepsilon R_0^2\right]^{1/2}, \quad (5)$$

$$z=\frac{sign(x)}{\sqrt{2}}\left[Q(x)+R^2(x)(2x^2-1)+\varepsilon R_0^2\right]^{1/2} \quad (6)$$

where $Q(x)=\left[R^4(x)+2\varepsilon R_0^2R^2(x)(2x^2-1)+\varepsilon^2R_0^4\right]^{1/2}$, εR_0^2 is the squared distance from the focus of Cassinian ovals to the origin and $-1\leq x\leq 1$. The function $R(x)$ is represented as a series in Legendre polynomials: $R(x)=R_0[1+\sum\alpha_mP_m(x)],\ \ m\geq 1$. The set of parameters (ε,α_m) determines the nuclear shape.

To define the potential, we consider the function

$$\Phi(\rho,z)=\left[(z^2+\rho^2)^2-2\varepsilon R_0^2(z^2-\rho^2)+\varepsilon^2R_0^4\right]^{1/4}-R_0\left[1+\sum_{m\geq 1}\alpha_mP_m(x)\right].$$

The equation $\Phi(\rho,z)=0$ represents the nuclear surface in cylindrical coordinates. The nuclear potential is given by

$$V=-V_0\left[1+\exp(\Theta/a_0)\right]^{-1} \quad (7)$$

where V_0 is the depth and a_0 the diffuseness. The quantity Θ is an approximation to the distance between a point and the nuclear surface, given by the expression: $\Theta(\rho, z) = \Phi/|\nabla\Phi|$ (see Ref.[5]). The spin-orbit interaction is taken proportional to the gradient of the potential (7) - see Ref.[6]. Applied to functions of the mentioned form (up + down) this gives rise to the additional terms in the Hamiltonians (3), (4). For charged particles, the potential contains also a Coulomb term (Ref.[5]).

THE ALGEBRAIC EIGENVALUE PROBLEM

For numerical treatment, the infinite physical domain should be limited to a finite one, which is discretized by a grid with the mesh points:

$$\rho_j = (j - \frac{1}{2})\Delta\rho,\ j = 1, \dots, N_\rho,\ \ z_k = k\Delta z,\ k = -N_z, \dots, N_z \tag{8}$$

(the singularity at $\rho = 0$ of the Hamiltonian was considered). At each point the partial derivatives appearing in $\mathcal{H}$ are approximated by finite difference formulae. The original eigenvalue problem is transformed to an **algebraic eigenvalue problem**, whose eigenvector is formed by the values of f_1 and f_2 at nodes. The matrix is sparse and contains the coefficients of derivative approximations, as well as the potential values at the mesh points. Its order is two times the number of nodes (at each point the solution has two components) and is usually large.

To solve the resulting eigenvalue problem (of the form $\boldsymbol{Ax} = \lambda\boldsymbol{x}$) we have used the Implicitly Restarted Arnoldi Method (Ref.[7]). It is able to efficiently find selected eigenpairs of a large sparse matrix. Thus only the portion of the spectrum with physical relevance is calculated. An important feature is that only the matrix-vector product $\boldsymbol{Ax}$ is needed. Hence, the matrix $\boldsymbol{A}$ should not be generated nor stored, which is a great advantage in computation, saving time and memory.

APPLICATION TO SCISSION NEUTRONS

Firstly, we consider the scission as a sudden transition between two different nuclear configurations. The emission probability of a neutron that before scission has occupied the state $|\Psi^i>$ is given by:

$$P^i_{em} = 1 - \sum_{bound\ states} |a_{if}|^2, \tag{9}$$

$$a_{if} = <\Psi^i|\Psi^f> = 2\pi \int (f^i_1 f^f_1 + f^i_2 f^f_2)\rho d\rho dz.$$

$|\Psi^f>$ are the eigenstates of the immediately-after-scission Hamiltonian. Summing over all occupied states one obtains the total number of scission neutrons per fission event

$$\nu_{sc} = \sum_i v^2_i P^i_{em}. \tag{10}$$

v_i^2 is the ground-state occupation probability of $|\Psi^i>$. To obtain it one can consider the neutrons either independent or pairing correlated. In the first case $v_i^2 = 1$ if the eigenenergy e_i is below some limit (around Fermi level) and 0 otherwise (a step function). In the second case

$$v_i^2 = \frac{1}{2}\left[1 - \frac{e_i - \lambda}{\sqrt{(e_i-\lambda)^2 + \Delta^2}}\right],$$

with the parameters Δ and λ deduced from Bardeen-Cooper-Schrieffer equations. Using the eigenstates, other characteristics of the process can be obtained, as the primary fragments excitation energy and the spatial distribution of the emission points.

We have applied the above formalism to study the emission of scission neutrons during the fission of ^{236}U. The numerical domain was: $\rho \in [\Delta\rho/2, 32+\Delta\rho/2], z \in [-32, 32]$, while $\Delta\rho = \Delta z = 1/8$. Number of grid points: $N \approx 1.3 \times 10^5$. The two deformations between which the sudden transition is supposed to occur are characterized by the parameters $\varepsilon_i = 0.985$ and $\varepsilon_f = 1.001$ in the Cassini description of the nuclear shapes. The fission can be symmetric (each fragment has the mass 118) or, more frequently, asymmetric. As an example we present the case when one of the fragments has the mass $A_L = 96$. Depending of this mass and of ε one more deformation parameter is obtained, namely α_1. The resulting values are: 0.0729 and 0.0711. There are 41 bound states for each configuration ($e_i(1) = -35.01095, e_i(41) = -0.12676$ and $e_f(1) = -35.03973, e_f(41) = -0.11423$). The depth of the nuclear potential was $V_0 = 40.22$ and $\Omega = 1/2$. The calculated scission-neutron multiplicity and excitation energy are given in Table 1.

An alternative approach is to consider the fission as a time dependent (diabatic) process. This implies the solution of the bi-dimensional time-dependent Schrödinger equation - TDSE. The wave functions before scission are propagated on some temporal interval $[0, T]$. The potential is also time dependent, its deformation parameters changing during propagation. The moment $t = 0$ corresponds to $\varepsilon = 0.985$, while $t = T$ corresponds to $\varepsilon = 1.001$. Then the propagated functions are used instead Ψ^i in the previous calculations. Thus, the physical quantities are evaluated in a more realistic model, since the fission is not a sudden process, but takes some time.

Practically, we consider a time step $\Delta t = 1/256 \times 10^{-22}$s and advance in time a number n of steps by numerically solving at each step the time-dependent Schrödinger equation ($T = n\Delta t$). As initial solution one takes each Ψ^i. TDSE has the form

$$i\hbar \frac{\partial \Psi(\rho,z,t)}{\partial t} = \mathscr{H}(\rho,z,t)\Psi(\rho,z,t). \tag{11}$$

Its approximate solution is obtained by the formula (of Crank-Nicolson type):

$$\left(1 + \frac{i\Delta t}{2\hbar}\mathscr{H} + \frac{i\Delta t^2}{4\hbar}V'\right)\Psi(t+\Delta t) = \left(1 - \frac{i\Delta t}{2\hbar}\mathscr{H} - \frac{i\Delta t^2}{4\hbar}V'\right)\Psi(t).$$

Using the same discretization as before we obtain the solution at each time step by solving a linear system (with the biconjugate gradient method [8]). In the next table we present multiplicities and excitation energies (in MeV) obtained with different numbers of time steps ($A_L = 96$). We include also values given by sudden approximation ($n_{ts} = 0$).

TABLE 1. SN multiplicity and excitation energy

n_{ts}	ν_{sc} - IP	ν_{sc} - PC	E_{sc} - IP	E_{sc} - PC
0	0.24902	0.23937	4.26008	0.97000
64	0.24500	0.23598	4.22364	0.93791
128	0.23475	0.22585	4.08930	0.81106
256	0.20598	0.19793	3.76628	0.50864

One can see that the values of ν_{sc} and E_{sc} descrease when n_{ts} increases (the sudden approximation is an upper limit). $n_{ts} = 256$ means a process duration of 10^{-22} s.

CONCLUSION

In order to numerically describe the emission of scission neutrons from fissioning nuclei we have used the sudden approximation and a time dependent approach. This implies the solution of the two-dimensional Schrödinger equation, both stationary and time-dependent. Applying adequate procedures, we were able to determine physical quantities like neutron multiplicity and excitation energy. The neutrons were considered independent as well as pairing correlated. These assumptions give close results for multiplicites, but different for energies. The time-dependent approach is more realistic, but require more calculations. If the duration of scission is short enough (of order $1/4 \times 10^{-22}$s) the sudden approximation is reasonable. For a complete investigation, one has to consider a wide range of masses (e.g. $A_L \in [70, 118]$) and several values of Ω $(1/2, 3/2, \ldots)$. Calculations of this type for a broad scale of fissioning nuclei can contribute to an improved estimation and characterization of neutrons emitted during the nuclear fission.

REFERENCES

1. N.V. Kornilov, A.B. Kagalenko, S.V. Poupko, P.A. Androsenko, F.J. Hambsch:*New evidence of an intense scission neutron source in the Cf-252 spontaneous fission*, Nucl. Phys. A **686** (2001) 187-203
2. G.V. Valsky:*Scission neutrons in the general systematics of the yields of light particles from ternary nuclear fission*, Physics of Atomic Nuclei **67**(7) (2004) 1264-1276
3. G.A. Petrov: *Current status of the search for scission neutrons in fission and estimation of their main characteristics*, Nuclear Fission and Fission-Product Spectroscopy, AIP Conf. Proc. **798** (2005) 205-212
4. N.V. Kornilov, F.J. Hambsch, A. S. Vorobyev: *Neutron emission in fission*, Nucl. Phys. A **789** (1-4) (2007) 55-72
5. V.V. Pashkevich, *On the asymmetric deformation of fissioning nuclei*, Nucl. Phys. A **169** (1971) 275-293
6. J. Damgaard, H.C. Pauli, V.V. Pashkevich, V.M. Strutinsky, *A method for solving the independent-particle Schrödinger equation with a deformed average field*, Nucl. Phys. A **135** (1969) 432-444
7. D. Sorensen, R. Lehoucq, Chao Yang, K. Maschhoff, *ARPACK - An implementation of the implicitly restarted Arnoldi method for computing a few selected eigenvalues and corresponding eigenvectors of a large sparse matrix*, **www.caam.rice.edu/software/ARPACK**, Rice University (2001)
8. M. Botchev, D. Fokkema, *zbcg2.f90 - Improved BiCGstab(l) iterative method*, **www.math.uu.nl/people/vorst/software.html**, Utrecht University (2001)

Update on radio detection of inclined air showers with LOPES-10

A. Săftoiu*

* *National Institute for Physics and Nuclear Engineering, Horia Hulubei, Bucharest, Romania*
allixme@gmail.com

Abstract. We investigate here inclined air showers observed using the radio technique. They are well detectable, from the radio point of view, and allow analyses of angular correlations over a much broader range in geomagnetic angle than near-vertical events. We present an updated analysis of highly inclined ($> 50°$ zenith angle), high energy ($N_\mu > 10^5$) air showers measured with LOPES-10 in coincidence with KASCADE-Grande. Data from the Grande rather than the KASCADE array are used for the reconstruction of the air shower events, giving us access to a broader range of core distances for an independent cross-check with an earlier analysis.

Keywords: air showers, radio detection, LOPES, cosmic rays
PACS: 98.70.Sa

INTRODUCTION

Air showers, produced through primary cosmic particle interaction with the atmosphere, generate radio waves while propagating through the atmosphere. Various mechanisms have been proposed so far. The last of them, dating back only to 2003, explains the radio emission from air showers as a coherent geosynchrotron emission due to the deflection of charged particles, electrons and positrons, in the Earth's magnetic field [1, 2]. In order to bring more light on the features of the radio emission and confirm dominant mechanisms responsible for the emission, we must investigate in detail correlations of the measured field strength with shower parameters. For this purpose we use the LOPES set-up, placed within the KASCADE-Grande array (KArlsruhe Shower Core and Array Detector and Grande array), and investigate shower radio emission in coincidence with shower parameters obtained from independent particle detectors in the energy range of $10^{16} - 10^{18}$ eV [3].

An air shower develops in the atmosphere and evolves due to multiple interactions. The electromagnetic component has a maximum height above the ground and decreases rapidly. Consequently, only a small number of electrons (comparing with the electron number at the shower maximum) reach ground level, in particular for inclined showers. But the total number of electrons is expected to be approximately proportional to the strength of the radio pulse detected by the antennas. Therefore, better knowledge of radio emission parameters may give extra information on shower development.

Radio emission can be observed for showers with different zenith angles from vertical to highly inclined with no limitations. The antenna sensitivity poses no problems when zenith angle increases in comparison with flat particle detectors for which the reconstruction becomes unreliable for zenith angles larger than $60°$.

CP972, *Exotic Nuclei and Nuclear/Particle Astrophysics (II)—Proceedings of the Carpathian Summer School of Physics 2007,* edited by L. Trache and S. Stoica

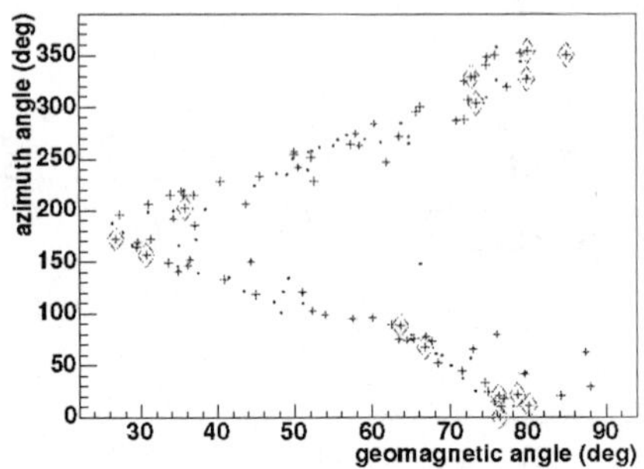

FIGURE 1. Azimuth angle as a function of geomagnetic angle. Red rhombs - clear radio signals and Grande reliable reconstruction; blue crosses - radio data and reliable Grande reconstruction; also processed with optimized beam forming procedure; black dots - all events recorded by Grande with zenith angle$> 50^0$ and $N_\mu > 10^5$.

DATA ANALYSIS

For the present study he have analyzed data recorded in 2004 by LOPES-10 which consists of 10 East-West polarization antennas placed inside KASCADE array [5].

The LOPES-10 data set undergo various selections in order to investigate different aspects. For this work we have selected only inclined events, zenith angle $> 50^0$, and high energy events, muon number $N_\mu > 10^5$. By analyzing inclined events we are able to investigate angular correlations over a much broader geomagnetic angle than vertical or near-vertical events. [1]

A previous work on inclined showers, [7], has reported correlations between measured field strength and shower parameters considering data in coincidence with the KASCADE array which covers an area on 200x200 m^2. The LOPES-10 antennas are also situated inside the KASCADE array, therefore only close events were processed. Now we extend the study and investigate events in coincidence with Grande data, keeping the zenith angle and energy cuts. This gives us the opportunity to investigate distant events of up to 800 m from the LOPES antennas. They have been detected by the Grande stations on an area of $\sim$0.5 km^2 [4]. For most large showers triggering KASCADE with at least 10/16 clusters fired there is data available simultaneously from the Grande array and the antennas. For 139 out of the 149 high energy inclined events, radio data was recorded with the LOPES array.

Because we use KASCADE-Grande data as a starting point in the digital beamforming of radio data we have to establish which of the detected radio events have reliable reconstructed parameters (core coordinates, azimuth and zenith angles).

Figure 1 shows the azimuth angle as a function of geomagnetic angle. Azimuth origin point (0°) is North. There are clear coherent events detected from North and South because the LOPES-10 antennas, measuring the East-West polarization only, are more

[1] The geomagnetic angle is the angle between the incident shower axis and Earth's magnetic field.

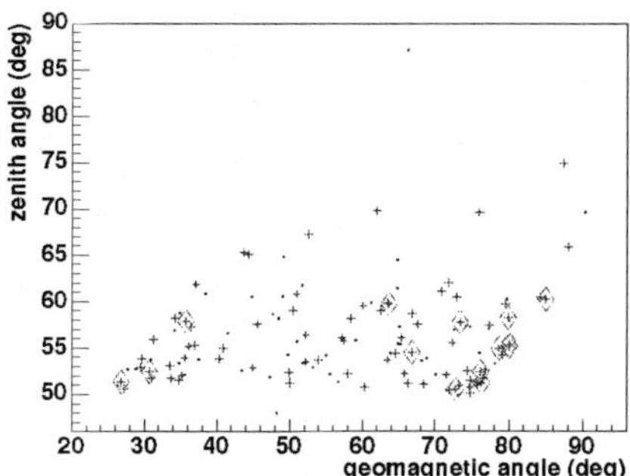

FIGURE 2. Zenith angle as a function of geomagnetic angle. Symbols are the same as in figure 1.

sensitive to such showers. For a shower coming from the East or West the emission is expected to be more North-South polarized, therefore such a shower wouldn't be detected with the LOPES-10 set-up. Also there are more events with azimuth angle corresponding to a northward direction than those coming from a southward direction. This effect is due to the fact that events coming from the North have larger geomagnetic angles, considering the orientation of the Northern Europe magnetic field.

Figure 2 shows the zenith angle as a function of geomagnetic angle for the same sets of data as figure 1. Most detected events are in the range of $50° - 60°$. Above this limit radio detection is possible but the shower parameters reconstruction becomes uncertain.

OPTIMIZED BEAM FORMING

The optimized beam forming procedure consists in varying the direction and core coordinates reconstructed for each event in the error range of Grande, i.e. $1.5°$ for the angle reconstruction and 20 m for the core coordinates reconstruction in order to improve beam parameters.

All the events marked with blue crosses in figures 1 and 2 were submitted to the optimized beam forming procedure. For each of these events sets of 50 randomly varied parameters were generated in the Grande error range.

For events with visible strong coherent signal the optimization lead to an important increase in the cross-correlation beam amplitude. For other 15 events the optimization improved the parameters only slightly, and did not clear reveal the radio signal. For the rest of the events the optimization process didn't make any difference.

CORRELATIONS

The dependence of the radio pulse height on different parameters can be separated into: A dependence on the geomagnetic angle, a dependence on the distance to the shower axis and a dependence on the primary energy ($\sim$muon number) [6, 1, 2, 7]. To separate

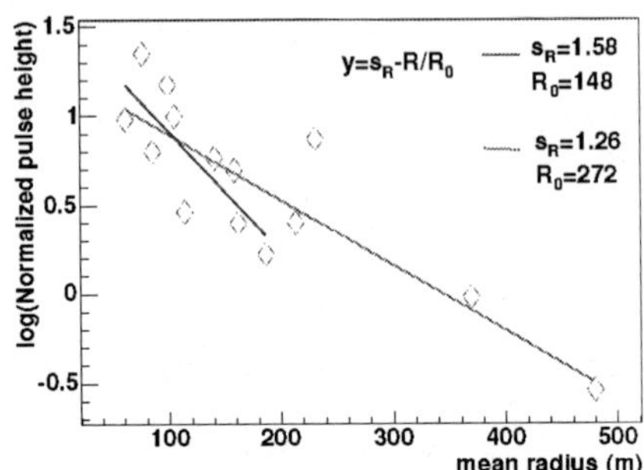

FIGURE 3. Dependence of normalized pulse height on mean distance from the shower axis to the antennas. The lines show results of a fit with a two parameter function: red line - parameter R_0=272 m, blue line - fit only for close events, mean distance<200 m, parameter R_0=148 m .

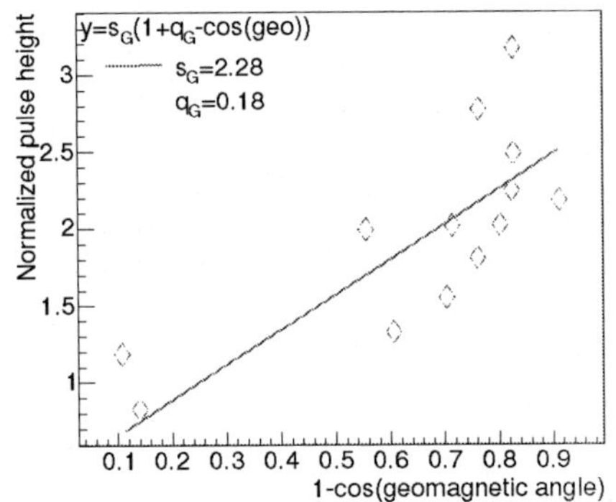

FIGURE 4. Dependence of the pulse height corrected for muon number and distance on the geomagnetic angle.

these parameters they are fitted separately in three iteration steps starting from the raw pulse height. In each fit the other dependencies are corrected for by dividing by the results of the fit from the previous iteration step.

Figure 3 shows the dependence of the pulse height, obtained after two iterations, corrected for muon number and geomagnetic angle, on the mean distance from the shower axis to the antenna set-up. There is a clear decrease with distance.

As shown in figure 4, there is a clear correlation of the pulse height, corrected for muon number and distance, with the geomagnetic angle which suggests that the emission is caused by the interaction of the shower with the Earth's magnetic field. The increase of the pulse height with geomagnetic angle supports the geosynchrotron theory.

Figure 5 shows a clear dependence of the pulse height, corrected for the geomagnetic angle and distance, on the muon number which is an approximate estimator of the primary particle energy. The dependence is fitted with a power law.

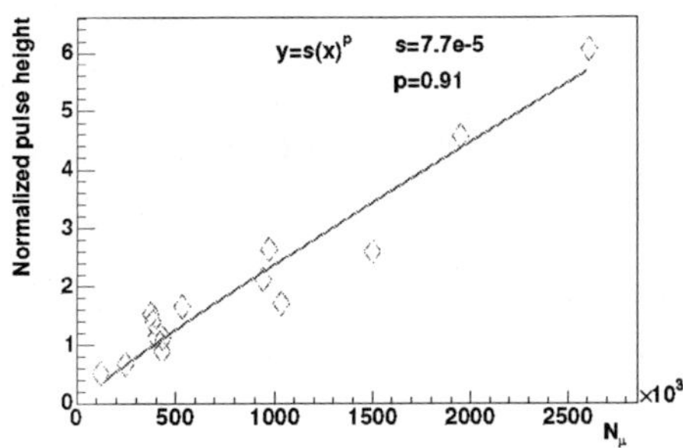

FIGURE 5. Dependence of the pulse height corrected for the geomagnetic angle and distance on muon number; fitted with a power law.

CONCLUSIONS

In this work we have extended previous studies of radio emission from inclined air showers to larger distances by analyzing LOPES-10 data in coincidence with KASCADE-Grande data. The predicted features, increase of the pulse height with the primary energy, decrease of the pulse height with distance and a dependence on the geomagnetic field are confirmed for LOPES-10 events analyzed in coincidence with the Grande array of the KASCADE-Grande experiment. In addition, the data proves that very inclined events can be detected in the radio domain at large distances showing the capabilities of the radio technique for measuring horizontal air showers.

ACKNOWLEDGMENTS

The author would like to thank the LOPES collaboration for the opportunity to perform this work and in particular to Tim Huege for the extended help. Thanks to the Carpathian Summer School organizers for the support and wonderful time.

REFERENCES

1. T. Huege and H. Falcke, Astron. Astrophys., 41, 19-34 (2003)
2. H. Falcke, P. Gorham, Astropart. Phys., 19, 477 (2003)
3. H. Falcke et al., Nature 435, 313 (2005)
4. A. Haungs et al., LOPES Coll., 30th ICRC Proc. Merida, Mexico, in press (2007)
5. A. Saftoiu et al., LOPES Coll.,30th ICRC Proc. Merida, Mexico, in press (2007)
6. H.R. Allan, Prog. in Elem. part. and Cos. Ray Phys., Vol. 10, 171 (1971)
7. J. Petrovic et al.,Astron. Astrophys., 462, 389 (2007)

Analysis of Shear Flows and Vortex Formation in Small-Scale Auroral Forms

Alexandra Spiridon[a]

[a]*Jacobs University Bremen, College Ring 1, D-28759 Bremen,Germany*

Abstract. Space weather is associated with and influences various processes in the Earth's atmosphere. Some of these influences can be seen as auroral emissions. From the large spectacular features such as arcs, rays and spirals to smaller but just as fascinating curls, folds and kinks, they all contribute to a better understanding of the mechanism and triggers behind such spectacular phenomena. For such purposes, the present study examines the optical data acquired in a campaign carried out by the Max-Planck-Institut für extraterrestrische Physik, Garching, Germany, which consists of 13 image sequences of auroral events. There has been a previous study based on the same set of data which analyzed one of the image sequences. The aim is to extend the analysis to all the image sequences with the purpose of obtaining a more statistical description of the results. The analysis is centered on the events that have not been analyzed before and looks to determine their spatial and temporal characteristics. Each curl system is described in terms of position in the auroral event, starting time of the event, duration, stability and brightness. The results consist of shear velocities, vorticities and brightness distributions, which are discussed in light of an electrostatic model of auroral acceleration. The aim is to check the validity of the model by looking at correlations or lack of correlation between the vorticities and brightness profiles.

Keywords: space physics, small-scale auroral forms, curl system
PACS: 92.60.hw

INTRODUCTION

Visible phenomena have always presented great interest. Auroral lights are considered to be some of the most fascinating of them and have been noted and researched for a long time. Auroral displays are produced when high-energy charged particles mostly electrons, impact the upper atmosphere and excite the atoms and molecules there. When they return to their initial states, they emit light. On the ground this is seen as Aurora Borealis.

Starting with the pioneering work of Hallinan and Davis [1], many studies and high-resolution observations have been carried out on small-scale spatially periodic distortions of auroral forms. This enabled a better understanding of their characteristics, dynamics and the processes behind their formation. Auroral curls can easily be recognized by their typical spiral form and in general have near-circular symmetry. They usually have a counterclockwise rotational motion as viewed from the ground in Northern Hemisphere and clockwise in the Southern Hemisphere. They are usually found in curl systems, groups of curls spaced quasi-periodically along an arc. They have wavelengths of less than 10 km, lifetimes of less than 2 sec and

CP972, *Exotic Nuclei and Nuclear/Particle Astrophysics (II)—Proceedings of the Carpathian Summer School of Physics 2007,* edited by L. Trache and S. Stoica

apparent horizontal velocities in the range 0 - 20 km/sec [1]. There can be more than one curl system in the same arc.

THEORY

Electrostatic Model of Auroral Acceleration

The parallel potential difference between the auroral acceleration region and the ionosphere below can be written as:

$$\Delta_{\parallel}\Phi = \int_{s_X}^{s_I} ds E_{\parallel} \tag{1}$$

s is the field-line coordinate and X corresponds to the auroral acceleration region and I corresponds to the ionosphere. If we consider the ambient magnetic field lines as converging field lines and apply magnetic field mapping [2] then the electric fields are:

$$m_1 E_{\perp X} = E_{\perp I} + \nabla_{\perp I}(\Delta_{\parallel}\Phi) \tag{2}$$

$m_1 = \sqrt{B_I / B_X}$ is the isotropic mapping factor. Neglecting the ionospheric electric field and with further calculations we get:

$$\nabla^2_{\perp I}(\Delta_{\parallel}\Phi) = -m_2 B_I \omega_{\parallel X} = -B_I \omega_{\parallel} \tag{3}$$

$m_2 = 1$ according to the isotropic mapping considered and $\omega_{\parallel}$ is the vorticity at ionospheric altitudes.

A local maximum of the parallel potential drop $\Delta_{\parallel}\Phi$ means $\nabla^2_{\perp I}(\Delta_{\parallel}\Phi) < 0$, which in turn corresponds to positive vorticity. Similarly a local minimum will be associated with negative vorticity. If the thermal flux is neglected for large accelerating potentials, the energy flux can be written as $\varepsilon \propto j_{\parallel} \cdot \Delta_{\parallel}\Phi$. Also assuming that brightness maxima corresponds to energy flux maxima, it is safe to say that there is the same correspondence between brightness and vorticity too. Thus, brightness maxima are associated with positive vorticity and brightness minima with negative vorticity. This is valid as long as the strength of the field-aligned-current does not decrease when the parallel potential drop increases [3].

Kelvin-Helmholtz Instability

The Kelvin - Helmholtz model of instability states that instabilities leading to the formation of auroral curls are due to fluid shear operating in the vicinity [1]. The Kelvin - Helmholtz dispersion relation relates the growth rate γ to the vorticity:

$$\gamma^2 = \frac{\omega_{\parallel}^2}{4}\left[e^{-4ka} - (1-2ka)^2\right] \tag{4}$$

Where, $\omega_{\parallel} = (\nabla \times v)$ is the vorticity component along the ambient magnetic field, k is the wave number, and $2a$ is the thickness of the shear layer in the initial phase of the instability. Under the assumption, that we considered before in the electrostatic model, for isotropic mapping, $\omega_{\parallel}$ and ka do not vary along magnetic field lines, thus removing the reference altitude assumption [3].

Throughout this paper we will denote by vorticity or velocity shear $\omega_{\parallel}$ and by rotational vorticity $\omega_{curl} = 4\pi/\tau$.

DATA COLLECTION

The data set was acquired during an auroral campaign carried out at the European Incoherent Scatter (EISCAT) site near Tromso, Norway, by the Max-Planck Institut für extraterrestrische Physik, Garching, Germany. The observations took place on 6th and 7th December, 1994 between 8 p.m. and 10 p.m. The instruments used were two image-intensified CCD cameras. There was an attempt to combine with data from the EISCAT radar however the latter was not positioned in the same field of view as the cameras, so it was not possible.

The cameras were operated at the standard PAL rate of 25 frames per second. One of the cameras was a wide-angled camera with an opening of 86° x 64° (190 km x 130 km). The other was a small-angled camera with an opening of 21° x 15° (40 km x 30 km). Time is expressed everywhere as UT and all spatial measurements are estimated considering a reference height of 110 km. For the study of small-scale auroral forms, data collected with this camera was used. Moreover, it was equipped with a filter to eliminate light from long-lived emissions (650 nm) from atomic oxygen which would render the small-scale structures hard to distinguish. For further information on the instruments, please refer to the work of Frey et al. [4].

ANALYSIS METHOD

The method used for the analysis of the image sequences is the "slice-stack-slopes" method which was introduced by Vogt et al. [3]. The three steps are described below. The end-result is a pair of profiles showing the velocity and vorticity distributions with relative brightness.

1. The first step in the method is to define cuts parallel to the flow and slice along those cuts through every frame in the sequence.
2. For each cut, the slices from every frame in the sequence are stacked in time to form time-cut diagrams. This method makes the transport phenomena visible through bright/dark oblique traces which indicate bright/dark substructures moving along the cut.
3. In these time-cut diagrams the slopes of the oblique traces can be measured and with appropriate geometry factors the velocities can be calculated. The result is then, an irregularly spaced time series of velocities for each cut.

The time series of velocities is then used to calculate vorticities and plot their distributions with brightness for each image sequence.

SUMMARIZED RESULTS AND CONCLUSION

The analysis results consist of numerical estimates of the general characteristics of curl systems and velocity and vorticity distributions with brightness for each image sequence. The distributions are too many to be shown here so only the discussion will be presented. We studied 53 small-scale distortions: 35 curl systems, 7 spinning patches and 11 arc decay events.

General Characteristics

The small-scale distortions studied had lifetimes in the range 0.4 s - 7.1 s with an average of 1.8 s and a preferred overall value of 0.7 s. For the 35 curl systems in particular, the average lifetime was 2 s and the same preferred value. The average wavelength for the curl systems was 3.9 km with a range of 2.1 km to 9.9 km and they had in general 3 curls. Velocities were in the range 0 to 100 km/s with a mean of 20 km/s and preferred values between 3 and 10 km/s. The average turnover time for the curls was 0.4 s and the average associated rotational vorticity was $\omega_{curl} = 31.4s^{-1}$. These results agree with what has been observed in previous studies although the velocities are somewhat higher.

Electrostatic Model of Auroral Acceleration

To check the validity of the model we looked at the vorticity distribution with brightness in each image sequence. We found that for 6 of the image sequences there is no correlation whatsoever between brightness and vorticity. For 4 of these it was expected. The dynamical activity varies on really short timescales. The few curl systems that we observed barely last a second and their vorticities are very low (less then $4s^{-1}$). It was however unexpected for the other two. The small-scale distortions last longer, 2.3 s on average, and the vorticities reach up to $10s^{-1}$. However brightness doesn't seem to vary so much, such that areas of minima and maxima are extremely small so it can be said the results are inconclusive.

There were 4 image sequences that showed an overall positive correlation. This wasn't particularly expected since in terms of small-scale structures and their characteristics, these sequences resemble some of the above. So it is surprising that they yield different results and makes it difficult to find the factor responsible. The dynamical activity varies rapidly however the brightness contrast is more clear and the vorticities are high, $8.5s^{-1}$. The highest vorticities were observed in this group $23.3s^{-1}$. So far it seems that positive correlation is associated with high vorticities and arc decay events or very short-lived curl systems.

The last 3 image sequences show an overall negative correlation. However, this doesn't seem to be correlated in some way to the formation of curl systems or other small-scale distortions. It seems to be somewhat random. It is clear that the electrostatic model of auroral acceleration is insufficient in most cases or maybe the wrong mechanism considered. It might be that there is a scale-length-mechanism involved as suggested by Vogt et al [3], however these results don't particularly support that theory either.

Kelvin-Helmholtz Instability Model

In this paper the analysis of the image sequences is based on a method introduced by Vogt et al. [3] which calculates the vorticities from distribution profiles, thus without making use of the Kelvin - Helmholtz dispersion relation. This makes it possible to test the validity of the Kelvin - Helmholtz model, having both the velocity shear and the spatial parameters estimated from the image sequences. The average values for velocity shear in the curl systems were much smaller, $\omega_{\square} = 10s^{-1}$ and thus inconsistent with the model expectation as calculated by Hallinan and Davis [1], $\omega_{\square} = 28s^{-1}$. However the rotational vorticities for individual curls were unexpectedly close to the values in the Kelvin - Helmholtz model $\omega_{\square} = 31.1s^{-1}$. The maximum growth rate occurs for $ka = 0.4$ as shown in the graph below (right) and the average value for ka is 0.7. Both values agree with results from previous studies.

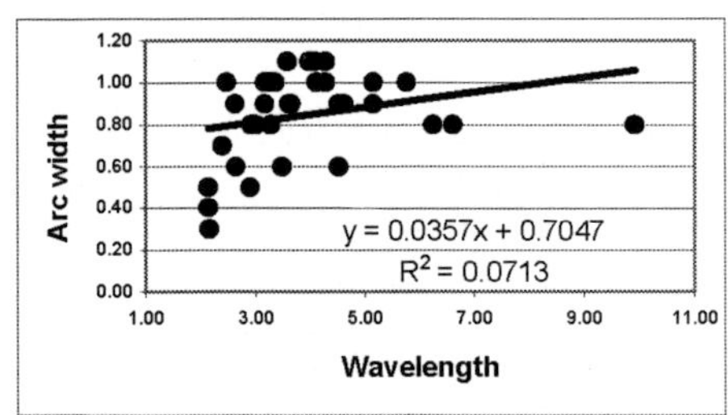

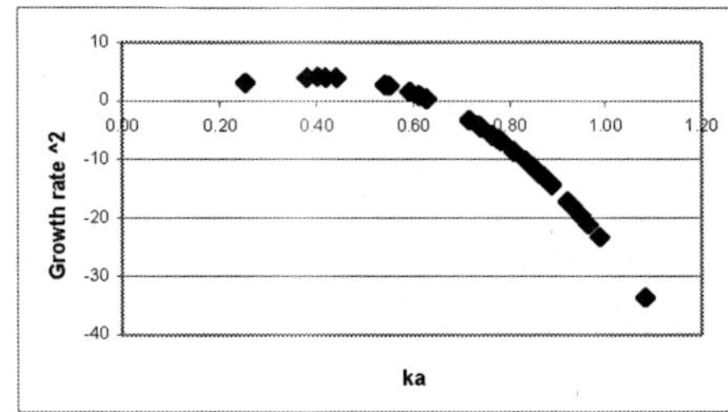

FIGURE 1. (left) Scatter plot, arc width as a function of wavelength, for curls systems. (right) Scatter plot, growth rate squared as a function of the *ka* parameter.

ACKNOWLEDGMENTS

I wish to thank Prof. Dr. Joachim Vogt for his support and advice. I also wish to thank the Max-Planck Institut für Extraterrestrische Physik, Garching, Germany for the optical data needed for this study.

REFERENCES

1. Hallinan, T. J. and T. N. Davis, "Small-scale auroral arc distortions", *Planet. Space Sci.*, 18, 1735, 1970.
2. Borovsky, J. E., Auroral arc thickness as predicted by various theories, *J. Geophys. Res.*, 98, 6101, 1993.
3. Vogt, J., Frey, H.U., Haerendel, G., Höfner, H., and Semeter, J.L.: 1999b, Shear velocity profiles associated with auroral curls, *J. Geophys. Res* 104, 17277.
4. Frey, H. U., Lieb W., Bauer O. H., Höfner H. and Haerendel G., CCD-camera system for stereoscopic optical observations of the aurora, *Proc. SPIE Int. Soc. Opt. Eng.*, 2863, 460, 1996.

No-Core Shell Model as an Effective Theory

Ionel Stetcu

Theoretical Division, Los Alamos National Laboratory, Los Alamos, NM 8754

Abstract. In the last years, much progress has been achived in the description of many-body systems. I discuss one possible way to improve upon the successes of the many-body approaches by employing general principles in order to construct more reliable interactions to be used in nuclear many-body calculations. Such approaches should give us more predictive power for the description of low-energy properties of nuclei far from stability.

Keywords: shell model, effective operators
PACS: 21.60.Cs

INTRODUCTION

More often than one would like, numerical approaches are used to describe quantum many-body systems. Of course, such approaches have, limitations. For example, in basis expansion methods, the infinite Hilbert space has to be truncated to a finite size, where is possible to obtain a numerical solution. Obtaining effective operators, which are designed to account for the correlations left out by the truncation, is an extremely difficult task [1].

At another level, the same issues appear already when one constructs the nucleon-nucleon (NN) and multinucleon interactions. Thus, the nucleons are not fundamental particles, and, in principle, the internucleon interactions are the direct consequences of the interactions between quarks and gluons inside nucleons. Such interactions are described by the QCD Lagrangian, but their consideration for the description of low-momentum observables of nuclei would be an overkill. Effective field theories (EFTs) provide a modern understanding of the theory of nuclear forces at low energies [2], even in the absence of exact solutions from QCD. The main idea behind EFT is that truncation to a model space generates all the terms allowed by the symmetries of the underlying theory, in the nuclear case the QCD. There is an infinite number of such terms, but they can be organized as an expansion of the relevant momentum scale over the nucleon mass, Q/M_N. Such an organization, ca lled power counting, allows for a consistent truncation of the relevant terms in the nuclear Hamiltonian. Moreover, it provides a systematic way to improve the results by including more terms in the calculation. The results are, thus, model independent and it is possible to estimate the errors, generated by neglecting higher order terms in the expansion.

The no-core shell model (NCSM) is a many-body technique that solves the non-relativistic Schröedinger equation for A nucleons in a truncated basis. It has been successfully applied to the description of nuclear properties of light nuclei, and progress is on the way to push the limits of applicability beyond mass 16 [4]. The method of choice to construct the effective interaction in the restricted model space is to perform a uni-

CP972, *Exotic Nuclei and Nuclear/Particle Astrophysics (II)—Proceedings of the Carpathian Summer School of Physics 2007,* edited by L. Trache and S. Stoica

tary transformation [3]. In this paper, I discuss an alternate method to renormalize the interaction for the finite space calculation. Such an approach has been proposed in some recent publications, and it has been successfully applied to describing states in ^{4}He and ^{6}Li [5], and harmonically trapped few-fermion systems [6].

NCSM AS AN EFFECTIVE THEORY: HOW TO CONSTRUCT AN INTERACTION

In NCSM, all particles are allowed to interact, and basis states are constructed using harmonic oscillator (HO) wave functions. The calculations can be done either in a Slater determinant (SD) basis [3], or in relative coordinates [7], where each Jacobi coordinate is described by a HO wave function. In a properly truncated SD basis (energy truncation), where one includes up to a maximum number of many-body excitations above a mean-field-like configuration, the center of mass can be removed exactly, and the results are identical with the ones obtained in the translationally invariant basis. Then, the intrinsic Hamiltonian of the A-body system is diagonalized numerically in the truncated basis.

Of course, any numerical calculation can be performed only by including a finite number of basis states. The space has to be truncated, and the correlations left out by this truncation have to be introduced not only in the effective Hamiltonian, but also in other effective operators defined in the model space. Thus, the usual approach is to start with an intrinsic Hamiltonian that includes, besides the relative kinetic energy, so called "realistic" two-body forces, which describe with high accuracy low-energy two-body observables, and theoretical three-body interaction adjusted to the three-body system. The method of choice to renormalize the interaction in the NCSM is a unitary transformation designed so that the effective Hamiltonian and general operators are energy dependent. (Note that in a HO basis another method avoiding the unitary transformation has been proposed, but only two- and three-particle results have been obtained in this framework [8].) However, c alculating the exact transformation is as hard as solving the infinite space A-body problem, so that a cluser approximation has to be employed. In this approximation, the transformation is obtained for $a < A$ particles, and then the effective operators are used in the A-body calculation. This approximation is not motivated *a priori*, but increasing the size of the model space reduces the associated errors, as by construction the effect of renormalization diminishes when one increases the model space.

For a wide range of nuclei, the agreement of the NCSM calculations with the experimental spectrum is, in general, remarkable. (Even more precise calculations are achieved in Green's function Monte Carlo framework [9], but this approach has other limitations that don't make the object of the present paper.) Other observables are, in general, more involved, as the renormalization is more complicated even in the lowest approximation ($a = 2$ or two-body cluster). While for short-range observables the unitary transformation renormalization in the lowest approximation produces effective operators that include well the effects of the truncation, and the rate of convergence with increasing the size of the model space is similar to the rate of convergence for the energy, much more work is required for the description of low-momentum observables [10].

The problems reported with the renormalization of long-range operators, as well as the need to produce more reliable effective interactions for heavier nuclei, have prompted us to investigate other avenues to produce effective operators in truncated spaces. Such methods treat the Hamiltonian and other observables on the same footing.

Because the truncation in HO basis can be treated as any other truncation, we have proposed recently a new approach based on the general principles of EFT. Thus, we construct internucleon forces directly in the truncated model spaces, using the power counting approach. That is, the interaction preserves the form in each model space, and the overall coupling constants of different terms in the Hamitonian are fixed to reproduce some observables. For example, in a theory where all the momenta above the pion mass are integrated out (pionless EFT), the leading order Hamiltonian writes as:

$$H = \frac{1}{2mA}\sum_{i,j}(\vec{p}_i - \vec{p}_j)^2 + C_0^0 \sum_{[i<j]^0} \delta(\vec{r}_i - \vec{r}_j) + C_0^1 \sum_{[i<j]^1} \delta(\vec{r}_i - \vec{r}_j) + D_0 \sum_{[i<j<k]} \delta(\vec{r}_i - \vec{r}_j)\delta(\vec{r}_j - \vec{r}_k), \tag{1}$$

where $[i<j]^s$ denotes pairs of particles in the S-wave NN channel of spin s –with corresponding parameters C_0^s– and $[i<j<k]$ triplets of particles in the spin-1/2 S-wave 3N channel —with parameter D_0. We have solved this Hamiltonian in a truncated HO basis, by adjusting C_0^1 in each model space so that we reproduce exacty the deuteron energy, and C_0^1 and D_0 so that one simultaneously reproduces the tritium and ^{4}He binding energies. This is somehow different from the usual approach in EFT, where the two-body forces are fixed by the scattering two-body data. However, because we work in a basis constructed from bound states, scatteting states are very difficult to describe when the number of states included is finite. Therefore, we used for fitting binding energies of ^{3}H, ^{4}He, and then we predicted the first excited state in ^{4}He within 10% of the experimental value, as well as the ^{6}Li binding energy within 70% of experiment [5]. While the LO results are not as precise as the ones obtained from phenomenological potentials, it has the advantage that it is consistent with QCD, with the information about QCD in the parameters. Moreover, neglecting operators beyond the LO is expected to produce an error of about 30%, so that the ^{6}Li result was the first successful application of the p ionless EFT to $A>4$ nuclei. In principle, adding next-to-leading order terms and beyond improves the precision, as it happens in continuum calculations. As a side note, it is interesting to find the limit of applicability of the pionless EFT for light nuclei, because by construction the theory breaks down if the processes described involve momentum transfer of order m_π or greater. When we have clear signal that it does fail, one can always apply the same principles to construct the interaction using the power counting in an EFT that includes pions (chiral EFT).

While the approach presented in Ref. [5] is perfectly valid, application beyond leading order EFT without pions (or even leading order in the theory with pions) becomes cumbersome because of the increased number of parameters which have to be adjusted to properties of light nuclei. This motivated us to devise in Ref. [6] a more flexible approach, to allow the adjustment of the two-body parameters to two-body data. The inspiration came from the more traditional NCSM. Thus, following Lipkin's idea [11], in the traditional no-core shell model one adds a center-of-mass HO term, which binds the nucleus, and all the $a<A$ clusters. The unitary transformation is constructed by the

requirement that in the truncated space the infinite space bound-state energies of a-body clusters are reproduced in the model space. The binding induced by the CM Hamiltonian is essential, as otherwize the clusters would not be always bound, since, for example, for $a = 2$ the NN interaction produces just one bound state, the deuteron. We follow the same approach of adding the CM Hamiltonian, and then, instead of constructing the unitary transformation, we use power counting to write down the interaction in a truncated HO basis. The coupling constants are then fixed to reproduce the "observables," which in this case are the energy levels of the cluster.

In the case of identical two-component fermion systems for which the range of the two-body interaction is neglijible, the power counting simplifies, and the Hamiltonian describing such systems includes just the first two terms in Eq. (1), plus the HO center-of-mass term. Two particles in a harmonic trap interacting via contact forces constitute the cluster, and the method outlined above can be used to obtain the renormalized Hamiltonian in a finite model space. In this case, the infinite space eigenvalues were obtained by Busch et. al. [12], and these levels are used to fix the coupling constants in each space. In particular, in LO, one fixes the strength of the two-body contact interaction to reproduce the ground-state energy of two trapped fermions. We have proved successfully that this approach reproduces the eigenvalues of the trapped three-fermion system governed by two-body interactions in the unitary limit [13]. We have further extended the approach away from unitary, and this allowed us to show an inversion of parity in the ground state of the system of three-fermions trapped in optical lattices. Thus, when the fermions interact only very weakly, the ground state has negative parity, but when the strength of the interaction is increased enough, the parity of the ground state becomes positive. Moreover, in the limit of untrapped system, we have shown that the three-body system is unbound and can be viewed as composed of a dimer and an additional particle.

CONCLUSION

In this paper, I have reviewed the first steps toward the construction of a new type of many-body theory, where the interaction is renormalized by means of the general principles of EFT, taking into account directly the basis and the model space where the many-body calculation would be performed. First applications were to the description of energy levels in light nuclei and identical fermions trapped in optical lattices. It is important to stress out that while the calculations have been carried out in a NCSM formalism and using an effective theory without pions, the same general principles can be applied using other many-body techniques, as well as to the chiral EFT. As a possible further application, we consider integration of nucleon degrees of freedom, which would allow us to perfom calculations similar to the more phenomenological shell model, where only a small number of nucleons outside a closed shell core, valence nucleons, are allowed to interact.

ACKNOWLEDGMENTS

I would like to thank B.R. Barrett, U. van Kolck, and J.P. Vary for enjoyable collaborations on the research reviewed here. This work was supported in part by the U.S. Department of Energy under grant number DE-AC52-06NA25396.

REFERENCES

1. B.R. Barrett, *Czech. J. Phys.* **49**, 1 (1998).
2. P.F. Bedaque and U. van Kolck, *Ann. Rev. Nucl. Part. Sci.* **52**, 339 (2002).
3. P. Navrátil, J.P. Vary, and B.R. Barrett, *Phys. Rev. Lett.* **84**, 5728 (2000); *Phys. Rev. C* **62**, 054311 (2000).
4. A. Lisetskyy and B. R. Barrett, talk at the INT worksop on New Approaches in Nuclear Many-Body Theory.
5. I. Stetcu, B.R. Barrett, and U. van Kolck, *Phys. Lett. B* **653** 358 (2007).
6. I. Stetcu, B.R. Barrett, U. van Kolck, and J.P. Vary, *Phys. Rev. A* (in press); arXiv:0705.4335.
7. P. Navrátil, G.P. Kamuntavicius, and B.R. Barrett, *Phys. Rev. C* **61**, 044001 (2000).
8. W. C. Haxton and T. C. Luu, *Phys. Rev. Lett.* **89**, 182503 (2002); W. C. Haxton, arXiv:0710.0289v1 (2007).
9. S. C. Pieper, R. B. Wiringa and J. Carlson, *Phys. Rev. C* **70**, 054325 (2004).
10. I. Stetcu, B.R. Barrett, P. Navrátil, and J.P. Vary, *Phys. Rev. C* **71**, 044325 (2005).
11. H.J. Lipkin, Phys. Rev. **109**, 2071 (1958).
12. T. Busch, B.-G. Englert, K. Rzążewski, and M. Wilkens, *Found. Phys.* **28**, 549 (1998).
13. F. Werner and Y. Castin, Phys. Rev. Lett. **97**, 150401 (2006).

The Upgrade Project at Texas A&M Cyclotron Institute

G. Tabacaru[a], J. Ärje[b], F. Abegglen[a], G.Chubaryan[a], G. Derrig[a], H.L. Clark[a], G. Kim[a], D. May[a], G. Souliotis[a] and R.E. Tribble[a]

[a]Cyclotron Institute Texas A&M University, College Station, 77843 TX, USA
[b]JYFL, Jyväskylä, Finland

Abstract. The Cyclotron Institute at Texas A&M University is currently under an upgrade process. The plan is to produce re-accelerated radioactive ion beams using the recommissioned 88" Cyclotron as a driver and two different devices. The technique for producing radioactive species will be presented.

Keywords: radioactive ion beams, sextupole ion guide.
PACS: 29.38.-c; 29.20.dg; 29.25.Rm; 29.27.Eg

PROJECT OVERVIEW

The Cyclotron Institute at Texas A&M University initiated [1] about 3 years ago a facility upgrade project. This project will extend the research capabilities as a stable beam facility with moderate rare beam capabilities. This will be achieved by re-activating the 88" Cyclotron to deliver high intensity light-particle and heavy-ion beams, to be used for production of rare isotopes for acceleration in the existing K500 Cyclotron. The plan is to produce radioactive species for re-acceleration by the existing K500 Cyclotron. The scientific program that drives this project is vary vast and here we'll just summarize the main items: nuclear astrophysics (the extension of the Asymptotic Normalization Coefficients method and study of the (^{3}He,d) reactions), nuclear structure (study of the Giant Monopole Resonances and the cluster structure of the nuclei using the radioactive beams), fundamental interactions and nuclear thermodynamics (multifragmentation). We are expecting also to gain valuable experience in the development of radioactive ion sources and different methods of diagnosis for weak beams.

The Project is divided in three tasks: a) recommission of the existing 88" Cyclotron and install new beam lines; b) construct light-ion and heavy-ion guides and produce and transport 1^+ radioactive ions; c) charge boost radioactive ions, transport and accelerate in the K500 Cyclotron.

The funding for this project comes from the U.S. Department of Energy, Texas A&M University, Cyclotron Institute and Welch Foundation. The Cyclotron Institute developed a testing facility: the Single Effect Events (SEE) facility. The SEE facility is designated for government and private companies for testing different types of electronic components that are implemented in satellites, aircrafts or other devices

CP972, *Exotic Nuclei and Nuclear/Particle Astrophysics (II)—Proceedings of the Carpathian Summer School of Physics 2007,* edited by L. Trache and S. Stoica

sensitive to radiation. The demands for using the SEE line at the Cyclotron Institute increasing each year and for the 2006-2007 year the SEE facility has been using about 50% from the total beam time [2].

RADIOACTIVE IONS PRODUCTION

Two techniques will be used to produce radioactive species: IGISOL (Ion Guide with a Separator On-Line) technique [3] – the Light-Ion guide and heavy ion deep inelastic reactions, fragmentation reactions in conjunction with a gas catcher – the Heavy-Ion guide.

Light-Ion Guide

The Light-Ion guide will produce radioactive species mainly from (p,n) reactions. The beam (a proton beam around 30 MeV) interacts with a production target (e.g. ^{27}Al) that's playing also the role of a separation foil for the gas cell. In the gas cell Helium gas is flowing continuously at constant pressure of 500 mbar maximum. The recoil ions (e.g. ^{27}Si from ^{27}Al(p,n)^{27}Si)) are trapped in the buffer gas and ejected at a 90° direction (with respect to the beam direction) through a small exit hole. The 1^+ ions are collected and transported by a Sextupole Ion Guide (SPIG): a non-resonant structure similar to the RFQ's from residual gas analyzers. A differential pumping system evacuates all the residual Helium gas. The 1^+ ions are injected then into a Charge Breeding ECR (CBECR) source which will further ionize them to higher charge states. The radioactive species are injected into the K500 Cyclotron and re-accelerated. The primary beam (proton beam) will exit the gas cell and will be stopped in the beam dump.

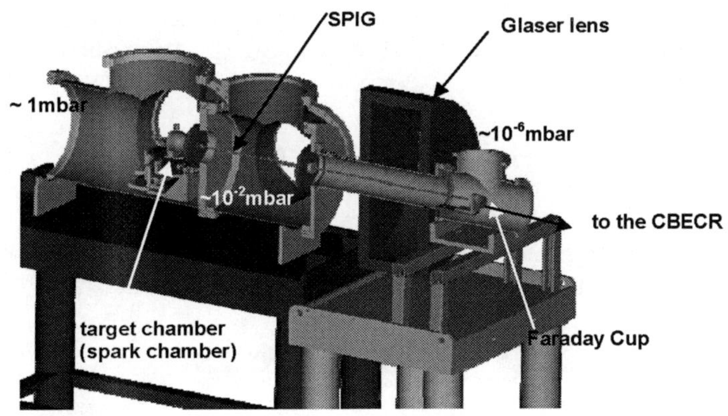

FIGURE 1. Detailed picture of the Light-Ion guide.

Off-line tests were performed using a spark discharge to create ions in the target chamber. Ions were transported 1 meter away through the SPIG and the current were measured on a Faraday cup (see Figure 1). The Glaser lens presented in Figure 1 is trying to reproduce the magnetic field from the injection side coil in the CBECR. No significant disturbance in the transport was observed. The maximum current intensity observed on the Faraday Cup was about 40 nA in certain conditions, which is sufficient for the future development and fine tuning. Our projected beams when the Light-Ion guide will be operated on-line are presented in Table 1.

TABLE 1. Projected beam intensities from the Light-Ion guide after the re-acceleration in the K500 Cyclotron.

Product (p,n)	Max Energy (MeV/A)	Intensity (particles/sec)
^{27}Si	57	4×10^4
^{50}Mn	45	1×10^5
^{54}Co	45	4×10^4
^{64}Ga	45	2×10^5
^{92}Tc	35	2×10^5
^{106}In	28	4×10^5
^{108}In	28	2×10^5
^{110}In	26	4×10^5

Heavy-Ion Guide

The second technique used for the production of the radioactive ion beams is based on a project developed at the Argonne National Laboratory [4]. A heavy ion beam will impinge a production target and the fragmentation products will be then separated with the superconducting solenoid BIGSOL. The radioactive products will be stopped in a gas catcher and transported to the CBECR. The gas catcher is a very complex structure where a combination of forces is present in order to obtain a fast extraction time over a big volume of gas. The forces that keep the radioactive species moving towards the exit orifice are related to the DC field, to the gas flow and to the RF electric field. Due to the complexity of this device extensive off-line tests are needed and currently a series of experiments were performed in order to characterize the functioning of the gas catcher. In the near future specific results will be presented for discussions.

ACKNOWLEDGMENTS

The authors would like to thank the staff of the Cyclotron Institute Texas A&M University for the commitment to their work. This work is supported by the U.S. Department of Energy under Grant number DE-FG02-93ER40773.

REFERENCES

1. "A proposed facility upgrade for the Texas A&M University Cyclotron Institute", http://cyclotron.tamu.edu/facility_upgrade.pdf .
2. H.L. Clark et al., "Progress in Research. April 1, 2006 – March 31, 2007", http://cyclotron.tamu.edu/2007%20Progress%20Report/cyclotron_progress2007.pdf .
3. J. Ärje et al., *Nucl. Instr. and Meth.* **A247**, 431 (1986).
4. G. Savard, "Gas Catcher Techniques", Working Group Meeting for the S3 Project at SPIRAL 2, Caen June 2006.

Investigation of the EAS Lateral Particle Density at 500 m Distance from Shower Core

G. Toma*

* *National Institute for Physics and Nuclear Engineering, Horia Hulubei, Bucharest, Romania*

Abstract. For the experimental conditions of the KASCADE-Grande experiment, the density of EAS charged particles at the distance of about 500 m from the shower core S(500) has been shown by detailed simulation studies to be an approximate energy estimator, being nearly independent of the mass of the primary particle. This report presents some experimentally observed features of the S(500) observable registered with the KASCADE-Grande array installed at the Forschungszentrum Karlsruhe, Germany. The measured energy deposits of particles in the 37 scintillation detector stations have been used to reconstruct the lateral charged particle distributions that are described by a Linsley parameterization (LDF). Among other features, the S(500) dependence from the EAS angle of incidence has been studied [1].

Keywords: air showers, cosmic rays, particle densities
PACS: 98.70.Sa

INTRODUCTION

It has been first shown by Hillas et al. [2] that the lateral charged particle density at a particular distance from the EAS core, dependent on the specific layout of the considered array, proves to be nearly independent of primary mass and maps only the primary energy. An estimate of the primary energy on this basis has been applied for various arrays [3]. KASCADE-Grande [4] consists as main component of an array of 37 detector stations of 10 m^2 scintillation counters, that covers a total area of 0.5 km^2. Dedicated simulation studies [5] have shown for the particular case of the layout of the KASCADE-Grande array that the charged particle density at the distance of around 500 m from shower centre appears most appropriate for the energy estimate.

RECONSTRUCTION OF S(500)

In context of the present studies, a software tool has been developed [6] and was used for analyzing the lateral particle density distributions event by event. The energy deposits of particles in detectors are converted into particle numbers [7] using appropriate lateral energy correction functions, dependent of the angle of incidence. The reconstructed particle numbers are converted into particle densities in the detectors. The next step of the analysis approximates the shape of the lateral particle density distribution by a Linsley LDF [8]. The LDF depends on three parameters, defining the total size and the lateral shape, which are determined by fitting the data. The results are used to deduce the value of the lateral particle density at 500 m from shower core, referred as S(500). The relation of S(500) to the primary energy, resulting from simulations is shown in

CP972, *Exotic Nuclei and Nuclear/Particle Astrophysics (II)—Proceedings of the Carpathian Summer School of Physics 2007,* edited by L. Trache and S. Stoica

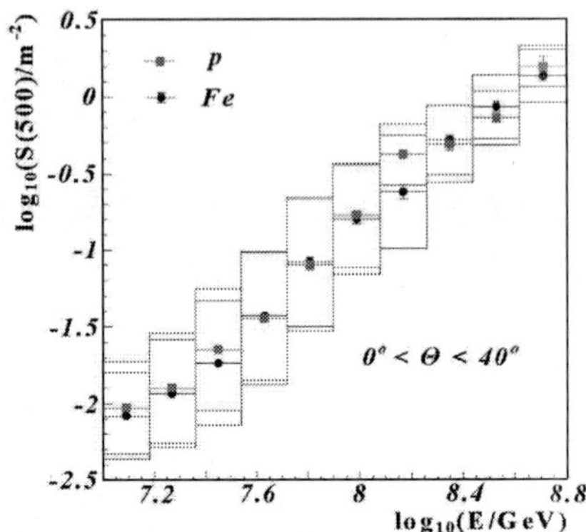

FIGURE 1. The relation of the mean values of S(500) to the primary energy E resulting from simulation studies for proton (squares) and Fe(dots) induced EAS: log S(500) vs. log E. In addition to the error of the mean (in many cases smaller than the dot size) the dispersion (standard deviation) is indicated by the error boxes.

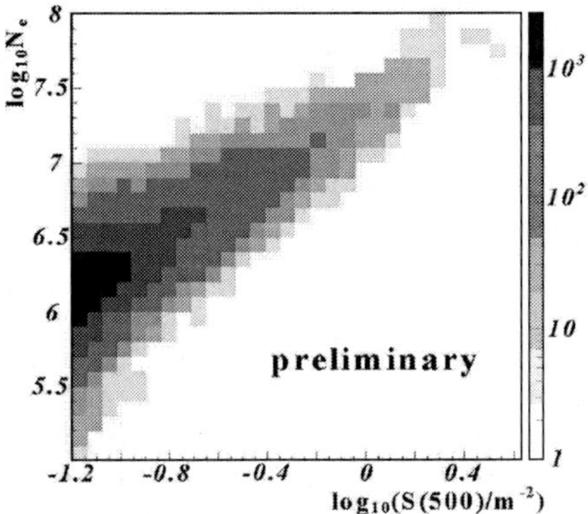

FIGURE 2. The dependence between electron shower size (N_e) and the S(500) for the given shower sample - a minimum of 20 triggered stations, showers inclined up to 45° degrees and shower cores reconstructed inside the array.

Fig.1 (simulations use the QGSJET II [9] model as high energy interaction model embeded in CORSIKA [10]; in this case, S(500) is evaluated for each shower with the same reconstruction procedure as for the experimental data).

We apply this energy estimator to the measured data in the primary energy range of about 10^{16}- 10^{18} eV. Fig. 2 shows a preliminary plot of the electron shower size (N_e) dependence with the S(500) for the given shower sample. The electron shower size is obtained using the standard reconstruction technique [11], while the S(500) is obtained with the described technique. The origin of the visible fluctuations is subject of further investigations. Fig. 3 shows for some S(500) ranges the measured charged particle distributions.

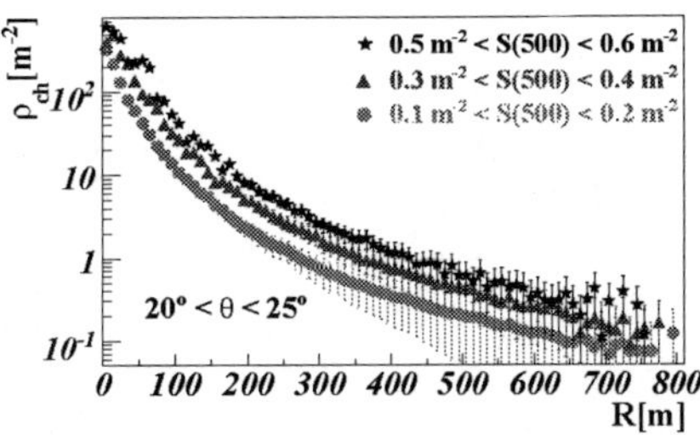

FIGURE 3. The measured charged particle lateral distributions for different S(500) ranges.

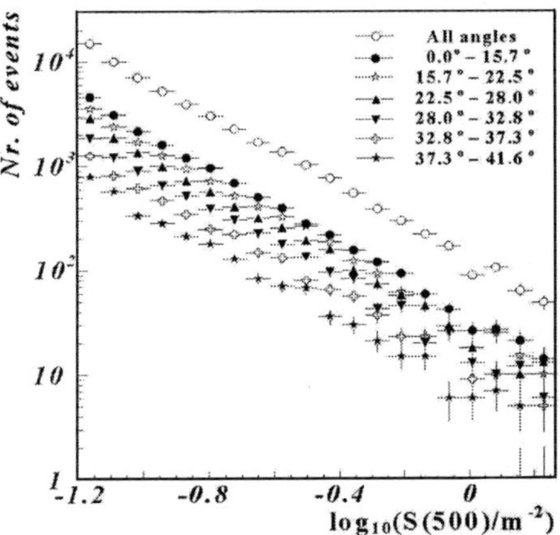

FIGURE 4. Differential spectra of the reconstructed S(500) observable for different EAS zenith angles opening equal solid angles and for the total shower sample.

DEPENDENCE OF S(500) ON THE ANGLE OF EAS INCIDENCE

A sample of the events detected by KASCADE-Grande has been studied in order to reconstruct the lateral distributions of charged particles and to determine S(500) distributions for EAS with different angles of incidence. In order to ensure a good quality of the shower sample, some restrictions have been applied with following requirements:(i)showers for which a minimum of 20 detector stations have been triggered,(ii) showers for which the reconstructed shower core falls inside the array and(iii) for which the Linsley LDF provided a good quality fit (as reported by MINUIT). Furthermore, only the showers for which the reconstructed zenith angle does not exceed 45° are considered. For values of S(500) > 0.6 m^2, the trigger threshold for all angular ranges is exceeded. Fig. 4 shows the distributions for all the showers and also specified for showers from different angles of EAS incidence in the full efficiency range of the KASCADE-Grande array.

In addition, the influence of attenuation in the Earth's atmosphere (reducing the value of S(500) for same number of events with increasing angle of EAS incidence), the spec-

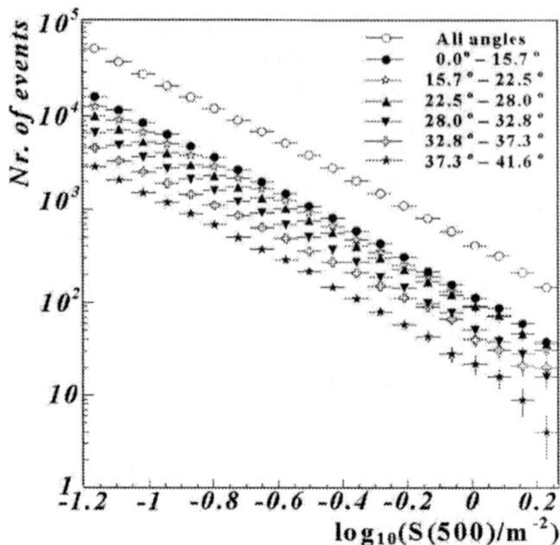

FIGURE 5. Integral spectra of the reconstructed S(500) observable for different zenith angles of EAS incidence as compared to that of the total shower sample.

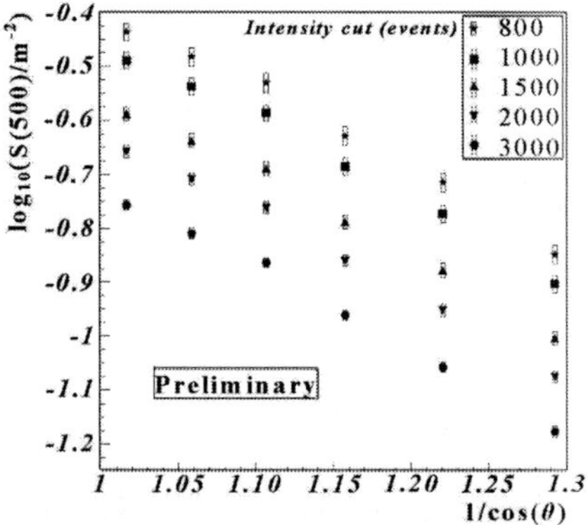

FIGURE 6. The S(500) dependence of the angle of incidence for various pre-chosen intensities (number of events). For a given angle of incidence, the $\log_{10}$S(500) value on the vertical axis is the corresponding $\log_{10}$S(500) in the integral spectrum for which we have an intensity equal to the one used for the constant intensity cut.

tra follow approximately the same slope for all angular intervals. This is a consequence of the isotropic incidence of the primary cosmic radiation. In Fig. 5 the integral S(500) spectrum is shown as derived from the data. Fig. 6 shows the S(500) dependence on the angle of EAS incidence for different pre-chosen constant intensities in the integral spectrum. After choosing different values for the intensity, the corresponding S(500) value has been calculated. Since the $1/\cos(\theta)$ value shows the atmospheric depth, the constant intensity cut provides a method for observing the S(500) attenuation with the atmospheric depth.

CONCLUSIONS

The value of particle density at 500 m from shower core was indicated by simulation studies to be a suitable energy estimator for EAS in the range of 10^{16}- 10^{18} eV. With this aspect, the experimental lateral density distribution of charged particles has been investigated for EAS events detected with the KASCADE-Grande array. The reconstructed experimental lateral density distributions have been approximated with a Linsley LDF and the values of particle density at 500 m distance from shower core S(500) are derived. The distribution of S(500), assumed to reflect the primary energy spectrum, and the dependence of S(500) on the angle of EAS incidence have been shown.

ACKNOWLEDGMENTS

The author would like to thank all colleagues from the KASCADE-Grande collaboration for their support and guidance during this study and also to the summer school organisers.

REFERENCES

1. G. Toma et. al. KASCADE-Grande collaboration, 30th ICRC Proc.,Merida,Mexico 2007, in press
2. A.M. Hillas et. al., Proc.12th ICRC, Hobart 3 (1971) 1001
3. D.M. Edge et. al., J. Phys. A: Math. Nucl. Gen.6 (1973) 1612; M. Nagano et. al., J. Phys. G:Nucl. Part. Phys.10(1984) 1295; Y. Dai et.al., J.Phys.G: Nucl. Part. Phys. 14(1998) 793; M. Roth et al.- Auger collaboration Proc. 28th ICRC, Tsukuba, Japan, vol. 2 (2003) 333
4. A. Haungs et al., KASCADE-Grande collaboration, Proc. 28th ICRC, Tsukuba, Japan, vol.2 (2003)985
5. H. Rebel and O. Sima et al. KASCADEGrande collaboration, Proc. 29th ICRC, Pune, India, vol.6 (2005)297 I.M. Brancus et al. KASCADE-Grande collaboration, Proc. 29th ICRC, Pune, India, vol.6 (2005)361
6. O. Sima et. al., Report FZKA 6985, Forschungszentrum Karlsruhe 2004
7. G. Toma et al., Proc.26th ECRS 2006, Lisbon, Portugal ; CERN program library, GEANT user's guide, (1997)
8. J. Linsley et. al., Journ. Phys. Soc. Japan 17 (1962) A-III
9. N.N. Kalmykov, S. Ostapchenko and A.I. Pavlov, Nucl. Phys. B (Proc. Suppl.) 52 (1997) 17v
10. D. Heck et al., CORSIKA: A Monte Carlo code to simulate extensive air showers, FZKA 6019, Forschungszentrum Karlsruhe, 1998
11. F. Cossavella et. al. KASCADE-Grande collaboration, 30th ICRC Proc.,Merida,Mexico 2007, in press

Neutral Current Neutrino-^{116}Cd Reaction Cross Sections at Low and Intermediate Energies

K.G. Balasi, T.S. Kosmas, P.C. Divari and V.C. Chasioti

Theoretical Physics Section, University of Ioannina, GR 45110 Ioannina, Greece

Abstract. Differential and integrated cross sections for inelastic scattering of electron neutrinos from a ^{116}Cd target at low and intermediate neutrino energies ($\varepsilon_i \leq 100$ MeV) are presented. The nuclear wave functions for the initial and final nuclear states are constructed in the context of quasi-particle random phase approximation (QRPA). By exploiting the results of these cross sections, we study the response of ^{116}Cd to supernova neutrino spectra by utilizing the folding procedure and employing a two-parameter Fermi-Dirac distribution for the supernova neutrino spectra. Our results show that, this isotope may play a significant role in supernova neutrino detection, in addition to its use in double-beta-decay experiments (COBRA, etc).

Keywords: Semi-leptonic electroweak interactions; Neutrino-nucleus reactions; Inelastic cross sections; Quasi-particle random phase approximation
PACS: 23.40.Bw;25.30.Pt;21.60.Jz;26.30.+k

INTRODUCTION

Neutrino-nucleus interactions have been proved to be of great importance in recent nuclear and astronuclear physics research. Such interactions enable us to study the neutrino intrinsic properties, the structure of the fundamental electroweak interactions and many astrophysical phenomena that take place in the interior of stars [1, 2, 3, 4].

In the last few decades several terrestrial experiments have been developed in order to investigate neutrino induced proccesses in nuclear and astroparticle physics [2, 4, 5]. On the other hand, neutrino-nucleus cross section calculations at low and intermediate neutrino-energies have been employed for studying the nuclear response of various targets to astrophysical neutrino spectra [3, 4, 6].

In the present work, we devote a special effort on this topic by performing realistic calculations for neutral-current neutrino-nucleus induced reactions within the quasi-particle random phase approximation (QRPA) [7, 8]. Particularly, we investigate the dependence on the scattering angle and initial neutrino-energy of the differential, integrated and total cross sections of the reaction $^{116}Cd(\nu,\nu')^{116}Cd^{*}$. We also evaluate the contribution to incoherent (inelastic) cross section of the various neutrino-induced transitions of this isotope. It is worth mentioning that, ^{116}Cd is already used by several recent experiments as a neutrinoless double-beta-decay target, such as COBRA [2], NEMO [5], etc., due to its high value of transition energy that makes ^{116}Cd a significant detector in avoiding background events in the double beta decay region.

One of our primary purposes in this work, is to investigate the response of the ^{116}Cd isotope to supernova neutrino spectra and explore its possibility to be used as a supernova neutrino detector as well.

CP972, *Exotic Nuclei and Nuclear/Particle Astrophysics (II)—Proceedings of the Carpathian Summer School of Physics 2007,* edited by L. Trache and S. Stoica

FORMALISM FOR NEUTRINO-NUCLEUS CROSS SECTIONS

In the neutral-current neutrino-nucleus processes we consider in the present work, low and intermediate energy neutrinos (or antineutrinos) are elastically or inelastically scattered from a nucleus (A,Z) via the exchange of neutral Z^0 bosons. Such a reaction is represented by

$$\nu + (A,Z) \longrightarrow \nu' + (A,Z)^* \tag{1}$$

where ν ($\overline{\nu}$) denote neutrinos (anti-neutrinos) of any flavor.

In our J-projected nuclear structure calculations method, the initial and final nuclear states are assumed to have well-defined spins and parities, $|J^{\pi}_{m}\rangle$ (the index m counts the multipole states from low to high energies) [8]. Then, within the Donnely-Walecka formalism the differential neutrino-nucleus reaction cross section is written as

$$\left(\frac{\mathrm{d}^2\sigma_{i\to f}}{\mathrm{d}\Omega \mathrm{d}\omega}\right)_{\nu/\overline{\nu}} = \frac{G_F^2}{\pi}\frac{|\mathbf{k}_f|\,\varepsilon_f}{(2J_i+1)}\left(\sum_{J=0}^{\infty}\sigma_{CL}^J + \sum_{J=1}^{\infty}\sigma_T^J\right) \tag{2}$$

where $\omega = \varepsilon_i - \varepsilon_f$ denotes the excitation energy of the nucleus, with ε_i the energy of the incoming neutrino, while ε_f ($\mathbf{k}_f$) represent the energy (momentum) of the outgoing lepton. In our convention ($\hbar = c = 1$) it holds: $|\mathbf{k}_f| = \varepsilon_f$. In the right-hand side of Eq. (2), the partial contributions σ_{CL}^J, for the Coulomb and Longitudinal part, and σ_T^J, for the transverse part of the multipole operators, are respectively written as

$$\begin{aligned}\sigma_{CL}^J &= (1+\cos\Phi)\left|\langle J_f||\widehat{\mathscr{M}}_J(q)||J_i\rangle\right|^2 + \left(1+\cos\Phi - 2b\sin^2\Phi\right)\left|\langle J_f||\widehat{\mathscr{L}}_J(q)||J_i\rangle\right|^2 \\ &+ \left[\frac{\omega}{q}(1+\cos\Phi)\right] 2\Re e\langle J_f||\widehat{\mathscr{L}}_J(q)||J_i\rangle\langle J_f||\widehat{\mathscr{M}}_J(q)||J_i\rangle^*\end{aligned} \tag{3}$$

$$\begin{aligned}\sigma_T^J &= \left(1-\cos\Phi + b\sin^2\Phi\right)\left[\left|\langle J_f||\widehat{\mathscr{T}}_J^{mag}(q)||J_i\rangle\right|^2 + \left|\langle J_f||\widehat{\mathscr{T}}_J^{el}(q)||J_i\rangle\right|^2\right] \\ &\mp \frac{(\varepsilon_i+\varepsilon_f)}{q}(1-\cos\Phi)2\Re e\langle J_f||\widehat{\mathscr{T}}_J^{mag}(q)||J_i\rangle\langle J_f||\widehat{\mathscr{T}}_J^{el}(q)||J_i\rangle^*\end{aligned} \tag{4}$$

where Φ is the lepton scattering angle and $b = \varepsilon_i\varepsilon_f/q^2$. The magnitude of the three momentum transfer q is given by the kinematics of the process (see e.g. Ref. [6, 8]). We note that, the summations in Eq. (2) contain contributions from both the polar-vector and the axial-vector components of the Coulomb $\widehat{\mathscr{M}}_J$, longitudinal $\widehat{\mathscr{L}}_J$, transverse electric $\widehat{\mathscr{T}}_J^{el}$ and transverse magnetic $\widehat{\mathscr{T}}_J^{mag}$ multipole operators of the nuclear current [6].

RESULTS AND DISCUSSION

In the first step of this work, using quasi-particle random phase approximation, we performed realistic state-by-state calculations of $\mathrm{d}^2\sigma_{i\to f}/\mathrm{d}\Omega\mathrm{d}\omega$ for all inelastic neutrino induced transitions (up to about 50 MeV) of ^{116}Cd. In order to derive these excitations, we used a rather rich model space which consists of twelve active single particle levels, up to N=$4\hbar\omega$ major harmonic oscillator quanta with core the ^{16}O nucleus.

At the BCS level, the values of the pairing parameters for protons (g^p_{pair}) and neutrons (g^n_{pair}), were fixed in the usual way [7] and we found $g^p_{pair} = 1.43$ and $g^n_{pair} = 1.37$, respectively. At the QRPA level, the strength parameters for the particle-hole channel, g_{ph}, and the particle-particle channel, g_{pp}, were fixed so as the energies obtained from the solutions of the QRPA-equations to reproduce the lowest lying spectrum of ^{116}Cd [2]. We mention that the determination of these strength parameters was achieved separately for each set of multipole states. The obtained values lie in the region $0.05 \ll g_{ph} \ll 0.90$ and $0.85 \ll g_{pp} \ll 1.20$, respectively.

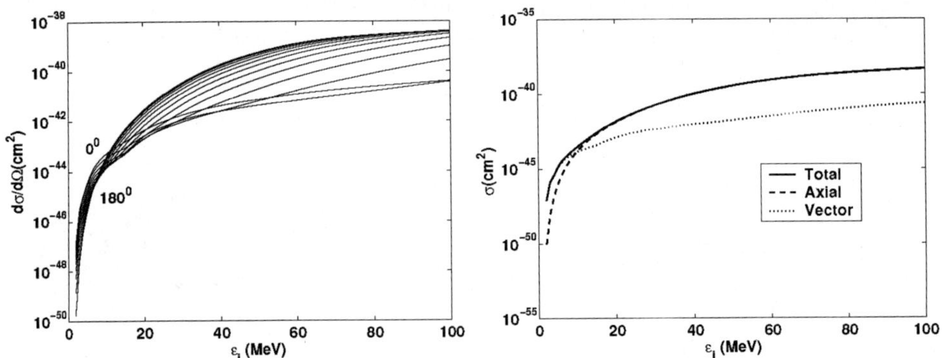

FIGURE 1. Integrated differential cross sections for the neutral-current reaction $^{116}Cd(\nu,\nu')^{116}Cd^*$. The curves correspond to the sum over partial rates for all J-projected states up to J^π=8^+ (left), and total cross section (separated in vector and axial vector parts) of this isotope (right).

In the next step, we studied systematically the angular and initial-neutrino energy dependence of the differential cross sections of the reaction ν-^{116}Cd, for initial neutrino energies $\varepsilon_i \leq 100$ MeV. This neutrino-energy range covers fully the supernova neutrino spectra.

In Fig. 1 (left-hand side), we show the differential cross sections obtained for various values of the scattering angle Φ (step $\Delta\Phi = 15^o$) of the reaction $^{116}Cd(\nu_e,\nu_e')^{116}Cd^*$. The dependance of $d\sigma/d\omega$ on the initial neutrino energy ε_i, shown in Fig. 1, is rather smooth. As can be seen, for low neutrino energies ($\varepsilon_i \leq 10-12$ MeV), i.e in the region of the discrete energy spectrum of ^{116}Cd, the differential cross section decreases as the scattering angle increases but for higher energies the behavior of the $d\sigma/d\omega(\varepsilon_i)$ is reversed.

By using numerical techniques, from the above results we obtained the total cross section $\sigma_{tot}(\varepsilon_i)$ shown in Fig. 1 (right-hand side) and its individual contributions coming from the vector and axial vector components of the nuclear current. We notice that, for energies greater than $\varepsilon_i \simeq 15$ MeV, the contribution of the axial vector component dominates, result in agreement with previous calculations obtained for many other nuclear systems [6].

Finally, by using the total cross sections shown in Fig. 1 (right-hand side), we examine the nuclear response of ^{116}Cd to supernova neutrino spectra by calculating the flux

TABLE 1. Total cross sections (in $10^{-40} cm^2$) of the neutral current reaction ^{116}Cd$(\nu_e, \nu_e')^{116}$Cd* averaged over supernova neutrino spectra described by a two parameter Fermi-Dirac distribution [1]

Quantity	$\alpha = 0$						$\alpha = 3$					
T (MeV)	2.75	3.50	4.00	5.00	6.40	8.00	2.75	3.50	4.00	5.00	6.40	8.00
$\langle \sigma_{tot} \rangle$	2.94	4.18	1.84	2.46	4.04	3.03	2.80	3.98	1.75	2.34	3.84	3.14

averaged cross sections for this isotope. To this aim, we assumed that a two-parameter Fermi-Dirac distribution offers a sufficiently good description for the supernova neutrino spectra [1, 6]. Our results are listed in Table 1 for various values of neutrino temperature T and chemical potential α. We see that the flux averaged cross sections are quite large compared to those of other nuclei like ^{56}Fe, ^{98}Mo [8]. From this point of view, our results suggest that, ^{116}Cd may be considered as a promising supernova detector.

SUMMARY AND CONCLUSIONS

In the present work, we performed realistic state-by-state calculations for incoherent processes of the reaction $^{116}Cd(\nu_e, \nu_e')^{116}Cd^*$ by using the QRPA. We investigated the dependence of the differential cross sections (they include both Fermi and Gamow-Teller type contributions) on the scattering angle and the initial neutrino energy.

Total cross-sections obtained subsequently offered the possibility to explore the response of the ^{116}Cd isotope as a supernova detector. On the basis of our neutral current flux averaged cross sections for this isotope, one may conclude that ^{116}Cd may be considered as important supernova detector.

ACKNOWLEDGMENTS

This research was supported by the ΠΕΝΕΔ No 03ΕΔ807 project of the General Secretariat for Research and Technology of the Hellenic Ministry of Development.

REFERENCES

1. H. Ejiri, *Phys. Rep.* **338**, 265–351 (2000), and references therein.
2. K. Zuber, *Phys. Lett. B* **519**, 1–7 (2001); *Prog. Part. Nucl. Phys.* **57**, 235–240 (2006); Private communication (2007).
3. T.S. Kosmas and E. Oset, *Phys. Rev.* **C 53**, 1409–1415 (1996).
4. E. Kolbe and T.S. Kosmas, *Springer Trac. Mod. Phys.* **163**, 199–225 (2000).
5. R. Arnold, *Phys. Rev. Lett.* **95**, 182302 (2005).
6. E. Kolbe, K. Langanke and P. Vogel, *Phys. Rev. D* **66**, 013007 (2002).
7. T.S. Kosmas,*Nucl. Phys. A* **683**, 443 (2001).
8. V.Ch. Chasioti, T.S. Kosmas and P.C. Divari, *Prog. Part. Nucl. Phys.*, **59**, 481–485 (2007).

Thermonuclear Reaction Rates for Solar Neutrino Processes

G.I. Karathanou*, V. Tsikoudi*, Th. Liolios† and T.S. Kosmas*

*Theoretical Physics Section, University of Ioannina, GR 45110 Ioannina, Greece
† Hellenic Army Academy, Department of Physical Sciences & Applications, Vari, GR 16673, Attica, GREECE

Abstract. Using the formalism for the effective cross-section factor S_{eff} we calculate the thermonuclear reaction rates and neutrino fluxes for the pp-chain processes. We adopt the recent LUNA experimental data for the nuclear cross-section S-factor and we compute improved and updated solar neutrino fluxes which we compare with those previously obtained.

Keywords: Thermonuclear reactions; Solar neutrinos; Helioseismology; Neutrino-nucleus reactions
PACS: 25.60.Pj, 26.65+t, 96.60.Ly

INTRODUCTION

Over the past four decades, precise solar and stellar models have been constructed and many of them are based on the thermonuclear reaction rates of Caughlan and Fowler [1] compilation. Several other nuclear physics groups have also worked on compilations of updated thermonuclear reaction rates of astrophysical interest, Adelberger et al.(1998) [2], the European Nuclear Astrophysics Compilation of REaction rates (Angulo et al. 1999, NACRE) [3] and several others.

In a main-sequence star burning hydrogen via the pp-chain, the energy spectrum of the neutrinos emitted from the core is determined by the relative rates of the various ^3He-burning reactions and the relative burning rates of any subsequent ^7Be-burning reactions. This spectrum determines both the fraction of the pp-chain energy which is removed from star's energy budget, escaping in the form of neutrinos, and the counting rates for various solar neutrino with differing thresholds and response functions.

Existing (LUNA) and planned solar neutrino experiments (Borexino), are expected to determine whether the energy spectrum of electron type neutrinos created in the center of the sun is modified by physics beyond standard electroweak theory. Moreover, these experiments have the capability of determining the mechanism, if any, by which new physics is manifested in solar neutrino experiments and thereby determining how the original neutrino spectrum is altered by the new physics. Once we reach this stage, solar neutrino experiments will provide precision tests of solar model predictions for the rates at which nuclear reactions occur in the sun.

Another probe of the deep solar interior is provided by helioseismology. The highly precise measurements of p and f oscillation modes enable us to extract both the properties of the convective envelope (depth, helium abundance) and the sound speed along the solar profile with high accuracy. The production of solar neutrinos is mainly sensitive

CP972, *Exotic Nuclei and Nuclear/Particle Astrophysics (II)—Proceedings of the Carpathian Summer School of Physics 2007,* edited by L. Trache and S. Stoica

to the solar temperature. On the other hand, from helioseismic observations one cannot determine directly the temperature of the solar interior. Neutrino and helioseismic information are complementary. We shall review the information on nuclear fusion in the sun provided by neutrinos and by helioseismology [4].

The purpose of this work is to study the solar thermonuclear reactions rates and neutrino fluxes. We focus on the solar neutrino-producing chain reactions, focusing on the $^3He(\alpha,\gamma)^7Be$ and $^3He(^3He,2p)^4He$ processes, and we evaluate the pp-chain reaction neutrino fluxes by using the recent LUNA experimental data [5].

FORMALISM OF THE THERMONUCLEAR REACTION RATES

The key step that allowed quantitative modelling of the Sun and other main-sequence stars was the description of the relevant thermonuclear reactions.

A typical thermonuclear reaction is represented by $\alpha + X \rightarrow Y + b$, where a nucleus X is bombarded by a uniform flux of particles of type α moving with relative velocity υ_r. Assuming that n_X and n_α are the number density of the interacting nuclei, the reaction rate per unit volume is $n_X\sigma$ and the flux of particles of type α is $n_\alpha \upsilon_r$ then the reaction rate is given by $r_{\alpha X} = n_\alpha n_x \upsilon_r \sigma(\upsilon_r)$, where σ is the cross section which depends on the relative speed υ_r of the nuclei.

If $f(\upsilon_r)d\upsilon_r$ denotes the probability that the relative speed is between υ_r and $\upsilon_r + d\upsilon_r$ then the total thermonuclear reaction rate per unit volume becomes

$$r_{\alpha X} = n_\alpha n_x \int_0^\infty \upsilon_r \sigma(\upsilon_r) f(\upsilon_r) d\upsilon_r = n_\alpha n_x \langle \sigma \upsilon_r \rangle, \quad (1)$$

In most astrophysical situations, especially under solar conditions, the nuclei form a classical, non-relativistic gas with a Maxwell-Boltzmann velocity distribution.

Assuming that the relative kinetic energy in the center-of-mass system of X and α nuclei is $E = \frac{1}{2}\mu\upsilon^2$, with μ the reduced mass, k the Boltzmann constant and T the temperature, the mean value $\langle\sigma\upsilon_r\rangle$ takes the form

$$\langle \sigma \upsilon_r \rangle = [\frac{8}{\pi\mu(kT)^3}]^{1/2} \int_0^\infty E\sigma(E) \exp(-\frac{E}{kT}) dE, \quad (2)$$

In order to evaluate $\langle\sigma\upsilon_r\rangle$, the energy dependence of the reaction cross section must be provided. Almost all of the nuclear reactions relevant to solar energy generation are non-resonant and charged-particle induced. For such reactions, the cross section can be expressed as

$$\sigma(E) = S(E) \exp(-2\pi\eta) \frac{1}{E}, \quad (3)$$

where $S(E)$ is the nuclear cross-section factor and η is the Sommerfeld parameter.

Substituting (3) into (2) and expanding $S(E)$ in a Taylor series, Eq. (2) becomes

$$\langle \sigma \upsilon_r \rangle \quad = \quad (\frac{2}{\mu kT})^{1/2} (\Delta E_0/kT) S_{eff} \exp(-3E_0/kT), \quad (4)$$

where the quantities S_{eff}, E_0 and ΔE_0 are given as follows below

$$S_{eff} = S(0)[1+\frac{5kT}{36E_0}] + S'(0)E_0[1+\frac{35kT}{36E_0}] + \frac{1}{2}S''(0)E_0^2[1+\frac{89kT}{36E_0}], \tag{5}$$

$$E_0/kT = (\pi Z_1 Z_2 \alpha/\sqrt{2})^{2/3}(\mu c^2/kT)^{1/3}, \qquad \Delta E_0/kT = 4(E_0/3kT)^{1/2}. \tag{6}$$

The S-factor is convenient for extrapolating measured cross sections down to lower astrophysical energies. When S(E) is assumed to be a constant, the integrand in Eq. (2) is peaked at the "most effective energy" E_0 [6].

Nuclear astrophysics experiments measure the components of the effective astrophysical factors S_{eff}, i.e. $S(0)$, $S'(0)$, etc. These data are, then, inserted into Eq. (5) in order to provide parametric expressions for the thermonuclear reaction rates like

$$r_{tot}(\alpha_1 ... \alpha_7; T_9) = \exp(\alpha_1 + \alpha_2 T_9^{-1} + \alpha_3 T_9^{-1/3} + \alpha_4 T_9^{1/3} + \alpha_5 T_9 + \alpha_6 T_9^{5/3} + \alpha_7 \ln T_9), \tag{7}$$

which are used in stellar evolution simulations.

DISCUSSION OF THE RESULTS AND CONCLUSIONS

The most important fluxes for solar neutrino experiments, that have been carried out so far, are those of the low energy neutrinos originating from the fundamental pp-reaction, $\Phi(pp)$, the intermediate energy ^7Be line neutrinos, $\Phi(^7Be)$, and the rare high-energy neutrinos from ^8B decay, $\Phi(^8B)$. The pp neutrinos are the most abundant experimentally-accessible solar neutrinos and the ^8B neutrinos have the smallest detectable flux, according to the predictions of standard solar model [7].

The LUNA collaboration has recently derived more accurate than before, zero-energy astrophysical factors $S_{33}(0) = (5.32 \pm 0.08) MeVb$ and $S_{34}(0) = (0.560 \pm 0.017) keVb$ for the reactions $^3He(^3He, 2p)^4He$ and $^3He(\alpha, \gamma)^7Be$. In the present work, we use these factors to update the solar neutrino fluxes (Φ). For each reaction channel the neutrino flux Φ reads

$$\Phi_{channel}(LUNA) \propto \frac{\left(S_{33}^{\alpha} S_{34}^{\beta}\right)_{LUNA}}{\left(S_{33}^{\alpha} S_{34}^{\beta}\right)_{model}} \Phi_{channel}(model) \tag{8}$$

In the latter expression, α and β denote the logarithmic partial derivatives of the neutrino fluxes which depend upon a number of solar input parameters [7].

In Table 1, we show the improved solar neutrino fluxes for *pp*, *pep*, ^{7}Be, ^{8}B neutrinos for four models (column 1): YA, the Young and Arnett model [8], A98, the Adelberger et al. model [2], N99, the NACRE compilation [3], and BPB, the Bahcall, Pinsonneault and Basu model [9]. In columns 2 – 5 we give the neutrino fluxes Φ for the different thermonuclear reaction of the *pp*-chain. The first value of Φ, in each of the columns 2 – 5, is the neutrino flux calculated by the model shown in column 1, while the second value (present results) is the neutrino-flux computed by using the LUNA-experiment S-factors [5].

TABLE 1. Improved solar neutrino fluxes for *pp, pep,* ^7Be, ^8B neutrinos calculated by employing the resent LUNA data [5] for the cross-section S-factors.

Model	$\Phi_{pp} \times 10^{10}(cm^2 s^{-1})$		$\Phi_{pep} \times 10^{8}(cm^2 s^{-1})$		$\Phi_{^7Be} \times 10^{9}(cm^2 s-1)$		$\Phi_{^8B} \times 10^{6}(cm^2 s-1)$	
YA	*5.95*	*5.93*	*1.42*	*1.41*	*4.83*	*5.08*	*5.51*	*5.77*
A98	*5.92*	*5.90*	*1.40*	*1.39*	*4.80*	*5.06*	*5.17*	*5.44*
N99	*5.91*	*5.90*	*1.40*	*1.39*	*4.90*	*4.99*	*5.68*	*5.79*
BPB	*5.96*	*5.92*	*1.39*	*1.38*	*4.81*	*5.27*	*5.44*	*5.93*

We notice that, the updated Φ_{pp} and Φ_{pep} are slightly improved compared to previous models of column 1. Nevertheless, they modify the energy balance, the chemical composition and the neutrino generation in the solar core. We also notice that the $\Phi_{^7Be}$, $\Phi_{^8B}$ fluxes are appreciably improved over the previous models. An extensive discussion and additional detailed results will be published elsewhere.

In the stellar problem, the nuclear reaction cross-section σ times the relative velocity υ of the interacting particles, i.e. $\langle \sigma \upsilon \rangle$, must be integrated over the Maxwell-Boltzmann velocity distribution for energy and temperature T which occurs in the stellar core.

ACKNOWLEDGMENTS

The present work was partially supported by the ΠΕΝΕΔ No 03ΕΔ807 project of the General Secretariat for Research and Technology of the Hellenic Ministry of Development.

REFERENCES

1. G.A. Caughlan and W.A. Fowler, *At. Data Nucl. Data Tables* **40**, 283 (1988).
2. E.G. Adelberger et al., *Rev. Mod. Phys.* **70**, 1265 (1998).
3. C. Angulo et al, *Nucl. Phys. A* **656**, 3 (1999).
4. J.Christensen-Dalsgaard, *Rev. Mod. Phys.* **74**, 1073 (2002).
5. G. Gyurky et al. (LUNA-experiment), *Phys. Rev. C* **75**, 035805 (2007).
6. W.C. Haxton, P.D. Parker and C.E. Rolfs, nucl-th/0501020v1.
7. J. Bahcall, Neutino Astrophysics (Cambridge University, Cambridge) (1989).
8. P.A. Young and D. Arnett, astro-ph/0409658.
9. J.N. Bahcall, M.H. Pinsonneault and S. Basu, *Ap. J.* **555**, 990 (2001).
10. T.E. Liolios and T.S. Kosmas, *Phys. Rev. C* **72**, 025803 (2005).

On the Nuclear Response of ^{56}Fe to Supernova Neutrino Spectra

V.A. Tsakstara*, T.S. Kosmas*, V.C. Chasioti* and J. Sinatkas †

*Theoretical Physics Section, University of Ioannina, GR 45110 Ioannina, Greece
†Department of Informatics and Computer Technology, TEI of Western Macedonia, GR-52100 Kastoria, Greece

Abstract. Neutral current cross sections of scattering of ν_e on the nuclear isotope ^{56}Fe, which play a significant role in astrophysical neutrino studies, are used in order to explore the response of this detector to supernova neutrino spectra. We assume a two-parameter Fermi-Dirac distribution for the supernova neutrino spectra and evaluate the flux averaged cross sections for various values of the neutrino temperature and chemical potential parameters.

Keywords: Supernova neutrinos; Neutrino-nucleus interactions; neutrino detection; nuclear response to neutrino spectra; convolution method
PACS: 26.30.+k, 25.30.Pt, 13.15.+g, 26.50.+x

INTRODUCTION

The detection of astrophysical neutrinos (solar, supernova, atmospheric, etc.) through the neutrino-nucleus reactions, i.e. by probing a nuclear target in terrestrial experiments, is an important research field in neutrino physics. In particular, the search of supernova neutrinos, has, recently, attracted much attention in nuclear, astro-nuclear physics and cosmology and supernova neutrinos provide a sensitive probe for studying the properties of neutrinos and stellar collapse mechanisms.

As it is well known, the central core of a massive star gets unstable when it uses up the nuclear fusion fuels and the gravitational core collapse occurs because there is no nuclear fusion energy to hold the star against the gravitational pressure [1, 2, 3]. During this process, most of the emitted energy is carried away by neutrinos of all flavors which interact with matter (neutron, proton, electron, heavier nuclei) through the neutral current interactions.

Historically, the observation of the SN1987A neutrinos has encouraged further experimental and theoretical studies of supernova neutrinos in order to investigate in detail supernova explosion mechanisms and neutrino properties.

In the present work, we focus on the investigation of supernova detection by terrestrial experiment. In particular, we are going to study the nuclear response to supernova neutrino spectra of the Fe isotopes. To this end, differential, integrated and total cross section are provided through realistic state-by-state calculations (performed within the context of the QRPA) and specific neutrino-energy distributions for the supernova neutrino spectra are employed for the convolution procedure.

CP972, *Exotic Nuclei and Nuclear/Particle Astrophysics (II)—Proceedings of the Carpathian Summer School of Physics 2007*, edited by L. Trache and S. Stoica

THE FOLDING METHOD IN SUPERNOVA NEUTRINO STUDIES

In general, in order to estimate the response of a nuclear target to a specific source of neutrinos, the calculated (differential or total) cross sections of the reaction of neutrinos with the nuclear detector [4, 5, 6, 7], must be convoluted with the neutrino energy distribution of the source in question [7].

In the case of differential neutrino-nucleus cross sections, $d\sigma(E_\nu,\omega)/d\omega$, the folding is defined by the expression [8]

$$\frac{d\sigma(\omega)}{d\omega} = \int_{\omega}^{\infty} \frac{d\sigma(E_\nu,\omega)}{d\omega} f_\nu(E_\nu) dE_\nu \,, \tag{1}$$

where $\omega = E_f - E_i = \varepsilon_i - \varepsilon_f$, represents the excitation energy of the nucleus. E_f (E_i) and ε_f (ε_i) denote the final (initial) nuclear-energy and neutrino-energy (or, in general, outgoing lepton-energy), respectively. $f_{\nu_i}(E_{\nu_i})$, with ν_i=ν_e, ν_μ, $\tilde{\nu}_e$, etc., describes the energy distribution of the neutrinos originating from the specific neutrino source [7].

In the case of total neutrino-nucleus cross section, $\sigma(E_\nu)$, the folding is rather simpler and it is defined as

$$\langle \sigma_\nu \rangle = \int_{E_{thres}}^{\infty} \sigma(E_\nu) f_\nu(E_\nu) dE_\nu \,, \tag{2}$$

where E_{thres} denotes the energy-threshold of the nuclear detector. The result obtained from the convolution of Eq. (2), is known as the flux averaged cross section, $\langle \sigma_\nu \rangle$, of the studied nucleus with respect to the assumed neutrino source.

During a supernova explosion, all six flavor neutrinos, ν_e, $\tilde{\nu}_e$, ν_μ, $\tilde{\nu}_\mu$, ν_τ, and $\tilde{\nu}_\tau$, are produced. Among them, ν_e decouples and escapes at the largest radius via the charged- and the neutral-current interactions with matter. Then, the temperature of ν_e is about 3.5 MeV. On the other hand, $\tilde{\nu}_e$ decouples at a little inner region than ν_e, since the proton density relevant to $\tilde{\nu}_e$ is less than the neutron one. Therefore, $\tilde{\nu}_e$ has the higher temperature of around 5 MeV. The other neutrinos decouple at the sphere with the highest temperature of around 8 MeV.

Numerical simulations on supernova neutrino dynamics have shown [1] that the energy-spectra of supernova neutrinos could be well described by a two-parameter Fermi-Dirac distribution of the form

$$f_\nu(E_\nu, T, \alpha) = N_2(\alpha) \frac{1}{T^3} \frac{E_\nu^2}{1 + e^{(E_\nu/T + \alpha)}} \,, \tag{3}$$

where T is the neutrino temperature and α parameter associated with the non-zero chemical potential. $N_2(\alpha)$ denotes the normalization factor given from

$$\frac{1}{N_n(\alpha)} = \int_0^{\infty} \frac{x^n}{e^{(x-\alpha)} + 1} dx \,, \tag{4}$$

for n=2. The average neutrino energy $\langle E_\nu \rangle$ can be written in terms of the functions of Eq. (4) as [2]

$$\langle E_\nu \rangle = \frac{N_2(\alpha)}{N_3(\alpha)} T = (3.1515 + 0.125\alpha + 0.0429\alpha^2 + \ldots) T \tag{5}$$

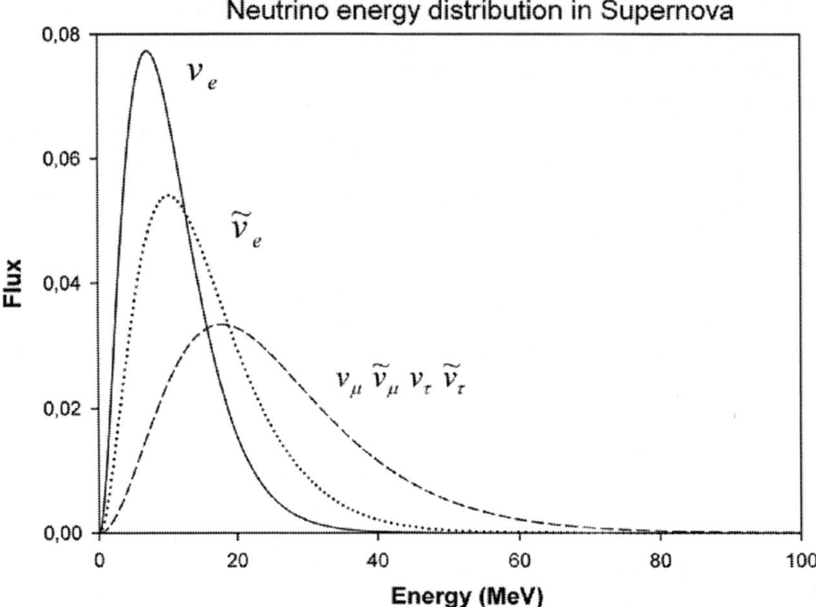

FIGURE 1. Energy-spectra of neutrinos emerging out of a core collapse supernova. They are approximately described by a two-parameter Fermi-Dirac distribution, $f_\nu(E_\nu, T, \alpha)$, of Eq. (3), where the values of the temperature T, and degeneracy parameter α, depend on the neutrino flavor [3].

Thus, for a given value of the degeneracy parameter α, the neutrino temperature T is fixed by using the relation between average neutrino energy $\langle E_\nu \rangle$ and T given in Eq. (5). The numerical simulation studies predict the average energies for various flavors of neutrinos to be $\langle E_{\nu_e} \rangle \approx 10 - 11 MeV$, $\langle E_{\tilde{\nu}_e} \rangle \approx 15 - 16 MeV$, $\langle E_{\nu_x} \rangle \approx 23 - 25 MeV$ [1, 2, 3]. The corresponding spectra to these values are illustrated in Fig. 1.

DISCUSSION OF THE RESULTS AND CONCLUSIONS

In the present work, we examine the response of ^{56}Fe to the supernova neutrino spectra of electron type (ν_e). To this aim, we use the results for the total cross sections of inelastic neutrino-nucleus scattering cross sections of Ref. [9].

The results obtained for various values of neutrino temperatures T and degeneracy parameter α of the assumed Fermi-Dirac distribution are shown in Table 1. In addition, in this Table we include the mean cross sections given for the individual polar-vector $\langle \sigma_V \rangle$ and axial-vector $\langle \sigma_A \rangle$ components of the hadronic current. Since, for the neutral current neutrino-nucleus process the coherent ($gs \rightarrow gs$) channel is also possible, the corresponding $\langle \sigma_{coh} \rangle$ are also presented in Table 1.

As can be seen, there is an increase of the flux averaged cross sections with the

TABLE 1. Flux averaged cross sections for the ^{56}Fe nucleus obtained for the supernova neutrino/antineutrino spectra. A two-parameter Fermi-Dirac distributions for various values of neutrino temperatures T and chemical potential α has been employed (see text).

	$\alpha = 0$						$\alpha = 3$					
T	2.75	3.5	4.0	5.0	6.4	8.0	2.75	3.5	4.0	5.0	6.4	8.0
$\langle\sigma_{coh}\rangle$	12.9	20.4	26.2	39.2	59.9	85.3	19.0	29.9	38.2	56.8	85.7	120.0
$\langle\sigma_{tot}\rangle$	0.7	1.6	2.4	5.2	12.4	27.9	1.2	2.7	4.1	8.7	21.0	47.7
$\langle\sigma_{A}\rangle$	0.6	1.4	2.2	4.9	12.0	27.5	1.0	2.4	3.8	8.3	20.5	47.3
$\langle\sigma_{V}\rangle$	0.1	0.1	0.1	0.2	0.4	0.7	0.1	0.2	0.2	0.4	0.6	1.0

neutrino temperature T and the chemical potential α for σ_{tot} and other individual components of the cross section. These results are in good agreement with the averaged cross sections obtain by using continuum random phase approximation [7].

In conclusion, we have evaluated averaged cross-sections for the neutral current $^{56}\mathrm{Fe}(\nu,\nu')^{56}\mathrm{Fe}^*$ reaction. We used the convolution method and employed Fermi-Dirac neutrino energy distribution which is appropriate for neutrinos produced by a Supernova explosion.

ACKNOWLEDGMENTS

This research was funded by the program ΠΕΝΕΔ No 03ΕΔ807 of the General Secretariat for Research and Technology of the Hellenic Ministry of Development.

REFERENCES

1. H.T. Janka and B. Muller, *Phys. Reports* **256**, 135 (1995); H.T. Janka and W. Hillebrand, *Astron. Astrophys.* **224**, 49 (1989).
2. M.S. Athar, S. Ahmad, and S.K. Sing, *Phys. Rev. C* **71**, 045501 (2005).
3. H. Ejiri, *Phys. Reports* **338**, 265–351 (2000).
4. J. Toivanen, E. Kolbe, G. Martinez-Pinedo, and P. Vogel, *Nucl. Phys. A* **694** 395–408 (2001).
5. K. Langanke, G. Martinez-Pinedo, B. Muler, H.Th. Janka, A. Marek, W.R. Hix, A. Juodagalvis, and J.M. Sampaio, astro-ph/07061687.
6. E. Kolbe and T.S. Kosmas, *Springer Trac. Mod. Phys.* **163**, 199–225 (2000).
7. T.S. Kosmas and E. Oset, *Phys. Rev.* **C 53**, 1409–1415 (1996); E. Kolbe, *Phys. Rev. C* **54**, 1741–1748 (1996); E. Kolbe, K. Langanke and P. Vogel, *Phys. Rev. D* **66**, 013007 (2002);E. Kolbe and K. Langanke, *Phys. Rev. C* **63**, 025802 2001.
8. V. Tsakstara, T.S. Kosmas, J. Sinatkas, V.C. Chasioti and P.C. Divari, Advances in Computational Methods in Shiences and Engineering, to appear.
9. V.Ch. Chasioti, P.C. Divari and T.S. Kosmas, *American J. Phys.*, in press.

PART III: OUTREACH SESSION AND CLOSING SESSION

Outreach Session: Science education and the Public Understanding of Science

Livius Trache[a] and Cristinel Diaconu[b]

[a]*Cyclotron Institute, Texas A&M University, College Station, TX 77843-3366*
[b]*Centre de Physique des Particules de Marseille, IN2P3, France*

CONVENORS REMARKS

In many ways, the organizers of this 22nd edition of the Carpathian Summer School of Physics 2007 (CSSP07), see it a direct continuation of its 20th edition CSSP05, in Mamaia. It was then the “International Year of Physics”, or the “Einstein year”, and we made particular efforts to add outreach sessions to that event. However, in spite of having excellent evening lectures on the life of Einstein, on dark matter and dark energy, and on the origin of chemical elements and stars, we had very little success in attracting audience from outside the school. This year we adopted a slightly different approach. Sir Arnold Wolfendale, who offered to talk about Public Understanding of Science (PUS), inspired the subject of the session. We added Science Education in the title, because we believe we cannot have good PUS in a society (national or wider, up to global) that does not have a healthy science education in place. Or, for that matter, a good science, a notion that includes, in our times, and even starts with, a healthy science management. After we chose the subject, we invited specifically people from Romania, from the government, universities, research institutes, high schools and media, which we thought would be interested, and made sure some will attend. We had a mixed response. In addition to school’s participants, we had in the audience one-day guests involved in science management and education in Romania, representatives of the media. Professor Mircea Miclea, ex-ministry of Education in the current government is the chair of a presidential commission that gave recently a blasting assessment of the status of education in Romania. No representatives of the current management of research institutes or national science agencies could or chose to attend. However, we had a good response from the people doing groundwork in science education. We have (had) here:

Prof. Ion Bararu, Colegiul National "Mircea cel Batran", Constanta
Ion Toma, Colegiul National "Mihai Viteazu", Bucharest
Seryl Talpalaru, Colegiul National "Emil Racovita", Iasi
Romulus Pop, Dept. for Secondary Schools, MEC, Bucharest.

These four leaders of high schools with interest and best results in physics were here, and the presentation of professor Bararu about the activities and successes of the group of students that he coaches in Constanta, was impressive for all in the audience.

CP972, *Exotic Nuclei and Nuclear/Particle Astrophysics (II)—Proceedings of the Carpathian Summer School of Physics 2007,* edited by L. Trache and S. Stoica

THE PUBLIC UNDERSTANDING OF SCIENCE: 'THE WOLFENDALE REPORT' AND OTHER MATTERS

A.W.Wolfendale

Physics Department, Durham University, South Road, Durham DH1 3LE

Abstract. 'The public understanding of Science' is a matter of interest and concern and one that we neglect at our peril. This paper deals with two aspects : the 'Wolfendale Report' of 1995, commissioned by the British Government, and some personal reflections on how to make our presentations to 'the public' more interesting.

Keywords: Public Understanding of Science.
PACS: 00, 01.10.Fv, 01.30-y.

THE WOLFENDALE REPORT: PREAMBLE

A number of factors and worries about the perception of the public of matters-scientific led to the formation of a Committee to:

"Review the contribution of scientists and engineers to the public understanding of science, engineering and technology".

It will be noted immediately that the scope was broadened from 'science' alone to the inclusion of engineering and technology. Such a move was welcomed.

The membership was as follows, the names and affiliations being included so as to give an appreciation of our scope.

Chairman

Sir Arnold Wolfendale FRS President of the Institute of Physics

Members

Mr Eric Hassall	Consulting Engineer
Dr Iain Murray	Lecturer in Computer Science, University of Dundee
Mr Ronald Paul	Retired headmaster of Currie High School, Edinburgh
Dr Geoffrey Thomas	Director, Department for Continuing Education, Oxford University; President of Kellogg College, Oxford
Ms Jenny Thurston	Deputy General Secretary, IPMS
Dr David Whitehouse	Science Correspondent, BBC News and Current Affairs
Dr Tom Wilkie	Science Editor, "The Independent"
Dr Monica Winstanley	Head of Public Relations, BBSRC

CP972, *Exotic Nuclei and Nuclear/Particle Astrophysics (II)—Proceedings of the Carpathian Summer School of Physics 2007,* edited by L. Trache and S. Stoica

At the start of our deliberations we were told that the objective of the Government's policy on public understanding are:

"To contribute to the economic wealth and quality of life on the Nation, particularly by drawing more of our best young people into careers in science, engineering and technology.

To strengthen the effectiveness of the democratic process through better informed public debate of issues of public concern arising in the fields of science, engineering and technology."

A number of 'perspectives' presented themselves, of which two can be given here:

The Nation's research output

In a changing world, the maintenance of research support, and hopefully its enhancement, and also the increased take-up of science and engineering subjects by people of all ages, will depend on public appreciation of science and engineering and their practitioners. This will have a direct effect on the economic wealth and quality of life of the Nation. This aspect is particularly important in the area of basic science where the tax-payer is the main provider.

Public attitudes to universities

One of the results of adopting the Committee's proposals would be a wider appreciation of the role of universities in a modern society and, not least, the further breaking down of barriers – real and imagined – between these institutions and the general public.

CONCLUSIONS AND RECOMMENDATIONS

The major items are presented here.

We hope that the work of the Committee and its recommendations will encourage all those in the professional scientific and engineering community, who have an interest in creating wealth and enhancing our quality of life, to participate in promoting the **public understanding of science, engineering and technology.**

The steps currently being taken in universities to equip and encourage professional scientists, engineers and research students to contribute to improved public understanding of science, engineering and technology vary considerably across the country. A few universities are making a major contribution. Most undertake some public understanding activities.

There is scope for improvement, however, particularly in the areas of training in communication skills, and of support for those in public understanding activities.

The teaching of science and technology in schools is a continuing source of concern.

- ❖ Scientists, engineers and research students in receipt of public funds have a duty to explain their work to the general public.
- ❖ Research Councils should develop further their grant procedures to make public understanding an integral feature and continue with their other schemes in the public understanding of science, engineering and technology field.
- ❖ In universities there should be increased emphasis on the acquisition of communication skills and their use for the benefit of the public.

The increased responsibility should bring with it increased recognition for both individual and institution.

The Funding Councils should take the lead in ensuring a continuing role for adult education provision in promoting the public understanding of science and engineering.

The Committee on the Public Understanding of Science (COPUS) is well placed to act as co-ordinator of public understanding activities across all sectors of society. If it is to fulfill that role, it should be expanded to include engineering and one or more senior engineering bodies should be invited to become permanent members of the new Committee.

- ❖ The Professional Institutes should develop further their work in the public understanding of science, engineering and technology.
- ❖ The OST should sponsor a study of the present degree of public understanding of science, engineering and technology.
- • Universities should develop cost-effective ways of measuring performance in public understanding.
- ❖ The OST should commission the production of a "best practice" guide.

Those marked with a star have been acted upon, to some extent at least. The biggest problem is still an inadequate recognition for those who undertake PUS work.

THE ROLE OF HUMOUR IN PUS

There is no doubt that humour can play a big part in getting a message across. Straightforward jokes are, in the author's view, always appreciated and it is well known that audiences remember the serious matters – in part – when jokes are inter-spaced.

A number of other aspects of humour are also relevant. Typographical errors ('typos') can also be fascinating. Some examples from the Author's stock are as follows:

Some typos' from 2006

In The William Herschel Society's journal we read

"... there is a distinct shift towards ***scared*** *and vocal music in the Bath years."*

In the instructions for Prince Philip's visit to Westminster Abbey to unveil the memorial to John Harrison:

*"... Prince Philip will move on to **sing** the visitors book."*

In the magazine 'The Historian' (2006) in an article by Arnold Wolfendale about the joys of lecturing on cruise ships he apparently writes:

*"Pre-history and modern science are blended in **cornets,** asteroids and things that go bump in the night".*

PREDICTIONS BY THE GREAT

Humour with a purpose is afforded by some of the predictions by the great scientists of the past. These illustrate the fallibility of even the greatest minds. Some examples are the following:

"Heavier than air flying machines are impossible."

(Lord Kelvin, 1885)

"I have not the smallest molecule of faith in aerial navigation other than ballooning."

(Lord Rayleigh, 1889)

"The abdomen, the chest and the brain will forever be shut from the intrusion of the wise and humane surgeon."

(Sir John Erickson, Surgeon Extraordinary to Queen Victoria, 1873)

"Anyone who expects a source of power from the transformation of [the nuclei of] atoms is talking moonshine."

(Lord Rutherford, 1933)

"A few decades hence, energy may be free, just like unmetered air."

(J. von Neumann, 1956)

"The possibility of travel in space seems at present to appeal to schoolboys more than to scientists."

(Sir George Paget Thomson, 1956)

"Space-travel is utter bilge."

(Sir Richard van der Riet Woolley, Astronomer Royal, 1956)

Cartoons, too have wide appeal and illustrate some of the more difficult concepts. One such is:

Another, germane to the popular topic of 'The Search for Intelligent Life' is:

Another aspect is Audience-participation. Often it is possible to enlist their support in solving some of the great problems of science. Two examples can be taken. Firstly, in a lecture on 'The Origin of the Universe' the audience are invited to vote for one or the following:

1. The Universe is closed.
2. The Universe is open
3. The universe is perfectly tuned.

It is fascinating to compare the answers from audiences of different average age. In 'The Search for Intelligent Life' the choice is simple:

1. We are alone
2. There are many intelligent civilization in our Galaxy.

CONTENTION IN SCIENCE

Everyone likes a good argument and it is often useful to draw attention to opposing views. A good example some years ago was the 'fight' between proponents of 'Continuous Creation' and 'the Big Bang'.

Another area is the quest for credit. A good example is the raw deal suffered by the 17th Century British physicist, Robert Hooke. His experiences can be summarized as follows:

- 1670 Hooke – Cutlerian lecture – motions of planets affected by the attractive power of the sun, Newton claimed precedence, and the inverse-square law.
- Hooke's idea of polar wandering (precession) – pre-Newton.
- Hooke postulated that the Earth is an oblate spheroid.
- 1664 – 5 Tracked a comet – said it had first appeared in 1618. Pre-dated Edmond Halley.
- Hooke described the interference colours from thin films – known now as 'Newton's Rings'.
- He designed the great dome of St. Paul's Cathedral – Wren got most of the credit.
- Hooke was the founder of scientific geology – Steno got the credit.
- Hooke's instruments, models and fossils – also his portrait – missing from the Royal Society (Newton destroyed them?)

CONCLUSION

'Public understanding ...' or, more properly. 'Public appreciation of science etc' has never been more necessary. Those who talk to the public about these matters can make it fun for their audiences, as well as for themselves.

REFERENCES

1. Wolfendale, A.W. 'Report of the Committee to review the contribution of scientists and engineers to the public understanding of science, engineering and technology.' October, 2005. UK Office of Science and Technology, Whitehall, London.

Research and Teaching across Borders

Peter L. Biermann[a,b,c]

[a]Max-Plank Institute for Radioastronomy, Bonn, Germany
[b]Department of Physics and astronomy, University of Bonn, Bonn, Germany
[c]Department of Physics and Astronomy, University of Alabama, Tuscaloosa, AL, USA

Abstract. Reflecting back to teaching and doing scientific research since the late 1960ies to 2007, I report here a very personal story and some advice. It mentions both my family and my research group.

Keywords: Teaching, Career in Science, Science Education, Public Understanding of Science
PACS: 01.40.-d, 01.40.Fk, 01.40.gb, 01.85.+f

MY CREDENTIALS: STUDENTS, POSTDOCS AND VISITING PROFESSORS

Over the years we have had as participants in our research students, postdocs and visiting professors from/in: Korea, Japan, Australia, India, Israel, Ruanda, Armenia, Iran, Russia, Turkey, Bulgaria, Romania, Hungary, Poland, Greece, Italy, Switzerland, Germany, Great Britain, Norway, France, Netherlands, Spain, Argentina, Mexico, Canada, USA.

We have been teaching by giving small series of lectures in: Korea, China, Australia, India, Bulgaria, Poland, Hungary, Slovenia, Romania, Italy, Germany, Spain, Ireland, Mexico, Argentina, Brazil, Canada, and the USA.

TEACHING THE NEXT GENERATION

Starting with the University of Bucharest we have had Erasmus/Sokrates contracts for teaching and education with universities in: Bucharest, Cluj-Napoca, Budapest, Cracow, Szeged, and Ljubljana.

We have (now twelve) former students and postdocs now involved in teaching (at the level of professor in the USA) in and/or from: China, Australia, India, Bulgaria, Romania, Germany, Netherlands, Spain, Mexico, Canada, USA.

The group has had 80 MSc or PhD students and postdocs. Of all the still active group members half are women.

CP972, *Exotic Nuclei and Nuclear/Particle Astrophysics (II)—Proceedings of the Carpathian Summer School of Physics 2007,* edited by L. Trache and S. Stoica

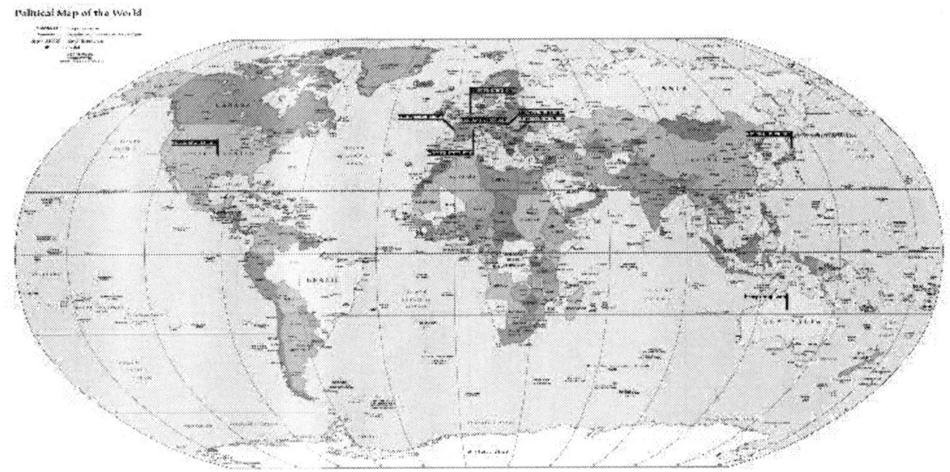

FIGURE 1. Where the Romanian students are.

CAREERS FOR SCIENTISTS

There are basically three ways today to start a career after studying physics:

A) Direct application of what you learn, and examples are: 1) The solid state physics of detectors, memories and microchips. 2) Optical pulsing and number theory (in the future perhaps quantum theory) as a source of code systems for transmission of confidential data. 3) Safe fuel cell and battery systems for electronics.

B) Quantum theory, relativity, dark matter, dark energy: good basis for perpetually fresh thinking in economics and finance: Big financial advisory companies hire few economists, mostly physicists and astro-physicicists (they just hired a PhD student away from us).

C) Teaching the open questions: reach the fresh young minds, and do research to keep them excited.

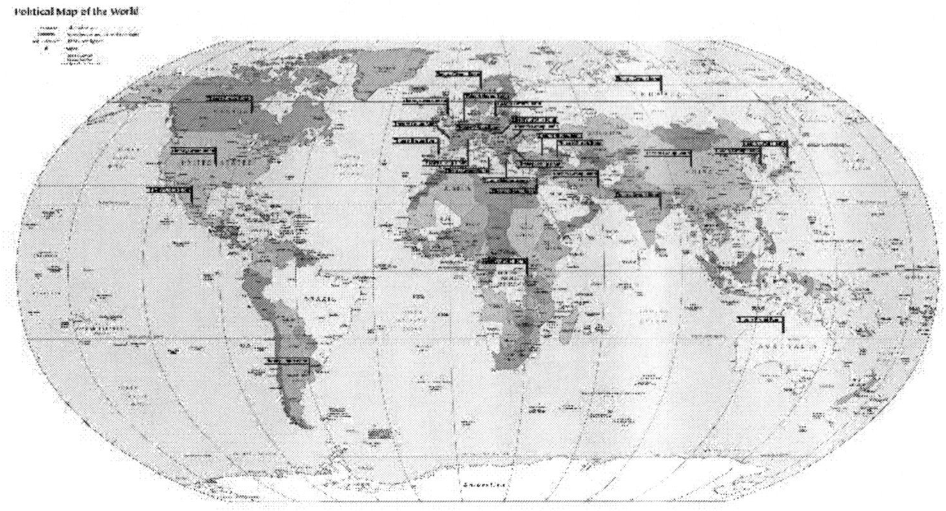

FIGURE 2. Where the students, postdocs and visiting professors come from/are.

SCIENCE EDUCATION INTERNATIONALLY

In my immediate family we cover four countries on three continents; in schools/universities as both students and faculty.

Career choices as a scientist are usually not clear to young people.

The combination of hands-on approach and theoretical understanding usually not well arranged.

Few students take science as a result.

PUBLIC UNDERSTANDING OF SCIENCE

There is lots of efforts in many countries: with TV shows, internet programs like ``Hands-on-universe", Children-University, advanced placement courses in schools, Science nights, Science years.

However, many such programs exist only in few languages.

A Science Honor Society in High School in USA is a pure initiative of students, and so focuses the young minds, as everything depends on their initiative.

Weekend schools and regular seminars organized by students for all faculties are another example of an encouraging student initiative.

However, lectures and presentations by faculty are often seen by children as high level entertainment, but with little thought for deciding a career choice.

CONCLUSIONS

We clearly need more effort in outreach, with a reality check on what such a career choice implies (work as a physicist, in a financial company, or as a teacher; international families, multi-language children).

More effort on languages is required, to communicate worldwide.

Advice for each country: Compete for the young talent worldwide.

Advice again for each country: Keep the young scientists happy and creative - so they stay.

Last advice for each country: Entice your own talent to stay or come back!

ACKNOWLEDGMENTS

Much of the inspiration for this lecture has come from Mircea V. Rusu. Support for PLB is coming from the AUGER membership and theory grant 05 CU 5PD 1/2 via DESY/BMBF.

Some Psychological Knowledge for Scientists' Use

Mircea Miclea

Abstract. Relying on empirical evidences our paper presents the most salient personality traits, developmental factors and cognitive characteristics of the scientists. We claim that a sound exploration of scientists' mind and patterns of behavior could improve public support for science and enhance scientists' mutual understanding.

Keywords: psychology of science, expertise, personality
PACS: 14.60.Pq, 98.80.-k, 98.80.Es

address=Babes-Bolyai University,Department of Psychology,Mihail Kogalniceanu no.1,Cluj-Napoca,Romania mirceamiclea@psychology.ro

INTRODUCTION

Psychology, as many other sciences, owns a lot to physics. Starting from physicists' models (and sometimes even trying to emulate them), psychology has succeeded to become, during the 20th century, a mature, experimental science of human mind and behavior. It is now its turn to contribute to the promotion of natural sciences, of physics in particular, if not by its theoretical and methodological achievements at least by its facts and knowledge about the mind, behavior and personality of the scientists. The aim of this paper is to draw some conclusions from psychological research which could be useful for scientists in mathematics or natural sciences, with special reference to physics. It is our belief that psychological knowledge about scientists' mind and patterns of behavior could have a threefold utility: (1) helping the scientists to understand themselves and to enhance their occupational identification, (2) offering some guidance when senior scientists are recruiting young scientists, (3) widening public understanding about science and, as a consequence, public support for scientific enterprises. In other words, we need to understand not only scientific theories but also scientists' behaviors in order to have a comprehensive image about science and supportive attitudes from the larger public. In the following, we shall focus on the most stable personality characteristics, developmental histories and scientist's expertise that could be useful in achieving our aim.

PERSONALITY CHARACTERISTICS

1.Be patient if you want to understand a (or another) scientist. He/she is introverted, independent and less sociable. [32] used Eysenck Personality Questionnaire

CP972, *Exotic Nuclei and Nuclear/Particle Astrophysics (II)—Proceedings of the Carpathian Summer School of Physics 2007,* edited by L. Trache and S. Stoica

(EPQ) to measure personality characteristics of 109 male and 133 female physicists and compared their scores to the population norms. They found that male physicists are almost one standard deviation (d = - 0.94) and female physicists three-fourths of a standard deviation (d = - 0.73), lower on the Sociable-Unsociable dimension than the general population.Moreover, highly creative scientists are less affiliative than low creative scientists. Being thing-oriented rather than people-oriented, scientists working in natural sciences and mathematics are even more independent and asocial than their colleagues in social sciences ([18]). [3] have administered a non-clinical measure of autism and found that mathematicians, physical scientists and! engineers score higher than social scientists. They also found a similar pattern when a scale of high functioning autism (Asperger's Syndrome) was used. Moreover, autistic children are more than twice as likely as non-autistic children to have a father or grandfather who was an engineer ([16]). Although highly functioning autism is an extreme form of asocial behavior and it is very rare, its increased incidence among mathematicians and natural scientists is significant.

2.Be cool when approaching a scientist. He/she is dominant, arrogant - even hostile - and self-confident. A large number of studies have documented the presence (but not the exclusivity) of these personality traits to scientists (e.g. [7]; [19]; [14, 15]. [14, 16] elaborated a structural equation model of scientific eminence which shows that eminence in science is indirectly (being mediated by productivity) but significantly related to arrogant working style, whereas the paths between observer-related hostility and eminence is direct and significant. The emergence of these personality traits among scientists could be explained by the existing highly competitive climate. In the world of natural sciences in particular, the winner takes all: best human resources, expensive equipments, exquisite laboratories. Characteristics as those mentioned above may b! e instrumental to increase a scientist's competitive advantage in a highly competitive environment.

3.Be trustful when you work with a scientist. He/she is conscientious, flexible and intrinsically motivated. Outstanding scientists are hard-working and driven. [5], who have studied the personality characteristics of 196 biologists, 201 chemists and 171 physicists, found that commitment to work (i.e. "the need to concentrate intensively over long periods of time on one's work") is the strongest predictor for scientific productivity. Similar findings reported by [1] show that the most creative and eminent scientists are the most persistent, hard-working and driven they proved to be. In the same vein, using the "Five Factor Model" of personality, [16] reported that, relative to non-scientists, scientists are roughly a half a standard deviation higher on conscientiousness and controlling of impulses. In another study, [21] have compared 326 science majors with 357 art majors a! nd found a significant difference on conscientiousness between the two groups of d = 1.59 (d is the effect size measured as difference between the two means divided by the average standard deviation). An analysis of laboratory meetings undertook by [9] has revealed the intellectual flexibility of the scientists (i.e. their ability to creatively generate but also modify experimental hypothesis). He also found that expert scientists are more willing to modify or discard hypotheses than novices. To conclude this short presentation of the most stable personality characteristics of the natural scientists, two cautionary remarks should be expressed. First, none of these characteristics is present only in scientists and there is no such a thing as linear determination of scientific behavior by the personality characteristics of the scientists. We just claim that the personality traits mentioned above are more frequent among scientists than non-scientists and, usually, they are enhanced in highly creative scientists than in less creative ones. But we cannot deny the possibility that some or all of them may characterize also other specific groups when arguably they are compared with the population norms. Second, between the scientists' behavior, personality traits and scientific environment, there are mutual and dynamic interactions. For example, an initially thing-oriented interest of a person can be enhanced if he/she is deciding to work in physics rather than in clinical psychology. On the other hand, the daily tasks in a physics lab elicit rather thing-oriented behavior than people-oriented. And it is difficult to say whether the driven and conscientiousness of the scientists make the scientific environment so competitive or, on the contrary, the highly competitive character of the scientific world has elicited the hard-working style and conscientiousness of the scientists. But the difficulty to isolate experimentally each variable (personality, environment and behavior) does not reduce the importance of identifying the most stable personality characteristics of the scientists which account for a large part of their behavior variance.

DEVELOPMENTAL HISTORIES

>From a developmental perspective, the psychology of science has accumulated salient results with regard to the following aspects: (1) life-span and scientific productivity and interests; (2) family socio-cultural background; (3) the role of teacher/mentor and attitude toward school; (4) gender issues. A short outline of the main results will be presented below.

1.Life-span and scientific productivity and interest. The available data are converging on the conclusion that the relationship between age and scientific productivity approximate an inverted U ([8, 27, 28, 29, 33]). If plotted against career age, productivity (expressed by the number of publications) peaks at about 20 years into one's career. This means that maximum productivity is reached in one's early 40's or late 30's, with a small variation between disciplines: earlier in mathematics and physics, later in biology and geology. The growth rate before this peak is steeper than the decline rate afterwards. We must however underline that the individual variations are rather large and that age only accounts for a relatively small percentage of the overall variability in productivity. Moreover, some recent data seem to support the idea of another increase in publication rate (a sort of "swan song! ") at the very end of one's scientific career ([16]). As for scientific interests, the available data show that eminent scientists express clear interest for research at early age, varying since 12 years old (Westinghouse finalists) to 18 years old (member of the National Academy of Science), Feist ([16]). Although the existent data are convergent, they could be undermined by the so-called hindsight bias (i.e. reconstruction of the past in order to make sense or to legitimate the present). In other words, the consecrated scientists are prone to reconstruct their past, including interest for research, from the perspective of their present achievements and to project their scientific interests earlier in their lives than it actually happened to be. There are no substantial evidence yet, for the opposite relationship, i.e., that one's interest to become a scientist at 12 or 18 years old is a strong predictor for a future scientific career, there! fore the data should be cautiously considered.

2.Family socio-cultural background. The most empirically sound data concerning family variables refer to immigrant status and religion. Research done in USA found that scientific elite is about three times more likely to be offspring of immigrant families than the population in general. About 40 percent of science majors, scientists and elite scientists had at least one parent who was an immigrant, whereas only 12 percent of US population in 2005 is first generation of Americans - foreign born ([25], [6]). However, we must strongly emphasize that these results cannot be generalized. If we consider other areas with low scientific immigration but emerging scientific performance as EU, Japan, China or India, the results are likely to be completely different. Moreover, recent OECD data show that academic performance of immigrants, especially in mathematics, is rather lower than the average mathematics performance of their host cou! ntry.If we consider also the religious background of the belonging family, not necessarily one's current religious behavior, we may reliably estimate that 20 to 30 percent of the eminent scientists come from Jewish families ([33, 13].

3.The influence of a teacher or mentor and attitude toward school. Good grades or high scores at aptitude testes do not predict reliably the retention rate for scientific career and even less, creative achievements ([31]). Longitudinal studies show that from a sample of very precocious children in mathematics only about 25 percent continue in science and math through graduate school and even fewer follow a scientific career ([4]). On the other hand, a large majority of eminent scientists strongly underline the importance of mentorships in stimulation of creativity. [13], for example, shows that 65 percent of the elite natural scientists reported having a charismatic mentor in high school and 80 percent reported having one in graduate school. In high school, the mentor role is played either by a teacher (29 percent) or a parent (26 percent), whereas in graduate school the mentor is embodied by one's PhD advisor ! (56 percent) or another professor (20 percent). The most reliable data concerning the impact of discipline-related, more precisely mathematics-related attitudes upon academic performance are provided by OECD indicators published in [24].The data show that an increase with one unit of the intrinsic motivation for mathematics is associated with 11.9 unit increase in mathematic performance, OECD average. Self-concept in mathematics (i.e. the extent to which students believe that they can handle difficulties in mathematics) produce even larger positive effects of 32.4 upon mathematics performance. On the contrary, an increase with one unit of the anxiety index in mathematics deteriorates with 35.3 units the mathematics performance. And finally, an improvement with one unit of the disciplinary climate (disciplinary climate refers to the level of disorder and disruption in the classroom) increases with 18.3 units the score at mathematics performance. Although the indices presented above emphasize the re! lationship between mathematics attitude and mathematics performance in school at the age of 15, it is very likely that the contribution of the same factors could be critical for a scientific career in mathematics and natural sciences.

4.Gender is a issue. On the average, in OECD countries, the males graduate ratio in science per 100,000 employed 25-34-year-olds is almost double than the rate of females. ([24]) However, there are large variations from country to country and gender gap is expected to be much smaller in the future, due to the recent feminization of higher education and significant changes in career prospect for women. Note also that not all the fields of science are equally gender biased in their distribution. An unequal distribution is most striking in physical sciences and least striking in social sciences. Only 17 percent of the engineering degrees and 35 percent of physical-earth sciences degrees are awarded to women, but in biological and social sciences women have 50 percent and 73 percent respectively, of the awarded degrees ([23]). Men produce more publications than women but besides numbers and ratios, once the number of publications is held constant, women pro! duce works of better quality than men ([22, 30]).

ACHIEVING EXPERTISE

Achieving expertise, in many areas but especially in sciences, requires critical changes in at least two cognitive areas: knowledge base and meta-cognition.

1.To be an expert, you need a large and conditional knowledge base. To be an expert, you need a large and conditional knowledge base. The knowledge base of an expert mind contains three types of knowledge: declarative knowledge (i.e. knowledge about facts, principles, and processes), procedural knowledge (i.e. knowledge about how to do things, from how to run an experiment to how to use equipment) and tacit knowledge (i.e. that knowledge you don't know that you know). Compared to a novice, an expert possesses not only a richer knowledge base, but he/she excels especially in tacit knowledge, achieved through extensive practice, innumerable interactions with other experts and incidental learning. According to some studies, you need about 10 years of practicing in a specific field and constant reflection upon your practice in order to acquire an expert knowledge base ([10, 11]).Besides its richness, an expert knowledge base is charac! terized by its conditional character: it is organized around relevant problems. If a particular problem arises then relevant adjacent knowledge are automatically activated. Thus, the expert is not the owner of an inert mass of knowledge but the user of a problem-related knowledge base. The novices, on the contrary, still pay tribute to their formal learning and use to organize information according to its source ("what I learned at that discipline, in that chapter", "what my PhD supervisor told me" etc.) or other shallow characteristics rather than to its relevance for the problems to be solved. According to the existent data, the conditional knowledge base of an expert not only improves the access to relevant information but it may even change the vector of problem solving process. When facing a problem, an expert in physics usually proceeds forward from what is known to what is asked for whereas a novice is primarily focused on what is to be found and then proceeds backwa! rd to what is known (DeMey, 1989). A reach and conditional knowledge b ase also allows an easier use of analogies and metaphors, which are strong tools in the process of scientific creation (Miller, 1989).

2.To become an expert, you should develop your meta-cognition.Meta-cognition refers to one's ability and knowledge to reflect upon own cognitions, to monitor and change it if necessary. From the available research, we may conclude that experts' meta-cognitions are much more powerful than those of novices ([2]). Compared to novices in science the experts can make much easier distinctions between theories and evidences and recalibrate the theories to account for the data. They are also much more aware of the underlying assumptions that should be reconsidered in the case of a paradigm shift. Extensive meta-cognitions also allow expert scientists to undertake more successful hypothesis-testing procedures as, for example, M. Faraday, who used the confirm-early-disconfirm-later heuristic: confirm until you have a well corroborated hypothesis than try to disconfirm it ([18]). To conclude, a reach and conditional knowledge base and powerful meta-cognitions are critical factors for achieving scientific expertise. They should be very seriously considered when someone wants to train or to evaluate a scientist.

CONCLUSIONS

We have tried to outline, in this paper, the most salient psychological results concerning personality traits, developmental histories and cognitive characteristics of the scientists. It is our hope that, by this kind of results, the psychological sciences can contribute to a better understanding of scientists by themselves, by their fellow colleagues and by the general public. In fact, the science is not an impersonal and abstract enterprise but the outcome of the scientists' behaviors, of their emerging personalities and developmental histories. There is so much humid life behind their dry theories.

ACKNOWLEDGMENTS

I would like to thank the organisers of the Carpathian Summer School of Physics for a most enjoyable meeting.

REFERENCES

1. T. Amabile,*Creativity in context*,New-York: Westview,(1996),
2. Y. Anzai, Learning and use of representation for physics expertise. In K.A. Ericsson & J. Smith (Eds.)."*Toward a general theory of expertise*. Cambridge University Press, England.(1991),
3. S. Baron-Cohen, S. Wheelwright, P. Bolton & I. Goodyear, Is there a link between engineering and autism?, *Autism*, 1, 101-109.
4. C.P. Benbow, D. Lubinski, Psychological profiles of the mathematically talented. Some sex differences and evidence supporting their biological basis. In G.R. Bock and K. Ackrill (Eds.). *The origins and development of high ability* (pp. 44-66). Chichester: UK, Wiley and Sons.S. S. Gershtein and Y. B. Zeldovich (1993),
5. T.V. Busse, R.S. Mansfield, Selected personality traits and achievement in male scientists. *The Journal of Psychology*, 116, 117-131.(1984),
6. D. Camarota, Immigrants at mid-decade: A snapshot of Americaâ20ac2122s foreign-born population in 2005. Retrieved on October 12, 2006, from http://www.cis.org/articles/2005/back1405.pdf, (2005),
7. J.A. Chambers, . Relating personality and biographical factors to scientific creativity. *Psychological Monographs: General and Applied*, 78, 1-2,(1964),
8. A.M. Diamond, The life-cycle research productivity of mathematicians and scientists. *Journal of Gerontology*, 41, 520-525,(1986),
9. K. Dunbar, How scientists really reason: Scientific reasoning in real-world laboratories, in R.J. Sternberg & J. Davidson (Eds.), *The nature of insight* (pp. 365-396). Cambridge, MA: MIT Press, (1995),
10. K.A. Ericsson, J. Smith, Eds., *Toward a general theory of expertise*, Cambridge University Press, England,(1991),

11. K.A. Ericsson, J. Smith, Prospects and limits of the empirical study of expertise: An introduction. In D.J. Levitin (Ed.). *Foundations of Cognitive Psychology*, MIT Press,(2002),
12. G.J. Feist, *The emotional and cognitive aspects and personality dimensions of artistic and scientific productivity*. Unpublished manuscript. University of California, Berkeley (cited apud Feist, G.J., 2006), (1987),
13. G.J. Feist, Unpublished doctoral dissertation (cited apud Feist, G.J., 2006),(1991),
14. G.J. Feist, A structural model of scientific eminence. *Psychological Science*, 4, 366-371,(1993),
15. G.J. Feist, A meta-analysis of the impact of personality on scientific and artistic creativity. *Personality and Social Psychological Review*, 2, 290-309,(1998),
16. G.J. Feist, *The psychology of science and the origins of the scientific mind*. New haven, CT: Yale University Press,(2006),
17. G.J. Feist, F. Barron, *Do hostile and arrogant scientists become eminent or are eminent scientists likely to become hostile and arrogant?*, Paper presented at the annual conference of Society for Social Studies of Science, Charlottesville, VA.,(1995)
18. G.J. Feist, M.E. Gorman, The Psychology of Science. Reviewed integration of a nascent discipline. *Review of General Psychology*, 2, 13-47,(1998),
19. R.L. Helmreich, J.T. Spence, R.S. Pred, Making it without losing it: Type A, achievement motivation and scientific attainment revisited. *Personality and Social Psychology Bulletin*, 14, 495-504,(1988),
20. R. Helson, Women mathematicians and the creative personality. *Journal of Consulting and Clinical Psychology*, 36, 210-220,(1971),
21. P. Kline, S.L. Lapham, Personality and faculty in British universities. *Personality and Individual Differences*, 13, 855-857.(1992),
22. J.S. Long, Measures of sex differences in scientific productivity. *Social Forces*, 71, 159-178,(1992),
23. National Science Foundation â20ac" NSF (1999). *Women, minorities, and persons with disabilities in science and engineering*: 1998 (NSF 99-87). Arlington, VA: National Science Foundation,
24. Organization for Economic Co-operation and Development - OECD (2007). *Education at a glance 2007*. OECD Indicators: OECD.,
25. A. Portes, R.G. Rumbaut, Legacies: *The story of the immigrant second generation*. Berkeley, CA: University of California Press,(2001),
26. A. Roe, A psychological study of eminent psychologists and anthropologists, and a comparison with biological and physical scientists. *Psychological Monographs: General and Applied*, 67, 1-55, (1953),
27. D.K. Simonton, *Scientific genius: A psychology of science*. Cambridge, England: Cambridge University Press,(1988),
28. D.K. Simonton, Age and outstanding achievement: What do we know after a century of research? *Psychological Bulletin*, 194, 251-267,(1988),
29. D.K. Simonton, Career landmarks in science: Individual differences and interdisciplinary contrasts. *Developmental Psychology*, 27, 119-130,(1991),
30. G. Sonnert, What makes a good scientist?: Determinants of peer evaluation among biologists. *Social Studies of Science*, 25, 35-55,(1995),
31. R.J. Sternberg, A three-facet model of creativity, in R.J. Sternberg (Ed.), *The nature of creativity* (pp. 125-147). Cambridge, England: Cambridge University Press,(1988),
32. G.D. Wilson, C. Jackson, The personality of physicists.*Personality and Individual Differences*, 16, 187-189,(1994),
33. H. Zuckerman, *Scientific elite: Nobel laureates in the United States*. (2nd edition). New Brunswick, NJ: Transaction Publishers,(1996),

Science Education through Cosmic Ray Experiments in High Schools

V. Popa

Institute for Space Sciences, RO 077125 Bucharest - Măgurele, Romania

Abstract.
We discuss the idea of using simple cosmic ray physics experiments to increase the motivation of both teachers and high school students in teaching and learning physics, and, more generally, technical sciences and mathematics. Many different such initiatives emerged in Northen America and in some European countries. An European network dedicated to such activities was recently formed.

Keywords: Science in secondary school, Research in physics education, Teacher in service training, Lab. experiments and apparatus, Cosmic rays
PACS: 01.40.Ek,01.40.Fk,01.40.jh,01.50.My,96.50.S-

INTRODUCTION

Cosmic Ray Physics is one of the most dynamic fields of modern physics; it yielded so far fundamental results, and its future development implies very large experimental facilities as well as proportional educated man power. It has, trough its "exotic" nature, a significant fascinating potential for the young students, if they have the opportunity to be involved in an up-to-date research activity, without need of complicated notions far beyond of the normal high school science curricula.

We discuss the involvement of high school students and teachers in a network of "simple minded", easy to build, maintain and use, and not in the end, cost effective, cosmic ray detectors. This allows to make efficient science education, but also valid research, once the network expanded at a large scale.

THE ALTA AND NALTA NETWORKS

The original idea of an educational - research network composed of cosmic ray detectors installed and operated in high schools, was implemented from 1998 by the Alberta University, Edmonton, Canada, under the acronym ALTA (**A**lberta **L**arge **T**ime coincidence **A**rray) [1]. An ALTA station (the basic unit of the network, operated in high schools) consists in 3 plastic scintillator detectors disposed in the angles of a equilateral triangle with the distance between two adjacent scintillators of about 10 m. The stations are completed by a GPS receiver (for precise time stamps), dedicated acquisition electronics, callibration LEDs and a personal computer used for control and interconnection with the rest of the network. Different other sensors may be added. The ALTA concept spread in all North America; different similar networks appeared, and their intercon-

CP972, *Exotic Nuclei and Nuclear/Particle Astrophysics (II)—Proceedings of the Carpathian Summer School of Physics 2007,* edited by L. Trache and S. Stoica

nections created a continental network known as NALTA (**N**orth **A**merican **L**arge **T**ime coincidence **A**rray), counting 12 member or affiliates networks.

SCIENTIFIC AND EDUCATIONAL GOALS OF SPARSE ARRAYS OF SIMPLE COSMIC RAY DETECTORS

The spread of the ALTA concept is based on its physical and educational goals. Avoiding large costs and infrastructure difficulties inherent to large "professional" experiments, such kind of sparse networks can still address significant scientific and educational targets. Single muon rates could be used to monitor seasonal variations, correlations with the atmospheric preassure, but could also give early warnings for large electromagnetic storms (trough the Forbush effect). Local time coincidences allow studying Extensive Air Showers, while coincidences between stations are sensitive to Giant Air Showers, which offer still subjects for scientific debates.

From the educational point of view, such initiatives respond the targets set in Lisbon 2000: "... to increase the number of students completing a scientific or technical study in 2010 by 15%". Teachers would have the opportunity to teach basic concepts of astrophysics, from the solar wind to exotic ultra high energy cosmic ray sources; the students will benefit of hands-on experience on a real cosmic ray experiment, would acquire notions about the construction and use of radiation detectors, data analysis, scientific internet use, will find out how modern science works and get the feeling to belong to a large comunity.

EUROCOSMICS

EuroCosmics is a recently created consortium of research institutes and universities from 15 European countries, in which similar initiatives are developed. The consortium aims to interconnect the national networks in a single and powerful continental net, to ensure standards for safe, robust, cost-effective and easy-to-handle instruments to be made and operated by high school students, to standardize the data formats, develop web-based tools for data storage and access, digital archives for technical and educational materials, and coordinate the knowledge transfer, including the EuroCosmics portal and the organization of a Summer Academy for physics teachers.

ACKNOWLEDGMENTS

We thank the organizers of the Carpathian Summer School for hosting the illuminating "Science education & Public understanding of Physics" round table.

REFERENCES

1. W. Brouwer et al., *Nucl. Instr. & Meths. in Phys. Res.* **A539** (2005) 595 and references therein.

CONCLUDING REMARKS—

NUCLEAR ASTROPHYSICS, NULCEAR STRUCTURE, PARTICLE ASTROPHYSICS

Concluding Remarks on the Physics at CSSP07

Livius Trache

Cyclotron Institute, Texas A&M University, College Station, TX 77843-3366

Abstract. A brief summary of the activities and of the nuclear astrophysics subjects presented in the two weeks of the Carpathian Summer shool of Physics 2007 [1] are presented. Emphasis is on the talks on nuclear astrophysics.

Keywords: nuclear astrophysics, nucleosynthesis
PACS: 20, 26

INTRODUCTION

Our conference is coming today to its conclusion. We have spent two weeks together, under the auspices of nuclei, cosmic rays, stars, and the mountains surrounding us. We have chosen an image of supernova SN1987 as part of school's logo, but we could as well have chosen an image representing our research, or our tools, or an image from the Carpathians, like that of "The Sphinx", high above the city of Sinaia (*on the cover of this volume).* We had very good conditions for our activities here. We had full days with over 8 hours of work, we had a well-rounded program and lively sessions! That is not only my own opinion, or that of the organizers, but that of so many guests, as directly or indirectly expressed to us! We were very glad to hear your appreciations! The conference, called a school for reasons of tradition, as you know, offered us the possibility to hear interesting lectures from accomplished and well known scientists (we had 52 invited lectures), as well as short communications (29 of them) from many future accomplished and famous scientists! I truly believe that, and we would be glad to know that some young physicists, in the future, will appreciate that their participation in this event meant something in their careers! I only wish the students had asked more questions, because we have to acknowledge that actually Sir Arnold Wolfendale was our most inquiring participant!

The organizers chose the title of the event due to the worldwide current interest in these apparently disjoint fields, but also in relation with the interests of our people from the organizing institutes. And we are very glad that we have had such a good response and so many invited lecturers attended. We have worked to arrange sessions with rather well defined subjects, such that the speakers will complement each other. And in large part we succeeded; where not, it was mostly for reasons of the very busy schedules of our guests. As such, we had in 10 working days, 16 regular sessions, two per day. We had also a special one-day symposium honoring Prof. Robert E Tribble on his 60th anniversary on Friday Aug. 24th. It was a simple celebration of one of our ones, during which past or current collaborators talked about the physics research they did or do together with Bob or under his leadership. And all ended with a dinner at

CP972, *Exotic Nuclei and Nuclear/Particle Astrophysics (II)—Proceedings of the Carpathian Summer School of Physics 2007,* edited by L. Trache and S. Stoica

that spectacular location which is "Cota 1400", anchored on the side of the mountain, half the way between the city and the top. We also had an outreach session in the afternoon of Monday Aug 27 entitled "Science education and the public understanding of science". It was, arguably, the liveliest session of all. In addition to school's participants we had one-day guests involved in science management and education in Romania, representatives of the media. It was most telling, probably, the fact that while we did not succeed to bring officials currently in charge of the science management, at institute or ministry level, we were able to attract people doing the groundwork. Four leaders of high schools with best results in physics were here, and the presentation of professor Bararu about the activities and successes of the students that he coaches in Constanta, was impressive for all in the audience. I want also to mention that it was in this session that the students in the audience were most active in asking questions and expressing own opinions.

In addition to our work, we all had many opportunities to interact, with one another, with our colleagues from Romanian institutes or universities, with our hosts in Sinaia, with this beautiful place and this beautiful country! We had these opportunities during our stay here at the "International" hotel, during dinners in the evenings, in the excursion last Saturday, during the visits in Sinaia or the surroundings, for those more courageous that took longer trips or hikes in the mountains or several cities around here. We respected the traditions of the school, like the football game Romania vs. the Rest of the World (the youth of the Romanian team won, but ROW was not offered a re-match, which makes some consider it a draw overall!). We had a nice banquet last night, with lots of food, music and lot of dancing, giving or reminding us the flavors of Romanian customs, cuisine, music and dances!

It may be that the length of that banquet did not allow for an extensive coordination among the four speakers about the material of today's summaries! We may miss some, we may overlap some, please bare with us! Because the material was so vast, it would have been difficult in any case to give a just representation of all subjects, and all speakers. Moreover, when we prepared the current volume of proceedings, we felt that the best way to summarize the school was to struggle to have all or most of the lectures included in the book, which we have achieved in large measure! Therefore, the printed considerations will be short!

NUCLEAR ASTROPHYSICS

It falls to me to say a few words about the sessions on "Nuclear Physics for Astrophysics" and "Nucleosynthesis". These were among the most consistent and well rounded sessions, as the organizers were lucky enough to be able to correlate several, if not all speakers. And lectures consistent with a "school" were presented, giving the students the opportunity for a rounded first contact with the subjects and to those more experienced in the audience the same feeling one gets while listening to a known piece in the interpretation of a brilliant pianist. The "Nuclear Physics for Astrophysics" day started with an excellent lecture by Claus Rolfs (*Editor's note: unfortunately absent from this volume)* who first introduced the field and presented its challenges, then continued to his most recent work, in particular that on the $^{12}C(\alpha,\gamma)^{16}O$ reaction using a magnetic spectrometer. This was for some time, and remains, the most important and challenging reaction in nuclear astrophysics. Michael Hass took us to another challenging realm: that of the solar neutrinos problem. He summarized the evolution and the current status of the solar neutrino research and then presented his and his group's work on the most important nuclear reactions behind it: $^{3}He(^{4}He,\gamma)^{7}Be$ and $^{7}Be(p,\gamma)^{8}B$. Or S_{34} and S_{17}, as we call them in "slang". If both these speakers referred to direct measurements, the next two went to talk about the use of indirect methods. Moshe Gai showed us how the same S_{17} problem was solved using Coulomb dissociation. Then Livius Trache presented four different indirect methods using radioactive nuclear beams (RNB): ANC from transfer reactions, breakup at intermediate energies, beta decay and beta-delayed proton decay. He chose to exemplify with four different experiments of the group at Texas A&M University. Later in the day, Yoneda has summarized for us the work done at RIKEN using Coulomb Dissociation of sd-shell RNB, and Junker took us back to direct measurement made in the low background environment of LUNA. F. Negoita presented a dream of a small Bucharest group: to have a small electrostatic accelerator installed in a salt mine in Romania, open to the European and international nuclear astrophysics community. Whether realistic or not, feasible or not, the existence of the project is remarkable.

Many other speakers treated nuclear astrophysics subjects or matters relevant to nuclear astrophysics during the rest of the school. In fact, many studies of the structure and decay of exotic nuclei, or of nuclear matter, are motivated by or relevant to nuclear astrophysics. Muhsin Harakeh and Phil Woods exemplified that in their lectures. Mihai Horoi tells more about this interplay of nuclear physics branches in his summary below.

A full batch of lectures and communications on NA were presented in the "Tribble symposium" by his past students or current collaborators. Xiaodong Tang presented new measurements on the beta-delayed alpha decay of ^{16}N with relevance to the $^{12}C(\alpha,\gamma)^{16}O$, while Vaclav Kroha, Tariq Al-Abdulah and Adriana Banu showed examples of the use of the ANC method using proton or neutron transfer reaction. Gianluca Pizzone has introduced us the Trajan Horse method, while insisting on the most recent results of the Catania group.

The session on "Nucleosynthesis", this Monday morning, was short, but had three excellent lectures, starting with that of Brian Field on Big Bang Nucleosynthesis and the new cosmology. In his, arguably the most captivating lecture, he has lifted our spirits and thoughts to the level of the whole Universe, to what we know and still have to learn about its beginning. And how scientific knowledge in apparently different fields can come together to give a coherent image of what our Universe is, and with it what we are. Further, Karl-Ludwig Kratz has shown us how, starting from the building blocks Brian has talked about, a considerable part of the chemical elements were produced in the r-process. He summarized the current knowledge after 50 years from the seminal B^2FH paper, including his own important contributions. Hendrik Schatz has then talked about explosive and super-explosive stellar scenarios and the importance of nuclear data in understanding them.

As I pass the baton to the next speaker, I thank you'll for coming, and wish you good and safe return home! See you next time!

ACKNOWLEDGEMENTS

This school went so smooth and well because of the visible and invisible efforts of many people, before and during the event. I must thank them all. I want to acknowledge the contribution of the management and staff of the hotel "International" that hosted us here in Sinaia. We need to congratulate Sabin for proposing this venue for our school! Good choice! I want to thank the members of the International Advisory Committee and in particular those that chose to attend the school they helped us organize, for their invaluable advice and support. Most, if not all of the members of the Organizing Committee were very active during the preparations for the school and during the school. In addition to them, I have to thank the very young ones that were de facto OC members: Oana, Andrei, Cristian and those I forget to mention, for their help with technical and other matters during the sessions.

Last, but not least, we have to acknowledge the support of our home institutions: IFIN-HH Bucharest and the Cyclotron Institute at Texas A&M University, and of our sponsor, Romanian National Authority for Scientific Research (ANCS).

REFERENCES

1. These Proceedings.

Concluding Remarks on Nuclear Structure and its Role in Nuclear Astrophysics

Mihai Horoi

Department of Physics, Central Michigan University, Mount Pleasant, Michigan 48859, USA

Abstract. Nuclear structure knowledge is essential for Nuclear Astrophysics. Many developments in this area of research that occurred recently were presented and hotly debated at the 2007 edition of the Carpathian Summer School in Sinaia. Here we give a short overview of what was discussed.

Keywords: Nuclear masses, Excitation energies, Spectroscopic factors, Gamow-Teller transition, Double beta decay
PACS: 21.10.Dr, 21.10.Pc, 21.10.Jx, 21.10.Ky, 21.10.Tg

Lectures discussing topics related to Nuclear Structure or/and Nuclear Astrophysics were addressing key issues in this area of research, such as nuclear masses of unstable nuclei, excitation energies, spectroscopic factors, Gamow-Teller transitions, beta decay, as well as the hotly debated topic of double beat decay.

Precise knowledge of the nuclear masses of known or recently discovered nuclides is essential for the understanding of the transitions rates in astrophysical processes. Several speakers, including C. Rolfs, H. Schatz, M. Junker, have discussed experimental aspects of measuring nuclear masses of unstable nuclei. On the theory side M. Horoi discussed the efforts in US to build a comprehensive Universal Energy Density Functional (UNEDF, see e.g. www.unedf.org) for nuclei, and use the Department of Energy supercomputing resources to calculate with increased precision masses and low energy properties of all nuclei. As an important aspect of nuclear masses, not yet fully understood, S. Lenzi analyzed the mirror symmetry breaking for *pf*-shell nuclei.

Low-lying levels, their excitation energies and quantum numbers are also essential to ascertain reaction rates of interest for nuclear astrophysics. Several talks, including those given by C. Rolfs, P. Woods, and S. Grevy presented selected topics and recent developments, including new understanding of the magic numbers for nuclei far away from the valley of stability. On the theory side, J, Vary presented in two talks recent efforts to use realistic nucleon-nucleon interactions and the machinery of the no-core shell model to describe the low energy structure of the *p*-shell nuclei. I. Stetcu discussed the effect of short-range correlations on the observables, within the no-core shell model approach. M. Horoi presented in his talk the efforts of traditional shell model community to describe heavier nuclei using core+valence space models with phenomenologically adjusted effective interactions.

CP972, *Exotic Nuclei and Nuclear/Particle Astrophysics (II)—Proceedings of the Carpathian Summer School of Physics 2007,* edited by L. Trache and S. Stoica

Spectroscopic factors are essential ingredients in the description of any reaction process. They represent the effect of nuclear structure seen in nuclear reactions. A. Buonacorso and F. Carstoiu discussed aspects of recent experiments with exotic radioactive beams and their implications to the nuclear structure of the reactants. One can also extract the spectroscopic factor from direct nuclear structure calculations. Comparison with the spectroscopic factors extracted from cross sections could provide new information about our understanding of nuclear forces and nuclear dynamics. An alternative approach is describing the reaction process using the Asymptotic Normalization Coefficient (ANC). This route was adopted in L. Trache's talk, in the contributions of A. Banu and T. Al-Abdullah. Selected topics on spectroscopic amplitudes in cold emission processes were discussed by D. Delion, C. Borcea and in the contribution of A. Neacsu.

Beta decay and Gamow-Teller transitions are very important processes in any stellar environment. Several recent experimental investigations of beta decays and Gamow-Teller transitions of interest were presented by H. Schatz, and H. Ejiri in their talks. Presentations of theoretical calculations for these processes were abundant, covering a wide range of techniques: Complex Vampir (A. Petrovici), QRPA (J. Suhonen), no-core shell model (J. Vary), shell model with effective interactions (M. Horoi). Special attention was given to cases with exotic nuclear structure, such as shape coexistence (A. Petrovici).

Double beta decay was highly debated at the school due to its potential of describing physics beyond the Standard Model, if neutrino turns out to be a Majorana particle. If observed, neutrinoless double beta decay offers a unique opportunity of determining the absolute masses of the neutrinos. The talks of H. Ejiri, and A. Bettini highlighted the present status of the experiments that are searching such rare events. S. Stoica presented a general overview of the theoretical aspects of double beta decay. J. Suhonen devoted special attention to the role of short-range correlations for the neutrinoless double beta decay lifetime. While no confirmed neutrinoless double beta decay results are know, there are about 10 cases for which 2-neutrino double beta decay lifetimes were determined. Calculations of these lifetimes are used to fine-tune the calculations of the neutrinoless double beta decay matrix elements. M. Horoi presented a new method to efficiently calculate these rates.

In conclusions, the 2007 edition of the Carpathian Summer School was a successful meeting of researchers with a large spectrum of interests in nuclear structure and nuclear astrophysics. The topics covered by the talks and contributions were attracting much attentions due to their novelty and originality. The discussions were lively and abundant. Last but not least, the large attendance of young researcher, as well as the opportunity for most of them to present their results, contributed to the overall success of this meeting.

Concluding Remarks on Neutrino Physics at CSSP07

Sabin Stoica

"Horia Hulubei" National Institute for Physics and Nuclear Engineering, 407 Atomistilor Street, P.O. Box MG-6, 077125 Magurele-Bucharest, Romania

Abstract. Neutrino physics presented at CSSP07 is reviewed.

Keywords: Neutrino, Double beta decay
PACS: 14.60.Pq, 98.80.-k, 98.80.Es

Neutrino physics is presently a very active field, both experimentally and theoretically and was included between the main topics of this School (CSSP07). Neutrinos are particles that play an important role in particle and nuclear physics, astrophysics and cosmology. The Standard Model incorporates neutrinos as massless particles which are left-handed partners of the charged bosons. However, in the recent past, neutrino oscillation experiments have convincingly demonstrated that neutrinos are massive and can transform from one flavor into another one. This is the first signal demonstrating that new physics beyond the SM exists. Neutrino physics is focused today on the crucial problem of the neutrino mass.

From the neutrino oscillation experiments one can determine only mass square differences of the neutrino types and thereby sets a lower bound on the largest of the neutrino masses. The results obtained by several atmospheric neutrino oscillation experiments are reviewed in Giacomelli's lecture. Also, the results of different long baseline neutrino experiments are considered (see G. Giacomelli, this volume and the references therein). Atmospheric neutrinos are electron and muon neutrinos and their antiparticles that are produced in the hadronic showers induced by primary cosmic rays in the earth's atmosphere. MACRO, SUDAN and Super-Kamiokande experiments detected deficits in neutrino fluxes with respect to the theoretical predictions and angular distributions distortions. These features were interpreted as $\nu_\mu \to \nu_\tau$ oscillations. Several long baseline experiments like K2K and MINOS have confirmed the neutrino oscillation results, and others, like CNGS or Opera and LVD are now under test or are accumulating data.

From neutrino oscillation experiments, one can measure only the mass square differences and the mixing angles. Therefore, the absolute scale of the neutrino masses as well as the hierarchy of them remains unknown. The absolute mass of electron neutrinos can be measured in tritium beta decay experiments. The decay results in a ^{3}He, an electron and an electron antineutrino. If neutrinos have non-zero mass, the spectrum of the electrons is deformed at the high-energy part, i.e. the neutrino mass

CP972, *Exotic Nuclei and Nuclear/Particle Astrophysics (II)—Proceedings of the Carpathian Summer School of Physics 2007,* edited by L. Trache and S. Stoica

determines the maximum energy of emitted electrons. The advantage of these experiments are the following: β decay is superallowed, the energy released and the half-life (12.5 years) are convenient quantities, precisely known, and the NME involved are constant; so, the electron spectrum is determined only by the phase spaces. Nevertheless, the inconvenient, is the reduced sensitivity. The current limits for $<m_\beta>$ come from Troisk and Mainz experiments and are 2.05 eV and 2.3 eV at a 95% CL, respectively. The future KATRIN experiment is planned to reach 0.2 eV at 90% CL and it expected to check the degenerate hierarchy ν-mass region. An up-to-date review of this experiment was given in Bonn's lecture (see Bonn, this volume and the references therein). Neutrino properties beyond the SM are studied by high-sensitivity neutrinoless double beta decay (0νββ) experiments. They are sensitive and realistic probes for studying the Majorana ν nature, the mass spectrum, the absolute mass scale, as suggested by ν-oscillation experiments, and even the Majorana CP phases. Moreover, the potential of 0νββ experiments to search for beyond-SM physics is broader. For example, from the non-observability of the 0νββ decay one may extract information on the lepton number non-conservation, existence of right-handed (RH) components in the weak interaction, L-R symmetric models 0νββ mediated by heavy RH bosons), Majoron existence (a light or massless boson) or leptoquark existence (bosons carrying both lepton and baryon number) - particles that may also mediate 0νββ decay, compositeness (possible substructure of quarks and leptons at an energy scale of about TeV), R-parity non-conservation and possibility that 0νββ occurs by exchange of SUSY particles between nucleons, etc. (see Stoica's lecture in this volume). The current running experiments use ββ isotopes of the order of 10 kg. Thus, the ν-masses to be studied are limited to the 200-300 meV. New-generation 0νββ experiments, both calorimetric experiments with ββ isotopes integrated into the detectors, like: Majorana/GERDA, CUORE, EXO and others, or spectroscopic experiments with ββ isotopes separated from the detectors like MOON, Super-NEMO and others, are planned to study the Majorana masses in the quasi-degenerate (QD 100 meV) and the inverted hierarchy (IH 30 meV) ν-mass regions. Nice reviews about the present and future of the 0νββ experiments were done in Ejiri's and Bettini's lectures. The measured 0νββ-decay half-life is connected with the effective neutrino mass $<m_{\beta\beta}>$, which, in turn, is connected to the relevant nuclear matrix elements (NME). Nuclear models of different type have been used to compute these NME. Accurate calculation of these NME is still one challenging theoretical issue in the study of ββ decay. There are essentially two competing families of nuclear models on the market, the nuclear shell model (ShM) and the proton-neutron quasiparticle random-phase approximation (pnQRPA). Reviews of them were done in Stoica and Suhonen lectures in this volume. Although pnQRPA and its extensions were more extensively used to perform calculations of the NME, much progress was also done in the ShM approaches. Therefore, one can now proceed to a systematic comparison between the two types of methods in view of reducing the uncertainties in the calculations. The importance of the short-range correlations in treating the neutrino potentials and the proper treatment of these correlations turns out to be an important issue related to the accuracy of the NME calculations (see Suhonen's lecture in this volume). In addition, there are other independent probes of checking the NME:

ordinary β–decay transitions, charge-exchange reactions, neutrino-nucleus charged-current scattering and ordinary muon capture, that were discussed in Suhonen's lecture.

Besides the 0νββ decay experiments, from precision measurements of the cosmic microwave background (for example WMAP) combined with observations of the large scale structure of the universe (SSDS) and analysis of the Lyman-α (Lya) spectra, one can derive stringent limits for the sum of the neutrino masses. The presence of neutrino mass is manifested in a suppression of power on small scales. Using this effect one can derive bounds on the sum of the neutrino masses. The mass bound depends very much on which data sets are used to produce it. In particular, one can obtain an extremely tight bound of $\sum m_\nu < 0.17$ eV if Lya absorption lines are included. However, the Lya data used are subject of unknown systematic effects, and show a 2σ discrepancy in the fluctuation amplitude when compared with CMB data. This discrepancy pushes the allowed parameter space to a ``compromise region'' and artificially pushes down the neutrino mass bound. Therefore, one should treat bounds from Lya with extreme caution. Also, it is worth mentioning that the neutrino mass bounds are model-dependent: adding to the model parameters that are degenerate with the neutrino mass tends to relax the mass bound. The bottom line is that precision cosmology can at present constrain the sum of the neutrino masses to the order of 1 eV. To obtain tighter bounds one would need to make assumptions about unknown factors such as the nature of dark energy and the initial conditions of the primordial density fluctuations. All these interesting issues were discussed in detail in Wong's lecture.

Finally, neutrino interactions with nuclei offer significant probes to detect or distinguish the neutrinos of several flavors and study the basic structure of the weak interaction. For computing the various ν-nucleus cross sections one also needs to calculate accurately NME. Therefore, the nuclear models (like QRPA) can also be tested in such calculations. The inelastic neutrino-nucleus reaction cross sections at low and intermediate neutrino energies were discussed in Kosmas' lecture. The results presented refer to the differential, integrated and total cross sections of the neutral-current induced reaction of ν_e with the ^{98}Mo and ^{40}Ar nuclei. Such reliable results are significant to explore the nuclear response of the studied isotopes to supernova neutrino spectra.

Concluding, one can say, it is very likely that our knowledge on the neutrino properties will be significantly improved in the next future by the expected progress in the ββ decay study (both theoretically and experimentally), in neutrino oscillation experiments (particularly the long baseline experiments), in the precision of cosmological data and in the accurate calculation of the ν-nucleus cross sections.

Astroparticle Physics at the CSSP07

Heinigerd Rebel

Institut für Kernphysik, Forschungszentrum Karlsruhe, 76021 Karlsruhe,Germany
rebel@ik.fzk.de

Abstract. With a tour along the energy spectrum of primary cosmic rays the scientific messages of the Carpathian Summer School of Physics 2007 about various energy regions and aspects of the current status of the research field are commented

Keywords: Astroparticle Physics, Cosmic Rays, Extensive Air Showers.
PACS: 96.50.sd ; 98.70.Sa; 95.55.Jz.

INTRODUCTION

Astroparticle physics is the endeavour to study astronomical objects in the light of elementary particles. Such a definition is a little ambiguous since in modern understanding only the neutrino and photon astronomy would be covered with this term Nuclear and most subnuclear particles, observed in cosmic rays investigations are nowadays understood as composite particles. Thus we must relax a rigorous limitation by saying "in the light of subatomic particles" with including cosmic rays, whose discovery has historically launched this branch of science [1] being a driving force for many actual investigations of problems of fundamental character. Most recently this has been shown by the communication of the Pierre Auger Collaboration [2] about the experimental demonstration of a correlation between the arrival directions of Ultra-High -Energy cosmic rays (UHECR) and the positions of active galactic nuclei (AGN) lying within a distance of about 75 mega parsecs. from Earth. When we follow the suggestions that UHECR are protons, we may call these exciting results as the advent of *proton astronomy.* The anisotropy observation of the UHECR arrival directions by the Pierre Auger collaboration provides a great push for a future observatory in the northern hemisphere for viewing the entire sky.

Though these results of the Pierre Auger Collaboration had not yet been published and officially communicated to the community at the time, when we were discussing various aspects of the present status of Astroparticle and Cosmic Ray Physics at the CSSP07, at the fine place of Sinaia, the horizon of the presentations covered the background and familiarized quite well about the scientific richness of the field. The present survey and comments are focused, in particular to the topics of the following lectures and seminar contributions:

CP972, *Exotic Nuclei and Nuclear/Particle Astrophysics (II)—Proceedings of the Carpathian Summer School of Physics 2007,* edited by L. Trache and S. Stoica

Lectures

Tina Suomijarvi	*Highest Energies-CR: AUGER*
Gianni Navarra	*Transition galactic to extragalactic CR: KASCADE – Grande*
Andreas Haungs	*EAS-Radioemissio: LOPES*
Heinigerd Rebel	*Lateral EAS Asymmetries - Muon Charge Ratio in EAS*
Masahiro Teshima	*Gamma Ray Astronomy*
Gregorz Wrochna	*Gamma Ray Bursts*
Arnold Wolfendale	*Cosmological Aspects*

Seminar contributions

Maria Garzelli	*FLUKA Code*
Gabriel Toma	*EAS Energy Determination*
Athina Meli	*Cosmic Ray Acceleration Mechanisms*
Gina Isar	*LOPES*
Alexandra Saftoiu	*LOPES*
Bogdan Mitrica	*WILLI Detector*
Eduardo Medicanelli	*Massive Particles: SLIM experiment*

In order to specify the kind of phenomena of Nature we are talking about, let us recall some basics of cosmic rays (CR). Cosmic rays are energetic particles and nuclei from the cosmos that strike the Earth's atmosphere. Their energy distribution ranges from a few 10^8 eV to beyond $E=10^{20}$ eV, with a rapidly decreasing flux along power laws ($\sim E^{-n}$ with changing values of the spectral index n). Figure 1 displays the energy spectrum as determined by various experiments and it shows different regions: (i) the region up to 10^{14}-10^{15} eV where some information about the spectrum and the mass composition can be provided by direct measurements with balloon or satellite borne detectors (ii) the region around the *knee* (the kink at about $3 \cdot 10^{15}$ eV) with the end of galactic and the transition to extragalactic cosmic rays (up to the *dip* at 10^{18} eV with the beginning of the *ankle*) (iii) the region around and beyond the predicted GKZ cut-off at $6 \cdot 10^{19}$ eV. As experimental method of CR studies above 10^{14} eV Extensive Air Showers (EAS) are observed whose various components (electromagnetic particles, muons, hadrons, Cherenkov and fluorescence light, radio emission) are analyzed with respect of the information on the primary energy and the variation of the CR mass composition (for a survey of various EAS components, see the contribution of *Maria Garzelli*).

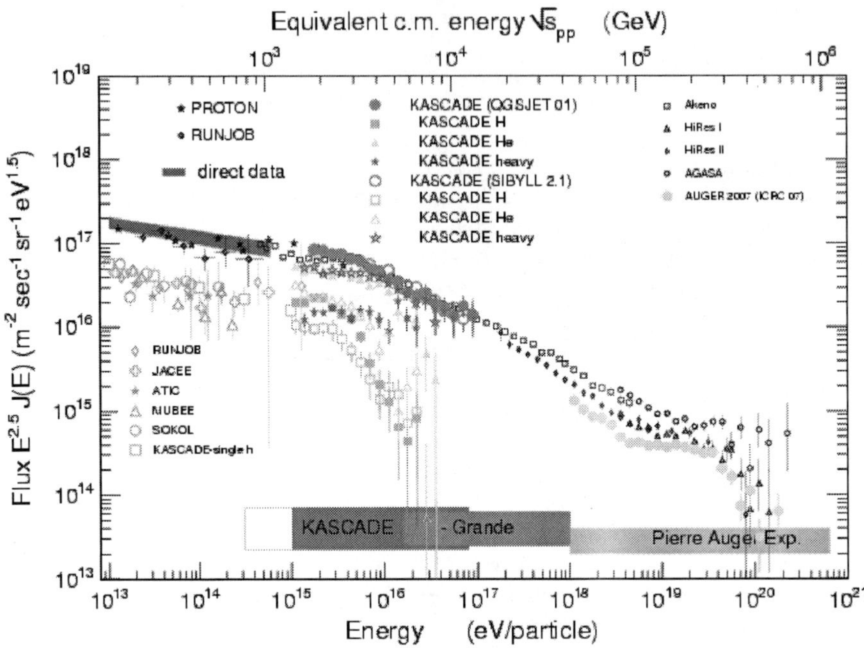

FIGURE 1: The energy spectrum of the cosmic ray flux (status: 2006). Note the flux is multiplied by $E^{-2.5}$ in order to enhance the peculiarities with changes of the spectral index. (Status: ICRC 2007, Merida, Mexico)

SPECTRAL REGION OF THE ANKLE AND ABOVE

A remarkable feature is arising from the interaction of CR with the 2.7 K microwave background radiation in the Universe. Above a certain energy threshold the particles interact with these photons, blue-shifted to high – energy γ- rays. For protons of an energy $> 6 \cdot 10^{19}$ eV photo production of π starts ($p + \gamma_{3K} \rightarrow \Delta^+ \rightarrow p + \pi^0$ ($n + \pi^+$)) and cools down the proton. This is the Greisen- Zatsepin-Kuz`min (GZK) spectral cut off [3]. As consequence the Universe gets rather opaque for UHECR, and their sources do not exceed the distance of 200 Mps ("GZK horizon"), unless rather exotic particles are envisaged. The latter possibility has been extensively discussed in the past, triggered by the message from AGASA detector and others (in 1999) that trans - GZK events would have been discovered. That situation has prompted the project of the Pierre Auger laboratory in order to clarify the situation. In fact the recent results demonstrating the anisotropy of the arrival directions of UHECR and the rapid decrease of the flux above $6 \cdot 10^{19}$ eV are now fairly consistent with the existence of the GZK cut off. The scientific scope and the layout oft the **Pierre Auger Southern Observatory**, this year nearing completion in Argentina, have been presented and discussed by *Tiina Sumjarvi* at the CSSP07. Her lecture introduced in the features of the hybrid detector combining two detection techniques: Sampling the lateral distributions of charged particles with a surface detector array extending over 3000 km^2 with 1600 water Cerenkov tanks. In addition the fluorescence light induced by the EAS in the atmosphere is registered by 24 telescopes at four sites of the array. Though the new results had already been at the horizon and could been partly inferred from preliminary communications, clearly the lecturer (member of the Auger collaboration) was rather reluctant to emphasize the most recent achievements in his oral presentation. However, *Arnold Wolfendale* whom the audience of the CSSP07 owes a lot of clarifying and enlightening comments, had already anticipated the existence of the anisotropies, when critically discussing in his lecture: **"Cosmic rays and cosmology",** the origin of extragalactic CR. With the observation of the anisotropies the origin is not clarified at all. How might AGNs accelerate protons to such mysterious energies? Sir Arnold concluded with the remark: *"It may be that the cosmos will protect its secret about the UHECR origin somewhat longer".*

TRANSITION FROM GALACTIC TO EXTRAGALALACTIC CR

One of the main conclusion of the multi-detector KASCADE experiment, set up in Forschungszentrum Karlsruhe for exploring the knee region of the all-particle spectrum, is the evidence that the kink in the spectrum is due to the escape of the light mass component from the galaxy (Fig.1). A change towards a heavier composition above the knee is expected in acceleration models (of the kind as discussed in the contribution of *Athina Meli*), with a rigidiy dependent position of a further knee, in the energy region just below where the transition to extragalactic CR may set in with the *ankle*. Following these arguments the KASCADE-experiment has been extended to **KASCADE-Grande** to cover the energy range up to 10^{18} eV. The arguments for this extension have been detailed in the lecture of *Gianni Navarra*, describing the experimental layout of KASCADE-Grande, the present status of the investigations and the open questions. Actual results of the analysis of the observed EAS have been shown (see also the contribution of *Gabriel Toma*).

There was a message about the analysis procedures of the lateral density distributions (PLD) of charged EAS particles, from which important EAS observables are usually deduced. It is by no means appropriate and may lead to systematic errors of the EAS observables when anticipating circular symmetry of the lateral distributions. PLD varies also azimuthally (i) due to geometrical effects with the projection from the observational plane to the normal shower plane, (ii) due to the attenuation of the density of inclined showers and (ii) due to the distortions by the geomagnetic field. *Heinigerd Rebel* has outlined these features, especially the peculiar geomagnetic influence on the EAS muon component, leading to a pronounced azimuthal variation of the charge ratio of the EAS muon density, being studied in context of the idea to measure this quantity in future with a dedicated detector installation (see contribution of *Bogdan Mitrica*).

RADIO EMISSSION FROM EAS

A long time neglected EAS feature, though known since the 60ties of the last century, is the radio emission accompanying EAS. Due to the advanced techniques with digital electronics the interest experienced a revival in our days with various activities, in particular by the LOPES experiment in Forschungszentrum Karlsruhe and the CODALEMA experiment in France. As process for the radio emission from EAS a coherent geosynchotron mechanism has been theoretically suggested: Electron-positron pairs generated in EAS gyrate in the geomagnetic field, and synchotron radiation pulses are emitted . In order to establish definitively the phenomenon dipole antenna have been set up within the detector array of KASCADE – Grande in order to observe radio emission correlated with EAS when registered by the surface detector array: Experiment **LOPES**. The definite proof of principle for detecting radio flashes from EAS has been communicated in 2005 [5], and the project has progressed with various studies of the features of radio emission towards an understanding in terms of EAS properties, in particular of the primary energy. Andreas Haungs reported in his lecture about the LOPES project, the scientific aims, the experimental procedures and current status. The lecture has been illustrated by supplementary contributions of *Gina Isar* and *Alexandra Saftoiu* about detail investigations. I would like to stress the aspect, that the technology of LOPES can be used in existing cosmic ray experiments and is actually forseen to be embedded in the Pierre Auger Observatory. Far

Radio Emission from Cosmic Ray Air Shower
LOPES

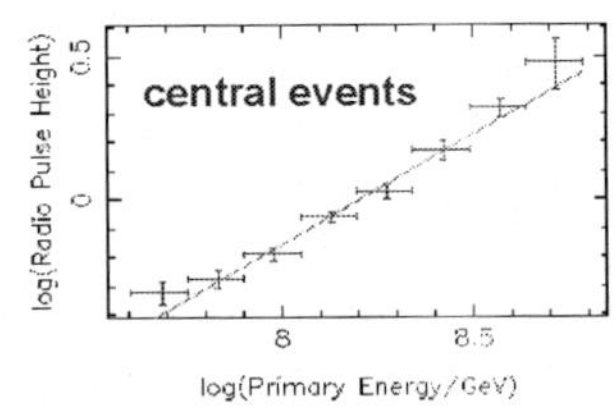

Radio signal (electric field) scales with primary energy:
$\varepsilon_v \sim E_0^{\sim 1}$

➔ Power of electric field scales approximately quadratically with primary energy !

FIGURE 2. LOPES

reaching perspectives are applications for detecting UHE neutrinos by showers induced in ice or salt. In dense media the Cherenkov-Askaryan effect plays the dominant role as emission mechanism. The lecturer concluded about this aspect: *"Presently a very active field, but yet no positive detection".*

GAMMA RAYS FROM COSMOS

Gamma Ray Astronomy, studying TeV γ-ray emission from astronomical objects has been started with the observation of the Whipple telescope in 1989 registering γ-rays from the Crab nebula, the remnants of a Supernova seen first in 1054 by Chinese astronomers. *Masahiro Teshima* introduced us with an extremely nice scholar lecture into the experimental techniques of observing the Cerenkov light of γ- ray induced air showers, discriminating efficiently hadron and γ-ray induced events. He presented an account of the current status of the research field. At present a larger number (ci.70) of sources is known, and the general approach is focused to study the γ-ray spectrum of these sources, which may in turn be interpreted by theoretical models in terms of the mass composition of the ejected hadronic charged particles. *Masahiro Teshima* concluded *"The physics in TeV - γ- ray astronomy is very rich and still there are many open questions. We definitely need large instruments like CTA for the development of TeV -γ- ray astronomy after HESS, MAGIC, VERITAS and CANGAROO"*. CTA (*Cerenkov Telescope Array*) stands for an initiated project to set up the next generation ground-based γ-ray instruments which will provide the deepest ever insight into the non-thermal high-energy universe.

Rapid astrophysical processes as source of γ- rays from the cosmos! *Greogoz Wrochna* made us familiar with the most luminous events in the universe since the Big Bang: **Gamma -ray bursts** (GRB). They are flashes of

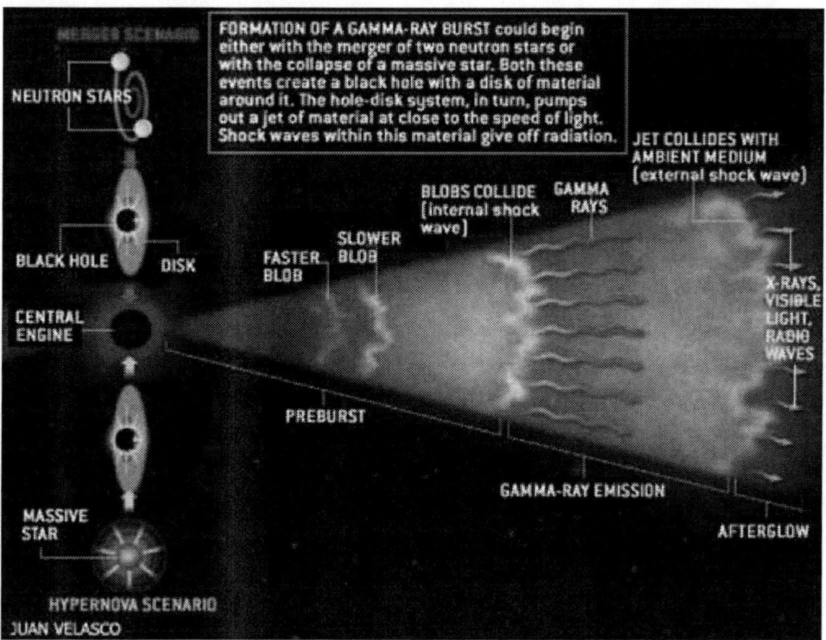

FIGURE 3. Formation of GRB

γ- rays from seemingly random places in the cosmos at random times, lasting milliseconds to minutes. We listened about the history of detection and the early debates about their origin since their discovery in 1990 by the US Vela nuclear test detection satellites and about the current understanding of their formation and perspectives of further observations. We may recall that for a short while GRBs had been speculated to be sources of UHECR. Of general interest may be the suggestion that nearby GRBs would possibly have had implications for historical mass extinctions on our planet with damages of the biosphere.

FINALE

The course of the interesting lectures and presented seminar contributions has displayed a lively and active research field. In a creative scientific atmosphere we had the chance and opportunity to learn and to get informed about the current problems and activities at the frontier of our science, by following the lectures and discussions in the audience, by interactions of researchers of different provenance, and in particular by the meetings of young scientists with experienced oldies, being inspired by the nice surroundings of the Prahova valley. It remains to thank for the excellent organisation of the CSSP07, the directors *Livius Trache* and *Sabin Stoica*, the Organisation Committee and the whole staff, in particular *Aurora Anitoaie, Alexandra Olteanu and Corina Simion* for kindness and various help. Finally let me express my hope for a future edition of the Carpathian Summer School of Physics, guided by a lively scientific spirit and revealing our joy in physics.

REFERENCES

1. Victor Hess, *Z.Phys.**12**, 998(1911); ibid**13**, 1084 (1912)*
2. Pierre Auger Collaboration, *Science, 9, November 2007, vol. 318, pp. 938-943*
3. K. Greisen, Phys.Rev.Lett.**16**, 748 (1966);
4. G.T.Zatsepin, V.A. Kuz`min, Sov.Phys.JETP Lett.**4**, 48 (1966)
5. H. Falcke et al.-LOPES Collaboration , Nature **435**, 313 (2005)

LIST OF PARTICIPANTS

Tariq Al-Abdullah
Cyclotron Institute
Texas A&M University
College Station, TX 77843-3366
USA
abdullatq@comp.tamu.edu

Cristian Andrei
National Institute of Physics and Nuclear Engineering – Horia Hulubei
Bucharest-Magurele, POB MG-6
Romania
acristian@niham.nipne.ro

Marilena Avrigeanu
National Institute of Physics and Nuclear Engineering – Horia Hulubei
Bucharest-Magurele, POB MG-6
Romania
mavrig@ifin.nipne.ro

Vlad Avrigeanu
National Institute of Physics and Nuclear Engineering – Horia Hulubei
Bucharest-Magurele, POB MG-6
Romania
vavrig@ifin.nipne.ro

Konstantia Balasi
University of Ioannina
GR 45110, Ioannina
Greece
dbalasi@hotmail.com

Adriana Banu
Cyclotron Institute
Texas A&M University
College Station, TX 77843-3366
USA
banu@comp.tamu.edu

Laurent Benoit
Laboratoire de Physique Corpusculaire
IN2P3-CNRS
& Université de Caen
F-14050 Caen cedex
France
laurent@lpccaen.in2p3.fr

Alessandro Bettini
University ofPadova
Padova, Italy
alessandro.bettini@pd.infn.it

Peter Biermann
Max Planck Institute for Radioastronomy
Postfach 2024
D-53010 Bonn
Germany
plbiermann@mpifr-bonn.mp.de

Jochen Bonn
Universitaet Mainz
Mainz
Germany
jbonn@uni-mainz.de

Angela Bonaccorso
Istituto Nazionale di Fisica
Sezione di Pisa
Italy
bonaccorso@pi.infn.it

Catalin Borcea
EC-JRC-IRMM NP Unit,
B-2440 Geel
Belgium
Catalin.BORCEA@ec.europa.eu

Iliana Magda Brancus
National Institute of Physics and Nuclear Engineering - Horia Hulubei
Bucharest-Magurele, POB MG-6
Romania
iliana@muon1.nipne.ro

Florin Carstoiu
National Institute of Physics and Nuclear Engineering - Horia Hulubei
Bucharest-Magurele, POB MG-6
Romania
fcarstoiu@ifin.nipne.ro

Niculina Chis
National Institute of Physics and Nuclear Engineering - Horia Hulubei
Bucharest-Magurele, POB MG-6
Romania
nchis@nipne.ro

Danka Chmielewska
Institute for Nuclear Studies, (IPJ)
05-400 Otwock-Swierk, Warsaw
Poland
danka@ipj.gov.pl

Dan Cozma
National Institute of Physics and Nuclear
Engineering - Horia Hulubei
Bucharest-Magurele, POB MG-6
Romania
cozma@niham.nipne.ro

Doru Delion
National Institute of Physics and Nuclear
Engineering - Horia Hulubei
Bucharest-Magurele, POB MG-6
Romania
ddelion@ifin.nipne.ro

Cristinel Diaconu
CPPM/IN2P3 Marseille, France and
DESY, Hamburg
Germany
diaconu@cppm.in2p3.fr

Hiro Ejiri
RCNP, Osaka University Ibaraki
Osaka 567-0047
Japan
ejiri@rcnp.osaka-u.ac.jp

Brian Fields
University of Illinois at Urbana-Champaign
Department of Astronomy 1002
Urbana, IL
USA
bfields@uiuc.edu

C.A. Gagliardi
Texas A&M University
College Station, TX 77843-3366
USA
gagliardi@comp.tamu.edu

Gai Moshe
University of Connecticut
Laboratory for Nuclear Science at Avery
Point and
Yale University, New Haven
Connecticut, USA
moshe.gai@yale.edu

Maria Garzelli
University of Milano, INFN
Milano, Italy
Maria.Garzelli@infn.it

Radu Gherghescu
National Institute of Physics and Nuclear
Engineering - Horia Hulubei
Bucharest-Magurele, POB MG-6
Romania
rgherg@ifin.nipne.ro

Giorgio Giacomelli
Dipartamento di Fisica dell Univ.
di Bologna, Bologna
Italy

Nicolae Grama
National Institute of Physics and Nuclear
Engineering - Horia Hulubei
Bucharest-Magurele, POB MG-6
Romania
grama@ifin.nipne.ro

Stephane Grevy
GANIL, IN2P3
Caen
France
grevy@in2p3.fr

Dan Gurban
National Institute of Physics and Nuclear
Engineering - Horia Hulubei
Bucharest-Magurele, POB MG-6
Romania
gurban_dan@yahoo.com

Michael Hass
Weizmann Institute
Department of Particle Physics
76100 Rehovot
Israel
fnhass@wisemail.weizmann.ac.il

Mushin Harakeh
KVI, University of Groningen
9747 AA Groningen
The Netherlands
m.n.harakeh@KVI.nl

John C.Hardy
Texas A&M University
College Station
TX 77843-3366
USA
hardy@comp.tamu.edu

Andreas Haungs
Institut fur Kernphysik Forschungszentrum
Karlsruhe, D-76021
Germany
Andreas.Haungs@ik.fzk.de

Dumitru Haseganu
Institute for Spatial Sciences
Bucharest
Romania
hasegan@venus.nipne.ro

Andrei Herghelegiu
National Institute of Physics and Nuclear
Engineering - Horia Hulubei
Bucharest-Magurele, POB MG-6
Romania
ehgreh@yahoo.com

Mihai Horoi
Department of Physics
Central Michigan University
Mt. Pleasant, MI 48859
USA
horoi@phys.cmich.edu

Laura Husti Tesileanu
Universita degli Studi di Torino
Torino
Italy
husti@ph.unito.it

Aurel Isar
National Institute of Physics and Nuclear
Engineering - Horia Hulubei
Bucharest-Magurele, POB MG-6
Romania
isar@ifin.nipne.ro

Paula Gina Isar
Max Planck Institute for Radioastronomy,
Postfach 2024
D-53010 Bonn
Germany
pgisar@mpifr-bonn.mpg.de

Leila Joulaeizadeh
KVI, Zernikelaan 25
NL-9747AA Groningen
The Netherlands
joulaei@kvi.nl

Matthias Junker
INFN
Gran Sasso National Laboratory
Italy
junker@lngs.infn.it

Georgia Karathanou
University of Ioannina
Physics Department
Gr 45110 Ioannina, Greece
me01306@cc.uoi.gr

Karl-Ludwig Kratz
Universitaet Mainz
Institut fuer Kernchemie, Mainz, D-55128
Germany
klkratz@uni-mainz.de

V.Kroha
Institute of Nuclear Physics
Prague-Rez.
Czechoslovakia Republic
kroha@ujf.cas.cz

Haris Kosmas
University of Ioannina
Physics Department
Greece
hkosmas@cc.uoi.gr

Vladimir Kuz'min
Joint Institute for Nuclear Researches,
Dubna
Russia
kuzmin@thsun.jinr.ru

Marco La Cognata
Universita di Catania
& INFN Lab. Nazionali del Sud,
Catania, Italy
Lacognata@lns.infn.it

Silvia Lenzi
INFN & Dipartimento di Fisica
Padova University
lenzi@pd.infn.it

Pawel Luczak
The A.Soltan Institute for Nuclear Studies
Department of Cosmic Ray Physics
Poland
luczak@ik.fzk.de

Matthew McCleskey
Cyclotron Institute
Texas A&M University
College Station, TX 77843-3366
USA
mmccleskey@gmail.com

Cristian Dumitru Manailescu
University of Bucharest
Faculty of Physics
Bucharest
Romania

Eduardo Medinaceli
University of Bologna
Italy
medinaceli@bo.infn.it

Athina Meli
University of Athens
Department of Physics
Athens
Greece
ameli@phys.uoa.gr

Saskia Mioduszewski
Texas A&M University
College Station
TX 77843-3366
USA
mio@comp.tamu.edu

Mihai Mirea
National Institute of Physics and Nuclear Engineering - Horia Hulubei
Bucharest-Magurele, POB MG-6
Romania
mirea@ifin.nipne.ro

Bogdan Mitrica
National Institute of Physics and Nuclear Engineering - Horia Hulubei
Bucharest-Magurele, POB MG-6
Romania
bmitrica2002@yahoo.com

Hossein Moeini
KVI Institute
University of Groningen
The Netherlands
moeini@kvi.nl

Claudia-Mihaela Morariu
University of Bucharest
Faculty of Physics
Bucharest
Romania
claudia.morariu@gmail.com

Szymon Myalski
Institute of Nuclear Physics
Polish Academy of Sciences
Krakow
Poland
szymon.myalski@ifj.edu.pl

Gianni Navarra
INFN
University of Torino
Torino, Italy
navarra@to.infn.it

Andrei Neacsu
National Institute of Physics and Nuclear Engineering - Horia Hulubei
Bucharest-Magurele, POB MG-6
Romania
nandrei@theory.nipne.ro

Florin Negoita
National Institute of Physics and Nuclear Engineering - Horia Hulubei
Bucharest-Magurele, POB MG-6
Romania
fnegoita@nipne.ro

Gabriela Emilia Pavalas
Institute of Spatial Sciences
Bucharest
Romania
gpavalas@venus.nipne.ro

Eduard Pauna
National Institute of Physics and Nuclear Engineering - Horia Hulubei
Bucharest-Magurele, POB MG-6
Romania
eduardp@theory.nipne.ro

Bogdan Pastrav
National Institute of Physics and Nuclear Engineering - Horia Hulubei
Bucharest-Magurele, POB MG-6
Romania

Sami Peltonen
University of Jyvaskyla
Teoreettinen Ydinfysiika
Tyohuone
Jyvaskyla
Finland
sjpelton@cc.jyu.fi

Marian Petre
National Institute of Physics and Nuclear Engineering - Horia Hulubei
Bucharest-Magurele, POB MG-6
Romania
petre_m@yahoo.com

Petrovici Alexandrina
National Institute of Physics and Nuclear Engineering - Horia Hulubei
Bucharest-Magurele, POB MG-6
Romania
spetro@ifin.nipne.ro

Mihai Petrovici
National Institute of Physics and Nuclear Engineering - Horia Hulubei
Bucharest-Magurele, POB MG-6
Romania
mpetro@ifin.nipne.ro

Gianluca Rosario Pizzone
Universita di Catania
& INFN Lab. Nazionali del Sud,
Catania, Italy
rgpizzone@lns.infn.it

Dorin Poenaru
National Institute of Physics and Nuclear Engineering - Horia Hulubei
Bucharest-Magurele, POB MG-6
Romania
poenaru@ifin.nipne.ro

Lucia Popa
Institute of Space Sciences
Bucharest, Romania

Vlad Popa
Institute of Space Sciences
Bucharest
Romania
Vlad.Popa@bo.infn.it

Alessio Porcelli
University of Pisa
Pisa, Italy
ale_led@yahoo.it

Oana Georgeta Radu
National Institute of Physics and Nuclear Engineering - Horia Hulubei
Bucharest-Magurele, POB MG-6
Romania
oanaradu@nipne.ro

Ralf Friedrich Rapp
Texas A&M University
College Station, TX 77843-3366
USA
rapp@comp.tamu.edu

Adriana Rodica Raduta
National Institute of Physics and Nuclear Engineering - Horia Hulubei
Bucharest-Magurele, POB MG-6
Romania
araduta@nipne.ro

Giuseppe Gabriele Rapisarda
INFN, LNS
Catania, Italy
grapisarda@lns.infn.it

Heinigerd Rebel
Institut fur Kernphysik
Forschungzentrum Karlshrue Germany
Heinrich.Rebel@ik3.fzk.de

Catherine Rigollet
Kernphysisch Versneller Institut
9747 AA Groningen
The Netherlands
C.E.Rigollet@rug.nl

Margarit Rizea
National Institute of Physics and Nuclear Engineering - Horia Hulubei
Bucharest-Magurele, POB MG-6
Romania
rizea@ifin.nipne.ro

Claus Rolfs
Institut fuer Physik mit Ionenstrahlen
Ruhr-Universitaet Bochum
Germany
rolfs@ep3.rub.de

Florin Rotaru
National Institute of Physics and Nuclear Engineering - Horia Hulubei
Bucharest-Magurele, POB MG-6
Romania
frotaru@tandem.nipne.ro

Alexandra Saftoiu
National Institute of Physics and Nuclear Engineering - Horia Hulubei
Bucharest-Magurele, POB MG-6
Romania
allixme@yahoo.com

Antti Saastamoinen
Department of Physics
University of Jyväskylä, PO Box 35
FIN-40351 Jyväskylä
Finland
ajsaasta@phys.jyu.fi

Nicu Sandulescu
National Institute of Physics and Nuclear Engineering - Horia Hulubei
Bucharest-Magurele, POB MG-6
Romania
nsand@nipne.ro

Adela-Consuela Scafes
National Institute of Physics and Nuclear Engineering - Horia Hulubei
Bucharest-Magurele, POB MG-6
Romania
ascafes@tandem.nipne.ro

Hendrik Schatz
National Superconducting Laboratory
Michigan State University
East Lansing. MI
USA
schatz@nscl.msu.edu

Shalom Shlomo
Cyclotron Institute, Texas A&M University
College Station, TX 77843-3366
USA
shlomo@comp.tamu.edu

Ion Silisteanu
National Institute of Physics and Nuclear Engineering - Horia Hulubei
Bucharest-Magurele, POB MG-6
Romania
silist@theory.nipne.ro

Corina Simion
National Institute of Physics and Nuclear Engineering - Horia Hulubei
Bucharest-Magurele, POB MG-6
Romania
corinasi@yahoo.com

Asym Soylu
Erciyes University
Turkey
asimsoylu@gmail.com

Alexandra Spiridon
Jacobs University Bremen
Bremen
Germany
a.spiridon@iu-bremen.de

Ionel Stetcu
Los Alamos National Laboratory
NM 87545
USA
istet@lanl.gov

Sabin Stoica
National Institute of Physics and Nuclear Engineering - Horia Hulubei
Bucharest-Magurele, POB MG-6
Romania
stoica@nipne.ro

Jouni Suhonen
Department of Physics
University of Jyväskylä, PO Box 35
FIN-40351 Jyväskylä
Finland
suhonen@phys.jyu.fi

Tiina Suomijarvi
Institut de Physique Nucleaire
91406 Orsay
France
ipno@in2p3.fr

Gabriel Tabacaru
Cyclotron Institute
Texas A&M University
College Station, TX 77843-3366
USA
tabacaru@comp.tamu.edu

Xiaodong Tang
Department of Physics
University of Notre Dame
Notre Dame, Indiana
USA
x.tang@nd.edu

Masahiro Teshima
Max Planck Institute for Physics
Munich
Germany
mteshima@mppmu.mpg.de

Ovidiu Tesileanu
Universita degli Studi di Torino
Torino
Italy
tesileanu@ph.unito.it

Gabriel Toma
National Institute of Physics and Nuclear Engineering - Horia Hulubei
Bucharest-Magurele, POB MG-6
Romania
gtoma@muon1.nipne.ro

Livius Trache
Cyclotron Institute
Texas A&M University
College Station, TX 77843-3366
USA
livius_trache@tamu.edu

Robert E. Tribble
Cyclotron Institute
Texas A&M University
College Station, TX 77843-3366
USA
tribble@comp.tamu.edu

Vaitsa Tsakstara
Physics Department
University of Ioanina
GR 45110 Ioannina
Greece
vtsakstara@cc.uoi.gr

Ana Vasile
Institute of Spatial Sciences
Bucharest
Romania
ana_ymail@yahoo.com

James Vary
Department of Physics and Astronomy
Iowa State University
Ames, IA 50011
USA
jvary@iastate.edu

Arnold Wolfendale
Physics Department
University of Durham
United Kingdom
ia.w.wolfendale@durham.ac.uk

Phillip J. Woods
University of Edinburgh
United Kingdom
pjw@ph.ed.ac.uk

Yvonne Wong
Max Planck Institute for Physics
Munich
Germany
ywong@mppmu.mpg.de

Gregorz Wrochna
Soltan Institute for Nuclear Studies
Otwock-Swierk, Warsaw
Poland
wrochna@ipj.gov.pl

K. Yoneda
Heavy Ion Nuclear Physics Laboratory, RIKEN
Tokyo
Japan
motobaya@riken.jp

Nicolae V. Zamfir
National Institute of Physics and Nuclear Engineering - Horia Hulubei
Bucharest-Magurele, POB MG-6
Romania
zamfir@tandem.nipne.ro

Author Index

A

B

C

D

E

F

G

H